THE SAUNDERS SERIES

Consulting Editor: John A. Thorpe, SUNY at Stony Brook

Howard Eves: AN INTRODUCTION TO THE HISTORY OF
MATHEMATICS, 5e   1983
Theodore W. Gamelin and Robert Everist Greene:
INTRODUCTION TO TOPOLOGY   1983
P. G. Kumpel and J. A. Thorpe: LINEAR ALGEBRA   1983
Martin Guterman and Zbigniew Nitecki: DIFFERENTIAL
EQUATIONS: A FIRST COURSE   1984
Zbigniew H. Nitecki and Martin M. Guterman: DIFFERENTIAL
EQUATIONS WITH LINEAR ALGEBRA   1986

 **SAUNDERS COLLEGE PUBLISHING**

Philadelphia   New York   Chicago
San Francisco   Montreal   Toronto
London   Sydney   Tokyo   Mexico City
Rio de Janeiro   Madrid

Zbigniew H. Nitecki
Martin M. Guterman
Tufts University, Medford, Massachusetts

DIFFERENTIAL EQUATIONS

WITH LINEAR ALGEBRA

THE SAUNDERS SERIES

Address orders to:
383 Madison Avenue
New York, NY 10017

Address editorial correspondence to:
West Washington Square
Philadelphia, PA 19105

Text Typeface: 10/12 Times Roman
Compositor: Waldman Graphics
Acquisitions Editor: Robert Stern
Developmental Editor: Jay Freedman
Project Editors: Joanne Fraser and Robin C. Bonner
Copyeditor: Rebecca Gruliow
Art Director: Carol C. Bleistine
Text Design: Lawrence R. Didona
Cover Design: Lawrence R. Didona
Text Artwork: J & R Technical Services
Production Manager: Tim Frelick
Assistant Production Manager: JoAnn Melody

Library of Congress Cataloging in Publication Data

Nitecki, Zbigniew.
  Differential equations with linear algebra.

  Bibliography: p.
  Includes index

    1. Differential equations. 2. Algebras, Linear.
I. Guterman, Martin M.   II. Title.
QA371.N54   1985   515.3′5   85-10746
ISBN 0-03-002719-5

DIFFERENTIAL EQUATIONS WITH LINEAR ALGEBRA   ISBN 0-03-002719-5

  6   032   98765432

CBS COLLEGE PUBLISHING
Saunders College Publishing
Holt, Rinehart and Winston
The Dryden Press

For

Alicia, Elizabeth,
Sonia, Lila, and Beth

with love and thanks

PREFACE

This book is an integrated introduction to standard topics in differential equations and basic linear algebra for engineering, science, and mathematics students at the sophomore level. The fundamental first three chapters develop the solution of linear differential equations and systems. This motivates two subsequent chapters treating vector spaces and linear transformations. Two final chapters give mutually independent treatments of Laplace transform and series methods; these chapters can be approached at any time after Chapter 2. The present book differs from our other text, *Differential Equations—A First Course,* in its deeper exploration of linear algebra (in Chapters 4 and 5); this replaces chapters on numerical methods and Fourier series methods for partial differential equations in the earlier book.

The present book is intended to serve as the basis for a combined introduction to differential equations and linear algebra following a standard two- or three-semester calculus sequence. Traditionally, a combined exposition of these subjects takes a purely deductive approach, beginning with linear algebra and then applying it to the solution of differential equations and systems. We have found, however, that for most students at this level, the abstract ideas of linear algebra are far harder to motivate than the basic "bread-and-butter" problem of solving differential equations that model physical systems. As a result, we have adopted an inductive approach, in which motivation is a key consideration. Thus, the introductory section at the beginning of each of the chapters on differential equations considers a class of physical models as a practical motivation for the mathematical discussion that follows. (These sections can be treated as independent reading when no class time is available for modeling or applications.) We have used differential equations to motivate linear algebra. Notions of linear algebra are first introduced as they arise in the process of studying differential equations. The more general topics concerning vector spaces and linear transformations come only after the solution of linear differ-

v

ential equations and systems has provided us with a wealth of concrete instances of linear phenomena.

Our exposition is aimed at the beginning user of differential equations; it guides the reader through the underlying ideas of the subject while maintaining a hands-on experience of specific problems. The discussion proceeds from the concrete to the abstract by means of many worked-out examples and observation of general patterns. Boxed summaries reiterate the main points to remember, including explicit problem-solving procedures. These serve as handy reference points for the reader. The notes that follow the summaries discuss technical fine points and specific shortcuts or difficulties that arise in practice. Exercises at the end of each section are arranged roughly in order of increasing difficulty and abstraction; especially involved problems are designated with an asterisk (*). Review problems at the end of each chapter provide an opportunity for the reader to check his or her understanding of the chapter as a whole. The answers to odd-numbered exercises appear at the end of the book.

The chapter-by-chapter contents are summarized as follows:

First-Order Equations (Chapter 1), motivated by population models, are handled primarily by separation of variables and variation of parameters. The discussion of graphing and exact equations can be skipped or deferred until later.

Linear Differential Equations (Chapter 2) are motivated by models involving damped springs and treated by the method of characteristic roots. The discussion of completeness tests for lists of solutions leads naturally to a discussion of determinants and to the notion of linear independence for functions. The methods of undetermined coefficients and variation of parameters are treated as natural extensions of earlier techniques. Some further consideration of physical models can be used or skipped at the user's discretion.

Our discussion of **Linear Systems of Differential Equations** (Chapter 3) is motivated by electrical circuit models and explores the analogies between the solution of a single differential equation and that of a system. Basic tools from matrix theory are developed as the need for them arises. Matrices and vectors first arise as notational conveniences for expressing a linear system of differential equations. Linear independence of vectors arises in the course of investigating the completeness of lists of solutions to a homogeneous linear system. Eigenvalues and eigenvectors occur when we attempt to carry over to systems the exponential solutions we have seen earlier for one differential equation. A fundamental computational tool in this chapter is row reduction of matrices (3.6), which in itself is one of the most useful techniques for a student of engineering or science to learn at this stage.

Our treatment of **Vector Spaces** in Chapter 4 looks back at phenomena encountered in previous chapters and puts them in a more unified, abstract framework. Basic notions treated here are those of a vector space, subspaces, spanning sets, linear independence and bases, dimension, and linear transformations. Examples include R^n and spaces of scalar- and vector-valued functions.

Chapter 5 explores the **Matrix Representation of Linear Transformations.** Sections covering the matrix of a product, the change-of-basis formula, and invertibility of matrices lead to a discussion of diagonal and block diagonal forms and their relation to the solution of differential systems with multiple eigenvalues. An optional section extends this discussion to Jordan normal form.

The motivation for Chapter 6, **The Laplace Transform,** comes from problems with discontinuous forcing terms. The chapter develops an operational calculus for handling initial value problems via the Laplace transform, including the adaptation to systems.

The motivation for Chapter 7, **Linear Equations with Variable Coefficients: Power Series,** comes from the Cauchy-Euler, Legendre, and Bessel equations as they arise in temperature distribution problems. After a review of power series, we develop power series solution of o.d.e.'s with polynomial coefficients, including the Frobenius method.

The present text draws heavily from our earlier book, and we remain indebted to the many people whom we have named there. In addition, we would like to register our gratitude to the following people for their additional help in making the present book possible: John Benfatto, Robin Bonner, Joni Fraser, Jay Freedman, Leslie Hawke, Harvey Keynes, Maggie Konner, Maria Maggio, Kate Pachuta, Bill Reynolds, Steve Schwarz, Bob Stern, Richard Weiss, and David Wend.

Z.N.
M.G.

CONTENTS

PREFACE   *vii*

ONE   FIRST-ORDER EQUATIONS   *1*

    1.1   Introduction   *1*
    1.2   Separation of Variables   *9*
    1.3   First-Order Linear Equations   *16*
    1.4   Graphing Solutions (Optional)   *26*
    1.5   Exact Differential Equations (Optional)   *31*

       Review Problems   *40*

TWO   LINEAR DIFFERENTIAL EQUATIONS   *43*

    2.1   Some Spring Models   *43*
    2.2   Linear Differential Equations: A Strategy   *48*
    2.3   Homogeneous Linear Equations: General
          Properties   *55*
    2.4   Determinants   *62*
    2.5   The Wronskian   *73*
    2.6   Linear Independence of Functions   *79*
    2.7   Homogeneous Linear Equations with Constant
          Coefficients: Real Roots   *85*
    2.8   Homogeneous Linear Equations with Constant
          Coefficients: Complex Roots   *93*
    2.9   Nonhomogeneous Linear Equations:
          Undetermined Coefficients   *99*
    2.10   Nonhomogeneous Linear Equations: Variation
          of Parameters   *110*
    2.11   Behavior of Spring Models (Optional)   *119*
    2.12   Rotational Models (Optional)   *127*

       Review Problems   *136*

THREE   **LINEAR SYSTEMS OF DIFFERENTIAL
EQUATIONS**   *139*

3.1   Some Electrical Circuit Models   *139*
3.2   Linear Systems, Matrices, and Vectors   *147*
3.3   Linear Systems of O.D.E.'s: General Properties   *161*
3.4   Linear Independence of Vectors   *172*
3.5   Homogeneous Systems, Eigenvalues, and
       Eigenvectors   *180*
3.6   Systems of Algebraic Equations: Row
       Reduction   *191*
3.7   Homogeneous Systems with Constant
       Coefficients: Real Roots   *203*
3.8   Homogeneous Systems with Constant
       Coefficients: Complex Roots   *214*
3.9   Double Roots and Matrix Products   *222*
3.10   Homogeneous Systems with Constant
       Coefficients: Multiple Roots   *236*
3.11   Nonhomogeneous Systems   *250*

       Review Problems   *261*

FOUR   **VECTOR SPACES**   *263*

4.1   Definition and Examples   *263*
4.2   Subspaces   *271*
4.3   Spanning Sets   *282*
4.4   Linear Independence and Bases   *292*
4.5   Dimension   *305*
4.6   Linear Transformations   *312*
4.7   Some Special Types of Linear Transformations
       (Optional)   *321*
4.8   Inner Products and Orthogonality (Optional)   *330*

       Review Problems   *343*

FIVE   **MATRIX REPRESENTATION OF LINEAR
TRANSFORMATIONS**   *349*

5.1   The Matrix of a Linear Transformation   *349*
5.2   Multiplication of Linear Transformations
       and Multiplication of Matrices   *362*
5.3   Change of Bases   *370*
5.4   Invertible Matrices   *377*
5.5   Diagonalizable Matrices and Block Diagonal
       Form   *386*

5.6   Jordan Form (Optional)   *396*

Review Problems   *408*

SIX   **THE LAPLACE TRANSFORM**   *411*

6.1   Old Models from a New Viewpoint   *411*
6.2   Definitions and Basic Calculations   *419*
6.3   The Laplace Transform and Initial Value
       Problems   *431*
6.4   Further Properties of the Laplace Transform
       and Inverse Transform   *442*
6.5   Functions Defined in Pieces   *451*
6.6   Convolution   *463*
6.7   Review: Laplace Transform Solution of Initial
       Value Problems   *475*

Review Problems   *483*
6.8   Laplace Transforms for Systems   *484*

SEVEN   **LINEAR EQUATIONS WITH VARIABLE
              COEFFICIENTS: POWER SERIES**   *493*

7.1   Temperature Models: O.D.E.'s from P.D.E.'s   *493*
7.2   Review of Power Series   *504*
7.3   Solutions about Ordinary Points   *516*
7.4   Power Series on Programmable Calculators
       (Optional)   *529*
7.5   The Cauchy-Euler Equation   *543*
7.6   Regular Singular Points: Frobenius Series   *551*
7.7   A Case Study: Bessel Functions of the First
       Kind   *564*
7.8   Regular Singular Points: Exceptional Cases   *576*

Review Problems   *593*

SUPPLEMENTARY READING LIST   *595*

ANSWERS TO ODD-NUMBERED EXERCISES   *A.1*

INDEX   *I.1*

First-order Equations

1.1 INTRODUCTION

In the late seventeenth century, Isaac Newton in England (1665, 1687) and Gottfried Wilhelm Leibniz in Germany (1673) synthesized several centuries of mathematical thought to create a language and method for describing and predicting the motion of bodies in various physical situations. The invention of the calculus was immediately followed by a period of intense mathematical activity, and the effect of these ideas on the development of mathematics, science, and technology makes this event surely one of the most important in the history of western thought. During the development of calculus, differential equations and their solutions played the central role. They arose as mathematical formulations of physical problems, and attempts at their solution motivated much of the mathematical development of calculus.

The role of differential equations in the modeling of physical phenomena is well illustrated by **Newton's second law** of motion, familiar to all physics students in the mnemonic form

$$F = ma.$$

In the situation that most interested Newton (gravity), the force F is the weight of the body, the constant m is its mass, and a is its acceleration. Although we are really interested in the position of the body, the equation tells us about neither the position nor its rate of change, but rather about the rate of change of the rate of change of the position. In the language of calculus, if $x = x(t)$ represents the position at time t, then the velocity is the derivative of x, $v = dx/dt$, and the acceleration is the derivative of velocity, $a = dv/dt = d^2x/dt^2$. Thus Newton's second law

$$F = m\,\frac{d^2x}{dt^2}$$

is an equation involving a derivative of the interesting variable—that is, it is a **differential equation.**

1

It was Newton's brilliant observation that in many physical situations the relation between rates of change of observable quantities is simpler than the relation between the quantities themselves. This is at the same time the source of the power of differential equations and the central problem in using them to predict physical phenomena. For if Newton's second law is to lead to useful physical predictions, we must translate this statement about the second derivative of position into a prediction of the position of the body at some time in the future; that is, we must express x as a function of time

$$x = \phi(t).$$

Any such prediction (function) that is consistent with a given law (differential equation) is called a **solution** of the differential equation. The problem of obtaining solutions to a given differential equation is a purely mathematical one and forms the subject of this book.

Some important features of the solution of differential equations can be illustrated by a special case of Newton's law. When the force is constant, it is easy to solve the equation by integrating both sides twice. We first recall that $v = dx/dt$ and write the law in the form

$$\frac{dv}{dt} = \frac{F}{m}.$$

Integrating both sides with respect to t

$$\int \frac{dv}{dt}\, dt = \int \frac{F}{m}\, dt$$

gives

$$\frac{dx}{dt} = v = \frac{F}{m} t + c_1.$$

Now, we integrate again

$$\int \frac{dx}{dt}\, dt = \int \left(\frac{F}{m} t + c_1 \right) dt$$

$$x = \frac{F}{2m} t^2 + c_1 t + c_2.$$

Note that the solution involves two "arbitrary constants," c_1 and c_2, which resulted from taking two indefinite integrals. The physical significance of these constants becomes clearer when we think of a specific instance of this equation, the motion of a falling ball under the force of gravity. Whereas the differential equation takes into account the Earth's gravity and the mass of the ball, this is hardly enough to predict the ball's position. We need to know where it started from and whether it was dropped or thrown. Without such information, we can make only a **general** prediction, vague enough to apply to all possible circumstances of the ball. To make a **specific** prediction without ambiguity, we need to know the initial position (the value of x when $t = 0$) and the initial velocity (the value of $v = dx/dt$ when $t = 0$). If we pick specific numerical values for the constants c_1 and c_2 in the general solution above, we find upon substituting $t = 0$ that the initial position is

$$x(0) = \frac{F}{2m}(0)^2 + c_1(0) + c_2 = c_2$$

while the initial velocity is

$$v(0) = \frac{F}{m}(0) + c_1 = c_1.$$

We see that in this case the values of the two "arbitrary constants" in our general solution are numerically equal to the **initial conditions** that determine a specific solution. We shall consider the role of initial conditions in determining specific solutions as we study various kinds of differential equations.

Of course, the process of finding the general solution in the preceding case was very easy. Some of the difficulties of the subject become clearer if we consider Newton's law in a context closer to the problems that motivated it in the first place. When the distance between bodies reaches interplanetary scale, the gravitational force depends on position according to an inverse-square law. Fixing masses and the constant g appropriately, this leads to the differential equation

$$\frac{-g}{x^2} = \frac{d^2x}{dt^2}.$$

If we try to solve this equation by integrating both sides, we run into trouble on the left. Remember, we want to take "$\int (\quad)dt$" of both sides. To evaluate

$$\int \frac{-g}{x^2}\, dt$$

we need to express x as a function of t. But if we knew *that*, we wouldn't need to integrate, since the equation would already be solved.

This shows that, even when our final goal is a practical one, we need a certain amount of theory to handle the differential equations that arise in physical models. We will consider specific theoretical questions as they come up in our study of solution methods for differential equations. For the moment we consider some population models, with an eye toward understanding the different kinds of differential equations that can arise in modeling various phenomena. Other physical models leading to similar differential equations are considered in the exercises that follow. We will solve the equations of these examples later in the text.

Example 1.1.1

A population grows at the rate of 5% per year. If $x = x(t)$ stands for the number of individuals in the population after t years, then the rate of change of x is numerically equal to 5% of x. Written as an equation this is

$$\frac{dx}{dt} = \frac{5}{100} x.$$

Whereas Newton's law involves the second derivative of the interesting variable, the equation in Example 1.1.1 involves only the first derivative. In general, we refer to the highest order of differentiation as the **order** of the equation. Thus, Newton's law is a second-order differential equation, while the population equation is of first order.

Example 1.1.2

Our first population model assumed a constant growth rate. This will not always be realistic. For example, the growth rate of the United States rose sharply after World War II but recently has been decreasing slowly. One function that exhibits this phenomenon (although it does not accurately portray the U. S. population) is

$$g(t) = \frac{t}{t^2 + 1}.$$

Some calculation shows that starting from $g(0) = 0$, $g(t)$ rises to a maximum value of $g(1) = 1/2$ (i.e., 50% annual growth rate), then falls off gradually, approaching 0 in the distant future (see Figure 1.1). A model based on this changing growth rate would give the equation

$$\frac{dx}{dt} = \frac{t}{t^2 + 1} x.$$

This equation, like the previous one, is of first order. Note that here the variable t appears explicitly in the coefficients of the equation.

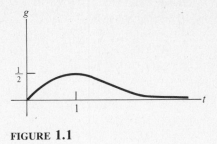

FIGURE **1.1**

Example 1.1.3

The rates of change of the populations in the previous examples were multiples of the populations. In some cases there may be an additional component (immigration) that depends only on time.

Suppose a disease-causing organism reproduces in its host by dividing once a day (on the average). Suppose also that its presence causes the host's resistance to deteriorate, so that on the tth day after the initial infection, t thousand organisms are able to enter the host from the surrounding environment. Let $x = x(t)$ be the number of organisms in the host, measured in thousands. Then the rate of change of x has two components: reproduction contributes x to dx/dt, and new organisms entering from the surrounding environment contribute t. The total rate of change is

$$\frac{dx}{dt} = x + t.$$

Example 1.1.4

Suppose a population consisting of $x = x(t)$ thousand organisms would, in an unlimited environment, have a growth rate of 5% per year. Assume the environment is limited and can support at most a population of 10 thousand. Then as the population approaches 10 thousand, we would expect the growth rate to decline. The simplest way to take account of this fact is to multiply the unlimited growth rate by a factor that approaches zero as x approaches 10; the simplest such factors are constant multiples of $10 - x$. Thus, we expect our population to satisfy

$$\frac{dx}{dt} = .05 \, \alpha(10 - x)x.$$

Since a small population experiences little competition, the limited-environment growth rate should approach the unlimited-environment growth rate as x approaches zero.

Then $\alpha = 1/10$. Our model has the equation

$$\frac{dx}{dt} = .005(10 - x)x.$$

Note that this equation, like our second gravitational example, involves an x^2 term.

Example 1.1.5

Suppose two neighboring countries, with populations $x_1(t)$ and $x_2(t)$, have natural growth rates (birth rate minus death rate) of 15% and 10%, respectively. Suppose that 4% of the first population moves to the second country each year, while 3% of the second population moves to the first country each year. Then the rate of change of each of the populations is made up of three components, the natural growth rate, emigration, and immigration:

$$\frac{dx_1}{dt} = .15\,x_1 - .04\,x_1 + .03\,x_2 = .11\,x_1 + .03\,x_2$$

$$\frac{dx_2}{dt} = .10\,x_2 - .03\,x_2 + .04\,x_1 = .04\,x_1 + .07\,x_2$$

In this case we have a **system** of two differential equations, each involving both variables x_1 and x_2 in an unavoidable way.

Each of our examples so far has involved only ordinary derivatives (as opposed to partial derivatives). We refer to such equations as **ordinary differential equations** (abbreviated **o.d.e.'s**). An equation like

$$\frac{\partial u}{\partial t} = c\,\frac{\partial^2 u}{\partial x^2}$$

which involves partial derivatives, is called a **partial differential equation (p.d.e.)**.

In this book we shall concentrate primarily on a special class of o.d.e.'s and systems (linear ones) for which a systematic solution procedure can be formulated and which are used in a broad variety of physical models. The precise delineation of this class will occur piecemeal as we study various specific instances.

In this chapter we will look at first-order equations, like the ones in Examples 1.1.1 through 1.1.4. These are the simplest from the point of view of calculus, since they involve only first derivatives. Yet a large variety of phenomena can be described by using such equations.

We close this section with a summary of the basic definitions that will play a large role in the first part of the book.

SOME BASIC DEFINITIONS

An **ordinary differential equation** (abbreviated **o.d.e.**) is an equation whose unknown x is a function of one independent variable t. The equation relates values of x and its derivatives to values of t.

The **order** of an o.d.e. is the highest order of differentiation of x appearing in the equation.

A **solution** of an nth-order o.d.e. is a function $x = \phi(t)$, with derivatives at least up to order n, which when substituted into the o.d.e. yields an identity on the domain of definition of $\phi(t)$.

The **general solution** of an o.d.e. of order n is a formula (usually involving n ''arbitrary constants'') that describes all **specific solutions** of the equation. A specific solution (or, equivalently, the value of each of the constants in the general solution) is determined by certain **initial conditions**, such as the starting point and the starting velocity.

EXERCISES

1. Determine the order of each of the following o.d.e.'s.

    a. $t^4 \dfrac{d^3x}{dt^3} + t \dfrac{dx}{dt} - x = t^7$ b. $\left(\dfrac{dx}{dt}\right)^5 + \dfrac{d^4x}{dt^4} - t^3x^7 + t^7 = 0$

    c. $x^8 \dfrac{dx}{dt} + \dfrac{d^7x}{dt^7} = x + t^9$ d. $(x')^2 x''' = x^4 x'' + t^5 x'$

In Exercises 2 to 6, check to see whether the given function $x = \phi(t)$ is a solution of the given o.d.e.

2. $\phi(t) = t^5$; $\dfrac{d^2x}{dt^2} - \dfrac{20x}{t^2} = t^3$ 3. $\phi(t) = e^{3t}$; $\dfrac{d^3x}{dt^3} - 9\dfrac{d^2x}{dt^2} = 0$

4. $\phi(t) = te^{3t}$; $\dfrac{d^2x}{dt^2} - 9\dfrac{dx}{dt} = 6e^{3t}$ 5. $\phi(t) = \ln(-t)$, $t < 0$; $tx' = 1$

6. $\phi(t) = \begin{cases} t^2, & t > 0 \\ 3t^2, & t < 0 \end{cases}$; $tx' - 2x = 0$

In Exercises 7 to 12, find all values of the constant k for which the given function $x = \phi(t)$ is a solution of the given o.d.e.

7. $\phi(t) = t^k$, $t > 0$; $t^2x'' - 6x = 0$

8. $\phi(t) = e^{kt}$; $\dfrac{d^2x}{dt^2} - x = 0$

9. $\phi(t) = k;$ $\dfrac{d^7x}{dt^7} + \dfrac{dx}{dt} - x = 7$

10. $\phi(t) = t^k,$ $t > 0;$ $16t^2xx'' + 3x^2 = 0$

11. $\phi(t) = ke^t;$ $\dfrac{d^2x}{dt^2} + 5\dfrac{dx}{dt} = 3e^t$

12. $\phi(t) = kte^{3t};$ $x'' - 3x' = e^{3t}$

In Exercises 13 to 17, find (a) the general solution of the o.d.e. and (b) the specific solution satisfying the given initial condition.

13. $\dfrac{d^2x}{dt^2} = 3t + 1;$ $x(0) = 2, x'(0) = 3$

14. $\dfrac{d^2x}{dt^2} = \dfrac{-1}{(t + 1)^2},$ $t > -1;$ $x(0) = 2, x'(0) = 3$

15. $x''' = 6;$ $x(1) = x'(1) = x''(1) = 0$

16. $\dfrac{d^3x}{dt^3} = e^{-t};$ $x(0) = x'(0) = x''(0) = 1$

17. $\dfrac{d^2x}{dt^2} = te^t;$ $x(0) = x'(0) = 0$

Exercises 18 to 25 describe some more models leading to first-order o.d.e.'s.

18. *Compound Interest:* A bank may compute interest by continuous compounding, that is, by treating the interest rate as the instantaneous rate of change of the principal. Set up a differential equation to model the principal $x = x(t)$ in an account accruing interest at 8% per year, compounded continuously.

19. *A savings account* pays 8% interest per year, compounded continuously. In addition, the income from another investment is credited to the account continuously, at the rate of $400 per year. Set up a differential equation to model this account.

20. *Radioactive Decay:* The atoms of a radioactive substance tend to decompose into atoms of a more stable substance at a rate proportional to the number $x = x(t)$ of unstable atoms present. Set up a differential equation for $x = x(t)$.

21. *Newton's Law of Cooling:* The simplest model for the rate at which a warm body in cooler surroundings loses heat assumes that the rate of change in temperature is proportional to the difference between the temperature of the body x and that of its environment y. Set up a differential equation to model the change in temperature.

22. *A Mixing Problem:* A tanker carrying 100,000 gallons of oil runs aground off Nantucket. Water pours into the tanker at one end at 1000 gallons per hour, while the polluted water-oil mixture pours out at the other end, also at 1000 gallons per hour. Set up a differential equation to predict the amount $x = x(t)$ of oil in the tanker. (*Hint:* What percentage of the polluted mixture is oil?)

23. *Memorization:* Empirical studies suggest that each person has a certain maximum number of symbols that he can remember at any one time; the rate at which new symbols can be memorized is proportional to the difference between this maximum and the number of symbols already memorized. Set up a differential equation to model the rate at which a person can memorize new symbols.

24. *Parachutes:* A parachutist falling to earth is subject to two forces: her weight, $32m$ (where m is her mass), and the drag created by the parachute, which we take to be proportional to her velocity. Use Newton's second law of motion to write down a differential equation predicting the parachutist's velocity.

25. *Prices:* Assume that the rise in the price $p = p(t)$ of a product is proportional to the difference between the demand $w(t)$ and the supply $s(t)$ (known as the excess demand) and that the demand depends on the price as a first-degree polynomial. Set up a differential equation for the price.

1.2 SEPARATION OF VARIABLES

As we saw in Section 1.1, certain differential equations can be solved by the simple device of "integrating both sides." In this section we explore an extension of this technique to a large class of first-order equations.

Let's try to codify the technique of integrating both sides by investigating the solution of the following simple equation:

$$\frac{dx}{dt} = t + 1.$$

Formally, to integrate both sides we do two things. We put a "dt" to the right of each side of the equation, and then we put an integral sign to the left of each side of the equation:

$$\frac{dx}{dt} \, dt = (t + 1) \, dt$$

$$\int \frac{dx}{dt} \, dt = \int (t + 1) \, dt.$$

The integral on the right is a simple calculus problem:

$$\int (t + 1) \, dt = \frac{t^2}{2} + t + c.$$

The integral on the left presents a slightly more abstract problem, since it involves an unknown function, $x = x(t)$. Fortunately, we know from the definition of the indefinite integral that whatever x may be, integrating its derivative gives us back x. Thus we get

$$x = \frac{t^2}{2} + t + c.$$

The procedure can be sped up if we use "differential" notation. We shall write dx as shorthand notation for $(dx/dt)dt$. We can then describe the process we used as follows. First multiply both sides of the original equation by dt to get

$$dx = (t + 1)\, dt.$$

Then integrate both sides. By definition, the left side integrates to x:

$$\int dx = \int (t + 1)\, dt$$

$$x = \frac{t^2}{2} + t + c.$$

Our manipulation of differential expressions may appear cavalier to readers who did not use differentials in calculus. The notes at the end of this section indicate some dangers as well as some justification of this procedure.

Let's look next at the equation

$$\frac{dx}{dt} = x^2.$$

We multiply both sides by dt to get

$$dx = x^2\, dt.$$

If we try to integrate both sides we run into a problem. Since x is a function of t, we can't integrate the right side unless we have an expression for x^2 in terms of t. Fortunately, we can play a little with the differential form before rushing in with an integral sign. If we divide both sides of the last equation by x^2, we obtain

$$\frac{1}{x^2}\, dx = dt$$

which *is* integrable, in a perfectly reasonable way:

$$\int \frac{1}{x^2}\, dx = \int dt$$

$$-\frac{1}{x} = t + c$$

$$x = \frac{-1}{t + c}.$$

When we divided by x^2 we implicitly assumed $x \neq 0$. It is easy to check that $x = 0$ is also a solution of the equation.

This example teaches us a lesson. Even if the differential form obtained by multiplying both sides of the equation by dt can't be integrated immediately, it may be possible to manipulate the expression to make it formally integrable. Basically, to integrate we must have all "x" expressions accompanied by a dx and all "t" expressions accompanied by a dt. In particular, we cannot have any mixed expressions; we must separate the variables. Finally, if our manipulations involved division by a function, we must check whether setting the function equal to zero yields any solutions that were overlooked in the process.

Let's explore some more examples.

Example 1.2.1

Solve

$$\frac{dx}{dt} = \lambda x$$

where λ is a constant.

We first multiply by dt to obtain the differential form

$$dx = \lambda x \, dt.$$

We next divide by x and integrate:

$$\int \frac{1}{x} \, dx = \int \lambda \, dt$$

$$\ln |x| = \lambda t + c.$$

We can exponentiate both sides of this equation to get

$$|x| = e^{\lambda t + c} = e^c e^{\lambda t}.$$

Since c represents an arbitrary constant, e^c represents an arbitrary positive constant. We can write

$$|x| = k e^{\lambda t}$$

so long as we remember that k is positive. Solving for x gives us an ambiguity

$$x = \pm k e^{\lambda t}.$$

that can be resolved by allowing k to be negative. Finally, noting that we divided by x, we check that $x = 0$ is a solution of our equation. This solution would be included in the preceding formula if we allowed $k = 0$. Thus the formula

$$x = ke^{\lambda t}$$

where k represents an arbitrary constant, constitutes a list of all solutions of the equation; it is the general solution of the equation. Note that when $t = 0$ we have

$$x(0) = ke^0 = k.$$

It might be more suggestive to call this value x_0 and to write the general solution in the form

$$x = x_0 e^{\lambda t}.$$

Example 1.2.2

Solve the equation

$$\frac{dx}{dt} = \frac{t}{t^2 + 1} x$$

which describes the model in Example 1.1.2.

We multiply by dt to obtain the differential form, separate the variables by dividing by x, and integrate:

$$dx = \frac{t}{t^2 + 1} x \, dt$$

$$\int \frac{1}{x} \, dx = \int \frac{t}{t^2 + 1} \, dt$$

$$\ln |x| = \frac{1}{2} \ln (t^2 + 1) + c.$$

We exponentiate the resulting equation and solve for x:

$$x = \pm e^c (t^2 + 1)^{1/2} = k(t^2 + 1)^{1/2}.$$

We check that $x = 0$ is a solution of our equation and note that it is included in our formula for x if we allow $k = 0$. When $t = 0$, $x(0) = k$, so we can write the general solution in the form

$$x = x_0 (t^2 + 1)^{1/2}.$$

Note that even though the growth *rate* is decreasing, the population *size* continues to increase indefinitely.

The method we have used is summarized below. It was first formulated by Johannes Bernoulli in 1694.

SEPARATION OF VARIABLES

Given a first-order o.d.e. of the form

$$A(x,t) \frac{dx}{dt} = B(x,t):$$

1. Multiply by dt to obtain the differential form

$$A(x,t)\, dx = B(x,t)\, dt.$$

2. If by multiplying both sides by expressions in x or t we can put all the x's on the left (with dx) and all the t's on the right (with dt), then the equation is called **separable;** we separate to obtain an equation of the form

$$f(x)\, dx = g(t)\, dt.$$

3. We integrate the separated form

$$\int f(x)\, dx = \int g(t)\, dt$$

to obtain an equation

$$F(x) = G(t) + c.$$

(Note that a constant of integration need only be introduced on one side.)

4. We can now try to solve the preceding equation for x in terms of t.

5. If when separating variables we divided by a function $r(x,t)$, then we check to see whether any solutions of $r(x,t) = 0$ are solutions of our o.d.e. that were overlooked in steps 1 to 4.

In practice, step 4 may be difficult or impossible to perform. In these cases we content ourselves with the equation obtained in step 3.

Notes

1. A warning

Differential notation does not make sense if we have either a $(dx/dt)^2$ or a higher derivative.

Blind formal manipulation of $(dx/dt)^2$ as if it were $(dx)^2/(dt)^2$ leads to nonsensical expressions like $\int A(x)\,(dx)^2$. Of course, sometimes an equation involving higher powers of dx/dt can be solved for dx/dt and then can be separable. Any such algebra should be performed *before* switching to differential notation.

It is a common mistake, but a serious one, to think that higher derivatives can be interpreted as quotients of differentials or powers of differentials. This also leads to total nonsense. *Separating variables is strictly a first-order method*, and even then it doesn't always work.

2. An explanation

Our formal manipulation of dx and dt can be justified quite rigorously. The basis for this justification is the chain rule: if $y = F(x)$ and $x = x(t)$, then

$$\frac{dy}{dt} = \frac{dy}{dx}\frac{dx}{dt} = F'(x)\frac{dx}{dt}.$$

This says that if $F(x)$ is a function whose derivative is $f(x)$, then substituting $x = x(t)$ will give a function of t whose derivative is $f(x(t))x'(t)$. This translates into the integral statement

$$\int f(x)\frac{dx}{dt}\,dt = \int f(x)\,dx.$$

Thus at any stage of our manipulations, a formal equation involving dx on one side and dt on the other can be interpreted rigorously by replacing the dx by $(dx/dt)\,dt$ and putting in integral signs. Our individual manipulations are justified by the fact that integrals of equal functions are equal (up to a constant of integration), and our various multiplications and divisions do not change the equality of the integrands on either side.

An advantage of the differential notation is its symmetry. Sometimes our final answer is more naturally interpreted as an expression for t in terms of x, and the prediction $x = x(t)$ is only given implicitly. The differential notation helps us ignore this problem until after we integrate.

3. On extensions

By making appropriate substitutions, certain equations that are not separable can be transformed into ones that are. An important example of such a substitution is noted in Exercise 15.

EXERCISES

In Exercises 1 to 8, solve the o.d.e. by separation of variables. (You may omit step 4 of the summary if necessary.)

1. $3\dfrac{dx}{dt} = 2x$

2. $3\dfrac{dx}{dt} = 2x + 1$

3. $e^{2t}x' + e^t = 1$

4. $\sin t \, dx + \cos t \, dt = 0$

5. $(x + 1)(t^2 + 1) = t\dfrac{dx}{dt}$

6. $(t^2 - 1)x' + 2x = 0$

7. $(10x^4 + 6)x' = x^5 + 3x + 2$

8. $x'(x - 1)(t^2 + 1)^{1/2} + t^3 + t = x(t^3 + t)$

In Exercises 9 to 14, find the specific solution that satisfies the given initial conditions.

9. $3\dfrac{dx}{dt} + 5x = 0;$ $x(0) = 3$

10. $(t + 1)x' + tx = 0;$ $x(0) = 2$

11. $t^2\dfrac{dx}{dt} = x^2;$ $x(1) = 2$

12. $x^2\dfrac{dx}{dt} = t^2;$ $x(1) = 2$

13. $t^2\dfrac{dx}{dt} = x^2 + 1;$ $x(1) = 0$

14. $e^{-t}\dfrac{dx}{dt} + tx^2 = 0;$ $x(0) = 1$

15. A first-order o.d.e. is said to have **homogeneous coefficients** if it can be written in the form

(*)
$$\frac{dx}{dt} = G\!\left(\frac{x}{t}\right).$$

For an o.d.e. of this type, the substitution $x = vt$ (and $\dfrac{dx}{dt} = t\dfrac{dv}{dt} + v$) changes the equation into one that is separable. For each of the following equations (i) write it in the form (*) and (ii) solve it using the substitution $x = vt$. Express your final answer in terms of x and t.

a. $x^2 t \dfrac{dx}{dt} + t^3 - x^3 = 0$

b. $(x + t)\dfrac{dx}{dt} = x$

c. $t^2 \dfrac{dx}{dt} = x^2 + 4xt$

Exercises 16 to 21 refer back to our models.

16. Solve the differential equation of Exercise 22, Section 1.1, modeling a mixing problem. How much oil is left in the tanker after 10 days (240 hours)?

17. a. Solve the differential equation of Exercise 20, Section 1.1, modeling radioactive decay.
     b. Find an expression for the constant of proportionality in terms of the *half-life* of the substance (the time it takes for half of a given sample to decay).
     c. The half-life of white lead is 22 years. Find a formula predicting the change in size of a sample of white lead over a long period of time.

18. a. Solve the differential equation of Exercise 18, Section 1.1, modeling compound interest.
     b. Calculate the percentage increase in such a deposit over one year. (This is called the effective annual rate.)

19. How long does it take for the population in Example 1.1.1 to double?

20. *Cramming for Exams:* Alex Smart memorizes symbols according to the model in Exercise 23, Section 1.1. Based on his experience of last term, he knows he can remember at most 10,000 symbols and, if t is measured in days, the constant of proportionality is $\gamma = 1/100$. He now knows 6500 symbols. One week from now Alex is going to have a math exam requiring 225 new symbols and an organic chemistry exam requiring 275 new symbols. Will he be able to memorize them all in time?

21. a. Solve the differential equation of Example 1.1.4.
     b. Show that any solution of the equation with $x(0) > 0$ satisfies $\lim\limits_{t \to \infty} x(t) = 10$.

1.3 FIRST-ORDER LINEAR EQUATIONS

In this section we consider a large class of first-order equations that, while not always separable themselves, can be solved by considering a related separable equation. These equations, called linear o.d.e.'s, are important because they and their higher-order cousins encompass many elementary physical models. The basic strategy we use for first-order equations in this section will apply as well to higher-order linear equations in later chapters.

Before formulating a general definition, let's consider the equation that describes the population model in Example 1.1.3,

$$\frac{dx}{dt} = x + t.$$

Note that this equation is *not* separable. Recall that dx/dt was made up of two parts: reproduction contributed x, and immigration contributed t. One way to study such a model is to concentrate first on the isolated behavior of the organism, that is, the growth caused only by reproduction. Once we understand this type of growth, we can try to see how to modify our prediction when immigration is present.

Mathematically, the first step involves solving the equation obtained from the preceding one by dropping the t term:

$$\frac{dx}{dt} = x.$$

This equation is easily separable. Its general solution is

$$x = ke^t$$

where $k = x(0)$.

Now, how does this prediction change when immigration is present? We note that if the immigration all occurred at one time t_1, then for $t \geq t_1$ the formula for x would be just like the one we got in the isolated case, except that the parameter k

would be set higher than the real initial population because of the immigration. By analogy, it might be feasible to use this modified prediction for our actual model as well, provided the parameter k is continuously readjusted to account for the continuing immigration. In mathematical terms we expect the parameter k, which was constant in the isolated model, now to be a function of t. Thus we expect a solution to the original equation to have the form

$$x = k(t)e^t.$$

The problem, of course, is to find the adjusted parameter $k(t)$. We do this by substituting this form of x back into the original differential equation. We calculate

$$\frac{dx}{dt} = k'(t)e^t + k(t)e^t$$

so the equation $dx/dt = x + t$ reads

$$k'(t)e^t + k(t)e^t = k(t)e^t + t$$

or

$$k'(t)e^t = t.$$

Dividing by e^t and integrating by parts (with $u = t$ and $dv = e^{-t}dt$), we have

$$k(t) = \int k'(t)\, dt = \int te^{-t}\, dt = -te^{-t} - e^{-t} + c.$$

Finally, we find the solution of our original equation by multiplying $k(t)$ by e^t:

$$x = k(t)e^t = -t - 1 + ce^t.$$

The method used in this example works just as well if the effect of immigration is a continuous function of t and if the terms dx/dt and x in the equation are multiplied by continuous functions of t. It is customary when dealing with equations of this type to rewrite them by gathering the terms describing the isolated system on one side of the equation and the extra term, involving the external effects (immigration in our example), on the other side.

Definition: *A first-order o.d.e. is **linear** if it can be written in the form*

$$a_1(t)\frac{dx}{dt} + a_0(t)x = E(t)$$

where $a_1(t)$, $a_0(t)$, and $E(t)$ are functions of t (possibly constant).

The first step in the solution of our example was to solve the equation we got by replacing $E(t)$ by zero (this corresponded to ignoring immigration). We give such equations a special name:

Definition: *A linear o.d.e. is* **homogeneous** *if it is of the form*

$$a_1(t) \frac{dx}{dt} + a_0(t)x = 0.$$

Otherwise, the o.d.e. is **nonhomogeneous.**

Our method had three basic steps. First, we used separation of variables to solve the homogeneous equation obtained by replacing $E(t)$ with zero. Second, we replaced the parameter k appearing in the solution of the homogeneous equation with a variable $k(t)$, substituted the resulting form for x into the original equation, and obtained a formula for $k'(t)$. Third, we integrated the formula for $k'(t)$ to obtain $k(t)$ and hence the desired solution x. In solving for $k'(t)$, the terms involving $k(t)$ canceled. If we could verify that this was part of a general pattern, it would shorten the work required by our method.

To work out the general pattern, it is convenient to divide the form in the definition of linearity

$$a_1(t) \frac{dx}{dt} + a_0(t)x = E(t)$$

by $a_1(t)$, to obtain what we shall call the **standard form** of the equation:

(N) $$\frac{dx}{dt} + r(t)x = q(t)$$

(here $r(t) = a_0(t)/a_1(t)$ and $q(t) = E(t)/a_1(t)$).

If we apply our method starting from standard form, we first consider the related homogeneous equation

(H) $$\frac{dx}{dt} + r(t)x = 0.$$

This is separable. Its general solution is of the form

$$x = kh(t)$$

where k is an arbitrary constant and $h(t) = e^{-\int r(t)\,dt}$. Note that $h(t)$ solves (H).

To solve the nonhomogeneous equation, we allow the parameter to vary and look for a solution of the form

$$x = k(t)h(t).$$

We substitute this into the left side of (N), using the product rule to differentiate:

$$\frac{dx}{dt} + r(t)x = k'(t)h(t) + k(t)h'(t) + r(t)k(t)h(t)$$

$$= k'(t)h(t) + k(t)[h'(t) + r(t)h(t)].$$

Since $h(t)$ solves the homogeneous equation, the quantity in brackets is zero. By setting the resulting expression for $dx/dt + r(t)x$ equal to $q(t)$, we obtain the simple formula

(V) $$k'(t)h(t) = q(t).$$

From here, we proceed as in the example: divide by $h(t)$, integrate to get $k(t)$, and multiply by $h(t)$ to find the solution x.

Example 1.3.1

Solve

$$t\frac{dx}{dt} + x = t^3, \qquad 0 < t < +\infty.$$

We first divide by t to put the equation in standard form,

(N) $$\frac{dx}{dt} + \frac{1}{t}x = t^2.$$

Next we solve the related homogeneous equation:

(H) $$\frac{dx}{dt} + \frac{1}{t}x = 0$$

$$\frac{1}{x}\,dx = -\frac{1}{t}\,dt$$

$$\ln|x| = -\ln|t| + c$$

$$= \ln\frac{1}{|t|} + c$$

$$x = \pm\,(e^c/t) = k\frac{1}{t}.$$

We now let the parameter vary and try for a solution of (N) in the form

$$x = k(t)\frac{1}{t}.$$

As a result of the analysis preceding this example, we know that substitution of x into (N) yields the equation

(V) $$k'(t)\frac{1}{t} = t^2.$$

We solve for $k'(t)$ and integrate to find $k(t)$:

$$k'(t) = t^3$$

$$k(t) = \frac{t^4}{4} + c.$$

Finally, we substitute into our formula $x = k(t)/t$ to obtain the general solution of (N):

$$x = \frac{c}{t} + \frac{t^3}{4}.$$

Example 1.3.2
Solve the equation

(N) $$\frac{dx}{dt} - \frac{t}{t^2 + 1}x = e^{-t}(t^2 + 1)^{1/2}.$$

Then find the specific solution that satisfies the initial condition $x(0) = 1$.

This equation is already in standard form, so we begin by solving the related homogeneous equation

(H) $$\frac{dx}{dt} - \frac{t}{t^2 + 1}x = 0.$$

In Example 1.2.2 we found that the general solution of this equation is

$$x = k(t^2 + 1)^{1/2}.$$

We now seek a solution to (N) in the form

$$x = k(t)(t^2 + 1)^{1/2}.$$

The varying parameter $k(t)$ has to satisfy

(V) $$k'(t) (t^2 + 1)^{1/2} = e^{-t}(t^2 + 1)^{1/2}.$$

We solve for $k'(t)$ and integrate to find $k(t)$:

$$k'(t) = e^{-t}$$

$$k(t) = \int e^{-t} \, dt = -e^{-t} + c.$$

Finally, we substitute this into our formula for x to obtain the general solution of (N):

$$x = c(t^2 + 1)^{1/2} - e^{-t}(t^2 + 1)^{1/2}.$$

To find the specific solution satisfying $x(0) = 1$, we substitute $t = 0$ into the general solution:

$$1 = x(0) = c - 1.$$

This gives $c = 2$, so the specific solution is

$$x = 2(t^2 + 1)^{1/2} - e^{-t}(t^2 + 1)^{1/2}.$$

The general pattern for solving first-order linear o.d.e.'s can be summarized in the following streamlined version of our method, which is called **variation of parameters.** It was first formalized by Johannes Bernoulli in 1697.

FIRST-ORDER LINEAR DIFFERENTIAL EQUATIONS

To solve the **linear** first-order o.d.e.

$$a_1(t) \frac{dx}{dt} + a_0(t)x = E(t):$$

0. Divide by $a_1(t)$ to obtain **standard form:**

(N) $$\frac{dx}{dt} + r(t)x = q(t).$$

1. Separate variables in the **related homogeneous equation**

    (H) $$\frac{dx}{dt} + r(t)x = 0$$

    to obtain the general homogeneous solution

    $$x = kh(t).$$

2. We expect solutions of (N) to be of the form

    $$x = k(t)h(t).$$

    Substituting this into (N) yields the equation

    (V) $$k'(t)h(t) = q(t)$$

    which we solve for $k'(t)$.

3. Integrate the formula for $k'(t)$ to obtain $k(t)$, being sure to include the constant of integration:

    $$k(t) = \int k'(t)\, dt + c.$$

    Multiply $k(t)$ by $h(t)$ to obtain x:

    $$x = k(t)h(t).$$

    The final solution always has the form

    $$x = ch(t) + p(t).$$

Notes

1. A warning

The division by $a_1(t)$ to obtain standard form is needed if the simple formula $k'(t)h(t) = q(t)$ for the substitution is to hold. If we don't divide, the result of the substitution on the left will be $a_1(t)k'(t)h(t)$ (see Exercise 23).

2. Points where $a_1(t)$ is zero

Since we divided by $a_1(t)$, we can only be sure our results hold on intervals that do not contain points at which $a_1(t)$ is zero. If we are interested in an interval that contains a point t_1 with $a_1(t_1) = 0$, then we must determine how the solutions for $t < t_1$ and for $t > t_1$ can be pieced together to give solutions valid on the whole interval (see Exercises 25 to 27).

3. An existence theorem

We have seen that we can list all solutions of a linear equation by a formula involving a single constant of integration (assuming $a_1(t)$ is never zero). If we are given numbers t_0 and α, there will be exactly one value of the constant for which the resulting solution has the value $x = \alpha$ when $t = t_0$. It turns out that, even in cases where we are not clever enough to display the solutions, it is possible to say something about the theoretical existence of solutions and their relation to initial conditions. The theorem will appear in various guises in this book. The first version was formulated by Cauchy (1820), and several basic modifications were made by Liouville (1838), Lipschitz (1876), and Picard (1890). We state here a very limited version.

Theorem: Existence and Uniqueness of Solutions. *Suppose that a first-order o.d.e. can be written in the form $dx/dt = f(t, x)$, with both $f(t, x)$ and $\partial f/\partial x$ continuous. Then for any real numbers t_0 and α, there is an open interval containing t_0, $a < t < b$, on which there exists precisely one solution satisfying the initial condition $x(t_0) = \alpha$.*

Note that this theorem guarantees both the existence and the uniqueness of solutions. That is, it tells us

1. There is a function $x = \phi(t)$, defined and continuous for $a < t < b$, that satisfies $x' = f(t, x)$ and $x(t_0) = \alpha$, and
2. If $x = \psi(t)$ is any function, defined and continuous for $a < t < b$, that also satisfies $x' = f(t, x)$ and $x(t_0) = \alpha$, then $\phi(t) = \psi(t)$ for $a < t < b$.

4. Extensions of the method

Some equations of first order that are not linear can be transformed to linear ones, either by dividing out a factor common to all terms or by a clever substitution (see Exercise 18). In addition, sometimes it is possible to reverse the roles of dependent and independent variable to obtain a linear equation (see Exercise 12).

EXERCISES

1. Which of the following o.d.e.'s are linear? Of the linear ones, which are homogeneous? (In answering this question, take x to be the dependent variable; $x' = dx/dt$.)
   a. $x' + x + t = 0$
   b. $x' + xt = 0$
   c. $x't + x = 0$
   d. $x'x + t = 0$
   e. $x' + x + t^2 = 0$
   f. $(x')^2 + x + t = 0$
   g. $x' + x^2 + t = 0$
   h. $x't + xt^2 + 1 = 0$

In Exercises 2 to 12, find the general solution on the indicated interval. (If no interval is given, take it to be $-\infty < t < \infty$.)

2. $\dfrac{dx}{dt} - x = e^t$

3. $3x' + 2x = 1$

4. $\dfrac{dx}{dt} + x = e^{-t}\sin 2t$

5. $\dfrac{dx}{dt} + 3x = t$

6. $t\dfrac{dx}{dt} - 3x = t^3 + 2t, \quad t > 0$

7. $x' + x\tan t = \cos t, \quad -\dfrac{\pi}{2} < t < \dfrac{\pi}{2}$

8. $x' + x \tan t = \sin t$, $-\dfrac{\pi}{2} < t < \dfrac{\pi}{2}$ 9. $dx + (3x + t^2 + 1)\, dt = 0$

10. $tx' + x = \sin t$, $t > 0$

11. $(t^2 - 1)\dfrac{dx}{dt} + x = (t - 1)^{1/2}$, $t > 1$

12. $(t + x)\, dx + dt = 0$ (*Hint:* Solve for t in terms of x.)

In Exercises 13 to 17, solve the initial-value problem.

13. $\dfrac{dx}{dt} - 2x = 8$, $x(0) = 1$ 14. $\dfrac{dx}{dt} + 3x = 8e^t$, $x(0) = 0$

15. $(t^2 + 1)x' - tx = 0$, $x(0) = 3$ 16. $tx' - x = t^3$, $x(1) = 0$

17. $\dfrac{dx}{dt} - tx = t$, $x(0) = \dfrac{1}{2}$

18. An equation of the form $x' + r(t)x = q(t)x^n$ is called a **Bernoulli equation.** When $n = 0$ or 1, the equation is linear. When $n \neq 1$, the substitution $y = x^{1-n}$ yields a linear equation in y. Use this substitution to solve the following o.d.e.'s.
    a. $x' + 3x = x^2$ b. $x' - 4x = x^{-1}$
    c. $\dfrac{dx}{dt} + 6x = 5x^{\frac{1}{2}}$ d. $t\dfrac{dx}{dt} - t^2x = t^2x^{-2}$
    e. $x^2\dfrac{dx}{dt} + x^3 = t$

Exercises 19 to 22 refer back to our models.

19. Find a formula for the velocity of the parachutist in Exercise 24, Section 1.1, given that she starts from rest ($v(0) = 0$).

20. Predict the growth of the savings account in Exercise 19, Section 1.1, starting with an initial deposit of $1000.

21. A pail of water at 70°F is placed outdoors when the temperature is 40°F. The water loses heat according to Newton's law of cooling (Exercise 21, Section 1.1) with constant of proportionality $\gamma = 1/10$. Suppose the temperature outdoors decreases steadily at the rate of 3°F per hour.
    a. What is the temperature outdoors at time t?
    b. Substitute the answer from (a) into the equation of Exercise 21, Section 1.1, to obtain an equation for the temperature of the water. Solve this equation.
    c. What is the temperature of the water after five hours?

22. Assume the price of Blue Mountain coffee behaves as described in Exercise 25, Section 1.1. Assume also that (i) the supply is constant at $s = 500{,}000$ pounds per year, (ii) if the coffee were free, the demand would be ten million pounds per year, (iii) if the price were to hit $10 per pound, no one would buy it, (iv) the price rises 5¢ for every 100,000 pounds per year of excess demand, and (v) the present price is $4.50 per pound. Predict the price after six months and after one year. (*Caution:* Watch your units.)

Exercises 23 to 29 involve some more theoretical considerations.

23. Show that if we perform variation of parameters for the first-order o.d.e. $a_1(t)x' + a_0(t)x = E(t)$ without first putting it in standard form, then the substitution $x = k(t)h(t)$,

where $h(t)$ solves the related homogeneous equation, leads to the equation $a_1(t)k'(t)h(t) = E(t)$.

24. Try the "variation of parameters" procedure on the nonlinear equation $x' + x^2 = t$ as follows: (a) find the general solution of $x' + x^2 = 0$ by separating variables, then (b) replace the parameter with a variable and substitute back. *What goes wrong?*

25. a. Show that the general solution on the interval $0 < t$ or on the interval $t < 0$ of the equation

    (*)
    $$tx' - 2x = 0$$

    is $x = ct^2$.

    b. Show that for any choice of c_1 and c_2,

    $$x = \begin{cases} c_1t^2 \text{ for } t \le 0 \\ c_2t^2 \text{ for } t > 0 \end{cases}$$

    is a continuous solution of (*) on $-\infty < t < \infty$. Note that when $c_1 \ne c_2$, this solution is *not* of the form $x = ct^2$ for any constant c.

26. Show that, when $a_1(t) = 0$, the existence and uniqueness theorem of Note 3 can break down, by proving that *every* solution of $tx' - 2x = 0$ must satisfy $x(0) = 0$.

27. Show that *no* solution of $t^3x' - 2t^2x = 1$ can be continuous at $t = 0$.

28. Show that all homogeneous linear o.d.e.'s of first order are separable. Which nonhomogeneous linear o.d.e.'s (of first order) are separable?

29. In the model solved at the beginning of this section, the growth *without* immigration is ke^t, while the growth *with* immigration is $ce^t - t - 1$. One expects immigration to increase the population, but here we are subtracting the positive quantity $t + 1$ from the growth without immigration. Explain this apparent contradiction. [*Hint:* Interpret k and c in terms of $x(0)$.]

30. *Another Approach.*
    a. Show that for any linear first-order o.d.e. in standard form

    (N)
    $$\frac{dx}{dt} + r(t)x = q(t)$$

    multiplication by the function

    $$\rho(t) = e^{\int r(t)\, dt}$$

    leads to an equation that can be rewritten in the form

    $$\frac{d}{dt}\left(e^{\int r(t)\, dt}x\right) = e^{\int r(t)\, dt}q(t).$$

    b. Integrate this equation and solve for x to find

    $$x = \left(\int e^{\int r(t)\, dt}q(t)\, dt + c\right)e^{-\int r(t)\, dt}.$$

c. Check that the integrals in this expression are the same integrals that arise in solving (N) by the method described in the text.

1.4 GRAPHING SOLUTIONS (Optional)

Our main approach to differential equations will be the one followed in the preceding two sections. We try to express a solution by a formula involving familiar functions (polynomials, exponentials, logarithms, and trigonometric functions). This is known as "closed-form solution." Unfortunately, many equations cannot be solved in closed form. We are forced to represent solutions of these equations in different ways.

Aside from a formula, there are three important ways of representing a function: a convergent power series, a table of values, and a graph. In this section we shall see how to sketch graphs of solutions to a first-order o.d.e., using only the information provided directly by the equation.

Let's see how this is done for the specific example

$$\frac{dx}{dt} = x.$$

What does the equation tell us about the graph of x as a function of t? The left-hand side, dx/dt, is the slope of the curve. Thus at any point on the graph of a solution, the slope of the curve is numerically equal to x, the height of the point above the t-axis. Conversely, any curve that satisfies this condition at all its points is the graph of a solution.

To represent this information geometrically, we choose some points (t,x) of the t-x plane and draw little line segments with slope $m = x$ to represent the tangents to solution curves at these points. In this example, solution curves will all cross a given horizontal line $x = k$ at the same slope, $m = k$. Thus it is convenient to pick a few values of k and then to draw many segments, all crossing $x = k$ with slope $m = k$. In Figure 1.2(a) we have plotted segments for the values $k = -2, -3/2, -1, -1/2, 0, 1/2, 1, 3/2,$ and 2.

In Figure 1.2(b) we have plotted segments for values of k in increments of $1/5$ rather than $1/2$. At this stage, the segments seem to form curves. The curves fitted to these segments (Figure 1.2(c)) give us a family of graphs of solutions to the original equation.

We have succeeded in graphing solutions without first solving the equation. Of course, our graphs are only approximately correct. Their accuracy depends in part on the number of "tangent" segments we have drawn. The more densely we scatter these segments, the more accurate our final graphs are likely to be. Unfortunately, the more segments we draw, the more likely we are to get writer's cramp (Figure 1.2(b), for example, required over 400 segments). Computers with plotting devices can be very helpful in drawing the segments.

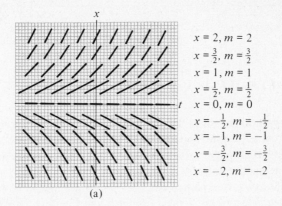

$$x = 2, m = 2$$
$$x = \frac{3}{2}, m = \frac{3}{2}$$
$$x = 1, m = 1$$
$$x = \frac{1}{2}, m = \frac{1}{2}$$
$$x = 0, m = 0$$
$$x = -\frac{1}{2}, m = -\frac{1}{2}$$
$$x = -1, m = -1$$
$$x = -\frac{3}{2}, m = -\frac{3}{2}$$
$$x = -2, m = -2$$

(a)

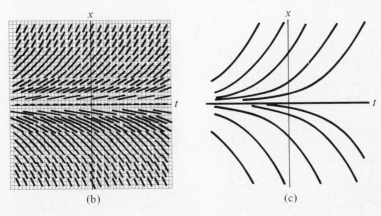

(b)   (c)

FIGURE 1.2 $x' = x$

In general, when drawing segments by hand it is useful to locate curves along which $m = dx/dt$ is constant. A single slope will serve for all the segments along such a curve. The curves along which dx/dt is constant are called **isoclines** of the equation. In the preceding example, the isoclines were the horizontal lines $x = k$. We will see in Example 1.4.3 that isoclines need not be straight lines.

It is important to remember that even when an isocline is a straight line, its slope is in general different from the slope m of the segments crossing it. In the preceding example, every isocline had slope 0 regardless of the value of m. However, it may happen that a *particular* isocline has slope coinciding with m at all its points (this occurred above for $k = 0$). The segments we draw on this isocline, instead of actually crossing it, will lie along it. This isocline will itself be a solution.

The isocline for $k = 0$ is of special interest even when it does not coincide with a solution curve. The segments crossing this isocline will always be horizontal. Any maxima or minima of solutions to the equation will lie on this isocline.

Example 1.4.1

$$\frac{dx}{dt} = t + 1$$

The isoclines in this example are the lines $t + 1 = k$. The segments crossing the isocline $t + 1 = k$ all have slope $m = k$. Since the isoclines are all vertical, no isocline can itself be a solution curve. The line $t = -1$ is the isocline for $k = 0$; segments cross this line horizontally.

In Figure 1.3(a) we have plotted isoclines for values of k in increments of $1/5$ and sketched in segments crossing these isoclines. In Figure 1.3(b) we have fitted solution curves to these segments.

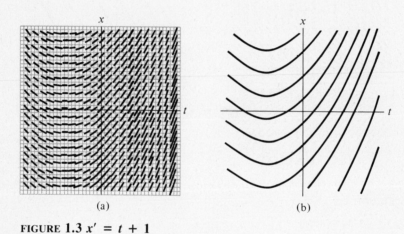

(a)   (b)

FIGURE 1.3 $x' = t + 1$

Example 1.4.2

$$\frac{dx}{dt} = x + t$$

The isoclines are the lines $x + t = k$. The segments crossing $x + t = k$ have slope $m = k$. Since every isocline is a line of slope -1, the particular isocline for $k = -1$ has slope coinciding with m at every point. Therefore $x + t = -1$ is a solution of the equation. The isocline for $k = 0$ is the line $x = -t$. We have plotted segments in Figure 1.4(a) and fitted curves to them in Figure 1.4(b).

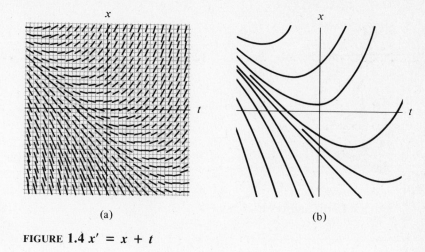

(a)   (b)

FIGURE 1.4 $x' = x + t$

Example 1.4.3

$$\frac{dx}{dt} = x^2 + t$$

The isoclines in this example are the parabolas $x^2 + t = k$. Segments are horizontal along the isocline $x^2 = -t$. We have sketched some isoclines and segments in Figure 1.5(a and b) and fitted some solution curves to these segments in Figure 1.5(c).

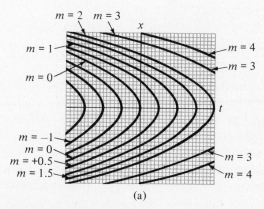

(a)

FIGURE 1.5(a) $x' = x^2 + t$

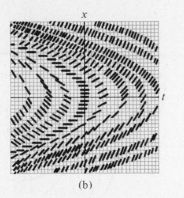

(b)

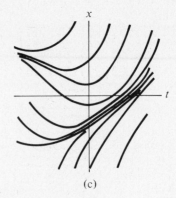

(c)

FIGURE 1.5(b) and (c) $x' = x^2 + t$

Our graphs can be used to obtain qualitative information about solutions of a given equation. Let's see, for example, what Figure 1.2(c) tells us about solutions of $dx/dt = x$. Solutions move away from the t-axis (i.e., $|x|$ increases) as t increases. The t-axis ($x = 0$) is itself a solution. No solution changes sign. We notice that the solution curves fill up the t-x plane and do not cross each other. In particular, each point on the x-axis ($t = 0$) is crossed by some solution, and no two solutions cross the x-axis at the same point. This last observation is a geometric restatement of the Existence and Uniqueness Theorem (Note 3, Section 1.3) for this equation.

From Figure 1.4(b) we can see that solutions of $dx/dt = x + t$ that lie above the line $x + t = -1$ approach that line as t decreases, have a minimum at $x = -t$ (why?), and rise to $+\infty$ as $t \to \infty$. Solutions below the line $x + t = -1$ also approach this line as t decreases; as t increases, x decreases at a progressively faster rate and tends to $-\infty$ as $t \to \infty$.

Let's summarize this technique.

GRAPHING SOLUTIONS OF $\dfrac{dx}{dt} = f(t,x)$

1. Try to sketch **isoclines** $f(t, x) = k$ for a few specific values of k. The value $k = 0$ is of special interest.

2. For each specific value of k, draw lots of line segments of slope $m = k$ along the corresponding isocline $f(t,x) = k$. These segments will be horizontal for $k = 0$.

3. Find any values of k for which the isocline $f(t,x) = k$ is a straight line of slope k. Any isocline of this type is itself a solution curve. Draw it!

4. Fit solution curves tangent to the segments you have drawn. Note that any maxima or minima occur along the isocline with $k = 0$.

Note

On Approximations

In situations where quantitative information about solutions is needed, graphs obtained by this method can be used to yield approximations. In practical situations the differential equations are based on measurements, which in turn are subject to experimental error. Thus, practically speaking, a reasonably good approximate solution to an o.d.e. should be all we need.

Of course, an approximation is not very useful unless we have a good idea of its accuracy. The analysis of approximation schemes from the point of view of accuracy is a delicate and difficult subject, which is beyond the scope of this book.

EXERCISES

Plot graphs of solutions for the following o.d.e.'s in the region $-2 \leq t \leq 2$, $-2 \leq x \leq 2$. Use at least five isoclines, and sketch at least three different solutions.

1. $x' = 3x + 2t$

2. $x' = x + 3t$

3. $\dfrac{dx}{dt} = xt$

4. $\dfrac{dx}{dt} = x + t^2$

5. $\dfrac{dx}{dt} = x - t^2$

6. $\dfrac{dx}{dt} = x^2 + t^2$

7. $tx' = x$

8. $tx' = 1$ (Note that $x = \ln t$ is by definition the solution of this equation satisfying $x(1) = 0$. Be sure to include the graph of this solution.)

1.5 EXACT DIFFERENTIAL EQUATIONS (Optional)

The solutions of the o.d.e.'s we dealt with in Sections 1.2 and 1.3 were given explicitly by formulas for x in terms of t or implicitly by formulas of the form $F(x) = G(t) + c$. In this section we deal with equations whose solutions are given by formulas of the form $F(t,x) = c$.

We first note that if x is a function of t, then so is $F(t,x)$. We can use the chain rule for partial derivatives to calculate dF/dt:

$$\frac{dF}{dt} = \frac{\partial F}{\partial t}\frac{dt}{dt} + \frac{\partial F}{\partial x}\frac{dx}{dt} = \frac{\partial F}{\partial t} + \frac{\partial F}{\partial x}\frac{dx}{dt}.$$

Then

$$dF = \frac{\partial F}{\partial t}\,dt + \frac{\partial F}{\partial x}\,dx.$$

Example 1.5.1

If $F(t,x) = t^3x^2 + e^{-t} - x^5$, then

$$\frac{\partial F}{\partial t} = 3t^2x^2 - e^{-t} \quad \text{and} \quad \frac{\partial F}{\partial x} = 2t^3x - 5x^4$$

so that

$$dF = (3t^2x^2 - e^{-t}) \, dt + (2t^3x - 5x^4) \, dx.$$

Suppose now that we are given a differential equation that can be written in the form

$$M(t,x) \, dt + N(t,x) \, dx = 0.$$

We will say the equation is **exact** if there is a function $F(t,x)$ so that

$$\frac{\partial F}{\partial t} = M \quad \text{and} \quad \frac{\partial F}{\partial x} = N.$$

If the equation is exact, it can be rewritten as

$$dF = 0.$$

The solution of this equation is

$$F = c.$$

In order to make effective use of our observation about exact equations, we need a simple way of recognizing when an equation is exact and, if it is, a way of finding $F(t,x)$. Note that if $M \, dt + N \, dx = 0$ is exact, then

$$\frac{\partial M}{\partial x} = \frac{\partial}{\partial x}\left(\frac{\partial F}{\partial t}\right) = \frac{\partial^2 F}{\partial x \partial t} = \frac{\partial^2 F}{\partial t \partial x} = \frac{\partial}{\partial t}\left(\frac{\partial F}{\partial x}\right) = \frac{\partial N}{\partial t}.$$

Suppose, conversely, that

$$\frac{\partial M}{\partial x} = \frac{\partial N}{\partial t}.$$

Let's look for a function $F(t,x)$ with

$$\frac{\partial F}{\partial t} = M \quad \text{and} \quad \frac{\partial F}{\partial x} = N.$$

We can integrate the first of these equations with respect to t, holding x constant (since x is constant, we have to allow the "constant" of integration to be a function of x):

$$F = \int M \, \partial t + g(x).$$

Then

$$\frac{\partial F}{\partial x} = \frac{\partial}{\partial x} \int M \, \partial t + \frac{dg(x)}{dx}.$$

We want this partial derivative to equal N:

$$\frac{\partial}{\partial x} \int M \, \partial t + \frac{dg(x)}{dx} = N.$$

Then

$$\frac{dg(x)}{dx} = N - \frac{\partial}{\partial x} \int M \, \partial t.$$

This last equation makes sense only if the right side is a function of x alone, that is, if the partial derivative with respect to t of the right side is 0. This is the case here:

$$\frac{\partial}{\partial t}\left(N - \frac{\partial}{\partial x} \int M \, \partial t \right) = \frac{\partial N}{\partial t} - \frac{\partial^2}{\partial t \partial x} \int M \, \partial t$$

$$= \frac{\partial N}{\partial t} - \frac{\partial^2}{\partial x \partial t} \int M \, \partial t = \frac{\partial N}{\partial t} - \frac{\partial M}{\partial x} = 0.$$

We can integrate our formula for dg/dx to find g and substitute into our formula for F to find F.

We have obtained a simple test for exactness.

Fact: *The equation*

$$M \, dt + N \, dx = 0$$

is exact if and only if

$$\frac{\partial M}{\partial x} = \frac{\partial N}{\partial t}.$$

What's more, our proof includes a method for finding F!

Example 1.5.2

Solve

$$(3t^2 \sin^2 x)\, dt + (2t^3 \sin x \cos x - 2e^{2x})\, dx = 0.$$

In this example,

$$M = 3t^2 \sin^2 x \qquad \text{and} \qquad N = 2t^3 \sin x \cos x - 2e^{2x}.$$

The differential equation is exact since

$$\frac{\partial M}{\partial x} = \frac{\partial}{\partial x}(3t^2 \sin^2 x) = 6t^2 \sin x \cos x = \frac{\partial}{\partial t}(2t^3 \sin x \cos x - 2e^{2x}) = \frac{\partial N}{\partial t}.$$

If we find a function $F(t,x)$ with

$$\frac{\partial F}{\partial t} = 3t^2 \sin^2 x \qquad \text{and} \qquad \frac{\partial F}{\partial x} = 2t^3 \sin x \cos x - 2e^{2x}$$

then the solution of the o.d.e. will be $F = c$.

We first integrate the equation $\partial F/\partial t = 3t^2 \sin^2 x$ with respect to t, holding x constant:

$$F = t^3 \sin^2 x + g(x).$$

We next set the partial with respect to x of the resulting expression for F equal to N:

$$\frac{\partial F}{\partial x} = 2t^3 \sin x \cos x + \frac{dg(x)}{dx} = 2t^3 \sin x \cos x - 2e^{2x}.$$

We solve for dg/dx:

$$\frac{dg(x)}{dx} = -2e^{2x}.$$

We integrate this equation to find $g(x)$ (since any function $g(x)$ that satisfies this equation will yield a function $F(t,x)$ with the required properties, we needn't include a constant of integration):

$$g(x) = -e^{2x}.$$

We substitute this value into our expression for F:

$$F = t^3 \sin^2 x - e^{2x}.$$

The solution of our o.d.e. is

$$t^3 \sin^2 x - e^{2x} = c.$$

Of course, equations are not always given in the form $Mdt + Ndx = 0$. In order to use the methods of this section, one must first rewrite the given equation in the proper form.

Example 1.5.3
Solve

$$(2t^3x - 5x^4) \frac{dx}{dt} = -(3t^2x^2 + 1).$$

We can rewrite the equation in the form $M\,dt + N\,dx = 0$:

$$(3t^2x^2 + 1)\,dt + (2t^3x - 5x^4)\,dx = 0.$$

This equation is exact since

$$\frac{\partial}{\partial x}(3t^2x^2 + 1) = 6t^2x = \frac{\partial}{\partial t}(2t^3x - 5x^4).$$

If we find a function $F(t,x)$ with

$$\frac{\partial F}{\partial t} = 3t^2x^2 + 1 \qquad \text{and} \qquad \frac{\partial F}{\partial x} = 2t^3x - 5x^4,$$

then the solution of the o.d.e. will be $F = c$.

We first integrate the equation $\partial F/\partial t = 3t^2x^2 + 1$ with respect to t, holding x constant:

$$F = t^3x^2 + t + g(x).$$

We set the partial with respect to x of the resulting expression for F equal to N:

$$2t^3x + \frac{dg(x)}{dx} = 2t^3x - 5x^4.$$

We solve for dg/dx and integrate to find $g(x)$:

$$\frac{dg(x)}{dx} = -5x^4$$

$$g(x) = -x^5.$$

We substitute this value into our expression for F:

$$F = t^3x^2 + t - x^5.$$

The solution of our o.d.e. is

$$t^3x^2 + t - x^5 = c.$$

In the next example we see what goes wrong if we attempt to find F for an equation that is not exact.

Example 1.5.4

The equation

$$2x \, dt + (2 + x)t \, dx = 0$$

is not exact since

$$\frac{\partial M}{\partial x} = 2 \quad \text{and} \quad \frac{\partial N}{\partial t} = 2 + x \neq 2.$$

Let's still try to find a function F with

$$\frac{\partial F}{\partial t} = 2x \quad \text{and} \quad \frac{\partial F}{\partial x} = (2 + x)t.$$

If we integrate the formula for $\partial F/\partial t$ with respect to t, holding x constant, we get

$$F = 2xt + g(x).$$

If we set the partial with respect to x of this expression for F equal to N, we get

$$2t + \frac{dg(x)}{dx} = (2 + x)t.$$

Then

$$\frac{dg(x)}{dx} = xt.$$

This equation is impossible to solve; the left side is a function of x only, whereas the right side is a function of both x and t.

Let's summarize.

EXACT DIFFERENTIAL EQUATIONS

A differential equation of the form

(E) $$M\,dt + N\,dx = 0$$

is **exact** if there is a function $F(t,x)$ so that

$$\frac{\partial F}{\partial t} = M \quad \text{and} \quad \frac{\partial F}{\partial x} = N.$$

Test for Exactness

The o.d.e. (E) is exact if and only if

$$\frac{\partial M}{\partial x} = \frac{\partial N}{\partial t}.$$

Solving Exact Equations

If the equation (E) is exact, we find the solutions as follows.

1. Integrate the equation $\partial F/\partial t = M$ with respect to t, holding x constant (we must allow a function of x as the "constant" of integration):

$$F = \int M \, \partial t + g(x).$$

2. Set the partial with respect to x of the resulting expression for F equal to N:

$$\frac{\partial F}{\partial x} = \frac{\partial}{\partial x} \int M \, \partial t + \frac{dg(x)}{dx} = N.$$

3. Solve for dg/dx:

$$\frac{dg(x)}{dx} = N - \frac{\partial}{\partial x} \int M \, \partial t.$$

4. Integrate the formula for dg/dx to find g. Since any function satisfying this formula yields a function F with the required properties, we needn't include a constant of integration:

$$g(x) = \int \left(N - \frac{\partial}{\partial x} \int M \, \partial t \right) dx.$$

5. Obtain F by substituting the value for g into the formula for F found in step 1.
6. The solution of the o.d.e. (E) is

$$F = c.$$

Notes
1. A technicality

We have used theorems from multidimensional calculus without worrying about whether the hypotheses of the theorems were satisfied. (For example, we used the "fact" that $\partial^2 F/\partial x \partial t = \partial^2 F/\partial t \partial x$.) The precise statement of the conditions under which our method works involves conditions on the region of the t-x plane in which we are working. A region of the t-x plane is said to be *simply connected* if "it has no holes." That is, a region is simply connected if it contains the interior of every simple closed curve lying in it. The region in Figure 1.6(a) is simply connected, whereas the region in Figure 1.6(b) is not. The method we have described works provided that M and N have continuous first partial derivatives in a simply connected region.

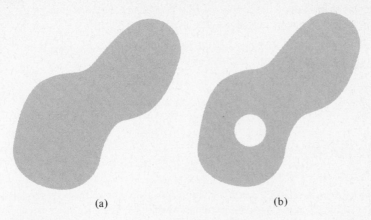

(a)   (b)

FIGURE 1.6

2. On finding F

The integration in step 1 is sometimes hard. In these cases it may be easier to work with the equation $\partial F/\partial x = N$ first. We can integrate this equation with respect to x, holding t constant (the "constant" of integration will be a function of t):

$$F = \int N \, \partial x + k(t).$$

Substitution of this expression into the equation $\partial F/\partial t = M$ will yield an equation for dk/dt. If the o.d.e. is exact, we can solve this equation for $k(t)$ and thereby find F.

EXERCISES

Test the equations in Exercises 1 to 11 for exactness, and solve those that are exact.

1. $t \, dt + x \, dx = 0$ 2. $x \, dt + t \, dx = 0$

3. $x \, dt - t \, dx = 0$ 4. $(2x^2 + 2t + 1) \, dt + (4x^3 + 4tx) \, dx = 0$

5. $(2xt^2 + 2x) \, dt + (2tx^2 + 2t) \, dx = 0$

6. $(t^3 + x \sin t) \, dt + (2x - \cos t) \, dx = 0$

7. $xe^{tx} \, dt + e^{tx} \, dx = 0$

8. $xe^{tx} \, dt + te^{tx} \, dx = 0$

9. $x^2 \cos t \, dt + (xt \cos tx + \sin tx) \, dx = 0$

10. $(2x \sin t \cos t + e^x + t) \, dt + (\sin^2 t + te^x) \, dx = 0$

11. $\dfrac{x(1 - t^2)}{(t^2 + 1)^2} \, dt + \dfrac{t}{t^2 + 1} \, dx = 0$

If an equation is not exact, it may be possible to multiply by a function $\rho(t,x)$ (called an **integrating factor**) so that the resulting equation $\rho M \, dt + \rho N \, dx = 0$ is exact.

12. Each of the following equations has an integrating factor of the form $\rho = t^\alpha$. Find α, multiply by ρ, and solve the resulting exact equation.

   a. $xt^2 \, dt + t^3 \, dx = 0$

   b. $(6x + 5t) \, dt + 2t \, dx = 0$

   c. $\dfrac{x}{2} \, dt + (t + t^{1/2} \sin x) \, dx = 0$

   d. $-x \, dt + t \, dx = 0$

REVIEW PROBLEMS

The methods we have discussed are just a few of the many known methods for solving first-order o.d.e.'s. For Exercises 1 to 15, determine whether the given equation can be solved by any of these methods. If so, (i) find the general solution and (ii) find the specific solution satisfying the given initial condition.

1. $(t^2 - 1) \dfrac{dx}{dt} + x = 0, t > 1;$ $x(2) = 1$

2. $x' = x + t^2;$ $x(0) = 1$

3. $x' = x + x^2;$ $x(0) = 1$

4. $x' + 5x = t;$ $x(0) = 0$

5. $t x' = -x, t > 0;$ $x(2) = 1$

6. $t x' + 4x = t^3, t > 0;$ $x(2) = 1$

7. $3t^2 x x' = t x + x + t + 1, t > 0;$ $x(2) = 1$

8. $\cos t \dfrac{dx}{dt} = x \sin t, -\dfrac{\pi}{2} < t < \dfrac{\pi}{2};$ $x(0) = 1$

9. $x' \cos t + x \sin t = \cos t, -\dfrac{\pi}{2} < t < \dfrac{\pi}{2};$ $x(0) = 1$

10. $3t \, dx + 4xt \, dt = 0, t > 0;$ $x(3) = 1$

11. $(t + x) \, dx + (x - t) \, dt = 0;$ $x(1) = 1$

12. $(e^t - 1) \, dx + e^{2t} \, dt = 0, t > 0;$ $x(\ln 2) = 1$

13. $t^2 x' - (t + 1)x + t = 0, t > 0;$ $x(1) = 1$

14. $x' + 5x = tx^3;$ $x(0) = 1$

15. $(x^2 t + t^2 x) \dfrac{dx}{dt} = (x^3 + t^3), t > 0;$ $x(2) = 1$

16. *Tea:* Boiling water (212°F) is poured into a teapot and cools according to Newton's law of cooling (Exercise 21, Section 1.1). The temperature of the room is 65°F. After 2 minutes, the tea has a temperature of 191°F.

   a. What will be the temperature of the tea 16 minutes after the water is added?

   b. When tea reaches a temperature below 100°F, it is tepid and tastes bad. How long can we talk to our guests before pouring the tea? (A calculator may be useful here.)

17. *House Heating:* (A calculator may be useful here.) Suppose the heating system of a house could, if the house were perfectly insulated, raise the temperature 10°F per hour. However,

the house is not perfectly insulated. It loses heat according to Newton's law of cooling (Exercise 21, Section 1.1).

a. If the heat is turned off when the house is at 70°F and the temperature outside is 30°F, then the house temperature reaches 60°F after 2 hours. How long does it take, from the time the heat is turned off, for the house to reach 50°F?

b. When the temperature inside reaches 50°F, the heat is turned back on. What will the temperature be 2 hours later?

c. When the temperature inside reaches 70°F, the temperature outside drops steadily at 15°F per hour (from the initial temperature of 30°). If the heating system remains on, what will the temperature of the house be when the temperature outside reaches 0°F?

18. *A Space Launch:* An object of mass m, shot upward so that it travels along a straight line through the center of the Earth, is subject to a gravitational force $F = -mgR^2/x^2$, where g is the gravitational constant, R is the radius of the Earth, and x is the distance of the object from the center of the Earth. Assume that this is the only force acting on the object (we are ignoring air resistance, the gravitational pull of the moon, and so on).

a. Using Newton's second law of motion, set up a differential equation for the distance x of the object from the center of the earth.

b. Use the fact that $a = d^2x/dt^2 = dv/dt = (dx/dt)(dv/dx) = v(dv/dx)$ to obtain a differential equation for velocity v in terms of x.

c. Solve the equation in (b). Express the arbitrary constant in terms of the initial velocity v_0, using $x(0) = R$.

d. If the velocity is 0 at some time, then the object falls back to Earth. If the velocity stays positive—that is, if it is never 0—then the object escapes into space. Show that the object escapes if and only if $v_0^2 \geq 2gR$. (*Hint:* What would x have to be when $v = 0$?)

e. Find a formula for x at time t, if the initial velocity is $v_0 = \sqrt{2gR}$.

Linear Differential Equations

2.1 SOME SPRING MODELS

In this section we consider some elementary models, involving damped spring systems, that lead to "linear" o.d.e.'s of order higher than one. The basic model for a spring is given by **Hooke's law** (Robert Hooke was a contemporary of Newton). This states that *a spring stretched or squeezed from its natural length L to length L + x will exert a force − kx, where k is a positive constant depending on the makeup of the spring*. The constant k is called the **spring constant.**

Example 2.1.1   Undamped Springs
Suppose a mass of m grams moves along a horizontal track but is attached to a wall at one end of the track by a spring with natural length L cm and spring constant k dynes/cm. In modeling the motion, it is most convenient to specify the position of the mass, *not* by the distance from the end of the track, but rather by the deviation x of the length of the spring from equilibrium (see Figure 2.1). Hooke's law tells us that at position x (that is, when the mass is $L + x$ from the wall), the spring exerts a force

$$F_{\text{spr}} = - kx.$$

Since $F = ma = m\, d^2x/dt^2$, we can rewrite this second-order o.d.e. in the form

$$m \frac{d^2x}{dt^2} + kx = 0.$$

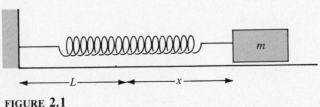

FIGURE 2.1

43

Example 2.1.2 Damped Springs

Let's now take friction into account in the preceding model. Friction depends on the velocity and opposes its direction. The simplest model for friction is **viscous damping**, in which the magnitude of the force is proportional to the velocity:

$$F_{\text{fric}} = -b \frac{dx}{dt}.$$

Here, the (positive) constant of proportionality b is called the **damping coefficient.**

When we take into account both the spring force and viscous damping, the total force on the mass is

$$F_{\text{tot}} = F_{\text{spr}} + F_{\text{fric}} = -kx - b \frac{dx}{dt}.$$

Newton's second law gives us the second-order o.d.e.

$$m \frac{d^2x}{dt^2} = -kx - b \frac{dx}{dt}$$

or

$$m \frac{d^2x}{dt^2} + b \frac{dx}{dt} + kx = 0.$$

Example 2.1.3 External Forces

The spring and damping forces in the preceding examples were internal, in the sense that they depended only on the position and velocity of the mass in the system. When an outside force is imposed on a system of this type, we add an "external" force term

$$F_{\text{ext}} = E(t).$$

In the presence of internal (spring and damping) and external forces, the total force is

$$F_{\text{tot}} = F_{\text{spr}} + F_{\text{fric}} + F_{\text{ext}}.$$

Newton's second law leads to the second-order equation

$$m \frac{d^2x}{dt^2} + b \frac{dx}{dt} + kx = E(t).$$

Two common forms of the external force term are constants $E(t) = E$ (e.g., gravity acting on a system hung vertically) and oscillating forces, such as the sinusoidal term $E(t) = E \sin \beta t$ with amplitude E and period $2\pi/\beta$.

Example 2.1.4   A Coupled Spring System

Suppose two masses of m_1 and m_2 grams, respectively, move along a frictionless horizontal track. Assume they are attached to each other by a spring with natural length L_2 and spring constant k_2, while the first mass is attached to a wall at the end of the track by a spring with constant k_1 and natural length L_1. A description of the configuration of this system requires two positions, one for each mass. When both springs are at equilibrium, the first mass is L_1 cm from the wall and the second is $L_1 + L_2$ cm from the wall. Let's specify the positions in general by x_1 and x_2 when the first mass is $L_1 + x_1$ cm from the wall and the second is $L_1 + L_2 + x_2$ cm from the wall (see Figure 2.2).

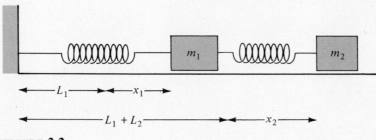

FIGURE 2.2

To apply Newton's second law, we must analyze the forces on each mass. The second mass is acted on only by the connecting spring. The length of this spring is the distance between the two masses,

$$(L_1 + L_2 + x_2) - (L_1 + x_1) = L_2 + (x_2 - x_1),$$

which represents a deviation of $(x_2 - x_1)$ cm from its natural length. Thus the force on the second mass is

$$F_2 = -k_2(x_2 - x_1).$$

The situation of the first mass is slightly more complicated. The spring attaching it to the wall is pulled to a length of $L_1 + x_1$ cm, so it exerts a force $F_1 = -k_1 x_1$. However, the connecting spring also acts on the first mass, pulling with a force equal in magnitude and opposite in direction to the force it exerts on the second mass, that is, $-F_2$. Thus the total force on the first mass is

$$F_1 - F_2 = -k_1 x_1 + k_2(x_2 - x_1) = -(k_1 + k_2)x_1 + k_2 x_2.$$

Applying Newton's law to each mass in turn leads to two equations

$$m_1 \frac{d^2x_1}{dt^2} = -(k_1 + k_2)x_1 + k_2x_2$$

$$m_2 \frac{d^2x_2}{dt^2} = k_2x_1 - k_2x_2.$$

Since each equation involves both x_1 and x_2, we cannot hope to solve either equation separately. Note, however, that we can solve the first equation for x_2 in terms of x_1 and d^2x_1/dt^2:

$$x_2 = \frac{m_1}{k_2} \frac{d^2x_1}{dt^2} + \frac{(k_1 + k_2)}{k_2} x_1.$$

We can differentiate this equation twice, to get

$$\frac{d^2x_2}{dt^2} = \frac{m_1}{k_2} \frac{d^4x_1}{dt^4} + \frac{(k_1 + k_2)}{k_2} \frac{d^2x_1}{dt^2}.$$

We can now substitute our expressions for x_2 and d^2x_2/dt^2 into the second equation to get

$$\frac{m_2m_1}{k_2} \frac{d^4x_1}{dt^4} + \frac{m_2(k_1 + k_2)}{k_2} \frac{d^2x_1}{dt^2} = k_2x_1 - m_1 \frac{d^2x_1}{dt^2} - (k_1 + k_2)x_1$$

or, multiplying by k_2 and collecting terms,

$$m_2m_1 \frac{d^4x_1}{dt^4} + (m_2k_1 + m_2k_2 + m_1k_2) \frac{d^2x_1}{dt^2} + k_2k_1x_1 = 0.$$

This fourth-order o.d.e. can be solved for x_1. We can then obtain x_2 by substitution into our expression for x_2 in terms of x_1 and d^2x_1/dt^2.

EXERCISES

1. *A Floating Box:* A cube of wood 1 foot on each side, weighing 16 lb (so that its mass is 16/32 slugs), floats in the Charles River with the bottom of the box $x < 1$ foot below the surface of the water. According to the *Principle of Archimedes*, the water buoys up the

box by a force equal to the weight of the water it displaces. Given that the density of water is 62.5 lb/ft³, write an o.d.e. for x

a. assuming no frictional resistance and

b. assuming resistance with damping constant b lb/(ft/sec).

Be sure to take into account both the gravitational force ($=$ weight) downward and the buoyant force upward.

2. *Supply and Demand:* Assume, as in Exercise 25, Section 1.1, that the price of a product increases at a rate proportional to the excess demand (demand minus supply) and that demand in turn depends on price as a first-degree polynomial. Assume further that the supply of this product is controlled so that it increases at a rate proportional to the price.

   a. Differentiate the equation of Exercise 25, Section 1.1, to obtain an expression for d^2p/dt^2 in terms of dp/dt and ds/dt.

   b. Use the fact that ds/dt is proportional to p to rewrite this expression as a second-order o.d.e. for $p(t)$.

3. *An LRC Circuit:* The current I in the circuit of Figure 2.3 is related to the charge Q on the capacitor by the two equations

$$\frac{dQ}{dt} = I$$

$$V(t) - RI - L\frac{dI}{dt} - \frac{1}{C}Q = 0.$$

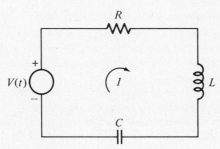

FIGURE **2.3**

Here, C, L, and R are constants, and $V(t)$ is an externally controlled voltage. (We discuss circuits in greater detail in Section 3.1.) Substitute $I = dQ/dt$ in the second equation to obtain a single second-order o.d.e. for Q.

4. *One Mass, Two Springs:* A mass of m grams moves along a horizontal track of length B cm and is attached to walls at each end of the track by springs with natural lengths L_1 and L_2 cm and spring constants k_1 and k_2 dynes/cm, respectively (see Figure 2.4).

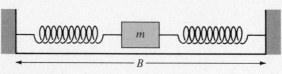

FIGURE **2.4**

The mass encounters frictional resistance with damping constant b.

   a. Set up a differential equation for the amount $x = x(t)$ by which the first spring is stretched. [*Hint:* If the first spring is stretched by x, then the second spring is stretched by $B - (L_2 + L_1 + x)$.]

   b. Check that we get the same equation if we assume the mass is attached to only one wall, by a spring of natural length $L = L_1$ with spring constant $k = k_1 + k_2$, and subject to an external force $E(t) = k_2(B - L_1 - L_2)$.

5. *A Moving Spring:* A 16-lb weight hangs suspended from the bottom of an elevator by a spring with natural length L feet and spring constant k lb/ft. The elevator rises at a constant rate of 2 ft/sec and the weight encounters frictional resistance with damping constant b. Let $x = x(t)$ denote the height of the weight above the ground.

   a. Use the fact that the forces acting on the weight are the gravitational force ($=$ weight), the restoring force of the spring, and friction, to obtain an expression for d^2x/dt^2 in terms of dx/dt, x, and the distance $y = y(t)$ from the bottom of the elevator to the ground. (*Hint:* When the mass is x ft from the ground, the spring is stretched $y - x - L$ ft.)

   b. Differentiate this expression and substitute $dy/dt = 2$ to obtain a third-order o.d.e. for x.

6. *An Anchored Floating Box:* Suppose the cube of Exercise 1 is anchored to the bottom of the river, at a point where the water is 3 feet deep, by a line that behaves like a spring of natural length $L = 2$ ft and spring constant $k = 22$ lb/ft. Set up an o.d.e. modeling the motion of the box,

   a. ignoring friction and

   b. assuming viscous damping with constant b lb/(ft/sec).

7. *The Box Adrift:* The box and anchor in Exercise 6 are towed out to deeper water so that the anchor hangs suspended from the box, but off the river bottom. The anchor weighs 16 lb and has a volume of $1/5$ ft^3. Obtain an o.d.e. modeling the motion of the box, ignoring friction, as follows:

   a. By analyzing the forces on the box, obtain an expression for the amount $y = y(t)$ by which the spring is stretched, in terms of x and d^2x/dt^2.

   b. By analyzing the forces on the anchor, obtain a second-order o.d.e. for y.

   c. Substitute the expression from (a) into the equation of (b) to obtain a single fourth-order o.d.e. for x.

*8. *Coupled Springs with Friction:* Write an o.d.e. modeling the motion of the first mass (x_1) in Example 2.1.4 when $L_1 = L_2 = L$, $m_1 = m_2 = m$, $k_1 = k_2 = k$, and each mass is subject to damping with constant b.

2.2 LINEAR DIFFERENTIAL EQUATIONS: A STRATEGY

   In Section 1.3 we found that the general solution of a first-order linear o.d.e. has the form

$$x = ch(t) + p(t).$$

Note that $x = p(t)$ is itself a particular solution of the o.d.e. and that $x = ch(t)$ is the general solution of the related homogeneous equation. In this section we will see that solutions of a class of higher-order o.d.e.'s follow a similar pattern.

Definition: *An nth-order o.d.e. is* **linear** *if it can be written in the form*

$$a_n(t) \frac{d^n x}{dt^n} + \ldots + a_1(t) \frac{dx}{dt} + a_0(t)x = E(t).$$

The functions $a_0(t)$, $a_1(t)$, . . ., $a_n(t)$ are called the **coefficients** *of the equation. The equation is* **homogeneous** *if $E(t) = 0$. The equation is* **normal** *on an interval I of the form $\alpha < t < \beta$ provided the functions $a_0(t)$, . . ., $a_n(t)$, and $E(t)$ are continuous on I and $a_n(t)$ is never zero on I.*

The o.d.e.'s in Examples 2.1.1 to 2.1.4 are all linear. Let's look at some other examples to see what the solution of a normal linear o.d.e. looks like.

Example 2.2.1
Solve

(N) $$\frac{d^3 x}{dt^3} = t.$$

This is a third-order linear o.d.e. with $a_3(t) = 1$, $a_2(t) = a_1(t) = a_0(t) = 0$, and $E(t) = t$. It is normal on $-\infty < t < +\infty$. The related homogeneous equation is

(H) $$\frac{d^3 x}{dt^3} = 0.$$

Both the nonhomogeneous and the homogeneous equations can be solved by integrating three times. The solution to (N) is

$$x = c_1 \frac{t^2}{2} + c_2 t + c_3 + \frac{t^4}{24}.$$

The solution to (H) is

$$x = c_1 \frac{t^2}{2} + c_2 t + c_3.$$

Example 2.2.2

Solve

(N)
$$t \frac{d^2x}{dt^2} - \frac{dx}{dt} = t^2, \qquad 0 < t < +\infty.$$

This is a second-order linear o.d.e. with $a_2(t) = t$, $a_1(t) = -1$, $a_0(t) = 0$, and $E(t) = t^2$. It is normal on the given interval. The related homogeneous equation is

(H)
$$t \frac{d^2x}{dt^2} - \frac{dx}{dt} = 0.$$

At first glance it looks as if we don't know enough to solve these equations. However, if we set $y = dx/dt$, then we can rewrite (N) as

(N′)
$$t \frac{dy}{dt} - y = t^2.$$

This is a linear first-order equation! Its solution is

$$y = c_1 t + t^2.$$

Since $dx/dt = y$, integration of the preceding equation gives

$$x = c_1 \frac{t^2}{2} + c_2 + \frac{t^3}{3}.$$

Similarly, one can check that the solution of (H) is

$$x = c_1 \frac{t^2}{2} + c_2.$$

The solutions in our examples have the form $x = H(t) + p(t)$ where $x = p(t)$ is a particular solution of the nonhomogeneous equation and $x = H(t)$ is the general solution of the related homogeneous equation. Let's see whether this pattern holds for all linear o.d.e.'s.

We start our discussion of the general situation by introducing some notation. In place of our usual notation for mth derivatives, we will use

$$D^m x = \frac{d^m x}{dt^m}.$$

We will extend this formalism to expressions that appear on the left-hand sides of linear o.d.e.'s as follows:

$$[a_n(t)D^n + \ldots + a_1(t)D + a_0(t)] x = a_n(t)D^n x + \ldots + a_1(t)Dx + a_0(t)x$$

$$= a_n(t)\frac{d^n x}{dt^n} + \ldots + a_1(t)\frac{dx}{dt} + a_0(t)x.$$

An expression of the form

$$L = a_n(t)D^n + \ldots + a_1(t)D + a_0(t)$$

is called a **linear differential operator.** When we apply an operator L to a function x we get a function Lx.

Example 2.2.3

Find $L(\sin 2t)$ and $L(e^{-4t})$ where $L = D^3 - tD^2 + t^3$.

$$L(\sin 2t) = D^3 \sin 2t - tD^2 \sin 2t + t^3 \sin 2t$$
$$= -8 \cos 2t + 4t \sin 2t + t^3 \sin 2t.$$
$$L(e^{-4t}) = D^3 e^{-4t} - tD^2 e^{-4t} + t^3 e^{-4t}$$
$$= (-64 - 16t + t^3)e^{-4t}.$$

Among the first properties we learned in calculus is that the derivative of a sum is the sum of the derivatives. This rule extends to an arbitrary linear differential operator

$$L = a_n(t)D^n + \ldots + a_1(t)D + a_0(t).$$

Here

$$L(x_1 + x_2) = a_n(t)D^n(x_1 + x_2) + \ldots + a_1(t)D(x_1 + x_2) + a_0(t)(x_1 + x_2)$$
$$= a_n(t)(D^n x_1 + D^n x_2) + \ldots + a_1(t)(Dx_1 + Dx_2) + a_0(t)(x_1 + x_2)$$
$$= a_n(t)D^n x_1 + \ldots + a_1(t)Dx_1 + a_0(t)x_1$$
$$+ a_n(t)D^n x_2 + \ldots + a_1(t)Dx_2 + a_0(t)x_2$$
$$= Lx_1 + Lx_2.$$

Fact: Principle of Superposition.   *If L is a linear differential operator, then*

$$L(x_1 + x_2) = Lx_1 + Lx_2.$$

Similarly, the fact that $D(cx) = c(Dx)$ for any constant c extends to linear differential operators.

Fact: Principle of Proportionality.   *If L is a linear differential operator and c is a constant, then*

$$L(cx) = c(Lx).$$

Let's now turn our attention to the linear o.d.e.

(N) $$a_n(t)\frac{d^n x}{dt^n} + \ldots + a_1(t)\frac{dx}{dt} + a_0(t)x = E(t).$$

If we set

$$L = a_n(t)D^n + \ldots + a_1(t)D + a_0(t)$$

then we can rewrite (N) as

(N) $$Lx = E(t).$$

The related homogeneous equation is

(H) $$Lx = 0.$$

Let $x = p(t)$ be a solution of (N) and let $x = h(t)$ be a solution of (H). Then

$$Lp(t) = E(t) \qquad \text{and} \qquad Lh(t) = 0.$$

The Principle of Superposition applies to give

$$L[h(t) + p(t)] = Lh(t) + Lp(t) = 0 + E(t) = E(t).$$

Thus $x = h(t) + p(t)$ is another solution of (N). If we let $h(t)$ range over all the solutions of (H), we get a list of solutions of (N).

Suppose now that $x = f(t)$ is also a solution of (N). Then $Lf(t) = E(t)$. Let $h_1(t) = f(t) - p(t)$. Then

$$f(t) = h_1(t) + p(t)$$

and

$$Lh_1(t) = L[f(t) - p(t)] = Lf(t) - Lp(t) = E(t) - E(t) = 0.$$

Thus $x = h_1(t)$ is a solution of (H) and $f(t)$ is on the list we described in the previous paragraph.

We have obtained powerful information about the form of the solution of a linear o.d.e.

Fact: *If we find a particular solution $x = p(t)$ of the nonhomogeneous linear o.d.e.*

(N) $$Lx = E(t)$$

and if $x = H(t)$ is a formula that describes all solutions of the related homogeneous equation

(H) $$Lx = 0$$

then

$$x = H(t) + p(t)$$

describes all solutions of (N).

We can now describe our strategy for solving a linear o.d.e. First find the general solution $x = H(t)$ of the related homogeneous equation. Then find a particular solution $x = p(t)$ of the nonhomogeneous equation. The general solution of the nonhomogeneous equation is $x = H(t) + p(t)$.

Let's summarize the major points of this section.

LINEAR DIFFERENTIAL EQUATIONS: A STRATEGY

An expression of the form

$$L = a_n(t)D^n + \ldots + a_1(t)D + a_0(t)$$

is called a **linear differential operator.** When L is applied to a function x, it gives a function Lx by the rule

$$Lx = a_n(t)\frac{d^n x}{dt^n} + \ldots + a_1(t)\frac{dx}{dt} + a_0(t)x.$$

Linear differential operators satisfy the following principles.

1. **Principle of Superposition:** $L(x_1 + x_2) = Lx_1 + Lx_2$.
2. **Principle of Proportionality:** $L(cx) = c(Lx)$.

A differential equation is **linear** if it can be written in the form $Lx = E(t)$ where L is a linear differential operator. The functions $a_0(t)$, $a_1(t)$, . . ., $a_n(t)$ in the description of the operator are called the **coefficients** of the equation. The equation is **homogeneous** if $E(t) = 0$. The equation is **normal** on an interval I if the coefficients and $E(t)$ are continuous on I and the highest coefficient $a_n(t)$ is never zero on I.

If $x = p(t)$ is a particular solution of the linear o.d.e.

(N) $$Lx = E(t)$$

and $x = H(t)$ is the general solution of the **related homogeneous equation**

(H) $$Lx = 0$$

then

$$x = H(t) + p(t)$$

is the general solution of (N).

EXERCISES

Which of the following o.d.e.'s are linear? Rewrite each linear o.d.e. in operator notation. Of the linear equations, which are homogeneous?

1. $\dfrac{d^2x}{dt^2} - 5tx = t\dfrac{dx}{dt} - 25$

2. $\dfrac{d^2x}{dt^2} + 2t^2x^2 = 0$

3. $\left(\dfrac{dx}{dt}\right)^2 - 2t^2 = 0$

4. $\dfrac{d^3x}{dt^3} = -x \sin t$

5. $\dfrac{d^2x}{dt^2} = -x\dfrac{dx}{dt}$

6. $\dfrac{d^3x}{dt^3} - 3x = t\dfrac{d^2x}{dt^2}$

7. $\dfrac{d^4x}{dt^4} + 5t^3\dfrac{dx}{dt} = \sqrt{t^2 - 1}$

8. $x\dfrac{d^2x}{dt^2} + t\dfrac{dx}{dt} = t$

In Exercises 9 to 13 find $Lf_i(t)$ for each given function $f_i(t)$.

9. $L = D^2 - 2D + 1$; $f_1(t) = e^t$, $f_2(t) = 3e^{2t}$

10. $L = D^2 - 2D + 1$; $f_1(t) = te^t$, $f_2(t) = \dfrac{5}{2}te^t$

11. $L = D^2 + 3D - 3$; $f_1(t) = e^{2t}$, $f_2(t) = t^3$, $f_3(t) = 5e^{2t} - 2t^3$

12. $L = D^3 - tD + 6$; $f_1(t) = \sin 2t$, $f_2(t) = 1$, $f_3(t) = 1 - \sin 2t$

13. $L = D$; $f_1(t) = t$, $f_2(t) = e^t$, $f_3(t) = te^t$

Note that in this problem $L[f_1(t)f_2(t)] \neq L[f_1(t)] L [f_2(t)]$.

Solve Exercises 14 to 20 by treating each as a first-order o.d.e. in a different variable, or by integrating several times.

14. $\dfrac{d^4x}{dt^4} = 3t - 1$

15. $\dfrac{d^2x}{dt^2} - \dfrac{dx}{dt} = e^t$

16. $\dfrac{d^2x}{dt^2} - \dfrac{dx}{dt} = 1$

17. $\dfrac{d^2x}{dt^2} = \dfrac{dx}{dt} + e^{2t}$

18. $\dfrac{d^3x}{dt^3} - 3\dfrac{d^2x}{dt^2} = 6$

19. $t\dfrac{d^4x}{dt^4} - \dfrac{d^3x}{dt^3} = 0, \quad t > 0$

20. $\dfrac{d^nx}{dt^n} = 0$

In Exercises 21 to 25 you are given a nonhomogeneous equation (N) $Lx = E(t)$, the general solution $x = H(t)$ of the related homogeneous equation, and an expression involving one or more constants for a particular solution $x = p(t)$ of (N). In each problem,

a. Find values of the constants for which $x = p(t)$ is a solution of (N).
b. Find the general solution of (N).

21. $(D^2 - D - 2)x = -t + 4$; $H(t) = c_1e^{2t} + c_2e^{-t}$; $p(t) = At + B$

22. $(D^2 - D - 2)x = 3 \sin t$; $H(t) = c_1e^{2t} + c_2e^{-t}$; $p(t) = A \sin t + B \cos t$

23. $(D^3 - D^2 - 2D)x = 4$; $H(t) = c_1 + c_2e^{2t} + c_3e^{-t}$; $p(t) = At$

24. $(D^2 + 4)x = e^t + 1$; $H(t) = c_1 \sin 2t + c_2 \cos 2t$; $p(t) = Ae^t + B$

25. $(D^2 - 1)x = t^2$; $H(t) = c_1e^t + c_2e^{-t}$; $p(t) = A + Bt + Ct^2$

2.3 HOMOGENEOUS LINEAR EQUATIONS: GENERAL PROPERTIES

In Section 1.3 we saw that the general solutions of first-order homogeneous linear o.d.e.'s have the form $x = ch(t)$. In Example 2.2.2 we determined that the general solution of a particular second-order homogeneous linear o.d.e. was of the form $x = c_1h_1(t) + c_2h_2(t)$. The general solution of the homogeneous third-order equation in Example 2.2.1 was of the form $x = c_1h_1(t) + x_2h_2(t) + c_3h_3(t)$. In this section we shall see that this pattern is typical of normal homogeneous linear o.d.e.'s.

We begin by noting an important consequence of the Principles of Superposition and Proportionality. Suppose $x = h_1(t), \ldots, x = h_n(t)$ are solutions of the homogeneous linear equation $Lx = 0$ and that $c_1, \ldots, c_n$ are constants. Then

$$L[c_1h_1(t) + c_2h_2(t) + \ldots + c_nh_n(t)]$$
$$= L[c_1h_1(t)] + L[c_2h_2(t)] + \ldots + L[c_nh_n(t)]$$
$$= c_1Lh_1(t) + c_2Lh_2(t) + \ldots + c_nL\,h_n(t)$$
$$= \quad 0 \quad + \quad 0 \quad + \ldots + \quad 0 \quad = 0.$$

Thus $x = c_1h_1(t) + \ldots + c_nh_n(t)$ is also a solution of $Lx = 0$. We refer to a function of the form $c_1h_1(t) + \ldots + c_nh_n(t)$ as a **linear combination** of $h_1(t), \ldots, h_n(t)$. Thus we have

Fact: *If $h_1(t), \ldots, h_n(t)$ are solutions of $Lx = 0$, then any linear combination of these functions is also a solution.*

If we can find some solutions of $Lx = 0$, then we can use the preceding fact to generate a list of solutions. We will apply the following theorem to decide whether the list we get is a *complete* list of solutions.

Theorem: Existence and Uniqueness of Solutions of Linear O.D.E.'s. *Let $Lx = E(t)$ be an nth-order linear o.d.e. that is normal on an interval I, and let t_0 be a fixed value of t in I. Then, given any real numbers $\alpha_0, \alpha_1, \ldots, \alpha_{n-1}$, there exists a solution $x = \phi(t)$ of the o.d.e. that is defined for all t in I and that satisfies the initial condition*

$$x(t_0) = \alpha_0, \quad x'(t_0) = \alpha_1, \ldots, \quad x^{(n-1)}(t_0) = \alpha_{n-1}.$$

Furthermore, if $x = \theta(t)$ is a solution of the o.d.e. that satisfies the same initial condition as $x = \phi(t)$, then $\theta(t) = \phi(t)$ for all t in I.

How do we use this theorem? To decide whether a list of solutions is complete, we try to match every initial condition at t_0 with a function on our list. *If there is an initial condition at t_0 that we can't match, then* the "existence" part of the theorem says *there is a solution that is not on our list.* On the other hand, *a list that matches every initial condition at t_0 must be complete,* since any solution satisfies some initial condition and, by "uniqueness," must theorefore agree with a function on the list.

Example 2.3.1
Solve the second-order equation

(H) $$(D^2 - 1)x = 0.$$

A lucky guess might lead us to look for solutions of the form $e^{\lambda t}$. On substitution of $e^{\lambda t}$ into (H) we find that the two functions e^t and e^{-t} are solutions. We can generate a list of solutions

(S) $$x = c_1e^t + c_2e^{-t}$$

by taking all linear combinations of these two functions.

Let's take $t_0 = 0$ and see whether we can match all initial conditions

$$x(0) = \alpha_0, \qquad x'(0) = \alpha_1.$$

For a function on our list (S),

$$x(0) = c_1 + c_2.$$

Also

$$x' = c_1 e^t - c_2 e^{-t}$$

so

$$x'(0) = c_1 - c_2.$$

If we try to match the initial conditions, we get two algebraic equations

(A)
$$c_1 + c_2 = \alpha_0$$
$$c_1 - c_2 = \alpha_1$$

in the two unknowns c_1 and c_2. Addition of these equations yields

$$c_1 = \frac{1}{2}(\alpha_0 + \alpha_1).$$

Subtraction yields

$$c_2 = \frac{1}{2}(\alpha_0 - \alpha_1).$$

Thus, by choosing c_1 and c_2 appropriately, we can match any initial condition at $t_0 = 0$. Formula (S) represents all solutions of (H).

Example 2.3.2

Solve the third-order equation

(H)
$$(t^2 D^3 + 2t D^2 - 2D)x = 0, \qquad 0 < t < +\infty.$$

A lucky guess might lead us to look for solutions of the form t^k. On substitution of t^k into (H) we find that the three functions t^2, $1/t$, and $1 (= t^0)$ are solutions.

These functions generate a list of solutions

(S)
$$x = c_1 t^2 + c_2 \frac{1}{t} + c_3.$$

Let's take $t_0 = 1$ and see whether we can match all initial conditions

$$x(1) = \alpha_0, \qquad x'(1) = \alpha_1, \qquad x''(1) = \alpha_2.$$

For a function on our list,

$$x' = 2c_1 t - c_2 \frac{1}{t^2}$$

$$x'' = 2c_1 + 2c_2 \frac{1}{t^3}.$$

If we try to match the initial conditions, we get three algebraic equations for c_1, c_2, and c_3:

(A)
$$
\begin{aligned}
c_1 + c_2 + c_3 &= \alpha_0 \\
2c_1 - c_2 \phantom{{}+ c_3} &= \alpha_1 \\
2c_1 + 2c_2 \phantom{{}+ c_3} &= \alpha_2.
\end{aligned}
$$

We can use the last two equations to get c_1 and c_2 in terms of α_1 and α_2:

$$c_1 = \frac{1}{6}(\alpha_2 + 2\alpha_1), \qquad c_2 = \frac{1}{3}(\alpha_2 - \alpha_1).$$

Substitution of these expressions into the first equation yields

$$c_3 = \frac{1}{2}(2\alpha_0 - \alpha_2).$$

Thus we can match any initial condition at $t_0 = 1$ by choosing c_1, c_2, and c_3 appropriately. Our formula (S) represents all solutions of (H).

Example 2.3.3

The three functions 1, t, and $2t - 3$ are solutions of the third-order equation

(H)
$$(D^3 - 2D^2)x = 0.$$

Then each function of the form

$$x = c_1 + c_2 t + c_3(2t - 3)$$

is also a solution of (H). For a function on this list,

$$x' = c_2 + 2c_3 \quad \text{and} \quad x'' = 0.$$

This list can't be complete, since we could never match an initial condition with $x''(t_0) \neq 0$.

We note that e^{2t} is also a solution of (H) and leave it to the reader to verify that

(S) $$x = c_1 + c_2 t + c_3 e^{2t}$$

is a complete list of solutions to (H).

Suppose now that we are given an nth-order homogeneous linear o.d.e. $Lx = 0$ that is normal on an interval I. Let t_0 be a fixed value of t in I. Then the Existence Theorem guarantees that there is a solution $x = h_1(t)$ that satisfies the initial condition

$$x(t_0) = 1, x'(t_0) = x''(t_0) = \ldots = x^{(n-1)}(t_0) = 0.$$

There is also a solution $x = h_2(t)$ that satisfies the initial condition

$$x(t_0) = 0, x'(t_0) = 1, x''(t_0) = \ldots = x^{(n-1)}(t_0) = 0.$$

Indeed, for each $i = 1, 2, \ldots, n$, there is a solution $x = h_i(t)$ that satisfies the initial condition

$$x(t_0) = \ldots = x^{(i-2)}(t_0) = 0, \quad x^{(i-1)}(t_0) = 1, \quad x^{(i)}(t_0) = \ldots = x^{(n-1)}(t_0) = 0.$$

These functions generate a list of solutions

(S) $$x = c_1 h_1(t) + \ldots + c_n h_n(t).$$

To see whether our list is complete, let's try to match all initial conditions

$$x(t_0) = \alpha_0, x'(t_0) = \alpha_1, \ldots, x^{(n-1)}(t_0) = \alpha_{n-1}.$$

For a function on our list,

$$
\begin{aligned}
x(t_0) &= c_1 h_1(t_0) + c_2 h_2(t_0) + \ldots + c_n h_n(t_0) = c_1 \\
x'(t_0) &= c_1 h_1'(t_0) + c_2 h_2'(t_0) + \ldots + c_n h_n'(t_0) = c_2 \\
&\quad\ \vdots \\
x^{(n-1)}(t_0) &= c_1 h_1^{(n-1)}(t_0) + c_2 h_2^{(n-1)}(t_0) + \ldots + c_n h_n^{(n-1)}(t_0) = c_n.
\end{aligned}
$$

We can match any initial condition by choosing $c_1 = \alpha_0, c_2 = \alpha_1, \ldots, c_n = \alpha_{n-1}$. Thus (S) is a complete list of solutions of $Lx = 0$.

We have verified the pattern for solutions of normal homogeneous linear equations.

Fact: *The general solution of the nth-order normal homogeneous linear equation $Lx = 0$ is of the form*

$$
x = c_1 h_1(t) + \ldots + c_n h_n(t)
$$

for a suitable choice of $h_1(t), \ldots, h_n(t)$.

Note that the functions $h_1(t), \ldots, h_n(t)$ in our examples did not satisfy the initial conditions we used to verify the pattern in general. In Sections 2.5 and 2.6 we determine conditions on solutions $h_1(t), \ldots, h_n(t)$ that guarantee that the list of solutions they generate is the general solution.

Let's summarize.

HOMOGENEOUS LINEAR EQUATIONS: GENERAL PROPERTIES

Given an nth-order homogeneous linear o.d.e.

(H) $$Lx = 0$$

that is normal on an interval I:

1. If $h_1(t), \ldots, h_k(t)$ are solutions of (H), then so is any **linear combination** $c_1 h_1(t) + \ldots + c_k h_k(t)$.
2. Let t_0 be a fixed value of t in I. A list of solutions of (H) is complete if and only if it matches all **initial conditions**

$$
x(t_0) = \alpha_0, \ x'(t_0) = \alpha_1, \ \ldots, \ x^{(n-1)}(t_0) = \alpha_{n-1}.
$$

---

3. The general solution of (H) is of the form

$$x = c_1 h_1(t) + \ldots + c_n h_n(t)$$

for a suitable choice of $h_1(t), \ldots, h_n(t)$.

Notes

1. A technicality

Our results depended heavily on the Existence and Uniqueness Theorem, which assumed we were dealing with an interval I on which the o.d.e. was normal and which provided information about solutions defined on I. *All statements in this and later sections about general solutions or, equivalently, about complete lists of solutions are valid only on such intervals.* In addition, *the values t_0 that are used to test for completeness of a list of solutions are always in the interval I.* Of course, when all the coefficients are constant, the equation is normal on $-\infty < t < \infty$.

2. On the existence of solutions of linear o.d.e.'s

In this section we concentrated on the implications for *homogeneous* linear equations of the Existence and Uniqueness Theorem. Note that the theorem holds for *any* normal linear o.d.e., homogeneous or not. In particular, *normal linear o.d.e.'s always have solutions.*

If we are interested in solutions of a linear o.d.e. on an interval where the equation is *not* normal, then we must proceed as we did in the similar case for first-order equations (see Note 2, Section 1.3). We must determine whether or not the solutions on the subintervals where the equation is normal can be pieced together to yield solutions on the whole interval.

EXERCISES

In Exercises 1 to 8, you are given an o.d.e., a list of solutions, and two initial conditions. Decide whether each initial condition can be matched with a function on the list.

1. $(D^2 - 9)x = 0; x = c_1 e^{3t} + c_2 e^{-3t}$
   a. $x(0) = 0, x'(0) = 1$ b. $x(0) = 1, x'(0) = 0$
2. $(D^2 - 2D + 1)x = 0; x = c_1 e^t + c_2 e^{t+1}$
   a. $x(0) = 1, x'(0) = 0$ b. $x(0) = x'(0) = 1$
3. $(D^2 + 4)x = 0; x = c_1 \sin 2t + c_2 \cos 2t$
   a. $x(0) = 1, x'(0) = 2$ b. $x(0) = x'(0) = 4$
4. $(D^2 + 1)x = 0; x = c_1 \sin t + c_2 \cos t$
   a. $x(\pi/2) = 1, x'(\pi/2) = 0$ b. $x(\pi/2) = 0, x'(\pi/2) = 1$
5. $D^3 x = 0; x = c_1(1 + t) + c_2(1 + t^2) + c_3(t^2 - t)$
   a. $x(1) = x'(1) = 0, x''(1) = 1$ b. $x(1) = 2, x'(1) = 1, x''(1) = 0$
6. $(D^3 + D)x = 0; x = c_1 \sin t + c_2 \cos t + c_3 \sin \left(t + \dfrac{\pi}{6} \right)$
   a. $x(0) = x'(0) = 0, x''(0) = 1$ b. $x(0) = 1, x'(0) = 0, x''(0) = -1$
7. $(D^3 - 2D^2 + D)x = -2e^{-t}; x = c_1 e^t + c_2 t e^t + c_3(t - 1)e^t - e^{-t}$
   a. $x(0) = -1, x'(0) = 1, x''(0) = 0$ b. $x(0) = -1, x'(0) = 2, x''(0) = 1$

8. $(D^3 - 4D)x = -3e^t$; $x = c_1e^{2t} + c_2e^{-2t} + c_3 + e^t$
    a. $x(0) = x'(0) = 0$, $x''(0) = 1$ b. $x(0) = x'(0) = 1$, $x''(0) = 0$.

In Exercises 9 to 16, a list of solutions is given. Determine whether the list is complete.

9. $(D^2 + 1)x = 0$; $x = c_1 \sin t + c_2 \cos t$
10. $(D^3 + D^2)x = 0$; $x = c_1e^{-t} + c_2$
11. $(D^3 + D^2)x = 0$; $x = c_1e^{-t} + c_2 + c_3t$
12. $(D^3 + D^2)x = 0$; $x = c_1e^{-t} + c_2(t + 2e^{-t}) + c_3t$
13. $(D^3 - 2D^2 + D)x = 0$; $x = c_1e^t + c_2te^t + c_3(2 + t)e^t$
14. $(D^3 - 2D^2 + D)x = 0$; $x = c_1 + c_2e^t + c_3te^t$
15. $(D^4 - 1)x = 0$; $x = c_1 \sin t + c_2 \cos t + c_3e^t$
16. $(D^2 - 4)x = 12t$; $x = c_1e^{2t} + c_2e^{-2t} - 3t$

In Exercises 17 to 23,
a. Find all solutions of the form $e^{\lambda t}$ or t^α.
b. Determine whether the solutions found in (a) generate a complete list of solutions.

17. $(D^2 - 1)x = 0$ 18. $(t^2D^2 - tD)x = 0$, $t > 0$
19. $(tD^2 + 2D)x = 0$, $t > 0$ 20. $(D^3 + D^2 - D + 2)x = 0$
21. $(D^3 - 6D^2 + 11D - 6)x = 0$ 22. $(D^4 - 1)x = 0$
23. $(D^4 - 3D^2 - 4)x = 0$

Two cautionary exercises:

24. Suppose $x = \phi(t)$ is a solution of the linear o.d.e. $Lx = E(t)$ where $E(t) \neq 0$. Show that $x = 2\phi(t) = \phi(t) + \phi(t)$ is *not* a solution of $Lx = E(t)$. Thus, linear combinations of solutions of *nonhomogeneous* equations need not be solutions.

25. Show that $x = \dfrac{1}{t^2 + 1}$ is a solution of $\dfrac{dx}{dt} + 2tx^2 = 0$, but $x = \dfrac{2}{t^2 + 1} = \dfrac{1}{t^2 + 1} + \dfrac{1}{t^2 + 1}$ is not. Note that this o.d.e. is *not* linear.

2.4 DETERMINANTS

In the examples of the last section, the question of whether a list of solutions of an nth-order homogeneous linear o.d.e. was complete reduced to the question of whether a set of n algebraic equations could always be solved for the "unknowns" $c_1, \ldots, c_n$. In this section we describe a test that will enable us to answer this last question without actually solving the equations.

Let's begin by considering the two algebraic equations in two unknowns

(A)
$$b_{11}u_1 + b_{12}u_2 = r_1$$
$$b_{21}u_1 + b_{22}u_2 = r_2.$$

Here the b_{ij}'s, r_1, and r_2 are given real numbers, and we wish to solve for the unknowns u_1 and u_2.

We multiply the first equation by b_{22} and the second by b_{12} to obtain

$$b_{22}b_{11}u_1 + b_{22}b_{12}u_2 = b_{22}r_1$$
$$b_{12}b_{21}u_1 + b_{12}b_{22}u_2 = b_{12}r_2.$$

Subtraction yields

$$(b_{11}b_{22} - b_{12}b_{21})u_1 = r_1b_{22} - b_{12}r_2.$$

Similarly, subtraction of the first equation multiplied by b_{21} from the second equation multiplied by b_{11} yields

$$(b_{11}b_{22} - b_{12}b_{21})u_2 = b_{11}r_2 - r_1b_{21}.$$

If $b_{11}b_{22} - b_{12}b_{21} \neq 0$, we can solve for u_1 and u_2:

$$u_1 = \frac{r_1b_{22} - r_2b_{12}}{b_{11}b_{22} - b_{12}b_{21}}, \qquad u_2 = \frac{b_{11}r_2 - r_1b_{21}}{b_{11}b_{22} - b_{12}b_{21}}.$$

We call $b_{11}b_{22} - b_{12}b_{21}$ the **determinant of coefficients** of (A) and write

$$\det \begin{bmatrix} b_{11} & b_{12} \\ b_{21} & b_{22} \end{bmatrix} = b_{11}b_{22} - b_{12}b_{21}.$$

The numerators of our expressions for u_1 and u_2 can also be expressed as determinants. We have

$$u_1 = \frac{\det \begin{bmatrix} r_1 & b_{12} \\ r_2 & b_{22} \end{bmatrix}}{\det \begin{bmatrix} b_{11} & b_{12} \\ b_{21} & b_{22} \end{bmatrix}}, \qquad u_2 = \frac{\det \begin{bmatrix} b_{11} & r_1 \\ b_{21} & r_2 \end{bmatrix}}{\det \begin{bmatrix} b_{11} & b_{12} \\ b_{21} & b_{22} \end{bmatrix}}.$$

Example 2.4.1

Solve

$$2u_1 - 3u_2 = 7$$
$$5u_1 + 6u_2 = -1.$$

The determinant of coefficients is

$$\det \begin{bmatrix} 2 & -3 \\ 5 & 6 \end{bmatrix} = (2)(6) - (-3)(5) = 27.$$

Then

$$u_1 = \frac{\det \begin{bmatrix} 7 & -3 \\ -1 & 6 \end{bmatrix}}{27} = \frac{39}{27} = \frac{13}{9}$$

$$u_2 = \frac{\det \begin{bmatrix} 2 & 7 \\ 5 & -1 \end{bmatrix}}{27} = \frac{-37}{27}.$$

What if the determinant of coefficients is zero? In this case the answer depends on the values of r_1 and r_2. For certain values of r_1 and r_2 there will be no solutions. For all other values, there will be infinitely many solutions.

Example 2.4.2

Consider the equations

(A)
$$u_1 + 4u_2 = r_1$$
$$2u_1 + 8u_2 = r_2.$$

The determinant of coefficients is

$$\det \begin{bmatrix} 1 & 4 \\ 2 & 8 \end{bmatrix} = (1)(8) - (4)(2) = 0.$$

If we subtract twice the first equation from the second, we get

$$0 = r_2 - 2r_1.$$

This is impossible if $r_2 \neq 2r_1$; there are no solutions in this case. If $r_2 = 2r_1$, then $u_1 = r_1 - 4k$ and $u_2 = k$ will be a solution for each value of k; there are infinitely many solutions in this case.

With the right interpretation of determinants, the preceding observations can be extended to systems of n algebraic equations in n unknowns.

Fact: Cramer's Determinant Test. *Given the system of n algebraic equations*

$$b_{11}u_1 + b_{12}u_2 + \ldots + b_{1n}u_n = r_1$$
$$b_{21}u_1 + b_{22}u_2 + \ldots + b_{2n}u_n = r_2$$

(A)

$$\cdot$$
$$\cdot$$
$$\cdot$$

$$b_{n1}u_1 + b_{n2}u_2 + \ldots + b_{nn}u_n = r_n$$

there is a number depending on the coefficients

$$\Delta = \det \begin{bmatrix} b_{11} & b_{12} & \cdots & b_{1n} \\ b_{21} & b_{22} & \cdots & b_{2n} \\ & & \cdot & \\ & & \cdot & \\ & & \cdot & \\ b_{n1} & b_{n2} & \cdots & b_{nn} \end{bmatrix}$$

*called the **determinant of coefficients** with the following properties:*

1. *If $\Delta \neq 0$, then the equations (A) have a unique solution.*
2. *If $\Delta = 0$, then the equations (A) have no solutions for some choices of $r_1, \ldots, r_n$ and have infinitely many solutions for all other choices of $r_1, \ldots, r_n$.*

In the case that $\Delta \neq 0$, the unique solution of (A) can be calculated explicitly, using determinants (see Note 3 of this section). However, in practice Cramer's determinant test for large n is primarily a theoretical tool.

To make use of Cramer's test, we need to know how to compute the $n \times n$ determinant Δ. The method we'll use is called **expansion by minors:**

1. Associated to each entry b_{ij} of the $n \times n$ determinant is a smaller $(n - 1) \times (n - 1)$ determinant, called the (i,j)th **minor.** It is obtained by crossing out the row and column containing the given entry b_{ij}.
2. Also associated to each entry b_{ij} is a plus or minus sign, obtained by forming a "checkerboard" pattern starting with $+$ in the upper left hand corner:

$$\begin{bmatrix} + & - & + & - & \cdot & \cdot & \cdot \\ - & + & - & + & \cdot & \cdot & \cdot \\ + & - & + & - & \cdot & \cdot & \cdot \\ \cdot & \cdot & \cdot & & & & \\ \cdot & \cdot & \cdot & & & & \\ \cdot & \cdot & \cdot & & & & \end{bmatrix}$$

This sign can also be described by a formula. The sign associated to b_{ij} is $(-1)^{i+j}$.

3. The determinant is obtained by expanding along a given row or column. One fixes a row (or column) and adds the entries of the row (or column) times the associated sign times the associated minor:

$$\det = \sum (-1)^{i+j} (b_{ij})\{(i,j)\text{th minor}\}.$$

There is a great deal of work in actually checking that calculations of a determinant using different rows or columns all give the same answer and that this answer is what we need to make Cramer's determinant test work. We'll take this on faith.

Example 2.4.3

Calculate

$$\Delta = \det \begin{bmatrix} 1 & 2 & 3 & 4 \\ 2 & 0 & 3 & 1 \\ 3 & 2 & 1 & 0 \\ 5 & -2 & 2 & 0 \end{bmatrix}$$

We expand Δ by minors along the last column:

i. The sign associated to the "4" in the top right-hand corner is "$-$". The minor associated to this "4" is obtained by crossing out the top row and the last column of Δ:

$$\det \begin{bmatrix} 2 & 0 & 3 \\ 3 & 2 & 1 \\ 5 & -2 & 2 \end{bmatrix}$$

The corresponding summand in the expansion of Δ is

$$-4 \det \begin{bmatrix} 2 & 0 & 3 \\ 3 & 2 & 1 \\ 5 & -2 & 2 \end{bmatrix}$$

ii. The sign associated to the "1" in the last column and second row of Δ is "$+$". The associated minor is obtained by crossing out the last column and the second row of Δ. The corresponding summand is

$$+1 \det \begin{bmatrix} 1 & 2 & 3 \\ 3 & 2 & 1 \\ 5 & -2 & 2 \end{bmatrix}$$

iii. The summands corresponding to the two "0"'s in the last column of Δ are both of the form

$$(\text{sign})(0)(\text{minor}) = 0.$$

Thus

$$\Delta = -4 \det \begin{bmatrix} 2 & 0 & 3 \\ 3 & 2 & 1 \\ 5 & -2 & 2 \end{bmatrix} + \det \begin{bmatrix} 1 & 2 & 3 \\ 3 & 2 & 1 \\ 5 & -2 & 2 \end{bmatrix} + 0 + 0.$$

We calculate each of the 3×3 determinants in our expansion by expanding along the first row:

$$\det \begin{bmatrix} 2 & 0 & 3 \\ 3 & 2 & 1 \\ 5 & -2 & 2 \end{bmatrix} = +2 \det \begin{bmatrix} 2 & 1 \\ -2 & 2 \end{bmatrix} + 0 + 3 \det \begin{bmatrix} 3 & 2 \\ 5 & -2 \end{bmatrix}$$

$$= 2[(2)(2) - (1)(-2)] + 3[(3)(-2) - (2)(5)]$$

$$= -36$$

$$\det \begin{bmatrix} 1 & 2 & 3 \\ 3 & 2 & 1 \\ 5 & -2 & 2 \end{bmatrix} = +1 \det \begin{bmatrix} 2 & 1 \\ -2 & 2 \end{bmatrix} - 2 \det \begin{bmatrix} 3 & 1 \\ 5 & 2 \end{bmatrix} + 3 \det \begin{bmatrix} 3 & 2 \\ 5 & -2 \end{bmatrix}$$

$$= 6 - 2 - 48 = -44.$$

Substitution of the values for the 3×3 determinants into our expression for Δ yields

$$\Delta = -4(-36) + (-44) = 100.$$

Example 2.4.4

Determine whether the equations

(A)
$$\begin{aligned} u_1 + 2u_2 + 3u_3 + 4u_4 &= r_1 \\ 2u_1 \qquad\ + 3u_3 +\ \ u_4 &= r_2 \\ 3u_1 + 2u_2 +\ \ u_3 \qquad &= r_3 \\ 5u_1 - 2u_2 + 2u_3 \qquad &= r_4 \end{aligned}$$

always have a solution.

The determinant of coefficients is

$$\Delta = \det \begin{bmatrix} 1 & 2 & 3 & 4 \\ 2 & 0 & 3 & 1 \\ 3 & 2 & 1 & 0 \\ 5 & -2 & 2 & 0 \end{bmatrix}$$

We found in Example 2.4.3 that

$$\Delta = 100.$$

Since $\Delta \neq 0$, the equations (A) always have a solution.

Example 2.4.5

Determine whether the equations

(A)
$$
\begin{aligned}
3u_1 - u_2 + u_3 + u_4 &= r_1 \\
2u_1 \quad\quad + u_3 \quad\quad &= r_2 \\
8u_1 - 2u_2 + 3u_3 + 2u_4 &= r_3 \\
u_1 \quad\quad - u_3 \quad\quad &= r_4
\end{aligned}
$$

always have a solution.

The determinant of coefficients is

$$
\Delta = \det \begin{bmatrix}
3 & -1 & 1 & 1 \\
2 & 0 & 1 & 0 \\
8 & -2 & 3 & 2 \\
1 & 0 & -1 & 0
\end{bmatrix}
$$

We expand Δ along the second column:

$$
\Delta = -(-1) \det \begin{bmatrix}
2 & 1 & 0 \\
8 & 3 & 2 \\
1 & -1 & 0
\end{bmatrix} + 0 - (-2) \det \begin{bmatrix}
3 & 1 & 1 \\
2 & 1 & 0 \\
1 & -1 & 0
\end{bmatrix} + 0.
$$

We expand each of the 3×3 determinants along the last column:

$$
\Delta = -(-1)[-2 \det \begin{bmatrix} 2 & 1 \\ 1 & -1 \end{bmatrix}] - (-2)[1 \det \begin{bmatrix} 2 & 1 \\ 1 & -1 \end{bmatrix}]
$$

$$
= (-2 + 2) \det \begin{bmatrix} 2 & 1 \\ 1 & -1 \end{bmatrix} = 0.
$$

The equations (A) have no solutions for some choices of $r_1, r_2, r_3,$ and r_4; they have infinitely many solutions for all other choices.

In general, the expansion of an $n \times n$ determinant involves n minors, each of which is an $(n - 1) \times (n - 1)$ determinant. Each of these minors can be expanded in terms of $(n - 2) \times (n - 2)$ determinants, and so on. Eventually, the whole problem reduces to 2×2 determinants. As we saw in the examples, expansion along a row or column with zeroes in it saves a lot of work, since we needn't calculate the corresponding minors. We list some other useful facts about determinants in the notes that follow our summary.

CALCULATING $n \times n$ DETERMINANTS

Calculation of 2×2 Determinants:

$$\det \begin{bmatrix} b_{11} & b_{12} \\ b_{21} & b_{22} \end{bmatrix} = b_{11}b_{22} - b_{12}b_{21}.$$

Calculation of $n \times n$ Determinants:

$$\det \begin{bmatrix} b_{11} & \cdots & b_{1n} \\ & \cdot & \\ & \cdot & \\ b_{n1} & \cdots & b_{nn} \end{bmatrix}$$

Choose a row or column. Associated to each entry b_{ij} of the row or column is a sign, $(-1)^{i+j}$, and a **minor,** the $(n - 1) \times (n - 1)$ determinant obtained by crossing out the row and column containing the entry. The determinant is the sum along the chosen row or column of the terms consisting of entry times sign times minor.

Cramer's Determinant Test

Given the system of n algebraic equations for the n unknowns $u_1, \ldots, u_n$

$$b_{11}u_1 + \ldots + b_{1n}u_n = r_1$$
$$\cdot$$
$$\cdot$$
$$\cdot$$
$$b_{n1}u_1 + \ldots + b_{nn}u_n = r_n,$$

let Δ be the **determinant of coefficients,**

$$\Delta = \det \begin{bmatrix} b_{11} & \cdots & b_{1n} \\ & \cdot & \\ & \cdot & \\ & \cdot & \\ b_{n1} & \cdots & b_{nn} \end{bmatrix}$$

1. If $\Delta \neq 0$, then the equations have a unique solution.
2. If $\Delta = 0$, then the equations have no solutions for some choices of $r_1, \ldots, r_n$ and have infinitely many solutions for all other choices.

Notes

1. A warning

Some readers may have seen a "diagonal" method for calculating 3×3 determinants. This method does *not* extend to $n \times n$ determinants with $n > 3$.

2. On calculating determinants

The following properties of determinants are useful in simplifying calculations (for proofs, see Exercises 34 to 36):

i. If all entries of a single row (or column) have a common factor, this factor can be pulled out in front of the determinant.

ii. If a determinant has two identical rows (or columns), then the determinant is 0.

iii. If $i \neq j$, then replacement of row i by [(row i) + k(row j)] doesn't change the determinant:

$$\det \begin{bmatrix} b_{11} & b_{12} & \cdots & b_{1n} \\ & & \cdot & \\ & & \cdot & \\ b_{i1} & b_{i2} & \cdots & b_{in} \\ & & \cdot & \\ & & \cdot & \\ b_{n1} & b_{n2} & \cdots & b_{nn} \end{bmatrix} = \det \begin{bmatrix} b_{11} & b_{12} & \cdots & b_{1n} \\ & & \cdot & \\ & & \cdot & \\ b_{i1}+kb_{j1} & b_{i2}+kb_{j2} & \cdots & b_{in}+kb_{jn} \\ & & \cdot & \\ & & \cdot & \\ b_{n1} & b_{n2} & \cdots & b_{nn} \end{bmatrix}.$$

The calculation of the 4×4 determinant in Example 2.4.5 can be simplified by the use of properties i and ii. We first factor out -1 from the second column; the resulting determinant is 0 since it has two identical columns:

$$\det \begin{bmatrix} 3 & -1 & 1 & 1 \\ 2 & 0 & 1 & 0 \\ 8 & -2 & 3 & 2 \\ 1 & 0 & -1 & 0 \end{bmatrix} = (-1) \det \begin{bmatrix} 3 & 1 & 1 & 1 \\ 2 & 0 & 1 & 0 \\ 8 & 2 & 3 & 2 \\ 1 & 0 & -1 & 0 \end{bmatrix} = (-1)(0) = 0.$$

We can simplify the calculation of the 4×4 determinant Δ in Example 2.4.3 by using property iii. We replace row 1 by [(row 1) $-$ 4 (row 2)] and expand by minors along the last column:

$$\Delta = \det \begin{bmatrix} 1 & 2 & 3 & 4 \\ 2 & 0 & 3 & 1 \\ 3 & 2 & 1 & 0 \\ 5 & -2 & 2 & 0 \end{bmatrix} = \det \begin{bmatrix} -7 & 2 & -9 & 0 \\ 2 & 0 & 3 & 1 \\ 3 & 2 & 1 & 0 \\ 5 & -2 & 2 & 0 \end{bmatrix} = \det \begin{bmatrix} -7 & 2 & -9 \\ 3 & 2 & 1 \\ 5 & -2 & 2 \end{bmatrix}.$$

To calculate the 3×3 determinant, replace row 2 by [(row 2) $-$ (row 1)] and then replace row 3 by [(row 3) $+$ (row 1)]:

$$\Delta = \det \begin{bmatrix} -7 & 2 & -9 \\ 10 & 0 & 10 \\ 5 & -2 & 2 \end{bmatrix} = \det \begin{bmatrix} -7 & 2 & -9 \\ 10 & 0 & 10 \\ -2 & 0 & -7 \end{bmatrix}.$$

We expand this last determinant by minors along the second column:

$$\Delta = -2 \det \begin{bmatrix} 10 & 10 \\ -2 & -7 \end{bmatrix} = (-2)(-50) = 100.$$

3. Cramer's rule

When the determinant of coefficients Δ is not zero, the unique solution of the equations (A) can be calculated explicitly from the coefficients and right-hand side, using determinants.

Fact: Cramer's Rule. *Suppose the determinant of coefficients Δ of the system (A) is not zero. Let Δ_i be the determinant obtained from Δ by replacing the entries of the ith column with the right-hand side $r_1, \ldots, r_n$:*

$$\Delta_i = \det \begin{bmatrix} b_{11} & \cdots & b_{1\,i-1} & r_1 & b_{1\,i+1} & \cdots & b_{1n} \\ b_{21} & \cdots & b_{2\,i-1} & r_2 & b_{2\,i+1} & \cdots & b_{2n} \\ & & & \cdot & & & \\ & & & \cdot & & & \\ & & & \cdot & & & \\ b_{n1} & \cdots & b_{n\,i-1} & r_n & b_{n\,i+1} & \cdots & b_{nn} \end{bmatrix}.$$

Then the unique solution of the system (A) is given by the formulas

$$u_1 = \frac{\Delta_1}{\Delta}, u_2 = \frac{\Delta_2}{\Delta}, \ldots, u_n = \frac{\Delta_n}{\Delta}.$$

Note that when $n = 2$, these are just the formulas we obtained at the beginning of this section.

EXERCISES

Calculate the determinants in Exercises 1 to 16.

1. $\det \begin{bmatrix} 1 & 2 \\ 3 & 4 \end{bmatrix}$ 2. $\det \begin{bmatrix} 1 & -1 \\ 2 & 5 \end{bmatrix}$ 3. $\det \begin{bmatrix} e^t & 1 \\ 1 & e^{-t} \end{bmatrix}$ 4. $\det \begin{bmatrix} \sin t & \cos t \\ \cos t & -\sin t \end{bmatrix}$

5. $\det \begin{bmatrix} 0 & 1 & 0 \\ 5 & 2 & 7 \\ 5 & 9 & 3 \end{bmatrix}$ 6. $\det \begin{bmatrix} 1 & 0 & -1 \\ 2 & 3 & 2 \\ 5 & 5 & 7 \end{bmatrix}$ 7. $\det \begin{bmatrix} 1 & -1 & 2 \\ 2 & 3 & 2 \\ 6 & 8 & 8 \end{bmatrix}$

8. $\det \begin{bmatrix} e^t & \sin t & \cos t \\ e^t & \cos t & -\sin t \\ e^t & -\sin t & -\cos t \end{bmatrix}$ 9. $\det \begin{bmatrix} e^t & e^{2t} & e^{-t} \\ e^t & 2e^{2t} & -e^{-t} \\ e^t & 4e^{2t} & e^{-t} \end{bmatrix}$

10. $\det \begin{bmatrix} 1 & 0 & 1 & 0 \\ 0 & -1 & -1 & 0 \\ 0 & 1 & 1 & -1 \\ 1 & 1 & 0 & -1 \end{bmatrix}$ 11. $\det \begin{bmatrix} 1 & 2 & 1 & 3 \\ 2 & 0 & 5 & 0 \\ 0 & 1 & 3 & 7 \\ 0 & 5 & 0 & 5 \end{bmatrix}$ 12. $\det \begin{bmatrix} 0 & -1 & 3 & 2 \\ 5 & 0 & 2 & 1 \\ 6 & 0 & 0 & 1 \\ 2 & 0 & 0 & 3 \end{bmatrix}$

13. $\det \begin{bmatrix} 1 & t & \cos 2t & \cos^2 t \\ 0 & 1 & -2 \sin 2t & -2 \cos t \sin t \\ 0 & 0 & -4 \cos 2t & -2 (\cos^2 t - \sin^2 t) \\ 0 & 0 & 8 \sin 2t & 8 \cos t \sin t \end{bmatrix}$ 14. $\det \begin{bmatrix} 1 & t & t^2 & t^3 \\ 0 & 1 & 2t & 3t^2 \\ 0 & 0 & 2 & 6t \\ 0 & 0 & 0 & 6 \end{bmatrix}$

15. $\det \begin{bmatrix} 1 & 7 & -8 & 2 & 3 \\ 0 & 2 & 3 & 6 & -4 \\ 0 & 0 & 3 & 5 & 8 \\ 0 & 0 & 0 & 4 & -9 \\ 0 & 0 & 0 & 0 & 5 \end{bmatrix}$ 16. $\det \begin{bmatrix} 0 & 0 & 0 & 1 & 0 \\ 1 & 3 & 2 & -3 & 3 \\ 0 & 2 & 0 & 2 & 0 \\ 1 & 5 & 3 & 5 & 4 \\ 1 & -2 & 2 & 6 & 5 \end{bmatrix}$

In each of Exercises 17 to 23, use Cramer's determinant test to determine whether the system has solutions for all choices of the right-hand side.

17. $\begin{aligned} x - y &= a \\ 3x - 3y &= b \end{aligned}$ 18. $\begin{aligned} x - y &= a \\ x + y &= b \end{aligned}$ 19. $\begin{aligned} x + y &= a \\ 3x - 3y &= b \end{aligned}$ 20. $\begin{aligned} x + 2y &= a \\ 2x + 4y &= b \end{aligned}$

21. $\begin{aligned} x - y + 3z &= a \\ x + y - 3z &= b \\ 2x \qquad - z &= c \end{aligned}$ 22. $\begin{aligned} x - y + 3z &= a \\ x + y - 3z &= b \\ 3x - y + 3z &= c \end{aligned}$ 23. $\begin{aligned} x - y + u + v &= a \\ 3x + 2y \qquad &= b \\ x \qquad - v &= c \\ y + 2u \qquad &= d \end{aligned}$

In Exercises 24 to 27, decide whether the given system of equations has no solutions, a unique solution, or infinitely many solutions.

24. $\begin{aligned} x - y &= 1 \\ 3x - 3y &= 3 \end{aligned}$ 25. $\begin{aligned} x - y &= 3 \\ 3x - 3y &= 1 \end{aligned}$ 26. $\begin{aligned} x - y &= 2 \\ x + y &= 4 \end{aligned}$ 27. $\begin{aligned} x + 2y &= 1 \\ 2x - y &= 1 \end{aligned}$

In Exercises 28 to 32, use Cramer's rule (Note 3) to solve for x only.

28. $\begin{aligned} 2x - 2y &= 1 \\ x + y &= 2 \end{aligned}$ 29 $\begin{aligned} 3x + 5y &= 7 \\ x + 4y &= 4 \end{aligned}$ 30. $\begin{aligned} x + 2y - 3z &= 1 \\ 2x + 2y + 3z &= 2 \\ x - y + z &= 0 \end{aligned}$

31. $\begin{aligned} u + v + x &= 2 \\ u - 3v + 2x &= -1 \\ 2u + 5v - 3x &= 2 \end{aligned}$ 32. $\begin{aligned} w + x + y - z &= 5 \\ 3w \quad + y \quad &= 1 \\ w + x \quad + z &= 2 \\ w \quad + y + z &= 0 \end{aligned}$

Some more abstract problems:

33. Show that

$$\det \begin{bmatrix} a_{11} & a_{12} & \cdots & a_{1n} \\ & a_{22} & \cdots & a_{2n} \\ & & \cdot & \cdot \\ & & & \cdot \\ 0\text{'s} & & & \cdot \\ & & & a_{nn} \end{bmatrix} = \det \begin{bmatrix} a_{11} & & & \\ a_{21} & a_{22} & & 0\text{'s} \\ \cdot & & \cdot & \\ \cdot & & & \cdot \\ \cdot & & & \cdot \\ a_{n1} & a_{n2} & \cdots & a_{nn} \end{bmatrix} = a_{11} a_{22} \cdots a_{nn}.$$

34. a. Show that a factor common to all the entries of a single row or column can be pulled out in front of the determinant. (*Hint:* Expand along this row or column.)
    b. Show that a determinant with a row or column of zeroes must equal zero.

35. a. Show that interchanging adjacent rows (say row i and row $i + 1$) in a determinant reverses the sign. [*Hint:* Expand the original determinant along the ith row and the new one along the $(i + 1)$st.]
    b. Show that interchanging any two rows in a determinant reverses the sign. [*Hint:* Use the result of (a) several times.]
    c. Show that a determinant with two equal rows is 0.

36. a. Suppose each entry of the ith row of a determinant is a sum, say $a_{ij} = b_{ij} + c_{ij}$ for all j. Show that the determinant is the sum of the two determinants obtained by replacing the a_{ij}'s in that row by the b_{ij}'s and the c_{ij}'s, respectively.
    b. Use (a) together with the results of earlier exercises to show that the addition of a multiple of one row to a different row does not change the determinant.

2.5 THE WRONSKIAN

In this section we will use Cramer's Determinant Test to obtain a test for checking whether n solutions of an nth-order homogeneous linear o.d.e. generate the general solution.

Suppose that $h_1(t), \ldots, h_n(t)$ are solutions of the nth-order homogeneous linear o.d.e. $Lx = 0$, which is normal on an interval I. Then every function of the form

$$x = c_1 h_1(t) + \ldots + c_n h_n(t)$$

is also a solution. Let t_0 be a value of t in I. Recall from Section 2.3 that to see whether the list of solutions generated in this way is complete, we must see whether we can match every initial condition

$$x(t_0) = \alpha_0, \; x'(t_0) = \alpha_1, \ldots, x^{(n-1)}(t_0) = \alpha_{n-1}.$$

Substitution of our expression for x into the equations describing the initial condition leads to the algebraic equations

$$c_1 h_1(t_0) + \ldots + c_n h_n(t_0) = \alpha_0$$
$$c_1 h_1'(t_0) + \ldots + c_n h_n'(t_0) = \alpha_1$$

$$\cdot$$
$$\cdot$$
$$\cdot$$

$$c_1 h_1^{(n-1)}(t_0) + \ldots + c_n h_n^{(n-1)}(t_0) = \alpha_{n-1}.$$

If the determinant of coefficients is not zero, then the equations always have a solution for the unknowns $c_1, \ldots, c_n$. If the determinant of coefficients is zero, then there are values of $\alpha_0, \ldots, \alpha_{n-1}$ for which the equations do not have a solution. Thus our list of solutions of $Lx = 0$ is complete if and only if the determinant of coefficients is not zero.

The determinant formed from the n solutions and their first $n - 1$ derivatives was introduced by the Polish mathematician H. Wronski (1811) and bears his name.

Definition: *Given a collection of n functions $h_1(t), \ldots, h_n(t)$ their **Wronskian** is the n × n determinant*

$$W[h_1, h_2, \ldots, h_n](t) = \det \begin{bmatrix} h_1(t) & h_2(t) & \ldots & h_n(t) \\ h_1'(t) & h_2'(t) & \ldots & h_n'(t) \\ & \cdot & & \\ & \cdot & & \\ & \cdot & & \\ h_1^{(n-1)}(t) & h_2^{(n-1)}(t) & \ldots & h_n^{(n-1)}(t) \end{bmatrix}$$

With this notation we can state our conclusion as follows:

Fact: *Let $Lx = 0$ be an nth-order homogeneous linear o.d.e. that is normal on an interval I and let t_0 be a value of t in I. The general solution of $Lx = 0$ is $x = c_1 h_1(t) + \ldots + c_n h_n(t)$ if and only if $h_1(t), \ldots, h_n(t)$ are solutions of the o.d.e. and*

$$W[h_1, \ldots, h_n](t_0) \neq 0.$$

Example 2.5.1

Show that the general solution of $(D^2 - 3D + 2)x = 0$ is

$$x = c_1 e^t + c_2 e^{2t}.$$

Substitution will show that the functions $h_1(t) = e^t$ and $h_2(t) = e^{2t}$ both solve the given o.d.e. Their Wronskian is

$$W[e^t, e^{2t}](t) = \det \begin{bmatrix} e^t & e^{2t} \\ e^t & 2e^{2t} \end{bmatrix}.$$

Thus,

$$W[e^t, e^{2t}](0) = \det \begin{bmatrix} 1 & 1 \\ 1 & 2 \end{bmatrix} = 1.$$

Since the Wronskian is not zero at $t_0 = 0$, the functions e^t and e^{2t} generate the general solution.

Example 2.5.2

Show that

$$x = c_1 \cos 2t + c_2 (\cos^2 t - 1/2)$$

is *not* the general solution of $(D^2 + 4)x = 0$.

Substitution will show that each of the functions $h_1(t) = \cos 2t$ and $h_2(t) = \cos^2 t - 1/2$ is a solution of the given o.d.e. The Wronskian of these functions is

$$W[\cos 2t, \cos^2 t - 1/2](t) = \det \begin{bmatrix} \cos 2t & \cos^2 t - 1/2 \\ -2 \sin 2t & -2 \cos t \sin t \end{bmatrix}.$$

Thus

$$W[\cos 2t, \cos^2 t - 1/2](0) = \det \begin{bmatrix} 1 & 1/2 \\ 0 & 0 \end{bmatrix} = 0.$$

Since the Wronskian is zero at $t_0 = 0$, the function $h_1(t)$ and $h_2(t)$ do *not* generate the general solution.

We note that $\sin 2t$ is also a solution of the o.d.e. and leave it to the reader to check that the general solution is

$$x = c_1 \cos 2t + c_2 \sin 2t.$$

Example 2.5.3

Show that the general solution of $(D^4 - 2D^3 + D^2)x = 0$ is

$$x = c_1 e^t + c_2 t e^t + c_3 + c_4 t.$$

Substitution will show that the functions $h_1(t) = e^t$, $h_2(t) = te^t$, $h_3(t) = 1$ and $h_4(t) = t$ are solutions of the o.d.e. Their Wronskian is

$$W[e^t, te^t, 1, t](t) = \det \begin{bmatrix} e^t & te^t & 1 & t \\ e^t & (t+1)e^t & 0 & 1 \\ e^t & (t+2)e^t & 0 & 0 \\ e^t & (t+3)e^t & 0 & 0 \end{bmatrix}.$$

We calculate the Wronskian at $t_0 = 0$ by expanding along the last column twice:

$$W[e^t, te^t, 1, t](0) = \det \begin{bmatrix} 1 & 0 & 1 & 0 \\ 1 & 1 & 0 & 1 \\ 1 & 2 & 0 & 0 \\ 1 & 3 & 0 & 0 \end{bmatrix}$$

$$= \det \begin{bmatrix} 1 & 0 & 1 \\ 1 & 2 & 0 \\ 1 & 3 & 0 \end{bmatrix}$$

$$= \det \begin{bmatrix} 1 & 2 \\ 1 & 3 \end{bmatrix} = 1.$$

Since the Wronskian is not zero at $t_0 = 0$, the functions e^t, te^t, 1, and t generate the general solution.

We close with a brief summary.

**THE WRONSKIAN TEST FOR SOLUTIONS
OF HOMOGENEOUS LINEAR O.D.E.'s**

Let $Lx = 0$ be an nth-order homogeneous linear o.d.e. that is normal on an interval I and let t_0 be a value of t in I. The general solution of $Lx = 0$ is $x = c_1 h_1(t) + \ldots + c_n h_n(t)$ if and only if $h_1(t), \ldots, h_n(t)$ are solutions of the o.d.e. and

$$W[h_1, \ldots, h_n](t_0) = \det \begin{bmatrix} h_1(t_0) & \cdots & h_n(t_0) \\ h_1'(t_0) & \cdots & h_n'(t_0) \\ & \vdots & \\ h_1^{(n-1)}(t_0) & \cdots & h_n^{(n-1)}(t_0) \end{bmatrix} \neq 0.$$

Notes

1.   On the Wronskian test for solutions of homogeneous o.d.e.'s

Let $h_1, \ldots, h_n$ be solutions of the nth-order homogeneous linear o.d.e. $Lx = 0$, which is normal on an interval I. Let t_0 and t_0' be two values of t in I. If $W[h_1, \ldots, h_n](t_0) \neq 0$, then $x = c_1h_1 + \ldots + c_nh_n$ is the general solution of $Lx = 0$. Then $W[h_1, \ldots, h_n](t_0') \neq 0$. Thus *the Wronskian is either always zero, or always nonzero, on* I.

2.   On the number of solutions needed to generate a general solution

Suppose that $h_1(t), \ldots, h_k(t)$ are solutions of an nth-order normal homogeneous linear o.d.e. $Lx = 0$, where $k < n$. Since the Wronskian test applies only if we have n solutions, let's throw in the extra solutions $h_{k+1}(t) = \ldots = h_n(t) = 0$. The Wronskian $W = W[h_1, \ldots, h_n](t)$ has at least one column (the last) consisting entirely of zeroes, so $W = 0$ (see Exercise 34, Section 2.4). Then the list of solutions $x = c_1h_1(t) + \ldots + c_nh_n(t)$ is not complete, so neither is the list $x = c_1h_1(t) + \ldots + c_kh_k(t)$. Thus *the general solution of an* n*th-order equation* $Lx = 0$ *cannot be generated by fewer than* n *solutions.*

Exercises

In Exercises 1 to 15 you are given an nth-order linear differential operator L and n solutions $h_1(t), \ldots, h_n(t)$ of $Lx = 0$. Use the Wronskian test to determine whether these solutions generate the general solution of $Lx = 0$.

1. $L = D^2 - a^2, a \neq 0; h_1(t) = e^{at}, h_2(t) = e^{-at}$

2. $L = D^2 - 2aD + a^2; h_1(t) = e^{at}, h_2(t) = ae^{at}$

3. $L = D^2 - 2aD + a^2; h_1(t) = e^{at}, h_2(t) = te^{at}$

4. $L = D^2 + 4; h_1(t) = \sin 2t, h_2(t) = \sin t \cos t$

5. $L = D^2 + a^2, a \neq 0; h_1(t) = \sin at, h_2(t) = \cos at$

6. $L = tD^2 - D, t > 0; h_1(t) = 2, h_2(t) = t^2$

7. $L = t^2D^2 + 4tD + 2, t > 0; h_1(t) = 1/t, h_2(t) = 1/t^2$

8. $L = D^3 + D^2; h_1(t) = e^{-1}, h_2(t) = t + 3e^{-t}, h_3(t) = t$

9. $L = D^3 - 4D; h_1(t) = e^{2t}, h_2(t) = e^{-2t}, h_3(t) = 1$

10. $L = D^4 + 5D^2 + 4; h_1(t) = \sin t, h_2(t) = \cos t, h_3(t) = \sin 2t,$

   $h_4(t) = \sin \left(\dfrac{\pi}{4} + t \right)$

11. $L = D^4 + 4D^3 + 6D^2 + 4D + 1; h_1(t) = e^{-t}, h_2(t) = te^{-t}, h_3(t) = t^2e^{-t},$

   $h_4(t) = t^3e^{-t}$

12. $L = D^4 - 1: h_1(t) = \sin t, h_2(t) = \cos t, h_3(t) = e^t, h_4(t) = e^{-t}$

13. $L = D^4; h_1(t) = t^3 + t^2, h_2(t) = t^2 + 1, h_3(t) = t^3 - 1, h_4(t) - t$

14. $L = D^4 + 2D^3 + D^2; h_1(t) = e^{-t}, h_2(t) = te^{-t}, h_3(t) = 1, h_4(t) = t$

15. $L = D^n; h_1(t) = 1, h_2(t) = t, \ldots, h_n(t) = t^{n-1}$

Some more abstract problems

16.   a. Show that $W[t, t^2](t_0)$ is 0 when $t_0 = 0$ and is not 0 when $t_0 = 1$.
   b. Why doesn't this contradict the observation in Note 1?

17. Suppose that $h_1(t)$ and $h_2(t)$ are solutions of the second-order homogeneous o.d.e.
    $x'' + b_1(t)x' + b_0(t)x = 0$. Set $W(t) = W[h_1,h_2](t) = h_1(t)h_2'(t) - h_1'(t)h_2(t)$.
    a. Show that $W'(t) = h_1(t)h_2''(t) - h_1''(t)h_2(t) = -b_1(t)W(t)$.
    b. By solving this first-order o.d.e. for $W(t)$, show that

$$W(t) = ce^{-\int b_1(t)dt}.$$

This formula is known as **Abel's formula.**

If $h_1(t)$ is a given solution of $x'' + b_1(t) + b_0(t)x = 0$, then its Wronskian with any other solution $h_2(t)$ must satisfy Abel's formula (Exercise 17b), for some constant c. In particular, if we set $c = 1$ and write $W(t)$ in terms of $h_2(t)$ and our known solution $h_1(t)$, we obtain a first-order o.d.e. for $h_2(t)$:

$$[h_1(t)]Dh_2(t) - [h_1'(t)]h_2(t) = e^{-\int b_1(t)dt}.$$

A solution $h_2(t)$ of this o.d.e. also solves the original o.d.e., and since we have cooked things up so that $W[h_1, h_2]$ is never zero, the solutions $h_1(t)$ and $h_2(t)$ generate the general solution of the second-order o.d.e. Use this idea in each of Exercises 18 to 20 to find the general solution of the given second-order o.d.e. from the single given solution $h_1(t)$.

18. $(D^2 - 2aD + a^2)x = 0$; $h_1(t) = e^{at}$
19. $(D^2 + 1)x = 0$; $h_1(t) = \cos t$
20. $(tD^2 - (t + 1)D + 1)x = 0$, $t > 0$; $h_1(t) = e^t$
*21. a. Show that

$$p_n(r) = \det \begin{bmatrix} 1 & 1 & & 1 & 1 \\ a_1 & a_2 & & a_n & r \\ a_1^2 & a_2^2 & & a_n^2 & r^2 \\ \cdot & \cdot & \cdots & \cdot & \cdot \\ \cdot & \cdot & & \cdot & \cdot \\ \cdot & \cdot & & \cdot & \cdot \\ a_1^n & a_2^n & & a_n^n & r^n \end{bmatrix}$$

is a polynomial in r of degree at most n, by expanding the determinant by minors along the last column.
    b. Show that $a_1, \ldots, a_n$ are roots of $p_n(r)$. (*Hint:* See Section 2.4, Note 2(ii).)
    c. Conclude that $p_n(r) = A_n(r - a_1) \cdots (r - a_n)$ for some constant A_n.
    d. Use mathematical induction to show that if $a_1, a_2, \ldots, a_n$ are distinct, then $A_n \neq 0$.
    e. By taking $r = a_{n+1}$, show that if $a_1, \ldots, a_{n+1}$ are distinct, then

$$\det \begin{bmatrix} 1 & 1 & & 1 & 1 \\ a_1 & a_2 & & a_n & a_{n+1} \\ a_1^2 & a_2^2 & & a_n^2 & a_{n+1}^2 \\ \cdot & \cdot & \cdots & \cdot & \cdot \\ \cdot & \cdot & & \cdot & \cdot \\ \cdot & \cdot & & \cdot & \cdot \\ a_1^n & a_2^n & & a_n^n & a_{n+1}^n \end{bmatrix} \neq 0.$$

This determinant is called the **Vandermonde determinant**.

f. Prove that if $a_1, \ldots, a_{n+1}$ are distinct, then $W[e^{a_1 t}, \ldots, e^{a_{n+1} t}](0) \neq 0$.

2.6 LINEAR INDEPENDENCE OF FUNCTIONS

In the last section we found that the Wronskian provides a test for determining whether solutions $h_1(t), \ldots, h_n(t)$ of a normal nth-order equation $Lx = 0$ generate the general solution. In this section we will use this test to obtain another characterization of solutions that generate the general solution. This new characterization provides an alternative approach in cases where computation of the Wronskian is cumbersome.

Let's begin by taking another look at Example 2.3.3. In that example the three functions $h_1(t) = 1$, $h_2(t) = t$, and $h_3(t) = 2t - 3$ were solutions of the third-order equation $(D^3 - 2D^2)x = 0$ but did *not* generate the general solution. Note that $h_3(t) = -3h_1(t) + 2h_2(t)$, so that

$$-3h_1(t) + 2h_2(t) - h_3(t) = 0.$$

We have found constants $c_1 = -3$, $c_2 = 2$, and $c_3 = -1$ so that

$$c_1 h_1(t) + c_2 h_2(t) + c_3 h_3(t) = 0$$

for all t. Of course, this last relationship would also hold if we took all the constants to be zero. The important thing is that we have found constants that are not all zero. We say these functions are linearly dependent.

Definition: *The functions $h_1(t), \ldots, h_n(t)$ are **linearly dependent** on the interval I if there exist constants $c_1, \ldots, c_n$, with at least one $c_i \neq 0$, so that*

$$c_1 h_1(t) + \cdots + c_n h_n(t) = 0$$

*for all t in I. The functions are **linearly independent** on I if the only constants for which the above relationship holds for all t in I are $c_1 = c_2 = \ldots = c_n = 0$.*

Example 2.6.1
Show that the functions 1, t, and e^{2t} are linearly independent on any interval I.
Suppose

$$c_1 + c_2 t + c_3 e^{2t} = 0$$

for all t in I. If we differentiate this relationship twice we see that

$$c_2 + 2c_3 e^{2t} = 0$$

and

$$4c_3 e^{2t} = 0$$

for all t in I. Since $e^{2t} \neq 0$, the last equation implies $c_3 = 0$. Substitution of this value into the second equation yields $c_2 = 0$. The first equation now reads $c_1 = 0$. Thus the functions are linearly independent on I.

Recall from Example 2.3.3 that these functions *do* generate the general solution of $(D^3 - 2D^2)x = 0$.

Example 2.6.2

Show that the functions 1, t, $\cos 2t$, and $\cos^2 t$ are linearly dependent on any interval I.

Since $\cos^2 t = (1 + \cos 2t)/2$,

$$\frac{1}{2}(1) + 0(t) + \frac{1}{2}\cos 2t + (-1)\cos^2 t = 0$$

for all t in I. Thus the functions are linearly dependent on I.

Example 2.6.3

Show that the functions t^3 and $|t^3|$ are linearly independent on $-\infty < t < +\infty$.

Suppose

$$c_1 t^3 + c_2 |t^3| = 0$$

for all t. Then the relationship certainly holds for $t = 1$ and $t = -1$:

$$c_1(1) + c_2(1) = 0$$
$$c_1(-1) + c_2(1) = 0.$$

Addition of these two equations yields $c_2 = 0$; subtraction yields $c_1 = 0$. Thus the functions are linearly independent on $-\infty < t < +\infty$.

In Example 2.6.1, repeated differentiation of the original relationship led to a system of equations, which we used to show that the functions were independent.

Let's try this in the general case. Suppose

$$c_1 h_1(t) + \ldots + c_n h_n(t) = 0$$

for all t in an interval I. Then

$$c_1 h_1'(t) + \ldots + c_n h_n'(t) \quad = 0$$

$$\cdot$$
$$\cdot$$
$$\cdot$$

$$c_1 h_1^{(n-1)}(t) + \ldots + c_n h_n^{(n-1)}(t) = 0$$

for all t in I. Substitution of a particular value of t, say $t = t_0$, yields the algebraic equations

$$c_1 h_1(t_0) + \ldots + c_n h_n(t_0) \quad = 0$$
$$c_1 h_1'(t_0) + \ldots + c_n h_n'(t_0) \quad = 0$$

(A)

$$\cdot$$
$$\cdot$$
$$\cdot$$

$$c_1 h_1^{(n-1)}(t_0) + \ldots + c_n h_n^{(n-1)}(t_0) = 0.$$

Note that these equations always have at least one solution, namely

$$c_1 = c_2 = \ldots = c_n = 0.$$

If the determinant of coefficients, $W[h_1, \ldots, h_n](t_0)$, is not zero, then Cramer's determinant test tells us that the equations (A) have a *unique* solution. In this case, $c_1 = c_2 = \ldots = c_n = 0$ is the *only* solution and the functions are independent.

Fact: Wronskian Test for Independence.   *If $W[h_1, \ldots, h_n](t_0) \neq 0$, then the functions $h_1(t), \ldots, h_n(t)$ are linearly independent on any interval I that contains t_0.*

Example 2.6.4

Show that the functions t and t^5 are linearly independent on $-\infty < t < +\infty$. The Wronskian of these functions is

$$W[t, t^5](t) = \det \begin{bmatrix} t & t^5 \\ 1 & 5t^4 \end{bmatrix} = 4t^5.$$

Since $W[t, t^5](1) = 4 \neq 0$, the functions are independent.

Unfortunately, $h_1(t), \ldots, h_n(t)$ may be independent even if the determinant of coefficients of (A) is zero. The functions in Example 2.6.4 are independent even though their Wronskian is zero at $t_0 = 0$. The functions in Example 2.6.3 are independent on $-\infty < t < +\infty$ even though their Wronskian is *always* zero.

Suppose, however, that $h_1(t), \ldots, h_n(t)$ are solutions of an nth-order homogeneous linear o.d.e. $Lx = 0$ that is normal on I. If $W[h_1, \ldots, h_n](t_0) = 0$, then the second part of Cramer's determinant test tells us that there are infinitely many solutions to the equations (A). Thus there are constants $c_1, \ldots, c_n$, with at least one $c_i \neq 0$, such that the equations (A) hold. Let $x = c_1 h_1(t) + \ldots + c_n h_n(t)$. Then x is a solution of $Lx = 0$. The equations (A) tell us that x satisfies the same initial condition as 0. The "uniqueness" part of the Existence and Uniqueness Theorem implies that $x = 0$—that is, $c_1 h_1(t) + \ldots + c_n h_n(t) = 0$ for all t in I. The functions are linearly dependent on I in this case.

If we combine the result of the preceding paragraph with the Wronskian Test for Independence, we see that the solutions $h_1(t), \ldots, h_n(t)$ of an nth-order homogeneous linear o.d.e. $Lx = 0$ that is normal on I are linearly independent if and only if their Wronskian is not zero at t_0. The Wronskian Test for Solutions tells us that this last condition holds if and only if these functions generate the general solution of $Lx = 0$. Thus we have the following:

Fact: *Let $h_1(t), \ldots, h_n(t)$ be solutions of the nth-order homogeneous linear o.d.e. $Lx = 0$, which is normal on I. The general solution of $Lx = 0$ is $x = c_1 h_1(t) + \ldots + c_n h_n(t)$ if and only if $h_1(t), \ldots, h_n(t)$ are linearly independent on I.*

Example 2.6.5

Show that the general solution of

(H) $$(D^4 + 4D^3 + 6D^2 + 4D + 1)x = 0$$

is

(S) $$x = c_1 e^{-t} + c_2 t e^{-t} + c_3 t^2 e^{-t} + c_4 t^3 e^{-t}.$$

Substitution will show that the four functions e^{-t}, te^{-t}, $t^2 e^{-t}$, and $t^3 e^{-t}$ are solutions of (H). Suppose

(A) $$c_1 e^{-t} + c_2 t e^{-t} + c_3 t^2 e^{-t} + c_4 t^3 e^{-t} = 0$$

for all t. Since $e^{-t} \neq 0$, we can divide to get

(B) $$c_1 + c_2 t + c_3 t^2 + c_4 t^3 = 0.$$

We differentiate (B) three times, obtaining three new relations:

(B′) $$c_2 + 2c_3t + 3c_4t^2 = 0$$

(B″) $$2c_3 + 6c_4t = 0$$

(B‴) $$6c_4 = 0.$$

Equation (B‴) gives $c_4 = 0$, which substituted into (B″) gives $c_3 = 0$. Substitution of these two values into (B′) gives $c_2 = 0$, and finally (B) gives us $c_1 = 0$. Thus the only way that (B), and hence (A), can hold is if

$$c_1 = c_2 = c_3 = c_4 = 0.$$

Our solutions of (H) are therefore independent, and (S) is the general solution.

Let's summarize.

LINEAR INDEPENDENCE OF FUNCTIONS

   The functions $h_1(t), \ldots, h_n(t)$ are **linearly dependent** on the interval I if there exist constants $c_1, \ldots, c_n$, with at least one $c_i \neq 0$, so that

$$c_1h_1(t) + \ldots + c_nh_n(t) = 0$$

for all t in I. The functions are **linearly independent** on I if the only constants for which this relationship holds for all t in I are $c_1 = c_2 = \ldots = c_n = 0$.

   Let $Lx = 0$ be an nth-order homogeneous linear o.d.e. that is normal on I and let t_0 be a value of t in I. Suppose $h_1(t), \ldots, h_n(t)$ are solutions of $Lx = 0$. Then the following are equivalent:

1. $h_1(t), \ldots, h_n(t)$ generate the general solution on I of $Lx = 0$.
2. $W[h_1, \ldots, h_n](t_0) \neq 0$.
3. $h_1(t), \ldots, h_n(t)$ are linearly independent on I.

EXERCISES

In Exercises 1 to 12, check the given functions for linear independence on $-\infty < t < \infty$.

1. $f_1(t) = 2t,$ $\qquad\qquad$ $f_2(t) = 3t$
2. $f_1(t) = e^{2t},$ $\qquad\qquad$ $f_2(t) = e^{3t}$

3. $f_1(t) = \sin 5t$, $f_2(t) = \cos 5t$

4. $f_1(t) = 1$, $f_2(t) = t$, $f_3(t) = t^2$

5. $f_1(t) = t + 1$, $f_2(t) = t^2 + 1$, $f_3(t) = t^2 - t$

6. $f_1(t) = t^2 + t$, $f_2(t) = t^2 + 1$, $f_3(t) = t^2 - 1$

7. $f_1(t) = te^t$, $f_2(t) = (t + 1)e^t$, $f_3(t) = te^{t+1}$

8. $f_1(t) = (t + 1)e^t$, $f_2(t) = (t - 1)e^t$, $f_3(t) = e^t$

9. $f_1(t) = \sin^2 t$, $f_2(t) = \cos^2 t$, $f_3(t) = \sin 2t$

10. $f_1(t) = \sin^2 t$, $f_2(t) = \cos^2 t$, $f_3(t) = \cos 2t$

11. $f_1(t) = e^t$, $f_2(t) = e^t \sin t$, $f_3(t) = e^t \cos t$

12. $f_1(t) = e^t$, $f_2(t) = te^t$, $f_3(t) = e^{2t}$, $f_4(t) = te^{2t}$

Some more abstract problems:

13. a. Show that if a, b, and c are distinct constants, then the functions e^{at}, e^{bt}, and e^{ct} are linearly independent.
    b. Show that the functions e^{at}, te^{at}, . . ., $t^{k-1} e^{at}$ are linearly independent.

14. a. Suppose we know that

$$f_1(0) = 1 \qquad f_1(1) = 0 \qquad f_1(2) = 0$$
$$f_2(0) = 0 \qquad f_2(1) = 1 \qquad f_2(2) = 0$$
$$f_3(0) = 0 \qquad f_3(1) = 0 \qquad f_3(2) = 1$$

Show that $f_1(t)$, $f_2(t)$, and $f_3(t)$ are linearly independent on the interval $-1 < t < 4$.
    b. Suppose we know that

$$g_1(0) = 0 \qquad g_1(1) \neq 0 \qquad g_1(2) \neq 0$$
$$g_2(0) \neq 0 \qquad g_2(1) = 0 \qquad g_2(2) \neq 0$$
$$g_3(0) \neq 0 \qquad g_3(1) \neq 0 \qquad g_3(2) = 0.$$

Are $g_1(t)$, $g_2(t)$, and $g_3(t)$ necessarily independent on the interval $-1 < t < 4$?

15. a. Suppose the functions $f_1(t)$, . . ., $f_n(t)$ are linearly independent on an interval I. Show that then $g_1(t) = e^t f_1(t)$, . . ., $g_n(t) = e^t f_n(t)$ are also linearly independent on I.
    b. Show that e^t in (a) can be replaced by any function that does not vanish on I.

16. a. Show that if $f_1(t)$, . . ., $f_n(t)$ are polynomials of degrees 1, 2, . . ., n, respectively, then they are linearly independent.
    b. Show that any collection of nonconstant polynomials of distinct degrees is independent.

*17. Show that a collection of more than $n + 1$ polynomials of degree at most n must be dependent.

18. Show that if $f_1(t)$, . . ., $f_n(t)$ are independent, then the functions $g_1(t) = a_1 f_1(t) + . . . + a_n f_n(t)$ and $g_2(t) = b_1 f_1(t) + . . . + b_n f_n(t)$ are different, unless $a_1 = b_1$, . . ., $a_n = b_n$.

19. a. Show that if $f_1(t)$, . . ., $f_n(t)$ are linearly dependent, then $f_1'(t)$, . . . $f_n'(t)$ are also linearly dependent.

b. Find an example of two linearly independent functions $f(t)$, $g(t)$ whose derivatives are linearly dependent.

2.7 HOMOGENEOUS LINEAR EQUATIONS WITH CONSTANT COEFFICIENTS: REAL ROOTS

We saw in the last section that the problem of finding the general solution of an nth-order homogeneous linear o.d.e. boils down to finding n linearly independent solutions. In this section and the next we will develop an algebraic way of finding such solutions for equations

(H)
$$(a_n D^n + \ldots + a_1 D + a_0)x = 0$$

with *constant coefficients*.

The operator in (H) is a polynomial expression in D. We will call the corresponding polynomial,

$$P(r) = a_n r^n + \ldots + a_1 r + a_0,$$

the **characteristic polynomial** of (H) and we will denote the operator by $P(D)$:

$$P(D) = a_n D^n + \ldots + a_1 D + a_0.$$

We can take our algebraic notation further by making sense of multiplication of operators. We define the product $Q(D)F(D)$ to be the operator whose value at a function x is obtained by first applying $F(D)$ to x, and then applying $Q(D)$ to $F(D)x$:

$$[Q(D)F(D)]x = Q(D)[F(D)x].$$

Example 2.7.1

$$[(D^2 + D)(D - 1)]x = (D^2 + D)[(D - 1)x] = (D^2 + D)[Dx - x]$$
$$= D^2[Dx - x] + D[Dx - x] = D^3x - D^2x + D^2x - Dx$$
$$= D^3x - Dx = (D^3 - D)x.$$

This example illustrates a remarkable fact about polynomial expressions in D. The effect of applying first $F(D)$ to x and then $Q(D)$ to the result is exactly the same as applying their product *as polynomials* to x.

Fact: *If $P(r) = Q(r)F(r)$, then for any function x, $P(D)x = Q(D)[F(D)x]$.*

Notice that if $P(r) = Q(r)F(r)$ and if $F(D)\phi(t) = 0$, then

$$P(D)\phi(t) = Q(D)[F(D)\phi(t)] = Q(D)[0] = 0.$$

Fact: *If $F(r)$ is a factor of $P(r)$, then any solution of $F(D)x = 0$ is also a solution of $P(D)x = 0$.*

We begin our search for solutions of $P(D)x = 0$ by looking for factors of $P(r)$. If the real number λ is a root of $P(r)$—that is, if $P(\lambda) = 0$—then $(r - \lambda)$ is a factor of $P(r)$. In this case, any solution of $(D - \lambda)x = 0$ will also be a solution of $P(D)x = 0$. In particular, $e^{\lambda t}$ solves $P(D)x = 0$.

Fact: *If the real number λ is a root of $P(r)$, then $e^{\lambda t}$ is a solution of $P(D)x = 0$.*

Example 2.7.2
Solve $(D^3 - D)x = 0$.
The characteristic polynomial factors easily:

$$r^3 - r = r(r^2 - 1) = r(r - 1)(r + 1).$$

Corresponding to its roots, 0, 1, and -1, we get solutions, $e^{0t} = 1$, e^t, and e^{-t}. Since these three solutions are independent (check this), the general solution of our third-order o.d.e. is

$$x = c_1 + c_2 e^t + c_3 e^{-t}.$$

Suppose now that λ is a root of multiplicity k, that is, suppose $(r - \lambda)^k$ is the highest power of $(r - \lambda)$ that is a factor of $P(r)$. We know that $e^{\lambda t}$ is a solution of $P(D)x = 0$. Can we find any other solutions corresponding to λ? Let's look for some, using a trick we first used when dealing with first-order equations. Let's look for solutions of the form $e^{\lambda t}y$ where y is a function of t.

The derivative of $e^{\lambda t}y$ is

$$D[e^{\lambda t}y] = e^{\lambda t}Dy + \lambda e^{\lambda t}y = e^{\lambda t}(D + \lambda)y.$$

Thus we can pull out the factor $e^{\lambda t}$, provided we replace D by $D + \lambda$. Then

$$D^2[e^{\lambda t}y] = D[e^{\lambda t}(D + \lambda)y] = e^{\lambda t}[(D + \lambda)(D + \lambda)y] = e^{\lambda t}[(D + \lambda)^2 y].$$

If we continue this way, we get the general formula

$$D^m[e^{\lambda t}y] = e^{\lambda t}[(D + \lambda)^m y].$$

This formula extends easily to arbitrary polynomials in D.

Fact: Exponential Shift. $P(D)[e^{\lambda t}y] = e^{\lambda t}P(D + \lambda)y.$

Example 2.7.3

$$(D^2 - 4D + 5)[e^{2t} \cos t] = e^{2t}[(D + 2)^2 - 4(D + 2) + 5] \cos t$$
$$= e^{2t}(D^2 + 1) \cos t = e^{2t}(-\cos t + \cos t) = 0.$$

Example 2.7.4

$$(D - 1)^3[c_1 e^{2t} + c_2 t e^{2t}] = e^{2t}(D + 2 - 1)^3[c_1 + c_2 t]$$
$$= e^{2t}(D + 1)^3[c_1 + c_2 t]$$
$$= e^{2t}(D^3 + 3D^2 + 3D + 1)[c_1 + c_2 t]$$
$$= e^{2t}[(3c_2 + c_1) + c_2 t]$$
$$= (3c_2 + c_1)e^{2t} + c_2 t e^{2t}.$$

Let's apply the Exponential Shift formula in the case where $P(r) = Q(r)(r - \lambda)^k$. Here

$$P(D)[e^{\lambda t}y] = Q(D)(D - \lambda)^k[e^{\lambda t}y] = e^{\lambda t}[Q(D + \lambda)(D + \lambda - \lambda)^k y]$$
$$= e^{\lambda t}[Q(D + \lambda) D^k y].$$

If we take y to be a function whose kth derivative is 0, then $e^{\lambda t}y$ will be a solution of $P(D)x = 0$. In particular, taking $y = 1, t, \ldots, t^{k-1}$, we see that $e^{\lambda t}, te^{\lambda t}, \ldots, t^{k-1}e^{\lambda t}$, are solutions of $P(D)x = 0$. What's more, an argument like the one in Example 2.6.5 will show they are independent (see Exercise 13, Section 2.6, and Exercise 27, Section 2.7).

Fact: *If the real number λ is a root of $P(r)$ of multiplicity k, then $e^{\lambda t}, te^{\lambda t}, \ldots, t^{k-1}e^{\lambda t}$ are linearly independent solutions of $P(D)x = 0$.*

Example 2.7.5

Solve $(D^3 - 3D^2 + 3D - 1)x = 0$.

The characteristic polynomial is $r^3 - 3r^2 + 3r - 1 = (r - 1)^3$. Corresponding to the triple root 1, we get three independent solutions, e^t, te^t, and t^2e^t, of our third-order o.d.e.. The general solution is

$$x = c_1e^t + c_2te^t + c_3t^2e^t.$$

Example 2.7.6

Solve $(D - 2)^2(D - 1)^3x = 0$.

The characteristic polynomial is $(r - 2)^2(r - 1)^3$. Corresponding to the double root 2, we get two independent solutions e^{2t} and te^{2t}. Corresponding to the triple root 1, we get three independent solutions e^t, te^t, and t^2e^t. We have found five solutions of our fifth-order o.d.e.

Are the solutions independent? Suppose

(A) $\qquad c_1e^{2t} + c_2te^{2t} + c_3e^t + c_4te^t + c_5t^2e^t = 0$

for all t. Note that e^t, te^t, and t^2e^t are solutions of $(D - 1)^3x = 0$. Application of $(D - 1)^3$ to (A) will eliminate the terms involving these functions and will affect the first two terms as in Example 2.7.4:

$$(3c_2 + c_1)e^{2t} + c_2te^{2t} = 0.$$

Since e^{2t} and te^{2t} are independent,

$$3c_2 + c_1 = 0 \qquad \text{and} \qquad c_2 = 0.$$

Then

$$c_1 = c_2 = 0.$$

Now (A) reads

$$c_3e^t + c_4te^t + c_5t^2e^t = 0.$$

Since e^t, te^t, and t^2e^t are independent,

$$c_3 = c_4 = c_5 = 0.$$

Our solutions are independent. The general solution of the o.d.e. is

$$x = c_1e^{2t} + c_2te^{2t} + c_3e^t + c_4te^t + c_5t^2e^t$$

The argument we used to show independence in the preceding example can be generalized to show that the solutions corresponding to different real roots are always independent.

Fact: *Associate functions to the polynomial $P(r)$ as follows: for each real root λ of $P(r)$, include the k functions $e^{\lambda t}$, $te^{\lambda t}$, . . ., $t^{k-1}e^{\lambda t}$, where k is the multiplicity of λ as a root. These functions are linearly independent solutions of $P(D)x = 0$.*

Example 2.7.7

Solve $(D^3 - D^2 - 8D + 12)x = 0$.

The characteristic polynomial is $P(r) = r^3 - r^2 - 8r + 12$. By trial and error, we note that ± 1 are not roots, but 2 is. This tells us that $r - 2$ is a factor of $P(r)$. Dividing, we see that $P(r) = (r - 2)(r^2 + r - 6)$. We can find the roots of the quadratic term by factoring or by the quadratic formula; these are 2 and -3. Thus $P(r)$ has a double root 2 and a single root -3. Corresponding to these roots, we get three independent solutions, e^{2t}, te^{2t}, and e^{-3t}, of our third-order o.d.e. The general solution is

$$x = c_1 e^{2t} + c_2 te^{2t} + c_3 e^{-3t}.$$

Example 2.7.8

Solve $(3D^5 - D^4 - 15D^3 + 5D^2 + 18D - 6)x = 0$.

The characteristic polynomial is

$$3r^5 - r^4 - 15r^3 + 5r^2 + 18r - 6 = 3(r - 1/3)(r^2 - 3)(r^2 - 2)$$

(see Note 2 for a description of how we factored this polynomial). Corresponding to its roots, $1/3$, $\pm\sqrt{3}$, and $\pm\sqrt{2}$, we get five linearly independent solutions, $e^{t/3}$, $e^{\sqrt{3}t}$, $e^{-\sqrt{3}t}$, $e^{\sqrt{2}t}$, and $e^{-\sqrt{2}t}$, of our fifth-order equation. The general solution is

$$x = c_1 e^{t/3} + c_2 e^{\sqrt{3}t} + c_3 e^{-\sqrt{3}t} + c_4 e^{\sqrt{2}t} + c_5 e^{-\sqrt{2}t}.$$

Example 2.7.9

Solve the initial value problem

$$(D^3 - 2D^2 + D)x = 0; \qquad x(0) = x'(0) = 0, \quad x''(0) = 1.$$

The characteristic polynomial is $r^3 - 2r^2 + r = r(r - 1)^2$, so the general solution of the o.d.e. is

$$x = c_1 + c_2 e^t + c_3 t e^t.$$

Any function on this list satisfies

$$x' = (c_2 + c_3)e^t + c_3 t e^t$$
$$x'' = (c_2 + 2c_3)e^t + c_3 t e^t.$$

Substitution of our initial condition yields the equations

$$0 = c_1 + c_2$$
$$0 = \quad\; c_2 + c_3$$
$$1 = \quad\; c_2 + 2c_3$$

which we solve to find

$$c_3 = 1, \qquad c_2 = -1, \qquad \text{and} \qquad c_1 = 1.$$

The solution of our initial value problem is

$$x = 1 - e^t + t e^t.$$

Let's summarize our facts about operators of the form $P(D)$.

HOMOGENEOUS EQUATIONS WITH CONSTANT COEFFICIENTS: REAL ROOTS

Associated to each polynomial $P(r) = a_n r^n + \ldots + a_1 r + a_0$ is an operator $P(D) = a_n D^n + \ldots + a_1 D + a_0$. Operators of this form satisfy the following:

1. If $P(r) = Q(r)F(r)$, then $P(D)x = Q(D)[F(D)x]$.
2. If $F(r)$ is a factor of $P(r)$, then any solution of $F(D)x = 0$ is also a solution of $P(D)x = 0$.
3. **Exponential Shift:** $P(D)[e^{\lambda t}y] = e^{\lambda t}[P(D + \lambda)y]$.
4. Associate functions to $P(r)$ as follows: for each real root λ of $P(r)$, include the k functions $e^{\lambda t}, te^{\lambda t}, \ldots, t^{k-1}e^{\lambda t}$, where k is the multiplicity of λ as a root. These functions are linearly independent solutions of $P(D)x = 0$.

Notes

1. A warning about variable-coefficient operators

The basis for our results was the observation that formal multiplication of polynomial expressions in D corresponds to successive application of these operators. In particular, the commutative law for multiplication holds for these operators: $Q(D)F(D)$ and $F(D)Q(D)$ multiply out to give the same operator.

We can still use the notion of successive application to define the product of two variable-coefficient operators, but the commutative law breaks down. For example,

$$[(D - t)D]x = (D - t)[Dx] = D^2x - tDx = (D^2 - tD)x,$$

but

$$[D(D - t)]x = D[(D - t)x] = D[Dx - tx] = D^2x - tDx - x = (D^2 - tD - 1)x.$$

Note that D is a factor of both $D(D - t)$ and $(D - t)D$ and that 1 is a solution of $Dx = 0$. Although 1 is also a solution of $[(D - t)D]x = 0$, it is *not* a solution of $[D(D - t)]x = 0$. This example illustrates why we can't expect a nice algebraic technique for solving general variable-coefficient equations.

2. Finding roots of polynomials

The problem of finding roots of polynomials is a deep one. For polynomials of degree two, there is a formula; the roots of $ar^2 + br + c$ are $(-b \pm \sqrt{b^2 - 4ac})/2a$. Similar, but messier, formulas exist for finding the roots of polynomials of degrees three and four. No such formula exists for polynomials of degree $n > 4$.

Although trial and error plays a part in finding roots, there are some tricks worth knowing. Among these is one for finding *rational* roots of polynomials $P(r)$ with *integer* coefficients. If $\lambda = a/b$ is a root of $P(r)$, where a and b are integers from which we've canceled all common factors, then a is a divisor of the constant coefficient a_0 of $P(r)$ and b is a divisor of the highest coefficient a_n. Let's use this fact to help find the roots of

$$P(r) = 3r^5 - r^4 - 15r^3 + 5r^2 + 18r - 6$$

(as needed in Example 2.7.8).

The divisors of the constant coefficient of $P(r)$ are ± 1, ± 2, ± 3, and ± 6. The divisors of the highest coefficient are ± 1 and ± 3. The only possibilities for rational roots are ± 1, $\pm 1/3$, ± 2, $\pm 2/3$, ± 3, $\pm 3/3$, ± 6, and $\pm 6/3$. Substitution will show that ± 1 are not roots, but $1/3$ is. Then $r - 1/3$ is a factor. We divide to get

$$P(r) = (r - 1/3)(3r^4 - 15r^2 + 18) = 3(r - 1/3)(r^4 - 5r^2 + 6).$$

The fact that the fourth-degree factor is a quadratic in $y = r^2$ helps us to spot a factorization:

$$P(r) = 3(r - 1/3)(r^2 - 3)(r^2 - 2).$$

The roots of $P(r)$ are $1/3$, $\pm\sqrt{3}$, and $\pm\sqrt{2}$.

EXERCISES

Find the general solution in Exercises 1 to 13.

1. $x'' - 5 = 0$

2. $x'' - 2x' - 15x = 0$

3. $2x'' - 5x' = 0$

4. $x'' + 4x' + 4x = 0$

5. $9x'' - 12x' + 4x = 0$

6. $x''' - 2x'' - x' + 2x = 0$

7. $x^{(5)} - 2x^{(4)} + x^{(3)} = 0$

8. $(D + 2)^3(D - 1)x = 0$

9. $D^2(D + 1)^3(D - 2)(3D + 5)(2D - 3)x = 0$

10. $(D^3 - 11D^2 + 31D - 21)x = 0$

11. $(D^2 - 2)(D^2 + D - 1)x = 0$

12. $(4D^3 - 24D^2 + 35D - 12)x = 0$

13. $(D^4 - 2D^2 + 1)x = 0$

Solve the initial value problems in Exercises 14 to 17.

14. $15x'' - 2x' - x = 0$; $x(0) = x'(0) = 1$

15. $x'' - 5x' = 0$; $x(0) = x'(0) = 10$

16. $x''' + 3x'' + 3x' + x = 0$; $x(0) = x'(0) = x''(0) = 0$

17. $(D - 1)^2(D + 2)x = 0$; $x(0) = x'(0) = 0, x''(0) = 9$

Exercises 18 to 21 refer back to our models.

18. *Damped Spring:*
    a. Find the general solution of the o.d.e. in Example 2.1.2, assuming $m = 5$ grams, $L = 3$ cm, $k = 60$ dynes/cm, and $b = 40$ dynes/(cm/sec).
    b. The mass in (a) starts at rest with the spring stretched to 5 cm. Find a formula for its position after t seconds.

19. *LRC Circuits:* Using the o.d.e. derived in Exercise 3, Section 2.1, find a formula for the charge Q at time t seconds on the capacitor of an *LRC* circuit with no external voltage ($V(t) = 0$), resistance $R = 2$ ohms, capacitance $C = 1$ farad, and inductance $L = 1$ henry, given that charge $Q = 0$ and current $I = 1$ ampere at time $t = 0$.

20. *Punk Rock:* Consider the supply-and-demand model in Exercise 2, Section 2.1, under the following assumptions concerning the new album, "Annihilation," by Dee and the Operators:
    a. If it were free, 5000 people per week would want a copy; every $1 in price reduces this number by 600.
    b. The contract that Dee and the Operators signed with Golden Groove Records contains an escalator clause: their commission (and hence the price of the record) is increased each week by 10¢ for every 100 fans who ask for, but can't get, the album.
    c. Every week, the record production plant hires enough new people to increase their weekly output of copies by 50 for every dollar in price of the record.
    Using the model of Exercise 2, Section 2.1, predict what will happen to the price of the record after one year (52 weeks), if its initial price is $p = \$4.00$ and $dp/dt = 0$ at $t = 0$.

*21. *Coupled Springs:*
  a. Find a formula for the roots of the polynomial $P(r) = (mr^2 + br + k)^2 + k(mr^2 + br)$ by means of the substitution $z = mr^2 + br$ and two uses of the quadratic formula.
  b. Find the general solution of the o.d.e. modeling the coupled spring system with friction in Exercise 8, Section 2.1, when $m = 1$, $k = 1$, and $b = 1 + \sqrt{5}$. [*Hint:* $(1 + \sqrt{5})^2 = 2(3 + \sqrt{5})$.]

Use the Exponential Shift to calculate $Lf_i(t)$ in Exercises 22 to 25:

22. $L = D^2 + 3D - 2;$ $f_1(t) = e^{2t},$ $f_2(t) = t^3 e^{2t}$

23. $L = (D + 1)^2;$ $f_1(t) = e^t,$ $f_2(t) = e^t \sin t,$ $f_3(t) = e^{-t} \sin t$

24. $L = D^2 + 4D + 5;$ $f_1(t) = te^{-2t} \cos t,$ $f_2(t) = te^{-2t} \sin t$

25. $L = [D^2 - 2\alpha D + (\alpha^2 + \beta^2)];$ $f_1(t) \, e^{\alpha t} \cos \beta t,$ $f_2(t) = e^{\alpha t} \sin \beta t$

In Exercises 26 and 27, you are asked to use an argument similar to the one used in Example 2.7.7 to show the independence of given sets of functions.

26. Show that if $m_1, \ldots, m_k$ are distinct, then $e^{m_1 t}, \ldots, e^{m_k t}$ are independent as follows: Suppose

$$c_1 e^{m_1 t} + \ldots + c_k e^{m_k t} = 0$$

for all t. Show that the operator

$$P_k(D) = (D - m_1)(D - m_2) \ldots (D - m_{k-1})$$

changes the equation above into

$$c_k(m_k - m_1)(m_k - m_2) \ldots (m_k - m_{k-1})e^{m_k t} = 0$$

from which we can conclude $c_k = 0$. What operator should one now use to show $c_{k-1} = 0$?

27. Show that $e^{\lambda t}, te^{\lambda t}, \ldots, t^{k-1}e^{\lambda t}$ are independent as follows: Suppose $c_1 e^{\lambda t} + c_2 te^{\lambda t} + \ldots + c_k t^{k-1} e^{\lambda t} = 0$. Show that the operator $(D - \lambda)^{k-1}$ changes this equation into $e^{\lambda t}(k - 1)! \, c_k = 0$, from which we can conclude $c_k = 0$. What operator should we now use to show $c_{k-1} = 0$?

2.8 HOMOGENEOUS EQUATIONS WITH CONSTANT COEFFICIENTS: COMPLEX ROOTS

In the last section we saw how to solve equations $P(D)x = 0$ when all the roots of the characteristic polynomial $P(r)$ are real. In this section we complete our treatment of homogeneous linear equations with constant coefficients by determining how to deal with complex roots.

Suppose $\alpha + \beta i$ is a root of $P(r)$, where α and β are real numbers and $\beta \neq 0$.

Then $\alpha - \beta i$ is also a root, so that

$$F(r) = r^2 - 2\alpha r + (\alpha^2 + \beta^2) = [r - (\alpha + \beta i)][r - (\alpha - \beta i)]$$

is a factor of $P(r)$. Any solution of $F(D)x = 0$ will also be a solution of $P(D)x = 0$.

For the moment, let's ignore the fact that the roots are complex. Then the general solution of $F(D)x = 0$ is

$$x = k_1 e^{(\alpha + \beta i)t} + k_2 e^{(\alpha - \beta i)t}.$$

We can use **Euler's formula,**

$$e^{u + vi} = e^u (\cos v + i \sin v),$$

to rewrite our formula for x:

$$\begin{aligned} x &= k_1 e^{\alpha t}(\cos \beta t + i \sin \beta t) + k_2 e^{\alpha t}(\cos \beta t - i \sin \beta t) \\ &= (k_1 + k_2)e^{\alpha t} \cos \beta t + (k_1 i - k_2 i)e^{\alpha t} \sin \beta t \\ &= c_1 e^{\alpha t} \cos \beta t + c_2 e^{\alpha t} \sin \beta t. \end{aligned}$$

This formula involves two real-valued candidates for solutions.

Substitution (aided by the Exponential Shift formula) will show that $e^{\alpha t} \cos \beta t$ and $e^{\alpha t} \sin \beta t$ are solutions of $F(D)x = 0$. Are they independent? Suppose

$$c_1 e^{\alpha t} \cos \beta t + c_2 e^{\alpha t} \sin \beta t = 0$$

for all t. Then, taking $t = 0$,

$$c_1 = 0.$$

Since $e^{\alpha t} \sin \beta t$ is not always 0, it now follows that

$$c_2 = 0.$$

We have found two independent solutions of $F(D)x = 0$.

Fact: *If $\alpha \pm \beta i$ are roots of $P(r)$, where α and β are real and $\beta \neq 0$, then $e^{\alpha t} \cos \beta t$ and $e^{\alpha t} \sin \beta t$ are linearly independent solutions of $P(D)x = 0$.*

Example 2.8.1

Solve $(D^2 + 4D + 5)x = 0$, which is the equation for the damped-spring system of Example 2.1.2, with $m = 1$, $b = 4$, $k = 5$.

The roots of the characteristic polynomial, $r^2 + 4r + 5$, are $-2 \pm i$. Corresponding to these roots, we get two linearly independent solutions, $e^{-2t} \cos t$ and $e^{-2t} \sin t$, of our second-order equation. The general solution is

$$x = c_1 e^{-2t} \cos t + c_2 e^{-2t} \sin t.$$

Suppose now that $\alpha + \beta i$ is a root of $P(r)$ of multiplicity k. Then $\alpha - \beta i$ is also. The polynomial

$$G(r) = [r^2 - 2\alpha + (\alpha^2 + \beta^2)]^k = [r - (\alpha + \beta i)]^k [r - (\alpha - \beta i)]^k$$

is a factor of $P(r)$. Any solution of $G(D)x = 0$ is also a solution of $P(D)x = 0$. Let's look at an example to see what solutions of $G(D)x = 0$ look like.

Example 2.8.2

Solve $(D^2 + 4D + 5)^2 x = 0$.

The roots of the characteristic polynomial are $-2 \pm i$, each of which has multiplicity 2. We know that $e^{-2t} \cos t$ and $e^{-2t} \sin t$ are linearly independent solutions. We need two more. By analogy with what worked when we had real roots of multiplicity 2, we guess that $te^{-2t} \cos t$ and $te^{-2t} \sin t$ are also solutions. We can verify this by substitution (aided, once more, by the Exponential Shift formula).

Are our four solutions independent? Suppose

$$c_1 e^{-2t} \cos t + c_2 e^{-2t} \sin t + c_3 te^{-2t} \cos t + c_4 te^{-2t} \sin t = 0$$

for all t. Application of $(D^2 + 4D + 5)$ to this equation yields

$$-2c_3 e^{-2t} \sin t + 2c_4 e^{-2t} \cos t = 0.$$

Since $e^{-2t} \sin t$ and $e^{-2t} \cos t$ are independent,

$$c_3 = c_4 = 0.$$

Now

$$c_1 e^{-2t} \cos t + c_2 e^{-2t} \sin t = 0,$$

so that

$$c_1 = c_2 = 0.$$

Our solutions are independent. The general solution is

$$x = c_1 e^{-2t} \cos t + c_2 e^{-2t} \sin t + c_3 t e^{-2t} \cos t + c_4 t e^{-2t} \sin t.$$

The general case follows a similar pattern.

Fact: *Suppose $\alpha \pm \beta i$ are roots of $P(r)$ of multiplicity k, where α and β are real numbers and $\beta \neq 0$. Then the $2k$ functions $e^{\alpha t} \cos \beta t$, $e^{\alpha t} \sin \beta t$, $t e^{\alpha t} \cos \beta t$, $t e^{\alpha t} \sin \beta t$, . . ., $t^{k-1} e^{\alpha t} \cos \beta t$, $t^{k-1} e^{\alpha t} \sin \beta t$ are linearly independent solutions of $P(D)x = 0$.*

Once again, the argument in Example 2.7.6 can be generalized to show that the solutions we get corresponding to distinct roots of $P(r)$ are independent. We can now solve any equation $P(D)x = 0$, *provided we can find the roots of $P(r)$.*

Fact: *Associate functions to the polynomial $P(r)$ as follows:*

1. *For each real root λ of $P(r)$, include the k functions $e^{\lambda t}$, $t e^{\lambda t}$, . . ., $t^{k-1} e^{\lambda t}$, where k is the multiplicity of λ as a root.*
2. *For each pair of complex roots $\alpha \pm \beta i$, include the $2k$ functions $e^{\alpha t} \cos \beta t$, $e^{\alpha t} \sin \beta t$, $t e^{\alpha t} \cos \beta t$, $t e^{\alpha t} \sin \beta t$, . . ., $t^{k-1} e^{\alpha t} \cos \beta t$, $t^{k-1} e^{\alpha t} \sin \beta t$, where k is the multiplicity of each of $\alpha \pm \beta i$ as a root.*

These functions are linearly independent solutions of $P(D)x = 0$.

Example 2.8.3

Solve $(D + 2)^3 D^2 (D^2 - D + 1)x = 0$.

The roots of the characteristic polynomial, $(r + 2)^3 r^2 (r^2 - r + 1)$, are -2 (multiplicity 3), 0 (multiplicity 2), and $\dfrac{1}{2} \pm \dfrac{i\sqrt{3}}{2}$ (multiplicity 1). Corresponding to the roots, we get seven linearly independent solutions, e^{-2t}, $t e^{-2t}$, $t^2 e^{-2t}$, 1, t, $e^{t/2} \cos \dfrac{\sqrt{3}}{2} t$, and $e^{t/2} \sin \dfrac{\sqrt{3}}{2} t$, of our seventh-order o.d.e. The general solution is

$$x = c_1 e^{-2t} + c_2 t e^{-2t} + c_3 t^2 e^{-2t} + c_4 + c_5 t$$
$$+ c_6 e^{t/2} \cos \frac{\sqrt{3}}{2} t + c_7 e^{t/2} \sin \frac{\sqrt{3}}{2} t.$$

Example 2.8.4

Solve $(D^8 - 8D^4 + 16)x = 0$.

The characteristic polynomial is $r^8 - 8r^4 + 16 = (r^4 - 4)^2 = [(r^2 - 2)(r^2 + 2)]^2$. Its roots are $\pm\sqrt{2}$, and $\pm\sqrt{2}\,i$, each of multiplicity 2. The general solution of our o.d.e. is

$$x = c_1 e^{\sqrt{2}\,t} + c_2 t e^{\sqrt{2}\,t} + c_3 e^{-\sqrt{2}\,t} + c_4 t e^{-\sqrt{2}\,t}$$
$$+ c_5 \cos\sqrt{2}\,t + c_6 \sin\sqrt{2}\,t + c_7 t \cos\sqrt{2}\,t + c_8 t \sin\sqrt{2}\,t.$$

In our next two examples we turn the tables. We are given a function $E(t)$ and are asked to find an operator $A(D)$ so that $A(D)\,E(t) = 0$. An operator that satisfies this condition is said to **annihilate** $E(t)$. Thus, saying $A(D)$ annihilates $E(t)$ is the same as saying $E(t)$ is a solution of $A(D)x = 0$.

Example 2.8.5

Find an operator $A(D)$ with constant coefficients that annihilates $E(t) = -3e^{3t} + 10te^{3t}$.

Any linear operator that annihilates both e^{3t} and te^{3t} will annihilate $E(t)$. We know that e^{3t} and te^{3t} are solutions of any homogeneous equation whose characteristic polynomial has 3 as a double root. The simplest such polynomial is $A(r) = (r - 3)^2$. $E(t)$ is annihilated by

$$A(D) = (D - 3)^2.$$

Example 2.8.6

Find an operator $A(D)$ with constant coefficients that annihilates $E(t) = 1 + 65e^t \cos 2t$.

We know that $1 = e^{0t}$ is a solution of any homogeneous equation whose characteristic polynomial has 0 as a root. Also, $e^t \cos 2t$ is a solution of any homogeneous equation whose characteristic polynomial has $1 \pm 2i$ as roots. The simplest polynomial that has all three numbers as roots is

$$A(r) = r[r - (1 + 2i)][r - (1 - 2i)] = r(r^2 - 2r + 5).$$

$E(t)$ is annihilated by

$$A(D) = D(D^2 - 2D + 5).$$

We close with a description of our method for solving homogeneous linear equations with constant coefficients. This method was published by Euler (1724).

**HOMOGENEOUS LINEAR EQUATIONS
WITH CONSTANT COEFFICIENTS**

To solve a homogeneous linear o.d.e. with constant coefficients

(H) $$(a_nD^n + \ldots + a_1D + a_0)x = 0:$$

1. Find all the roots of the characteristic polynomial

$$P(r) = a_nr^n + \ldots + a_1r + a_0.$$

2. Obtain functions $h_1(t), \ldots, h_n(t)$ as follows:
   a. For each real root λ of $P(r)$, include on the list the k functions

$$e^{\lambda t}, te^{\lambda t}, \ldots, t^{k-1}e^{\lambda t}$$

   where k is the multiplicity of λ as a root.
   b. For each pair of complex roots $\alpha \pm \beta i$, include on the list the $2k$ functions

$$e^{\alpha t} \cos \beta t, e^{\alpha t} \sin \beta t, te^{\alpha t} \cos \beta t, te^{\alpha t} \sin \beta t,$$
$$\ldots, t^{k-1}e^{\alpha t} \cos \beta t, t^{k-1}e^{\alpha t} \sin \beta t$$

   where k is the multiplicity of each of $\alpha \pm \beta i$ as a root.

3. The general solution of (H) is generated by these functions:

$$x = c_1h_1(t) + \ldots + c_nh_n(t).$$

EXERCISES

Find the general solution in Exercises 1 to 12.

1. $9x'' + x = 0$
2. $x'' + 2x' + 5x = 0$
3. $x^{(4)} - 16x = 0$
4. $(4D^2 + 1)(D^2 + 2D + 2)x = 0$
5. $(D^2 + 1)(D^2 + D + 1)x = 0$
6. $(D^4 + 2D^2 + 1)x = 0$
7. $(D^3 - D^2 + 9D - 9)x = 0$
8. $(D^4 + 9D^2 + 20)x = 0$

9. $(D^4 - 1)x = 0$

10. $(D^4 - 2D^3 + 2D^2 - 2D + 1)x = 0$

11. $(D^2 + 1)^4 x = 0$

*12. $(D^4 + 2D^3 + 3D^2 + 2D + 1)x = 0$

Solve the initial value problems in Exercises 13 to 18.

13. $16x'' + x = 0$; $x(0) = 2, x'(0) = 9$

14. $x^{(4)} - 81x = 0$; $x(0) = -2, x'(0) = 9, x''(0) = 18, x'''(0) = -27$

15. $(D^2 + 1)x = 0$; $x(\pi) = 0, x'(\pi) = 0$

16. $(D^2 + 1)x = 0$; $x(\pi) = 1, x'(\pi) = 1$

17. $(5D^2 + 2D + 1)x = 0$; $x(0) = 0, x'(0) = 1$

18. $(D^3 - 2D^2 + 2D - 4)x = 0$; $x(0) = 0, x'(0) = 4, x''(0) = 12$

Find an annihilator of smallest possible order for the given function in Exercises 19 to 24.

19. $e^{3t} - 5e^t$

20. $3 + te^t$

21. $t^2 e^{2t} - e^t + te^t$

22. $e^t + \sin 2t - 3$

23. $t \sin 2t$

24. $t^2 + e^t \sin 3t$

Exercises 25 to 29 refer back to our models.

25. *Damped Spring:*
    a. Find the general solution for the o.d.e. of Example 2.1.2, assuming $m = 2$ grams, $L = 7$ cm, $k = 5$ dynes/cm, and $b = 6$ dynes/(cm/sec).
    b. Find the specific solution that corresponds to a starting position at which the spring is 5 cm long and becoming longer at 1 cm/sec.

26. *Coupled Springs:* Find the general solution of the o.d.e. in Example 2.1.4, assuming $m_1 = m_2 = 1$ gram, $L_1 = 10$ cm, $L_2 = 3$ cm, $k_1 = 3$ dynes/cm, and $k_2 = 2$ dynes/cm.

27. *Circuits:* Find a formula, in terms of the inductance L and capacitance C, for the charge Q in the circuit of Exercise 3, Section 2.1, assuming $R = 0$, $V(t) = 0$, $Q(0) = Q_0$, and $I(0) = 0$.

28. *Anchored Floating Box:*
    a. Find the general solution of the homogeneous equation related to the model of the anchored floating box in Exercise 6, Section 2.1, assuming damping constant $b = 5$.
    b. Find a constant solution of the nonhomogeneous equation (this is the equilibrium position of the box).
    c. Write a formula for the motion of the box in terms of its initial position x_0 and velocity v_0.

29. *Two Springs:* Find the general solution of the o.d.e. in Exercise 4, Section 2.1, assuming $m = 5$ grams, $k_1 = k_2 = 1/2$ dyne/cm, $L_1 + L_2 = B$, and $b = 2$ dynes/(cm/sec).

2.9 NONHOMOGENEOUS LINEAR EQUATIONS: UNDETERMINED COEFFICIENTS

Now that we have a method for solving homogeneous linear equations with constant coefficients, it is time to turn to the behavior of "forced" models, that is, to nonhomogeneous equations. Recall that if $x = p(t)$ is a particular solution of

(N) $Lx = E(t)$

and $x = H(t)$ is the general solution of the related homogeneous equation

(H) $Lx = 0$

then the general solution of (N) is

$$x = H(t) + p(t).$$

Once we can solve the homogeneous equation, it suffices to find just one function solving the nonhomogeneous equation.

In this section we will describe a method for finding a particular solution of (N) when $L = P(D)$ has constant coefficients and $E(t)$ is itself the solution of a homogeneous equation with constant coefficients. If

$$A(D)E(t) = 0$$

we say that $A(D)$ **annihilates** $E(t)$. In this case, any function that solves

(N) $P(D)x = E(t)$

also solves

(H*) $A(D)P(D)x = A(D)E(t) = 0.$

Thus, we should look for our particular solution $x = p(t)$ among the solutions of the *homogeneous* equation (H*).

Since (H*) is homogeneous, we know how to get a complete list, $x = k_1h_1(t) + \ldots + k_sh_s(t)$, of its solutions. Our particular solution is on this list. That is,

$$p(t) = k_1h_1(t) + \ldots + k_sh_s(t)$$

for some choice of the coefficients k_i. We can *treat this as a guess for $x = p(t)$, substitute into (N), and determine the coefficients.* However, we want to minimize our work. Note that if $h_i(t)$ is a solution of (H), then the term $k_ih_i(t)$ will vanish upon substitution into the left side of (N); this term will contribute nothing toward attaining the right side of (N). We therefore *obtain a simpler guess for $p(t)$ by dropping terms that solve (H).*

Example 2.9.1

Solve

(N)
$$(D^2 - 2D - 3)x = 6 - 8e^t.$$

The characteristic polynomial of the related homogeneous equation

(H)
$$(D^2 - 2D - 3)x = 0$$

is $(r - 3)(r + 1)$. Thus the general solution of (H) is $x = H(t)$ where

$$H(t) = c_1 e^{3t} + c_2 e^{-t}.$$

The right side of (N) is a solution of any homogeneous equation whose characteristic polynomial has roots 0 and 1. In particular, it is annihilated by $A(D) = D(D - 1)$. Thus any function $x = p(t)$ that solves (N) will also solve

(H*)
$$D(D - 1)(D^2 - 2D - 3)x = 0.$$

The characteristic polynomial of (H*) is $r(r - 1)(r - 3)(r + 1)$, so $p(t)$ has to be of the form

$$p(t) = k_1 + k_2 e^t + k_3 e^{3t} + k_4 e^{-t}.$$

We get a simpler guess for $p(t)$ by crossing out the terms that solve (H):

$$p(t) = k_1 + k_2 e^t.$$

If we substitute our simpler guess for $x = p(t)$ into the left side of (N), we find

$$(D^2 - 2D - 3)(k_1 + k_2 e^t) = -3k_1 - 4k_2 e^t.$$

We want this to equal the right side of (N), $6 - 8e^t$. Thus

$$k_1 = -2 \quad \text{and} \quad k_2 = 2.$$

Our particular solution is

$$p(t) = -2 + 2e^t.$$

The general solution of (N) is

$$x = H(t) + p(t) = c_1 e^{3t} + c_2 e^{-t} - 2 + 2e^t.$$

Example 2.9.2

Solve

(N)
$$(D^2 - 4)x = -3e^{3t} + 10te^{3t}.$$

The general solution of the related homogeneous equation

(H)
$$(D^2 - 4)x = 0$$

is $x = H(t)$ where

$$H(t) = c_1 e^{2t} + c_2 e^{-2t}.$$

The right side of (N) is annihilated by $(D - 3)^2$ (see Example 2.8.5), so any function $x = p(t)$ that solves (N) also solves

(H*)
$$(D - 3)^2(D^2 - 4)x = 0.$$

The characteristic polynomial of (H*) is $(r - 3)^2(r - 2)(r + 2)$, so $p(t)$ has to be of the form

$$p(t) = k_1 e^{3t} + k_2 te^{3t} + k_3 e^{2t} + k_4 e^{-2t}.$$

We get a simpler guess for $p(t)$ by dropping the terms that solve (H):

$$p(t) = k_1 e^{3t} + k_2 te^{3t}.$$

If we substitute our simpler guess for $x = p(t)$ into the left side of (N) and use the Exponential Shift, we find

$$(D^2 - 4)(k_1 e^{3t} + k_2 te^{3t}) = e^{3t}[(D + 3)^2 - 4] (k_1 + k_2 t)$$
$$= e^{3t}[D^2 + 6D + 5] (k_1 + k_2 t)$$
$$= (5k_1 + 6k_2)e^{3t} + 5k_2 te^{3t}.$$

We want this to equal $-3e^{3t} + 10te^{3t}$, or

$$5k_1 + 6k_2 = -3, \qquad 5k_2 = 10.$$

Then

$$k_1 = -3, \qquad k_2 = 2.$$

Our particular solution is

$$p(t) = -3e^{3t} + 2te^{3t}.$$

The general solution of (N) is

$$x = H(t) + p(t) = c_1e^{2t} + c_2e^{-2t} - 3e^{3t} + 2te^{3t}.$$

In each of the preceding examples, the simplified guess for $x = p(t)$ was of the same form as the right side of (N), with specific constants replaced by undetermined coefficients. The following examples show that this is not always the case. The form of the simplified guess is sometimes more complicated than the right side of (N).

Example 2.9.3
Solve

(N) $$(D^2 - 4)x = e^t + 2e^{2t}.$$

The general solution of the related homogeneous equation

(H) $$(D^2 - 4)x = 0$$

is $x = H(t)$, where

$$H(t) = c_1e^{2t} + c_2e^{-2t}.$$

The right side of (N) is annihilated by $(D - 1)(D - 2)$, so a particular solution $x = p(t)$ of (N) will also solve

(H*) $$(D - 1)(D - 2)(D^2 - 4)x = 0.$$

The characteristic polynomial of (H*) is $(r - 1)(r - 2)^2(r + 2)$, so our particular solution has to be of the form

$$p(t) = k_1e^t + k_2e^{2t} + k_3te^{2t} + k_4e^{-2t}.$$

We obtain a simpler guess for $p(t)$ by crossing out the terms that solve (H):

$$p(t) = k_1e^t + k_3te^{2t}.$$

We substitute the simpler guess for $p(t)$ into the left side of (N):

$$\begin{aligned}
(D^2 - 4)p(t) &= (D^2 - 4)k_1e^t + (D^2 - 4)k_3te^{2t} \\
&= -3k_1e^t + e^{2t}[(D + 2)^2 - 4]k_3t \\
&= -3k_1e^t + e^{2t}(D^2 + 4D)k_3t \\
&= -3k_1e^t + 4k_3e^{2t}.
\end{aligned}$$

We want this to equal $e^t + 2e^{2t}$, so

$$k_1 = -1/3, \qquad k_3 = 1/2.$$

Our particular solution is $x = p(t)$ where

$$p(t) = -\frac{1}{3}e^t + \frac{1}{2}te^{2t}.$$

The general solution of (N) is

$$x = H(t) + p(t) = c_1e^{2t} + c_2e^{-2t} - \frac{1}{3}e^t + \frac{1}{2}te^{2t}.$$

Example 2.9.4

Find the general solution of

(N) $$(D^2 + 4)x = \sin 2t$$

which describes a forced undamped spring system as in Example 2.1.3, with $m = 1$, $b = 0$, $k = 4$, and $E(t) = \sin 2t$. Find the specific solution satisfying the initial conditions

$$x(0) = 0, \qquad x'(0) = 0.$$

The related homogeneous equation is

(H) $$(D^2 + 4)x = 0.$$

Its characteristic polynomial, $r^2 + 4$, has roots $\pm 2i$. The general solution of (H) is

$x = H(t)$ where

$$H(t) = c_1 \cos 2t + c_2 \sin 2t.$$

The right side of (N) is annihilated by $(D^2 + 4)$, so a particular solution $x = p(t)$ of (N) will also solve

(H*) $$(D^2 + 4)^2 x = 0.$$

The roots, $\pm 2i$, of the characteristic polynomial of (H*) both have multiplicity 2. Our particular solution has to be of the form

$$p(t) = k_1 \cos 2t + k_2 \sin 2t + k_3 t \cos 2t + k_4 t \sin 2t.$$

We obtain a simpler guess by crossing off the terms that solve (H):

$$p(t) = k_3 t \cos 2t + k_4 t \sin 2t.$$

We substitute the simpler guess into the left side of (N):

$$(D^2 + 4)(k_3 t \cos 2t + k_4 t \sin 2t) = -4k_3 \sin 2t + 4k_4 \cos 2t.$$

We want this to equal $\sin 2t$, so

$$k_3 = -1/4, \qquad k_4 = 0.$$

Our particular solution is $x = p(t)$ where

$$p(t) = -\frac{t}{4} \cos 2t.$$

The general solution of (N) is

$$x = H(t) + p(t) = c_1 \cos 2t + c_2 \sin 2t - \frac{t}{4} \cos 2t.$$

To find the specific solution of (N) satisfying the given initial conditions, we first differentiate the general solution to find

$$x' = -2c_1 \sin 2t + 2c_2 \cos 2t - \frac{1}{4} \cos 2t + \frac{t}{2} \sin 2t$$

so that

$$x(0) = c_1 \quad \text{and} \quad x'(0) = 2c_2 - \frac{1}{4}.$$

Matching this to the initial conditions, we find

$$c_1 = 0 \text{ and } c_2 = \frac{1}{8}.$$

The desired specific solution is

$$x = \frac{1}{8} \sin 2t - \frac{t}{4} \cos 2t.$$

Example 2.9.5

Solve

(N) $$(D^2 - 4)x = 1 + 65e^t \cos 2t.$$

The general solution of the related homogeneous equation

(H) $$(D^2 - 4)x = 0$$

is $x = H(t)$, where

$$H(t) = c_1 e^{2t} + c_2 e^{-2t}.$$

Since the right side of (N) is annihilated by $D(D^2 - 2D + 5)$ (see Example 2.8.6), a particular solution $x = p(t)$ of (N) will also solve

(H*) $$D(D^2 - 2D + 5)(D^2 - 4)x = 0.$$

The characteristic polynomial of (H*) is $r(r^2 - 2r + 5)(r - 2)(r + 2)$, so $p(t)$ has to be of the form

$$p(t) = k_1 + k_2 e^t \cos 2t + k_3 e^t \sin 2t + k_4 e^{2t} + k_5 e^{-2t}.$$

We get a simpler guess by crossing out the terms that solve (H):

$$p(t) = k_1 + k_2 e^t \cos 2t + k_3 e^t \sin 2t.$$

We substitute the simpler guess into the left side of (N):

$$(D^2 - 4)p(t) = (D^2 - 4)k_1 + (D^2 - 4)(k_2e^t \cos 2t + k_3e^t \sin 2t)$$
$$= -4k_1 + e^t(D^2 + 2D - 3)(k_2 \cos 2t + k_3 \sin 2t)$$
$$= -4k_1 + (-7k_2 + 4k_3)e^t \cos 2t + (-4k_2 - 7k_3)e^t \sin 2t.$$

We want this to equal $1 + 65e^t \cos 2t$, or

$$-4k_1 \qquad\qquad\qquad = 1$$
$$-7k_2 + 4k_3 = 65$$
$$-4k_2 - 7k_3 = 0.$$

Then

$$k_1 = -\frac{1}{4}, \qquad k_2 = -7, \qquad k_3 = 4.$$

Our particular solution is $x = p(t)$ where

$$p(t) = -\frac{1}{4} - 7e^t \cos 2t + 4e^t \sin 2t.$$

The general solution of (N) is

$$x = H(t) + p(t) = c_1e^{2t} + c_2e^{-2t} - \frac{1}{4} - 7e^t \cos 2t + 4e^t \sin 2t.$$

In order for the method we used in the examples to work, the right side of the o.d.e. must be a solution of a constant-coefficient homogeneous equation. But we know what all such solutions are. They are linear combinations of functions of the following types:

i.   constants
ii. exponentials: $e^{\lambda t}$
iii. trig functions: $\sin \beta t$, $\cos \beta t$
iv. exponentials times trig functions: $e^{\alpha t} \cos \beta t$, $e^{\alpha t} \sin \beta t$
v.   positive integer powers of t times any of the above.

In brief, our method requires the right side of the o.d.e. to be a linear combination of nonnegative integer powers of t times exponentials times sines or cosines. We cannot use our method otherwise, because we will not be able to find a constant-coefficient annihilator for $E(t)$ if, say, $E(t)$ is sec t or ln t or even $1/(1 + e^t)$. We can summarize the method as follows.

UNDETERMINED COEFFICIENTS

To solve a constant-coefficient equation

(N) $$P(D)x = E(t)$$

where $E(t)$ is a linear combination of nonnegative integer powers of t times exponentials times sines or cosines:

1. Solve the related homogeneous equation

    (H) $$P(D)x = 0.$$

2. Find an **annihilator** $A(D)$ for the right side of (N):

    $$A(D)E(t) = 0.$$

3. Since a particular solution $x = p(t)$ of (N) also solves

    (H*) $$A(D)P(D)x = 0,$$

    we obtain a description of the form of $p(t)$ from the general solution of (H*).
4. Obtain a simpler guess for $p(t)$ by dropping those terms that solve (H).
5. Substitute the simpler guess into the left side of (N). (The Exponential Shift is often helpful in this step.)
6. Determine the coefficients that yield the required right side, $E(t)$.
7. The general solution of (N) is the function obtained in step 6 plus the general homogeneous solution from step 1.

EXERCISES

Find the general solution in Exercises 1 to 10.

1. $(D^2 + 2D + 1)x = 5 + t$

2. $(D - 1)^4x = 2t + e^{-t}$

3. $(D^4 + 4D^3 + 4D^2)x = 2t$

4. $(3D^2 + 2D - 1)x = 2 \sin t$

5. $(D^2 + 1)x = \cos t$

6. $(4D^2 + 3D - 1)x = 25 - t^2$

7. $(D^2 + 6D + 10)x = 80e^t \sin t$

8. $(D^2 - D + 2)x = t^2 - 8e^{2t}$

9. $(9D^2 - 1)x = t \sin t$

10. $(D^5 - 4D^4 + 4D^3)x = 240 \, t^2 + 4e^{2t}$

Solve the initial value problems in Exercises 11 to 13.

11. $(D^2 + 1)x = \sin 3t$; $x(0) = x'(0) = 0$

12. $(D^3 + 5D^2 - 6D)x = 3e^t$; $x(0) = 1, x'(0) = 3/7, x''(0) = 6/7$

13. $(5D^2 + 2D + 1)x = 5t$; $x(0) = -1, x'(0) = 0$

Make a *simplified* guess for a particular solution in Exercises 14 to 16. Do *not* solve for the coefficients.

14. $(D - 1)^2(D^2 + 1)^3(D + 2)x = t^2 e^{3t} + e^t + e^{-t} \sin 3t + t^4$

15. $(D - 3)(D^2 + D + 1)(D + 1)^3 x = 3 - t^2 + t e^{-t/2} \sin (t \sqrt{5}) + e^{3t}$

16. $(D + 2)^7(D^2 + 1)^6 x = t e^{-2t} + \sin t$

Exercises 17 to 21 refer back to our models.

17. *Floating Boxes:* Solve the o.d.e. modeling the floating box in Exercise 1, Section 2.1, assuming (a) no frictional resistance, (b) viscous damping with $b = 10$, and (c) viscous damping with $b = 15$.

18. *Circuits:* Solve the o.d.e. modeling the *LRC* circuit in Exercise 3, Section 2.1, assuming $C = 1$, $V(t) = 10 \sin 2t$, and (a) $L = 1$, $R = 1$; (b) $L = 4$, $R = 4$; (c) $L = 4$, $R = 0$.

19. *Two Springs:* Show that in Exercise 4, Section 2.1, the distance from the mass to the left wall is the same as if the mass were attached to a single spring with constant $k = k_1 + k_2$ and natural length $L = \dfrac{k_1 L_1 + k_2(B - L_2)}{k_1 + k_2}$, subject to no external forces.

20. *Moving Spring:* Find the general solution of the o.d.e. modeling the moving spring in Exercise 5, Section 2.1, assuming (a) $b = 5$ and $k = 8$; (b) $b = 2$ and $k = 2$; (c) $b = 0$ and $k = 2$.

21. *Box Adrift:* Find the general solution of the o.d.e. in Exercise 7c, Section 2.1.

Another useful substitution:

22. An o.d.e. of the form

   (N) $(a_n t^n D^n + a_{n-1}t^{n-1} D^{n-1} + \ldots + a_1 t D + a_0)x = E(t), t > 0,$

   where $a_0, \ldots, a_n$ are constants, is known as an **equidimensional** or **Cauchy-Euler** equation. The substitution $s = \ln t$ (a change in time scale) transforms an equidimensional equation into a linear o.d.e. with constant coefficients.
   a. Show that if $s = \ln t$, then

$$\frac{dx}{dt} = \frac{1}{t}\frac{dx}{ds} \quad \text{and} \quad \frac{d^2x}{dt^2} = \frac{1}{t^2}\left[\frac{d^2x}{ds^2} - \frac{dx}{ds}\right].$$

b. Use (a) to show that, when $n = 2$, the substitution $s = \ln t$ transforms (N) into the o.d.e.

$$a_2 \frac{d^2x}{ds^2} + (a_1 - a_2) \frac{dx}{ds} + a_0 x = E(e^s).$$

Use the substitution described in Exercise 22 to solve the following o.d.e.'s. Express your final answers in terms of t.

23. $t^2x'' - 2tx' + 2x = 8 + t$, $t > 0$

24. $t^2x'' + 5tx' + 3x = \ln t$, $t > 0$

25. $t^2x'' + 5tx' + 4x = t^2 - 3t$, $t > 0$

26. $t^2x'' + x = 3 - \ln t$, $t > 0$

2.10 NONHOMOGENEOUS LINEAR EQUATIONS: VARIATION OF PARAMETERS

In the last section we discussed a method of finding a particular solution $x = p(t)$ of $Lx = E(t)$ that works whenever the equation has constant coefficients and $E(t)$ is of a special form. In this section we consider a method that does not require $E(t)$ to be of a special form and does not assume the equation has constant coefficients. It requires only that we know the general solution of the related homogeneous equation, $Lx = 0$. This method is therefore much more general than the preceding one. Unfortunately, it requires us to solve n equations in n unknown functions and can lead to hard integrals. The method is an extension of the method of varying parameters that we used in Section 1.3.

Let's begin by considering the second-order equation

$$[a_2(t)D^2 + a_1(t)D + a_0(t)]x = E(t).$$

We will assume that the equation is normal on an interval I and will look for solutions valid on I. We first divide by $a_2(t)$ to obtain an equation in **standard form:**

(N) $$[D^2 + b_1(t)D + b_0(t)]x = q(t).$$

Suppose we know that the general solution of the related homogeneous equation,

(H) $$[D^2 + b_1(t)D + b_0(t)]x = 0,$$

is

$$x = c_1h_1(t) + c_2h_2(t).$$

We look for a particular solution of (N) in the form $x = p(t)$ with

$$p(t) = c_1(t)h_1(t) + c_2(t)h_2(t).$$

In order to substitute our expression for x into (N), we must calculate $Dp(t)$ and $D^2p(t)$. We begin with $Dp(t)$:

$$Dp(t) = c_1(t)h_1'(t) + c_2(t)h_2'(t) + [c_1'(t)h_1(t) + c_2'(t)h_2(t)].$$

This formula is quite messy; $D^2p(t)$ will be even worse. Keep in mind, however, that we're not really interested in $c_1(t)$ and $c_2(t)$ for their own sake, but only as a means of finding $x = p(t)$. In general, many different choices of $c_1(t)$ and $c_2(t)$ will yield the same function $p(t)$. Thus we are allowed some freedom in picking $c_1(t)$ and $c_2(t)$. Let's make a simplifying assumption. Let's assume that the term in brackets in our expression for Dx is zero:

(A) $$c_1'(t)h_1(t) + c_2'(t)h_2(t) = 0.$$

Our formula for $Dp(t)$ now reads

$$Dp(t) = c_1(t)h_1'(t) + c_2(t)h_2'(t),$$

so that

$$D^2p(t) = c_1(t)h_1''(t) + c_2(t)h_2''(t) + c_1'(t)h_1'(t) + c_2'(t)h_2'(t).$$

Substitution of our expressions for $p(t)$, $Dp(t)$, and $D^2p(t)$ into (N) yields

$$c_1(t)[h_1''(t) + b_1(t)h_1'(t) + b_0(t)h_1(t)] + c_2(t)[h_2''(t) + b_1(t)h_2'(t) + b_0(t)h_2(t)]$$
$$+ c_1'(t)h_1'(t) + c_2'(t)h_2'(t) = q(t).$$

Since $h_1(t)$ and $h_2(t)$ solve (H), the terms in brackets are zero. Thus the last equation reads

(B) $$c_1'(t)h_1'(t) + c_2'(t)h_2'(t) = q(t).$$

Formulas (A) and (B) provide us with two equations for two unknown functions, $c_1'(t)$ and $c_2'(t)$:

(V) $$c_1'(t)h_1(t) + c_2'(t)h_2(t) = 0$$
$$c_1'(t)h_1'(t) + c_2'(t)h_2'(t) = q(t).$$

The determinant of coefficients of these equations is

$$\det \begin{bmatrix} h_1(t) & h_2(t) \\ h_1'(t) & h_2'(t) \end{bmatrix}.$$

This is just the Wronskian of $h_1(t)$ and $h_2(t)$. Since $h_1(t)$ and $h_2(t)$ generate the general solution of (H), the Wronskian test for solutions tells us that their Wronskian is never zero on I. Then Cramer's determinant test tells us that we can always solve for $c_1'(t)$ and $c_2'(t)$ (Cramer's rule, Note 3 of Section 2.4, gives a formula for the solution). We can solve and integrate to find $c_1(t)$ and $c_2(t)$. These in turn determine $p(t) = c_1(t)h_1(t) + c_2(t)h_2(t)$.

Example 2.10.1

Solve $(4D^2 - 4D + 1)x = t^{1/2}e^{t/2}$, $0 < t < +\infty$.

We first put the equation in standard form:

(N) $(D^2 - D + 1/4)x = (t^{1/2}e^{t/2})/4.$

The characteristic polynomial of the related homogeneous equation,

(H) $(D^2 - D + 1/4)x = 0,$

is $(r - 1/2)^2$. The general solution of (H) is $x = H(t)$ with

$$H(t) = c_1 e^{t/2} + c_2 t e^{t/2}.$$

We look for a particular solution of (N) in the form $x = p(t)$ with

$$p(t) = c_1(t)e^{t/2} + c_2(t)te^{t/2}.$$

In this example, $h_1(t) = e^{t/2}$ and $h_2(t) = te^{t/2}$, so that $h_1'(t) = (e^{t/2})/2$ and $h_2'(t) = (2 + t)(e^{t/2})/2$. The equations (V) read

(V) $c_1'(t)\, e^{t/2} + c_2'(t)\, te^{t/2} = 0$

$$c_1'(t)\frac{e^{t/2}}{2} + c_2'(t)\frac{(2 + t)e^{t/2}}{2} = \frac{t^{1/2}e^{t/2}}{4}.$$

If we subtract twice the second equation from the first, we get $-2c_2'(t)e^{t/2} = -(t^{1/2}e^{t/2})/2$, or

$$c_2'(t) = \frac{1}{4} t^{1/2}.$$

Substitution of this value into the first equation yields

$$c_1'(t) = -\frac{1}{4} t^{3/2}.$$

We integrate these formulas to obtain specific values of $c_1(t)$ and $c_2(t)$:

$$c_1(t) = -\frac{1}{10} t^{5/2}, \qquad c_2(t) = \frac{1}{6} t^{3/2}.$$

Our particular solution is $x = p(t)$ with

$$p(t) = -\frac{1}{10} t^{5/2} e^{t/2} + \frac{1}{6} t^{3/2} t e^{t/2} = \frac{1}{15} t^{5/2} e^{t/2}.$$

The general solution of (N) is

$$x = H(t) + p(t) = c_1 e^{t/2} + c_2 t e^{t/2} + \frac{1}{15} t^{5/2} e^{t/2}.$$

Example 2.10.2

  Solve

(N) $$(D^2 + 1)x = \sec t, \qquad -\frac{\pi}{2} < t < +\frac{\pi}{2}.$$

  The equation is already in standard form. The general solution of the related homogeneous equation is $x = H(t)$ with

$$H(t) = c_1 \cos t + c_2 \sin t.$$

  We look for a particular solution in the form $x = p(t)$ with

$$p(t) = c_1(t) \cos t + c_2(t) \sin t.$$

In this case the equations (V) read

(V) $$c_1'(t) \cos t + c_2'(t) \sin t = 0$$
$$-c_1'(t) \sin t + c_2'(t) \cos t = \sec t.$$

Cramer's rule provides formulas for $c_1'(t)$ and $c_2'(t)$:

$$c_1'(t) = \frac{\det \begin{bmatrix} 0 & \sin t \\ \sec t & \cos t \end{bmatrix}}{\det \begin{bmatrix} \cos t & \sin t \\ -\sin t & \cos t \end{bmatrix}} = \frac{-\sin t \sec t}{\cos^2 t + \sin^2 t} = -\frac{\sin t}{\cos t}$$

$$c_2'(t) = \frac{\det \begin{bmatrix} \cos t & 0 \\ -\sin t & \sec t \end{bmatrix}}{\det \begin{bmatrix} \cos t & \sin t \\ -\sin t & \cos t \end{bmatrix}} = \frac{\cos t \sec t}{\cos^2 t + \sin^2 t} = 1.$$

We integrate these formulas to get

$$c_1(t) = \ln |\cos t|, \qquad c_2(t) = t.$$

Our particular solution is $x = p(t)$ with

$$p(t) = (\ln |\cos t|) \cos t + t \sin t.$$

The general solution is

$$x = H(t) + p(t) = c_1 \cos t + c_2 \sin t + (\ln |\cos t|) \cos t + t \sin t.$$

This method, which we have worked out for second-order equations, can be extended to apply to nth-order equations. Instead of making a single simplifying assumption, we make $n - 1$ simplifying assumptions. Substitution into the o.d.e. provides another equation. The details are quite messy, so we will content ourselves with a statement of the result.

Fact: *Suppose we are given an nth-order o.d.e. in* **standard form**

(N) $[D^n + b_{n-1}(t)D^{n-1} + \ldots + b_1(t)D + b_0(t)]x = q(t)$

which is normal on an interval I. Suppose we know the general solution of the related homogeneous equation is $x = H(t)$ with

$$H(t) = c_1 h_1(t) + \ldots + c_n h_n(t).$$

If $c_1(t), \ldots, c_n(t)$ are functions that satisfy the equations

$$
\begin{aligned}
c_1'(t)h_1(t) \quad + \ldots + c_n'(t)h_n(t) \quad &= 0 \\
c_1'(t)h_1'(t) \quad + \ldots + c_n'(t)h_n'(t) \quad &= 0
\end{aligned}
$$

(V)

$$
\begin{aligned}
c_1'(t)h_1^{(n-2)}(t) + \ldots + c_n'(t)h_n^{(n-2)}(t) &= 0 \\
c_1'(t)h_1^{(n-1)}(t) + \ldots + c_n'(t)h_n^{(n-1)}(t) &= q(t)
\end{aligned}
$$

then

$$
p(x) = c_1(t)h_1(t) + \ldots + c_n(t)h_n(t)
$$

is a particular solution of (N).

The equations (V) are surprisingly easy to remember. They are equations for the unknowns $c_1'(t), \ldots, c_n'(t)$. The coefficients are just the entries of the Wronskian of $h_1(t), \ldots, h_n(t)$. The right sides are all zero, except the last, which is the right side of (N). Note, however, that *this assumes the o.d.e. is in standard form.*

Example 2.10.3

Solve $(t^2D^3 + 2tD^2 - 2D)x = t^3$, $0 < t < +\infty$.
We divide by t^2 to obtain standard form:

(N)
$$
\left(D^3 + \frac{2}{t}D^2 - \frac{2}{t^2}D \right)x = t.
$$

We obtained the general solution of the related homogeneous equation in Example 2.3.2; it is $x = H(t)$ with

$$
H(t) = c_1t^2 + c_2\frac{1}{t} + c_3.
$$

We look for a particular solution in the form $x = p(t)$ with

$$
p(t) = c_1(t)t^2 + c_2(t)\frac{1}{t} + c_3(t).
$$

In this case the equations (V) read

$$t^2 c_1'(t) + \frac{1}{t} c_2'(t) + c_3'(t) = 0$$

(V)
$$2t c_1'(t) - \frac{1}{t^2} c_2'(t) \qquad = 0$$

$$2 c_1'(t) + \frac{2}{t^3} c_2'(t) \qquad = t$$

If we multiply the last equation by t and subtract the second equation, we get $(3/t^2)c_2'(t) = t^2$, or

$$c_2'(t) = t^4/3.$$

Then

$$c_1'(t) = t/6,$$

so that

$$c_3'(t) = -t^3/2.$$

We integrate to get specific values for the $c_i(t)$'s:

$$c_1(t) = t^2/12, \qquad c_2(t) = t^5/15, \qquad c_3(t) = -t^4/8.$$

The function

$$p(t) = \frac{t^4}{12} + \frac{t^4}{15} - \frac{t^4}{8} = \frac{t^4}{40}$$

is a particular solution of (N). The general solution is

$$x = H(t) + p(t) = c_1 t^2 + c_2 \frac{1}{t} + c_3 + \frac{t^4}{40}.$$

A limited version of the method of Variation of Parameters was treated by Euler (1743). The systematic application for linear equations of all orders was worked out by Lagrange (1774).

VARIATION OF PARAMETERS

To solve an nth-order nonhomogeneous linear equation

$$[a_n(t)D^n + \ldots + a_1(t)D + a_0(t)]x = E(t):$$

0.   Divide by $a_n(t)$ to obtain an equation in **standard form:**

(N) $[D^n + \ldots + b_1(t)D + b_0(t)]x = q(t).$

1.   Find the general solution of the related homogeneous equation

(H) $[D^n + \ldots + b_1(t)D + b_0(t)]x = 0.$

This has the form $x = H(t)$ with

$$H(t) = c_1 h_1(t) + \ldots + c_n h_n(t).$$

2.   Solve the equations

$$c_1'(t)h_1(t) + \ldots + c_n'(t)h_n(t) \quad = 0$$
$$c_1'(t)h_1'(t) + \ldots + c_n'(t)h_n'(t) \quad = 0$$

(V)

$$\cdot$$
$$\cdot$$
$$\cdot$$

$$c_1'(t)h_1^{(n-2)}(t) + \ldots + c_n'(t)h_n^{(n-2)}(t) = 0$$
$$c_1'(t)h_1^{(n-1)}(t) + \ldots + c_n'(t)h_n^{(n-1)}(t) = q(t)$$

for the unknowns $c_1'(t), \ldots, c_n'(t).$

3.   Integrate to find specific values for $c_1(t), \ldots, c_n(t).$
4.   The function $p(t) = c_1(t)h_1(t) + \ldots + c_n(t)h_n(t)$ is a particular solution of (N).
5.   The general solution of (N) is obtained by adding the particular solution found in step 4 to the general homogeneous solution found in step 1.

Note
A shortcut

In our examples we obtained specific values for each $c_i(t)$ by integrating the formulas for $c_i'(t)$, taking the constant of integration to be zero. If we include arbitrary constants c_i when we integrate, then substitution of the resulting functions into our formula for x yields the general solution of (N).

EXERCISES

Find the general solution in Exercises 1 to 8 using Variation of Parameters.

1. $2x'' - 6x' + 4x = 6e^{2t}$

2. $4x'' - 4x' + x = \dfrac{8}{t^2}e^{t/2}, \quad t > 0$

3. $x'' + x = \tan t, \quad -\dfrac{\pi}{2} < t < \dfrac{\pi}{2}$

4. $x'' - 2x' + x = e^t \ln t, \quad t > 0$

5. $9x'' - 6x' + x = (te^t)^{1/3}$

6. $(2D - 1)(2D + 1)(D - 2)x = 15e^{2t}$

7. $(D - 1)^3 x = e^t/t^3, \quad t > 0$

8. $(D^3 + D)x = \sec^2 t, \quad -\dfrac{\pi}{2} < t < \dfrac{\pi}{2}$

Solve the initial value problems in Exercises 9 to 11.

9. $x'' + x = \sec t, \quad -\dfrac{\pi}{2} < t < \dfrac{\pi}{2}; \qquad x(0) = 1, x'(0) = 2$

10. $5x'' - 10x' + 5x = t^{1/5}e^t; \qquad x(0) = x'(0) = 0$

11. $5x'' - 10x' + 5x = 60t^{-5}e^t; \qquad x(\ln 2) = x'(\ln 2) = 0$

In Exercises 12 to 15, you are given the general solution $x = H(t)$ of the related homogeneous equation. Find the general solution of the given (nonhomogeneous) equation.

12. $[(t - 1)D^2 - tD + 1]x = (t - 1)^2 e^t, \quad t > 1; \qquad H(t) = c_1 t + c_2 e^t$

13. $[(t^2 + t)D^2 + (2 - t^2)D - (2 + t)]x = (t + 1)^2, \quad t > 0; \qquad H(t) = c_1 t^{-1} + c_2 e^t$

14. $(tD^3 + 3D^2)x = t^{-1/2}, \quad t > 0; \qquad H(t) = c_1 + c_2 t + c_3 t^{-1}$

15. $(t^3 D^3 + t^2 D^2 - 2tD + 2)x = t^{-1}, \quad t > 0; \qquad H(t) = c_1 t + c_2 t^2 + c_3 t^{-1}$

In Exercises 16 to 19, the general solution of the related homogeneous equation is generated by functions of the form t^α. Find the general solution of the given (nonhomogeneous) equation.

16. $(tD^2 - D)x = t^2, \quad t > 0$

17. $(t^2 D^2 + 4tD + 2)x = t^5, \quad t > 0$

18. $(t^3 D^3 - 3t^2 D^2 + 6tD - 6)x = t^{-1}, \quad t > 0$

19. $(tD^2 + 2D)x = \sqrt{t}, \quad t > 0$

Sometimes a linear o.d.e. with variable coefficients, $Lx = E(t)$, can be solved by a modified Variation-of-Parameters technique known as **reduction of order.** The method requires that we know one nonzero solution $x = h(t)$ of the related homogeneous o.d.e., $Lx = 0$. By analogy with Variation of Parameters, we guess that the general solution of $Lx = E(t)$ should have the form

$$x = k(t)h(t).$$

Substituting this guess for x into $Lx = E(t)$ leads to a linear equation of order $n - 1$ for $y = k'(t)$. With luck we can solve for y, integrate to find $k(t)$, and multiply by $h(t)$ to find x. Use this trick to solve the o.d.e.'s in Exercises 20 to 24, given that each of the related homogeneous equations has a solution either of the form $x = t^\alpha$ or of the form $x = e^{\lambda t}$.

20. $[t^2D^2 + 5tD + 4]x = 0, \quad t > 0$

21. $[t^3D^2 + tD - 1]x = 0, \quad t > 0$

22. $[tD^2 - (t + 1)D + 1]x = 0, \quad t > 0$

23. $[tD^2 - (t + 1)D + 1]x = 4, \quad t > 0$

24. $[tD^2 + (2t - 1)D + (t - 1)]x = e^{-t}, \quad t > 0$

Exercises 25 to 27 refer back to our models.

25. *Springs:* Find the general solution of the o.d.e. modeling the spring of Example 2.1.3 if
   a. $m = 1$ gram, $b = 4$ dynes/(cm/sec), $k = 3$ dynes/cm, and $E(t) = e^{-t}$ dynes.
   b. $m = 2$ grams, $b = 0$, $k = 8$ dynes/cm, and $E(t) = 8 \sin^2 2t$ dynes.

26. *Floating Box:* The floating box of Exercises 1 and 6, Section 2.1, has the anchor unhooked from the line (with $L = 2$ and $k = 22$), which remains attached to the box. The other end is pulled down so that at time t its depth is $2 + \sec 13t$ feet, $0 < t < \pi/2$. Set up an equation to model the motion of the box, assuming no friction, and find the general solution via variation of parameters.

27. *Moving Spring:* Modify Exercise 5, Section 2.1, by assuming simply that at time t the elevator is $h(t)$ feet off the ground.
   a. Set up a second-order o.d.e. modeling the height x of the weight, assuming $b = 3$ and $k = 4$.
   b. Using Variation of Parameters, derive a formula (involving some integrals including $h(t)$, L, and two arbitrary constants) to predict the height x at any time $t > 0$.

2.11 BEHAVIOR OF SPRING MODELS (Optional)

In this section we consider again the first three models of Section 2.1. We find the general solutions of their differential equations and interpret them in physical terms.

Example 2.11.1   Undamped Springs Revisited

The equation (from Example 2.1.1) for a mass m on a spring with constant k and no damping can be written in operator form as

$$(mD^2 + k)x = 0.$$

The characteristic polynomial, $mr^2 + k$, has roots $r = \pm i\sqrt{k/m}$ (remember that, for physical reasons, $k > 0$ and $m > 0$). If we set

$$\omega = \sqrt{k/m}$$

then the general solution is

$$x(t) = c_1 \cos \omega t + c_2 \sin \omega t.$$

To determine a specific solution (that is, one particular motion of the mass), we must specify the initial position and velocity:

$$x(0) = x_0 \quad \text{and} \quad x'(0) = v_0.$$

The solution of this initial value problem, obtained by substituting the general solution into these initial conditions and solving for c_1 and c_2, is

$$x(t) = x_0 \cos \omega t + \frac{v_0}{\omega} \sin \omega t.$$

If the mass starts at rest ($v_0 = 0$) from the equilibrium position ($x_0 = 0$), then the solution is $x(t) = 0$, expressing the fact that the mass remains at rest. If x_0 and v_0 are not both zero, we can rewrite the solution in **phase-amplitude** form as

$$x(t) = A \cos (\omega t - \alpha)$$

where

$$A = \sqrt{x_0^2 + (v_0/\omega)^2}$$

and α satisfies

$$\cos \alpha = x_0/A \quad \text{and} \quad \sin \alpha = v_0/\omega A$$

(see Exercise 1). The phase-amplitude form exhibits the solution as a regular oscillation about the equilibrium position (see Figure 2.5).

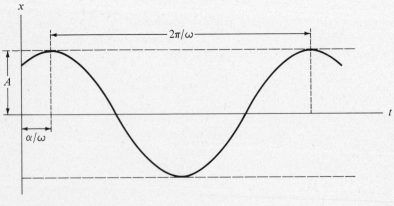

FIGURE 2.5

Example 2.11.2   Damped Springs Revisited

The equation of Example 2.1.2, modeling the motion of a mass m under the influence of a spring with constant k and damping with coefficient b, has the operator form

$$(mD^2 + bD + k)x = 0.$$

The characteristic polynomial, $mr^2 + br + k$, has roots

$$r = \frac{-b \pm \sqrt{b^2 - 4mk}}{2m}.$$

The nature of the solutions is determined by the sign of the discriminant, $b^2 - 4mk$.

If $b^2 - 4mk < 0$, we say the motion is **underdamped.** In this case, the roots are complex. If we set

$$\sigma = \frac{b}{2m} \quad \text{and} \quad \omega = \frac{\sqrt{4mk - b^2}}{2m},$$

then the roots are $r = -\sigma \pm \omega i$, so the general solution is

$$x(t) = c_1 e^{-\sigma t} \cos \omega t + c_2 e^{-\sigma t} \sin \omega t.$$

Substituting our formula for $x(t)$ into the initial value problem

$$x(0) = x_0, \qquad x'(0) = v_0,$$

we find

$$x(t) = x_0 e^{-\sigma t} \cos \omega t + \frac{(v_0 + \sigma x_0)}{\omega} e^{-\sigma t} \sin \omega t.$$

We can rewrite this in phase-amplitude form as

$$x(t) = A e^{-\sigma t} \cos (\omega t - \alpha)$$

where

$$A = \sqrt{x_0^2 + \frac{(v_0 + \sigma x_0)^2}{\omega^2}}$$

and α satisfies

$$\cos \alpha = x_0/A \quad \text{and} \quad \sin \alpha = (v_0 + \sigma x_0)/\omega A.$$

The solution represents an oscillation about the equilibrium position whose amplitude dies down as $t \to \infty$ (see Figure 2.6). The number σ determines the rate at which the amplitude dies down.

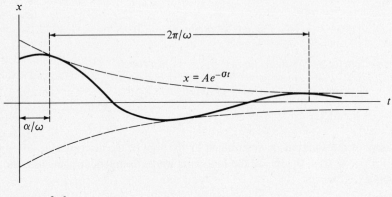

FIGURE 2.6

If $b^2 - 4mk > 0$, we say the motion is **overdamped.** In this case, the characteristic roots are real and distinct. If we set

$$\sigma_1 = \frac{b - \sqrt{b^2 - 4mk}}{2m} \quad \text{and} \quad \sigma_2 = \frac{b + \sqrt{b^2 - 4mk}}{2m},$$

then the roots are $-\sigma_1$ and $-\sigma_2$, so the general solution is

$$x(t) = c_1 e^{-\sigma_1 t} + c_2 e^{-\sigma_2 t}.$$

Note that $0 < \sigma_1 < \sigma_2$, so that both terms die down to 0 as $t \to \infty$. Substitution into the initial value problem

$$x(0) = x_0, \qquad x'(0) = v_0$$

gives

$$c_1 = \frac{\sigma_2 x_0 + v_0}{\sigma_2 - \sigma_1} \quad \text{and} \quad c_2 = -\frac{\sigma_1 x_0 + v_0}{\sigma_2 - \sigma_1}.$$

The solution crosses the equilibrium position at most once, at time

$$t_1 = \frac{1}{\sigma_2 - \sigma_1} \ln\left(-\frac{c_2}{c_1}\right) = \frac{1}{\sigma_2 - \sigma_1} \ln\left(\frac{\sigma_1 x_0 + v_0}{\sigma_2 x_0 + v_0}\right).$$

and has at most one relative extremum, at time

$$t_2 = \frac{1}{\sigma_2 - \sigma_1} \ln\left(-\frac{c_2}{c_1}\frac{\sigma_2}{\sigma_1}\right) = \frac{1}{\sigma_2 - \sigma_1} \ln\left[\left(\frac{\sigma_1 x_0 + v_0}{\sigma_2 x_0 + v_0}\right)\frac{\sigma_2}{\sigma_1}\right].$$

Note that in general $t_2 > t_1$. We sketch three typical forms for the graph of the solution. Figure 2.7(a) depicts a solution with $x_0 > 0$ and $v_0 > 0$ (so that $t_1 < 0 < t_2$). Figure 2.7(b) depicts a solution with $x_0 > 0$ and $v_0 < -\sigma_2 x_0$ (so that $0 < t_1 < t_2$). Figure 2.7(c) depicts a solution with $x_0 > 0$ and $-\sigma_2 x_0 \le v_0 < 0$ (so that t_1 and t_2 are both negative or both undefined). Physically, the first solution moves away from equilib-

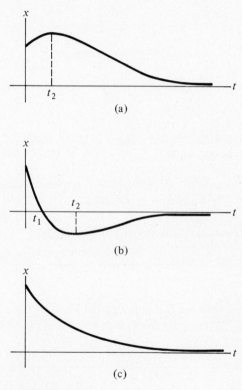

(a)

(b)

(c)

FIGURE 2.7

rium, turns back at time t_2, and then approaches equilibrium asymptotically from the right; the second initially moves toward equilibrium, overshoots it (at time t_1), then turns around and approaches equilibrium asymptotically from the left; the third solution simply approaches equilibrium asymptotically from the right.

If $b^2 - 4mk = 0$, we say the motion is **critically damped.** In this case, there is a real double root $r = -b/2m$. If we set

$$\sigma = \frac{b}{2m},$$

then the general solution is

$$x = c_1 e^{-\sigma t} + c_2 t e^{-\sigma t}.$$

The solution that satisfies the initial conditions

$$x(0) = x_0 \quad \text{and} \quad x'(0) = v_0$$

is

$$x = x_0 e^{-\sigma t} + (v_0 + \sigma x_0) t e^{-\sigma t}.$$

This solution, again, has at most one zero and at most one local extremum; these occur at times

$$t_1 = \frac{-x_0}{(v_0 + \sigma x_0)} \quad \text{and} \quad t_2 = \frac{v_0}{\sigma(v_0 + \sigma x_0)}$$

respectively. If $x_0 > 0$ and $v_0 > 0$, then $t_1 < 0 < t_2$ and the solution behaves roughly like Figure 2.7(a). If $x_0 > 0$ and $v_0 < -\sigma x_0$, then $t_2 > t_1 > 0$ and the graph resembles Figure 2.7(b). If $x_0 > 0$ and $-\sigma x_0 \leq v_0 < 0$, then t_1 and t_2 are both undefined or both negative and the solution behaves as in Figure 2.7(c).

The preceding discussion has indicated the kinds of unforced motion possible for a damped spring. We consider briefly two kinds of forced motion.

Example 2.11.3 Constant Forcing

When a mass of m grams, hooked to a spring with constant k dynes/cm and damping with coefficient b dynes/(cm/sec), is subject to a constant external force E dynes (in the direction of increasing x)—for example, when the system hangs vertically and is acted on by gravity—the operator form of the equation is

$$(mD^2 + bD + k)x = E.$$

Applying the method of undetermined coefficients, we see that solutions have the form

$$x = \frac{E}{k} + h(t)$$

where $h(t)$ is a solution of the homogeneous (unforced) equation treated in Example 2.11.2. We saw that in the unforced case the motion either oscillated about or tended asymptotically to the equilibrium. In the present case the graphs of all solutions are translated up by the constant amount E/k. This is the same effect as if the spring had natural length $L + E/k$. Thus we can describe the effect of constant forcing by saying that *in the presence of a constant external force E, a damped spring acts the same as an unforced system with the same mass m, damping coefficient b, and spring constant k, but with the equilibrium length L replaced by L + E/k.*

Example 2.11.4   Periodic Forcing—Resonance

When a mass of m grams on a spring with constant k and no damping is subjected to an oscillating force $E(t) = E \sin \beta t$, the equation, in operator form, is

$$(mD^2 + k)x = E \sin \beta t.$$

We know from Example 2.11.1 that the general solution of the homogeneous (unforced) equation has the form $x = H(t)$, where

$$H(t) = c_1 \cos \omega t + c_2 \sin \omega t$$

with $\omega = \sqrt{k/m}$. We can use the method of undetermined coefficients to find a particular solution, and hence the general solution, of the forced equation. Note, however, that we have to handle the cases $\beta \neq \omega$ and $\beta = \omega$ separately. (Why?)

If $\beta \neq \omega$, the general (forced) solution is

$$x = \frac{-E}{m(\beta^2 - \omega^2)} \sin \beta t + c_1 \cos \omega t + c_2 \sin \omega t.$$

This is a superposition of regular oscillations with different amplitudes and frequencies, and in general it can be quite complicated. Note, however, that the solution is bounded by $|c_1| + |c_2| + |E/m(\beta^2 - \omega^2)|$ and that if β and ω are both integer multiples of some number γ, say $\beta = B\gamma$ and $\omega = W\gamma$, then x will be periodic, repeating itself after $2\pi/\gamma$.

When $\beta = \omega$, a particular solution is

$$p(t) = \frac{-E}{2\omega m} t \cos \omega t.$$

Note that this function oscillates with increasing amplitude and becomes unbounded as $t \to \infty$. Since any solution of the homogeneous equation is bounded, this means that *when* $\beta = \omega$, *every motion oscillates with amplitude increasing without bound as* $t \to \infty$. This phenomenon is known as **resonance.** If this phenomenon were achieved physically, the spring would snap once the oscillations became too big, and our model would not apply. In practice, of course, there will be some friction present, and we will never have β *exactly* matching ω; it can be shown in these cases that resonant forcing leads to behavior with bounded but relatively large amplitude.

EXERCISES

Exercises 1 to 6 refer to the models discussed in the text.

1. Derive the values of the constants A and α in the phase-amplitude form of the solution of Example 2.11.1, using the trigonometric identity $\cos (\theta_1 - \theta_2) = \cos \theta_1 \cos \theta_2 + \sin \theta_1 \sin \theta_2$.

2. a. Verify the values of A and α in the phase-amplitude form of the solution in Example 2.11.2.
   b. Check the values of the times t_1 and t_2 in the overdamped case of Example 2.11.2.

3. Consider an unforced mass-spring system with $m = 1$ gram and $k = 100$ dynes/cm. For each of the following values of the damping constant, find the solution starting from equilibrium with velocity 1 cm/sec (increasing x), and graph it carefully for $0 \leq t \leq 1$ (you will need a calculator to do the numerical work).
   a. $b = 0$ b. $b = 12$ c. $b = 16$ d. $b = 20$ e. $b = 52$

4. In Example 2.11.3, the equilibrium position of a system with constant forcing is independent of the mass. Yet if we hang heavier objects on a single spring, the equilibrium position changes. Why?

5. What behavior should we expect in Example 2.11.4, as β approaches ω?

6. Reinterpret the results of Example 2.11.2 in terms of the charge Q in the *LRC* circuit of Exercise 3, Section 2.1. What can you say about the behavior of the *current* in the overdamped and underdamped cases? (*Hint:* Use phase-amplitude form for Q.)

Exercises 7 to 10 ask you to examine specific models. A calculator will help you carry out the numerical work.

7. *Springs:* Find the general solution of the o.d.e. modeling a 1-gram mass, attached to a spring of natural length 10 cm and spring constant 1 dyne/cm, with no external forcing, for each of the following damping constants.
   a. $b = 1$ dynes/(cm/sec)   b. $b = 10$ dynes/(cm/sec)   c. $b = 2$ dynes/(cm/sec)

8. *Quantitative Predictions:* For each of the models in Exercise 7, predict the length of the spring after 10 seconds if
     i.   the spring is initially 10 cm long and stretching at 5 cm/sec.
     ii.   the spring is initially 9 cm long and compressing at 1 cm/sec.
     iii.   the spring is initially 11 cm long and at rest.

9. *Qualitative Predictions:* For each of the models in Exercise 7, decide which of the following statements apply.
     i.   Unless the mass starts at the equilibrium position with zero velocity, it will move back and forth across equilibrium indefinitely.
     ii.   Given sufficient initial velocity at the equilibrium position, the spring will stretch indefinitely (until it breaks).
     iii.   The mass will, after sufficient time, stay forever within one angstrom (10^{-8} cm) of the equilibrium position.

10. *Floating Box:* A cubic box with 1-foot sides, of weight w pounds (and mass $w/32$ slugs), floats in the water as in Exercise 1, Section 2.1.
    a. Assuming no friction, find a general formula predicting the motion of the box.
    b. Find a formula to determine w from the frequency with which the box bobs up and down. [*Hint:* $x = A \cos(\lambda t - \alpha)$ has frequency $\lambda/2\pi$.]

2.12 ROTATIONAL MODELS (Optional)

In this section we consider some mechanical models in which a body turns or bends instead of moving back and forth along a straight line. In these contexts it is convenient to reformulate the physical quantities of position, mass, and force into their rotational equivalents: angular displacement, moment of inertia, and moment.

Suppose a body of mass m is constrained to move in a circle of radius r (Figure 2.8). Its position is naturally described by an angular displacement θ from some reference position. An analysis of the motion based on forces has to take account of the forces that keep the body moving in a circle. These forces are taken care of automatically if we calculate the equations of motion in terms of moments and moment of inertia. These are defined as follows.

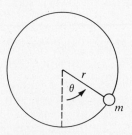

FIGURE 2.8

1. **Moment of inertia:** A point mass m traveling in a circle of radius r has moment of inertia

$$J = mr^2.$$

The moment of inertia for a continuously distributed mass is the integral of density times the square of the radius.

2. **Moment:** A force acting at radius r exerts a moment

$$M = F_{\text{tan}}r$$

where F_{tan} is the component of force tangent to the circle in the direction of increasing θ.

Newton's second law is expressed in terms of these quantities as

$$M = J\frac{d^2\theta}{dt^2}.$$

Example 2.12.1   A Twisted Shaft

A mass is attached to one end of a flexible shaft, the other end of which is fixed, and the contraption is mounted so that the mass can rotate (Figure 2.9). The mass is subject to two moments. The resistance of the shaft to twisting is reflected in a moment

$$M_{\text{shaft}} = -\kappa\theta$$

where θ is the angular displacement from an equilibrium position, while friction with the mounting mechanism or surrounding medium contributes

$$M_{\text{fric}} = -\beta\frac{d\theta}{dt}.$$

Combining these into Newton's law (in moment form), we obtain

$$J\frac{d^2\theta}{dt^2} = M_{\text{shaft}} + M_{\text{fric}} = -\kappa\theta - \beta\frac{d\theta}{dt}$$

or

$$(JD^2 + \beta D + \kappa)\theta = 0.$$

This is identical in form to the equation of an unforced damped spring, and the analysis of Examples 2.11.1 and 2.11.2 applies.

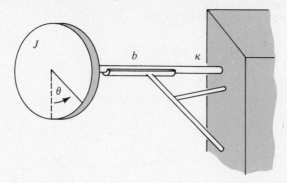

FIGURE 2.9

Example 2.12.2   The Simple Pendulum

A bob of mass m grams swings at the end of a rigid shaft of negligible weight that is L centimeters long; the other end of the shaft is attached to a support that allows it to rotate without friction (Figure 2.10). The force here is that of gravity, $G = gm$, where g is a constant ($g \approx 32$ ft/sec^2). If we measure angular displacement θ so that $\theta = 0$ when the shaft is hanging vertically, then the tangential component of the gravitational force has magnitude $G \sin \theta$. Thus, remembering that $r = L$ and gravity points down, we have

$$M_{\text{grav}} = -Lgm \sin \theta.$$

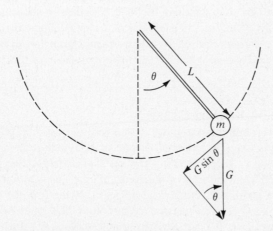

FIGURE 2.10

On the other hand, the moment of inertia is

$$J = mL^2.$$

Thus, the equation of motion is

$$mL^2 \frac{d^2\theta}{dt^2} = -Lgm \sin \theta$$

or

$$D^2\theta + \frac{g}{L} \sin \theta = 0.$$

This equation is nonlinear, and in fact the solution cannot be expressed in terms of elementary functions. However, for *small oscillations*—that is, when the bob is not very far from the equilibrium $\theta = 0$—we can approximate with a linear equation. Recall from calculus that

$$\lim_{\theta \to 0} \frac{\sin \theta}{\theta} = 1$$

which means, when $|\theta|$ is small, that $\sin \theta$ is approximately equal to the numerical value of θ (in radians). Hence, we can replace $\sin \theta$ with θ in the o.d.e. to obtain the approximate, but linear, o.d.e.

$$(D^2 + g/L)\theta = 0.$$

This is identical to the undamped spring equation of Example 2.11.1. Setting $\omega = \sqrt{g/L}$, we see that solutions oscillate periodically about equilibrium with a frequency of $\omega/2\pi$ cycles/sec. In particular, note that our approximation hypothesis—that $|\theta|$ is small—remains true in the approximate model for all time, provided the initial angular velocity is also small. (Compare Exercise 5 of this section.)

Example 2.12.3   Bending Beams

When a long, thin piece of metal (a thin wire several centimeters long or a steel beam many yards long) is supported at one or both ends and has various weights attached to it along its length, it bends (Figure 2.11). Suppose the beam is supported so that its undeflected shape is a horizontal line; then its deflected shape can be described by the graph of a function $y = -s(x)$, where x is the distance from one end of the beam and y is the vertical deflection of that point from its equilibrium position. Note that here, by contrast with our other examples, the independent variable is x, a distance, instead of time t.

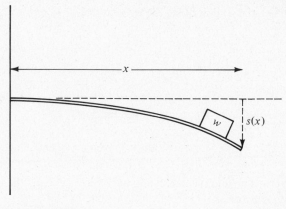

FIGURE 2.11

The equations of elasticity theory are formulated in terms of the **turning moment** $m(x)$ of a given force relative to the point x. This is defined as the sum of the moments, taking x as a center of rotation, of all forces acting to the right of x. If we have a weight distribution described by a function $w(x)$, then it can be shown that $w(x)$ and $m(x)$ are related by

$$\frac{d^2 m(x)}{dx^2} = w(x).$$

On the other hand, elasticity theory says that the turning moment is proportional to the curvature of the beam. Recall from analytic geometry that the curvature of $s(x)$ is given by $s''(x)[1 + s'(x)^2]^{-3/2}$. Thus the equation of elasticity theory is

$$m(x) = \frac{EI\, s''(x)}{[1 + s'(x)^2]^{3/2}}$$

where E and I are constants (E depends on the material of which the beam is composed, while I depends on the size and shape of a cross-section).

We have two second-order o.d.e.'s: one for $m(x)$ given $w(x)$, and the other for $s(x)$ given $m(x)$. The second equation, however, is nonlinear; trying to solve for $s(x)$ in this form would lead to a horrible mess. But for *small deflections*, the values of $s(x)$ and $s'(x)$ will be small, and we can approximate the quantity $[1 + s'(x)^2]^{3/2}$ by the constant 1. This simplifies the second equation to

$$m(x) = EI\, \frac{d^2 s}{dx^2}.$$

If we differentiate this twice and substitute into the relation between $m(x)$ and $w(x)$, we can eliminate the intermediate variable $m(x)$, which doesn't really interest us. This yields a fourth-order o.d.e. describing the deflection $s(x)$ in terms of the weight distribution $w(x)$:

$$EI \frac{d^4s}{dx^4} = w(x).$$

The general solution of this equation has the form

$$s(x) = H(x) + p(x)$$

where $p(x)$ is a particular solution depending on $w(x)$ and $H(x)$ is the general solution of the homogeneous equation

$$EI \frac{d^4H}{dx^4} = 0.$$

Thus

$$s(x) = c_1 + c_2x + c_3x^2 + c_4x^3 + p(x).$$

If we have an evenly distributed weight,

$$w(x) = w$$

then by undetermined coefficients

$$p(x) = \frac{w}{24EI} x^4$$

so

$$s(x) = c_1 + c_2x + c_3x^2 + c_4x^3 + \frac{w}{24EI} x^4.$$

In contrast to the earlier examples, for which the data naturally led to an initial-value problem, the natural data for the beam problem are values for $s(x)$ and/or its derivatives at each end of the beam. This is a **boundary-value problem.** For example, if the left end of the beam ($x = 0$) is embedded in the wall, $s(x)$ and $s'(x)$ are fixed at zero. If the right end ($x = L$) rests on a fulcrum support, we have $s(x) = 0$, $s'(x)$ unrestrained (because the beam is free to tilt), but $s''(x) = 0$ (because nothing is acting at the fulcrum to bend the beam). These give four conditions: two at $x = 0$ ($s(0) = 0$, $s'(0) = 0$) and two at $x = L$ ($s(L) = 0$, $s''(L) = 0$). For the evenly distributed weight, these boundary conditions give us

$$0 = s(0) = c_1$$

$$0 = s'(0) = c_2$$

$$0 = s(L) = c_1 + c_2L + c_3L^2 + c_4L^3 + \frac{w}{24EI}L^4$$

$$0 = s''(L) = 2c_3 + 6c_4L + \frac{w}{2EI}L^2$$

which we solve to find

$$s(x) = \frac{w}{48EI}[3L^2x^2 - 5Lx^3 + 2x^4].$$

In particular, if we set $L = 1$ and choose $w = 48EI$, we obtain

$$s(x) = 3x^2 - 5x^3 + 2x^4$$

for which the beam has the approximate shape in Figure 2.12.

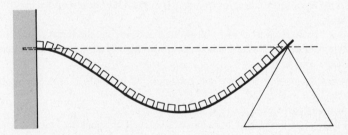

FIGURE 2.12

EXERCISES

1. *The Twisted Shaft:* What do the graphs in Figure 2.7 tell us about physical behavior when the equations model the shaft of Example 2.12.1?
2. *A beam* 6 ft long, with constants $EI = 1$, has both ends embedded in walls. Find the shape of the beam if the weight distribution is $w(x) = 4$. In particular, how far below the horizontal will the beam be at its lowest point?
3. *Another Beam:* Repeat Exercise 2 under the assumption that each end of the beam rests on a fulcrum support.
4. *A Simple Pendulum:* Solve the linear o.d.e. in Example 2.12.2 to predict the motion of a 2-ft pendulum under each of the following initial conditions:
   a. $\theta(0) = 0$, $\theta'(0) = 2$ radians/sec
   b. $\theta(0) = 1/20$ radian, $\theta'(0) = 0$
   c. $\theta(0) = 1/20$ radian, $\theta'(0) = -2$ radians/sec

Express your answers in phase-amplitude form (see Example 2.11.1) and, in each case, determine the maximum value of θ and the frequency of the oscillation.

5. *A Balanced Stick:* Consider the variation of Example 2.12.2 obtained by turning the pendulum upside down (Figure 2.13).

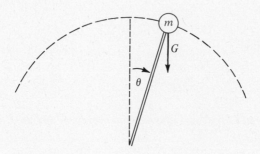

FIGURE 2.13

a. Find the appropriate linear approximation to the o.d.e. near the "balanced" equilibrium position.
b. Is this linearization as useful, on physical grounds, as the linear approximation to the equation modeling the original pendulum? (*Hint:* Does the approximation hypothesis continue to hold?)

6. *A Damped Pendulum:* Suppose the pendulum of Example 2.12.2 encounters frictional resistance with moment $M_{\text{fric}} = -b \, d\theta/dt$.
a. Obtain an approximate, but linear, o.d.e. for θ.
b. What does the discussion in Example 2.11.2 tell us about the physical behavior of the pendulum in case $b^2 < 4gL^3m$?

7. *The Shaking Finger:* Consider the following simplified model of a muscle controlling a limb (see Figure 2.14). An arm of length $\sqrt{2}$ ft (and negligible weight) pivots at one end from the bottom of a vertical rod. The free end of the arm supports a weight of w pounds and is attached to the vertical rod by a horizontal spring with constant $k = 5$ slugs/ft and natural length 1 ft. To simplify the mathematics, assume the spring is attached to a ring that can travel up and down the rod, so the spring remains horizontal.
a. Add the tangential components of the gravitational and spring forces to find F_{tan}.
b. Check that if no weight were supported ($w = 0$), the arm would rest at an angle of $\pi/4$ radians with the rod.
c. Use the result of (a) to obtain an o.d.e. for θ.
d. When θ is near $\pi/4$ radians, $\cos \theta \approx \dfrac{1}{\sqrt{2}}\left[1 - \left(\theta - \dfrac{\pi}{4}\right)\right]$, $\sin \theta \approx \dfrac{1}{\sqrt{2}}\left[1 + \left(\theta - \dfrac{\pi}{4}\right)\right]$, and $\cos \theta \sin \theta \approx \dfrac{1}{2}$. Use these approximations to obtain an approximate linear equation modeling the arm.
e. Predict the motion of the arm if the weight is $w = 1$ and the arm starts from rest with $\theta(0) = \pi/4$.

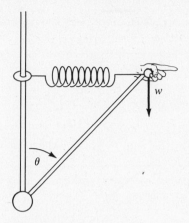

FIGURE 2.14

In the following exercise, which assumes some knowledge of geometric vectors, we derive the rotational form of Newton's second law of motion.

8. If a point moves counterclockwise on a circle of radius r centered at the origin, then its position at time t is described by the vector from the origin to the point:

$$\mathbf{R}(t) = r \cos \theta(t) \, \mathbf{i} + r \sin \theta(t) \, \mathbf{j}.$$

The vector

$$\mathbf{T}(t) = -\sin \theta(t) \, \mathbf{i} + \cos \theta(t) \, \mathbf{j}$$

is a unit vector, tangent to the circle, and points counterclockwise.

a. Differentiate $\mathbf{R}(t)$ twice (remembering that r is constant) to obtain a formula for the acceleration, $\mathbf{a}(t) = \mathbf{R}''(t)$.

b. Show that the force on the point, $\mathbf{F}(t) = m\mathbf{a}(t)$, is the sum of a force pointing toward the origin, $\mathbf{F}_1(t) = -m[\theta'(t)]^2 \, \mathbf{R}(t)$, and a tangential force, $\mathbf{F}_2(t) = mr\theta''(t)\mathbf{T}(t)$.

c. Conclude that $F_{\tan} = mr\theta''(t)$.

d. Use the expression for $F_{\tan}$ to check that $M = F_{\tan} \, r$ is equal to $J\theta''(t) = mr^2\theta''(t)$.

9. *A Derivation for the Bending Beam:*

a. Verify that if finitely many weights w_i, $i = 1, \ldots, k$, are placed at positions p_i, $i = 1, \ldots, k$, along a beam, then the turning moment relative to x (see Example 2.12.3) is

$$m(x) = \sum_{p_i > x} (p_i - x)w_i.$$

b. If the weight is continuously distributed along the beam according to the density function $w(p)$ (that is, the weight acting between points p and $q > p$ is given by

$\int_p^q w(t)\,dt$), then use a partition of the beam into portions of length Δp to obtain the approximation

$$m(x) \approx \sum_{p_i > x} (p_i - x)w(p_i)\Delta p$$

and, taking the limit as $\Delta p \to 0$,

$$m(x) = \int_x^L (p - x)w(p)\,dp.$$

c. Differentiate this formula twice with respect to x, using the fundamental theorem of calculus, to obtain the formulas

$$\frac{dm}{dx} = -\int_x^L w(p)\,dp \quad \text{and} \quad \frac{d^2m}{dx^2} = w(x).$$

REVIEW PROBLEMS

1. Let $L = tD^3 + 3D^2$.
   a. Find $L(1)$, $L(t)$, $L(1/t)$, and $L(t^3)$.
   b. Show that 1, t, $1/t$ are linearly independent on $0 < t$.
   c. Find the general solution of $Lx = 24t$ on $0 < t$.

Exercises 2 to 11 can be solved using the general methods discussed in the text. In Exercises 2 to 8, find the general solution.

2. $\dfrac{d^2x}{dt^2} + 2\dfrac{dx}{dt} - 3x = 3t + \sin 3t$

3. $9\dfrac{d^2x}{dt^2} - 6\dfrac{dx}{dt} + x = \dfrac{18}{t^3}e^{t/3}, \quad t > 0$

4. $4\dfrac{d^3x}{dt^3} - 8\dfrac{d^2x}{dt^2} - \dfrac{dx}{dt} + 2x = 0$

5. $\dfrac{d^3x}{dt^3} + 3\dfrac{d^2x}{dt^2} + 3\dfrac{dx}{dt} + x = \dfrac{2}{t}e^{-t}, \quad t > 0$

6. $(D - 7)^2(D + 1)x = e^{-t}$

7. $D^3(D^2 + 1)x = 3 + 2e^{-t}$

8. $(D - 1)(D^2 + 2D + 2)^3x = 3 + 2e^{-t}$

Solve the initial value problems in Exercises 9 to 11.

9. $(2D^2 + 2D + 5)x = 0; \quad x(0) = 2, x'(0) = -2$

10. $(D - 1)(D + 1)(D - 2)x = 6; \quad x(0) = 1, x'(0) = -1, x''(0) = 1$

11. $x'' - 2x' + x = 8e^{-t}; \quad x(0) = x'(0) = 1$

In Exercises 12 to 23, find the general solution. Some of these problems require the use of special techniques discussed in the exercises of this chapter.

12. $2x'' - 4x' + 10x = 4e^t \sec 2t, \quad -\dfrac{\pi}{4} < t < \dfrac{\pi}{4}$

13. $t^2x'' + tx' + x = 3/t, \quad t > 0$ 14. $tx'' - x' - (t + 1)x = 0, \quad t > 0$

15. $x'' + x = \sin t$ 16. $t\dfrac{d^3x}{dt^3} - \dfrac{d^2x}{dt^2} = 6t^2, \quad t > 0$

17. $t^2x'' - tx' + x = \dfrac{1}{t}, \quad t > 0$

18. $tx'' - (2t + 1)x' + 2x = t^2e^{2t}, \quad t > 0$

19. $4x'' - x = -6te^t + 8e^{t/2}$ 20. $t\dfrac{d^3x}{dt^3} + 2\dfrac{d^2x}{dt^2} = 0$

Linear Systems of Differential Equations

THREE

3.1 SOME ELECTRICAL CIRCUIT MODELS

In this section we formulate some models of electrical circuits. As in Example 2.1.4, the description of these models involves several interrelated quantities. In the following sections we will learn to handle the resulting systems of o.d.e.'s directly, instead of reducing everything to a single o.d.e. This direct approach is particularly useful when dealing with more than two unknowns.

The quantities used to describe the state of an electrical circuit are currents (measured in amperes), charges (measured in coulombs), and voltage changes (measured in volts). Our variables will be currents and charges. The differential equations of our models will be derived from a voltage analysis, using the fundamental laws first formulated in the mid-nineteenth century by the leading German physicist G. R. Kirchhoff. **Kirchhoff's current law** states that *the total current entering any point of a circuit equals the total current leaving it.* **Kirchhoff's voltage law** states that *the sum of the voltage changes around any loop in a circuit is zero.*

The circuits we consider involve four kinds of elements, denoted symbolically as in Figure 3.1. For each element in a given circuit, we choose a positive direction for measuring the current through the element (we will indicate our choices by arrows, and for a voltage source we will always choose the positive direction so that the arrow

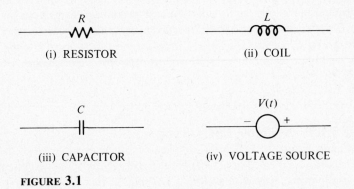

(i) RESISTOR   (ii) COIL

(iii) CAPACITOR   (iv) VOLTAGE SOURCE

FIGURE 3.1

points from the $-$ in the symbol to the $+$). As current flows through an element there is a voltage drop or rise (negative drop):

i.   In a **resistor,** the voltage drop is proportional to the current:

$$V_{res} = RI_{res}.$$

The positive constant R is called the **resistance** and is measured in ohms.

ii. In a **coil,** the voltage drop is proportional to the rate of change of the current:

$$V_{coil} = L \frac{dI_{coil}}{dt}.$$

The positive constant L is the **inductance** of the coil and is measured in henrys.

iii. In a **capacitor,** the voltage drop is proportional to the charge difference between two plates:

$$V_{cap} = \frac{1}{C} Q_{cap}$$

and the charge difference is related to the current by

$$I_{cap} = \frac{dQ_{cap}}{dt}.$$

The positive constant C is **capacitance** and is measured in farads.

iv. A **voltage source** imposes an externally controlled (possibly varying) voltage drop

$$V_{ext} = -V(t).$$

In order to apply Kirchhoff's voltage law correctly, we have to measure the voltage changes in the loop consistently. We add the voltage changes clockwise around the loop. *If the positive direction for the current through an element agrees with the clockwise direction, we add the voltage drop as given in (i) to (iv); if the directions disagree, we adjust the formula for the voltage drop by a sign change.*

Example 3.1.1   A Simple *LRC* Circuit

In Figure 3.2(a) we sketch the simplest circuit involving each of the four elements. The arrows indicate our choices for the positive directions of currents through the various elements. Kirchhoff's current law tells us that the current I is the same

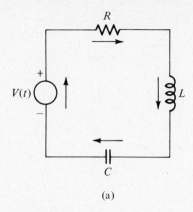
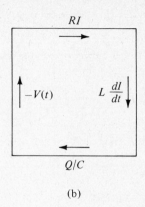

(a)   (b)

FIGURE 3.2

throughout the circuit:

$$I_{res} = I_{coil} = I_{cap} = I.$$

In Figure 3.2(b) we have sketched what relations (i) to (iv) tell us about the voltage drops across the elements. If we apply Kirchhoff's voltage law, we obtain

$$0 = V_{res} + V_{coil} + V_{cap} + V_{ext}$$

$$= RI + L\frac{dI}{dt} + \frac{1}{C}Q - V(t).$$

We solve this for $DI = dI/dt$ and combine the resulting equation with the relation of Q to I from (iii), to get the system of two o.d.e.'s

$$DQ = \qquad\qquad I$$

$$DI = \frac{-1}{LC}Q - \frac{R}{L}I + \frac{V(t)}{L}.$$

(We could, of course, substitute $I = DQ$ into the second equation to reduce this system to a single second-order o.d.e. We shall see that we can solve the system without doing this.)

Example 3.1.2   A Two-Loop Circuit

The wiring diagram in Figure 3.3(a) shows a circuit with two loops. The current I_1 through the left-hand resistor may differ from the current I_2 through the right-hand resistor. Kirchhoff's current law tells us that the current through the coil is $I_{coil} = I_2$ and that the current through the capacitor satisfies $I_1 = I_2 + I_{cap}$, so $I_{cap} = I_1 - I_2$.

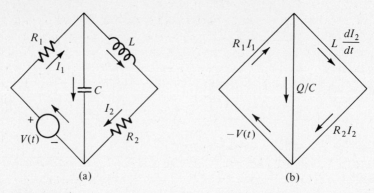

(a)   (b)

FIGURE 3.3

The voltage analysis, sketched in Figure 3.3(b), leads to two equations (one for each loop):

$$R_1 I_1 + \frac{1}{C} Q - V(t) = 0$$

$$L \, DI_2 + R_2 I_2 - \frac{1}{C} Q = 0.$$

As before, Q is related to I_{cap} by differentiation:

$$DQ = I_{\text{cap}} = I_1 - I_2.$$

The second and third equations describe DQ and DI_2 in terms of I_1, I_2, and Q. We can solve the first equation for I_1:

$$I_1 = \frac{-1}{R_1 C} Q + \frac{V(t)}{R_1}.$$

Substituting this into the equation for DQ, we obtain the system of two o.d.e.'s

$$DQ = -\frac{1}{R_1 C} Q - I_2 + \frac{V(t)}{R_1}$$

$$DI_2 = \frac{1}{LC} Q - \frac{R_2}{L} I_2.$$

When there are several loops in a circuit, the analysis of currents can be aided by imagining a **mesh current** flowing clockwise through each loop. In the last example, we could imagine mesh currents I_1 and I_2 flowing through the left-hand and right-hand loops, respectively. *The actual current flowing through an element is the sum of the signed mesh currents for the loops containing the element; the sign is $+$ if the mesh current flows in the positive direction and $-$ otherwise.* (The reader should check that this works in the last example.)

Example 3.1.3   A Three-Loop Circuit

To handle the circuit sketched in Figure 3.4(a), we imagine a mesh current in each loop. The currents through the coils L_2 and L_3 and the resistor R_1 are equal to the mesh currents of the loops in which they lie. To obtain the currents through the elements common to two loops, we must add signed mesh currents; the current through the resistor R_2 is $I_2 - I_3$ while the current through the capacitor is $I_1 - I_2$.

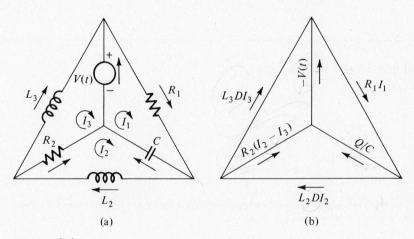

(a)   (b)

FIGURE 3.4

The voltage analysis sketched in Figure 3.4(b) leads to three equations:

$$-V(t) + R_1 I_1 + \frac{1}{C} Q \qquad\qquad = 0$$

$$\frac{-1}{C} Q + L_2\,DI_2 + R_2(I_2 - I_3) = 0$$

$$V(t) - R_2(I_2 - I_3) + L_3\,DI_3 = 0.$$

We also have the equation for DQ

$$DQ = I_1 - I_2.$$

The first voltage relation can be solved for I_1 to yield

$$I_1 = \frac{-1}{R_1 C} Q + \frac{V(t)}{R_1}.$$

Substituting this into the equation for DQ and combining with the second and third voltage relations, we come up with the system of three o.d.e.'s in Q, I_2, and I_3:

$$DQ = \frac{-1}{R_1 C} Q - I_2 \qquad\qquad + \frac{V(t)}{R_1}$$

$$DI_2 = \frac{1}{L_2 C} Q - \frac{R_2}{L_2} I_2 + \frac{R_2}{L_2} I_3$$

$$DI_3 = \qquad\qquad \frac{R_2}{L_3} I_2 - \frac{R_2}{L_3} I_3 - \frac{V(t)}{L_3}.$$

Example 3.1.4   Four Loops

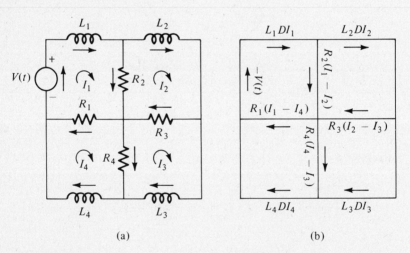

(a)   (b)

FIGURE 3.5

The complicated circuit of Figure 3.5(a) can be decomposed into four loops, with clockwise mesh currents I_j, $j = 1, 2, 3, 4$. Here there are no capacitors, hence no charges. The voltage analysis, sketched in Figure 3.5(b), leads directly to the system of four o.d.e.'s

$$L_1 DI_1 = -(R_1 + R_2)I_1 + R_2 I_2 + R_1 I_4 + V(t)$$
$$L_2 DI_2 = R_2 I_1 - (R_2 + R_3)I_2 + R_3 I_3$$

$$L_3 \, DI_3 = \qquad\qquad\qquad R_3 \, I_2 - (R_3 + R_4)I_3 + \qquad\qquad R_4 \, I_4$$
$$L_4 \, DI_4 = \qquad\quad R_1 \, I_1 + \qquad\qquad\qquad\qquad R_4 \, I_3 - (R_1 + R_4)I_4.$$

EXERCISES

1. *Supply and Demand:* The model in Exercise 2, Section 2.1, is most naturally modeled as a system of two first-order o.d.e.'s for price and supply. Find this system.

2. *Radioactive Decay:* A radioactive substance, A, decays at the rate of k_A percent annually into substance B. Substance B, while more stable than A, nevertheless decays at the rate of k_B percent annually into the stable substance C.
   a. Write a system of three first-order o.d.e.'s modeling the amounts of substances A, B, and C.
   b. Do the same under the additional assumption that α grams per year of substance A are steadily added to the mixture, while γ grams per year of substance C are steadily extracted.
   (*Note:* Assume that 1 gram of substance A decomposes into 1 gram of B, and 1 gram of B gives 1 gram of C.)

3. *Diffusion I:* When two solutions of a substance are separated by a permeable membrane, the *amount* of the substance that crosses the membrane in a given time is proportional to the difference between the *concentrations* of the neighboring solutions. The constant of proportionality is called the *permeability* of the membrane. Write a system of two first-order equations modeling the change in concentrations x_1 and x_2 of two saline solutions separated by a membrane with permeability P, assuming
   a. there are equal volumes $V_1 = V_2 = V$ of liquid on the two sides of the membrane.
   b. the two unequal volumes are V_1 and V_2, respectively.

4. *Diffusion II:* An organ has a double wall. The outer wall has a permeability P_1 to glucose, while the permeability of the inner wall is $0 < P_2 < P_1$. Let v and V denote, respectively, the (constant) volume between the two walls and the volume inside the organ. Denote the concentration of glucose in volume v by x, and in volume V by y. The blood surrounding the organ has a steady concentration G of glucose. Write two first-order equations in x and y to model the distribution of glucose in the organ.

5. *Political Coexistence:* Two neighboring countries, Camponesa and Mandachuva, have strikingly different economic and social structures. Camponesa is ruled by a socialist military front that is nationalizing all large corporations and developing various social welfare programs. Mandachuva, ruled by a right-wing junta, is experiencing rapid economic growth, although unemployment and inflation are high. There is a net annual birth rate in Camponesa of 15 percent, but the middle class is becoming uncomfortable with political developments and 4 percent of the population moves to Mandachuva each year. On the other hand, Mandachuva has an annual net birth rate of 10 percent, but 3 percent of the population (unemployed laborers and leftist intellectuals) annually move to Camponesa. Write a system of two differential equations modeling the demographic changes in Mandachuva and Camponesa. (Compare Example 1.1.5.)

More circuits: In Exercises 6 to 10, find a system of two first-order o.d.e.'s modeling the circuit of the indicated figure.

6. Figure 3.6; $R = 2$ ohms, $L_1 = L_2 = 1$ henry, and $V(t) = 4e^{-4t}$ volts.

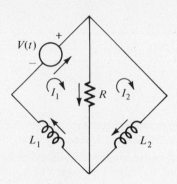

FIGURE 3.6

7. Figure 3.7; $R_1 = R_2 = 4$ ohms, $L = 2$ henrys, $C = 2$ farads, and $V(t) = 4e^{-t/2}$ volts.

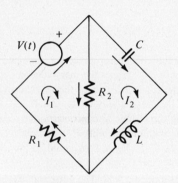

FIGURE 3.7

8. Figure 3.7; $R_1 = R_2 = 4$ ohms, $L = 2$ henrys, $C = 1$ farad, and $V(t) = 4e^{-t/2}$ volts.
9. Figure 3.8; $R_1 = R_2 = 1$ ohm, $L = 2$ henrys, $C = 1$ farad, and $V(t) = 4e^{-t/2}$ volts.

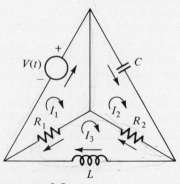

FIGURE 3.8

10.   Figure 3.9; $R_1 = R_2 = 1$ ohm, $C_1 = 3$ and $C_2 = 2$ farads, and $V(t) = 5$ volts.

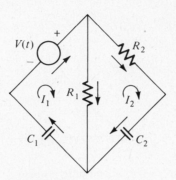

FIGURE 3.9

3.2 LINEAR SYSTEMS, MATRICES, AND VECTORS

In this section we introduce notation that helps us recognize parallels between the solutions of a single linear o.d.e. and the solutions of a linear system of o.d.e.'s. By the end of the chapter we hope you will agree that in this case a good notation, like a picture, is worth a thousand words.

A system of o.d.e.'s is **linear** if it can be written in the form

(S)

$$x_1' = a_{11}x_1 + a_{12}x_2 + \ldots + a_{1n}x_n + E_1(t)$$
$$x_2' = a_{21}x_1 + a_{22}x_2 + \ldots + a_{2n}x_n + E_2(t)$$
$$\cdot$$
$$\cdot$$
$$\cdot$$
$$x_n' = a_{n1}x_1 + a_{n2}x_2 + \ldots + a_{nn}x_n + E_n(t).$$

The **coefficients** a_{ij} may be constants or functions. (As with a single o.d.e., we will pay most attention to the constant-coefficient case.) The number n of unknowns is the **order** of the system. We will call the system **homogeneous** if $E_1(t) = E_2(t) = \ldots = E_n(t) = 0$.

The systems in Examples 3.1.1 to 3.1.4 were all linear. We can obtain other examples of linear systems from single linear o.d.e.'s.

Example 3.2.1
Consider the second-order equation

(N) $$(D^2 - 2D + 1)x = -t.$$

Suppose $x = x(t)$ is a solution of (N), and set

$$x_1 = x, \qquad x_2 = x'.$$

Then we have

$$x_1' = x' = x_2$$

and, since x solves (N),

$$x_2' = x'' = 2x' - x - t = 2x_2 - x_1 - t.$$

In other words, x_1 and x_2 solve the second-order system

(S$_N$)
$$\begin{aligned} x_1' &= & x_2 \\ x_2' &= -x_1 + 2x_2 - t. \end{aligned}$$

Conversely, if x_1 and x_2 solve (S$_N$), then

$$(D^2 - 2D + 1)x_1 = x_1'' - 2x_1' + x_1 = x_2' - 2x_2 + x_1 = -t.$$

That is, $x = x_1$ solves (N). Thus, solutions of (N) yield solutions of (S$_N$) and vice versa.

To solve (N), we first solve the related homogeneous o.d.e.

(H)
$$(D^2 - 2D + 1)x = 0$$

which, in the same way, is equivalent to the homogeneous system

(S$_H$)
$$\begin{aligned} x_1' &= & x_2 \\ x_2' &= -x_1 + 2x_2. \end{aligned}$$

The characteristic polynomial of (H) is $(r - 1)^2$, so the general solution of (H) is

$$x = c_1 e^t + c_2 t e^t$$

and the corresponding general solution of (S$_H$) is

$$\begin{aligned} x_1 &= x = c_1 e^t + c_2 t e^t \\ x_2 &= x' = c_1 e^t + c_2(t + 1)e^t. \end{aligned}$$

We next use undetermined coefficients to find a particular solution, $x = -2 - t$, of (N). The general solution of (N) is

$$x = c_1 e^t + c_2 t e^t - 2 - t$$

which corresponds to the general solution of (S_N)

$$x_1 = x = c_1 e^t + c_2 t e^t - 2 - t$$
$$x_2 = x' = c_1 e^t + c_2(t + 1)e^t - 1.$$

Example 3.2.2

The process we used in the preceding example can be used to replace the nth-order linear equation

(N) $$(D^n + a_{n-1}D^{n-1} + \ldots + a_1 D + a_0)x = E(t)$$

by an equivalent system. In this case, we introduce n unknowns

$$x_1 = x, \qquad x_2 = x', \qquad x_3 = x'', \ldots, x_n = x^{(n-1)}.$$

The single equation (N) is equivalent to the system

$$x_1' = x_2$$
$$x_2' = x_3$$

(S_N)

$$x_{n-1}' = x_n$$
$$x_n' = -a_0 x_1 - a_1 x_2 - \ldots - a_{n-1}x_n + E(t).$$

Note that the order of (S_N) is the same as the order of (N) and that (S_N) is homogeneous precisely when (N) is.

The right side of a linear system (S) is determined by its coefficients and the n functions $E_1(t), \ldots, E_n(t)$. A solution of (S) consists of n functions $x_1(t), x_2(t), \ldots, x_n(t)$. Thus we will be dealing with "arrays" of constants and functions. We need suitable notation and terminology for these arrays.

An **$n \times m$ matrix** is an array of the form

$$B = \begin{bmatrix} b_{11} & b_{12} & \ldots & b_{1m} \\ b_{21} & b_{22} & \ldots & b_{2m} \\ & & \cdot & \\ & & \cdot & \\ & & \cdot & \\ b_{n1} & b_{n2} & \ldots & b_{nm} \end{bmatrix}.$$

Note that an $n \times m$ matrix has n rows and m columns. The location of each **entry** b_{ij} is specified by two subscripts; the first indicates the row and the second the column in which b_{ij} appears. The entries may be constants or functions.

We will refer to $n \times 1$ matrices with constant entries as **n-vectors.** Vectors will be indicated by boldface letters. We will use **0** to denote the n-vector all of whose entries are 0.

An $n \times 1$ matrix whose entries are functions is an **n-vector valued function.** A vector valued function assigns a vector to each specific value of t. For example, the vector valued function

$$\mathbf{x}(t) = \begin{bmatrix} \sin t \\ 3t - 5 \\ t^2 + 2 \end{bmatrix}$$

assigns the 3-vector

$$\mathbf{x}(0) = \begin{bmatrix} 0 \\ -5 \\ 2 \end{bmatrix}$$

to the value $t = 0$.

We say that two $n \times m$ matrices B and C are **equal** provided each entry b_{ij} of B is equal to the corresponding entry c_{ij} of C. For example, the matrix equality

$$\begin{bmatrix} x_1 \\ x_2 \\ x_3 \end{bmatrix} = \begin{bmatrix} \sin t \\ 3t - 5 \\ t^2 + 2 \end{bmatrix}$$

means

$$x_1 = \sin t, \qquad x_2 = 3t - 5, \qquad \text{and} \qquad x_3 = t^2 + 2.$$

The matrix equality

$$\begin{bmatrix} y_1 \\ y_2 \\ y_3 \end{bmatrix} = \mathbf{0}$$

means

$$y_1 = y_2 = y_3 = 0.$$

We obtain the **sum of two $n \times m$ matrices** by adding corresponding entries:

$$
\begin{bmatrix} a_{11} & a_{12} & \cdots & a_{1m} \\ a_{21} & a_{22} & \cdots & a_{2m} \\ \cdot & \cdot & & \cdot \\ \cdot & \cdot & & \cdot \\ \cdot & \cdot & & \cdot \\ a_{n1} & a_{n2} & \cdots & a_{nm} \end{bmatrix} + \begin{bmatrix} b_{11} & b_{12} & \cdots & b_{1m} \\ b_{21} & b_{22} & \cdots & b_{2m} \\ \cdot & \cdot & & \cdot \\ \cdot & \cdot & & \cdot \\ \cdot & \cdot & & \cdot \\ b_{n1} & b_{n2} & \cdots & b_{nm} \end{bmatrix}
$$

$$
= \begin{bmatrix} (a_{11} + b_{11}) & (a_{12} + b_{12}) & \cdots & (a_{1m} + b_{1m}) \\ (a_{21} + b_{21}) & (a_{22} + b_{22}) & \cdots & (a_{2m} + b_{2m}) \\ \cdot & \cdot & & \cdot \\ \cdot & \cdot & & \cdot \\ \cdot & \cdot & & \cdot \\ (a_{n1} + b_{n1}) & (a_{n2} + b_{n2}) & \cdots & (a_{nm} + b_{nm}) \end{bmatrix} .
$$

The **product of a number and an $n \times m$ matrix** is obtained by multiplying each entry of the matrix by the number:

$$
c \begin{bmatrix} a_{11} & a_{12} & \cdots & a_{1m} \\ a_{21} & a_{22} & \cdots & a_{2m} \\ \cdot & \cdot & & \cdot \\ \cdot & \cdot & & \cdot \\ \cdot & \cdot & & \cdot \\ a_{n1} & a_{n2} & \cdots & a_{nm} \end{bmatrix} = \begin{bmatrix} ca_{11} & ca_{12} & \cdots & ca_{1m} \\ ca_{21} & ca_{22} & \cdots & ca_{2m} \\ \cdot & \cdot & & \cdot \\ \cdot & \cdot & & \cdot \\ \cdot & \cdot & & \cdot \\ ca_{n1} & ca_{n2} & \cdots & ca_{nm} \end{bmatrix} .
$$

For example

$$
-2 \begin{bmatrix} 1 & 2 \\ 0 & 3 \end{bmatrix} + 3 \begin{bmatrix} 4 & -1 \\ 1 & 0 \end{bmatrix} = \begin{bmatrix} -2 & -4 \\ 0 & -6 \end{bmatrix} + \begin{bmatrix} 12 & -3 \\ 3 & 0 \end{bmatrix} = \begin{bmatrix} 10 & -7 \\ 3 & -6 \end{bmatrix}
$$

and

$$
3 \begin{bmatrix} 0 \\ 1 \\ 2 \end{bmatrix} + \begin{bmatrix} 1 \\ 3 \\ 1 \end{bmatrix} - 2 \begin{bmatrix} -2 \\ 0 \\ 1 \end{bmatrix} = \begin{bmatrix} 0 \\ 3 \\ 6 \end{bmatrix} + \begin{bmatrix} 1 \\ 3 \\ 1 \end{bmatrix} + \begin{bmatrix} 4 \\ 0 \\ -2 \end{bmatrix} = \begin{bmatrix} 5 \\ 6 \\ 5 \end{bmatrix} .
$$

We define the **product of an $n \times m$ matrix and an m-vector** by the rule

$$
\begin{bmatrix} a_{11} & a_{12} & \cdots & a_{1m} \\ a_{21} & a_{22} & \cdots & a_{2m} \\ & & \cdot & \\ & & \cdot & \\ & & \cdot & \\ a_{n1} & a_{n2} & \cdots & a_{nm} \end{bmatrix} \cdot \begin{bmatrix} x_1 \\ x_2 \\ \cdot \\ \cdot \\ \cdot \\ x_m \end{bmatrix} = \begin{bmatrix} a_{11}x_1 + a_{12}x_2 + \ldots + a_{1m}x_m \\ a_{21}x_1 + a_{22}x_2 + \ldots + a_{2m}x_m \\ \cdot \\ \cdot \\ \cdot \\ a_{n1}x_1 + a_{n2}x_2 + \ldots + a_{nm}x_m \end{bmatrix} .
$$

Note that the product is an n-vector. The first (top) entry of the product is obtained by multiplying the entries of the first row of the matrix by the corresponding entries of the vector and then summing the products. Similarly, the ith entry of the product is obtained by "multiplying" the ith row of the matrix by the vector. For example,

$$
\begin{bmatrix} 3 & 2 & -1 \\ 4 & 3 & 5 \\ 0 & 1 & 0 \end{bmatrix} \begin{bmatrix} 1 \\ 2 \\ -2 \end{bmatrix} = \begin{bmatrix} 3(1) + 2(2) + (-1)(-2) \\ 4(1) + 3(2) + 5(-2) \\ 0(1) + 1(2) + 0(-2) \end{bmatrix} = \begin{bmatrix} 9 \\ 0 \\ 2 \end{bmatrix}.
$$

We define the **derivative of a vector valued function** by the rule

$$
D \begin{bmatrix} x_1(t) \\ x_2(t) \\ \cdot \\ \cdot \\ \cdot \\ x_n(t) \end{bmatrix} = \begin{bmatrix} x_1'(t) \\ x_2'(t) \\ \cdot \\ \cdot \\ \cdot \\ x_n'(t) \end{bmatrix}.
$$

For example,

$$
D \begin{bmatrix} \sin t \\ 3t - 5 \\ t^2 + 2 \end{bmatrix} = \begin{bmatrix} \cos t \\ 3 \\ 2t \end{bmatrix}.
$$

The equations that made up our system (S) can be replaced by an equivalent matrix equality, which we can simplify using the notions of vector addition and matrix multiplication:

$$
\begin{bmatrix} x_1' \\ x_2' \\ \cdot \\ \cdot \\ \cdot \\ x_n' \end{bmatrix} = \begin{bmatrix} a_{11}x_1 + a_{12}x_2 + \ldots + a_{1n}x_n + E_1(t) \\ a_{21}x_1 + a_{22}x_2 + \ldots + a_{2n}x_n + E_2(t) \\ \cdot \\ \cdot \\ \cdot \\ a_{n1}x_1 + a_{n2}x_2 + \ldots + a_{nn}x_n + E_n(t) \end{bmatrix}
$$

$$
= \begin{bmatrix} a_{11}x_1 + \ldots + a_{1n}x_n \\ a_{21}x_1 + \ldots + a_{2n}x_n \\ \cdot \\ \cdot \\ \cdot \\ a_{n1}x_1 + \ldots + a_{nn}x_n \end{bmatrix} + \begin{bmatrix} E_1(t) \\ E_2(t) \\ \cdot \\ \cdot \\ \cdot \\ E_n(t) \end{bmatrix}
$$

$$
= \begin{bmatrix} a_{11} & a_{12} & \cdots & a_{1n} \\ a_{21} & a_{22} & \cdots & a_{2n} \\ & & \cdot & \\ & & \cdot & \\ & & \cdot & \\ a_{n1} & a_{n2} & \cdots & a_{nn} \end{bmatrix} \begin{bmatrix} x_1 \\ x_2 \\ \cdot \\ \cdot \\ \cdot \\ x_n \end{bmatrix} + \begin{bmatrix} E_1(t) \\ E_2(t) \\ \cdot \\ \cdot \\ \cdot \\ E_n(t) \end{bmatrix}.
$$

If we set

$$\mathbf{x} = \begin{bmatrix} x_1 \\ x_2 \\ \cdot \\ \cdot \\ \cdot \\ x_n \end{bmatrix}, \qquad A = \begin{bmatrix} a_{11} & a_{12} & \cdots & a_{1n} \\ a_{21} & a_{22} & \cdots & a_{2n} \\ & & \cdot \\ & & \cdot \\ & & \cdot \\ a_{n1} & a_{n2} & \cdots & a_{nn} \end{bmatrix}, \qquad \text{and} \qquad \mathbf{E}(t) = \begin{bmatrix} E_1(t) \\ E_2(t) \\ \cdot \\ \cdot \\ \cdot \\ E_n(t) \end{bmatrix}$$

then our system (S) can be written in matrix form as

$$D\mathbf{x} = A\mathbf{x} + \mathbf{E}(t).$$

Example 3.2.3

The matrix form of the system

$$(S_N) \qquad \begin{aligned} x_1' &= x_2 \\ x_2' &= -x_1 + 2x_2 - t \end{aligned}$$

from Example 3.2.1 is $D\mathbf{x} = A\mathbf{x} + \mathbf{E}(t)$, where

$$\mathbf{x} = \begin{bmatrix} x_1 \\ x_2 \end{bmatrix}, \qquad A = \begin{bmatrix} 0 & 1 \\ -1 & 2 \end{bmatrix}, \qquad \text{and} \qquad \mathbf{E}(t) = \begin{bmatrix} 0 \\ -t \end{bmatrix}.$$

The general solution of (S_N) can be given in matrix form as

$$\begin{aligned} \mathbf{x} &= \begin{bmatrix} c_1 e^t + c_2 t e^t - 2 - t \\ c_1 e^t + c_2 (t + 1) e^t - 1 \end{bmatrix} = \begin{bmatrix} c_1 e^t \\ c_1 e^t \end{bmatrix} + \begin{bmatrix} c_2 t e^t \\ c_2 (t + 1) e^t \end{bmatrix} + \begin{bmatrix} -2 - t \\ -1 \end{bmatrix} \\ &= c_1 \begin{bmatrix} e^t \\ e^t \end{bmatrix} + c_2 \begin{bmatrix} t e^t \\ (t + 1) e^t \end{bmatrix} + \begin{bmatrix} -2 - t \\ -1 \end{bmatrix}. \end{aligned}$$

If we set

$$\mathbf{h}_1(t) = \begin{bmatrix} e^t \\ e^t \end{bmatrix}, \qquad \mathbf{h}_2(t) = \begin{bmatrix} t e^t \\ (t + 1) e^t \end{bmatrix}, \qquad \text{and} \qquad \mathbf{p}(t) = \begin{bmatrix} -2 - t \\ -1 \end{bmatrix}$$

then the general solution of (S_N) is

$$\mathbf{x} = c_1 \mathbf{h}_1(t) + c_2 \mathbf{h}_2(t) + \mathbf{p}(t).$$

Note that $\mathbf{x} = \mathbf{p}(t)$ is one particular solution of (S_N), while the general solution of (S_H) in Example 3.2.1 can be written

$$\mathbf{x} = c_1 \mathbf{h}_1(t) + c_2 \mathbf{h}_2(t).$$

Note all the parallels between the solution of the system and the solution of the o.d.e.!

Example 3.2.4

The matrix form of the system (S_N) in Example 3.2.2 is $D\mathbf{x} = A\mathbf{x} + \mathbf{E}(t)$, where

$$
\mathbf{x} = \begin{bmatrix} x_1 \\ x_2 \\ \cdot \\ \cdot \\ \cdot \\ x_{n-1} \\ x_n \end{bmatrix}, A = \begin{bmatrix} 0 & 1 & 0 & \ldots & 0 \\ 0 & 0 & 1 & \ldots & 0 \\ & & \cdot & & \\ & & \cdot & & \\ & & \cdot & & \\ 0 & 0 & 0 & \ldots & 1 \\ -a_0 & -a_1 & -a_2 & \ldots & -a_{n-1} \end{bmatrix}, \text{ and } \mathbf{E}(t) = \begin{bmatrix} 0 \\ 0 \\ \cdot \\ \cdot \\ \cdot \\ 0 \\ E(t) \end{bmatrix}.
$$

Example 3.2.5

The matrix form of the system in Example 3.1.3 is $D\mathbf{x} = A\mathbf{x} + \mathbf{E}(t)$, where

$$
\mathbf{x} = \begin{bmatrix} Q \\ I_2 \\ I_3 \end{bmatrix}, \quad A = \begin{bmatrix} \dfrac{-1}{R_1 C} & -1 & 0 \\ \dfrac{1}{L_2 C} & -\dfrac{R_2}{L_2} & \dfrac{R_2}{L_2} \\ 0 & \dfrac{R_2}{L_3} & -\dfrac{R_2}{L_3} \end{bmatrix}, \quad \text{and} \quad \mathbf{E}(t) = \begin{bmatrix} \dfrac{V(t)}{R_1} \\ 0 \\ -\dfrac{V(t)}{L_3} \end{bmatrix}.
$$

The following summary and notes include lists of important rules of matrix algebra and of differentiation of vector valued functions. We invite you to verify these rules (see Note 3 and Exercises 25 to 27).

LINEAR SYSTEMS, MATRICES, AND VECTORS

An $n \times m$ **matrix** is an array of the form

$$
B = \begin{bmatrix} b_{11} & \ldots & b_{1m} \\ & \cdot & \\ & \cdot & \\ & \cdot & \\ b_{n1} & \ldots & b_{nm} \end{bmatrix}
$$

The b_{ij}'s are called the **entries** of B. An $n \times 1$ matrix with constant entries is called an ***n*-vector.** We denote the n-vector all of whose entries are 0 by **0.** An $n \times 1$ matrix whose entries are functions is an ***n*-vector valued function.**

The **sum of two $n \times m$ matrices** is the $n \times m$ matrix whose entries are obtained by adding the corresponding entries of the two matrices. The **product of a number and an $n \times m$ matrix** is the new $n \times m$ matrix obtained by multiplying each entry of the old matrix by the number.

The **product of an $n \times m$ matrix and an *m*-vector** is the new n-vector given by the rule

$$
\begin{bmatrix} a_{11} & \cdots & a_{1m} \\ & \cdot & \\ & \cdot & \\ & \cdot & \\ a_{n1} & \cdots & a_{nm} \end{bmatrix}
\begin{bmatrix} x_1 \\ \cdot \\ \cdot \\ \cdot \\ x_m \end{bmatrix}
=
\begin{bmatrix} a_{11}x_1 + \ldots + a_{1m}x_m \\ \cdot \\ \cdot \\ \cdot \\ a_{n1}x_1 + \ldots + a_{nm}x_m \end{bmatrix}.
$$

If **v** and **w** are m-vectors, A is an $n \times m$ matrix, and c is a number, then

(M1) $\qquad\qquad\qquad A(\mathbf{v} + \mathbf{w}) = A\mathbf{v} + A\mathbf{w}$

(M2) $\qquad\qquad\qquad A(c\mathbf{v}) = c(A\mathbf{v}) = (cA)\mathbf{v}.$

The **derivative of an *n*-vector valued function** is the new n-vector valued function obtained by differentiating each of the entries of the old one. If **x** and **y** are n-vector valued functions and c is a number, then

(D1) $\qquad\qquad\qquad D(\mathbf{x} + \mathbf{y}) = D\mathbf{x} + D\mathbf{y}$

(D2) $\qquad\qquad\qquad D(c\mathbf{x}) = c(D\mathbf{x}).$

A system of o.d.e.'s is **linear** if it can be written in the form

$$x_1' = a_{11}x_1 + \ldots + a_{1n}x_n + E_1(t)$$

$$\cdot$$
$$\cdot$$
$$\cdot$$

$$x_n' = a_{n1}x_1 + \ldots + a_{nn}x_n + E_n(t).$$

The matrix form of this system is

$$D\mathbf{x} = A\mathbf{x} + \mathbf{E}(t)$$

where

$$\mathbf{x} = \begin{bmatrix} x_1 \\ \cdot \\ \cdot \\ \cdot \\ x_n \end{bmatrix}, \qquad A = \begin{bmatrix} a_{11} & \cdots & a_{1n} \\ & \cdot & \\ & \cdot & \\ a_{n1} & \cdots & a_{nn} \end{bmatrix}, \qquad \text{and} \qquad \mathbf{E}(t) = \begin{bmatrix} E_1(t) \\ \cdot \\ \cdot \\ \cdot \\ E_n(t) \end{bmatrix}.$$

The **order** of the system is n. The **coefficients** are the entries a_{ij} of A. The system is **homogeneous** if $\mathbf{E}(t) = \mathbf{0}$.

Notes

1. Geometry and vectors

You may have encountered vectors in the plane and in space before as representations of forces or velocities in a physics course, or as an analytic device in multivariate calculus. Usually, we think of these as arrows that are free to move in position, so long as they maintain the same direction and length. In the plane, such arrows can be described by their x and y components and are often written in the form $x\mathbf{i} + y\mathbf{j}$. If we identify $\mathbf{i}$ and $\mathbf{j}$, respectively, with the 2×1 matrices

$$\mathbf{i} = \begin{bmatrix} 1 \\ 0 \end{bmatrix}, \qquad \mathbf{j} = \begin{bmatrix} 0 \\ 1 \end{bmatrix}$$

then any 2-vector in our sense, written

$$\mathbf{v} = \begin{bmatrix} x \\ y \end{bmatrix},$$

corresponds uniquely to a free arrow in the plane, written

$$x\mathbf{i} + y\mathbf{j},$$

and vice versa. The pair of numbers in the matrix $\mathbf{v}$ can also be regarded as the coordinates of a point $P(x,y)$. These three interpretations are all related, as follows: The matrix $\mathbf{v}$ corresponds to the free arrow $x\mathbf{i} + y\mathbf{j}$, which, when it sits with its tail at the origin, has its head at the point $P(x,y)$. Much of what we do in this chapter can be given a geometric interpretation in terms of free arrows or in terms of points.

2. The laws of matrix algebra

Properties (M1) and (M2) in the summary show some analogies between the algebra of matrices and the algebra of numbers. In this note, we list some further properties of matrices.

We begin with a list of laws satisfied by matrix addition and by multiplication of a number times a matrix. In this list, O represents the $n \times m$ matrix all of whose entries are 0:

$$O = \begin{bmatrix} 0 & \dots & 0 \\ \cdot & & \cdot \\ \cdot & & \cdot \\ \cdot & & \cdot \\ 0 & \dots & 0 \end{bmatrix}.$$

This matrix is called the $n \times m$ **zero matrix.**

Fact: *If A, B, and C are $n \times m$ matrices, and if c and d are numbers, then*

(A1) $(A + B) + C = A + (B + C)$

(A2) $A + B = B + A$

(A3) $A + O = A$

(A4) $A + (-1)A = O$

(A5) $c(A + B) = cA + cB$

(A6) $(c + d)A = cA + dA$

(A7) $c(dA) = (cd)A$

(A8) $1A = A.$

Multiplication of a matrix times a vector satisfies the following list of laws. In this list, I denotes the $m \times m$ matrix whose entries a_{ij} are 0 if $i \neq j$ and 1 if $i = j$:

$$I = \begin{bmatrix} 1 & 0 & \dots & 0 \\ 0 & 1 & \dots & 0 \\ \cdot & \cdot & & \cdot \\ \cdot & \cdot & & \cdot \\ \cdot & \cdot & & \cdot \\ 0 & 0 & \dots & 1 \end{bmatrix}.$$

This matrix is called the $m \times m$ **identity matrix.**

Fact: *Suppose A and B are $n \times m$ matrices, v and w are m-vectors, and c is a number. Then*

(M1) $A(\mathbf{v} + \mathbf{w}) = A\mathbf{v} + A\mathbf{w}$

(M2) $A(c\mathbf{v}) = c(A\mathbf{v}) = (cA)\mathbf{v}$

(M3) $(A + B)\mathbf{v} = A\mathbf{v} + B\mathbf{v}$

(M4) $I\mathbf{v} = \mathbf{v}$

(M5) $A\mathbf{0} = \mathbf{0}$

(M6) $O\mathbf{v} = \mathbf{0}.$

3. Verifying the properties of matrix operations

The verification of properties (M1) to (M6), (A1) to (A8), and (D1) and (D2) follows a pattern, which we illustrate by verifying (M1). Let

$$
A = \begin{bmatrix} a_{11} & \cdots & a_{1m} \\ & & \\ & \cdot & \\ & \cdot & \\ & \cdot & \\ a_{n1} & \cdots & a_{nm} \end{bmatrix}, \quad \mathbf{v} = \begin{bmatrix} v_1 \\ \cdot \\ \cdot \\ \cdot \\ v_m \end{bmatrix}, \quad \text{and} \quad \mathbf{w} = \begin{bmatrix} w_1 \\ \cdot \\ \cdot \\ \cdot \\ w_m \end{bmatrix}.
$$

Then

$$
A(\mathbf{v} + \mathbf{w}) = \begin{bmatrix} a_{11} & \cdots & a_{1m} \\ & \cdot & \\ & \cdot & \\ & \cdot & \\ a_{n1} & \cdots & a_{nm} \end{bmatrix} \begin{bmatrix} v_1 + w_1 \\ \cdot \\ \cdot \\ \cdot \\ v_m + w_m \end{bmatrix}
$$

$$
= \begin{bmatrix} a_{11}(v_1 + w_1) + \ldots + a_{1m}(v_m + w_m) \\ \cdot \\ \cdot \\ \cdot \\ a_{n1}(v_1 + w_1) + \ldots + a_{nm}(v_m + w_m) \end{bmatrix}
$$

$$
= \begin{bmatrix} (a_{11}v_1 + \ldots + a_{1m}v_m) + (a_{11}w_1 + \ldots + a_{1m}w_m) \\ \cdot \\ \cdot \\ \cdot \\ (a_{n1}v_1 + \ldots + a_{nm}v_m) + (a_{n1}w_1 + \ldots + a_{nm}w_m) \end{bmatrix}
$$

$$
= \begin{bmatrix} a_{11}v_1 + \ldots + a_{1m}v_m \\ \cdot \\ \cdot \\ \cdot \\ a_{n1}v_1 + \ldots + a_{nm}v_m \end{bmatrix} + \begin{bmatrix} a_{11}w_1 + \ldots + a_{1m}w_m \\ \cdot \\ \cdot \\ \cdot \\ a_{n1}w_1 + \ldots + a_{nm}w_m \end{bmatrix}
$$

$$
= A\mathbf{v} + A\mathbf{w}.
$$

EXERCISES

1. Given $\mathbf{v} = \begin{bmatrix} 1 \\ 2 \end{bmatrix}$, $\mathbf{w} = \begin{bmatrix} -1 \\ 0 \end{bmatrix}$, $\mathbf{u} = \begin{bmatrix} 0 \\ 3 \end{bmatrix}$, $A = \begin{bmatrix} 1 & -1 \\ 2 & 3 \end{bmatrix}$, and $B = \begin{bmatrix} 1 & 2 \\ 0 & 1 \end{bmatrix}$, find

a. $\mathbf{v} + \mathbf{w}$ b. $3\mathbf{u}$ c. $2\mathbf{v} - 5\mathbf{w} + \mathbf{u}$
d. $A\mathbf{v}$ e. $A\mathbf{w}$ f. $A(3\mathbf{v} - \mathbf{w})$
g. $-3A$ h. $A - 2B$ i. $(A + B)\mathbf{v}$

2. Given $v = \begin{bmatrix} 1 \\ 2 \\ 0 \\ 0 \end{bmatrix}$, $w = \begin{bmatrix} -3 \\ 3 \\ 1 \\ -1 \end{bmatrix}$, and $A = \begin{bmatrix} 1 & 2 & 0 & 3 \\ 0 & 1 & -1 & 2 \\ 0 & 0 & 1 & 1 \\ 1 & -1 & 1 & 1 \end{bmatrix}$, find

   a. $3v$ b. $v - 2w$ c. Av d. $Av - 2Aw$

   e. $A - 3I$, where I is the 4×4 identity matrix (see Note 2)

Determine which of the systems in Exercises 3 to 9 are linear. For each linear system,

a. determine whether the system is homogeneous,

b. find its order, and

c. write it in matrix form.

3. $x' = -x + t + y$
    $y' = t - 2x$

4. $x' = 2x - 3xy$
    $y' = 3xy - 4y$

5. $x' = 5x - 6y$
    $y' = 2x + y$

6. $x' = -y - z$
    $y' = -x - z$
    $z' = -x - y$

7. $x' = -ty - z + t$
    $y' = -\dfrac{x}{t} - \dfrac{z}{t} + 1$
    $z' = x - ty$

8. $x' = y^2 + z$
    $y' = x + z$
    $z' = x - z$

9. $x' = x + 3y + t^2$
    $y' = 2x + y + t$
    $z' = x - t^2 + 3ty$

In Exercises 10 to 17, you are given an o.d.e. (N). For each,

a. write the equivalent system (S_N) in matrix form, and

b. find the general solution of (N) and use it to obtain the general solution of (S_N) in the form
    $x = c_1\mathbf{h}_1(t) + \ldots + c_n\mathbf{h}_n(t) + \mathbf{p}(t)$.

10. $(D^2 - 1)x = 0$

11. $(D^2 - 1)x = t$

12. $(D^2 + 1)x = 0$

13. $(D^2 + 1)x = 1$

14. $(D - 1)^2(D + 1)x = 0$

15. $(D - 1)^2(D + 1)x = 4$

16. $(D^3 - D)x = 0$

17. $(D^3 - D)x = 1$

In Exercises 18 to 22, you are given A, $\mathbf{E}(t)$, and $\mathbf{x}_i(t)$.

a. Find $D\mathbf{x}_i(t)$.

b. Find $A\mathbf{x}_i(t) + \mathbf{E}(t)$.

c. Determine whether $\mathbf{x}_i(t)$ is a solution of $D\mathbf{x} = A\mathbf{x} + \mathbf{E}(t)$.

18. $A = \begin{bmatrix} 1 & 2 \\ 0 & 3 \end{bmatrix}$, $\mathbf{E}(t) = 0$, $\mathbf{x}_1(t) = \begin{bmatrix} e^t \\ e^{3t} \end{bmatrix}$, $\mathbf{x}_2(t) = \begin{bmatrix} e^{3t} \\ e^{3t} \end{bmatrix}$

19. $A = \begin{bmatrix} 0 & 1 \\ -1 & 0 \end{bmatrix}$, $\mathbf{E}(t) = \begin{bmatrix} t \\ -1 \end{bmatrix}$, $\mathbf{x}_1(t) = \begin{bmatrix} \sin t \\ \cos t \end{bmatrix}$, $\mathbf{x}_2(t) = \begin{bmatrix} 0 \\ -t \end{bmatrix}$

20. $A = \begin{bmatrix} 0 & 1 \\ 0 & 0 \end{bmatrix}$, $\quad$ $\mathbf{E}(t) = \mathbf{0}$, $\quad$ $\mathbf{x}_1(t) = \begin{bmatrix} t \\ 1 \end{bmatrix}$, $\quad$ $\mathbf{x}_2(t) = \begin{bmatrix} 1 \\ 0 \end{bmatrix}$

21. $A = \begin{bmatrix} 1 & 2 & 0 \\ 0 & -1 & 0 \\ 0 & 0 & 2 \end{bmatrix}$, $\quad$ $\mathbf{E}(t) = \mathbf{0}$, $\quad$ $\mathbf{x}_1(t) = \begin{bmatrix} e^t \\ 0 \\ e^{2t} \end{bmatrix}$, $\quad$ $\mathbf{x}_2(t) = 2\mathbf{x}_1(t)$

22. $A = \begin{bmatrix} 1 & 2 & 0 \\ 0 & -1 & 0 \\ 0 & 0 & 2 \end{bmatrix}$, $\quad$ $\mathbf{E}(t) = \begin{bmatrix} -1 \\ 2 \\ -4 \end{bmatrix}$, $\quad$ $\mathbf{x}_1(t) = \begin{bmatrix} e^t - 3 \\ 2 \\ 2 \end{bmatrix}$, $\quad$ $\mathbf{x}_2(t) = 2\mathbf{x}_1(t)$

23. Several of the models in Chapter 2 led naturally to systems of o.d.e.'s involving second derivatives. We dealt with each of them by replacing the system by a single higher-order o.d.e. An alternative approach is to proceed as in Example 3.2.1. If an equation involves the second derivative of a variable, we introduce the first derivative as a new variable and replace the equation by two equations involving only first derivatives. Use this idea to obtain a linear system of o.d.e.'s for the given variables:
   a. The model of Example 2.1.4; x_1, $v_1 = x_1'$, x_2, and $v_2 = x_2'$.
   b. The model of Exercise 5, Section 2.1; the distance x of the weight from the ground, $v = x'$, and the distance y from the bottom of the elevator to the ground.

Some more abstract problems:

24. Suppose A is an $n \times n$ matrix, I is the $n \times n$ identity matrix (see Note 2), $\mathbf{v}$ is an n-vector, and λ is a number.
   a. Show that $A - \lambda I$ is the matrix obtained from A by subtracting λ from each of the diagonal entries a_{ii} while leaving the other entries alone.
   b. Show that $A\mathbf{v} - \lambda\mathbf{v} = (A - \lambda I)\mathbf{v}$.

25. Verify properties (A1) to (A8) of Note 2.

26. Verify properties (M2) to (M5) of Note 2.

27. Verify properties (D1) and (D2) of the summary.

28. a. Use some of the properties (M1), (M2), (D1), and (D2) to show that if $\mathbf{x} = \mathbf{h}_1(t)$ and $\mathbf{x} = \mathbf{h}_2(t)$ are solutions of the homogeneous system $D\mathbf{x} = A\mathbf{x}$ and c_1 and c_2 are constant scalars, then $\mathbf{x} = c_1\mathbf{h}_1(t) + c_2\mathbf{h}_2(t)$ is also a solution.
   b. Does (a) remain true if c_1 and c_2 are replaced by scalar valued functions?
   c. Does (a) remain true if the system there is replaced by the nonhomogeneous system $D\mathbf{x} = A\mathbf{x} + \mathbf{E}(t)$?

29. Use some of the properties in the summary to show that if $\mathbf{x} = \mathbf{h}(t)$ is a solution of the homogeneous system $D\mathbf{x} = A\mathbf{x}$, while $\mathbf{x} = \mathbf{p}(t)$ is a solution of the related nonhomogeneous system $D\mathbf{x} = A\mathbf{x} + \mathbf{E}(t)$, then $\mathbf{x} = \mathbf{p}(t) + \mathbf{h}(t)$ is also a solution of the nonhomogeneous system.

30. Use some of the properties in the summary to show that if $\mathbf{x} = \mathbf{p}(t)$ and $\mathbf{x} = \mathbf{q}(t)$ are both solutions of the same nonhomogeneous system $D\mathbf{x} = A\mathbf{x} + \mathbf{E}(t)$, then $\mathbf{x} = \mathbf{p}(t) - \mathbf{q}(t)$ is a solution of the related homogeneous system $D\mathbf{x} = A\mathbf{x}$.

31. Suppose A and B are $n \times m$ matrices with the property that $A\mathbf{v} = B\mathbf{v}$ for every m-vector $\mathbf{v}$. Show that the matrices A and B are equal. (*Hint:* What are $A\mathbf{v}$ and $B\mathbf{v}$ in the special case that all the entries of $\mathbf{v}$ except the jth are 0, and the jth entry of $\mathbf{v}$ is 1?)

3.3 LINEAR SYSTEMS OF O.D.E.'s: GENERAL PROPERTIES

We saw in Examples 3.2.1 and 3.2.2 that single linear o.d.e.'s can be replaced by equivalent systems. In Example 3.2.1 we solved such a system by first solving the single equation. If we learn to solve systems directly, we will be able to solve single equations by first solving the equivalent system. Thus our theory of systems includes the results of Chapter 2 as a special case. It should not surprise us that many of the features we see in solutions of systems will look familiar.

Suppose we are given an nth-order linear system of o.d.e.'s whose matrix form is

(S) $$D\mathbf{x} = A\mathbf{x} + \mathbf{E}(t)$$

where

$$\mathbf{x} = \begin{bmatrix} x_1 \\ \cdot \\ \cdot \\ \cdot \\ x_n \end{bmatrix}, \quad A = \begin{bmatrix} a_{11} & \cdots & a_{1n} \\ & \cdot & \\ & \cdot & \\ & \cdot & \\ a_{n1} & \cdots & a_{nn} \end{bmatrix}, \quad \text{and} \quad \mathbf{E}(t) = \begin{bmatrix} E_1(t) \\ \cdot \\ \cdot \\ \cdot \\ E_n(t) \end{bmatrix}.$$

Let $\mathbf{x} = \mathbf{p}(t)$ be a solution of (S) and let $\mathbf{x} = \mathbf{h}(t)$ be a solution of the **related homogeneous system**

(H) $$D\mathbf{x} = A\mathbf{x}.$$

Then

$$D[\mathbf{h}(t) + \mathbf{p}(t)] = D\mathbf{h}(t) + D\mathbf{p}(t)$$
$$= A\mathbf{h}(t) + [A\mathbf{p}(t) + \mathbf{E}(t)] = A[\mathbf{h}(t) + \mathbf{p}(t)] + \mathbf{E}(t).$$

Thus $\mathbf{x} = \mathbf{h}(t) + \mathbf{p}(t)$ is a solution of (S). If we let $\mathbf{h}(t)$ range over all solutions of (H), we get a list of solutions of (S).

Suppose now that $\mathbf{x} = \mathbf{f}(t)$ is also a solution of (S). Let $\mathbf{h}_1(t) = \mathbf{f}(t) - \mathbf{p}(t)$. Then

$$\mathbf{f}(t) = \mathbf{h}_1(t) + \mathbf{p}(t)$$

and

$$D\mathbf{h}_1(t) = D[\mathbf{f}(t) - \mathbf{p}(t)] = D\mathbf{f}(t) - D\mathbf{p}(t)$$
$$= [A\mathbf{f}(t) + \mathbf{E}(t)] - [A\mathbf{p}(t) + \mathbf{E}(t)] = A[\mathbf{f}(t) - \mathbf{p}(t)]$$
$$= A\mathbf{h}_1(t).$$

Thus $x = h_1(t)$ is a solution of (H) and $f(t)$ is on the list we described in the previous paragraph.

We have shown the following:

Fact: *If we find a particular solution* $x = p(t)$ *of the nonhomogeneous linear system (S), and if* $x = H(t)$ *is a formula that describes all solutions of the homogeneous system (H), then*

$$x = H(t) + p(t)$$

describes all solutions of (S).

Our strategy for solving linear systems will be the same as our strategy for solving single linear equations (see Section 2.2). First we will find the general solution $x = H(t)$ of the related homogeneous system (H). Then we will look for a particular solution $x = p(t)$ of the nonhomogeneous system (S). The general solution of (S) will be $x = H(t) + p(t)$.

Let's now take a look at what the general solution of (H) looks like. (You should look back at Sections 2.3 and 2.4, noting all of the parallels between the properties of one homogeneous o.d.e. and the present situation.) Suppose we know that $x = h_1(t), \ldots, x = h_n(t)$ are solutions of (H) and that $c_1, \ldots, c_n$ are constants. Then

$$D[c_1h_1(t) + \ldots + c_nh_n(t)] = c_1Dh_1(t) + \ldots + c_nDh_n(t)$$
$$= c_1Ah_1(t) + \ldots + c_nAh_n(t)$$
$$= A[c_1h_1(t) + \ldots + c_nh_n(t)].$$

Thus $x = c_1h_1(t) + \ldots + c_nh_n(t)$ is a solution of (H). We refer to a vector valued function of this form as a **linear combination** of $h_1(t), \ldots, h_n(t)$. Thus we have the following:

Fact: *If* $h_1(t), \ldots, h_n(t)$ *are solutions of (H), then any linear combination of these vector valued functions is also a solution of (H).*

If we find some solutions of (H), then we can use this fact to generate a list of solutions. The following theorem will provide us with a criterion for deciding whether the list we get is a *complete* list of solutions.

Theorem: Existence and Uniqueness of Solutions of Linear Systems.

Let $Dx = Ax + E(t)$ *be an nth-order linear system of o.d.e.'s. Suppose the entries* $E_i(t)$ *of* $E(t)$ *and the coefficients* a_{ij} *are continuous on an interval I. Let* t_0 *be a fixed value of t in I. Then, given any n-vector* v, *there exists a solution* $x = \phi(t)$

of the system, which is defined for all t in I and which satisfies the initial condition

$$\mathbf{x}(t_0) = \mathbf{v}.$$

Furthermore, if $\mathbf{x} = \boldsymbol{\psi}(t)$ *is a solution of the system that satisfies the same initial condition as* $\mathbf{x} = \boldsymbol{\phi}(t)$, *then* $\boldsymbol{\phi}(t) = \boldsymbol{\psi}(t)$ *for all t in I.*

    If a list of solutions fails to match an initial condition at t_0, *then* the "existence" part of the theorem says that *at least one solution has been excluded from the list.* On the other hand, *a list that matches all initial conditions at* t_0 *must be complete,* since any solution satisfies some initial condition at t_0 and, by uniqueness, must therefore agree with a vector valued function on the list.

    To determine whether a list, $\mathbf{x} = c_1\mathbf{h}_1(t) + \ldots + c_n\mathbf{h}_n(t)$, of solutions of the nth-order homogeneous system (H) is complete, we must determine whether we can match each initial condition. That is, we must determine whether the equation

(M) $$c_1\mathbf{h}_1(t) + c_2\mathbf{h}_2(t_0) + \ldots + c_n\mathbf{h}_n(t_0) = \mathbf{v}$$

can be solved for $c_1, \ldots, c_n$ for all choices of $\mathbf{v}$. Now each $\mathbf{h}_i(t)$ is an n-vector valued function:

$$\mathbf{h}_1(t) = \begin{bmatrix} h_{11}(t) \\ h_{21}(t) \\ \cdot \\ \cdot \\ \cdot \\ h_{n1}(t) \end{bmatrix}, \quad \mathbf{h}_2(t) = \begin{bmatrix} h_{12}(t) \\ h_{22}(t) \\ \cdot \\ \cdot \\ \cdot \\ h_{n2}(t) \end{bmatrix}, \ldots, \mathbf{h}_n(t) = \begin{bmatrix} h_{1n}(t) \\ h_{2n}(t) \\ \cdot \\ \cdot \\ \cdot \\ h_{nn}(t) \end{bmatrix}.$$

The vector $\mathbf{v}$ is an n-vector:

$$\mathbf{v} = \begin{bmatrix} v_1 \\ v_2 \\ \cdot \\ \cdot \\ \cdot \\ v_n \end{bmatrix}.$$

Equation (M) is merely a matrix formulation of the system of algebraic equations

$$c_1 h_{11}(t_0) + \ldots + c_n h_{1n}(t_0) = v_1$$
$$c_1 h_{21}(t_0) + \ldots + c_n h_{2n}(t_0) = v_2$$

(M')

$$\cdot$$
$$\cdot$$
$$\cdot$$

$$c_1 h_{n1}(t_0) + \ldots + c_n h_{nn}(t_0) = v_n.$$

Cramer's determinant test tells us that we can solve this system for all choices of $v_1, \ldots, v_n$ if and only if the determinant of coefficients is not zero.

We define the **Wronskian** of $h_1(t), \ldots, h_n(t)$ to be the determinant of the $n \times n$ matrix whose columns are $h_1(t), \ldots, h_n(t)$:

$$W[\mathbf{h}_1, \ldots, \mathbf{h}_n](t) = \det \begin{bmatrix} h_{11}(t) & \cdots & h_{1n}(t) \\ & \cdot & \\ & \cdot & \\ & \cdot & \\ h_{n1}(t) & \cdots & h_{nn}(t) \end{bmatrix}$$

The determinant of coefficients of (M') is the Wronskian evaluated at t_0. Thus we can state the conclusion of the preceding paragraph as follows:

Fact: *Suppose the coefficients a_{ij} of the nth-order linear homogeneous system (H) are continuous on an interval I. Let $\mathbf{h}_1(t), \ldots, \mathbf{h}_n(t)$ be solutions of (H) and let t_0 be a fixed value of t in I. The general solution of (H) is*

$$\mathbf{x} = c_1\mathbf{h}_1(t) + \ldots + c_n\mathbf{h}_n(t)$$

if and only if

$$W[\mathbf{h}_1, \ldots, \mathbf{h}_n](t_0) \neq 0.$$

Example 3.3.1

The equations of the simple *LRC* circuit of Example 3.1.1 with inductance $L = 1$ henry, resistance $R = 3$ ohms, capacitance $C = 1/2$ farad, and no voltage source ($V(t) = 0$) can be written in the form $D\mathbf{x} = A\mathbf{x}$, where

$$\mathbf{x} = \begin{bmatrix} Q \\ I \end{bmatrix} \quad \text{and} \quad A = \begin{bmatrix} 0 & 1 \\ -2 & -3 \end{bmatrix}.$$

Let

$$\mathbf{h}_1(t) = \begin{bmatrix} e^{-t} \\ -e^{-t} \end{bmatrix} \quad \text{and} \quad \mathbf{h}_2(t) = \begin{bmatrix} e^{-2t} \\ -2e^{-2t} \end{bmatrix}.$$

Then

$$A\mathbf{h}_1(t) = \begin{bmatrix} 0 & 1 \\ -2 & -3 \end{bmatrix} \begin{bmatrix} e^{-t} \\ -e^{-t} \end{bmatrix} = \begin{bmatrix} -e^{-t} \\ e^{-t} \end{bmatrix} = D\mathbf{h}_1(t)$$

and

$$Ah_2(t) = \begin{bmatrix} 0 & 1 \\ -2 & -3 \end{bmatrix} \begin{bmatrix} e^{-2t} \\ -2e^{-2t} \end{bmatrix} = \begin{bmatrix} -2e^{-2t} \\ 4e^{-2t} \end{bmatrix} = Dh_2(t).$$

Thus, $h_1(t)$ and $h_2(t)$ are solutions of $Dx = Ax$.
The Wronskian of $h_1(t)$ and $h_2(t)$ is

$$W[h_1,h_2](t) = \det \begin{bmatrix} e^{-t} & e^{-2t} \\ -e^{-t} & -2e^{-2t} \end{bmatrix}.$$

Thus

$$W[h_1,h_2](0) = \det \begin{bmatrix} 1 & 1 \\ -1 & -2 \end{bmatrix} = -1 \neq 0.$$

Since the Wronskian is not zero at $t_0 = 0$, $h_1(t)$ and $h_2(t)$ generate the general solution of $Dx = Ax$,

$$x = c_1 \begin{bmatrix} e^{-t} \\ -e^{-t} \end{bmatrix} + c_2 \begin{bmatrix} e^{-2t} \\ -2e^{-2t} \end{bmatrix}.$$

In terms of the charge and current in the circuit model, this reads

$$Q = c_1 e^{-t} + c_2 e^{-2t}$$
$$I = -c_1 e^{-t} - 2c_2 e^{-2t}.$$

Example 3.3.2

The equations of the three-loop circuit of Example 3.1.3 with $V(t) = 0$, $R_1 = R_2 = 1$, $L_2 = L_3 = 1$, and $C = 1$ can be written $Dx = Ax$, where

$$x = \begin{bmatrix} Q \\ I_2 \\ I_3 \end{bmatrix} \quad \text{and} \quad A = \begin{bmatrix} -1 & -1 & 0 \\ 1 & -1 & 1 \\ 0 & 1 & -1 \end{bmatrix}.$$

Direct substitution will show that

$$h_1(t) = \begin{bmatrix} (1+t)e^{-t} \\ -e^{-t} \\ -(1+t)e^{-t} \end{bmatrix}, \quad h_2(t) = \begin{bmatrix} (1-t)e^{-t} \\ e^{-t} \\ -(1-t)e^{-t} \end{bmatrix}, \quad h_3(t) = \begin{bmatrix} (3+t)e^{-t} \\ -e^{-t} \\ -(3+t)e^{-t} \end{bmatrix}$$

are solutions of this system. If we evaluate the Wronskian of these functions at $t = 0$, we get

$$W[\mathbf{h}_1, \mathbf{h}_2, \mathbf{h}_3](0) = \det \begin{bmatrix} 1 & 1 & 3 \\ -1 & 1 & -1 \\ -1 & -1 & -3 \end{bmatrix} = 0$$

so that $\mathbf{h}_1$, $\mathbf{h}_2$, $\mathbf{h}_3$ do not generate the general solution.

On the other hand, we leave it to the reader to check that

$$\mathbf{k}(t) = \begin{bmatrix} (2 - t^2)e^{-t} \\ 2te^{-t} \\ t^2 e^{-t} \end{bmatrix}$$

is a solution of $D\mathbf{x} = A\mathbf{x}$ and that $\mathbf{h}_1(t)$, $\mathbf{h}_2(t)$, and $\mathbf{k}(t)$ generate the general solution,

$$\mathbf{x} = c_1 \begin{bmatrix} (1 + t)e^{-t} \\ -e^{-t} \\ -(1 + t)e^{-t} \end{bmatrix} + c_2 \begin{bmatrix} (1 - t)e^{-t} \\ e^{-t} \\ -(1 - t)e^{-t} \end{bmatrix} + c_3 \begin{bmatrix} (2 - t^2)e^{-t} \\ 2te^{-t} \\ t^2 e^{-t} \end{bmatrix}.$$

In terms of charge and current, this reads

$$Q = (c_1 + c_2 + 2c_3)e^{-t} + (c_1 - c_2)te^{-t} - c_3 t^2 e^{-t}$$
$$I_2 = (-c_1 + c_2)e^{-t} + 2c_3 te^{-t}$$
$$I_3 = (-c_1 - c_2)e^{-t} + (-c_1 + c_2)te^{-t} + c_3 t^2 e^{-t}.$$

Example 3.3.3

If we add to the circuit of Example 3.3.1 a voltage source with

$$V(t) = 10 \cos t$$

then the equations read $D\mathbf{x} = A\mathbf{x} + \mathbf{E}(t)$, where

$$\mathbf{x} = \begin{bmatrix} Q \\ I \end{bmatrix}, \qquad A = \begin{bmatrix} 0 & 1 \\ -2 & -3 \end{bmatrix}, \qquad \text{and} \qquad \mathbf{E}(t) = \begin{bmatrix} 0 \\ 10 \cos t \end{bmatrix}.$$

We saw in Example 3.3.1 that the general solution of the homogeneous system $D\mathbf{x} = A\mathbf{x}$ (no voltage source) is $\mathbf{x} = \mathbf{H}(t)$, where

$$\mathbf{H}(t) = c_1 \begin{bmatrix} e^{-t} \\ -e^{-t} \end{bmatrix} + c_2 \begin{bmatrix} e^{-2t} \\ -2e^{-2t} \end{bmatrix}.$$

Let

$$\mathbf{p}(t) = \begin{bmatrix} \cos t + 3 \sin t \\ 3 \cos t - \sin t \end{bmatrix}.$$

Then

$$A\mathbf{p}(t) + \mathbf{E}(t) = \begin{bmatrix} 0 & 1 \\ -2 & -3 \end{bmatrix} \begin{bmatrix} \cos t + 3 \sin t \\ 3 \cos t - \sin t \end{bmatrix} + \begin{bmatrix} 0 \\ 10 \cos t \end{bmatrix}$$

$$= \begin{bmatrix} 3 \cos t - \sin t \\ -11 \cos t - 3 \sin t \end{bmatrix} + \begin{bmatrix} 0 \\ 10 \cos t \end{bmatrix} = \begin{bmatrix} 3 \cos t - \sin t \\ -\cos t - 3 \sin t \end{bmatrix}$$

$$= D\mathbf{p}(t).$$

Thus, $\mathbf{p}(t)$ solves $D\mathbf{x} = A\mathbf{x} + \mathbf{E}(t)$, and the general solution is

$$\mathbf{x} = \mathbf{H}(t) + \mathbf{p}(t) = c_1 \begin{bmatrix} e^{-t} \\ -e^{-t} \end{bmatrix} + c_2 \begin{bmatrix} e^{-2t} \\ -2e^{-2t} \end{bmatrix} + \begin{bmatrix} \cos t + 3 \sin t \\ 3 \cos t - \sin t \end{bmatrix}.$$

In terms of charge and current,

$$Q = c_1 e^{-t} + c_2 e^{-2t} + \cos t + 3 \sin t$$
$$I = -c_1 e^{-t} - 2c_2 e^{-2t} + 3 \cos t - \sin t.$$

In each of the examples we have considered, the general solution of an nth-order homogeneous system has been of the form $\mathbf{x} = c_1 \mathbf{h}_1(t) + \ldots + c_n \mathbf{h}_n(t)$. Let's now check that this is always the case.

The Existence theorem tells us that an nth-order homogeneous system has solutions that match each initial condition at t_0. In particular, there are solutions $\mathbf{h}_1(t), \ldots, \mathbf{h}_n(t)$ that match the initial conditions

$$\mathbf{h}_1(t_0) = \begin{bmatrix} 1 \\ 0 \\ 0 \\ \cdot \\ \cdot \\ \cdot \\ 0 \end{bmatrix}, \quad \mathbf{h}_2(t_0) = \begin{bmatrix} 0 \\ 1 \\ 0 \\ \cdot \\ \cdot \\ \cdot \\ 0 \end{bmatrix}, \ldots, \mathbf{h}_n(t_0) = \begin{bmatrix} 0 \\ 0 \\ 0 \\ \cdot \\ \cdot \\ \cdot \\ 1 \end{bmatrix}.$$

Then

$$W[\mathbf{h}_1, \ldots, \mathbf{h}_n](t_0) = \det \begin{bmatrix} 1 & 0 & \ldots & 0 \\ 0 & 1 & \ldots & 0 \\ 0 & 0 & \ldots & 0 \\ & & \cdot & \\ & & \cdot & \\ & & \cdot & \\ 0 & 0 & \ldots & 1 \end{bmatrix} = 1 \neq 0.$$

These solutions generate a complete list of solutions.

Fact: *Suppose the coefficients of the nth-order linear homogeneous system (H) are continuous on an interval I. Then the general solution of (H) has the form*

$$\mathbf{x} = c_1 \mathbf{h}_1(t) + \ldots + c_n \mathbf{h}_n(t)$$

for a suitable choice of $\mathbf{h}_1(t), \ldots, \mathbf{h}_n(t)$.

Let's summarize.

LINEAR SYSTEMS OF O.D.E.'s: GENERAL PROPERTIES

Given an nth-order linear system of o.d.e.'s

(S) $D\mathbf{x} = A\mathbf{x} + \mathbf{E}(t).$

The **related homogeneous system** is

(H) $D\mathbf{x} = A\mathbf{x}$

Suppose the entries a_{ij} of A and $E_i(t)$ of $\mathbf{E}(t)$ are continuous on an interval I. Then the following are true:

1. The general solution of (S) is of the form

    $$\mathbf{x} = \mathbf{H}(t) + \mathbf{p}(t)$$

    where $\mathbf{x} = \mathbf{H}(t)$ is the general solution of (H) and $\mathbf{x} = \mathbf{p}(t)$ is a particular solution of (S).
2. The general solution of (H) is of the form $\mathbf{x} = \mathbf{H}(t)$ where

    $$\mathbf{H}(t) = c_1 \mathbf{h}_1(t) + \ldots + c_n \mathbf{h}_n(t)$$

for a suitable choice of $\mathbf{h}_1(t), \ldots, \mathbf{h}_n(t)$.

3. Let t_0 be a fixed value of t in I. A list of solutions of (H) or (S) is a complete list if and only if it matches every initial condition $\mathbf{x}(t_0) = \mathbf{v}$.

4. Let t_0 be a fixed value of t in I and let $\mathbf{h}_1(t), \ldots, \mathbf{h}_n(t)$ be solutions of (H). The general solution of (H) is $\mathbf{x} = c_1\mathbf{h}_1(t) + \ldots + c_n\mathbf{h}_n(t)$ if and only if $W[\mathbf{h}_1, \ldots, \mathbf{h}_n](t_0) \neq 0$. Here $W[\mathbf{h}_1, \ldots, \mathbf{h}_n](t)$, the **Wronskian** of $\mathbf{h}_1, \ldots,$ $\mathbf{h}_n$, is the determinant of the $n \times n$ matrix whose columns are $\mathbf{h}_1, \ldots, \mathbf{h}_n$.

Notes

1. A technicality

Our results depend heavily on the Existence and Uniqueness Theorem, which assumes that the coefficients a_{ij} and $E_i(t)$ are continuous on an interval I and which provides information about solutions defined on I. *All statements in this and later sections about general solutions are valid only on such intervals. The values t_0 that we use to test for completeness of a list of solutions will always be in the interval I.*

2. On the new and old Wronskians

If $x = h(t)$ is a solution of the single nth-order linear homogeneous equation $(D^n + \ldots + a_0)x = 0$, then

$$
\mathbf{h}(t) = \begin{bmatrix} h(t) \\ h'(t) \\ \cdot \\ \cdot \\ \cdot \\ h^{(n-1)}(t) \end{bmatrix}
$$

is a solution of the equivalent system (see Example 3.2.2). If we start with n solutions $h_1(t), \ldots, h_n(t)$ of the o.d.e., then we get n solutions $\mathbf{h}_1(t), \ldots, \mathbf{h}_n(t)$ of the system. The Wronskian of $\mathbf{h}_1(t), \ldots, \mathbf{h}_n(t)$ is

$$
W[\mathbf{h}_1, \ldots, \mathbf{h}_n](t) = \det \begin{bmatrix} h_1(t) & \cdots & h_n(t) \\ \cdot & & \cdot \\ \cdot & & \cdot \\ \cdot & & \cdot \\ h_1^{(n-1)}(t) & \cdots & h_n^{(n-1)}(t) \end{bmatrix}.
$$

This is the same as $W[h_1, \ldots, h_n](t)$ as we defined it in Chapter 2.

3. Fundamental matrices

Suppose the vector valued functions

$$
\mathbf{h}_1(t) = \begin{bmatrix} h_{11}(t) \\ \cdot \\ \cdot \\ \cdot \\ h_{n1}(t) \end{bmatrix}, \quad \mathbf{h}_2(t) = \begin{bmatrix} h_{12}(t) \\ \cdot \\ \cdot \\ \cdot \\ h_{n2}(t) \end{bmatrix}, \ldots, \mathbf{h}_n(t) = \begin{bmatrix} h_{1n}(t) \\ \cdot \\ \cdot \\ \cdot \\ h_{nn}(t) \end{bmatrix}
$$

generate all solutions of the homogeneous system $D\mathbf{x} = A\mathbf{x}$. Then every solution can be written in the form

$$\mathbf{x} = c_1\mathbf{h}_1(t) + c_2\mathbf{h}_2(t) + \ldots + c_n\mathbf{h}_n(t)$$

$$= \begin{bmatrix} c_1h_{11}(t) + c_2h_{12}(t) + \ldots + c_nh_{1n}(t) \\ \cdot \\ \cdot \\ \cdot \\ c_1h_{n1}(t) + c_2h_{n2}(t) + \ldots + c_nh_{nn}(t) \end{bmatrix}.$$

Note that this last expression can be written in the matrix form

$$\mathbf{x} = \boldsymbol{\phi}(t)\ \mathbf{C}$$

where

$$\boldsymbol{\phi}(t) = \begin{bmatrix} h_{11}(t) & h_{12}(t) & \ldots & h_{1n}(t) \\ & \cdot \\ & \cdot \\ & \cdot \\ h_{n1}(t) & h_{n2}(t) & \ldots & h_{nn}(t) \end{bmatrix}$$

is the $n \times n$ matrix whose columns are $\mathbf{h}_1(t), \ldots, \mathbf{h}_n(t)$, and

$$\mathbf{C} = \begin{bmatrix} c_1 \\ \cdot \\ \cdot \\ \cdot \\ c_n \end{bmatrix}$$

is a constant n-vector whose entries are the coefficients of the linear combination. The matrix $\boldsymbol{\phi}(t)$ is called a **fundamental matrix** of $D\mathbf{x} = A\mathbf{x}$. Note that $\boldsymbol{\phi}(t)$ is the matrix whose determinant we compute to find the Wronskian of $\mathbf{h}_1(t), \ldots, \mathbf{h}_n(t)$:

$$W[\mathbf{h}_1, \ldots, \mathbf{h}_n](t) = \det \boldsymbol{\phi}(t).$$

4. A warning

Statement 3 in the summary applies to homogeneous and nonhomogeneous systems alike: a list of solutions is complete if and only if it satisfies all possible initial conditions. However, the Wronskian test (statement 4), based on the specific form of solutions as linear combinations, applies only to solutions of homogeneous equations. Of course, in the expression $\mathbf{x} = \mathbf{H}(t) + \mathbf{p}(t)$, one can use the Wronskian to check that $\mathbf{H}(t)$ represents a general solution of (H) and then use statement 1 to conclude that $\mathbf{H}(t) + \mathbf{p}(t)$ is the general solution of (S).

EXERCISES

In Exercises 1 to 10, you are given a matrix A and vector valued functions $\mathbf{h}_1(t), \ldots, \mathbf{h}_k(t)$. For each problem, decide (a) whether the functions $\mathbf{x} = \mathbf{h}_i(t)$ are solutions of the homogeneous system (H) $D\mathbf{x} = A\mathbf{x}$ and (b) whether the ones that are solutions generate the general solution of (H).

1. $A = \begin{bmatrix} -3 & -2 \\ 1 & 0 \end{bmatrix}$; $\quad \mathbf{h}_1(t) = \begin{bmatrix} 2e^{-2t} \\ -e^{-2t} \end{bmatrix}$, $\quad \mathbf{h}_2(t) = \begin{bmatrix} e^{-t} \\ -e^{-t} \end{bmatrix}$

2. $A = \begin{bmatrix} 0 & -1 \\ 4 & 0 \end{bmatrix}$; $\quad \mathbf{h}_1(t) = \begin{bmatrix} \cos 2t \\ 2 \sin 2t \end{bmatrix}$, $\quad \mathbf{h}_2(t) = \begin{bmatrix} \sin 2t \\ -2 \cos 2t \end{bmatrix}$

3. $A = \begin{bmatrix} 0 & 1 \\ -4 & 0 \end{bmatrix}$; $\quad \mathbf{h}_1(t) = \begin{bmatrix} \cos 2t \\ -2 \sin 2t \end{bmatrix}$, $\quad \mathbf{h}_2(t) = \begin{bmatrix} \sin 2t \\ -2 \cos 2t \end{bmatrix}$

4. $A = \begin{bmatrix} 5 & -3 \\ 3 & -5 \end{bmatrix}$; $\quad \mathbf{h}_1(t) = \begin{bmatrix} 3e^{4t} + e^{-4t} \\ e^{4t} + 3e^{-4t} \end{bmatrix}$, $\quad \mathbf{h}_2(t) = \begin{bmatrix} 3e^{4t} - e^{-4t} \\ e^{4t} - 3e^{-4t} \end{bmatrix}$

5. $A = \begin{bmatrix} 5 & -3 \\ 3 & -5 \end{bmatrix}$; $\quad \mathbf{h}_1(t) = \begin{bmatrix} 3e^{4t} \\ e^{4t} \end{bmatrix}$, $\quad \mathbf{h}_2(t) = \begin{bmatrix} 3e^{-4t} \\ e^{-4t} \end{bmatrix}$

6. $A = \begin{bmatrix} 1 & 1 \\ 1 & 1 \end{bmatrix}$; $\quad \mathbf{h}_1(t) = \begin{bmatrix} 1 \\ 1 \end{bmatrix}$, $\quad \mathbf{h}_2(t) = \begin{bmatrix} 1 \\ -1 \end{bmatrix}$, $\quad \mathbf{h}_3(t) = \begin{bmatrix} 1 + e^{2t} \\ -1 + e^{2t} \end{bmatrix}$

7. $A = \begin{bmatrix} -1 & -1 & 0 \\ 0 & -1 & 0 \\ 0 & 1 & -1 \end{bmatrix}$; $\quad \mathbf{h}_1(t) = \begin{bmatrix} e^{-t} \\ 0 \\ -e^{-t} \end{bmatrix}$, $\quad \mathbf{h}_2(t) = \begin{bmatrix} te^{-t} \\ -e^{-t} \\ -te^{-t} \end{bmatrix}$,

$$\mathbf{h}_3(t) = \begin{bmatrix} (3t - 2)e^{-t} \\ -3e^{-t} \\ -(3t - 2)e^{-t} \end{bmatrix}$$

8. $A = \begin{bmatrix} 2 & 0 & 4 \\ 0 & 2 & 0 \\ -1 & 0 & 2 \end{bmatrix}$; $\quad \mathbf{h}_1(t) = \begin{bmatrix} 2e^{2t} \cos 2t \\ e^{2t} \\ -e^{2t} \sin 2t \end{bmatrix}$, $\quad \mathbf{h}_2(t) = \begin{bmatrix} 2e^{2t} \sin 2t \\ e^{2t} \\ e^{2t} \cos 2t \end{bmatrix}$,

$$\mathbf{h}_3(t) = \begin{bmatrix} 2e^{2t} (\sin 2t + \cos 2t) \\ 2e^{2t} \\ e^{2t} (\cos 2t - \sin 2t) \end{bmatrix}$$

9. $A = \begin{bmatrix} 2 & 0 & 4 \\ 0 & 2 & 0 \\ -1 & 0 & 2 \end{bmatrix}$; $\quad \mathbf{h}_1(t) = \begin{bmatrix} 2e^{2t} \cos 2t \\ e^{2t} \\ -e^{2t} \sin 2t \end{bmatrix}$, $\quad \mathbf{h}_2(t) = \begin{bmatrix} 2e^{2t} \sin 2t \\ e^{2t} \\ e^{2t} \cos 2t \end{bmatrix}$,

$$\mathbf{h}_3(t) = \begin{bmatrix} 0 \\ e^{2t} \\ 0 \end{bmatrix}$$

10. $A = \begin{bmatrix} 0 & 8 & 0 & 0 \\ -2 & 0 & 0 & 0 \\ 0 & 0 & 0 & 2 \\ 0 & 0 & -8 & 0 \end{bmatrix}$; $\mathbf{h}_1(t) = \begin{bmatrix} 2 \sin 4t \\ \cos 4t \\ 0 \\ 0 \end{bmatrix}$, $\mathbf{h}_2(t) = \begin{bmatrix} 2 \cos 4t \\ -\sin 4t \\ 0 \\ 0 \end{bmatrix}$,

$$\mathbf{h}_3(t) = \begin{bmatrix} 0 \\ 0 \\ \sin 4t \\ 2 \cos 4t \end{bmatrix}, \quad \mathbf{h}_4(t) = \begin{bmatrix} 0 \\ 0 \\ \cos 4t \\ -2 \sin 4t \end{bmatrix}$$

In Exercises 11 to 15, you are given a matrix A, a vector valued function $\mathbf{E}(t)$, and a list of solutions of the nonhomogeneous system $D\mathbf{x} = A\mathbf{x} + \mathbf{E}(t)$. In each case, decide whether the list is complete.

11. $A = \begin{bmatrix} -3 & -2 \\ 1 & 0 \end{bmatrix}$, $\mathbf{E}(t) = \begin{bmatrix} 2e^{-t} \\ -e^{-t} \end{bmatrix}$; $\begin{cases} x_1 = 2c_1 e^{-2t} + c_2 e^{-t} \\ x_2 = -c_1 e^{-2t} - c_2 e^{-t} + e^{-t} \end{cases}$

12. $A = \begin{bmatrix} 0 & -1 \\ 4 & 0 \end{bmatrix}$, $\mathbf{E}(t) = \begin{bmatrix} 0 \\ -5e^t \end{bmatrix}$; $\begin{cases} x_1 = c_1 \cos 2t + c_2 \sin 2t + e^t \\ x_2 = -2c_2 \cos 2t + 2c_1 \sin 2t - e^t \end{cases}$

13. $A = \begin{bmatrix} 5 & -3 & 0 \\ 3 & -5 & 0 \\ 0 & 1 & 2 \end{bmatrix}$, $\mathbf{E}(t) = \begin{bmatrix} 0 \\ 0 \\ 4 \end{bmatrix}$; $\begin{cases} x_1 = 6c_1 e^{4t} - 2c_2 e^{-4t} \\ x_2 = 2c_1 e^{4t} - 6c_2 e^{-4t} \\ x_3 = c_1 e^{4t} + c_2 e^{-4t} - 2 \end{cases}$

14. $A = \begin{bmatrix} 5 & -3 & 0 \\ 3 & -5 & 0 \\ 0 & 1 & 2 \end{bmatrix}$, $\mathbf{E}(t) = \begin{bmatrix} 0 \\ 0 \\ 4 \end{bmatrix}$; $\begin{cases} x_1 = (6c_1 + 6c_3)e^{4t} + (-2c_2 + 2c_3)e^{-4t} \\ x_2 = (2c_1 + 2c_3)e^{4t} + (-6c_2 + 6c_3)e^{-4t} \\ x_3 = (c_1 + c_3)e^{4t} + (c_2 - c_3)e^{-4t} - 2 \end{cases}$

15. $A = \begin{bmatrix} 5 & -3 & 0 \\ 3 & -5 & 0 \\ 0 & 1 & 2 \end{bmatrix}$, $\mathbf{E}(t) = \begin{bmatrix} 0 \\ 0 \\ 4 \end{bmatrix}$; $\begin{cases} x_1 = 6c_1 e^{4t} - 2c_2 e^{-4t} \\ x_2 = 2c_1 e^{4t} - 6c_2 e^{-4t} \\ x_3 = c_1 e^{4t} + c_2 e^{-4t} + c_3 e^{2t} - 2 \end{cases}$

3.4 LINEAR INDEPENDENCE OF VECTORS

In the last section we obtained a Wronskian test for determining whether solutions $\mathbf{h}_1(t), \ldots, \mathbf{h}_n(t)$ of an nth-order homogeneous linear system (H) generate the general solution. In this section we obtain another test. The new test involves a concept that will be important when we try to build generating sets of solutions.

Recall that the solutions $\mathbf{h}_1(t), \ldots, \mathbf{h}_n(t)$ of (H) generate the general solution provided we can always solve the equation

$$c_1 \mathbf{h}_1(t_0) + \ldots + c_n \mathbf{h}_n(t_0) = \mathbf{v}$$

for $c_1, \ldots, c_n$. This is really a statement about the *initial vectors* $\mathbf{h}_1(t_0), \ldots, \mathbf{h}_n(t_0)$. Let's take a look at the vectors that arise in this way in Example 3.3.2. In that example, $\mathbf{h}_1(t)$, $\mathbf{h}_2(t)$, and $\mathbf{h}_3(t)$ do not generate the general solution. Since t_0 is 0, the vectors in question are

$$\mathbf{h}_1(0) = \begin{bmatrix} 1 \\ -1 \\ -1 \end{bmatrix}, \qquad \mathbf{h}_2(0) = \begin{bmatrix} 1 \\ 1 \\ -1 \end{bmatrix}, \qquad \text{and} \qquad \mathbf{h}_3(0) = \begin{bmatrix} 3 \\ -1 \\ -3 \end{bmatrix}.$$

Note that $\mathbf{h}_3(0) = 2\mathbf{h}_1(0) + \mathbf{h}_2(0)$ so that

$$2\mathbf{h}_1(0) + \mathbf{h}_2(0) - \mathbf{h}_3(0) = \mathbf{0}.$$

We have found constants $c_1 = 2$, $c_2 = 1$, and $c_3 = -1$ so that

$$c_1\mathbf{h}_1(0) + c_2\mathbf{h}_2(0) + c_3\mathbf{h}_3(0) = \mathbf{0}.$$

Of course, this last relationship would also be true if we took all the constants to be zero. The important thing is that we have found constants that are not all zero. We say these vectors are linearly dependent.

Definition: *The n-vectors* $\mathbf{v}_1, \ldots, \mathbf{v}_k$ *are **linearly dependent** if there exist constants* $c_1, \ldots, c_k$ *with at least one* $c_i \neq 0$ *so that*

$$c_1\mathbf{v}_1 + \ldots + c_k\mathbf{v}_k = \mathbf{0}.$$

*The vectors are **linearly independent** if the only constants for which this relationship holds are*

$$c_1 = c_2 = \ldots = c_k = 0.$$

Note that the vectors $\mathbf{v}_1, \ldots, \mathbf{v}_k$ all have to be *n*-vectors (otherwise we couldn't form the linear combination $c_1\mathbf{v}_1 + \ldots + c_k\mathbf{v}_k$). We do *not* require that $k = n$.

Example 3.4.1
Check for independence:

$$\mathbf{v}_1 = \begin{bmatrix} 1 \\ -1 \\ -1 \end{bmatrix}, \qquad \mathbf{v}_2 = \begin{bmatrix} 1 \\ 1 \\ -1 \end{bmatrix}, \qquad \mathbf{v}_3 = \begin{bmatrix} 2 \\ 0 \\ 0 \end{bmatrix}.$$

A typical linear combination of $\mathbf{v}_1$, $\mathbf{v}_2$, and $\mathbf{v}_3$ looks like

$$c_1\mathbf{v}_1 + c_2\mathbf{v}_2 + c_3\mathbf{v}_3 = \begin{bmatrix} c_1 + c_2 + 2c_3 \\ -c_1 + c_2 \\ -c_1 - c_2 \end{bmatrix}.$$

We wish to know which values of c_1, c_2, and c_3 give $\mathbf{0}$. Equating corresponding entries of the linear combination and $\mathbf{0}$ gives us the equations

$$
\begin{aligned}
c_1 + c_2 + 2c_3 &= 0 \\
-c_1 + c_2 &= 0 \\
-c_1 - c_2 &= 0.
\end{aligned}
$$

Adding the first and last equations, we obtain $2c_3 = 0$, so $c_3 = 0$. The last two equations force $c_1 = c_2 = 0$. Thus, the only values of c_1, c_2, and c_3 for which these equations hold are

$$
c_1 = c_2 = c_3 = 0.
$$

Hence, the vectors $\mathbf{v}_1$, $\mathbf{v}_2$, and $\mathbf{v}_3$ are independent.

Note that $\mathbf{v}_1$, $\mathbf{v}_2$, and $\mathbf{v}_3$ are the initial values at $t = 0$ of the vector valued functions $\mathbf{h}_1(t)$, $\mathbf{h}_2(t)$, and $\mathbf{k}(t)$ in Example 3.3.2. These functions *do* generate the general solution of the homogeneous system (H) in that example.

Example 3.4.2

Check for independence:

$$
\mathbf{v}_1 = \begin{bmatrix} 1 \\ 0 \\ 0 \\ 0 \end{bmatrix}, \quad
\mathbf{v}_2 = \begin{bmatrix} 0 \\ -1 \\ 0 \\ 0 \end{bmatrix}, \quad
\mathbf{v}_3 = \begin{bmatrix} 2 \\ 3 \\ 0 \\ 0 \end{bmatrix}.
$$

A typical linear combination of these vectors looks like

$$
c_1\mathbf{v}_1 + c_2\mathbf{v}_2 + c_3\mathbf{v}_3 = \begin{bmatrix} c_1 + 2c_3 \\ -c_2 + 3c_3 \\ 0 \\ 0 \end{bmatrix}.
$$

We wish to know which values of c_1, c_2, and c_3 give $\mathbf{0}$. Equating corresponding entries of the linear combination and $\mathbf{0}$ gives us the equations

$$
\begin{aligned}
c_1 + 2c_3 &= 0 \\
-c_2 + 3c_3 &= 0 \\
0 &= 0 \\
0 &= 0.
\end{aligned}
$$

The constants c_1, c_2, and c_3 will solve these equations if and only if

$$c_1 = -2c_3 \quad \text{and} \quad c_2 = 3c_3.$$

We *can* find constants that are not all 0 and that solve these equations. For example, we could take

$$c_3 = 1, \quad c_1 = -2, \quad c_2 = 3.$$

The vectors are linearly dependent.

Example 3.4.3

Check for independence:

$$\mathbf{v}_1 = \begin{bmatrix} 1 \\ 0 \\ 0 \end{bmatrix}, \quad \mathbf{v}_2 = \begin{bmatrix} 1 \\ 1 \\ 0 \end{bmatrix}, \quad \mathbf{v}_3 = \begin{bmatrix} 1 \\ 2 \\ 0 \end{bmatrix}, \quad \mathbf{v}_4 = \begin{bmatrix} 2 \\ -2 \\ 2 \end{bmatrix}.$$

A typical linear combination looks like

$$c_1\mathbf{v}_1 + c_2\mathbf{v}_2 + c_3\mathbf{v}_3 + c_4\mathbf{v}_4 = \begin{bmatrix} c_1 + c_2 + c_3 + 2c_4 \\ c_2 + 2c_3 - 2c_4 \\ 2c_4 \end{bmatrix}.$$

Equating corresponding entries of the linear combination and $\mathbf{0}$ gives us the equations

$$c_1 + c_2 + c_3 + 2c_4 = 0$$
$$c_2 + 2c_3 - 2c_4 = 0$$
$$2c_4 = 0.$$

The constants c_1, c_2, c_3, and c_4 solve these equations if and only if

$$c_4 = 0$$
$$c_2 = -2c_3 + 2c_4 \qquad = -2c_3$$
$$c_1 = -c_2 - c_3 - 2c_4 = 2c_3 - c_3 = c_3.$$

We *can* find constants that are not all zero and that solve these equations. For example, we could take

$$c_4 = 0, \quad c_3 = 1, \quad c_2 = -2, \quad c_1 = 1.$$

The vectors are linearly dependent.

Example 3.4.4

Check for independence:

$$\mathbf{v}_1 = \begin{bmatrix} 2 \\ 0 \\ 0 \\ 2 \end{bmatrix}, \quad \mathbf{v}_2 = \begin{bmatrix} 2 \\ 0 \\ 3 \\ 0 \end{bmatrix}, \quad \mathbf{v}_3 = \begin{bmatrix} 1 \\ 1 \\ 0 \\ 0 \end{bmatrix}.$$

A typical linear combination looks like

$$c_1\mathbf{v}_1 + c_2\mathbf{v}_2 + c_3\mathbf{v}_3 = \begin{bmatrix} 2c_1 + 2c_2 + c_3 \\ c_3 \\ 3c_2 \\ 2c_1 \end{bmatrix}.$$

Equating corresponding second, third, and fourth entries of the linear combination and $\mathbf{0}$ yields $c_3 = 0$, $3c_2 = 0$, and $2c_1 = 0$. The only solution is

$$c_1 = c_2 = c_3 = 0.$$

The vectors are linearly independent.

In general, to determine whether a given collection $\mathbf{v}_1, \ldots, \mathbf{v}_k$ of n-vectors is linearly independent, we look for values of $c_1, \ldots, c_k$ for which the linear combination $c_1\mathbf{v}_1 + \ldots + c_k\mathbf{v}_k$ is $\mathbf{0}$. If we set corresponding entries of the linear combination and $\mathbf{0}$ equal, we get n algebraic equations in the k unknowns $c_1, \ldots, c_k$. The coefficients of these equations are just the entries of the vectors $\mathbf{v}_1, \ldots, \mathbf{v}_k$. These equations always have at least one solution, $c_1 = \ldots = c_k = 0$. The vectors are independent if and only if this is the *only* solution.

In Section 3.6 we discuss a systematic method for solving systems of algebraic equations. For now, let's recall that *if* $k = n$ (that is, if the number of unknowns is equal to the number of equations), then Cramer's determinant test (Section 2.4) applies. If the determinant of coefficients is not zero, then $c_1 = c_2 = \ldots = c_n = 0$ is the only solution; the vectors $\mathbf{v}_1, \ldots, \mathbf{v}_n$ are linearly independent in this case. If the determinant of coefficients is zero, then there will be infinitely many other solutions; the vectors are linearly dependent in this case. Thus we have the following:

Fact: *The n-vectors* $\mathbf{v}_1, \ldots, \mathbf{v}_n$ *are linearly independent if and only if* $\det V \neq 0$ *where V is the* $n \times n$ *matrix whose columns are* $\mathbf{v}_1, \ldots, \mathbf{v}_n$.

If the vectors $\mathbf{v}_1, \ldots, \mathbf{v}_n$ are initial vectors,

$$\mathbf{v}_1 = \mathbf{h}_1(t_0), \ldots, \mathbf{v}_n = \mathbf{h}_n(t_0)$$

then det V is just $W[\mathbf{h}_1, \ldots, \mathbf{h}_n](t_0)$. Thus $\mathbf{v}_1, \ldots, \mathbf{v}_n$ are linearly independent if and only if this Wronskian is not zero. If we combine this with the Wronskian Test from the previous section, we get the following:

Fact: *Suppose $\mathbf{h}_1(t), \ldots, \mathbf{h}_n(t)$ are solutions of the nth-order homogeneous linear system of o.d.e.'s $D\mathbf{x} = A\mathbf{x}$. Suppose the coefficients a_{ij} are continuous on an interval I and t_0 is a fixed value of t in I. The general solution of $D\mathbf{x} = A\mathbf{x}$ is $\mathbf{x} = c_1\mathbf{h}_1(t) + \ldots + c_n\mathbf{h}_n(t)$ if and only if the initial vectors $\mathbf{h}_1(t_0), \ldots, \mathbf{h}_n(t_0)$ are linearly independent.*

This new test determines the strategy we will follow when we look for the general solution of a homogeneous system with constant coefficients. We will start by looking for solutions with linearly independent initial vectors. If we find n of these immediately, then they will generate the general solution. If we find fewer than n, then we will try to enlarge our collection of solutions to a larger set of solutions with linearly independent initial vectors. We will continue until we get n linearly independent solutions.

Let's summarize.

LINEAR INDEPENDENCE OF VECTORS

The n-vectors $\mathbf{v}_1, \ldots, \mathbf{v}_k$ are **linearly dependent** if there exist constants $c_1, \ldots, c_k$ with at least one $c_i \neq 0$ so that

$$c_1\mathbf{v}_1 + \ldots + c_k\mathbf{v}_k = \mathbf{0}.$$

The vectors are **linearly independent** if the only constants for which this relationship holds are

$$c_1 = \ldots = c_k = 0.$$

To check whether a given set of vectors

$$\mathbf{v}_1 = \begin{bmatrix} v_{11} \\ \cdot \\ \cdot \\ \cdot \\ v_{n1} \end{bmatrix}, \quad \mathbf{v}_2 = \begin{bmatrix} v_{12} \\ \cdot \\ \cdot \\ \cdot \\ v_{n2} \end{bmatrix}, \ldots, \mathbf{v}_k = \begin{bmatrix} v_{1k} \\ \cdot \\ \cdot \\ \cdot \\ v_{nk} \end{bmatrix}$$

is independent, we set corresponding coefficients of $c_1\mathbf{v}_1 + \ldots + c_k\mathbf{v}_k$ and $\mathbf{0}$ equal to obtain a system of algebraic equations:

$$c_1v_{11} + c_2v_{12} + \ldots + c_kv_{1k} = 0$$

$$.$$
$$.$$
$$.$$

$$c_1v_{n1} + c_2v_{n2} + \ldots + c_kv_{nk} = 0.$$

If $c_1 = \ldots = c_k = 0$ is the only solution of these equations, then the vectors are linearly independent. Otherwise, they are linearly dependent.

Suppose $\mathbf{h}_1(t), \ldots, \mathbf{h}_n(t)$ are solutions of the nth-order linear system of o.d.e.'s $D\mathbf{x} = A\mathbf{x}$. Suppose the coefficients a_{ij} are continuous on an interval I and t_0 is a fixed value in I. Then the general solution of $D\mathbf{x} = A\mathbf{x}$ is $\mathbf{x} = c_1\mathbf{h}_1(t) + \ldots + c_n\mathbf{h}_n(t)$ if and only if the initial vectors $\mathbf{h}_1(t_0), \ldots, \mathbf{h}_n(t_0)$ are linearly independent.

Note
On the number of solutions needed to generate a general solution

By an argument just like the one in Note 3, Section 2.4, we can show that *the general solution of an nth-order homogeneous system* $D\mathbf{x} = A\mathbf{x}$ *cannot be generated by fewer than n solutions.*

Suppose, on the other hand, that we are given $k > n$ solutions $\mathbf{h}_1(t), \ldots, \mathbf{h}_k(t)$ that generate the general solution of $D\mathbf{x} = A\mathbf{x}$. Then every initial condition $\mathbf{x}(t_0) = \mathbf{v}$ can be matched by a function on the list $\mathbf{x} = c_1\mathbf{h}_1(t) + \ldots + c_k\mathbf{h}_k(t)$. This says that every n-vector $\mathbf{v}$ can be expressed as a linear combination of the vectors $\mathbf{v}_1 = \mathbf{h}_1(t_0), \ldots, \mathbf{v}_k = \mathbf{h}_k(t_0)$. Since $k > n$, there is a j so that every n-vector $\mathbf{v}$ can still be expressed as a linear combination of $\mathbf{v}_1, \ldots, \mathbf{v}_{j-1}, \mathbf{v}_{j+1}, \ldots, \mathbf{v}_k$ (see Exercise 17). The corresponding solutions $\mathbf{h}_1(t), \ldots, \mathbf{h}_{j-1}(t), \mathbf{h}_{j+1}(t), \ldots, \mathbf{h}_k(t)$ still generate the general solution of $D\mathbf{x} = A\mathbf{x}$, since every initial condition can be matched by some linear combination of these functions. This means that *a generating set of k > n solutions is redundant,* since we can generate the general solution without using all of them.

EXERCISES

In Exercises 1 to 10, check the given set of vectors for linear independence.

1. $\mathbf{v}_1 = \begin{bmatrix} 1 \\ 2 \end{bmatrix}$, $\mathbf{v}_2 = \begin{bmatrix} 2 \\ 1 \end{bmatrix}$

   2. $\mathbf{v}_1 = \begin{bmatrix} 3 \\ -1 \end{bmatrix}$, $\mathbf{v}_2 = \begin{bmatrix} 12 \\ -4 \end{bmatrix}$

3. $\mathbf{v}_1 = \begin{bmatrix} 3 \\ -2 \end{bmatrix}$, $\mathbf{v}_2 = \begin{bmatrix} 1 \\ 1 \end{bmatrix}$, $\mathbf{v}_3 = \begin{bmatrix} 2 \\ 2 \end{bmatrix}$

4. $\mathbf{v}_1 = \begin{bmatrix} 1 \\ 0 \\ 1 \end{bmatrix}$, $\quad \mathbf{v}_2 = \begin{bmatrix} 1 \\ 1 \\ -1 \end{bmatrix}$, $\quad \mathbf{v}_3 = \begin{bmatrix} 3 \\ 1 \\ 1 \end{bmatrix}$

5. $\mathbf{v}_1 = \begin{bmatrix} 1 \\ 0 \\ 1 \end{bmatrix}$, $\quad \mathbf{v}_2 = \begin{bmatrix} 1 \\ 1 \\ -1 \end{bmatrix}$, $\quad \mathbf{v}_3 = \begin{bmatrix} 1 \\ 0 \\ 0 \end{bmatrix}$

6. $\mathbf{v}_1 = \begin{bmatrix} 1 \\ 0 \\ 1 \\ 1 \end{bmatrix}$, $\quad \mathbf{v}_2 = \begin{bmatrix} 1 \\ 0 \\ 0 \\ 3 \end{bmatrix}$, $\quad \mathbf{v}_3 = \begin{bmatrix} 1 \\ 1 \\ 0 \\ 0 \end{bmatrix}$

7. $\mathbf{v}_1 = \begin{bmatrix} 1 \\ 0 \\ 1 \\ 1 \end{bmatrix}$, $\quad \mathbf{v}_2 = \begin{bmatrix} 1 \\ 0 \\ 0 \\ 3 \end{bmatrix}$, $\quad \mathbf{v}_3 = \begin{bmatrix} 1 \\ 0 \\ 2 \\ -2 \end{bmatrix}$

8. $\mathbf{v}_1 = \begin{bmatrix} 1 \\ 2 \\ 3 \\ 4 \\ 1 \end{bmatrix}$, $\quad \mathbf{v}_2 = \begin{bmatrix} 2 \\ 1 \\ 4 \\ 3 \\ 2 \end{bmatrix}$, $\quad \mathbf{v}_3 = \begin{bmatrix} -1 \\ 1 \\ -1 \\ 1 \\ -1 \end{bmatrix}$

9. $\mathbf{v}_1 = \begin{bmatrix} 1 \\ 1 \\ 1 \\ 1 \\ 1 \end{bmatrix}$, $\quad \mathbf{v}_2 = \begin{bmatrix} 0 \\ 1 \\ 1 \\ 1 \\ 1 \end{bmatrix}$, $\quad \mathbf{v}_3 = \begin{bmatrix} 0 \\ 0 \\ 1 \\ 1 \\ 1 \end{bmatrix}$, $\quad \mathbf{v}_4 = \begin{bmatrix} 0 \\ 0 \\ 0 \\ 1 \\ 1 \end{bmatrix}$, $\quad \mathbf{v}_5 = \begin{bmatrix} 0 \\ 0 \\ 0 \\ 0 \\ 1 \end{bmatrix}$

10. $\mathbf{v}_1 = \begin{bmatrix} 1 \\ 1 \\ 1 \\ 1 \\ 0 \end{bmatrix}$, $\quad \mathbf{v}_2 = \begin{bmatrix} 2 \\ 0 \\ 3 \\ 2 \\ 1 \end{bmatrix}$, $\quad \mathbf{v}_3 = \begin{bmatrix} 2 \\ 0 \\ 4 \\ 2 \\ 0 \end{bmatrix}$, $\quad \mathbf{v}_4 = \begin{bmatrix} 1 \\ 0 \\ 2 \\ 1 \\ 0 \end{bmatrix}$

Some more abstract questions:

11. Show that any set of n-vectors that includes $\mathbf{0}$ is dependent.

12. Show that if two vectors in a set are equal, then the vectors are dependent.

13. a. Show that if $\mathbf{v}_j = a_1\mathbf{v}_1 + \ldots + a_{j-1}\mathbf{v}_{j-1}$, then $\mathbf{v}_1, \ldots, \mathbf{v}_j, \ldots, \mathbf{v}_k$ are dependent.
    b. Show that if $\mathbf{v}_1, \ldots, \mathbf{v}_k$ are dependent, then some $\mathbf{v}_j$ can be written as a linear combination of the preceding ones, $\mathbf{v}_j = a_1\mathbf{v}_1 + \ldots + a_{j-1}\mathbf{v}_{j-1}$.

14. Show that if $\mathbf{v}_1, \ldots, \mathbf{v}_k$ are independent, then the set obtained by deleting $\mathbf{v}_i$ is also independent.

15. Suppose $\mathbf{v}_1, \ldots, \mathbf{v}_k$ are n-vectors and, for each i, the vector $\mathbf{v}_i$ has a nonzero entry in a position where all the other vectors $\mathbf{v}_1, \ldots, \mathbf{v}_{i-1}, \mathbf{v}_{i+1}, \ldots, \mathbf{v}_k$ have zero entries. Show that $\mathbf{v}_1, \ldots, \mathbf{v}_k$ are linearly independent.

16. a. Suppose $\mathbf{v}_1, \ldots, \mathbf{v}_n$ are n linearly independent n-vectors. Show that every n-vector $\mathbf{v}$ can be written as a linear combination of $\mathbf{v}_1, \ldots, \mathbf{v}_n$:

$$\mathbf{v} = a_1\mathbf{v}_1 + \ldots + a_n\mathbf{v}_n.$$

(*Hint:* View this as a system of equations for the unknowns $a_1, \ldots, a_n$. What can you say about the determinant of coefficients?)

   b. Prove that any set $\mathbf{v}_1, \ldots, \mathbf{v}_n, \mathbf{v}$ consisting of $n + 1$ different n-vectors must be linearly dependent.

   c. Does a set of $k \leq n$ different n-vectors have to be independent?

17. Suppose that every n-vector $\mathbf{v}$ can be expressed as a linear combination $\mathbf{v} = c_1\mathbf{v}_1 + \ldots + c_k\mathbf{v}_k$ of the vectors $\mathbf{v}_1, \ldots, \mathbf{v}_k$, where $k > n$. Use the results of Exercises 16(b) and 13(b) to show that for some j, every n-vector $\mathbf{v}$ can be expressed as a linear combination $\mathbf{v} = b_1\mathbf{v}_1 + \ldots + b_{j-1}\mathbf{v}_{j-1} + b_{j+1}\mathbf{v}_{j+1} + \ldots + b_k\mathbf{v}_k$ of the vectors $\mathbf{v}_1, \ldots, \mathbf{v}_{j-1}, \mathbf{v}_{j+1}, \ldots, \mathbf{v}_k$, that is, without using $\mathbf{v}_j$.

3.5 HOMOGENEOUS SYSTEMS, EIGENVALUES, AND EIGENVECTORS

We now have methods for determining whether given solutions of a homogeneous linear system

(H) $$D\mathbf{x} = A\mathbf{x}$$

generate the general solution. In this section we begin our discussion of a method for finding solutions in case the system has *constant coefficients*.

In Chapter 2 we saw that if λ is a root of the polynomial $P(r)$, then $x = e^{\lambda t}$ is a solution of the o.d.e. $P(D)x = 0$. The equivalent system (see Example 3.2.2) has a solution

$$\mathbf{x} = \begin{bmatrix} x \\ x' \\ \cdot \\ \cdot \\ \cdot \\ x^{(n-1)} \end{bmatrix} = \begin{bmatrix} e^{\lambda t} \\ \lambda e^{\lambda t} \\ \cdot \\ \cdot \\ \cdot \\ \lambda^{n-1}e^{\lambda t} \end{bmatrix} = e^{\lambda t}\begin{bmatrix} 1 \\ \lambda \\ \cdot \\ \cdot \\ \cdot \\ \lambda^{n-1} \end{bmatrix}.$$

Let's look for solutions to (H) of the form

$$\mathbf{x} = e^{\lambda t}\mathbf{v}.$$

Note that the initial vector at $t = 0$ of such a solution will be $\mathbf{x}(0) = \mathbf{v}$. Since we want, eventually, to have solutions with linearly independent initial vectors, we certainly want $\mathbf{v} \neq \mathbf{0}$. If we substitute $\mathbf{x} = e^{\lambda t}\mathbf{v}$ into (H), we get

$$\lambda e^{\lambda t}\mathbf{v} = e^{\lambda t}A\mathbf{v}.$$

Cancellation of $e^{\lambda t}$ yields

$$\lambda \mathbf{v} = A\mathbf{v}.$$

If we find a constant λ and a vector $\mathbf{v}$ with this property, then $\mathbf{x} = e^{\lambda t}\mathbf{v}$ will be a solution of (H).

Definition: *Let A be an n $\times$ n matrix with constant entries. We say the number λ is an **eigenvalue** of A provided there exists a nonzero vector $\mathbf{v}$ such that*

$$A\mathbf{v} = \lambda \mathbf{v}.$$

*Any nonzero vector with this property is then called an **eigenvector** of A corresponding to λ.*

Note that an eigen*value* λ can be zero, but $\mathbf{0}$ is *not* allowed as an eigen*vector*.

With this terminology, we can state our observation about solutions of (H) as follows.

Fact: *If λ is an eigenvalue of A and $\mathbf{v}$ is an eigenvector of A corresponding to λ, then $\mathbf{x} = e^{\lambda t}\mathbf{v}$ is a solution of $D\mathbf{x} = A\mathbf{x}$.*

Example 3.5.1

Let

$$A = \begin{bmatrix} 2 & 1 \\ 0 & 2 \end{bmatrix} \quad \text{and} \quad \mathbf{v} = \begin{bmatrix} 1 \\ 0 \end{bmatrix}.$$

Then

$$A\mathbf{v} = \begin{bmatrix} 2 \\ 0 \end{bmatrix} = 2\mathbf{v}.$$

Thus 2 is an eigenvalue of A, and $\mathbf{v}$ is an eigenvector of A corresponding to the eigenvalue 2. The vector valued function

$$\mathbf{x} = e^{2t}\mathbf{v} = \begin{bmatrix} e^{2t} \\ 0 \end{bmatrix}$$

is a solution of $D\mathbf{x} = A\mathbf{x}$.

In order to make use of the preceding observation, we need a procedure for calculating the eigenvalues and eigenvectors of a given matrix

$$A = \begin{bmatrix} a_{11} & \cdots & a_{1n} \\ & \cdot & \\ & \cdot & \\ & \cdot & \\ a_{n1} & \cdots & a_{nn} \end{bmatrix}.$$

An eigenvalue of A is a number λ for which the equation

$$A\mathbf{v} = \lambda\mathbf{v}$$

has a nonzero solution

$$\mathbf{v} = \begin{bmatrix} v_1 \\ \cdot \\ \cdot \\ \cdot \\ v_n \end{bmatrix}.$$

Now

$$A\mathbf{v} = \begin{bmatrix} a_{11}v_1 + \ldots + a_{1n}v_n \\ \cdot \\ \cdot \\ \cdot \\ a_{11}v_1 + \ldots + a_{nn}v_n \end{bmatrix} \quad \text{and} \quad \lambda\mathbf{v} = \begin{bmatrix} \lambda v_1 \\ \cdot \\ \cdot \\ \cdot \\ \lambda v_n \end{bmatrix}.$$

Equating corresponding entries and regrouping terms, we get the system of algebraic equations

$$\begin{aligned} (a_{11} - \lambda)v_1 + \quad a_{12}v_2 + \ldots + \quad a_{1n}v_n &= 0 \\ a_{21}v_1 + (a_{22} - \lambda)v_2 + \ldots + \quad a_{2n}v_n &= 0 \\ &\vdots \\ a_{n1}v_1 + \quad a_{n2}v_2 + \ldots + (a_{nn} - \lambda)v_n &= 0. \end{aligned}$$

(C)

An eigenvalue of A is a number λ for which the equations (C) have a solution other than the trivial solution $v_1 = \ldots = v_n = 0$.

The system (C) can be rewritten in matrix form as

(C) $$(A - \lambda I)\mathbf{v} = \mathbf{0}$$

where $A - \lambda I$ is the matrix obtained from A by subtracting λ from each of the diagonal entries a_{ii} of A:

$$A - \lambda I = \begin{bmatrix} a_{11}-\lambda & a_{12} & \cdots & a_{1n} \\ a_{21} & a_{22}-\lambda & \cdots & a_{2n} \\ & & \cdot & \\ & & \cdot & \\ & & \cdot & \\ a_{n1} & a_{n2} & \cdots & a_{nn}-\lambda \end{bmatrix}.$$

Note that this matrix is the same as the matrix obtained by subtracting from A the product of the number λ with the $n \times n$ identity matrix I, which we introduced in Note 2 of Section 3.2 (see Exercise 24 in Section 3.2). Cramer's determinant test tells us that (C) has nontrivial solutions if and only if the determinant of coefficients, $\det(A - \lambda I)$, is zero. Thus we have the following:

Fact: *The number λ is an eigenvalue of A if and only if*

$$det(A - \lambda I) = 0.$$

Example 3.5.2

Find the eigenvalues of

$$A = \begin{bmatrix} -1 & -1 & 0 \\ 1 & -\frac{3}{2} & \frac{3}{2} \\ 0 & 1 & -1 \end{bmatrix}.$$

Here

$$\det(A - \lambda I) = \det \begin{bmatrix} -1-\lambda & -1 & 0 \\ 1 & -\frac{3}{2}-\lambda & \frac{3}{2} \\ 0 & 1 & -1-\lambda \end{bmatrix}$$

$$= -(1+\lambda) \det \begin{bmatrix} -\frac{3}{2}-\lambda & \frac{3}{2} \\ 1 & -1-\lambda \end{bmatrix} - \det \begin{bmatrix} -1 & 0 \\ 1 & -1-\lambda \end{bmatrix}$$

$$= -(1+\lambda)(\lambda^2 + \tfrac{5}{2}\lambda + 1)$$

$$= -(1+\lambda)(2+\lambda)(\tfrac{1}{2}+\lambda).$$

The eigenvalues of A are the values that make this zero,

$$\lambda = -1, \qquad \lambda = -2, \qquad \text{and} \qquad \lambda = -\tfrac{1}{2}.$$

Example 3.5.3

Find the eigenvalues of

$$A = \begin{bmatrix} 2 & 1 & 0 \\ 0 & 2 & 0 \\ 0 & 0 & 2 \end{bmatrix}.$$

Here

$$\det(A - \lambda I) = \det \begin{bmatrix} 2 - \lambda & 1 & 0 \\ 0 & 2 - \lambda & 0 \\ 0 & 0 & 2 - \lambda \end{bmatrix} = (2 - \lambda)^3.$$

The only eigenvalue of A is

$$\lambda = 2.$$

Example 3.5.4

Find the eigenvalues of

$$A = \begin{bmatrix} 1 & 0 & 4 \\ 0 & 3 & 0 \\ 1 & 0 & 1 \end{bmatrix}.$$

Here

$$\det(A - \lambda I) = \det \begin{bmatrix} 1 - \lambda & 0 & 4 \\ 0 & 3 - \lambda & 0 \\ 1 & 0 & 1 - \lambda \end{bmatrix} = (3 - \lambda)\det \begin{bmatrix} 1 - \lambda & 4 \\ 1 & 1 - \lambda \end{bmatrix}$$

$$= (3 - \lambda)(\lambda^2 - 2\lambda - 3) = (3 - \lambda)(\lambda - 3)(\lambda + 1).$$

The eigenvalues of A are

$$\lambda = 3 \quad \text{and} \quad \lambda = -1.$$

In each of our examples, $\det(A - \lambda I)$ was a polynominal in λ. This is true in general (see Exercises 19 and 20).

Fact: *If A is an $n \times n$ matrix, then $\det(A - \lambda I)$ is a polynominal in λ of degree n.*

We refer to $\det(A - \lambda I)$ as the **characteristic polynomial** of A. The eigenvalues of A are the roots of the characteristic polynomial.

Once we have found a specific eigenvalue λ, we find eigenvectors corresponding to λ by finding nonzero solutions of $(A - \lambda I)\mathbf{v} = \mathbf{0}$. Note that this system will have infinitely many solutions (why?), so there will be infinitely many eigenvectors corresponding to λ.

Example 3.5.5

The eigenvalues of

$$A = \begin{bmatrix} 1 & 0 & 4 \\ 0 & 3 & 0 \\ 1 & 0 & 1 \end{bmatrix}$$

are 3 and -1 (see Example 3.5.4). Find an eigenvector corresponding to each eigenvalue.

To find an eigenvector corresponding to -1, we must solve $(A - (-1)I)\mathbf{v} = \mathbf{0}$, which is just

$$\begin{bmatrix} 2 & 0 & 4 \\ 0 & 4 & 0 \\ 1 & 0 & 2 \end{bmatrix} \mathbf{v} = \mathbf{0}.$$

Written out in terms of the entries v_1, v_2, and v_3 of $\mathbf{v}$, this reads

$$\begin{aligned} 2v_1 \quad\quad + 4v_3 &= 0 \\ 4v_2 \quad\quad &= 0 \\ v_1 \quad\quad + 2v_3 &= 0. \end{aligned}$$

If we solve these equations for v_1 and v_2, we get

$$v_1 = -2v_3, \quad\quad v_2 = 0.$$

Any choice of v_3, say $v_3 = a$, will lead to a solution:

$$v_1 = -2a, \quad\quad v_2 = 0, \quad\quad v_3 = a.$$

In vector terms, the solutions are the vectors of the form

$$\mathbf{v} = \begin{bmatrix} -2a \\ 0 \\ a \end{bmatrix} = a \begin{bmatrix} -2 \\ 0 \\ 1 \end{bmatrix}.$$

Any nonzero vector of this form will be an eigenvector. In particular, taking $a = 1$,

$$\mathbf{v} = \begin{bmatrix} -2 \\ 0 \\ 1 \end{bmatrix}$$

is an eigenvector corresponding to -1.

To find an eigenvector corresponding to 3, we must solve $(A - 3I)\mathbf{w} = \mathbf{0}$. Written out in terms of the entries of $\mathbf{w}$, this reads

$$\begin{aligned} -2w_1 \quad + 4w_3 &= 0 \\ 0 &= 0 \\ w_1 \quad - 2w_3 &= 0. \end{aligned}$$

If we solve for w_1, we get

$$w_1 = 2w_3.$$

Any choices of w_2 and w_3, say $w_2 = a$ and $w_3 = b$, will lead to a solution:

$$w_1 = 2b, \qquad w_2 = a, \qquad w_3 = b.$$

In vector terms, the solutions are the vectors of the form

$$\mathbf{w} = \begin{bmatrix} 2b \\ a \\ b \end{bmatrix} = a \begin{bmatrix} 0 \\ 1 \\ 0 \end{bmatrix} + b \begin{bmatrix} 2 \\ 0 \\ 1 \end{bmatrix}.$$

Any nonzero vector of this form is an eigenvector. In particular, the vectors

$$\mathbf{w}_1 = \begin{bmatrix} 0 \\ 1 \\ 0 \end{bmatrix} \qquad \text{(take } a = 1 \text{ and } b = 0\text{)}$$

and

$$\mathbf{w}_2 = \begin{bmatrix} 2 \\ 0 \\ 1 \end{bmatrix} \qquad \text{(take } a = 0 \text{ and } b = 1\text{)}$$

are eigenvectors corresponding to 3. Note that these two vectors are linearly independent.

In the next section we describe a systematic approach to solving systems of algebraic equations. This approach will be extremely useful in finding eigenvectors. In Sections 3.7 to 3.10 we elaborate on our observation about the connection between eigenvalues, eigenvectors, and solutions of homogeneous systems of o.d.e.'s to find special solutions that generate the general solution.

Example 3.5.6

Solve $D\mathbf{x} = A\mathbf{x}$, where

$$A = \begin{bmatrix} -3 & -1 & 0 \\ 2 & 0 & 0 \\ 0 & 0 & 0 \end{bmatrix}.$$

This system models the three-loop circuit of Example 3.1.3 with $V(t) = 0$, $R_2 = 0$, $R_1 = 1/3$, $L_2 = L_3 = 1/2$, and $C = 1$.

The characteristic polynomial, $\det(A - \lambda I) = -\lambda(\lambda^2 + 3\lambda + 2)$, has three roots

$$\lambda = -2, -1, 0.$$

To find an eigenvector of A for $\lambda = -2$, we must solve $[A - (-2)I]\mathbf{v} = \mathbf{0}$, or

$$
\begin{aligned}
-v_1 - v_2 &= 0 \\
2v_1 + 2v_2 &= 0 \\
2v_3 &= 0.
\end{aligned}
$$

The last equation forces $v_3 = 0$, while the other two give $v_1 = -v_2$. Any nonzero choice of v_2 leads to an eigenvector; for example, $v_2 = 1$ gives the eigenvector

$$\mathbf{v} = \begin{bmatrix} -1 \\ 1 \\ 0 \end{bmatrix}$$

and a corresponding solution of $D\mathbf{x} = A\mathbf{x}$

$$\mathbf{h}_1(t) = e^{-2t}\mathbf{v} = \begin{bmatrix} -e^{-2t} \\ e^{-2t} \\ 0 \end{bmatrix}.$$

To find an eigenvector corresponding to $\lambda = -1$, we solve $[A - (-1)I]\mathbf{w} = \mathbf{0}$, or

$$-2w_1 - w_2 \quad = 0$$
$$2w_1 + w_2 \quad = 0$$
$$w_3 = 0.$$

Here the last equation forces $w_3 = 0$, and the other two give $w_1 = w_2/2$. Any nonzero choice of w_2 leads to an eigenvector; for example, $w_2 = 2$ gives

$$\mathbf{w} = \begin{bmatrix} -1 \\ 2 \\ 0 \end{bmatrix}$$

and a corresponding solution to $D\mathbf{x} = A\mathbf{x}$

$$\mathbf{h}_2(t) = e^{-t}\mathbf{w} = \begin{bmatrix} -e^{-t} \\ 2e^{-t} \\ 0 \end{bmatrix}.$$

Finally, we look for an eigenvector corresponding to $\lambda = 0$, that is, a nonzero solution of $A\mathbf{u} = \mathbf{0}$. This amounts to

$$-3u_1 - u_2 \quad = 0$$
$$2u_1 \quad = 0$$
$$0 = 0.$$

The second equation forces $u_1 = 0$, which then forces $u_2 = 0$ in the first; there is no restriction on u_3. The choice $u_3 = 1$ gives the eigenvector

$$\mathbf{u} = \begin{bmatrix} 0 \\ 0 \\ 1 \end{bmatrix}$$

and a corresponding solution of $D\mathbf{x} = A\mathbf{x}$

$$\mathbf{h}_3(t) = e^{0t}\mathbf{u} = \mathbf{u} = \begin{bmatrix} 0 \\ 0 \\ 1 \end{bmatrix}.$$

The reader can check that the vectors $\mathbf{h}_1(0) = \mathbf{v}$, $\mathbf{h}_2(0) = \mathbf{w}$, and $\mathbf{h}_3(0) = \mathbf{u}$ are linearly independent. Thus, the general solution of our third-order system $D\mathbf{x} = A\mathbf{x}$ is

$$\mathbf{x} = c_1\mathbf{h}_1(t) + c_2\mathbf{h}_2(t) + c_3\mathbf{h}_3(t) = c_1 \begin{bmatrix} -e^{-2t} \\ e^{-2t} \\ 0 \end{bmatrix} + c_2 \begin{bmatrix} -e^{-t} \\ 2e^{-t} \\ 0 \end{bmatrix} + c_3 \begin{bmatrix} 0 \\ 0 \\ 1 \end{bmatrix}.$$

In terms of currents and charges, this is

$$Q = -c_1e^{-2t} - c_2e^{-t}$$
$$I_2 = c_1e^{-2t} + 2c_2e^{-t}$$
$$I_3 = c_3.$$

The absence of the resistance R_2 in this example means that the equation for DI_3 does not involve Q or I_2, and the equations for DQ and DI_2 do not involve I_3. In effect, the one-loop subcircuit carrying the current I_3 does not interact with the two-loop subcircuit carrying Q and I_2 (and hence I_1). The solutions $c_3\mathbf{h}_3(t)$, corresponding to the eigenvalue $\lambda = 0$, are precisely the solutions for which the two-loop subcircuit is dormant ($Q = I_2 = 0$). The solutions $c_1\mathbf{h}_1(t) + c_2\mathbf{h}_2(t)$ are the solutions for which the one-loop subcircuit is dormant ($I_3 = 0$).

The currents in the two-loop subcircuit do interact, via the capacitor. However, there is a different way of measuring the state of the circuit via variables (related to our eigenvectors) whose action is independent. Indeed, the differential equations describing our circuit can be rewritten in terms of the variables $J_1 = 2Q + I_2$, $J_2 = Q + I_2$, and I_3 as

$$DJ_1 = -2J_1, \qquad DJ_2 = -J_2, \qquad DI_3 = 0.$$

The solutions $c_1\mathbf{h}_1(t)$, corresponding to the eigenvalue $\lambda = -2$, are the solutions for which $J_2 = I_3 = 0$. The solutions $c_2\mathbf{h}_2(t)$, corresponding to $\lambda = -1$, are the solutions for which $J_1 = I_3 = 0$.

HOMOGENEOUS SYSTEMS, EIGENVALUES, AND EIGENVECTORS

Let A be an $n \times n$ matrix with constant entries. We say the number λ is an **eigenvalue** of A provided there exists a nonzero vector $\mathbf{v}$ such that $A\mathbf{v} = \lambda\mathbf{v}$. Any nonzero vector with this property is then called an **eigenvector** of A corresponding to λ.

Let $A - \lambda I$ be the matrix obtained from A by subtracting λ from each of the diagonal entries a_{ii} of A. The eigenvalues of A are the roots of the **characteristic polynomial,** $\det(A - \lambda I)$. The eigenvectors of A corresponding to λ are the nonzero solutions of

(C) $$(A - \lambda I)\mathbf{v} = \mathbf{0}.$$

If λ is an eigenvalue of A and $\mathbf{v}$ is an eigenvector of A corresponding to λ, then $\mathbf{x} = e^{\lambda t}\mathbf{v}$ is a solution of $D\mathbf{x} = A\mathbf{x}$.

Note
On the old and new characteristic polynomial

In Section 2.7 we referred to the polynomial $P(r) = r^n + \ldots + a_0$ as the characteristic polynomial of the equation $(D^n + \ldots + a_0)x = 0$. The matrix of coefficients of the equivalent system is

$$A = \begin{bmatrix} 0 & 1 & 0 & \ldots & 0 \\ 0 & 0 & 1 & \ldots & 0 \\ & & \cdot & & \\ & & \cdot & & \\ & & \cdot & & \\ 0 & 0 & 0 & \ldots & 1 \\ -a_0 & -a_1 & -a_2 & \ldots & -a_{n-1} \end{bmatrix}.$$

The characteristic polynomial of this matrix is $(-1)^n P(\lambda)$.

EXERCISES

In Exercises 1 to 6, find (a) the characteristic polynomial and (b) the eigenvalues of the given matrix A.

1. $A = \begin{bmatrix} 0 & 2 \\ -1 & 3 \end{bmatrix}$ 2. $A = \begin{bmatrix} 1 & 1 \\ 1 & 1 \end{bmatrix}$

3. $A = \begin{bmatrix} 1 & 1 \\ 3 & 1 \end{bmatrix}$ 4. $A = \begin{bmatrix} 1 & -1 \\ 1 & 3 \end{bmatrix}$

5. $A = \begin{bmatrix} 2 & 1 & -2 \\ -3 & 0 & 4 \\ -2 & -1 & 4 \end{bmatrix}$ 6. $A = \begin{bmatrix} 0 & -2 & 2 \\ 1 & 3 & -2 \\ 2 & 4 & -3 \end{bmatrix}$

In Exercises 7 to 11, find the eigenvalues of A and for each eigenvalue a corresponding eigenvector.

7. $A = \begin{bmatrix} 1 & 1 \\ 3 & -1 \end{bmatrix}$ 8. $A = \begin{bmatrix} -3 & 1 \\ -1 & -1 \end{bmatrix}$ 9. $A = \begin{bmatrix} 1 & 1 & 1 \\ 0 & 2 & -1 \\ 0 & 0 & -3 \end{bmatrix}$

10. $A = \begin{bmatrix} 1 & -1 & -1 \\ 0 & 0 & -1 \\ 0 & 0 & 1 \end{bmatrix}$ 11. $A = \begin{bmatrix} 1 & -1 & -1 \\ 0 & -1 & -1 \\ 0 & 0 & 1 \end{bmatrix}$

In Exercises 12 to 14, find the general solution of $Dx = Ax$ (which is generated by solutions corresponding to eigenvectors of A).

12. A as in Exercise 7   13. A as in Exercise 9

14. A as in Exercise 10

In Exercises 15 and 16, (a) find all solutions of $Dx = Ax$ corresponding to eigenvectors of A, and (b) show that these do *not* generate the general solution, by exhibiting an initial condition not satisfied by the list of solutions these generate.

15.   *A* as in Exercise 8   16.   *A* as in Exercise 11

Some more abstract questions:

17.   Suppose $\mathbf{x} = f(t)\mathbf{v}$ is a solution of the homogeneous system $D\mathbf{x} = A\mathbf{x}$ with $\mathbf{v} \neq \mathbf{0}$ and $f(t)$ a real-valued function that is not identically zero. Show that $f(t)$ must have the form $f(t) = ce^{\lambda t}$, where $c \neq 0$, λ is an eigenvalue of A, and $\mathbf{v}$ is a corresponding eigenvector. [*Hint:* If $\mathbf{v} \neq \mathbf{0}$ and $g(t)\mathbf{v} = h(t)\mathbf{u}$, then $\mathbf{u} = \lambda\mathbf{v}$ for some λ and $g(t) = \lambda h(t)$.]

18.   Suppose A is an $n \times n$ matrix in one of the two forms that follow:

$$\text{upper triangular: } A = \begin{bmatrix} a_{11} & a_{12} & \cdots & a_{1n} \\ & a_{22} & \cdots & a_{2n} \\ & & \ddots & \vdots \\ 0\text{'s} & & & \ddots & \vdots \\ & & & & a_{nn} \end{bmatrix}$$

$$\text{lower triangular: } A = \begin{bmatrix} a_{11} & & & & \\ a_{21} & a_{22} & & 0\text{'s} & \\ \vdots & & \ddots & & \\ \vdots & & & \ddots & \\ a_{n1} & \cdots & & & a_{nn} \end{bmatrix}$$

   a. Show that the determinant of A equals the product of its diagonal entries: $\det A = a_{11} a_{22} \cdots a_{nn}$.
   b. Show that the eigenvalues of A equal its diagonal entries: $\lambda = a_{11}$, $\lambda = a_{22}$, ..., $\lambda = a_{nn}$.

19.   Show that the characteristic polynomial of a 3×3 matrix is a polynomial of degree 3.

*20. a. Suppose B is a $k \times k$ matrix, each of whose entries is either a number or a term of the form $a_{ij} - \lambda$ (λ a variable). By induction on k, show that if the number of entries of B involving λ is $p \leq k$, then $\det B$ is a polynomial in λ of degree at most k. (*Hint:* Expand along a row containing an entry that involves λ, and count.)
   b. Use (a) to show that the characteristic polynomial of an $n \times n$ matrix is a polynomial of degree n.

3.6 SYSTEMS OF ALGEBRAIC EQUATIONS: ROW REDUCTION

Many of our methods require us to solve algebraic systems of equations. If a system consists of n equations in n unknowns and if the determinant of coefficients of the system is not zero, then Cramer's rule (Note 3, Section 2.4) provides us with a formula for the unique solution. The systems that arose in the last two sections often failed to satisfy these conditions. Indeed the systems $(A - \lambda I)\mathbf{v} = \mathbf{0}$ that we have to solve to find eigenvectors are precisely the ones whose determinant of coefficients is zero. In this section we will describe a systematic approach to solving algebraic systems of equations. Even when Cramer's rule can be used, our new approach is usually more efficient.

We will be dealing with systems of the form

$$b_{11}u_1 + b_{12}u_2 + \ldots + b_{1m}u_m = k_1$$
$$b_{21}u_1 + b_{22}u_2 + \ldots + b_{2m}u_m = k_2$$

(E)

$$\cdot$$
$$\cdot$$
$$\cdot$$

$$b_{n1}u_1 + b_{n2}u_2 + \ldots + b_{nm}u_m = k_n$$

The coefficients b_{ij} and the k_i's will be given, and we will wish to solve for the unknowns $u_1, \ldots, u_m$.

To begin with, we note that (E) can be rewritten in matrix form as

$$B\mathbf{u} = \mathbf{k}$$

where

$$B = \begin{bmatrix} b_{11} & b_{12} & \cdots & b_{1m} \\ b_{21} & b_{22} & \cdots & b_{2m} \\ & & \cdot & \\ & & \cdot & \\ & & \cdot & \\ b_{n1} & b_{n2} & \cdots & b_{nm} \end{bmatrix}, \quad \mathbf{u} = \begin{bmatrix} u_1 \\ u_2 \\ \cdot \\ \cdot \\ \cdot \\ u_m \end{bmatrix}, \quad \text{and} \quad \mathbf{k} = \begin{bmatrix} k_1 \\ k_2 \\ \cdot \\ \cdot \\ \cdot \\ k_n \end{bmatrix}.$$

The information essential for describing the system is contained in the **augmented matrix**

$$[B|\mathbf{k}] = \begin{bmatrix} b_{11} & b_{12} & \cdots & b_{1m} & k_1 \\ b_{21} & b_{22} & \cdots & b_{2m} & k_2 \\ & & \cdot & & \\ & & \cdot & & \\ & & \cdot & & \\ b_{n1} & b_{n2} & \cdots & b_{nm} & k_n \end{bmatrix}.$$

The vertical line before the last column of this matrix is there just as a reminder that this is the augmented matrix and *not* the coefficient matrix B.

Our general approach to solving linear algebraic systems will be to replace the given system by a simpler system that has the same solutions. We will do this by systematically eliminating variables from the equations of the system. Since the augmented matrix carries all the information needed to describe the system, we will work with it rather than with the equations themselves.

If we start with a system (E) and perform any of the following operations on the equations, then we get a system that has the same solutions as (E):

1. Adding multiples of one fixed equation to one or more of the other equations.
2. Multiplying an equation by a nonzero number.
3. Rearranging the order of the equations.

Each of these operations affects the augmented matrix of (E). The corresponding matrix operations are called **row operations:**

1. Adding multiples of a fixed row to one or more of the other rows.
2. Multiplying a row by a nonzero constant.
3. Rearranging the order of the rows.

We will say that two matrices are **row equivalent** if we can get from one to the other by a sequence of row operations. Since the operations do not affect the solutions of the corresponding systems, we have the following:

Fact: *If $[B|\mathbf{k}]$ and $[B'|\mathbf{k}']$ are row equivalent, then the systems $B\mathbf{u} = \mathbf{k}$ and $B'\mathbf{u} = \mathbf{k}'$ have the same solutions.*

Let's look at some examples. In each example we will start with the augmented matrix of the system and perform a sequence of row operations to obtain a row-equivalent matrix whose system is easier to solve. We aim for a matrix that has a lot of zeroes. In particular, we want as many zeroes as possible in the lower left. We use the notation $[B|\mathbf{k}] \xrightarrow[R_j \to R_j + cR_i]{} [B'|\mathbf{k}']$ to indicate that we get $[B'|\mathbf{k}']$ from $[B|\mathbf{k}]$ by replacing (row j) by $\{(\text{row } j) + c(\text{row } i)\}$, and $[B|\mathbf{k}] \xrightarrow[R_i \to cR_i]{} [B'|\mathbf{k}']$ to indicate that we get $[B'|\mathbf{k}']$ by multiplying (row i) by c. $[B|\mathbf{k}] \xrightarrow[\text{rearr}]{} [B'|\mathbf{k}']$ indicates that we get $[B'|\mathbf{k}']$ by rearranging the order of the rows of $[B|\mathbf{k}]$.

Example 3.6.1
Solve

$$u_1 + u_2 \quad + u_4 = 3$$
(E)
$$u_1 + 2u_2 \qquad = 1$$
$$u_3 + u_4 = 2$$
$$2u_1 + 2u_2 \quad + 3u_4 = 3.$$

Starting with the augmented matrix, we perform a sequence of row operations to introduce zeroes:

$$\left[\begin{array}{cccc|c} 1 & 1 & 0 & 1 & 3 \\ 1 & 2 & 0 & 0 & 1 \\ 0 & 0 & 1 & 1 & 2 \\ 2 & 2 & 0 & 3 & 3 \end{array}\right] \xrightarrow[R_4 \to R_4 - 2R_1]{R_2 \to R_2 - R_1} \left[\begin{array}{cccc|c} 1 & 1 & 0 & 1 & 3 \\ 0 & 1 & 0 & -1 & -2 \\ 0 & 0 & 1 & 1 & 2 \\ 0 & 0 & 0 & 1 & -3 \end{array}\right]$$

$$\xrightarrow[\substack{R_1 \to R_1 - R_4 \\ R_2 \to R_2 + R_4 \\ R_3 \to R_3 - R_4}]{} \left[\begin{array}{cccc|c} 1 & 1 & 0 & 0 & 6 \\ 0 & 1 & 0 & 0 & -5 \\ 0 & 0 & 1 & 0 & 5 \\ 0 & 0 & 0 & 1 & -3 \end{array}\right] \xrightarrow[R_1 \to R_1 - R_2]{} \left[\begin{array}{cccc|c} 1 & 0 & 0 & 0 & 11 \\ 0 & 1 & 0 & 0 & -5 \\ 0 & 0 & 1 & 0 & 5 \\ 0 & 0 & 0 & 1 & -3 \end{array}\right].$$

The last matrix is the augmented matrix of the system

$$
\begin{aligned}
u_1 &&&= 11 \\
&u_2 &&= -5 \\
&&u_3 &= 5 \\
&&&u_4 = -3.
\end{aligned}
$$

These equations are simply the statement of the (unique) solution of (E), which can be written in vector form as

$$\mathbf{u} = \left[\begin{array}{c} 11 \\ -5 \\ 5 \\ -3 \end{array}\right].$$

Example 3.6.2

Solve

(E)
$$
\begin{aligned}
2u_1 + u_2 + 4u_3 + 3u_4 &= 3 \\
2u_2 + 2u_4 + 5u_5 &= 8 \\
-u_1 + 4u_2 - 2u_3 + 3u_4 + 6u_5 &= 6 \\
u_1 + 2u_3 + u_4 - 2u_5 &= -2.
\end{aligned}
$$

We start with the augmented matrix and perform row operations to introduce zeroes:

$$\begin{bmatrix} 2 & 1 & 4 & 3 & 0 & \bigm| & 3 \\ 0 & 2 & 0 & 2 & 5 & \bigm| & 8 \\ -1 & 4 & -2 & 3 & 6 & \bigm| & 6 \\ 1 & 0 & 2 & 1 & -2 & \bigm| & -2 \end{bmatrix} \xrightarrow[R_3 \to R_3 + R_4]{R_1 \to R_1 - 2R_4} \begin{bmatrix} 0 & 1 & 0 & 1 & 4 & \bigm| & 7 \\ 0 & 2 & 0 & 2 & 5 & \bigm| & 8 \\ 0 & 4 & 0 & 4 & 4 & \bigm| & 4 \\ 1 & 0 & 2 & 1 & -2 & \bigm| & -2 \end{bmatrix}$$

$$\xrightarrow[R_3 \to R_3 - 4R_1]{R_2 \to R_2 - 2R_1} \begin{bmatrix} 0 & 1 & 0 & 1 & 4 & \bigm| & 7 \\ 0 & 0 & 0 & 0 & -3 & \bigm| & -6 \\ 0 & 0 & 0 & 0 & -12 & \bigm| & -24 \\ 1 & 0 & 2 & 1 & -2 & \bigm| & -2 \end{bmatrix} \xrightarrow[R_3 \to -\frac{1}{12}R_3]{R_2 \to -\frac{1}{3}R_2} \begin{bmatrix} 0 & 1 & 0 & 1 & 4 & \bigm| & 7 \\ 0 & 0 & 0 & 0 & 1 & \bigm| & 2 \\ 0 & 0 & 0 & 0 & 1 & \bigm| & 2 \\ 1 & 0 & 2 & 1 & -2 & \bigm| & -2 \end{bmatrix}$$

$$\xrightarrow[\substack{R_3 \to R_3 - R_2 \\ R_4 \to R_4 + 2R_2}]{R_1 \to R_1 - 4R_2} \begin{bmatrix} 0 & 1 & 0 & 1 & 0 & \bigm| & -1 \\ 0 & 0 & 0 & 0 & 1 & \bigm| & 2 \\ 0 & 0 & 0 & 0 & 0 & \bigm| & 0 \\ 1 & 0 & 2 & 1 & 0 & \bigm| & 2 \end{bmatrix} \xrightarrow{\text{rearr}} \begin{bmatrix} 1 & 0 & 2 & 1 & 0 & \bigm| & 2 \\ 0 & 1 & 0 & 1 & 0 & \bigm| & -1 \\ 0 & 0 & 0 & 0 & 1 & \bigm| & 2 \\ 0 & 0 & 0 & 0 & 0 & \bigm| & 0 \end{bmatrix}.$$

The last matrix represents the equations

$$\begin{aligned} u_1 \quad\ + 2u_3 + u_4 \ &= \ 2 \\ u_2 \qquad\quad + u_4 \ &= \ -1 \\ u_5 \ &= \ 2 \\ 0 \ &= \ 0. \end{aligned}$$

We solve these for u_1, u_2, and u_5 in terms of u_3 and u_4 and the constants on the right:

$$u_1 = 2 - 2u_3 - u_4, \qquad u_2 = -1 - u_4, \qquad u_5 = 2.$$

Any choice of u_3 and u_4, say $u_3 = a$ and $u_4 = b$, will lead to a solution of (E):

$$u_1 = 2 - 2a - b, \qquad u_2 = -1 - b, \qquad u_3 = a, \qquad u_4 = b, \qquad u_5 = 2.$$

In vector form, this reads

$$\mathbf{u} = \begin{bmatrix} 2 - 2a - b \\ -1 - b \\ a \\ b \\ 2 \end{bmatrix} = \begin{bmatrix} 2 \\ -1 \\ 0 \\ 0 \\ 2 \end{bmatrix} + a \begin{bmatrix} -2 \\ 0 \\ 1 \\ 0 \\ 0 \end{bmatrix} + b \begin{bmatrix} -1 \\ -1 \\ 0 \\ 1 \\ 0 \end{bmatrix}.$$

Example 3.6.3

Solve

$$u_1 + 2u_2 + 2u_3 = 0$$

(E) $$2u_1 + 5u_2 + 5u_3 = 0$$

$$2u_1 + 2u_2 + 2u_3 = 1.$$

We perform a sequence of row operations:

$$\begin{bmatrix} 1 & 2 & 2 & | & 0 \\ 2 & 5 & 5 & | & 0 \\ 2 & 2 & 2 & | & 1 \end{bmatrix} \xrightarrow[\substack{R_2 \to R_2 - 2R_1 \\ R_3 \to R_3 - 2R_1}]{} \begin{bmatrix} 1 & 2 & 2 & | & 0 \\ 0 & 1 & 1 & | & 0 \\ 0 & -2 & -2 & | & 1 \end{bmatrix} \xrightarrow[\substack{R_1 \to R_1 - 2R_2 \\ R_3 \to R_3 + 2R_2}]{} \begin{bmatrix} 1 & 0 & 0 & | & 0 \\ 0 & 1 & 1 & | & 0 \\ 0 & 0 & 0 & | & 1 \end{bmatrix}.$$

The last matrix is the augmented matrix of the system

$$u_1 \qquad\quad = 0$$

$$u_2 + u_3 = 0$$

$$0 = 1.$$

Since the last equation of this system is impossible, the system, and hence also (E), has no solutions.

In each of these examples, the matrix at the end of the sequence of row operations has the following properties:

1. Any row consisting entirely of zeroes is at the bottom.
2. The first nonzero entry of each nonzero row is 1—we refer to these as **corner** (or pivot) entries.
3. The corner entry of each nonzero row is further to the right than the corner entries of the preceding rows.
4. The corner entries are the only nonzero entries in their columns.

Any matrix with these four properties is said to be **reduced.**

Schematically, the nonzero entries of a reduced matrix lie in a kind of stair-step pattern. The entry in the corner of each step is 1. Below and to the left of the corners, all the entries are zero. The entries above and to the right of the corners can be anything, except that the corner entries are the only nonzero entries in their columns.

$$\begin{bmatrix} \boxed{1} & \text{anything} & 0 & \text{anything} & 0 & \text{anything} & 0 & \text{anything} \\ & & 1 & & 0 & & 0 & \\ & & & & 1 & & 0 & \\ & & 0\text{'s} & & & & 1 & \end{bmatrix}$$

Starting with the augmented matrix $[B|\mathbf{k}]$ of (E), we can always obtain a reduced matrix $[B^*|\mathbf{k}^*]$ by a sequence of row operations. In fact, $[B^*|\mathbf{k}^*]$ is uniquely determined by $[B|\mathbf{k}]$.

Fact: *Each matrix* $[B|\mathbf{k}]$ *is row equivalent to exactly one reduced matrix* $[B^*|\mathbf{k}^*]$.

The process of finding $[B^*|\mathbf{k}^*]$ is called **reduction.** The fact that $[B^*|\mathbf{k}^*]$ is unique means we needn't perform the row operations in any particular order. At each stage, any row operation that is convenient can be performed without affecting the final answer. The system $B^*\mathbf{u} = \mathbf{k}^*$ has the same solutions as (E). What's more, it is easy to solve.

Example 3.6.4

The eigenvalues of

$$A = \begin{bmatrix} -1 & -1 & 0 \\ 1 & -\frac{3}{2} & \frac{3}{2} \\ 0 & 1 & -1 \end{bmatrix}$$

are $\lambda = -1, -2, -1/2$ (see Example 3.5.2). Find the eigenvectors of A corresponding to each of these eigenvalues.

To find eigenvectors of A corresponding to $\lambda = -1$, we solve $[A - (-1)I]\,\mathbf{v} = \mathbf{0}$. We reduce the augmented matrix, $[A + I|\mathbf{0}]$:

$$\begin{bmatrix} 0 & -1 & 0 & | & 0 \\ 1 & -\frac{1}{2} & \frac{3}{2} & | & 0 \\ 0 & 1 & 0 & | & 0 \end{bmatrix} \xrightarrow[\substack{R_2 \to R_2 - \frac{1}{2}R_1 \\ R_3 \to R_3 + R_1}]{} \begin{bmatrix} 0 & -1 & 0 & | & 0 \\ 1 & 0 & \frac{3}{2} & | & 0 \\ 0 & 0 & 0 & | & 0 \end{bmatrix} \xrightarrow[\substack{R_1 \to -R_1 \\ \text{and rearr}}]{} \begin{bmatrix} 1 & 0 & \frac{3}{2} & | & 0 \\ 0 & 1 & 0 & | & 0 \\ 0 & 0 & 0 & | & 0 \end{bmatrix}.$$

The equations of the corresponding system,

$$v_1 \quad + \frac{3}{2} v_3 = 0$$
$$v_2 \qquad\quad = 0$$
$$0 \quad = 0$$

can be solved for the corner variables, v_1 and v_2, in terms of the noncorner variable v_3:

$$v_1 = -\tfrac{3}{2} v_3, \qquad v_2 = 0.$$

Any choice of v_3, say $v_3 = 2a$, leads to a solution. In vector terms, the solutions are the vectors of the form

$$\mathbf{v} = \begin{bmatrix} -3a \\ 0 \\ 2a \end{bmatrix} = a \begin{bmatrix} -3 \\ 0 \\ 2 \end{bmatrix}.$$

The nonzero vectors of this form are the eigenvectors for $\lambda = -1$.

To find the eigenvectors for $\lambda = -2$, we solve $[A - (-2)I]\mathbf{w} = \mathbf{0}$. The augmented matrix $[A + 2I|\mathbf{0}]$ can be reduced as follows:

$$\begin{bmatrix} 1 & -1 & 0 & | & 0 \\ 1 & \tfrac{1}{2} & \tfrac{3}{2} & | & 0 \\ 0 & 1 & 1 & | & 0 \end{bmatrix} \xrightarrow[\substack{R_1 \to R_1 + R_3 \\ R_2 \to R_2 - \tfrac{1}{2}R_3}]{} \begin{bmatrix} 1 & 0 & 1 & | & 0 \\ 1 & 0 & 1 & | & 0 \\ 0 & 1 & 1 & | & 0 \end{bmatrix} \xrightarrow[\substack{R_2 \to R_2 - R_1 \\ \text{and rearr}}]{} \begin{bmatrix} 1 & 0 & 1 & | & 0 \\ 0 & 1 & 1 & | & 0 \\ 0 & 0 & 0 & | & 0 \end{bmatrix}.$$

The equations corresponding to the last matrix are

$$w_1 \quad\ \ + w_3 = 0$$
$$w_2 + w_3 = 0$$
$$0 = 0.$$

Any choice of the noncorner variable, w_3, say $w_3 = a$, leads to a solution. These solutions can be written in the form

$$\mathbf{w} = \begin{bmatrix} -a \\ -a \\ a \end{bmatrix} = a \begin{bmatrix} -1 \\ -1 \\ 1 \end{bmatrix}.$$

The eigenvectors for $\lambda = -2$ are the nonzero vectors of this form.

To find eigenvectors corresponding to $-1/2$, we solve $[A - (-\tfrac{1}{2})I]\mathbf{u} = \mathbf{0}$ by reducing the augmented matrix $[A + \tfrac{1}{2}I \mid \mathbf{0}]$:

$$\begin{bmatrix} -\tfrac{1}{2} & -1 & 0 & | & 0 \\ 1 & -1 & \tfrac{3}{2} & | & 0 \\ 0 & 1 & -\tfrac{1}{2} & | & 0 \end{bmatrix} \xrightarrow[\substack{R_1 \to R_1 + R_3 \\ R_2 \to R_2 + R_3}]{} \begin{bmatrix} -\tfrac{1}{2} & 0 & -\tfrac{1}{2} & | & 0 \\ 1 & 0 & 1 & | & 0 \\ 0 & 1 & -\tfrac{1}{2} & | & 0 \end{bmatrix} \xrightarrow[\substack{R_1 \to R_1 + \tfrac{1}{2}R_2 \\ \text{and rearr}}]{} \begin{bmatrix} 1 & 0 & 1 & | & 0 \\ 0 & 1 & -\tfrac{1}{2} & | & 0 \\ 0 & 0 & 0 & | & 0 \end{bmatrix}.$$

The corresponding equations, solved for the corner variables u_1 and u_2 in terms of the noncorner variable u_3, give

$$u_1 = -u_3, \qquad u_2 = \tfrac{1}{2} u_3.$$

The eigenvectors of A with eigenvalue $\lambda = -1/2$ are the nonzero vectors of the form

$$\mathbf{u} = \begin{bmatrix} -2a \\ a \\ 2a \end{bmatrix} = a \begin{bmatrix} -2 \\ 1 \\ 2 \end{bmatrix}.$$

Example 3.6.5

Given that 2 is an eigenvalue of

$$A = \begin{bmatrix} 2 & 0 & 0 \\ -6 & 8 & -2 \\ -9 & 9 & -1 \end{bmatrix}$$

describe the eigenvectors of A corresponding to $\lambda = 2$.

We reduce the augmented matrix of $(A - 2I)\mathbf{v} = \mathbf{0}$:

$$\left[\begin{array}{ccc|c} 0 & 0 & 0 & 0 \\ -6 & 6 & -2 & 0 \\ -9 & 9 & -3 & 0 \end{array}\right] \xrightarrow[\substack{R_2 \to -\frac{1}{6}R_2 \\ \text{and rearr}}]{} \left[\begin{array}{ccc|c} 1 & -1 & \frac{1}{3} & 0 \\ -9 & 9 & -3 & 0 \\ 0 & 0 & 0 & 0 \end{array}\right] \xrightarrow[R_2 \to R_2 + 9R_1]{} \left[\begin{array}{ccc|c} 1 & -1 & \frac{1}{3} & 0 \\ 0 & 0 & 0 & 0 \\ 0 & 0 & 0 & 0 \end{array}\right].$$

If we solve the nontrivial equation of the corresponding system for the corner variable v_1 in terms of the noncorner variables, v_2 and v_3, we get

$$v_1 = v_2 - \tfrac{1}{3} v_3.$$

Any choice of the noncorner variables, say $v_2 = a$ and $v_3 = 3b$, leads to a solution. The eigenvectors are the nonzero vectors of the form

$$\mathbf{v} = \begin{bmatrix} a - b \\ a \\ 3b \end{bmatrix} = a \begin{bmatrix} 1 \\ 1 \\ 0 \end{bmatrix} + b \begin{bmatrix} -1 \\ 0 \\ 3 \end{bmatrix}.$$

Let's summarize.

SYSTEMS OF ALGEBRAIC EQUATIONS: ROW REDUCTION

Two matrices are **row equivalent** if we can get from one to the other by a sequence of **row operations:**

1. Adding multiples of one fixed row to one or more of the other rows.
2. Multiplying a row by a nonzero constant.
3. Rearranging the order of the rows.

A matrix is **reduced** provided:

1. Any row consisting entirely of zeroes is at the bottom.
2. The first nonzero entry, or **corner** entry, of each nonzero row is 1.
3. The corner entry of each nonzero row is further to the right than the corner entries of the preceding rows.
4. The corner entries are the only nonzero entries in their columns.

To solve $B\mathbf{u} = \mathbf{k}$, we first **reduce** the **augmented matrix** $[B|\mathbf{k}]$—that is, we perform a sequence of row operations to obtain a reduced matrix $[B^*|\mathbf{k}^*]$. We then solve $B^*\mathbf{u} = \mathbf{k}^*$. The solutions of $B\mathbf{u} = \mathbf{k}$ are the same as the solutions of $B^*\mathbf{u} = \mathbf{k}^*$.

Notes

1. A warning

The row operation of type 1 allows us to alter several rows at once using a single row R_i, by adding or subtracting multiples of R_i to each of the other rows. We urge you to be careful when doing this. It is tempting to try to do several such operations at once, using a different row to alter each of the original rows of the matrix simultaneously. However, if you are not careful, this can lead to subtracting a row from itself or some other operation equivalent to multiplying a row by zero, which is *not* an allowable operation.

2. On the solutions of $B^*\mathbf{u} = \mathbf{k}^*$

If $[B^*|\mathbf{k}^*]$ is reduced, then the location of the corners determines whether $B^*\mathbf{u} = \mathbf{k}^*$ has no solutions, a unique solution, or infinitely many solutions. If there is a corner in the last column, as in Example 3.6.3, then there are no solutions. If each column except the last has a corner, as in Example 3.6.1, then $\mathbf{u} = \mathbf{k}^*$ is the unique solution. If the last column does not have a corner and there are i other columns without corners, as in Example 3.6.2, then it is possible to find vectors $\mathbf{w}, \mathbf{v}_1, \ldots, \mathbf{v}_i$ so that the solutions are the vectors of the form $\mathbf{u} = \mathbf{w} + a_1\mathbf{v}_1 + \ldots + a_i\mathbf{v}_i$. There are infinitely many solutions in this case.

It is often possible to determine how many solutions a system $B\mathbf{u} = \mathbf{k}$ has without completing the reduction. If, for example, one of the rows of an intermediate matrix is $[0 \quad 0 \quad \ldots \quad 0 | 1]$, then this will also be true of the reduced matrix—there are no solutions in this case.

3. Row operations and determinants

Row operations can be useful when calculating the determinant of an $n \times n$ matrix M (see Note 2 in Section 2.4). However, it is *not* true that row-equivalent matrices necessarily have the same determinant. The following table shows the effect on the determinant of each of our row operations:

$$M \xrightarrow[R_j \to R_j + cR_i]{} M' \qquad \det M' = \det M$$

$$M \xrightarrow[R_j \to cR_j]{} M' \qquad \det M' = c(\det M)$$

$$M \xrightarrow[\text{rearr}]{} M' \qquad \det M' = (-1)^j(\det M) \text{ where } j \text{ is the}$$
number of interchanges of rows it would take to arrive at the rearrangement.

EXERCISES

In Exercises 1 to 4, reduce the given matrix.

1. $\begin{bmatrix} 1 & 3 & -1 & 1 \\ 0 & 1 & 0 & 0 \\ 1 & 1 & 1 & 1 \\ 2 & 2 & 2 & 2 \end{bmatrix}$

2. $\begin{bmatrix} 1 & 1 & 1 & 1 \\ 1 & 1 & 0 & 1 \\ 0 & 1 & 1 & 1 \\ 1 & 1 & 1 & 1 \end{bmatrix}$

3. $\begin{bmatrix} 1 & 0 & 2 & 3 \\ 2 & 1 & 4 & 5 \\ 1 & 1 & 2 & 2 \\ 0 & 1 & 0 & 1 \end{bmatrix}$

4. $\begin{bmatrix} 0 & 2 & 2 & 1 \\ 0 & 1 & 1 & 3 \\ 0 & 4 & 4 & -1 \\ 0 & 0 & 0 & 6 \end{bmatrix}$

In Exercises 5 to 9, find all solutions of the given system of equations (if any exist). Express your answer (a) as separate parametric equations for the variables and (b) as a linear combination of vectors.

5.
$$x + 2y - z + 4w - u = 0$$
$$2x - y + 3z - w + u = 0$$
$$-x + y - z + 2w + u = 0$$
$$x + y - 4z + 2w = 0$$
$$-x + 3z - u = 0$$

6.
$$x_1 + 2x_2 + 3x_3 - 2x_4 = 0$$
$$3x_1 - 7x_2 - 4x_3 + 7x_4 = 0$$
$$4x_1 - 3x_2 + x_3 + 3x_4 = 0$$
$$x_1 + 3x_2 + 4x_3 - 3x_4 = 0$$

7.
$$x - y + z = 1$$
$$2x + y + 2z = 5$$
$$x + 2y + z = 4$$

8.
$$x_1 + 2x_2 + x_3 - x_4 - x_5 = 2$$
$$2x_1 + 2x_2 + 2x_3 - 3x_4 - 2x_5 = 1$$
$$-x_1 - x_3 + 2x_4 + x_5 = 0$$

9.
$$x_1 + x_2 - x_3 + x_4 = 1$$
$$x_1 - x_2 + x_3 + x_4 = 0$$
$$x_1 + x_2 - x_3 - x_4 = 1$$
$$x_1 + x_2 + x_3 + x_4 = 0$$

In Exercises 10 to 13, solve $B\mathbf{u} = \mathbf{k}$ for $\mathbf{u}$ (if possible):

10. $B = \begin{bmatrix} 1 & -1 & 1 \\ 0 & 1 & 1 \\ 1 & 1 & 1 \end{bmatrix}$, $\quad \mathbf{k} = \begin{bmatrix} 1 \\ 2 \\ 3 \end{bmatrix}$

11. $B = \begin{bmatrix} 1 & 2 & 3 & 0 \\ 1 & -1 & -3 & 1 \\ 2 & 1 & 0 & 1 \end{bmatrix}$, $\quad \mathbf{k} = \begin{bmatrix} 1 \\ 1 \\ 1 \end{bmatrix}$

12. $B = \begin{bmatrix} 1 & 2 & 3 & 0 \\ 1 & -1 & -3 & 1 \\ 2 & 1 & 0 & 1 \end{bmatrix}$, $\quad \mathbf{k} = \begin{bmatrix} 1 \\ 1 \\ 2 \end{bmatrix}$

13. $B = \begin{bmatrix} 1 & 2 & 3 \\ 1 & -1 & 2 \\ 2 & 1 & 5 \\ 0 & 3 & 1 \end{bmatrix}$, $\quad \mathbf{k} = \begin{bmatrix} 1 \\ 2 \\ 3 \\ -1 \end{bmatrix}$

In Exercises 14 to 19, find the eigenvectors corresponding to each of the eigenvalues of A, where A is the matrix in the indicated exercise from Section 3.5:

14. Exercise 1   15. Exercise 2   16. Exercise 3

17. Exercise 4   18. Exercise 5   19. Exercise 6

We have encountered algebraic systems of equations before. In Exercises 20 and 21, use row reduction to help solve the algebraic systems that arise in solving the problems.

20. Check for independence:

a. $\mathbf{v}_1 = \begin{bmatrix} 1 \\ 2 \\ -1 \end{bmatrix}$, $\quad \mathbf{v}_2 = \begin{bmatrix} 0 \\ -1 \\ 3 \end{bmatrix}$, $\quad \mathbf{v}_3 = \begin{bmatrix} 1 \\ 4 \\ 5 \end{bmatrix}$

b. $\mathbf{v}_1 = \begin{bmatrix} 1 \\ 3 \\ -1 \\ 1 \end{bmatrix}$, $\quad \mathbf{v}_2 = \begin{bmatrix} -1 \\ 2 \\ 2 \\ 1 \end{bmatrix}$, $\quad \mathbf{v}_3 = \begin{bmatrix} 1 \\ 4 \\ 1 \\ 3 \end{bmatrix}$, $\quad \mathbf{v}_4 = \begin{bmatrix} 1 \\ -1 \\ 0 \\ 1 \end{bmatrix}$

c. $\mathbf{v}_1 = \begin{bmatrix} 1 \\ 0 \\ 1 \\ 2 \end{bmatrix}$, $\quad \mathbf{v}_2 = \begin{bmatrix} 1 \\ -1 \\ 0 \\ 1 \end{bmatrix}$, $\quad \mathbf{v}_3 = \begin{bmatrix} 1 \\ 2 \\ 3 \\ 4 \end{bmatrix}$

21. Find the general solution and the specific solution satisfying the given initial conditions.
    a. $(D - 1)(D + 1)(D^2 + 1)x = 0$; $x(0) = x'(0) = 1, x''(0) = x'''(0) = 2$
    b. $(D - 1)^2(D + 1)^2 x = 0$; $x(0) = x'(0) = 1, x''(0) = x'''(0) = 0$

Some more abstract questions:

22. Show that a reduced matrix with fewer rows than columns ($n \times m$, $n < m$) must have some columns without corners.

23. Use the fact proved in Exercise 22 to show that a system of n equations in $m > n$ unknowns, and with zero right-hand side, $Bu = 0$, always has nonzero solutions.

24. Use the fact proved in Exercise 23 to show that any set of $m > n$ different n-vectors must be linearly dependent.

25. Give an example to show that the zero right-hand-side assumption in Exercise 23 is necessary; that is, find a system of n equations in $m > n$ unknowns, $Bu = v$, with no solutions.

col

3.7 HOMOGENEOUS SYSTEMS WITH CONSTANT COEFFICIENTS: REAL ROOTS

In Section 3.4 we saw that the problem of solving an nth-order homogeneous system $Dx = Ax$ boils down to finding n solutions with linearly independent initial vectors. In Section 3.5 we saw that if v is an eigenvector of A corresponding to the eigenvalue λ, then $h(t) = e^{\lambda t}v$ is a solution with initial vector $h(0) = v$. If we find n *linearly independent* eigenvectors $v_1, \ldots v_n$ corresponding to the eigenvalues $\lambda_1, \ldots, \lambda_n$, respectively, then the associated solutions

$$h_1(t) = e^{\lambda_1 t}v_1, \ldots, h_n(t) = e^{\lambda_n t}v_n$$

will generate the general solution of $Dx = Ax$:

$$x = c_1 h_1(t) + \ldots + c_n h_n(t).$$

Example 3.7.1
Solve $Dx = Ax$, where

$$A = \begin{bmatrix} 1 & 0 & 4 \\ 0 & 3 & 0 \\ 1 & 0 & 1 \end{bmatrix}.$$

We found the eigenvalues and corresponding eigenvectors of A in Examples 3.5.4 and 3.5.5. The vector

$$v = \begin{bmatrix} -2 \\ 0 \\ 1 \end{bmatrix}$$

is an eigenvector of A corresponding to -1. Associated to this vector is a solution of $D\mathbf{x} = A\mathbf{x}$:

$$\mathbf{h}_1(t) = e^{-t}\mathbf{v} = \begin{bmatrix} -2e^{-t} \\ 0 \\ e^{-t} \end{bmatrix}.$$

The vectors

$$\mathbf{w}_1 = \begin{bmatrix} 0 \\ 1 \\ 0 \end{bmatrix} \quad \text{and} \quad \mathbf{w}_2 = \begin{bmatrix} 2 \\ 0 \\ 1 \end{bmatrix}$$

are linearly independent eigenvectors corresponding to 3. Associated to these vectors are two more solutions:

$$\mathbf{h}_2(t) = e^{3t}\mathbf{w}_1 = \begin{bmatrix} 0 \\ e^{3t} \\ 0 \end{bmatrix} \quad \text{and} \quad \mathbf{h}_3(t) = e^{3t}\mathbf{w}_2 = \begin{bmatrix} 2e^{3t} \\ 0 \\ e^{3t} \end{bmatrix}.$$

If the initial vectors $\mathbf{v}$, $\mathbf{w}_1$, and $\mathbf{w}_2$ are linearly independent, then $\mathbf{h}_1(t)$, $\mathbf{h}_2(t)$, and $\mathbf{h}_3(t)$ generate the general solution of our third-order system.

To determine whether $\mathbf{v}$, $\mathbf{w}_1$, and $\mathbf{w}_2$ are independent, we look for the values of c_1, c_2, and c_3 for which

(I) $$c_1\mathbf{v} + c_2\mathbf{w}_1 + c_3\mathbf{w}_2 = \mathbf{0}.$$

Since $\mathbf{v}$, $\mathbf{w}_1$, and $\mathbf{w}_2$ are eigenvectors of A corresponding to -1, 3, and 3, respectively, then

$$A\mathbf{v} = -\mathbf{v}, \quad A\mathbf{w}_1 = 3\mathbf{w}_1, \quad \text{and} \quad A\mathbf{w}_2 = 3\mathbf{w}_2.$$

Multiplication of (I) by A yields

(II) $$-c_1\mathbf{v} + 3c_2\mathbf{w}_1 + 3c_3\mathbf{w}_2 = \mathbf{0}.$$

If we subtract 3 times (I) from (II), we get $-4c_1\mathbf{v} = \mathbf{0}$. Since $\mathbf{v} \neq \mathbf{0}$,

$$c_1 = 0.$$

Substitution into (I) yields $c_2\mathbf{w}_1 + c_3\mathbf{w}_2 = \mathbf{0}$. Since $\mathbf{w}_1$ and $\mathbf{w}_2$ are linearly independent,

$$c_2 = c_3 = 0.$$

Thus $\mathbf{v}$, $\mathbf{w}_1$, and $\mathbf{w}_2$ are linearly independent.

The general solution of $D\mathbf{x} = A\mathbf{x}$ is

$$\mathbf{x} = c_1\mathbf{h}_1(t) + c_2\mathbf{h}_2(t) + c_3\mathbf{h}_3(t)$$

$$= c_1\begin{bmatrix} -2e^{-t} \\ 0 \\ e^{-t} \end{bmatrix} + c_2\begin{bmatrix} 0 \\ e^{3t} \\ 0 \end{bmatrix} + c_3\begin{bmatrix} 2e^{3t} \\ 0 \\ e^{3t} \end{bmatrix}.$$

The argument we used in this example to show independence can be extended to show that a list of eigenvectors of A corresponding to several eigenvalues will be independent provided the eigenvectors corresponding to each single eigenvalue are independent. Combining this with our earlier observations, we have the following:

Fact: *Let A be an $n \times n$ matrix with constant entries. Associate a list of vector valued functions to A as follows: for each eigenvalue λ of A, find as many linearly independent eigenvectors corresponding to λ as possible, and associate to each such eigenvector $\mathbf{v}$ the function $e^{\lambda t}\mathbf{v}$. These vector valued functions are solutions of $D\mathbf{x} = A\mathbf{x}$ with linearly independent initial vectors. In particular, if we find n such functions $\mathbf{h}_1(t), \ldots, \mathbf{h}_n(t)$, then the general solution of $D\mathbf{x} = A\mathbf{x}$ is $\mathbf{x} = c_1\mathbf{h}_1(t) + \ldots + c_n\mathbf{h}_n(t)$.*

One case in which we can be sure that this process will yield n functions occurs when A has n *distinct* eigenvalues. In this case, each eigenvalue will contribute one solution toward the n we need.

Example 3.7.2

Solve $D\mathbf{x} = A\mathbf{x}$, where

$$A = \begin{bmatrix} -1 & -1 & 0 \\ 1 & -\frac{3}{2} & \frac{3}{2} \\ 0 & 1 & -1 \end{bmatrix}.$$

This represents the three-loop circuit of Example 3.1.3 with $V(t) = 0$, $R_1 = 2$, $R_2 = 3$, $L_2 = 2$, $L_3 = 3$, and $C = 1/2$.

We found the eigenvalues and eigenvectors of A in Examples 3.5.2 and 3.6.4. The vectors

$$\mathbf{v} = \begin{bmatrix} -3 \\ 0 \\ 2 \end{bmatrix}, \qquad \mathbf{w} = \begin{bmatrix} -1 \\ -1 \\ 1 \end{bmatrix}, \qquad \text{and} \qquad \mathbf{u} = \begin{bmatrix} -2 \\ 1 \\ 2 \end{bmatrix}$$

are eigenvectors corresponding to the eigenvalues -1, -2, and $-1/2$, respectively. Associated to these eigenvectors are the three solutions

$$\mathbf{h}_1(t) = e^{-t}\mathbf{v}, \qquad \mathbf{h}_2(t) = e^{-2t}\mathbf{w}, \qquad \mathbf{h}_3(t) = e^{-t/2}\mathbf{u}$$

with independent initial vectors. The general solution of our third-order system is therefore

$$\mathbf{x} = c_1\mathbf{h}_1(t) + c_2\mathbf{h}_2(t) + c_3\mathbf{h}_3(t)$$

$$= c_1 \begin{bmatrix} -3e^{-t} \\ 0 \\ 2e^{-t} \end{bmatrix} + c_2 \begin{bmatrix} -e^{-2t} \\ -e^{-2t} \\ e^{-2t} \end{bmatrix} + c_3 \begin{bmatrix} -2e^{-t/2} \\ e^{-t/2} \\ 2e^{-t/2} \end{bmatrix}.$$

In terms of the currents and charges in the circuit of Example 3.1.3, this is

$$Q(t) = -3c_1e^{-t} - c_2e^{-2t} - 2c_3e^{-t/2}$$
$$I_2(t) = \qquad\quad - c_2e^{-2t} + c_3e^{-t/2}$$
$$I_3(t) = \quad 2c_1e^{-t} + c_2e^{-2t} + 2c_3e^{-t/2}.$$

Example 3.7.3

Find the solution of

$$
(H) \qquad
\begin{aligned}
x_1' &= x_1 - x_2 \\
x_2' &= - x_2 + 3x_3 \\
x_3' &= -x_1 + x_2
\end{aligned}
$$

that satisfies the initial condition

$$x_1(0) = x_2(0) = 0, \qquad x_3(0) = 1.$$

This system can be written in matrix terms as $D\mathbf{x} = A\mathbf{x}$ where

$$\mathbf{x} = \begin{bmatrix} x_1 \\ x_2 \\ x_3 \end{bmatrix} \quad \text{and} \quad A = \begin{bmatrix} 1 & -1 & 0 \\ 0 & -1 & 3 \\ -1 & 1 & 0 \end{bmatrix}.$$

The characteristic polynomial of A is

$$\det(A - \lambda I) = \det \begin{bmatrix} 1 - \lambda & -1 & 0 \\ 0 & -1 - \lambda & 3 \\ -1 & 1 & -\lambda \end{bmatrix}$$

$$= -\lambda^3 + 4\lambda = -\lambda(\lambda - 2)(\lambda + 2).$$

The eigenvalues of A are

$$\lambda = 0, 2, -2.$$

To find eigenvectors of A, we reduce $[A - 0I|0]$, $[A - 2I|0]$, and $[A - (-2)I|0]$. The row-equivalent reduced matrices are

$$\begin{bmatrix} 1 & 0 & -3 & | & 0 \\ 0 & 1 & -3 & | & 0 \\ 0 & 0 & 0 & | & 0 \end{bmatrix}, \quad \begin{bmatrix} 1 & 0 & 1 & | & 0 \\ 0 & 1 & -1 & | & 0 \\ 0 & 0 & 0 & | & 0 \end{bmatrix}, \quad \text{and} \quad \begin{bmatrix} 1 & 0 & 1 & | & 0 \\ 0 & 1 & 3 & | & 0 \\ 0 & 0 & 0 & | & 0 \end{bmatrix}$$

respectively. Each of the algebraic systems corresponding to the reduced matrices has a single noncorner variable. If we take the noncorner variable to be 1, we obtain eigenvectors

$$\mathbf{v} = \begin{bmatrix} 3 \\ 3 \\ 1 \end{bmatrix}, \quad \mathbf{w} = \begin{bmatrix} -1 \\ 1 \\ 1 \end{bmatrix}, \quad \text{and} \quad \mathbf{u} = \begin{bmatrix} -1 \\ -3 \\ 1 \end{bmatrix}$$

corresponding to 0, 2, and -2, respectively. The three solutions associated to these vectors generate the general solution of our third-order system (H). In matrix terms, the general solution is

$$\mathbf{x} = c_1 \mathbf{v} + c_2 e^{2t} \mathbf{w} + c_3 e^{-2t} \mathbf{u}$$

$$= c_1 \begin{bmatrix} 3 \\ 3 \\ 1 \end{bmatrix} + c_2 \begin{bmatrix} -e^{2t} \\ e^{2t} \\ e^{2t} \end{bmatrix} + c_3 \begin{bmatrix} -e^{-2t} \\ -3e^{-2t} \\ e^{-2t} \end{bmatrix}.$$

In matrix terms, the initial condition is

$$\mathbf{x}(0) = \begin{bmatrix} 0 \\ 0 \\ 1 \end{bmatrix}.$$

If we equate corresponding entries of $\mathbf{x}(0)$ and the given initial vector, we obtain the algebraic system of equations

$$3c_1 - c_2 - c_3 = 0$$
$$3c_1 + c_2 - 3c_3 = 0$$
$$c_1 + c_2 + c_3 = 1.$$

We reduce the augmented matrix of this algebraic system

$$\begin{bmatrix} 3 & -1 & -1 & | & 0 \\ 3 & 1 & -3 & | & 0 \\ 1 & 1 & 1 & | & 1 \end{bmatrix} \rightarrow \begin{bmatrix} 1 & 0 & 0 & | & \frac{1}{4} \\ 0 & 1 & 0 & | & \frac{3}{8} \\ 0 & 0 & 1 & | & \frac{3}{8} \end{bmatrix}$$

to determine the solution

$$c_1 = \tfrac{1}{4}, \qquad c_2 = \tfrac{3}{8}, \qquad c_3 = \tfrac{3}{8}.$$

The solution of (H) satisfying the given initial condition is

$$\mathbf{x} = (\tfrac{1}{4}) \begin{bmatrix} 3 \\ 3 \\ 1 \end{bmatrix} + (\tfrac{3}{8}) \begin{bmatrix} -e^{2t} \\ e^{2t} \\ e^{2t} \end{bmatrix} + (\tfrac{3}{8}) \begin{bmatrix} -e^{-2t} \\ -3e^{-2t} \\ e^{-2t} \end{bmatrix}.$$

Written parametrically, the solution is

$$x_1 = \tfrac{3}{4} - \tfrac{3}{8}e^{2t} - \tfrac{3}{8}e^{-2t}$$
$$x_2 = \tfrac{3}{4} + \tfrac{3}{8}e^{2t} - \tfrac{9}{8}e^{-2t}$$
$$x_3 = \tfrac{1}{4} + \tfrac{3}{8}e^{2t} + \tfrac{3}{8}e^{-2t}.$$

In Example 3.7.1 we had fewer than n distinct eigenvalues, but we still found n solutions with independent initial vectors. This was because we were able to find two linearly independent eigenvectors corresponding to a single eigenvalue. Let's take another look at our method for finding eigenvectors to see how many independent eigenvectors we can expect to find corresponding to a single eigenvalue.

The eigenvectors of A for the eigenvalue λ are the nonzero solutions of the algebraic system

(C) $(A - \lambda I)\mathbf{v} = \mathbf{0}.$

We solve (C) by reducing the augmented matrix and expressing the corner variables in terms of the noncorner variables. Suppose there are k noncorner variables. Then each choice of values $a_1, \ldots, a_k$ for these variables leads to a solution of (C) in the form

$$\mathbf{v} = a_1\mathbf{v}_1 + \ldots + a_k\mathbf{v}_k$$

where $\mathbf{v}_i$ is the solution obtained by setting the ith noncorner variable equal to 1 and the other noncorner variables equal to zero. The eigenvectors $\mathbf{v}_1, \ldots, \mathbf{v}_k$ are linearly independent (Exercise 15, Section 3.4) and cannot be increased to a larger independent set of eigenvectors for λ.

Although the process just described is by no means the only way of finding k linearly independent eigenvectors corresponding to λ, it results in as large a set of independent eigenvectors as one can find for λ. Furthermore, the number k is at most equal to the multiplicity of λ as a root of the characteristic polynomial.

Fact: *Let λ be a root of $det(A - \lambda I)$ with multiplicity m_λ. If we reduce the augmented matrix $[A - \lambda I | \mathbf{0}]$ of the algebraic system*

(C) $$\qquad\qquad\qquad\qquad (A - \lambda I)\mathbf{v} = \mathbf{0}$$

we obtain a matrix whose system has k_λ noncorner variables. For $j = 1, \ldots, k_\lambda$, let $\mathbf{v}_j$ be the solution of (C) obtained by taking the jth noncorner variable to be 1 while taking all the other noncorner variables to be 0. Then

i. $\mathbf{v}_1, \ldots, \mathbf{v}_{k_\lambda}$ are linearly independent eigenvectors for λ,

ii.   every eigenvector $\mathbf{v}$ for λ is a linear combination of $\mathbf{v}_1, \ldots, \mathbf{v}_{k_\lambda}$,

iii. k_λ is the largest possible number of independent eigenvectors for λ, and

iv. $1 \leq k_\lambda \leq m_\lambda$.

Example 3.7.4
Solve $D\mathbf{x} = A\mathbf{x}$, where

$$A = \begin{bmatrix} -2 & 1 & 0 & 1 \\ 1 & -2 & 1 & 0 \\ 0 & 1 & -2 & 1 \\ 1 & 0 & 1 & -2 \end{bmatrix}.$$

This represents the four-loop circuit of Example 3.1.4 with $V(t) = 0$, every $L_i = 1$, and every $R_i = 1$.

The characteristic polynomial of A is $\det(A - \lambda I) = (\lambda + 2)^2(\lambda + 4)\lambda$. There are three eigenvalues,

$$\lambda = 0, \qquad \lambda = -4, \qquad \text{and} \qquad \lambda = -2.$$

Note that $\lambda = -2$ is a double root.

To find eigenvectors for $\lambda = 0$, we reduce $[A|\mathbf{0}]$ to obtain

$$\left[\begin{array}{cccc|c} 1 & 0 & 0 & -1 & 0 \\ 0 & 1 & 0 & -1 & 0 \\ 0 & 0 & 1 & -1 & 0 \\ 0 & 0 & 0 & 0 & 0 \end{array}\right].$$

Thus, an eigenvector for $\lambda = 0$ is

$$\mathbf{v} = \begin{bmatrix} 1 \\ 1 \\ 1 \\ 1 \end{bmatrix}.$$

The corresponding solution (remember $e^0 = 1$) is

$$\mathbf{h}_1(t) = \begin{bmatrix} 1 \\ 1 \\ 1 \\ 1 \end{bmatrix}.$$

To find an eigenvector for $\lambda = -4$, we reduce $[A + 4I|\mathbf{0}]$ to get

$$\left[\begin{array}{cccc|c} 1 & 0 & 0 & 1 & 0 \\ 0 & 1 & 0 & -1 & 0 \\ 0 & 0 & 1 & 1 & 0 \\ 0 & 0 & 0 & 0 & 0 \end{array}\right].$$

An eigenvector for $\lambda = -4$ is

$$\mathbf{w} = \begin{bmatrix} -1 \\ 1 \\ -1 \\ 1 \end{bmatrix}.$$

The associated solution of $D\mathbf{x} = A\mathbf{x}$ is

$$\mathbf{h}_2(t) = \begin{bmatrix} -e^{-4t} \\ e^{-4t} \\ -e^{-4t} \\ e^{-4t} \end{bmatrix}.$$

Finally, we look for eigenvectors for $\lambda = -2$. We reduce $[A + 2I|\mathbf{0}]$ to get

$$\left[\begin{array}{cccc|c} 1 & 0 & 1 & 0 & 0 \\ 0 & 1 & 0 & 1 & 0 \\ 0 & 0 & 0 & 0 & 0 \\ 0 & 0 & 0 & 0 & 0 \end{array} \right].$$

The corresponding equations have two noncorner variables (the last two). If in turn we set each noncorner variable equal to 1 while setting the other equal to zero, we get two independent eigenvectors,

$$\mathbf{u}_1 = \begin{bmatrix} -1 \\ 0 \\ 1 \\ 0 \end{bmatrix} \quad \text{and} \quad \mathbf{u}_2 = \begin{bmatrix} 0 \\ -1 \\ 0 \\ 1 \end{bmatrix}.$$

The associated solutions of $D\mathbf{x} = A\mathbf{x}$,

$$\mathbf{h}_3(t) = \begin{bmatrix} -e^{-2t} \\ 0 \\ e^{-2t} \\ 0 \end{bmatrix} \quad \text{and} \quad \mathbf{h}_4(t) = \begin{bmatrix} 0 \\ -e^{-2t} \\ 0 \\ e^{-2t} \end{bmatrix}$$

have independent initial vectors.

The general solution of our fourth-order system is

$$\mathbf{x} = c_1\mathbf{h}_1(t) + c_2\mathbf{h}_2(t) + c_3\mathbf{h}_3(t) + c_4\mathbf{h}_4(t)$$

$$= c_1 \begin{bmatrix} 1 \\ 1 \\ 1 \\ 1 \end{bmatrix} + c_2 \begin{bmatrix} -e^{-4t} \\ e^{-4t} \\ -e^{-4t} \\ e^{-4t} \end{bmatrix} + c_3 \begin{bmatrix} -e^{-2t} \\ 0 \\ e^{-2t} \\ 0 \end{bmatrix} + c_4 \begin{bmatrix} 0 \\ -e^{-2t} \\ 0 \\ e^{-2t} \end{bmatrix}.$$

In terms of currents in our model, we have

$$I_1 = c_1 - c_2 e^{-4t} - c_3 e^{-2t}$$
$$I_2 = c_1 + c_2 e^{-4t} - c_4 e^{-2t}$$
$$I_3 = c_1 - c_2 e^{-4t} + c_3 e^{-2t}$$
$$I_4 = c_1 + c_2 e^{-4t} + c_4 e^{-2t}.$$

The method described in the summary will yield the general solution of $D\mathbf{x} = A\mathbf{x}$ provided the following are true:

1. Each of the roots of the characteristic polynomial of A is a real number.
2. For each eigenvalue λ of A, we can find as many independent eigenvectors corresponding to λ as the multiplicity of λ as a root of the characteristic polynomial.

In the next two sections, we discuss methods for solving systems for which one or both of these conditions fail.

HOMOGENEOUS SYSTEMS WITH CONSTANT COEFFICIENTS: REAL ROOTS

To solve $D\mathbf{x} = A\mathbf{x}$ where A is an $n \times n$ matrix:

1. Find the roots of the characteristic polynomial $\det(A - \lambda I)$.
2. For each real root λ, find as many linearly independent eigenvectors corresponding to λ as possible. (One way to do this is to reduce $[A - \lambda I | \mathbf{0}]$ and then list the solutions of the corresponding algebraic system obtained by in turn taking each of the noncorner variables to be 1 while taking all other noncorner variables to be 0.) Associate to each of these eigenvectors $\mathbf{v}$ the solution $e^{\lambda t}\mathbf{v}$ of $D\mathbf{x} = A\mathbf{x}$.
3. If by combining the solutions of $D\mathbf{x} = A\mathbf{x}$ associated to the various roots we obtain n vector valued functions $\mathbf{h}_1(t), \ldots, \mathbf{h}_n(t)$, then the general solution of $D\mathbf{x} = A\mathbf{x}$ is $\mathbf{x} = c_1\mathbf{h}_1(t) + \ldots + c_n\mathbf{h}_n(t)$.

EXERCISES

In Exercises 1 to 5, find the general solution of the system $D\mathbf{x} = A\mathbf{x}$, where A is the matrix considered earlier in the exercises indicated.

1. Exercise 1, Section 3.5 and Exercise 14, Section 3.6

2. Exercise 2, Section 3.5 and Exercise 15, Section 3.6

3. Exercise 3, Section 3.5 and Exercise 16, Section 3.6

4. Exercise 5, Section 3.5 and Exercise 18, Section 3.6

5. Exercise 6, Section 3.5 and Exercise 19, Section 3.6

In Exercises 6 to 13, find (a) the general solution of the given system of o.d.e.'s, and (b) the specific solution satisfying the given initial conditions.

6. $Dx = Ax$, where $A = \begin{bmatrix} -3 & 1 \\ -2 & 0 \end{bmatrix}$; $\quad x(0) = \begin{bmatrix} 0 \\ 1 \end{bmatrix}$

7. $\begin{aligned} x_1' &= x_1 + 2x_2 \\ x_2' &= 2x_1 + x_2 \end{aligned}$ $\quad x_1(0) = 1, x_2(0) = 3$

8. $\begin{aligned} x_1' &= -x_2 \\ x_2' &= 2x_1 + 3x_2 \end{aligned}$ $\quad x_1(0) = 1, x_2(0) = 3$

9. $Dx = Ax$, where $A = \begin{bmatrix} 1 & 1 & 0 \\ 1 & 1 & 0 \\ 0 & 0 & -1 \end{bmatrix}$; $\quad x(0) = \begin{bmatrix} 2 \\ 4 \\ 2 \end{bmatrix}$

10. $Dx = Ax$, where $A = \begin{bmatrix} 1 & 1 & 0 \\ 1 & 1 & 0 \\ 1 & 1 & 0 \end{bmatrix}$; $\quad x(0) = \begin{bmatrix} 2 \\ 4 \\ 2 \end{bmatrix}$

11. $\begin{aligned} x_1' &= -x_1 + x_2 \\ x_2' &= -6x_1 + 4x_2 \\ x_3' &= x_2 - x_3 \end{aligned}$ $\quad x_1(0) = 1, x_2(0) = 2, x_3(0) = 3$

12. $Dx = Ax$, where $A = \begin{bmatrix} 1 & 0 & 1 & 0 \\ 0 & 2 & 0 & 2 \\ 0 & 0 & -1 & 1 \\ 0 & 0 & 0 & 4 \end{bmatrix}$; $\quad x(0) = \begin{bmatrix} 2 \\ 16 \\ 5 \\ 15 \end{bmatrix}$

13. $\begin{aligned} x_1' &= -4x_1 + 6x_4 \\ x_2' &= 2x_2 \\ x_3' &= -2x_3 + 3x_4 \\ x_4' &= -5x_1 + 7x_4 \end{aligned}$ $\quad x_1(0) = x_2(0) = x_3(0) = 1, x_4(0) = 5$

Exercises 14 to 18 refer back to our earlier models.

14. *LRC Circuit:* Consider the *LRC* circuit of Example 3.1.1 with no external voltage, resistance 5 ohms, inductance 1/2 henry, and capacitance 1/8 farad. (a) Write down the system of o.d.e's modeling the circuit, (b) find the general solution, and (c) predict the charge Q after 1 second, given that initially there is no current and $Q = 1$ (use a table of exponentials or a calculator for this last part).

15. *Two-Loop Circuit:* Find (a) the general solution and (b) the solution with $I_2 = 1$, $I_1 = -3$ at time $t = 0$, for the system modeling the two-loop circuit of Example 3.1.2 with no external voltage, $R_1 = 2$ ohms, $R_2 = 3$ ohms, $L = 6$ henrys, and $C = 1/6$ farad.

16. *Radioactive Decay:*
    a. Find the general solution of the system modeling the radioactive decay chain of Exercise 2, Section 3.1, when no extra material is added or extracted.
    b. Recall from Exercise 17, Section 1.2, the relation between half-life and rate of decay. Suppose that Zigonium has a half-life of 3 years and decomposes into Martinium-123. This in turn has a half-life of 15 years and decomposes into carbon. Starting from an initial sample of 1 kg ($= 10^3$ gram) of Zigonium, find the amount of each substance present after 15 years.

17. *Diffusion I:* Find the general solution of the system modeling the diffusion problem of Exercise 3, Section 3.1, in case
    a. $V_1 = V_2 = 5, P = 0.1$
    b. $V_1 = 5, V_2 = 10, P = 0.1$

18. *Political Coexistence:*
    a. Find the general solution of the system modeling the problem of Exercise 5, Section 3.1.
    b. Assume that the population of Camponesa is now 12 million and the population of Mandachuva is 16 million.
    c. Will there ever be a time when the two populations are equal?

Some more abstract problems:

19. Suppose that $\mathbf{v}_1, \ldots, \mathbf{v}_k$ are independent eigenvectors for A corresponding to the eigenvalues $\lambda_1, \ldots, \lambda_k$, respectively. Suppose that $\mathbf{v} = c_1\mathbf{v}_1 + \ldots + c_k\mathbf{v}_k$ is also an eigenvector for A, corresponding to λ. By applying A to the expression for $\mathbf{v}$ and regrouping terms, show that

$$c_1(\lambda - \lambda_1)\mathbf{v}_1 + c_2(\lambda - \lambda_2)\mathbf{v}_2 + \ldots + c_k(\lambda - \lambda_k)\mathbf{v}_k = \mathbf{0}.$$

Use the independence of the $\mathbf{v}_i$'s to conclude that for each i, either $c_i = 0$ or $\lambda_i = \lambda$.

20. Use the result of Exercise 19 to show that if $\mathbf{v}_1, \ldots, \mathbf{v}_n$ is a collection of eigenvectors for A such that the eigenvectors corresponding to each particular eigenvalue are independent, then the whole collection is independent.

3.8 HOMOGENEOUS SYSTEMS WITH CONSTANT COEFFICIENTS: COMPLEX ROOTS

If the characteristic polynomial of A has complex roots, then the methods of the last section do not provide enough solutions to generate the general solution of $D\mathbf{x} = A\mathbf{x}$. We faced a similar situation in Section 2.8. In that case, we proceeded formally with the method we had developed for real roots, ignoring the fact that the roots were complex. We then rewrote our solution using Euler's formula,

$$e^{u+iv} = e^u(\cos v + i \sin v)$$

and found real-valued solutions that were part of our generating set for the general solution. In this section we describe a similar approach for systems.

By analogy with real roots, we refer to complex roots of the characteristic polynomial of A as **complex eigenvalues** of A. Since our matrices A have *real* entries, their characteristic polynomials have *real* coefficients. Thus complex eigenvalues come in conjugate pairs, $\alpha \pm \beta i$.

We refer to a nonzero vector with complex entries that satisfies $[A - (\alpha + \beta i)I]\mathbf{v} = \mathbf{0}$ as an **eigenvector** of A corresponding to $\alpha + \beta i$. Note that while the examples we saw in Section 3.6 all involved real coefficients, the methods of that section are perfectly valid for systems of algebraic equations with complex coefficients. Thus, we can use those methods to find eigenvectors corresponding to complex eigenvalues.

Let's see how proceeding formally, ignoring the fact that our roots are complex, works for a system of order 2.

Example 3.8.1

Solve $D\mathbf{x} = A\mathbf{x}$, where

$$A = \begin{bmatrix} -1 & -1 \\ 4 & -1 \end{bmatrix}.$$

This represents the two-loop circuit of Example 3.1.2 with $V(t) = 0$, $C = 1/4$, $L = 1$, $R_1 = 4$, and $R_2 = 1$.

The characteristic polynomial of A is $\det(A - \lambda I) = \lambda^2 + 2\lambda + 5$, so that A has a pair of complex eigenvalues,

$$\lambda = -1 + 2i \qquad \text{and} \qquad \lambda = -1 - 2i.$$

To find an eigenvector corresponding to $\lambda = -1 + 2i$, we reduce the augmented matrix $[A - (-1 + 2i)I \,|\, \mathbf{0}]$:

$$\begin{bmatrix} -2i & -1 & 0 \\ 4 & -2i & 0 \end{bmatrix} \xrightarrow[\substack{R_2 \to \frac{1}{4}R_2 \\ \text{and rearr}}]{} \begin{bmatrix} 1 & -i/2 & 0 \\ -2i & -1 & 0 \end{bmatrix} \xrightarrow[R_2 \to R_2 + 2iR_1]{} \begin{bmatrix} 1 & -i/2 & 0 \\ 0 & 0 & 0 \end{bmatrix}.$$

If we set the noncorner variable of the corresponding algebraic system to be 2, we obtain the eigenvector

$$\mathbf{v} = \begin{bmatrix} i \\ 2 \end{bmatrix}.$$

Associated to this vector is a complex "solution" of $D\mathbf{x} = A\mathbf{x}$,

$$\mathbf{g}_1(t) = e^{(-1 + 2i)t}\mathbf{v}.$$

We can use Euler's formula to rewrite $\mathbf{g}_1(t)$ in terms of its real and imaginary parts:

$$\mathbf{g}_1(t) = e^{-t}(\cos 2t + i \sin 2t)\begin{bmatrix} i \\ 2 \end{bmatrix} = \begin{bmatrix} e^{-t}(-\sin 2t + i \cos 2t) \\ e^{-t}(2 \cos 2t + 2i \sin 2t) \end{bmatrix}$$

$$= \begin{bmatrix} -e^{-t}\sin 2t \\ 2e^{-t}\cos 2t \end{bmatrix} + i\begin{bmatrix} e^{-t}\cos 2t \\ 2e^{-t}\sin 2t \end{bmatrix}.$$

To find an eigenvector for the second complex eigenvalue, we reduce $[A - (-1 - 2i)I|\mathbf{0}]$ to get

$$\begin{bmatrix} 1 & i/2 & | & 0 \\ 0 & 0 & | & 0 \end{bmatrix}.$$

We see that

$$\mathbf{w} = \begin{bmatrix} -i \\ 2 \end{bmatrix}$$

is an eigenvector of A corresponding to $\lambda = -1 - 2i$. Associated to $\mathbf{w}$ is another complex "solution,"

$$\mathbf{g}_2(t) = e^{(-1 - 2i)t}\mathbf{w} = \begin{bmatrix} -e^{-t}\sin 2t \\ 2e^{-t}\cos 2t \end{bmatrix} - i\begin{bmatrix} e^{-t}\cos 2t \\ 2e^{-t}\sin 2t \end{bmatrix}.$$

Proceeding formally, we obtain the "general solution" of $D\mathbf{x} = A\mathbf{x}$:

$$\mathbf{x} = k_1\mathbf{g}_1(t) + k_2\mathbf{g}_2(t)$$

$$= (k_1 + k_2)\begin{bmatrix} -e^{-t}\sin 2t \\ 2e^{-t}\cos 2t \end{bmatrix} + i(k_1 - k_2)\begin{bmatrix} e^{-t}\cos 2t \\ 2e^{-t}\sin 2t \end{bmatrix}$$

$$= c_1\begin{bmatrix} -e^{-t}\sin 2t \\ 2e^{-t}\cos 2t \end{bmatrix} + c_2\begin{bmatrix} e^{-t}\cos 2t \\ 2e^{-t}\sin 2t \end{bmatrix}.$$

This expression involves two real vector valued candidates for solutions,

$$\mathbf{h}_1(t) = \begin{bmatrix} -e^{-t}\sin 2t \\ 2e^{-t}\cos 2t \end{bmatrix} \quad \text{and} \quad \mathbf{h}_2(t) = \begin{bmatrix} e^{-t}\cos 2t \\ 2e^{-t}\sin 2t \end{bmatrix}.$$

Direct substitution will show that $\mathbf{h}_1(t)$ and $\mathbf{h}_2(t)$ are indeed solutions of $D\mathbf{x} = A\mathbf{x}$. Their initial vectors,

$$\mathbf{h}_1(0) = \begin{bmatrix} 0 \\ 2 \end{bmatrix} \quad \text{and} \quad \mathbf{h}_2(0) = \begin{bmatrix} 1 \\ 0 \end{bmatrix}$$

are linearly independent, so the real general solution of our second-order system is

$$\mathbf{x} = c_1\mathbf{h}_1(t) + c_2\mathbf{h}_2(t) = c_1 \begin{bmatrix} -e^{-t}\sin 2t \\ 2e^{-t}\cos 2t \end{bmatrix} + c_2 \begin{bmatrix} e^{-t}\cos 2t \\ 2e^{-t}\sin 2t \end{bmatrix}.$$

In terms of current and charge in our model, the solution is

$$Q = -c_1 e^{-t}\sin 2t + c_2 e^{-t}\cos 2t$$
$$I = 2c_1 e^{-t}\cos 2t + 2c_2 e^{-t}\sin 2t.$$

Note that both $\mathbf{h}_1(t)$ and $\mathbf{h}_2(t)$ appear in our expression for the first complex "solution" $\mathbf{g}_1(t)$, obtained from $\lambda = -1 + 2i$; $\mathbf{h}_1(t)$ is the "real part" and $\mathbf{h}_2(t)$ the "imaginary part" of $\mathbf{g}_1(t)$.

The major features of this example are quite typical.

Fact: *Let $\alpha \pm \beta i$ be complex eigenvalues of the real $n \times n$ matrix A. Let $\mathbf{v}$ be an eigenvector of A corresponding to $\alpha + \beta i$. Then $\mathbf{h}_1(t) = Re(e^{(\alpha + \beta i)t}\mathbf{v})$ and $\mathbf{h}_2(t) = Im(e^{(\alpha + \beta i)t}\mathbf{v})$ are solutions of $D\mathbf{x} = A\mathbf{x}$ with linearly independent initial vectors.*

Here we have used the convention that if r and s are real, then $\text{Re}(r + is) = r$ and $\text{Im}(r + is) = s$. Note that *we need only work with one of the two eigenvalues $\alpha \pm \beta i$ in order to find two solutions.*

In general, the characteristic polynomial of A could have mixed real and complex roots, any of which could be multiple roots. A modification of our previous methods, along the line suggested by the preceding fact, helps us build toward the general solution of $D\mathbf{x} = A\mathbf{x}$.

Fact: *Let A be an $n \times n$ matrix with real constant entries. Associate real vector valued functions to A as follows:*

1. *For each real eigenvalue λ, find as many linearly independent eigenvectors corresponding to λ as possible. To each such eigenvector $\mathbf{v}$, associate $e^{\lambda t}\mathbf{v}$.*
2. *For each pair of complex eigenvalues $\lambda = \alpha \pm \beta i$, find as many linearly independent eigenvectors corresponding to one of the eigenvalues, $\alpha + \beta i$, as possible. To each such eigenvector $\mathbf{v}$, associate the two functions $Re(e^{(\alpha + \beta i)t}\mathbf{v})$ and $Im(e^{(\alpha + \beta i)t}\mathbf{v})$.*

These vector valued functions are solutions of $D\mathbf{x} = A\mathbf{x}$ with linearly independent initial vectors. In particular, if we find n such functions $\mathbf{h}_1(t)$, . . ., $\mathbf{h}_n(t)$, then the general solution of $D\mathbf{x} = A\mathbf{x}$ is $\mathbf{x} = c_1\mathbf{h}_1(t) + \ldots + c_n\mathbf{h}_n(t)$.

Example 3.8.2

Solve $D\mathbf{x} = A\mathbf{x}$, where

$$A = \begin{bmatrix} -1 & -1 & 0 \\ 2 & -1 & 1 \\ 0 & 1 & -1 \end{bmatrix}.$$

This represents the three-loop circuit of Example 3.1.3 with $V(t) = 0$, $C = 1/2$, $L_2 = L_3 = 1$, $R_1 = 2$, and $R_2 = 1$.

The characteristic polynomial of A is $\det(A - \lambda I) = -(1 + \lambda)(\lambda^2 + 2\lambda + 2)$. The eigenvalues of A are

$$\lambda = -1, \qquad \lambda = -1 \pm i.$$

We find eigenvectors for $\lambda = -1$ by reducing $[A - (-1)I|\mathbf{0}]$:

$$\begin{bmatrix} 0 & -1 & 0 & | & 0 \\ 2 & 0 & 1 & | & 0 \\ 0 & 1 & 0 & | & 0 \end{bmatrix} \rightarrow \begin{bmatrix} 1 & 0 & \frac{1}{2} & | & 0 \\ 0 & 1 & 0 & | & 0 \\ 0 & 0 & 0 & | & 0 \end{bmatrix}.$$

The choice $v_3 = -2$ leads to an eigenvector for $\lambda = -1$,

$$\mathbf{v} = \begin{bmatrix} 1 \\ 0 \\ -2 \end{bmatrix}$$

and the associated solution of $D\mathbf{x} = A\mathbf{x}$,

$$\mathbf{h}_1(t) = e^{-t}\mathbf{v} = \begin{bmatrix} e^{-t} \\ 0 \\ -2e^{-t} \end{bmatrix}.$$

We choose *one* of the pair of complex eigenvalues, say $\lambda = -1 + i$, and look for the corresponding eigenvectors by reducing $[A - (-1 + i)I|\mathbf{0}]$:

$$\begin{bmatrix} -i & -1 & 0 & | & 0 \\ 2 & -i & 1 & | & 0 \\ 0 & 1 & -i & | & 0 \end{bmatrix} \rightarrow \begin{bmatrix} 1 & 0 & 1 & | & 0 \\ 0 & 1 & -i & | & 0 \\ 0 & 0 & 0 & | & 0 \end{bmatrix}.$$

The complex vector

$$\begin{bmatrix} -1 \\ i \\ 1 \end{bmatrix}$$

is an eigenvector for $\lambda = -1 + i$. The corresponding complex solution of $D\mathbf{x} = A\mathbf{x}$ is

$$e^{(-1 + i)t} \begin{bmatrix} -1 \\ i \\ 1 \end{bmatrix} = e^{-t}(\cos t + i \sin t) \begin{bmatrix} -1 \\ i \\ 1 \end{bmatrix} = \begin{bmatrix} -e^{-t}\cos t \\ -e^{-t}\sin t \\ e^{-t}\cos t \end{bmatrix} + i \begin{bmatrix} -e^{-t}\sin t \\ e^{-t}\cos t \\ e^{-t}\sin t \end{bmatrix}.$$

The real and imaginary parts of this complex solution,

$$\mathbf{h}_2(t) = \begin{bmatrix} -e^{-t}\cos t \\ -e^{-t}\sin t \\ e^{-t}\cos t \end{bmatrix} \quad \text{and} \quad \mathbf{h}_3(t) = \begin{bmatrix} -e^{-t}\sin t \\ e^{-t}\cos t \\ e^{-t}\sin t \end{bmatrix}$$

are solutions with independent initial vectors.

The general solution of our third-order system is

$$\mathbf{x} = c_1\mathbf{h}_1(t) + c_2\mathbf{h}_2(t) + c_3\mathbf{h}_3(t)$$

$$= c_1 \begin{bmatrix} e^{-t} \\ 0 \\ -2e^{-t} \end{bmatrix} + c_2 \begin{bmatrix} -e^{-t}\cos t \\ -e^{-t}\sin t \\ e^{-t}\cos t \end{bmatrix} + c_3 \begin{bmatrix} -e^{-t}\sin t \\ e^{-t}\cos t \\ e^{-t}\sin t \end{bmatrix}.$$

In terms of charges and currents, this reads

$$Q = c_1e^{-t} - c_2e^{-t}\cos t - c_3e^{-t}\sin t$$

$$I_2 = - c_2e^{-t}\sin t + c_3e^{-t}\cos t$$

$$I_3 = -2c_1e^{-t} + c_2e^{-t}\cos t + c_3e^{-t}\sin t.$$

Example 3.8.3

Solve $D\mathbf{x} = A\mathbf{x}$ where

$$A = \begin{bmatrix} 0 & -2 & 0 & 0 \\ 2 & 0 & 0 & 0 \\ 0 & 0 & 0 & -4 \\ 0 & 0 & 1 & 0 \end{bmatrix}.$$

The characteristic polynomial of A is $\det(A - \lambda I) = (\lambda^2 + 4)^2$. The roots are $\pm 2i$, each of which has multiplicity 2.

We choose *one* of the pair of complex eigenvalues, say $2i$, and look for corresponding eigenvectors by reducing $[A - 2iI|\mathbf{0}]$:

$$\begin{bmatrix} -2i & -2 & 0 & 0 & | & 0 \\ 2 & -2i & 0 & 0 & | & 0 \\ 0 & 0 & -2i & -4 & | & 0 \\ 0 & 0 & 1 & -2i & | & 0 \end{bmatrix} \rightarrow \begin{bmatrix} 1 & -i & 0 & 0 & | & 0 \\ 0 & 0 & 1 & -2i & | & 0 \\ 0 & 0 & 0 & 0 & | & 0 \\ 0 & 0 & 0 & 0 & | & 0 \end{bmatrix}.$$

The algebraic system corresponding to this matrix has two noncorner variables (the second and fourth). If in turn we set each of these to be 1 while setting the other to be zero, we obtain two independent eigenvectors corresponding to $2i$:

$$\mathbf{v}_1 = \begin{bmatrix} i \\ 1 \\ 0 \\ 0 \end{bmatrix} \quad \text{and} \quad \mathbf{v}_2 = \begin{bmatrix} 0 \\ 0 \\ 2i \\ 1 \end{bmatrix}.$$

Associated to these vectors are two complex solutions:

$$e^{2it}\mathbf{v}_1 = (\cos 2t + i \sin 2t) \begin{bmatrix} i \\ 1 \\ 0 \\ 0 \end{bmatrix} = \begin{bmatrix} -\sin 2t \\ \cos 2t \\ 0 \\ 0 \end{bmatrix} + i \begin{bmatrix} \cos 2t \\ \sin 2t \\ 0 \\ 0 \end{bmatrix}$$

$$e^{2it}\mathbf{v}_2 = (\cos 2t + i \sin 2t) \begin{bmatrix} 0 \\ 0 \\ 2i \\ 1 \end{bmatrix} = \begin{bmatrix} 0 \\ 0 \\ -2 \sin 2t \\ \cos 2t \end{bmatrix} + i \begin{bmatrix} 0 \\ 0 \\ 2 \cos 2t \\ \sin 2t \end{bmatrix}.$$

The real and imaginary parts of these complex solutions provide us with four solutions with independent initial vectors.

The general solution of our fourth-order system is

$$
\mathbf{x} = c_1 \begin{bmatrix} -\sin 2t \\ \cos 2t \\ 0 \\ 0 \end{bmatrix} + c_2 \begin{bmatrix} \cos 2t \\ \sin 2t \\ 0 \\ 0 \end{bmatrix} + c_3 \begin{bmatrix} 0 \\ 0 \\ -2 \sin 2t \\ \cos 2t \end{bmatrix} + c_4 \begin{bmatrix} 0 \\ 0 \\ 2 \cos 2t \\ \sin 2t \end{bmatrix}.
$$

The method described in the summary will yield the general solution of $D\mathbf{x} = A\mathbf{x}$ provided that for each eigenvalue (real or complex) we can find as many independent eigenvectors as the multiplicity of that eigenvalue as a root of the characteristic polynomial. In the next two sections, we discuss systems for which this condition fails to hold.

HOMOGENEOUS SYSTEMS WITH CONSTANT COEFFICIENTS: COMPLEX ROOTS

To solve $D\mathbf{x} = A\mathbf{x}$ where A is an $n \times n$ matrix:

1. Find the roots of the characteristic polynomial $\det(A - \lambda I)$.
2. For each real root λ, find as many linearly independent eigenvectors corresponding to λ as possible. Assign to each of these eigenvectors $\mathbf{v}$ the solution $e^{\lambda t}\mathbf{v}$.
3. For each pair of complex roots $\alpha \pm \beta i$, work with one of the roots as if it were real to obtain complex solutions. Associate to the pair of eigenvalues the real and imaginary parts of these complex solutions.

If by combining the solutions associated to the various roots we obtain n vector valued functions $\mathbf{h}_1(t), \ldots, \mathbf{h}_n(t)$, then the general solution of $D\mathbf{x} = A\mathbf{x}$ is $\mathbf{x} = c_1\mathbf{h}_1(t) + \ldots + c_n\mathbf{h}_n(t)$.

EXERCISES

In Exercises 1 to 8, find the general solution of $D\mathbf{x} = A\mathbf{x}$.

1. $A = \begin{bmatrix} 0 & 1 \\ -1 & 0 \end{bmatrix}$

2. $A = \begin{bmatrix} 3 & -2 \\ 2 & 3 \end{bmatrix}$

3. $A = \begin{bmatrix} 1 & 0 & -1 \\ 0 & 2 & 0 \\ 1 & 0 & 1 \end{bmatrix}$

4. $A = \begin{bmatrix} 1 & -2 & 2 \\ 0 & 1 & 1 \\ -1 & -2 & 4 \end{bmatrix}$

5. $A = \begin{bmatrix} 0 & 1 & -2 \\ -1 & -1 & 2 \\ 0 & 0 & -1 \end{bmatrix}$

6. $A = \begin{bmatrix} 1 & 1 & 0 & 0 \\ 3 & -1 & 0 & 0 \\ 0 & 0 & 0 & 1 \\ 0 & 0 & -1 & 0 \end{bmatrix}$.

7. $A = \begin{bmatrix} 0 & 1 & 1 & -1 \\ 0 & -1 & 0 & 1 \\ -5 & -5 & -2 & 0 \\ 0 & -4 & 0 & -1 \end{bmatrix}$

8. $A = \begin{bmatrix} 0 & 1 & 0 & 0 & 0 & 0 \\ -1 & 0 & 0 & 0 & 0 & 0 \\ 0 & 0 & 0 & -1 & 0 & 0 \\ 0 & 0 & 1 & 0 & 0 & 0 \\ 0 & 0 & 0 & 0 & 0 & -1 \\ 0 & 0 & 0 & 0 & -1 & 0 \end{bmatrix}$

In Exercises 9 to 12, find (a) the general solution and (b) the specific solution satisfying the given initial conditions.

9. $D\mathbf{x} = A\mathbf{x}$, where $A = \begin{bmatrix} 1 & -1 \\ 1 & 1 \end{bmatrix}$; $\mathbf{x}(0) = \begin{bmatrix} 2 \\ 3 \end{bmatrix}$.

10. $x_1' = 5x_1 - 4x_2$
    $x_2' = 10x_1 - 7x_2$ $x_1(0) = 70, x_2(0) = 10$

11. $x_1' = -x_2$
    $x_2' = 3x_1 + 2x_2$ $x_1(0) = 2, x_2(0) = 1$

12. $x_1' = x_1 + 2x_2 + x_4$
    $x_2' = -x_1 - x_2 + x_3$
    $x_3' = x_3 + x_4$ $x_1(0) = x_2(0) = x_3(0) = 0, x_4(0) = 4$
    $x_4' = -2x_3 - x_4$

Exercises 13 to 15 refer back to our earlier models.

13. *LRC Circuit:* Find the general solution of the system modeling the *LRC* circuit of Example 3.1.1 with $V(t) = 0$, $R = 1$ ohm, $C = 1$ farad, and $L = 1$ henry.

14. *Two-Loop Circuit:*
    a. Find the general solution of the system modeling the two-loop circuit of Example 3.1.2 with $V(t) = 0$, $R_1 = R_2 = 1$ ohm, $L = 1$ henry, and $C = 1$ farad.
    b. Find a formula for the current I_{cap} through the capacitor at time t, given that $I_2 = 1$ and $I_1 = 2$ amperes when $t = 0$.

15. *Coupled Springs:* Find the general solution of the system of Exercise 23a, Section 3.2, modeling two 1-gram masses for which the connecting spring has length $L_2 = 3$ cm and constant $k_2 = 2$ dynes/cm, while the spring connecting the first mass to the end of the track has length $L_1 = 10$ cm and constant $k_1 = 3$ dynes/cm. (Compare this to Exercise 26, Section 2.8.)

3.9 DOUBLE ROOTS AND MATRIX PRODUCTS

We have seen how to calculate the general solution of $D\mathbf{x} = A\mathbf{x}$ provided that for each eigenvalue λ the number of independent eigenvectors matches the multiplicity

m of λ as a root of the characteristic polynomial of A. In this section and the next, we discuss a method for associating m solutions with independent initial vectors to the eigenvalue λ. This method works even when there are fewer than m independent eigenvectors. The discussion in this section focuses on the simplest case, when λ is a real root of multiplicity two. Once the machinery for this case is set up, the extension to complex roots and higher multiplicities (in the next section) is quite straightforward.

Let's start by investigating a system of order two that is equivalent to a single o.d.e.

Example 3.9.1

Recall from Section 3.2 that the o.d.e.

(H) $$(D^2 - 2D + 1)x = 0$$

can be replaced by an equivalent system. Introducing two unknowns

$$x_1 = x \quad \text{and} \quad x_2 = x',$$

we obtain the system

(S_H) $$Dx = Ax$$

where

$$\mathbf{x} = \begin{bmatrix} x_1 \\ x_2 \end{bmatrix} \quad \text{and} \quad A = \begin{bmatrix} 0 & 1 \\ -1 & 2 \end{bmatrix}.$$

We leave it to you to check that the characteristic polynomial of A is $(\lambda - 1)^2$ and that all eigenvectors are multiples of a single eigenvector.

While the method of the preceding two sections will not yield the general solution of (S_H), we can find it by first solving (H). Since the general solution of (H) is $x = c_1 e^t + c_2 t e^t$, the general solution of (S_H) can be written

$$\mathbf{x} = \begin{bmatrix} x \\ x' \end{bmatrix} = \begin{bmatrix} c_1 e^t + c_2 t e^t \\ c_1 e^t + c_2 (t + 1) e^t \end{bmatrix}$$

$$= c_1 \mathbf{h}_1(t) + c_2 \mathbf{h}_2(t)$$

where

$$\mathbf{h}_1(t) = e^t \left(\begin{bmatrix} 1 \\ 1 \end{bmatrix} + t \begin{bmatrix} 0 \\ 0 \end{bmatrix} \right) \quad \text{and} \quad \mathbf{h}_2(t) = e^t \left(\begin{bmatrix} 0 \\ 1 \end{bmatrix} + t \begin{bmatrix} 1 \\ 1 \end{bmatrix} \right).$$

Based on this example, we guess that whenever λ is a double root of $\det(A - \lambda I)$, we can find solutions to $D\mathbf{x} = A\mathbf{x}$ of the form

$$(1) \qquad\qquad \mathbf{h}(t) = e^{\lambda t}(\mathbf{v}_0 + t\mathbf{v}_1).$$

Substitution of $\mathbf{x} = \mathbf{h}(t)$ into $D\mathbf{x} = A\mathbf{x}$ yields the equation

$$\lambda e^{\lambda t}(\mathbf{v}_0 + t\mathbf{v}_1) + e^{\lambda t}\mathbf{v}_1 = e^{\lambda t}(A\mathbf{v}_0 + tA\mathbf{v}_1).$$

If we divide by $e^{\lambda t}$ and subtract $\lambda(\mathbf{v}_0 + t\mathbf{v}_1)$ from both sides, we get

$$\mathbf{v}_1 = (A\mathbf{v}_0 - \lambda\mathbf{v}_0) + t(A\mathbf{v}_1 - \lambda\mathbf{v}_1).$$

This can be rewritten (see Exercise 24b, Section 3.2) as

$$(2) \qquad\qquad \mathbf{v}_1 = (A - \lambda I)\mathbf{v}_0 + t(A - \lambda I)\mathbf{v}_1.$$

When $t = 0$, this reads

$$(3) \qquad\qquad \mathbf{v}_1 = (A - \lambda I)\mathbf{v}_0$$

so that $\mathbf{v}_1$ will be determined as soon as we find $\mathbf{v}_0$. If we subtract (3) from (2) and take $t = 1$, we get

$$(A - \lambda I)\mathbf{v}_1 = \mathbf{0}.$$

By substituting (3) into this equation, we eliminate $\mathbf{v}_1$ and obtain

$$(4) \qquad\qquad (A - \lambda I)[(A - \lambda I)\mathbf{v}_0] = \mathbf{0}.$$

This tells us where to look for $\mathbf{v}_0$. We want a vector with the property that when we multiply it by $A - \lambda I$ and then multiply the resulting vector by $A - \lambda I$, we get the zero vector.

If we were able to rewrite (4) as a problem involving a single known matrix M times the unknown vector $\mathbf{v}_0$,

$$M\mathbf{v}_0 = \mathbf{0},$$

then we could solve it by reduction. Luckily it is possible to do this in a very general setting.

It turns out that the effect of multiplying a k-vector $\mathbf{v}$ by an $m \times k$ matrix C, and then multiplying the resulting m-vector $C\mathbf{v}$ by an $n \times m$ matrix B—that is, the calculation of the n-vector $B(C\mathbf{v})$—can be rewritten as multiplication of $\mathbf{v}$ by a single $n \times k$ matrix BC called the **product** of B and C. To calculate BC, we think of C as made up from its columns

$$\mathbf{c}_1 = \begin{bmatrix} c_{11} \\ \cdot \\ \cdot \\ \cdot \\ c_{m1} \end{bmatrix}, \quad \mathbf{c}_2 = \begin{bmatrix} c_{12} \\ \cdot \\ \cdot \\ c_{m2} \end{bmatrix}, \ldots, \mathbf{c}_k = \begin{bmatrix} c_{1k} \\ \cdot \\ \cdot \\ c_{mk} \end{bmatrix}.$$

We multiply the $n \times m$ matrix B times each of these m-vectors to obtain n-vectors

$$B\mathbf{c}_1, \quad B\mathbf{c}_2, \ldots, B\mathbf{c}_k.$$

These n-vectors form the columns of the product BC:

$$BC = B[\mathbf{c}_1 \ldots \mathbf{c}_k] = [B\mathbf{c}_1 \ldots B\mathbf{c}_k].$$

With this definition, we have the following (see Exercise 18).

Fact: *Suppose B is an $n \times m$ matrix, C is an $m \times k$ matrix, and $\mathbf{v}$ is a k-vector. Then*

$$(BC)\mathbf{v} = B(C\mathbf{v}).$$

Example 3.9.2
   Let

$$B = \begin{bmatrix} 1 & 2 & 3 & 0 \\ 1 & 0 & -1 & 2 \\ 0 & 1 & 0 & 1 \end{bmatrix} \quad \text{and} \quad C = \begin{bmatrix} 1 & 2 \\ 0 & 1 \\ 1 & 0 \\ -1 & 0 \end{bmatrix}.$$

Then C has two columns

$$\mathbf{c}_1 = \begin{bmatrix} 1 \\ 0 \\ 1 \\ -1 \end{bmatrix} \quad \text{and} \quad \mathbf{c}_2 = \begin{bmatrix} 2 \\ 1 \\ 0 \\ 0 \end{bmatrix}.$$

To obtain the columns of BC, we multiply B times the columns of C:

$$B\mathbf{c}_1 = \begin{bmatrix} 1 & 2 & 3 & 0 \\ 1 & 0 & -1 & 2 \\ 0 & 1 & 0 & 1 \end{bmatrix} \begin{bmatrix} 1 \\ 0 \\ 1 \\ -1 \end{bmatrix} = \begin{bmatrix} 4 \\ -2 \\ -1 \end{bmatrix},$$

$$B\mathbf{c}_2 = \begin{bmatrix} 1 & 2 & 3 & 0 \\ 1 & 0 & -1 & 2 \\ 0 & 1 & 0 & 1 \end{bmatrix} \begin{bmatrix} 2 \\ 1 \\ 0 \\ 0 \end{bmatrix} = \begin{bmatrix} 4 \\ 2 \\ 1 \end{bmatrix}.$$

Thus

$$BC = \begin{bmatrix} 4 & 4 \\ -2 & 2 \\ -1 & 1 \end{bmatrix}.$$

Note that *the product of two matrices is only defined if the number of columns of the first is the same as the number of rows of the second.* Since C has 2 columns and B has 3 rows, the product in the opposite order CB is not defined for the matrices in this example.

Example 3.9.3

Let

$$B = \begin{bmatrix} 2 & 1 \\ 1 & 0 \end{bmatrix} \quad \text{and} \quad C = \begin{bmatrix} 1 & 2 \\ 0 & -1 \end{bmatrix}.$$

To obtain the columns of BC, we multiply B times the columns $\mathbf{c}_1$ and $\mathbf{c}_2$ of C:

$$B\mathbf{c}_1 = \begin{bmatrix} 2 & 1 \\ 1 & 0 \end{bmatrix} \begin{bmatrix} 1 \\ 0 \end{bmatrix} = \begin{bmatrix} 2 \\ 1 \end{bmatrix}, \quad B\mathbf{c}_2 = \begin{bmatrix} 2 & 1 \\ 1 & 0 \end{bmatrix} \begin{bmatrix} 2 \\ -1 \end{bmatrix} = \begin{bmatrix} 3 \\ 2 \end{bmatrix}.$$

Thus

$$BC = \begin{bmatrix} 2 & 3 \\ 1 & 2 \end{bmatrix}.$$

These matrices can also be multiplied in the opposite order. The columns of CB are obtained by multiplying C times the columns $\mathbf{b}_1$ and $\mathbf{b}_2$ of B:

$$C\mathbf{b}_1 = \begin{bmatrix} 1 & 2 \\ 0 & -1 \end{bmatrix} \begin{bmatrix} 2 \\ 1 \end{bmatrix} = \begin{bmatrix} 4 \\ -1 \end{bmatrix}, \quad C\mathbf{b}_2 = \begin{bmatrix} 1 & 2 \\ 0 & -1 \end{bmatrix} \begin{bmatrix} 1 \\ 0 \end{bmatrix} = \begin{bmatrix} 1 \\ 0 \end{bmatrix}.$$

Thus

$$CB = \begin{bmatrix} 4 & 1 \\ -1 & 0 \end{bmatrix}.$$

Note *that CB $\neq$ BC*. Thus, *even if the products CB and BC are both defined, they need not be equal.*

In order to rewrite equation (4), we need to form the product of the $n \times n$ matrix $B = (A - \lambda I)$ with itself. Just as with numbers, we write the product of an $n \times n$ matrix with itself as a **square:**

$$B^2 = BB.$$

Higher powers are obtained by successively multiplying by B:

$$B^3 = BB^2, B^4 = BB^3, \text{ etc.}$$

Example 3.9.4
Let

$$B = \begin{bmatrix} -1 & 1 \\ -1 & 1 \end{bmatrix}.$$

The columns of $B^2 = BB$ are obtained by multiplying B times the columns of B:

$$\begin{bmatrix} -1 & 1 \\ -1 & 1 \end{bmatrix}\begin{bmatrix} -1 \\ -1 \end{bmatrix} = \begin{bmatrix} 0 \\ 0 \end{bmatrix}, \begin{bmatrix} -1 & 1 \\ -1 & 1 \end{bmatrix}\begin{bmatrix} 1 \\ 1 \end{bmatrix} = \begin{bmatrix} 0 \\ 0 \end{bmatrix}.$$

Thus

$$B^2 = \begin{bmatrix} 0 & 0 \\ 0 & 0 \end{bmatrix}.$$

Note that all the entries of the product BB are zero, even though none of the entries of the factors are zero.

Example 3.9.5

Let

$$B = \begin{bmatrix} 0 & 0 & 0 & 0 \\ 1 & 0 & 0 & 0 \\ 0 & 0 & 0 & 1 \\ 0 & 0 & 0 & 2 \end{bmatrix}.$$

Then

$$B^2 = BB = \begin{bmatrix} 0 & 0 & 0 & 0 \\ 1 & 0 & 0 & 0 \\ 0 & 0 & 0 & 1 \\ 0 & 0 & 0 & 2 \end{bmatrix} \begin{bmatrix} 0 & 0 & 0 & 0 \\ 1 & 0 & 0 & 0 \\ 0 & 0 & 0 & 1 \\ 0 & 0 & 0 & 2 \end{bmatrix} = \begin{bmatrix} 0 & 0 & 0 & 0 \\ 0 & 0 & 0 & 0 \\ 0 & 0 & 0 & 2 \\ 0 & 0 & 0 & 4 \end{bmatrix}$$

and

$$B^3 = BB^2 = \begin{bmatrix} 0 & 0 & 0 & 0 \\ 1 & 0 & 0 & 0 \\ 0 & 0 & 0 & 1 \\ 0 & 0 & 0 & 2 \end{bmatrix} \begin{bmatrix} 0 & 0 & 0 & 0 \\ 0 & 0 & 0 & 0 \\ 0 & 0 & 0 & 2 \\ 0 & 0 & 0 & 4 \end{bmatrix} = \begin{bmatrix} 0 & 0 & 0 & 0 \\ 0 & 0 & 0 & 0 \\ 0 & 0 & 0 & 4 \\ 0 & 0 & 0 & 8 \end{bmatrix}.$$

Using the notation of matrix powers, equation (4) in our earlier discussion can be rewritten

$$(A - \lambda I)^2 \mathbf{v}_0 = \mathbf{0}.$$

If $\mathbf{v}_0$ is a solution of this equation, and if $\mathbf{v}_1 = (A - \lambda I)\mathbf{v}_0$ (as in equation (3)), then substitution of these vectors into equation (1) will yield a solution of $D\mathbf{x} = A\mathbf{x}$.

Fact: *If* $(A - \lambda I)^2 \mathbf{v} = \mathbf{0}$, *then*

$$\mathbf{h}(t) = e^{\lambda t}(\mathbf{v} + t[A - \lambda I]\mathbf{v})$$

is a solution of $D\mathbf{x} = A\mathbf{x}$ *with initial vector* $\mathbf{h}(0) = \mathbf{v}$.

We refer to nonzero solutions of $(A - \lambda I)^2 \mathbf{v} = \mathbf{0}$ as **generalized eigenvectors** of A corresponding to the eigenvalue λ. If we find independent generalized eigenvectors, then we can use the preceding fact to obtain solutions of $D\mathbf{x} = A\mathbf{x}$ with independent initial vectors.

Let's see how this works for the system of Example 3.9.1.

Example 3.9.6

The characteristic polynomial of

$$A = \begin{bmatrix} 0 & 1 \\ -1 & 2 \end{bmatrix}$$

has a double root $\lambda = 1$. With this choice of λ,

$$A - \lambda I = \begin{bmatrix} -1 & 1 \\ -1 & 1 \end{bmatrix}.$$

We calculated the square of this matrix in Example 3.9.3:

$$(A - \lambda I)^2 = \begin{bmatrix} 0 & 0 \\ 0 & 0 \end{bmatrix}.$$

In this case, *every* vector is a solution of $(A - \lambda I)^2 \mathbf{v} = \mathbf{0}.$ In particular, the vectors

$$\mathbf{v} = \begin{bmatrix} 1 \\ 1 \end{bmatrix} \quad \text{and} \quad \mathbf{w} = \begin{bmatrix} 0 \\ 1 \end{bmatrix}$$

are independent generalized eigenvectors. The solutions associated to these vectors,

$$\mathbf{h}_1(t) = e^{\lambda}t(\mathbf{v} + t[A - \lambda I]\mathbf{v})$$

$$= e^{\lambda t}\left(\begin{bmatrix} 1 \\ 1 \end{bmatrix} + t\begin{bmatrix} -1 & 1 \\ -1 & 1 \end{bmatrix}\begin{bmatrix} 1 \\ 1 \end{bmatrix} \right)$$

$$= e^{\lambda t}\left(\begin{bmatrix} 1 \\ 1 \end{bmatrix} + t\begin{bmatrix} 0 \\ 0 \end{bmatrix} \right)$$

$$\mathbf{h}_2(t) = e^{\lambda t}(\mathbf{w} + t[A - \lambda I]\mathbf{w})$$

$$= e^{\lambda t}\left(\begin{bmatrix} 0 \\ 1 \end{bmatrix} + t\begin{bmatrix} -1 & 1 \\ -1 & 1 \end{bmatrix}\begin{bmatrix} 0 \\ 1 \end{bmatrix} \right)$$

$$= e^{\lambda t}\left(\begin{bmatrix} 0 \\ 1 \end{bmatrix} + t\begin{bmatrix} 1 \\ 1 \end{bmatrix} \right)$$

are the solutions that generated the general solution in Example 3.9.1.

If we have a system with more than one eigenvalue, then, as in Sections 3.7 and 3.8, we work with each individually and combine the resulting solutions.

Example 3.9.7

Find the general solution to $D\mathbf{x} = A\mathbf{x}$, where

$$A = \begin{bmatrix} -1 & 1 & 4 \\ -2 & 2 & 4 \\ -1 & 0 & 4 \end{bmatrix}.$$

The characteristic polynomial of A is $(2 - \lambda)^2(1 - \lambda)$, so $\lambda = 2$ is a double root and $\lambda = 1$ is a simple root.

To find the generalized eigenvectors corresponding to the double root $\lambda = 2$, we must solve

$$(A - 2I)^2\mathbf{v} = \mathbf{0}.$$

We first square $A - 2I$:

$$(A - 2I)^2 = \begin{bmatrix} -3 & 1 & 4 \\ -2 & 0 & 4 \\ -1 & 0 & 2 \end{bmatrix}\begin{bmatrix} -3 & 1 & 4 \\ -2 & 0 & 4 \\ -1 & 0 & 2 \end{bmatrix} = \begin{bmatrix} 3 & -3 & 0 \\ 2 & -2 & 0 \\ 1 & -1 & 0 \end{bmatrix}.$$

We then reduce the augmented matrix $[(A - 2I)^2|\mathbf{0}]$:

$$\begin{bmatrix} 3 & -3 & 0 & | & 0 \\ 2 & -2 & 0 & | & 0 \\ 1 & -1 & 0 & | & 0 \end{bmatrix} \rightarrow \begin{bmatrix} 1 & -1 & 0 & | & 0 \\ 0 & 0 & 0 & | & 0 \\ 0 & 0 & 0 & | & 0 \end{bmatrix}.$$

The vectors

$$\mathbf{v} = \begin{bmatrix} 0 \\ 0 \\ 1 \end{bmatrix} \quad \text{and} \quad \mathbf{w} = \begin{bmatrix} 1 \\ 1 \\ 0 \end{bmatrix}$$

are independent solutions of the corresponding algebraic system of equations. Associated to these generalized eigenvectors are two solutions of $D\mathbf{x} = A\mathbf{x}$:

$$\begin{aligned}
\mathbf{h}_1(t) &= e^{2t}(\mathbf{v} + t[A - 2I]\mathbf{v}) \\
&= e^{2t}\left(\begin{bmatrix} 0 \\ 0 \\ 1 \end{bmatrix} + t\begin{bmatrix} -3 & 1 & 4 \\ -2 & 0 & 4 \\ -1 & 0 & 2 \end{bmatrix}\begin{bmatrix} 0 \\ 0 \\ 1 \end{bmatrix}\right) \\
&= e^{2t}\left(\begin{bmatrix} 0 \\ 0 \\ 1 \end{bmatrix} + t\begin{bmatrix} 4 \\ 4 \\ 2 \end{bmatrix}\right) = \begin{bmatrix} 4te^{2t} \\ 4te^{2t} \\ (1 + 2t)e^{2t} \end{bmatrix}
\end{aligned}$$

$$\mathbf{h}_2(t) = e^{2t}(\mathbf{w} + t[A - 2I]\mathbf{w})$$

$$= e^{2t}\left(\begin{bmatrix} 1 \\ 1 \\ 0 \end{bmatrix} + t\begin{bmatrix} -3 & 1 & 4 \\ -2 & 0 & 4 \\ -1 & 0 & 2 \end{bmatrix}\begin{bmatrix} 1 \\ 1 \\ 0 \end{bmatrix}\right)$$

$$= e^{2t}\left(\begin{bmatrix} 1 \\ 1 \\ 0 \end{bmatrix} + t\begin{bmatrix} -2 \\ -2 \\ -1 \end{bmatrix}\right) = \begin{bmatrix} (1 - 2t)e^{2t} \\ (1 - 2t)e^{2t} \\ -te^{2t} \end{bmatrix}.$$

The eigenvectors for $\lambda = 1$ can be found by reducing $[A - I|\mathbf{0}]$:

$$\begin{bmatrix} -2 & 1 & 4 & | & 0 \\ -2 & 1 & 4 & | & 0 \\ -1 & 0 & 3 & | & 0 \end{bmatrix} \rightarrow \begin{bmatrix} 1 & 0 & -3 & | & 0 \\ 0 & 1 & -2 & | & 0 \\ 0 & 0 & 0 & | & 0 \end{bmatrix}.$$

Every eigenvector for $\lambda = 1$ is a multiple of

$$\mathbf{u} = \begin{bmatrix} 3 \\ 2 \\ 1 \end{bmatrix},$$

and a solution associated to $\lambda = 1$ is

$$\mathbf{h}_3(t) = e^t\mathbf{u} = \begin{bmatrix} 3e^t \\ 2e^t \\ e^t \end{bmatrix}.$$

Since the initial vectors

$$\mathbf{h}_1(0) = \begin{bmatrix} 3 \\ 2 \\ 1 \end{bmatrix}, \quad \mathbf{h}_2(0) = \begin{bmatrix} 0 \\ 0 \\ 1 \end{bmatrix}, \quad \mathbf{h}_3(0) = \begin{bmatrix} 1 \\ 1 \\ 0 \end{bmatrix}$$

are independent (check this), the general solution of our third-order system $D\mathbf{x} = A\mathbf{x}$ is

$$\mathbf{x} = c_1\mathbf{h}_1(t) + c_2\mathbf{h}_2(t) + c_3\mathbf{h}_3(t)$$

$$= c_1\begin{bmatrix} 3e^t \\ 2e^t \\ e^t \end{bmatrix} + c_2\begin{bmatrix} 4te^{2t} \\ 4te^{2t} \\ (1 + 2t)e^{2t} \end{bmatrix} + c_3\begin{bmatrix} (1 - 2t)e^{2t} \\ (1 - 2t)e^{2t} \\ -te^{2t} \end{bmatrix}.$$

The method we have outlined relies on our being able to find two linearly independent generalized eigenvectors for each eigenvalue of multiplicity two. This is guaranteed by a general theorem of linear algebra, which we state more explicitly in the next section.

MATRIX PRODUCTS AND DOUBLE ROOTS

The **product** of an $n \times m$ matrix B and an $m \times k$ matrix C is the $n \times k$ matrix whose columns are obtained by multiplying B times the columns of C:

$$BC = B[\mathbf{c}_1 \dots \mathbf{c}_k] = [B\mathbf{c}_1 \dots B\mathbf{c}_k].$$

The effect of multiplying a k-vector by BC is the same as multiplying the vector by C and then multiplying the resulting vector by B:

$$(BC)\mathbf{v} = B(C\mathbf{v}).$$

The **square** of an $n \times n$ matrix is its product with itself, and **higher powers** are obtained by successively multiplying by the matrix:

$$B^2 = BB, B^3 = BB^2, B^4 = BB^3, \text{etc.}$$

If λ is a double root of the characteristic polynomial $\det(A - \lambda I)$, then a **generalized eigenvector** corresponding to the eigenvalue λ is a vector $\mathbf{v} \neq \mathbf{0}$ satisfying

$$(A - \lambda I)^2\mathbf{v} = \mathbf{0}.$$

There are always two linearly independent generalized eigenvectors corresponding to a double eigenvalue of A. Associated to each of these generalized eigenvectors is a solution of $D\mathbf{x} = A\mathbf{x}$,

$$\mathbf{h}(t) = e^{\lambda t}(\mathbf{v} + t[A - \lambda I]\mathbf{v}).$$

Notes

1. On the entries of the product of two matrices

We have described the product BC of an $n \times m$ matrix B and an $m \times k$ matrix C by describing its columns. Sometimes it helps to have a description of the entry in the ith row and jth column of BC.

If $\mathbf{c}_j$ is the jth column of C, then $B\mathbf{c}_j$ is the jth column of BC. The entry in the ith row and jth column of BC is the same as the ith entry down in the vector $B\mathbf{c}_j$. To get this, we multiply the entries in the ith row of B by the corresponding entries in $\mathbf{c}_j$ and sum the products:

$$\begin{bmatrix} b_{11} & \cdots & b_{1n} \\ & & \\ & & \\ b_{i1} & \cdots & b_{in} \\ & & \\ & & \\ b_{m1} & \cdots & b_{mn} \end{bmatrix} \begin{bmatrix} c_{11} & \cdots & c_{1j} & \cdots & c_{1k} \\ & & & & \\ c_{n1} & \cdots & c_{nj} & \cdots & c_{nk} \end{bmatrix} = \begin{bmatrix} & & \\ & & \\ & (b_{i1}c_{1j} + \ldots + b_{in}c_{nj}) & \\ & & \\ & & \end{bmatrix} \begin{matrix} \\ \\ i\text{th row} \\ \longleftarrow \\ \\ \end{matrix}$$

$\uparrow$ *j*th column

Thus BC is the $n \times k$ matrix whose entry in the ith row and jth column is

$$b_{i1}c_{1j} + b_{i2}c_{2j} + \ldots + b_{in}c_{nj}.$$

2. On the properties of matrix multiplication

Examples 3.9.2 and 3.9.3 illustrate ways in which matrix multiplication differs from multiplication of numbers. The following list of basic properties of matrix multiplication demonstrates some of the similarities between these operations. In this list, we use I_n to denote the $n \times n$ identity matrix and $O_{m \times k}$ to denote the $m \times k$ zero matrix (see Note 2 in Section 3.2).

Fact: *Suppose that A and A' are $n \times m$ matrices, B and B' are $m \times k$ matrices, C is a $k \times r$ matrix, and c is a number. Then*

1. $\qquad\qquad\qquad (AB)C = A(BC)$

2. $\qquad\qquad\qquad I_n A = A \text{ and } AI_m = A$

3. $\qquad\qquad\qquad (cA)B = c(AB) = A(cB)$

4. $\qquad\qquad\qquad AO_{m \times k} = O_{n \times k} \text{ and } O_{r \times n}A = O_{r \times m}$

5. $\qquad\qquad\qquad A(B + B') = AB + AB'$

6. $\qquad\qquad\qquad (A + A')B = AB + A'B.$

3. Eigenvectors are generalized eigenvectors

Suppose that λ is a double eigenvalue of A. If $\mathbf{v}$ is an eigenvector of A corresponding to λ, then

$$(A - \lambda I)^2\mathbf{v} = (A - \lambda I)[(A - \lambda I)\mathbf{v}] = (A - \lambda I)\mathbf{0} = \mathbf{0}.$$

Thus $\mathbf{v}$ is also a generalized eigenvector of A. The solution of $D\mathbf{x} = A\mathbf{x}$ associated to $\mathbf{v}$ by virtue of the fact that $\mathbf{v}$ is a generalized eigenvector is

$$\mathbf{h}(t) = e^{\lambda t}(\mathbf{v} + t[A - \lambda I]\mathbf{v}) = e^{\lambda t}(\mathbf{v} + t\mathbf{0}) = e^{\lambda t}\mathbf{v}$$

which is the same as the solution associated to $\mathbf{v}$ by virtue of the fact that $\mathbf{v}$ is an eigenvector.

EXERCISES

1. Let $B = \begin{bmatrix} 3 & -1 & 0 \\ 1 & 0 & 1 \\ -1 & 0 & 0 \end{bmatrix}$, $C = \begin{bmatrix} 1 & 1 & 0 \\ 0 & 1 & -1 \\ 1 & 0 & 1 \end{bmatrix}$, and $v = \begin{bmatrix} 1 \\ 2 \\ -1 \end{bmatrix}$.

Find the indicated matrices.
a. BC b. $(BC)v$ c. $B(Cv)$
d. CB e. B^2 f. B^3
g. $(BC)^2$ h. B^2C^2 i. $(B + C)^2$

2. Let $A = \begin{bmatrix} 0 & 0 & 3 & 0 \\ 1 & 3 & 0 & -1 \\ 0 & 0 & 3 & 0 \\ 1 & 0 & -1 & 3 \end{bmatrix}$, $B = \begin{bmatrix} 0 & 1 \\ 1 & 0 \\ 0 & 1 \\ 0 & 1 \end{bmatrix}$, $C = \begin{bmatrix} 1 & 3 \\ 0 & -1 \\ 1 & 0 \\ 0 & 0 \end{bmatrix}$, and

$F = \begin{bmatrix} 1 & 0 & -1 & 0 \\ 0 & 1 & 0 & 0 \end{bmatrix}$.

Find the indicated matrices.
a. AB b. AC c. $A(B - C)$
d. FA e. BF f. FB
g. $(FA)B$ h. $F(AB)$ i. $(AB)F$

In Exercises 3 to 8, find the general solution of $Dx = Ax$.

3. A as in Exercise 4, Section 3.5 (and Exercise 17, Section 3.6)
4. A as in Exercise 8, Section 3.5
5. A as in Exercise 11, Section 3.5

6. $A = \begin{bmatrix} 2 & -1 & -4 \\ 0 & 2 & -4 \\ 0 & 1 & -2 \end{bmatrix}$ 7. $A = \begin{bmatrix} 2 & 0 & 1 & 0 \\ 0 & 1 & 0 & 1 \\ 0 & 0 & 2 & 1 \\ 0 & -1 & 0 & 1 \end{bmatrix}$

8. $A = \begin{bmatrix} -2 & 0 & 0 & 0 \\ 0 & -2 & 2 & -1 \\ 0 & 0 & -4 & 9 \\ 0 & 0 & -4 & 8 \end{bmatrix}$

In Exercises 9 to 12, find (a) the general solution and (b) the specific solution satisfying the given initial condition.

9. $Dx = Ax$, $A = \begin{bmatrix} 0 & 1 \\ -4 & 4 \end{bmatrix}$; $x(0) = \begin{bmatrix} 3 \\ 4 \end{bmatrix}$

10. $\begin{aligned} x_1' &= x_2 \\ x_2' &= -x_1 - 2x_2 \end{aligned}$ $x_1(0) = 3, x_2(0) = 4$

11. $x_1' = 2x_1 - x_2$ $x_1(0) = 3, x_2(0) = 4$
    $x_2' = x_1 + 4x_2$

12. $x_1' = 3x_1 - x_2 \quad - 4x_4$
    $x_2' = \qquad 3x_2 \quad - 4x_4$
    $x_3' = \qquad\quad 2x_3$ $x_1(0) = x_2(0) = x_3(0) = 1, x_4(0) = 0$
    $x_4' = \qquad x_2 \quad\; - x_4$

Exercises 13 and 14 refer back to our earlier models.

13. *LRC Circuit:* Find the general solution of the system modeling the *LRC* circuit of Example 3.1.1 with $V(t) = 0$, $R = 2$ ohms, $C = 1$ farad, and $L = 1$ henry.

14. *Two-Loop Circuit:*
    a. Find the general solution of the system modeling the two-loop circuit of Example 3.1.2 with $V(t) = 0$, $R_1 = 1$, $R_2 = 3$ ohms, $C = 1$ farad, and $L = 1$ henry.
    b. Find the specific solution satisfying $Q(0) = 1$ and $I_2(0) = 2$ amperes.

Some more abstract problems:

15. Suppose that A is an $n \times m$ matrix.
    a. Show that if X is a matrix for which AX and XA are both defined, then X must be $m \times n$.
    b. Show that if $A^2 = AA$ is defined, then $m = n$.

16. Let A and B be $n \times n$ matrices.
    a. Use properties (5) and (6) from Note 2 to show that

$$(A + B)^2 = A^2 + AB + BA + B^2.$$

    b. Find an example of 2×2 matrices A and B for which $(A + B)^2 \neq A^2 + 2AB + B^2$.

17. a. Show that any 2×2 matrix of the form $X = \begin{bmatrix} 0 & a \\ 0 & 0 \end{bmatrix}$ or $X = \begin{bmatrix} 0 & 0 \\ a & 0 \end{bmatrix}$ satisfies

$$X^2 = \begin{bmatrix} 0 & 0 \\ 0 & 0 \end{bmatrix}.$$

    b. Are there any other 2×2 matrices that satisfy this equation?

18. Suppose B is an $n \times m$ matrix, C is an $m \times k$ matrix, and $\mathbf{v}$ is a k-vector. Show that $(BC)\mathbf{v} = B(C\mathbf{v})$. (*Hint:* Use the description of BC given in Note 1.)

19. Verify properties (1) through (6) from Note 2. (*Hint:* You may use properties (M1) to (M6) from Note 2, Section 3.2, as well as Exercise 18 with B and C replaced by A and B.)

20. Suppose A is an $n \times n$ matrix. Show that if $(A - \lambda I)^2\mathbf{v} = \mathbf{0}$ but $(A - \lambda I)\mathbf{v} \neq \mathbf{0}$, then $\mathbf{v}_1 = (A - \lambda I)\mathbf{v}$ is an eigenvector of A.

21. An important fact about matrix multiplication is that the determinant of a product of $n \times n$ matrices is the same as the product of the determinants:

$$\det(AB) = (\det A)(\det B).$$

Verify this formula in the case where

a. $A = \begin{bmatrix} 1 & 2 \\ 0 & 3 \end{bmatrix}, B = \begin{bmatrix} -1 & 0 \\ 1 & 3 \end{bmatrix}$

b. $A = \begin{bmatrix} 1 & 0 & 3 \\ -1 & 2 & 0 \\ 1 & 1 & 0 \end{bmatrix}, B = \begin{bmatrix} 1 & 1 & 1 \\ 3 & 0 & 0 \\ -1 & 2 & 5 \end{bmatrix}$

c. $A = \begin{bmatrix} a_{11} & a_{12} \\ a_{21} & a_{22} \end{bmatrix}, B = \begin{bmatrix} b_{11} & b_{12} \\ b_{21} & b_{22} \end{bmatrix}$

22. The **Cayley-Hamilton Theorem** says that if the $n \times n$ matrix A has characteristic polynomial $p(\lambda) = \det(A - \lambda I) = b_n\lambda^n + \ldots + b_1\lambda + b_0$, then the matrix $p(A) = b_nA^n + \ldots + b_1A + b_0I_n$ is equal to the $n \times n$ zero matrix. Verify this fact in the case where

a. $A = \begin{bmatrix} 1 & 2 \\ -1 & 3 \end{bmatrix}$ b. $A = \begin{bmatrix} 1 & 0 & 3 \\ -1 & 2 & 0 \\ 1 & 1 & 0 \end{bmatrix}$ c. $A = \begin{bmatrix} a_{11} & a_{12} \\ a_{21} & a_{22} \end{bmatrix}$

3.10 HOMOGENEOUS SYSTEMS WITH CONSTANT COEFFICIENTS: MULTIPLE ROOTS

In this section we complete our treatment of homogeneous systems with constant coefficients by discussing (real or complex) eigenvalues of arbitrary multiplicity. Our method for dealing with multiple roots is a straightforward generalization of the method we used for real double roots.

In Section 3.9 we saw how to associate to a double root λ solutions of $D\mathbf{x} = A\mathbf{x}$ of the form $\mathbf{h}(t) = e^{\lambda t}(\mathbf{v}_0 + t\mathbf{v}_1)$. In general, if λ is a root of multiplicity m, we look for solutions of the form

(1) $$\mathbf{h}(t) = e^{\lambda t}(\mathbf{v}_0 + t\mathbf{v}_1 + t^2\mathbf{v}_2 + \ldots + t^{m-1}\mathbf{v}_{m-1}).$$

Substitution of $\mathbf{x} = \mathbf{h}(t)$ into $D\mathbf{x} = A\mathbf{x}$ yields the equation

$$\lambda e^{\lambda t}(\mathbf{v}_0 + t\mathbf{v}_1 + \ldots + t^{m-1}\mathbf{v}_{m-1}) + e^{\lambda t}(\mathbf{v}_1 + 2t\mathbf{v}_2 + \ldots + (m-1)t^{m-2}\mathbf{v}_{m-1})$$
$$= e^{\lambda t}(A\mathbf{v}_0 + tA\mathbf{v}_1 + \ldots + t^{m-1}A\mathbf{v}_{m-1}).$$

If we divide by $e^{\lambda t}$ and subtract $\lambda(\mathbf{v}_0 + \ldots + t^{m-1}\mathbf{v}_{m-1})$ from both sides, we get

$$\mathbf{v}_1 + 2t\mathbf{v}_2 + \ldots + (m-1)t^{m-2}\mathbf{v}_{m-1}$$
$$= (A\mathbf{v}_0 - \lambda\mathbf{v}_0) + t(A\mathbf{v}_1 - \lambda\mathbf{v}_1) + \ldots + t^{m-1}(A\mathbf{v}_{m-1} - \lambda\mathbf{v}_{m-1})$$
$$= (A - \lambda I)\mathbf{v}_0 + t(A - \lambda I)\mathbf{v}_1 + \ldots + t^{m-1}(A - \lambda I)\mathbf{v}_{m-1}.$$

Comparing corresponding powers of t on both sides, we obtain the equations

$$(A - \lambda I)\mathbf{v}_0 = \mathbf{v}_1$$

$$(A - \lambda I)\mathbf{v}_1 = 2\mathbf{v}_2$$

.

(2)   .

.

$$(A - \lambda I)\mathbf{v}_{m-2} = (m - 1)\mathbf{v}_{m-1}$$

$$(A - \lambda I)\mathbf{v}_{m-1} = \mathbf{0}.$$

Starting from the first equation of (2) and successively multiplying by $A - \lambda I$ yields

$$(A - \lambda I)\mathbf{v}_0 = \mathbf{v}_1$$

$$(A) - \lambda I)^2\mathbf{v}_0 = (A - \lambda I)\mathbf{v}_1 = 2\mathbf{v}_2$$

$$(A - \lambda I)^3\mathbf{v}_0 = 2(A - \lambda I)\mathbf{v}_2 = 3 \cdot 2\mathbf{v}_3$$

(3)   .

.

.

$$(A - \lambda I)^{m-1}\mathbf{v}_0 = (m - 1)!\mathbf{v}_{m-1}$$

$$(A - \lambda I)^m\mathbf{v}_0 = (m - 1)!(A - \lambda I)\mathbf{v}_{m-1} = \mathbf{0}.$$

The last equation of (3) tells us where to look for $\mathbf{v}_0$. Once we find $\mathbf{v}_0$, we can use the earlier equations to find the other $\mathbf{v}_j$'s; you should check that

$$\mathbf{v}_j = \frac{1}{j!} (A - \lambda I)^j\mathbf{v}_0.$$

Substitution of these vectors into (1) will yield a solution of $D\mathbf{x} = A\mathbf{x}$.

Fact: *If $(A = \lambda I)^m\mathbf{v} = \mathbf{0}$, then*

$$\mathbf{h}(t) = e^{\lambda t}(\mathbf{v} + t[A - \lambda I]\mathbf{v} + \frac{1}{2} t^2[A - \lambda I]^2\mathbf{v}$$

$$+ \ldots + \frac{1}{(m - 1)!} t^{m-1}[A - \lambda I]^{m-1}\mathbf{v})$$

is a solution of $D\mathbf{x} = A\mathbf{x}$ with initial vector $\mathbf{h}(0) = \mathbf{v}$.

Just as when the multiplicity is 2, we refer to nonzero solutions of $(A - \lambda I)^m \mathbf{v} = \mathbf{0}$ as **generalized eigenvectors** of A corresponding to λ. Independent generalized eigenvectors yield solutions with independent initial vectors.

Example 3.10.1

Solve $D\mathbf{x} = A\mathbf{x}$, where

$$A = \begin{bmatrix} -1 & -1 & 0 \\ 1 & -1 & 1 \\ 0 & 1 & -1 \end{bmatrix}.$$

This represents the three-loop circuit of Example 3.1.3 with $V(t) = 0$, $R_1 = R_2 = 1$, $L_2 = L_3 = 1$, and $C = 1$.

The characteristic polynomial of A is $-(\lambda + 1)^3$, so -1 is a root of multiplicity 3. To find solutions associated to $\lambda = -1$, we first calculate

$$A - (-1)I = \begin{bmatrix} 0 & -1 & 0 \\ 1 & 0 & 1 \\ 0 & 1 & 0 \end{bmatrix}, \quad [A - (-1)I]^2 = \begin{bmatrix} -1 & 0 & -1 \\ 0 & 0 & 0 \\ 1 & 0 & 1 \end{bmatrix}$$

and

$$[A - (-1)I]^3 = \begin{bmatrix} 0 & 0 & 0 \\ 0 & 0 & 0 \\ 0 & 0 & 0 \end{bmatrix}.$$

Every 3-vector satisfies $[A - (-1)I]^3 \mathbf{v} = \mathbf{0}$. In particular, the vectors

$$\mathbf{v} = \begin{bmatrix} 1 \\ 0 \\ 0 \end{bmatrix}, \quad \mathbf{w} = \begin{bmatrix} 0 \\ 1 \\ 0 \end{bmatrix}, \quad \text{and} \quad \mathbf{u} = \begin{bmatrix} 0 \\ 0 \\ 1 \end{bmatrix}$$

are independent generalized eigenvectors. Associated to these vectors are three solutions of $D\mathbf{x} = A\mathbf{x}$ with independent initial vectors:

$$\mathbf{h}_1(t) = e^{-t}(\mathbf{v} + t[A - (-1)I]\mathbf{v} + \frac{1}{2}t^2[A - (-1)I]^2\mathbf{v})$$

$$= e^{-t}\left(\begin{bmatrix} 1 \\ 0 \\ 0 \end{bmatrix} + t\begin{bmatrix} 0 \\ 1 \\ 0 \end{bmatrix} + \frac{1}{2}t^2\begin{bmatrix} -1 \\ 0 \\ 1 \end{bmatrix}\right) = \begin{bmatrix} (1 - \frac{1}{2}t^2)e^{-t} \\ te^{-t} \\ \frac{1}{2}t^2e^{-t} \end{bmatrix}$$

$$\mathbf{h}_2(t) = e^{-t}(\mathbf{w} + t[A - (-1)I]\mathbf{w} + \frac{1}{2}t^2[A - (-1)I]^2\mathbf{w})$$

$$= e^{-t}\left(\begin{bmatrix} 0 \\ 1 \\ 0 \end{bmatrix} + t\begin{bmatrix} -1 \\ 0 \\ 1 \end{bmatrix} + \frac{1}{2}t^2\begin{bmatrix} 0 \\ 0 \\ 0 \end{bmatrix}\right) = \begin{bmatrix} -te^{-t} \\ e^{-t} \\ te^{-t} \end{bmatrix}$$

$$\mathbf{h}_3(t) = e^{-t}(\mathbf{u} + t[A - (-1)I]\mathbf{u} + \frac{1}{2}t^2[A - (-1)I]^2\mathbf{u})$$

$$= e^{-t}\left(\begin{bmatrix} 0 \\ 0 \\ 1 \end{bmatrix} + t\begin{bmatrix} 0 \\ 1 \\ 0 \end{bmatrix} + \frac{1}{2}t^2\begin{bmatrix} -1 \\ 0 \\ 1 \end{bmatrix}\right) = \begin{bmatrix} \frac{1}{2}t^2e^{-t} \\ te^{-t} \\ (1 + \frac{1}{2}t^2)e^{-t} \end{bmatrix}.$$

The general solution of $D\mathbf{x} = A\mathbf{x}$ is

$$\mathbf{x} = c_1\mathbf{h}_1(t) + c_2\mathbf{h}_2(t) + c_3\mathbf{h}_3(t)$$

which, in terms of charge and current, is

$$Q = c_1(1 - \tfrac{1}{2}t^2)e^{-t} - c_2te^{-t} - \tfrac{1}{2}c_3t^2e^{-t}$$

$$I_2 = c_1te^{-t} \qquad\quad + c_2e^{-t} + c_3te^{-t}$$

$$I_3 = \tfrac{1}{2}c_1t^2e^{-t} \qquad\quad + c_2te^{-t} + c_3(1 + \tfrac{1}{2}t^2)e^{-t}.$$

Note that this system of o.d.e.'s is the same as that of Example 3.3.2. However, the generating solutions we have found here are different from those given there. As a result, the two expressions for the general solution look different. They are equivalent, however, in the sense that each list yields every solution exactly once as the parameters c_1, c_2, and c_3 range over all possible real values.

If $\alpha \pm \beta_i$ are complex roots of the characteristic polynomial, then we can work with one of the roots as if it were real to obtain complex solutions of $D\mathbf{x} = A\mathbf{x}$. The real and imaginary parts of these complex solutions are real solutions.

Fact: *Suppose that λ is a complex root of $\det(A - \lambda I)$ and $\mathbf{w}_1, \ldots, \mathbf{w}_j$ are independent complex solutions of $(A - \lambda I)^m\mathbf{w} = \mathbf{0}$. For $i = 1, 2, \ldots, j$ let*

$$\mathbf{h}_i(t) = e^{\lambda t}(\mathbf{w}_i + t[A - \lambda I]\mathbf{w}_i + \frac{1}{2}t^2[A - \lambda I]^2\mathbf{w}_i + \ldots$$

$$+ \frac{1}{(m-1)!}t^{m-1}[A - \lambda I]^{m-1}\mathbf{w}_i)$$

Then Re($\mathbf{h}_1(t)$), Im($\mathbf{h}_1(t)$), . . ., Re($\mathbf{h}_j(t)$), Im($\mathbf{h}_j(t)$) are solutions of $D\mathbf{x} = A\mathbf{x}$ *with independent initial vectors.*

Example 3.10.2

Solve $D\mathbf{x} = A\mathbf{x}$, where

$$A = \begin{bmatrix} 0 & -2 & 0 & 0 \\ 2 & 0 & 4 & 0 \\ 0 & 0 & 0 & -4 \\ 0 & 0 & 1 & 0 \end{bmatrix}.$$

The characteristic polynomial of A is $(\lambda^2 + 4)^2$. The roots are $\pm 2i$, each of which has multiplicity 2.

To find solutions associated to $\lambda = 2i$, we first calculate

$$A - 2iI = \begin{bmatrix} -2i & -2 & 0 & 0 \\ 2 & -2i & 4 & 0 \\ 0 & 0 & -2i & -4 \\ 0 & 0 & 1 & -2i \end{bmatrix}$$

and

$$(A - 2iI)^2 = \begin{bmatrix} -8 & 8i & -8 & 0 \\ -8i & -8 & -16i & -16 \\ 0 & 0 & -8 & 16i \\ 0 & 0 & -4i & -8 \end{bmatrix}.$$

We solve $(A - 2iI)^2\mathbf{w} = \mathbf{0}$ by reducing the augmented matrix:

$$[(A - 2iI)^2|\mathbf{0}] \rightarrow \left[\begin{array}{cccc|c} 1 & -i & 0 & 2i & 0 \\ 0 & 0 & 1 & -2i & 0 \\ 0 & 0 & 0 & 0 & 0 \\ 0 & 0 & 0 & 0 & 0 \end{array} \right].$$

The vectors

$$\mathbf{w}_1 = \begin{bmatrix} i \\ 0 \\ 0 \\ 0 \end{bmatrix} \quad \text{and} \quad \mathbf{w}_2 = \begin{bmatrix} -2i \\ 0 \\ 2i \\ 1 \end{bmatrix}$$

are independent complex generalized eigenvectors. Associated to these vectors are two complex solutions:

$$\mathbf{h}_1(t) = e^{2it}(\mathbf{w}_1 + t[A - 2iI]\mathbf{w}_1)$$

$$= (\cos 2t + i \sin 2t)\left(\begin{bmatrix} i \\ 1 \\ 0 \\ 0 \end{bmatrix} + t \begin{bmatrix} 0 \\ 0 \\ 0 \\ 0 \end{bmatrix}\right)$$

$$= \begin{bmatrix} -\sin 2t \\ \cos 2t \\ 0 \\ 0 \end{bmatrix} + i \begin{bmatrix} \cos 2t \\ \sin 2t \\ 0 \\ 0 \end{bmatrix}$$

$$\mathbf{h}_2(t) = e^{2it}(\mathbf{w}_2 + t[A - 2iI]\mathbf{w}_2)$$

$$= (\cos 2t + i \sin 2t)\left(\begin{bmatrix} -2i \\ 0 \\ 2i \\ 1 \end{bmatrix} + t \begin{bmatrix} -4 \\ 4i \\ 0 \\ 0 \end{bmatrix}\right)$$

$$= \begin{bmatrix} 2 \sin 2t - 4t \cos 2t \\ -4t \sin 2t \\ -2 \sin 2t \\ \cos 2t \end{bmatrix} + i \begin{bmatrix} -2 \cos 2t - 4t \sin 2t \\ 4t \cos 2t \\ 2 \cos 2t \\ \sin 2t \end{bmatrix}.$$

The real and imaginary parts of these solutions provide us with four solutions with independent initial vectors. The general solution of our fourth-order system $D\mathbf{x} = A\mathbf{x}$ is

$$\mathbf{x} = c_1 \begin{bmatrix} -\sin 2t \\ \cos 2t \\ 0 \\ 0 \end{bmatrix} + c_2 \begin{bmatrix} \cos 2t \\ \sin 2t \\ 0 \\ 0 \end{bmatrix} + c_3 \begin{bmatrix} 2 \sin 2t - 4t \cos 2t \\ -4t \sin 2t \\ -2 \sin 2t \\ \cos 2t \end{bmatrix}$$

$$+ c_4 \begin{bmatrix} -2 \cos 2t - 4t \sin 2t \\ 4t \cos 2t \\ 2 \cos 2t \\ \sin 2t \end{bmatrix}.$$

As usual, if we have a system with more than one eigenvalue, then we work with each individually and combine the resulting solutions.

Fact: *The solutions of $D\mathbf{x} = A\mathbf{x}$ that we get by combining solutions associated to various eigenvalues will have independent initial vectors as long as the solutions associated to each single eigenvalue or pair of complex eigenvalues have independent initial vectors.*

Example 3.10.3

Solve $D\mathbf{x} = A\mathbf{x}$, where

$$A = \begin{bmatrix} 2 & 0 & 0 & 0 \\ 1 & 2 & 0 & 0 \\ 0 & 0 & 2 & 1 \\ 0 & 0 & 0 & 4 \end{bmatrix}.$$

The characteristic polynomial is $(2 - \lambda)^3(4 - \lambda)$. The root $\lambda = 2$ has multiplicity 3, while $\lambda = 4$ has multiplicity 1.

We calculated the first three powers of $A - 2I$ in Example 3.9.4:

$$A - 2I = \begin{bmatrix} 0 & 0 & 0 & 0 \\ 1 & 0 & 0 & 0 \\ 0 & 0 & 0 & 1 \\ 0 & 0 & 0 & 2 \end{bmatrix}, \quad (A - 2I)^2 = \begin{bmatrix} 0 & 0 & 0 & 0 \\ 0 & 0 & 0 & 0 \\ 0 & 0 & 0 & 2 \\ 0 & 0 & 0 & 4 \end{bmatrix},$$

$$(A - 2I)^3 = \begin{bmatrix} 0 & 0 & 0 & 0 \\ 0 & 0 & 0 & 0 \\ 0 & 0 & 0 & 4 \\ 0 & 0 & 0 & 8 \end{bmatrix}.$$

We reduce $[(A - 2I)^3 | \mathbf{0}]$

$$[(A - 2I)^3 | \mathbf{0}] \rightarrow \begin{bmatrix} 0 & 0 & 0 & 0 & 0 \\ 0 & 0 & 0 & 0 & 0 \\ 0 & 0 & 0 & 0 & 0 \\ 0 & 0 & 0 & 1 & 0 \end{bmatrix}$$

and read off three independent generalized eigenvectors,

$$\begin{bmatrix} 1 \\ 0 \\ 0 \\ 0 \end{bmatrix}, \quad \begin{bmatrix} 0 \\ 1 \\ 0 \\ 0 \end{bmatrix}, \quad \text{and} \quad \begin{bmatrix} 0 \\ 0 \\ 1 \\ 0 \end{bmatrix}.$$

The associated solutions are

$$\mathbf{h}_1(t) = e^{2t} \left(\begin{bmatrix} 1 \\ 0 \\ 0 \\ 0 \end{bmatrix} + t \begin{bmatrix} 0 \\ 1 \\ 0 \\ 0 \end{bmatrix} + \frac{1}{2} t^2 \begin{bmatrix} 0 \\ 0 \\ 0 \\ 0 \end{bmatrix} \right) = \begin{bmatrix} e^{2t} \\ te^{2t} \\ 0 \\ 0 \end{bmatrix}$$

$$\mathbf{h}_2(t) = e^{2t} \left(\begin{bmatrix} 0 \\ 1 \\ 0 \\ 0 \end{bmatrix} + t \begin{bmatrix} 0 \\ 0 \\ 0 \\ 0 \end{bmatrix} + \frac{1}{2} t^2 \begin{bmatrix} 0 \\ 0 \\ 0 \\ 0 \end{bmatrix} \right) = \begin{bmatrix} 0 \\ e^{2t} \\ 0 \\ 0 \end{bmatrix}$$

$$\mathbf{h}_3(t) = e^{2t} \left(\begin{bmatrix} 0 \\ 0 \\ 1 \\ 0 \end{bmatrix} + t \begin{bmatrix} 0 \\ 0 \\ 0 \\ 0 \end{bmatrix} + \frac{1}{2} t^2 \begin{bmatrix} 0 \\ 0 \\ 0 \\ 0 \end{bmatrix} \right) = \begin{bmatrix} 0 \\ 0 \\ e^{2t} \\ 0 \end{bmatrix}.$$

To find a solution associated to $\lambda = 4$, we reduce $[A - 4I|\mathbf{0}]$:

$$\begin{bmatrix} -2 & 0 & 0 & 0 & | & 0 \\ 1 & -2 & 0 & 0 & | & 0 \\ 0 & 0 & -2 & 1 & | & 0 \\ 0 & 0 & 0 & 0 & | & 0 \end{bmatrix} \rightarrow \begin{bmatrix} 1 & 0 & 0 & 0 & | & 0 \\ 0 & 1 & 0 & 0 & | & 0 \\ 0 & 0 & 1 & 0 & | & 0 \\ 0 & 0 & 0 & -\frac{1}{2} & | & 0 \end{bmatrix}.$$

The vector

$$\begin{bmatrix} 0 \\ 0 \\ 1 \\ 2 \end{bmatrix}$$

is an eigenvector with associated solution

$$\mathbf{h}_4(t) = e^{4t} \begin{bmatrix} 0 \\ 0 \\ 1 \\ 2 \end{bmatrix} = \begin{bmatrix} 0 \\ 0 \\ e^{4t} \\ 2e^{4t} \end{bmatrix}.$$

If we combine the solutions associated to $\lambda = 2$ and $\lambda = 4$, we obtain four solutions with independent initial vectors. Since $D\mathbf{x} = A\mathbf{x}$ has order four, the general solution is

$$\mathbf{x} = c_1\mathbf{h}_1(t) + c_2\mathbf{h}_2(t) + c_3\mathbf{h}_3(t) + c_4\mathbf{h}_4(t)$$

$$= c_1 \begin{bmatrix} e^{2t} \\ te^{2t} \\ 0 \\ 0 \end{bmatrix} + c_2 \begin{bmatrix} 0 \\ e^{2t} \\ 0 \\ 0 \end{bmatrix} + c_2 \begin{bmatrix} 0 \\ 0 \\ e^{2t} \\ 0 \end{bmatrix} + c_4 \begin{bmatrix} 0 \\ 0 \\ e^{4t} \\ 2e^{4t} \end{bmatrix}.$$

Example 3.10.4

Solve $D\mathbf{x} = A\mathbf{x}$, where

$$A = \begin{bmatrix} 1 & 0 & 0 & 0 & 0 & 0 \\ 0 & 1 & 0 & 0 & 1 & 0 \\ 0 & 0 & 0 & 0 & 0 & 1 \\ 1 & 0 & 0 & 1 & 0 & 0 \\ 0 & 0 & 0 & 1 & 1 & 0 \\ 0 & 0 & 0 & 0 & 0 & 0 \end{bmatrix}.$$

The characteristic polynomial is $(1 - \lambda)^4\lambda^2$. The root $\lambda = 1$ has multiplicity 4 and $\lambda = 0$ has multiplicity 2.

The first four powers of $A - I$ are

$$(A - I) = \begin{bmatrix} 0 & 0 & 0 & 0 & 0 & 0 \\ 0 & 0 & 0 & 0 & 1 & 0 \\ 0 & 0 & -1 & 0 & 0 & 1 \\ 1 & 0 & 0 & 0 & 0 & 0 \\ 0 & 0 & 0 & 1 & 0 & 0 \\ 0 & 0 & 0 & 0 & 0 & -1 \end{bmatrix},$$

$$(A - I)^2 = \begin{bmatrix} 0 & 0 & 0 & 0 & 0 & 0 \\ 0 & 0 & 0 & 1 & 0 & 0 \\ 0 & 0 & 1 & 0 & 0 & -2 \\ 0 & 0 & 0 & 0 & 0 & 0 \\ 1 & 0 & 0 & 0 & 0 & 0 \\ 0 & 0 & 0 & 0 & 0 & 1 \end{bmatrix},$$

$$(A - I)^3 = \begin{bmatrix} 0 & 0 & 0 & 0 & 0 & 0 \\ 1 & 0 & 0 & 0 & 0 & 0 \\ 0 & 0 & -1 & 0 & 0 & 3 \\ 0 & 0 & 0 & 0 & 0 & 0 \\ 0 & 0 & 0 & 0 & 0 & 0 \\ 0 & 0 & 0 & 0 & 0 & -1 \end{bmatrix},$$

$$(A - I)^4 = \begin{bmatrix} 0 & 0 & 0 & 0 & 0 & 0 \\ 0 & 0 & 0 & 0 & 0 & 0 \\ 0 & 0 & 1 & 0 & 0 & -4 \\ 0 & 0 & 0 & 0 & 0 & 0 \\ 0 & 0 & 0 & 0 & 0 & 0 \\ 0 & 0 & 0 & 0 & 0 & 1 \end{bmatrix}.$$

The vectors

$$\begin{bmatrix} 1 \\ 0 \\ 0 \\ 0 \\ 0 \\ 0 \end{bmatrix}, \begin{bmatrix} 0 \\ 1 \\ 0 \\ 0 \\ 0 \\ 0 \end{bmatrix}, \begin{bmatrix} 0 \\ 0 \\ 0 \\ 1 \\ 0 \\ 0 \end{bmatrix}, \begin{bmatrix} 0 \\ 0 \\ 0 \\ 0 \\ 1 \\ 0 \end{bmatrix}$$

are independent solutions of $(A - I)^4\mathbf{v} = \mathbf{0}$. The solution associated to the first of these generalized eigenvectors is

$$\mathbf{h}_1(t) = e^t \left(\begin{bmatrix} 1 \\ 0 \\ 0 \\ 0 \\ 0 \\ 0 \end{bmatrix} + t \begin{bmatrix} 0 \\ 0 \\ 0 \\ 1 \\ 0 \\ 0 \end{bmatrix} + \frac{1}{2}t^2 \begin{bmatrix} 0 \\ 0 \\ 0 \\ 0 \\ 1 \\ 0 \end{bmatrix} + \frac{1}{3 \cdot 2}t^3 \begin{bmatrix} 0 \\ 1 \\ 0 \\ 0 \\ 0 \\ 0 \end{bmatrix} \right) = \begin{bmatrix} e^t \\ \frac{1}{6}t^3e^t \\ 0 \\ te^t \\ \frac{1}{2}t^2e^t \\ 0 \end{bmatrix}.$$

We leave it to you to check that the solutions associated to the other generalized eigenvectors are

$$\mathbf{h}_2(t) = \begin{bmatrix} 0 \\ e^t \\ 0 \\ 0 \\ 0 \\ 0 \end{bmatrix}, \quad \mathbf{h}_3(t) = \begin{bmatrix} 0 \\ \frac{1}{2}t^2e^t \\ 0 \\ e^t \\ te^t \\ 0 \end{bmatrix}, \quad \mathbf{h}_4(t) = \begin{bmatrix} 0 \\ te^t \\ 0 \\ 0 \\ e^t \\ 0 \end{bmatrix}.$$

To find solutions associated to $\lambda = 0$, we calculate

$$(A - 0I)^2 = A^2 = \begin{bmatrix} 1 & 0 & 0 & 0 & 0 & 0 \\ 0 & 1 & 0 & 1 & 2 & 0 \\ 0 & 0 & 0 & 0 & 0 & 0 \\ 2 & 0 & 0 & 1 & 0 & 0 \\ 1 & 0 & 0 & 2 & 1 & 0 \\ 0 & 0 & 0 & 0 & 0 & 0 \end{bmatrix}$$

and reduce $[A^2|\mathbf{0}]$:

$$[A^2|\mathbf{0}] \rightarrow \left[\begin{array}{cccccc|c} 1 & 0 & 0 & 0 & 0 & 0 & 0 \\ 0 & 1 & 0 & 0 & 0 & 0 & 0 \\ 0 & 0 & 0 & 1 & 0 & 0 & 0 \\ 0 & 0 & 0 & 0 & 1 & 0 & 0 \\ 0 & 0 & 0 & 0 & 0 & 0 & 0 \\ 0 & 0 & 0 & 0 & 0 & 0 & 0 \end{array} \right].$$

The vectors

$$
\begin{bmatrix} 0 \\ 0 \\ 1 \\ 0 \\ 0 \\ 0 \end{bmatrix}
\quad \text{and} \quad
\begin{bmatrix} 0 \\ 0 \\ 0 \\ 0 \\ 0 \\ 1 \end{bmatrix}
$$

are generalized eigenvectors with associated solutions (remember $e^{0t} = 1$)

$$
\mathbf{h}_5(t) = \begin{bmatrix} 0 \\ 0 \\ 1 \\ 0 \\ 0 \\ 0 \end{bmatrix}
\quad \text{and} \quad
\mathbf{h}_6(t) = \begin{bmatrix} 0 \\ 0 \\ t \\ 0 \\ 0 \\ 1 \end{bmatrix}.
$$

The general solution of $D\mathbf{x} = A\mathbf{x}$ is

$$
\mathbf{x} = c_1 \mathbf{h}_1(t) + c_2 \mathbf{h}_2(t) + c_3 \mathbf{h}_3(t) + c_4 \mathbf{h}_4(t) + c_5 \mathbf{h}_5(t) + c_6 \mathbf{h}_6(t)
$$

$$
= c_1 \begin{bmatrix} e^t \\ \frac{1}{6}t^3 e^t \\ 0 \\ te^t \\ \frac{1}{2}t^2 e^t \\ 0 \end{bmatrix}
+ c_2 \begin{bmatrix} 0 \\ e^t \\ 0 \\ 0 \\ 0 \\ 0 \end{bmatrix}
+ c_3 \begin{bmatrix} 0 \\ \frac{1}{2}t^2 e^t \\ 0 \\ e^t \\ te^t \\ 0 \end{bmatrix}
+ c_4 \begin{bmatrix} 0 \\ te^t \\ 0 \\ 0 \\ e^t \\ 0 \end{bmatrix}
+ c_5 \begin{bmatrix} 0 \\ 0 \\ 1 \\ 0 \\ 0 \\ 0 \end{bmatrix}
+ c_6 \begin{bmatrix} 0 \\ 0 \\ t \\ 0 \\ 0 \\ 1 \end{bmatrix}.
$$

We can now find the general solution of $D\mathbf{x} = A\mathbf{x}$ provided that we can find m independent generalized eigenvectors corresponding to each root of $\det(A - \lambda I)$ with multiplicity m. The following result from linear algebra guarantees that this is always possible.

Theorem: *If λ is a root of $\det(A - \lambda I)$ with multiplicity m, then the equation $(A - \lambda I)^m \mathbf{v} = \mathbf{0}$ has m linearly independent solutions for $\mathbf{v}$.*

HOMOGENEOUS SYSTEMS WITH CONSTANT COEFFICIENTS

To solve $D\mathbf{x} = A\mathbf{x}$ where A is an $n \times n$ matrix:

1. Find the roots of the characteristic polynomial $\det(A - \lambda I)$ and their multiplicities.

2. For each real root λ of multiplicity m, find m linearly independent solutions of $(A - \lambda I)^m \mathbf{v} = \mathbf{0}$. Associate to each of these **generalized eigenvectors** the solution of $D\mathbf{x} = A\mathbf{x}$

$$\mathbf{h}(t) = e^{\lambda t}(\mathbf{v} + t[A - \lambda I]\mathbf{v} + \frac{1}{2} t^2[A - \lambda I]^2\mathbf{v}$$

$$+ \ldots + \frac{1}{(m - 1)!} t^{m-1}[A - \lambda I]^{m-1}\mathbf{v}).$$

3. For each pair of complex roots $\alpha \pm \beta i$, work with one of the roots as if it were real to obtain complex solutions. Associate to the pair of eigenvalues the real and imaginary parts of these complex solutions.

4. Combine the solutions associated to the various roots to obtain n-vector valued functions $\mathbf{h}_1(t), \ldots, \mathbf{h}_n(t)$ that generate the general solution

$$\mathbf{x} = c_1\mathbf{h}_1(t) + \ldots + c_n\mathbf{h}_n(t).$$

Note

On finding generalized eigenvectors

If λ is a root of $\det(A - \lambda I)$ with multiplicity m, then we can always find m independent generalized eigenvectors by calculating $(A - \lambda I)^m$ and reducing $[(A - \lambda I)^m \mid \mathbf{0}]$. If A is a large matrix and λ is a root of high multiplicity, then this may be a messy calculation. In this case, we can try to find the generalized eigenvectors without calculating the mth power of $A - \lambda I$.

We've already seen (in Section 3.7) that if there are m independent eigenvectors, the associated solutions can be obtained without calculating any powers of $A - \lambda I$ at all. If $A - \lambda I$ has j independent eigenvectors, where $1 < j < m$, then the following theorem tells us that we can stop short of calculating the mth power of $A - \lambda I$.

Theorem: *Suppose λ is a root of $\det(A - \lambda I)$ with multiplicity m and that there are j independent eigenvectors corresponding to λ. Then every generalized eigenvector corresponding to λ is a solution of the equation*

$$(A - \lambda I)^{m-j+1}\mathbf{v} = \mathbf{0}.$$

Note that if the number j of independent eigenvectors is relatively high, this theorem tells us that to find the generalized eigenvectors we need only calculate a relatively low power of $A - \lambda I$. If, on the other hand, $j = 1$, the theorem tells us that every generalized eigenvector is a solution of $(A - \lambda I)^m\mathbf{v} = \mathbf{0}$; in this case, there is nothing gained by first finding an eigenvector.

Exercise 14 describes another approach that is effective in case the number of independent eigenvectors is relatively high. Exercises 15 and 16 are useful when dealing with systems for which this number is relatively low.

EXERCISES

In Exercises 1 to 10, find the general solution of $D\mathbf{x} = A\mathbf{x}$.

1. $A = \begin{bmatrix} 0 & 1 & 0 \\ 0 & 0 & 1 \\ 1 & -3 & 3 \end{bmatrix}$

2. $A = \begin{bmatrix} 1 & 0 & 0 \\ 1 & 0 & 1 \\ 1 & -1 & 2 \end{bmatrix}$

3. $A = \begin{bmatrix} 2 & 0 & -1 & 1 \\ 0 & 0 & -1 & 0 \\ 0 & -1 & 0 & 0 \\ -1 & -1 & 0 & 0 \end{bmatrix}$

4. $A = \begin{bmatrix} 1 & 1 & 0 & 0 \\ 0 & 1 & 1 & 0 \\ 0 & 0 & 1 & -1 \\ 0 & 0 & 0 & 2 \end{bmatrix}$

5. $A = \begin{bmatrix} 0 & 2 & 0 & 0 \\ -2 & 0 & 1 & 0 \\ 0 & 0 & 0 & 2 \\ 0 & 0 & -2 & 0 \end{bmatrix}$

6. $A = \begin{bmatrix} 2 & 1 & 0 & 0 \\ 0 & 1 & 1 & 0 \\ -1 & -1 & 0 & 0 \\ 0 & 0 & 0 & 1 \end{bmatrix}$

7. $A = \begin{bmatrix} 2 & 0 & 0 & 0 \\ 0 & 2 & 0 & 0 \\ 1 & 0 & 1 & 1 \\ 1 & 0 & -1 & 3 \end{bmatrix}$

8. $A = \begin{bmatrix} 1 & 1 & 0 & 0 \\ -1 & 1 & 1 & 0 \\ 0 & 0 & 1 & -1 \\ 0 & 0 & 1 & 1 \end{bmatrix}$

9. $A = \begin{bmatrix} 0 & 0 & 0 & 0 & 1 \\ 1 & -1 & 0 & -1 & 1 \\ 1 & 0 & -1 & -1 & 0 \\ 0 & 0 & 0 & 0 & 1 \\ 1 & 0 & 0 & -1 & 0 \end{bmatrix}$

10. $A = \begin{bmatrix} 2 & 1 & 0 & 0 & 0 & 0 \\ 0 & 2 & 1 & 0 & 0 & 0 \\ 0 & 0 & 2 & 1 & 0 & 0 \\ 0 & 0 & 0 & 2 & 0 & 0 \\ 0 & 0 & 0 & 0 & 0 & 1 \\ 0 & 0 & 0 & 0 & -2 & -2 \end{bmatrix}$

In Exercises 11 and 12, find (a) the general solution and (b) the specific solution satisfying the given initial condition.

11. $\begin{aligned} x_1' &= -x_1 + x_2 \\ x_2' &= \quad\ -x_2 \\ x_3' &= \qquad\qquad -x_3 + x_4 \\ x_4' &= \qquad\qquad\quad -x_4 \end{aligned}$ $x_1(0) = x_2(0) = 1, x_3(0) = -1, x_4(0) = 0$

12. $\begin{aligned} x_1' &= -x_1 + x_2 \\ x_2' &= \quad\ -x_2 + x_3 \\ x_3' &= \qquad\qquad -x_3 + z_4 \\ x_4' &= \qquad\qquad\quad -x_4 \end{aligned}$ $x_1(0) = 1, x_2(0) = x_3(0) = 0, x_4(0) = -1$

Some more abstract problems:

13. Suppose λ is a root of $\det(A - \lambda I)$ with multiplicity m and that $\mathbf{v}$ is a solution of $(A - \lambda I)^k \mathbf{v} = \mathbf{0}$ for some $k \leq m$. Show that $\mathbf{v}$ is a generalized eigenvector for A with associated solution

$$\mathbf{h}(t) = e^{\lambda t}(\mathbf{v} + t[A - \lambda I]\mathbf{v} + \frac{1}{2} t^2[A - \lambda I]^2\mathbf{v} + \dots + \frac{1}{(k-1)!} t^{k-1}[A - \lambda I]^{k-1}\mathbf{v})$$

Note, as a special case, that any eigenvector $\mathbf{v}$ is a generalized eigenvector with associated solution $\mathbf{h}(t) = e^{\lambda t}\mathbf{v}$.

14. If $\mathbf{v}$ is a solution of $(A - \lambda I)^k\mathbf{v} = \mathbf{0}$ for a low power k, then the associated solution of $D\mathbf{x} = A\mathbf{x}$ is easy to calculate (see Exercise 13). One approach to finding generalized eigenvectors is to start by finding as many independent eigenvectors as possible. If this is not enough, then we combine these eigenvectors with solutions of $(A - \lambda I)^2\mathbf{v} = \mathbf{0}$, making sure that the expanded list is independent. We continue this way, solving $(A - \lambda I)^k\mathbf{v} = \mathbf{0}$ for $k = 3, 4$, etc., until we have found enough independent generalized eigenvectors. Use this approach to solve the system of the indicated exercise.

    a. Exercise 2   b. Exercise 3   c. Exercise 7

15. Suppose λ is a root of $\det(A - \lambda I)$ with multiplicity m and that there is an integer $k \le m$ so that

$$(A - \lambda I)^k\mathbf{v} = \mathbf{0} \text{ but } (A - \lambda I)^{k-1}\mathbf{v} \ne \mathbf{0}$$

Then we refer to the vectors

$$\mathbf{w}_1 = \mathbf{v}, \mathbf{w}_2 = (A - \lambda I)\mathbf{v}, \ldots, \mathbf{w}_k = (A - \lambda I)^{k-1}\mathbf{v}$$

as a **string of generalized eigenvectors**.

    a. Show that $\mathbf{w}_1, \ldots, \mathbf{w}_k$ are indeed generalized eigenvectors and that the associated solutions are

$$\mathbf{h}_1(t) = e^{\lambda t}(\mathbf{w}_1 + t\mathbf{w}_2 + \ldots + \frac{1}{(k-2)!} t^{k-2}\mathbf{w}_{k-1} + \frac{1}{(k-1)!} t^{k-1}\mathbf{w}_k)$$

$$\mathbf{h}_2(t) = e^{\lambda t}(\mathbf{w}_2 + t\mathbf{w}_3 + \ldots + \frac{1}{(k-2)!} t^{k-2}\mathbf{w}_k)$$

$$\cdot$$
$$\cdot$$
$$\cdot$$

$$\mathbf{h}_k(t) = e^{\lambda t}\mathbf{w}_k.$$

    b. Show that $\mathbf{w}_1, \ldots, \mathbf{w}_k$ are independent as follows. Suppose $c_1\mathbf{w}_1 + \ldots + c_k\mathbf{w}_k = \mathbf{0}$. Multiply this equation by $(A - \lambda I)^{k-1}$ to obtain the equation $c_1\mathbf{w}_1 = \mathbf{0}$ from which you can conclude $c_1 = 0$. Now multiply the original equation by $(A - \lambda I)^{k-2}$ to show $c_2 = 0$, etc.

16. A theorem from linear algebra states that *if λ is a root of $\det(A - \lambda I)$ of multiplicity m, and if every eigenvector corresponding to λ is a multiple of one fixed eigenvector, then it is possible to find a single string of m generalized eigenvectors* (see Exercise 15). The solutions associated to this string are all determined once we have calculated the vectors in the string. Use this fact to obtain the general solution of the system in the indicated exercise.

    a. Exercise 4   b. Exercise 9   c. Exercise 10

3.11 NONHOMOGENEOUS SYSTEMS

We turn now to the problem of solving a nonhomogeneous system

$$(S) \qquad\qquad D\mathbf{x} = A\mathbf{x} + \mathbf{E}(t).$$

Recall that the general solution of (S) will be of the form

$$\mathbf{x} = \mathbf{H}(t) + \mathbf{p}(t)$$

where $\mathbf{x} = \mathbf{H}(t)$ is the general solution of the related homogeneous system

$$(H) \qquad\qquad D\mathbf{x} = A\mathbf{x}$$

and $\mathbf{x} = \mathbf{p}(t)$ is a particular solution of (S). Our technique of solution will rely on having $\mathbf{H}(t)$. We will find a particular solution by extending to systems the method of Variation of Parameters that we used in Section 2.10.

If A is an $n \times n$ matrix, then the general solution of (H) is of the form $\mathbf{x} = \mathbf{H}(t)$ where

$$\mathbf{H}(t) = c_1 \mathbf{h}_1(t) + \ldots + c_n \mathbf{h}_n(t)$$

for a suitable choice of $\mathbf{h}_1(t), \ldots, \mathbf{h}_n(t)$. By analogy with what we did in Section 2.10, we look for a solution $\mathbf{x} = \mathbf{p}(t)$ with

$$\mathbf{p}(t) = c_1(t)\mathbf{h}_1(t) + \ldots + c_n(t)\mathbf{h}_n(t).$$

When we substitute this formula into (S), the left side, $D\mathbf{p}(t)$, is a sum of terms of the form

$$D(c_i(t)\mathbf{h}_i(t)) = c_i'(t)\mathbf{h}_i(t) + c_i(t)\mathbf{h}_i'(t).$$

The first part of the right-hand side, $A\mathbf{p}(t)$, is a sum of terms of the form

$$A(c_i(t)\mathbf{h}_i(t)) = c_i(t)A\mathbf{h}_i(t).$$

Since $\mathbf{h}_i(t)$ solves (H), we know that $A\mathbf{h}_i(t) = \mathbf{h}_i'(t)$. Then

$$A(c_i(t)\mathbf{h}_i(t)) = c_i(t)\mathbf{h}_i'(t).$$

These terms appear on both sides. Cancellation leaves

$$(V) \qquad\qquad c_1'(t)\mathbf{h}_1(t) + \ldots + c_n'(t)\mathbf{h}_n(t) = \mathbf{E}(t).$$

Thus we have the following:

Fact: *Let* $\mathbf{x} = c_1\mathbf{h}_1(t) + \ldots + c_n\mathbf{h}_n(t)$ *be the general solution of (H). If* $c_1'(t), \ldots, c_n'(t)$ *satisfy*

(V) $$c_1'(t)\mathbf{h}_1(t) + \ldots + c_n'(t)\mathbf{h}_n(t) = \mathbf{E}(t)$$

then $\mathbf{p}(t) = c_1(t)\mathbf{h}_1(t) + \ldots + c_n(t)\mathbf{h}_n(t)$ *is a solution of (S).*

If we equate corresponding entries of the left and right sides of (V), we obtain an algebraic system of n equations in the n unknowns $c_1'(t), \ldots, c_n'(t)$. The matrix of coefficients of this system is just the matrix whose columns are $\mathbf{h}_1(t), \ldots, \mathbf{h}_n(t)$. The determinant of this matrix is the Wronskian of these functions, $W(t) = W[\mathbf{h}_1, \ldots, \mathbf{h}_n](t)$. Since $\mathbf{h}_1(t), \ldots, \mathbf{h}_n(t)$ generate the general solution of (H), $W(t_0) \neq 0$ for all t_0. Cramer's determinant test tells us we can always solve the system (V)—Cramer's rule gives a formula for the solution. Once we find $c_1'(t), \ldots, c_n'(t)$, we can integrate to find $c_1(t), \ldots, c_n(t)$. These in turn determine $\mathbf{p}(t)$.

Cramer's rule is one way of solving (V). We can also find the solution by reducing the augmented matrix of the system. This augmented matrix will be $[\Phi(t)|\mathbf{E}(t)]$, where $\Phi(t)$ is the matrix whose columns are $\mathbf{h}_1(t), \ldots, \mathbf{h}_n(t)$.

Example 3.11.1
    Solve

(S) $$D\mathbf{x} = A\mathbf{x} + \mathbf{E}(t)$$

where

$$A = \begin{bmatrix} 1 & 0 & 4 \\ 0 & 3 & 0 \\ 1 & 0 & 1 \end{bmatrix} \quad \text{and} \quad \mathbf{E}(t) = \begin{bmatrix} -2e^t \\ 9t \\ e^t \end{bmatrix}.$$

We found the general solution of the related homogeneous system in Example 3.7.1. It is $\mathbf{x} = \mathbf{H}(t)$ where

$$\mathbf{H}(t) = c_1 \begin{bmatrix} -2e^{-t} \\ 0 \\ e^{-t} \end{bmatrix} + c_2 \begin{bmatrix} 0 \\ e^{3t} \\ 0 \end{bmatrix} + c_3 \begin{bmatrix} 2e^{3t} \\ 0 \\ e^{3t} \end{bmatrix}.$$

We look for a particular solution of (S) of the form $\mathbf{x} = \mathbf{p}(t)$ with

$$\mathbf{p}(t) = c_1(t) \begin{bmatrix} -2e^{-t} \\ 0 \\ e^{-t} \end{bmatrix} + c_2(t) \begin{bmatrix} 0 \\ e^{3t} \\ 0 \end{bmatrix} + c_3(t) \begin{bmatrix} 2e^{3t} \\ 0 \\ e^{3t} \end{bmatrix}.$$

To find functions $c_1(t)$, $c_2(t)$, and $c_3(t)$ that work, we first solve

(V)
$$c_1'(t) \begin{bmatrix} -2e^{-t} \\ 0 \\ e^{-t} \end{bmatrix} + c_2'(t) \begin{bmatrix} 0 \\ e^{3t} \\ 0 \end{bmatrix} + c_3'(t) \begin{bmatrix} 2e^{3t} \\ 0 \\ e^{3t} \end{bmatrix} = \begin{bmatrix} -2e^t \\ 9t \\ e^t \end{bmatrix}$$

for $c_1'(t)$, $c_2'(t)$, and $c_3'(t)$, by reducing the augmented matrix:

$$\begin{bmatrix} -2e^{-t} & 0 & 2e^{3t} & -2e^t \\ 0 & e^{3t} & 0 & 9t \\ e^{-t} & 0 & e^{3t} & e^t \end{bmatrix} \xrightarrow{R_1 \to R_1 + 2R_3} \begin{bmatrix} 0 & 0 & 4e^{3t} & 0 \\ 0 & e^{3t} & 0 & 9t \\ e^{-t} & 0 & e^{3t} & e^t \end{bmatrix}$$

$$\xrightarrow{R_3 \to R_3 - \frac{1}{4}R_1} \begin{bmatrix} 0 & 0 & 4e^{3t} & 0 \\ 0 & e^{3t} & 0 & 9t \\ e^{-t} & 0 & 0 & e^t \end{bmatrix}$$

$$\xrightarrow[\substack{R_1 \to (e^{-3t}/4)R_1 \\ R_2 \to e^{-3t}R_2 \\ R_3 \to e^t R_3}]{} \begin{bmatrix} 0 & 0 & 1 & 0 \\ 0 & 1 & 0 & 9te^{-3t} \\ 1 & 0 & 0 & e^{2t} \end{bmatrix} \xrightarrow{\text{rearr}} \begin{bmatrix} 1 & 0 & 0 & e^{2t} \\ 0 & 1 & 0 & 9te^{-3t} \\ 0 & 0 & 1 & 0 \end{bmatrix}.$$

The solution of the algebraic system corresponding to the reduced matrix is

$$c_1'(t) = e^{2t}, \qquad c_2'(t) = 9te^{-3t}, \qquad c_3'(t) = 0.$$

Any functions with these derivatives will yield a particular solution. We can take

$$c_1(t) = e^{2t}/2, \qquad c_2(t) = -3te^{-3t} - e^{-3t}, \qquad c_3(t) = 0$$

to obtain the particular solution

$$\mathbf{p}(t) = (e^{2t}/2) \begin{bmatrix} -2e^{-t} \\ 0 \\ e^{-t} \end{bmatrix} + (-3te^{-3t} - e^{-3t}) \begin{bmatrix} 0 \\ e^{3t} \\ 0 \end{bmatrix} = \begin{bmatrix} -e^t \\ -3t - 1 \\ e^t/2 \end{bmatrix}.$$

The general solution of (S) is

$$\mathbf{x} = \mathbf{H}(t) + \mathbf{p}(t) = c_1 \begin{bmatrix} -2e^{-t} \\ 0 \\ e^{-t} \end{bmatrix} + c_2 \begin{bmatrix} 0 \\ e^{3t} \\ 0 \end{bmatrix} + c_3 \begin{bmatrix} 2e^{3t} \\ 0 \\ e^{3t} \end{bmatrix} + \begin{bmatrix} -e^t \\ -3t - 1 \\ e^t/2 \end{bmatrix}.$$

Example 3.11.2

Solve

(S) $$D\mathbf{x} = A\mathbf{x} + \mathbf{E}(t)$$

where

$$A = \begin{bmatrix} -1 & -1 \\ 4 & -1 \end{bmatrix} \quad \text{and} \quad \mathbf{E}(t) = \begin{bmatrix} 2e^{-t} \\ 0 \end{bmatrix}.$$

This system describes the two-loop circuit of Example 3.1.2 with $C = 1/4$, $L = 1$, $R_1 = 4$, $R_2 = 1$, and a varying voltage source with $V(t) = 8e^{-t}$.

We solved the related homogeneous system, corresponding to no voltage source, in Example 3.8.1. The general solution was $\mathbf{x} = \mathbf{H}(t)$ where

$$\mathbf{H}(t) = c_1 \begin{bmatrix} -e^{-t} \sin 2t \\ 2e^{-t} \cos 2t \end{bmatrix} + c_2 \begin{bmatrix} e^{-t} \cos 2t \\ 2e^{-t} \sin 2t \end{bmatrix}.$$

We look for a particular solution of (S) in the form

$$\mathbf{p}(t) = c_1(t) \begin{bmatrix} -e^{-t} \sin 2t \\ 2e^{-t} \cos 2t \end{bmatrix} + c_2(t) \begin{bmatrix} e^{-t} \cos 2t \\ 2e^{-t} \sin 2t \end{bmatrix}.$$

To find functions $c_1(t)$ and $c_2(t)$ that work, we solve

(V) $$c_1'(t) \begin{bmatrix} -e^{-t} \sin 2t \\ 2e^{-t} \cos 2t \end{bmatrix} + c_2'(t) \begin{bmatrix} e^{-t} \cos 2t \\ 2e^{-t} \sin 2t \end{bmatrix} = \begin{bmatrix} 2e^{-t} \\ 0 \end{bmatrix}$$

for $c_1'(t)$ and $c_2'(t)$, using Cramer's rule:

$$c_1'(t) = \det \begin{bmatrix} 2e^{-t} & e^{-t} \cos 2t \\ 0 & 2e^{-t} \sin 2t \end{bmatrix} \Big/ \det \begin{bmatrix} -e^{-t} \sin 2t & e^{-t} \cos 2t \\ 2e^{-t} \cos 2t & 2e^{-t} \sin 2t \end{bmatrix}.$$

$$= 4e^{-2t} \sin 2t / (-2e^{-2t} \sin^2 2t - 2e^{-2t} \cos^2 2t)$$

$$= 4e^{-2t} \sin 2t / (-2e^{-2t}) = -2 \sin 2t$$

$$c_2'(t) = \det \begin{bmatrix} -e^{-t} \sin 2t & 2e^{-t} \\ 2e^{-t} \cos 2t & 0 \end{bmatrix} \Big/ \det \begin{bmatrix} -e^{-t} \sin 2t & e^{-t} \cos 2t \\ 2e^{-t} \cos 2t & 2e^{-t} \sin 2t \end{bmatrix}$$

$$= -4e^{-2t} \cos 2t / (-2e^{-2t}) = 2 \cos 2t.$$

We can take

$$c_1(t) = \cos 2t \quad \text{and} \quad c_2(t) = \sin 2t$$

to obtain the particular solution

$$\mathbf{p}(t) = \cos 2t \begin{bmatrix} -e^{-t} \sin 2t \\ 2e^{-t} \cos 2t \end{bmatrix} + \sin 2t \begin{bmatrix} e^{-t} \cos 2t \\ 2e^{-t} \sin 2t \end{bmatrix} = \begin{bmatrix} 0 \\ 2e^{-t} \end{bmatrix}.$$

The general solution of (S) is

$$\mathbf{x} = \mathbf{H}(t) + \mathbf{p}(t) = c_1 \begin{bmatrix} -e^{-t} \sin 2t \\ 2e^{-t} \cos 2t \end{bmatrix} + c_2 \begin{bmatrix} e^{-t} \cos 2t \\ 2e^{-t} \sin 2t \end{bmatrix} + \begin{bmatrix} 0 \\ 2e^{-t} \end{bmatrix}.$$

Written in terms of currents and charges in the circuit, this reads

$$Q = -c_1 e^{-t} \sin 2t + c_2 e^{-t} \cos 2t$$
$$I_2 = 2c_1 e^{-t} \cos 2t + 2c_2 e^{-t} \sin 2t + 2e^{-t}.$$

Note that in our particular solution $\mathbf{p}(t)$ (i.e., $c_1 = c_2 = 0$), the initial condition of an uncharged capacitor and current of 2 amperes in the second loop leads to a perpetually uncharged capacitor, and currents I_1 and I_2 moving in opposite directions with equal (diminishing) magnitudes, so that they cancel out at the capacitor.

Example 3.11.3

Find the solution of the system

$$
\begin{aligned}
DQ &= -2Q - I_2 &&+ e^{-3t} \\
DI_2 &= 2Q - I_2 + I_3 \\
DI_3 &= 2I_2 - 2I_3 - 2e^{-3t}
\end{aligned}
$$

(S)

satisfying the initial conditions

$$Q(0) = 1, \qquad I_2(0) = 0, \qquad I_3(0) = 0.$$

This represents the circuit of Example 3.1.3 with $R_1 = R_2 = 2$, $L_2 = 2$, $L_3 = 1$, $C = 1/4$, and $V(t) = 2e^{-3t}$.

In matrix notation, (S) is $D\mathbf{x} = A\mathbf{x} + \mathbf{E}(t)$, where

$$\mathbf{x} = \begin{bmatrix} Q \\ I_2 \\ I_3 \end{bmatrix}, \qquad A = \begin{bmatrix} -2 & -1 & 0 \\ 2 & -1 & 1 \\ 0 & 2 & -2 \end{bmatrix}, \qquad \text{and} \qquad \mathbf{E}(t) = \begin{bmatrix} e^{-3t} \\ 0 \\ -2e^{-3t} \end{bmatrix}.$$

To solve the related homogeneous system (H), we find the characteristic polynomial of A:

$$\det(A - \lambda I) = \det \begin{bmatrix} -2 - \lambda & -1 & 0 \\ 2 & -1 - \lambda & 1 \\ 0 & 2 & -2 - \lambda \end{bmatrix} = -(\lambda + 2)^2(\lambda + 1).$$

Thus, the eigenvalues of A are $\lambda = -1$ and $\lambda = -2$.

To find the eigenvectors for $\lambda = -1$, we reduce $[A + I | \mathbf{0}]$:

$$\begin{bmatrix} -1 & -1 & 0 & | & 0 \\ 2 & 0 & 1 & | & 0 \\ 0 & 2 & -1 & | & 0 \end{bmatrix} \rightarrow \begin{bmatrix} 1 & 0 & \frac{1}{2} & | & 0 \\ 0 & 1 & -\frac{1}{2} & | & 0 \\ 0 & 0 & 0 & | & 0 \end{bmatrix}.$$

This gives the eigenvector

$$\mathbf{v} = \begin{bmatrix} 1 \\ -1 \\ -2 \end{bmatrix}$$

and corresponding solution of (H)

$$\mathbf{h}_1(t) = e^{-t}\mathbf{v} = \begin{bmatrix} e^{-t} \\ -e^{-t} \\ -2e^{-t} \end{bmatrix}.$$

To find the generalized eigenvectors for $\lambda = -2$, we reduce $[(A + 2I)^2 | \mathbf{0}]$:

$$[(A + 2I)^2 | \mathbf{0}] = \begin{bmatrix} -2 & -1 & 1 & | & 0 \\ 2 & 1 & 1 & | & 0 \\ 4 & 2 & 2 & | & 0 \end{bmatrix} \rightarrow \begin{bmatrix} 1 & \frac{1}{2} & \frac{1}{2} & | & 0 \\ 0 & 0 & 0 & | & 0 \\ 0 & 0 & 0 & | & 0 \end{bmatrix}.$$

We read off two independent solutions of $(A + 2I)^2\mathbf{w} = \mathbf{0}$,

$$\mathbf{w}_1 = \begin{bmatrix} -1 \\ 0 \\ 2 \end{bmatrix} \quad \text{and} \quad \mathbf{w}_2 = \begin{bmatrix} -1 \\ 2 \\ 0 \end{bmatrix},$$

and obtain the associated solutions of (H):

$$\mathbf{h}_2(t) = e^{-2t}(\mathbf{w}_1 + t[A + 2I]\mathbf{w}_1) = e^{-2t}\begin{bmatrix} -1 \\ 0 \\ 2 \end{bmatrix} = \begin{bmatrix} -e^{-2t} \\ 0 \\ 2e^{-2t} \end{bmatrix}$$

$$\mathbf{h}_3(t) = e^{-2t}(\mathbf{w}_2 + t[A + 2I]\mathbf{w}_2) = e^{-2t}\left(\begin{bmatrix} -1 \\ 2 \\ 0 \end{bmatrix} + t\begin{bmatrix} -2 \\ 0 \\ 4 \end{bmatrix} \right)$$

$$= \begin{bmatrix} -(2t + 1)e^{-2t} \\ 2e^{-2t} \\ 4te^{-2t} \end{bmatrix}.$$

The general solution of the related homogeneous system is $\mathbf{x} = \mathbf{H}(t)$, where

$$\mathbf{H}(t) = c_1\mathbf{h}_1(t) + c_2\mathbf{h}_2(t) + c_3\mathbf{h}_3(t)$$

$$= c_1\begin{bmatrix} e^{-t} \\ -e^{-t} \\ -2e^{-t} \end{bmatrix} + c_2\begin{bmatrix} -e^{-2t} \\ 0 \\ 2e^{-2t} \end{bmatrix} + c_3\begin{bmatrix} -(2t + 1)e^{-2t} \\ 2e^{-2t} \\ 4te^{-2t} \end{bmatrix}.$$

We look for a particular solution of (S) in the form

$$\mathbf{p}(t) = c_1(t)\mathbf{h}_1(t) + c_2(t)\mathbf{h}_2(t) + c_3(t)\mathbf{h}_3(t).$$

To find functions $c_1(t)$, $c_2(t)$, $c_3(t)$ that work, we solve

(V)
$$\begin{aligned} e^{-t}c_1'(t) - e^{-2t}c_2'(t) - (2t + 1)e^{-2t}c_3'(t) &= e^{-3t} \\ -e^{-t}c_1'(t) + 2e^{-2t}c_3'(t) &= 0 \\ -2e^{-t}c_1'(t) + 2e^{-2t}c_2'(t) + 4te^{-2t}c_3'(t) &= -2e^{-3t}. \end{aligned}$$

Using row reduction or Cramer's rule, we find

$$c_1'(t) = 0, \qquad c_2'(t) = -e^{-t}, \qquad c_3'(t) = 0.$$

Integration gives

$$c_1(t) = 0, \qquad c_2(t) = e^{-t}, \qquad c_3(t) = 0$$

so a particular solution of (S) is

$$\mathbf{p}(t) = 0\mathbf{h}_1(t) + e^{-t}\mathbf{h}_2(t) + 0\mathbf{h}_3(t) = \begin{bmatrix} -e^{-3t} \\ 0 \\ 2e^{-3t} \end{bmatrix}.$$

The general solution of (S) is

$$\mathbf{x} = \mathbf{H}(t) + \mathbf{p}(t)$$

$$= c_1 \begin{bmatrix} e^{-t} \\ -e^{-t} \\ -2e^{-t} \end{bmatrix} + c_2 \begin{bmatrix} -e^{-2t} \\ 0 \\ 2e^{-2t} \end{bmatrix} + c_3 \begin{bmatrix} -(2t+1)e^{-2t} \\ 2e^{-2t} \\ 4te^{-2t} \end{bmatrix} + \begin{bmatrix} -e^{-3t} \\ 0 \\ 2e^{-3t} \end{bmatrix}.$$

To find the specific solution satisfying the desired initial condition, we substitute $t = 0$ and set the resulting vector equal to the required initial vector:

$$\mathbf{x}(0) = c_1 \begin{bmatrix} 1 \\ -1 \\ -2 \end{bmatrix} + c_2 \begin{bmatrix} -1 \\ 0 \\ 2 \end{bmatrix} + c_3 \begin{bmatrix} -1 \\ 2 \\ 0 \end{bmatrix} + \begin{bmatrix} -1 \\ 0 \\ 2 \end{bmatrix} = \begin{bmatrix} 1 \\ 0 \\ 0 \end{bmatrix}.$$

This leads to the equations

$$\begin{aligned} c_1 - c_2 - c_3 &= 2 \\ -c_1 \qquad + 2c_3 &= 0 \\ -2c_1 + 2c_2 \qquad &= -2 \end{aligned}$$

with solutions

$$c_1 = -2, \quad c_2 = -3, \quad c_3 = -1.$$

The solution of the initial-value problem is

$$\mathbf{x} = -2\mathbf{h}_1(t) - 3\mathbf{h}_2(t) - \mathbf{h}_3(t) + \mathbf{p}(t).$$

Written in terms of charges and currents, this is

$$\begin{aligned} Q &= -2e^{-t} + (4 + 2t)e^{-2t} - e^{-3t} \\ I_2 &= 2e^{-t} - 2e^{-2t} \\ I_3 &= 4e^{-t} - (6 + 4t)e^{-2t} + 2e^{-3t}. \end{aligned}$$

Let's summarize.

NONHOMOGENEOUS SYSTEMS

To solve the *n*th-order system

(S) $$D\mathbf{x} = A\mathbf{x} + \mathbf{E}(t)$$

given that the general solution of the related homogeneous system $Dx = Ax$ is $x = H(t)$ with

$$H(t) = c_1 h_1(t) + \ldots + c_n h_n(t):$$

We look for a particular solution of (S) of the form $x = p(t)$ with

(P) $$p(t) = c_1(t) h_1(t) + \ldots + c_n(t) h_n(t).$$

To find functions $c_1(t), \ldots, c_n(t)$ that work, we solve

(V) $$c_1'(t) h_1(t) + \ldots + c_n'(t) h_n(t) = E(t)$$

for $c_1', \ldots, c_n'(t)$. We then integrate to find $c_1(t), \ldots, c_n(t)$. Substitution into (P) yields a particular solution of (S). The general solution of (S) is

$$x = H(t) + p(t).$$

Notes

1. A shortcut

In our examples we obtained specific values for each $c_i(t)$ by integrating the formulas for $c_i'(t)$, taking the constants of integration to be zero. If we include arbitrary constants c_i when we integrate, then substitution of the resulting functions into formula (P) yields the general solution of (S).

2. On the old and new methods of variation of parameters

We have seen (Example 3.2.2) that an nth-order o.d.e. $Lx = E(t)$ can be replaced by an equivalent system $Dx = Ax + E(t)$. The forcing term of this system is just

$$E(t) = \begin{bmatrix} 0 \\ \cdot \\ \cdot \\ \cdot \\ 0 \\ E(t) \end{bmatrix}.$$

If the general solution of the homogeneous equation $Lx = 0$ is $x = H(t)$, where $H(t) = c_1 h_1(t) + \ldots + c_n h_n(t)$, then the general solution of the homogeneous system $Dx = Ax$ is $x = H(t)$, where

$$H(t) = c_1 \begin{bmatrix} h_1(t) \\ h_1'(t) \\ \cdot \\ \cdot \\ \cdot \\ h_1^{(n-1)}(t) \end{bmatrix} + \ldots + c_n \begin{bmatrix} h_n(t) \\ h_n'(t) \\ \cdot \\ \cdot \\ \cdot \\ h_n^{(n-1)}(t) \end{bmatrix}.$$

If we use variation of parameters to look for a particular solution of $D\mathbf{x} = A\mathbf{x} + \mathbf{E}(t)$, then we must solve the algebraic system of equations

$$
c_1'(t)\begin{bmatrix} h_1(t) \\ h_1'(t) \\ \cdot \\ \cdot \\ \cdot \\ h_1^{(n-1)}(t) \end{bmatrix} + \ldots + c_n'(t)\begin{bmatrix} h_n(t) \\ h_n'(t) \\ \cdot \\ \cdot \\ \cdot \\ h_n^{(n-1)}(t) \end{bmatrix} = \begin{bmatrix} 0 \\ 0 \\ \cdot \\ \cdot \\ \cdot \\ E(t) \end{bmatrix}.
$$

This system of equations is *exactly the same* as the system we would have to solve if we used variation of parameters to look for a particular solution of $Lx = E(t)$.

EXERCISES

In Exercises 1 to 12, find the general solution of $D\mathbf{x} = A\mathbf{x} + \mathbf{E}(t)$. (Note: you solved $D\mathbf{x} = A\mathbf{x}$ in the exercise indicated.)

1. $A = \begin{bmatrix} 0 & 2 \\ -1 & 3 \end{bmatrix}$, $\mathbf{E}(t) = \begin{bmatrix} e^t \\ e^t \end{bmatrix}$ (Exercise 1, Section 3.7)

2. $A = \begin{bmatrix} 1 & 1 \\ 1 & 1 \end{bmatrix}$, $\mathbf{E}(t) = \begin{bmatrix} 0 \\ t \end{bmatrix}$ (Exercise 2, Section 3.7)

3. $A = \begin{bmatrix} 1 & 1 \\ 3 & 1 \end{bmatrix}$, $\mathbf{E}(t) = \begin{bmatrix} 1 \\ 1 \end{bmatrix}$ (Exercise 3, Section 3.7)

4. $A = \begin{bmatrix} 1 & -1 \\ 1 & 3 \end{bmatrix}$, $\mathbf{E}(t) = \begin{bmatrix} e^{2t} \\ 0 \end{bmatrix}$ (Exercise 3, Section 3.9)

5. $A = \begin{bmatrix} 2 & 1 & -2 \\ -3 & 0 & 4 \\ -2 & -1 & 4 \end{bmatrix}$, $\mathbf{E}(t) = \begin{bmatrix} e^t \\ e^t \\ e^t \end{bmatrix}$ (Exercise 4, Section 3.7)

6. $A = \begin{bmatrix} 0 & -2 & 2 \\ 1 & 3 & -2 \\ 2 & 4 & -3 \end{bmatrix}$, $\mathbf{E}(t) = \begin{bmatrix} 2e^t - 2t \\ t \\ e^t \end{bmatrix}$ (Exercise 5, Section 3.7)

7. $A = \begin{bmatrix} 1 & -2 & 2 \\ 0 & 1 & 1 \\ -1 & -2 & 4 \end{bmatrix}$, $\mathbf{E}(t) = \begin{bmatrix} 0 \\ e^{2t} \\ 0 \end{bmatrix}$ (Exercise 4, Section 3.8)

8. $A = \begin{bmatrix} 1 & 0 & -1 \\ 0 & 2 & 0 \\ 1 & 0 & 1 \end{bmatrix}$, $\mathbf{E}(t) = \begin{bmatrix} e^t \\ e^{2t}\sin t \\ 0 \end{bmatrix}$ (Exercise 3, Section 3.8)

9. $A = \begin{bmatrix} -3 & 1 \\ -1 & -1 \end{bmatrix}$, $E(t) = \begin{bmatrix} 2te^{-2t} \\ te^{-2t} \end{bmatrix}$ (Exercise 4, Section 3.9)

10. $A = \begin{bmatrix} 3 & -2 \\ 2 & 3 \end{bmatrix}$, $E(t) = \begin{bmatrix} 2e^{3t} \\ 2e^{3t} \end{bmatrix}$ (Exercise 2, Section 3.8)

11. $A = \begin{bmatrix} 1 & 0 & 0 \\ 1 & 0 & 1 \\ 1 & -1 & 2 \end{bmatrix}$, $E(t) = \begin{bmatrix} 0 \\ 0 \\ 1 \end{bmatrix}$ (Exercise 2, Section 3.10)

12. $A = \begin{bmatrix} 0 & 0 & 0 & 0 & 1 \\ 1 & -1 & 0 & -1 & 1 \\ 1 & 0 & -1 & -1 & 0 \\ 0 & 0 & 0 & 0 & 1 \\ 1 & 0 & 0 & -1 & 0 \end{bmatrix}$, $E(t) = \begin{bmatrix} t \\ 1 \\ 0 \\ t \\ 1 \end{bmatrix}$ (Exercise 9, Section 3.10)

In Exercises 13 to 16, find (a) the general solution and (b) the specific solution satisfying the given initial conditions.

13. $x_1' = -2x_1 + 2x_2 + e^{3t}$
    $x_2' = -4x_1 + 4x_2$ $x_1(0) = x_2(0) = \frac{2}{3}$

14. $Dx = Ax + E(t)$, $A = \begin{bmatrix} 2 & 1 & 0 \\ 0 & 2 & 0 \\ 1 & 0 & 2 \end{bmatrix}$, $E(t) = \begin{bmatrix} e^{2t} \\ 0 \\ 0 \end{bmatrix}$, $x(0) = \begin{bmatrix} 1 \\ 2 \\ 3 \end{bmatrix}$

15. $x_1' = -3x_2 + 2x_3 + e^{2t}$
    $x_2' = 3x_1 - 3x_3 + 3$ $x_1(0) = 1, x_2(0) = 2, x_3(0) = 3$
    $x_3' = 2x_3 + e^{2t}$

16. $Dx = Ax + E(t)$, $A = \begin{bmatrix} 2 & 0 & 0 \\ -1 & 5 & -2 \\ -1 & 3 & 0 \end{bmatrix}$, $E(t) = \begin{bmatrix} e^t \\ 0 \\ e^t \end{bmatrix}$, $x(0) = \begin{bmatrix} 1 \\ 2 \\ 1 \end{bmatrix}$

Exercises 17 to 21 refer back to our models.

17. *Two-Loop Circuit:* Find the general solution of the system modeling the two-loop circuit of Example 3.1.2 with $V(t) = e^{-t}$ volts, $R_1 = R_2 = 1$ ohm, $L = 1$ henry, and $C = 1$ farad. (*Note:* See Exercise 14, Section 3.8.)

18. *Four-Loop Circuit:* Find the general solution of the system modeling the four-loop circuit of Example 3.1.4 with each $R_i = 1$ ohm, each $L_i = 1$ henry, and $V(t) = 4$ volts. (*Note:* See Example 3.7.4.)

*19. *Radioactive Decay:* Professor Adam D. Kay is performing an experiment involving the radioactive chain of Exercise 2, Section 3.1. There are now 500 grams of each substance, and substances A and B decay at the rates of 50 percent and 10 percent per year, respectively. The experiment requires the removal of substance C at the rate of 100 grams per year and allows the addition of substance A at a rate of α grams per year. It is crucial that α be chosen so that the amounts of substances A, B, and C never hit 0. Prof. Kay guesses that $\alpha = 250$ is big enough.
    a. Find formulas for the amounts of substances A, B, and C at time t. (*Note:* See Exercise 16a, Section 3.7.)

b. By investigating the minima of the functions found in (a), decide whether Prof. Kay's guess will work.

20. *Diffusion II:* Find the general solution of the system in Exercise 4, Section 3.1, with $P_1 = 0.2$, $P_2 = 0.1$, $v = 20$, $V = 10$, and $G = 0.15$. What happens to x and y as $t \to \infty$?

21. *Supply and Demand:*
   a. Find the general solution of the system in Exercise 1, Section 3.1, under the following assumptions:
      i. If the product were free, 5000 people would buy it; every \$1 in price reduces this number by 600.
      ii. The price increases 10¢ per week for every 100 people who want the product but can't get it.
      iii. The weekly production is increased by 50 for every dollar in price of the product.
   b. Suppose that initially the price is $p_0 = \$4$ and the supply and demand are equal ($s_0 = w_0$). Predict the price, supply, and demand after 52 weeks. (Compare this to Exercise 20, Section 2.7.)

REVIEW PROBLEMS

In Exercises 1 to 12, find (a) the general solution and (b) the specific solution satisfying the given initial condition.

1. $\begin{aligned} x_1' &= 5x_1 - 2x_2 + 4e^{3t} \\ x_2' &= 4x_1 + x_2 + 4e^{3t} \end{aligned}$ $x_1(0) = -1$, $x_2(0) = 6$

2. $\begin{aligned} x_1' &= 4x_1 - 5x_2 - 4e^t \\ x_2' &= x_1 - 2x_2 + 12e^t \end{aligned}$ $x_1(0) = 11$, $x_2(0) = 7$

3. $\begin{aligned} x_1' &= 3x_1 - 4x_2 + e^{-t} + 1 \\ x_2' &= 4x_1 - 5x_2 \qquad\quad + 1 \end{aligned}$ $x_1(0) = 2$, $x_2(0) = 1$

4. $\begin{aligned} x_1' &= 2x_1 + x_2 + e^{3t} \\ x_2' &= \qquad\quad x_2 + e^{2t} \end{aligned}$ $x_1(0) = 3$, $x_2(0) = 2$

5. $\begin{aligned} x_1' &= 4x_1 \\ x_2' &= 2x_1 + 2x_2 - 4x_3 + 5e^{3t} \\ x_3' &= x_1 \qquad\quad - 2x_3 + 5e^{3t} \end{aligned}$ $x_1(0) = 6$, $x_2(0) = 0$, $x_3(0) = 1$

6. $\begin{aligned} x_1' &= 3x_1 - x_2 + x_3 \\ x_2' &= 2x_1 + x_2 + x_3 \\ x_3' &= x_1 \qquad + 2x_3 \end{aligned}$ $x_1(0) = 2$, $x_2(0) = 8$, $x_3(0) = 4$

7. $\begin{aligned} x_1' &= 2x_1 \qquad\qquad\quad + 2e^{2t} \\ x_2' &= x_1 + 2x_2 + 2x_3 + 2e^{4t} \\ x_3' &= x_1 \qquad\quad + 4x_3 - e^{2t} \end{aligned}$ $x_1(0) = -2$, $x_2(0) = 7$, $x_3(0) = -1$

8. $\begin{aligned} x_1' &= -2x_1 + 2x_2 - 2x_3 + 2 \\ x_2' &= -4x_1 + 2x_2 \qquad\quad + e^{2t} \\ x_3' &= \qquad\qquad\qquad\quad 2x_3 + e^{2t} \end{aligned}$ $x_1(0) = 3$, $x_2(0) = x_3(0) = 1$

9. $\begin{aligned} x_1' &= 2x_1 &&+ x_3 + te^{-2t} \\ x_2' &= &&2x_2 + x_3 \\ x_3' &= &&2x_3 + e^{-2t} \end{aligned}$ $x_1(0) = 0, x_2(0) = x_3(0) = 1$

10. $\begin{aligned} x_1' &= x_1 &&+ x_3 \\ x_2' &= &&x_2 &&+ x_4 \\ x_3' &= &&&&x_3 \\ x_4' &= x_1 &&&&+ x_4 \end{aligned}$ $x_1(0) = x_2(0) = 0, x_3(0) = x_4(0) = 1$

11. $\begin{aligned} x_1' &= x_1 + x_2 - x_3 + x_4 + 4e^{2t} \\ x_2' &= -x_2 + 2e^{2t} \\ x_3' &= -2x_2 + x_3 - 2x_4 + 4e^{2t} \\ x_4' &= -x_4 + 2e^{2t} \end{aligned}$ $x_1(0) = 4, x_2(0) = 5, x_3(0) = 6, x_4(0) = 1$

12. $\begin{aligned} x_1' &= 3x_2 \\ x_2' &= -3x_1 + 3x_3 \\ x_3' &= 3x_4 \\ x_4' &= -3x_3 \end{aligned}$ $x_1(0) = 0, x_2(0) = 3, x_3(0) = 2, x_4(0) = 4$

13. *Circuits:* Find the general solution of the system modeling the circuit of the indicated exercise from Section 3.1.

   a. Exercise 6   b. Exercise 7   c. Exercise 8   d. Exercise 9   e. Exercise 10

14. *Moving Springs:* Find the general solution of the system of Exercise 23b, Section 3.2, assuming

   a. $b = 5$ and $k = 8$ b. $b = 2$ and $k = 2$
   c. $b = 0$ and $k = 2$
   (Compare this to Exercise 20, Section 2.9.)

Vector Spaces

4.1 DEFINITION AND EXAMPLES

Our primary concern in the preceding chapters was to solve linear differential equations and systems. In the process we also dealt with the problem of solving algebraic systems of equations. From the point of view of linear algebra (the study of vector spaces) these problems are really examples of the same problem. In this chapter we take a look back at these problems from this point of view.

In Chapter 3 we introduced n-vectors as part of a convenient notation for dealing with systems. For our present purpose, we consider the n-vectors with real entries as points of a space, which we denote R^n.

When $n = 2$, this space is actually quite familiar. Geometrically we can think of the 2-vector

$$\mathbf{v} = \begin{bmatrix} x \\ y \end{bmatrix}$$

as an arrow in the plane from the origin to the point (x, y) (see Figure 4.1). This sets up a correspondence between 2-vectors, arrows in the plane starting from the origin, and points in the plane. (Those of you who have read Note 1 of Section 3.2 will already be aware of this correspondence.) Using this correspondence, we can picture

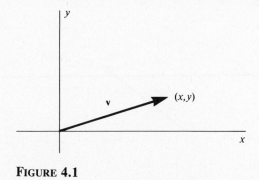

FIGURE 4.1

263

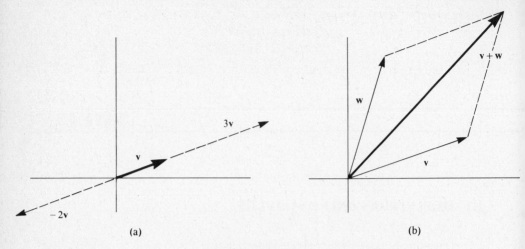

FIGURE 4.2

the set R^2 of all 2-vectors as the set of arrows from the origin or as the points in the plane. This visualization can be applied to vector operations: $c\mathbf{v}$ is an arrow with the same or opposite direction as $\mathbf{v}$ (depending on the sign of the number c) and with length $|c|$ times that of $\mathbf{v}$ (Figure 4.2a), while $\mathbf{v} + \mathbf{w}$ is the diagonal of the parallelogram determined by $\mathbf{v}$ and $\mathbf{w}$ (Figure 4.2b).

A similar correspondence can be set up between 3-vectors

$$\mathbf{v} = \begin{bmatrix} x \\ y \\ z \end{bmatrix},$$

arrows, and points (x, y, z) in 3-space (Figure 4.3). The visualization of vector operations in 3-space is just like that for 2-vectors.

If $n > 3$, then we can't draw pictures to represent an n-vector. But we can still think of the collection $V = R^n$ as a space. We can add elements of V and multiply

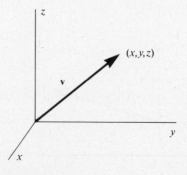

FIGURE 4.3

by real numbers, which we call **scalars** (to distinguish them from vectors). These operations obey many laws, including the following:

For all vectors **u**, **v**, *and* **w** *in V and all scalars b and c,*

(VS0)   **v** + **w** *and* c**v** *are in V.*

(VS1)   **v** + **w** = **w** + **v**.

(VS2)   **u** + (**v** + **w**) = (**u** + **v**) + **w**.

(VS3)   *There is a vector* **0** *in V, called the* **zero vector**, *so that* **v** + **0** = **v**.

(VS4)   **v** + (−1)**v** = **0**.

(VS5)   1**v** = **v**.

(VS6)   (bc)**v** = b(c**v**).

(VS7)   c(**v** + **w**) = c**v** + c**w**.

(VS8)   (b + c)**v** = b**v** + c**v**.

Each of the spaces R^n satisfies (VS0) to (VS8). So do many other spaces. In order to stress the similarities between them, such spaces are called *vector spaces* and their elements are called *vectors*.

Definition: *A **real vector space** consists of a collection V of objects called **vectors** together with two operations or rules: an operation called **addition**, which assigns to any two vectors* **v** *and* **w** *in V a third vector* **v** + **w**, *and an operation called **multiplication by scalars**, which assigns to any scalar (real number) c and any vector* **v** *in V another vector c**v**. These operations are required to satisfy properties (VS0) to (VS8) listed above.*

While our terminology highlights the similarities between general vector spaces and R^n, there are many examples of vector spaces that are quite different. Among these are the following examples, each of which is important in our discussion of linear o.d.e.'s and systems.

Example 4.1.1

Let C_∞ be the set of all those real-valued functions $x = x(t)$ that are defined for all real numbers t and that have derivatives of all orders. We add functions and multiply functions by scalars according to the rules

$$(x + y)(t) = x(t) + y(t)$$
$$(cx)(t) = c[x(t)].$$

The zero function is the function that assigns the value 0 to each real number t:

$$0(t) = 0.$$

If x and y are in C_∞, then the sum $x + y$ and scalar product cx are certainly defined for all values of t. What's more, since $D^k(x + y) = D^k x + D^k y$ and $D^k(cx) = c(D^k x)$, these functions have derivatives of all orders. Thus $x + y$ and cx are in C_∞—that is, property (VS0) holds for elements of C_∞.

In order to verify property (VS1), we must show that the two functions $x + y$ and $y + x$ are equal. To do this, we must check that they agree for all values of t. Now

$$
\begin{aligned}
(x + y)(t) &= x(t) + y(t) && \text{(definition of } x + y) \\
&= y(t) + x(t) && \text{(property of real numbers)} \\
&= (y + x)(t) && \text{(definition of } y + x).
\end{aligned}
$$

Thus $x + y = y + x$, and property (VS1) holds for functions in C_∞.

We leave it to you (Exercise 6) to verify that properties (VS2) to (VS8) also hold for functions in C_∞. Thus C_∞ is a real vector space.

Example 4.1.2

Let C_∞^n be the set of all those n-vector valued functions

$$
\mathbf{x}(t) = \begin{bmatrix} x_1(t) \\ \cdot \\ \cdot \\ \cdot \\ x_n(t) \end{bmatrix}
$$

whose entries $x_i(t)$ are real-valued functions defined for all real values of t and with derivatives of all orders. We add and multiply by scalars componentwise:

$$
\begin{bmatrix} x_1(t) \\ \cdot \\ \cdot \\ \cdot \\ x_n(t) \end{bmatrix}
+
\begin{bmatrix} y_1(t) \\ \cdot \\ \cdot \\ \cdot \\ y_n(t) \end{bmatrix}
=
\begin{bmatrix} x_1(t) + y_1(t) \\ \cdot \\ \cdot \\ \cdot \\ x_n(t) + y_n(t) \end{bmatrix}
$$

$$
c\begin{bmatrix} x_1(t) \\ \cdot \\ \cdot \\ \cdot \\ x_{n(t)} \end{bmatrix}
=
\begin{bmatrix} cx_1(t) \\ \cdot \\ \cdot \\ \cdot \\ cx_{n(t)} \end{bmatrix}.
$$

The zero vector in C_∞^n is the n-vector valued function that assigns the zero n-vector to each t:

$$\mathbf{0}(t) = \begin{bmatrix} 0 \\ \cdot \\ \cdot \\ \cdot \\ 0 \end{bmatrix}$$

We leave it to you to verify (Exercise 7) that C_∞^n is a real vector space.

If n is 2 or 3, then we can use our visualization of n-vectors as arrows to help picture elements $\mathbf{x} = \mathbf{x}(t)$ of C_∞^n. We regard $\mathbf{x}$ as a function that assigns to each time t an arrow $\mathbf{x}(t)$ from the origin. As time varies, the tip of the arrow will move along a path (see Figure 4.4). We can think of $\mathbf{x} = \mathbf{x}(t)$ as a parametrized curve describing the path.

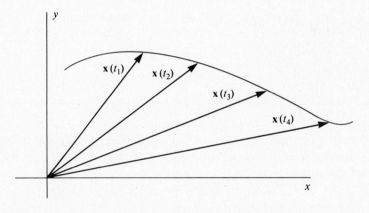

FIGURE 4.4

Because of the differences between these examples and R^n, we must be cautious not to assume that every property of R^n is automatically shared by every vector space. Among the properties that are shared by all vector spaces is the following description of the ways in which the product of a scalar and a vector can yield the zero vector.

Fact: *Let V be a vector space, let* **v** *be a vector in V, and let c be a scalar. Then*

1. $0\mathbf{v} = \mathbf{0}$.
2. $c\mathbf{0} = \mathbf{0}$.
3. *If* $c\mathbf{v} = \mathbf{0}$, *then* $c = 0$ *or* $\mathbf{v} = \mathbf{0}$.

As an example of how such properties are verified, we prove the first of these. For outlines of proofs for the others, see Exercises 12 and 13.

Example 4.1.3

Prove that if $\mathbf{v}$ is a vector in the vector space V, then $0\mathbf{v} = \mathbf{0}$.

We first use the fact that $0 + 0 = 0$, together with (VS8) to conclude that

$$0\mathbf{v} = (0 + 0)\mathbf{v} = 0\mathbf{v} + 0\mathbf{v}.$$

We add $(-1)[0\mathbf{v}]$ to both sides and use property (VS2) on the right to get

$$0\mathbf{v} + (-1)[0\mathbf{v}] = (0\mathbf{v} + 0\mathbf{v}) + (-1)[0\mathbf{v}]$$
$$= 0\mathbf{v} + (0\mathbf{v} + (-1)[0\mathbf{v}]).$$

Properties (VS4) and (VS3) now give

$$0 = 0\mathbf{v} + \mathbf{0}$$
$$= 0\mathbf{v}.$$

Exercises 14 through 17 are concerned with other properties that are shared by all vector spaces. In later sections we make use of these properties, as well as the following notational conventions:

$$-\mathbf{v} = (-1)\mathbf{v}$$
$$\mathbf{u} - \mathbf{v} = \mathbf{u} + (-1)\mathbf{v}$$
$$\mathbf{u} + \mathbf{v} + \mathbf{w} = (\mathbf{u} + \mathbf{v}) + \mathbf{w} = \mathbf{u} + (\mathbf{v} + \mathbf{w}).$$

We also write $\mathbf{u} + \mathbf{v} + \mathbf{w} + \mathbf{x}$ to denote the vector $(\mathbf{u} + \mathbf{v} + \mathbf{w}) + \mathbf{x}$. Continuing this way, we define the sum of k vectors by the rule

$$\mathbf{v}_1 + \ldots + \mathbf{v}_k = (\mathbf{v}_1 + \ldots + \mathbf{v}_{k-1}) + \mathbf{v}_k.$$

If we allow complex scalars in our definition of a vector space, then we obtain the definition of a **complex vector space**. In a complex vector space, the product of a vector by a *complex* number will always yield another vector. (In a real vector space, we are only guaranteed that we can multiply vectors by *real* numbers.) Also in a complex vector space, properties (VS6) to (VS8) hold for all *complex* numbers b and c.

Our most important example of a complex vector space is the space of n-vectors with complex entries.

Example 4.1.4

Let C^n be the set of all n-vectors with complex entries. We add vectors in C^n and multiply them by complex numbers componentwise, as usual. The zero vector is the n-vector all of whose entries are zero. Then C^n is a complex vector space.

In more advanced treatments of linear algebra, the definition of a vector space is broadened to allow multiplication of vectors by scalars that are elements in more general number systems called *fields*. The real and complex numbers are but two examples of fields. Each of the properties of vector spaces discussed in this chapter holds for any vector space, regardless of the field of scalars.

DEFINITION AND EXAMPLES

A **real** (respectively **complex**) **vector space** consists of a collection V of **vectors** together with two operations: **addition**, which assigns to any two vectors $\mathbf{v}$ and $\mathbf{w}$ a third vector $\mathbf{v} + \mathbf{w}$, and **multiplication by scalars**, which assigns to any real (respectively complex) **scalar** (number) c and any vector $\mathbf{v}$ another vector $c\mathbf{v}$. These operations are required to satisfy the following laws:

(VS0) $\mathbf{v} + \mathbf{w}$ and $c\mathbf{v}$ are in V.

(VS1) $\mathbf{v} + \mathbf{w} = \mathbf{w} + \mathbf{v}$.

(VS2) $\mathbf{u} + (\mathbf{v} + \mathbf{w}) = (\mathbf{u} + \mathbf{v}) + \mathbf{w}$.

(VS3)   There is a vector $\mathbf{0}$ in V with the property that $\mathbf{v} + \mathbf{0} = \mathbf{v}$ for all vectors $\mathbf{v}$.

(VS4) $\mathbf{v} + (-1)\mathbf{v} = \mathbf{0}$.

(VS5) $1\mathbf{v} = \mathbf{v}$.

(VS6) $(bc)\mathbf{v} = b(c\mathbf{v})$.

(VS7) $c(\mathbf{v} + \mathbf{w}) = c\mathbf{v} + c\mathbf{w}$.

(VS8) $(b + c)\mathbf{v} = b\mathbf{v} + c\mathbf{v}$.

Among the examples of real vector spaces are

R^n = the collection of n-vectors with real entries.

C_∞ = the collection of those real-valued functions $x(t)$ that are defined for all values of t and that have derivatives of all orders.

C_∞^n = the collection of those n-vector valued functions $\mathbf{x}(t)$ whose entries are real-valued functions that are defined for all real values of t and that have derivatives of all orders.

Our most important example of a complex vector space is

C^n = the collection of n-vectors with complex entries.

EXERCISES

1. Determine whether the given set V is a real vector space with respect to the usual addition and scalar multiplication of 2-vectors.

   a. The set V of all those 2-vectors $\begin{bmatrix} 0 \\ y \end{bmatrix}$ that have first entry 0.

   b. The set V of all those 2-vectors $\begin{bmatrix} 1 \\ y \end{bmatrix}$ that have first entry 1.

   c. The set V of all those 2-vectors $\begin{bmatrix} n \\ m \end{bmatrix}$ that have integer entries.

   d. The set V of all those 2-vectors $\begin{bmatrix} x \\ x \end{bmatrix}$ that have equal entries.

   e. The set V of all those 2-vectors $\begin{bmatrix} x \\ y \end{bmatrix}$ for which $y \geq x$.

2. Determine whether the given set V is a real vector space with respect to the usual addition and scalar multiplication of functions in C_∞.
   a. The set V of all those functions $x(t)$ in C_∞ for which $x(0) = 0$.
   b. The set V of all those functions $x(t)$ in C_∞ for which $x(0) = 1$.
   c. The set V of all those functions $x(t)$ in C_∞ for which $(D + 1)x(t) = 0$.
   d. The set V of all those functions $x(t)$ in C_∞ for which $(D + 1)x(t) = e^t$.
   e. The set V of all those functions $x(t)$ in C_∞ that are of the form $x(t) = a_0 + a_1 t$—that is, the set of all polynomials of degree at most 1, together with the zero function.

3. Check that the set $R_{m \times n}$ consisting of all $m \times n$ matrices with real entries is a real vector space with respect to the usual addition and scalar multiplication of matrices.

4. Check that the set $C_{m \times n}$ consisting of all $m \times n$ matrices with complex entries is a complex vector space with respect to the usual addition and scalar multiplication of matrices.

5. Let V be the set of all complex numbers $r + si$. We add elements of V and multiply by real numbers as usual:

   $$(r + si) + (r' + s'i) = (r + r') + (s + s')i \quad \text{and} \quad c(r + si) = cr + csi.$$

   Show that V is a real vector space.

6. Complete the verification that C_∞ is a real vector space.

7. Complete the verification that C_∞^n is a real vector space.

8. Let P be the set of all polynomials with real coefficients $a_0 + a_1t + \ldots + a_nt^n$. Show that P is a vector space with respect to the usual addition and scalar multiplication of polynomials.

9. Let S be the set of all sequences of real numbers $\{a_n\}$. We add and multiply by scalars according to the following rules:

$$\{a_n\} + \{b_n\} = \{(a_n + b_n)\} \quad \text{and} \quad c\{a_n\} = \{ca_n\}.$$

Show that S is a real vector space.

10. Let $C[a, b]$ be the set consisting of all those real-valued functions that are defined and continuous for $a \leq t \leq b$. Show that $C[a, b]$ is a real vector space with respect to the usual addition and scalar product of functions:

$$[x + y](t) = x(t) + y(t), \quad [cx](t) = c[x(t)].$$

11. Determine whether the set V of all real 2-vectors is a real vector space with respect to the following unusual addition and scalar multiplication:

$$\begin{bmatrix} x \\ y \end{bmatrix} \dotplus \begin{bmatrix} u \\ v \end{bmatrix} = \begin{bmatrix} x + u \\ 0 \end{bmatrix}, \quad c \cdot \begin{bmatrix} x \\ y \end{bmatrix} = \begin{bmatrix} cx \\ 0 \end{bmatrix}.$$

More abstract problems:

12. Prove that in a vector space $c0 = 0$ as follows. First use the fact that $0 + 0 = 0$ (property (VS3) with $v = 0$) together with (VS7) to show that $c0 = c0 + c0$. Now add $(-1)[c0]$ to both sides and use properties (VS2), (VS4), and (VS3).

13. Prove that in a vector space $cv = 0$ implies $c = 0$ or $v = 0$ as follows. Assume $c \neq 0$. Multiply by $1/c$ and use (VS6) and (VS5) to prove $v = 0$.

14. Prove that if u, v, and w are vectors in a vector space V, and if $u + v = w$, then $u = w - v$. (*Hint*: add $-v$ to both sides of the given equation.)

15. Use the preceding exercise to prove the following properties of a vector space.
    a. If $u + v = x + v$ for some vector v, then $u = x$.
    b. If $u + v = v$ for some vector v, then $u = 0$.
    c. If $u + v = 0$ for some vector v, then $u = -v$.

16. Use mathematical induction to prove the following properties of a vector space.
    a. $(c_1 + \ldots + c_k)v = c_1v + \ldots + c_kv$.
    b. $c(v_1 + \ldots + v_k) = cv_1 + \ldots + cv_k$.

17. Prove that in a vector space

$$(u + v + w) + x = u + (v + w + x) = (u + v) + (w + x).$$

4.2 SUBSPACES

In this section we shall look for more examples of vector spaces among the subsets of the spaces we have considered in Section 4.1. These examples will include the spaces most relevant to the solution of linear o.d.e.'s, linear systems of o.d.e.'s, and algebraic systems of equations.

In a vector space V, each of the laws (VS0) through (VS8) listed in Section 4.1 is satisfied. Most of these properties are given as equations that hold for all vectors in V and thus continue to hold if we restrict our attention to those vectors that lie in a given subset W of V. To determine whether W is itself a vector space, we need to check whether the remaining properties are also satisfied in W. We need to check (VS0), that sums and scalar multiples of vectors in W lie in W, and to complete (VS3), we need to check that $\mathbf{0}$ lies in W. Subsets of V that satisfy these special properties are called subspaces.

Definition: *A subset W of a vector space V is called a **subspace** of V provided*

(S1) $\mathbf{0}$ *is in W.*

(S2)   *If $\mathbf{w}_1$ and $\mathbf{w}_2$ are in W, then so is $\mathbf{w}_1 + \mathbf{w}_2$.*

(S3)   *If $\mathbf{w}$ is in W and c is a scalar, then $c\mathbf{w}$ is in W.*

A subset of V that satisfies (S2) is said to be **closed under addition**, while a subset satisfying (S3) is said to be **closed under multiplication by scalars**. Thus a subspace of V is a subset that contains $\mathbf{0}$ and is closed under both addition and multiplication by scalars.

Using this definition, we can rephrase our earlier observation as follows.

Fact: *If W is a subspace of a vector space V, then W is itself a vector space.*

Any vector space V has two subsets that obviously satisfy conditions (S1) to (S3)—namely, the entire vector space $W = V$ and the subset consisting of only the zero vector, $W = \{\mathbf{0}\}$. These are referred to as **trivial subspaces** of V. Let's consider some other examples.

Example 4.2.1

Let W be the set of all those 4-vectors that can be written in the form

$$\begin{bmatrix} v_1 \\ v_1 - v_4 \\ 2v_1 \\ v_4 \end{bmatrix}$$

for some choice of v_1 and v_4. Since the choice $v_1 = v_4 = 0$ yields

$$\begin{bmatrix} 0 \\ 0 - 0 \\ 2 \cdot 0 \\ 0 \end{bmatrix} = \begin{bmatrix} 0 \\ 0 \\ 0 \\ 0 \end{bmatrix} = \mathbf{0},$$

the zero vector belongs to W. If $\mathbf{v}$ and $\mathbf{w}$ belong to W, then we can write them in the required form:

$$\mathbf{v} = \begin{bmatrix} v_1 \\ v_1 - v_4 \\ 2v_1 \\ v_4 \end{bmatrix}, \qquad \mathbf{w} = \begin{bmatrix} w_1 \\ w_1 - w_4 \\ 2w_1 \\ w_4 \end{bmatrix}.$$

Then

$$\mathbf{v} + \mathbf{w} = \begin{bmatrix} v_1 + w_1 \\ (v_1 - v_4) + (w_1 - w_4) \\ 2v_1 + 2w_1 \\ v_4 + w_4 \end{bmatrix} = \begin{bmatrix} (v_1 + w_1) \\ (v_1 + w_1) - (v_4 + w_4) \\ 2(v_1 + w_1) \\ (v_4 + w_4) \end{bmatrix}.$$

This has the required form, so $\mathbf{v} + \mathbf{w}$ also belongs to W. If c is any scalar, then

$$c\mathbf{v} = c\begin{bmatrix} v_1 \\ v_1 - v_4 \\ 2v_1 \\ v_4 \end{bmatrix} = \begin{bmatrix} (cv_1) \\ (cv_1) - (cv_4) \\ 2(cv_1) \\ (cv_4) \end{bmatrix}$$

so $c\mathbf{v}$ belongs to W. Thus, since W contains $\mathbf{0}$ and is closed under addition and multiplication by scalars, it is a subspace of R^4.

The following examples are somewhat more general in flavor. The arguments used to verify that the next two examples are subspaces should be familiar from Sections 2.3 and 3.3.

Example 4.2.2

Let L be a linear differential operator with constant coefficients, and let W be the set of all those functions x in C_∞ that satisfy $Lx = 0$. Since

$$L0 = 0,$$

the zero function is in W. If x and y are in W and if c is a scalar, then

$$L(x + y) = Lx + Ly = 0 + 0 = 0$$

$$L(cx) = c(Lx) = c(0) = 0$$

so that $x + y$ and cx are also in W. Thus W is a subspace of C_∞.

Note that by definition W is the set of those solutions of $Lx = 0$ that happen to lie in C_∞. Our description of the solutions of constant-coefficient homogeneous o.d.e.'s in Section 2.7 and 2.8 shows that any solution of $Lx = 0$ is automatically in C_∞. Thus W consists of all solutions of $Lx = 0$.

Example 4.2.3

Let A be an $n \times n$ matrix with real entries, and let W be the set of all those n-vector valued functions $\mathbf{x}$ in C_∞^n that satisfy $D\mathbf{x} = A\mathbf{x}$. Since

$$D\mathbf{0} = \mathbf{0} = A\mathbf{0},$$

the zero-vector function is in W. If $\mathbf{x}$ and $\mathbf{y}$ are in W and if c is a scalar, then

$$D(\mathbf{x} + \mathbf{y}) = D\mathbf{x} + D\mathbf{y} = A\mathbf{x} + A\mathbf{y} = A(\mathbf{x} + \mathbf{y})$$

$$D(c\mathbf{x}) = c(D\mathbf{x}) = c(A\mathbf{x}) = A(c\mathbf{x})$$

so that $\mathbf{x} + \mathbf{y}$ and $c\mathbf{x}$ also belong to W. Thus W is a subspace of C_∞^n.

Here again we note that W is defined to consist of those solutions of $D\mathbf{x} = A\mathbf{x}$ that happen to lie in C_∞^n, but our description of the solutions of constant-coefficient systems (Sections 3.7 to 3.10) shows that W is the set of all solutions of $D\mathbf{x} = A\mathbf{x}$.

Example 4.2.4

Let B be an $m \times n$ matrix with real entries and define W to be the set of all vectors $\mathbf{v}$ in R^n satisfying $B\mathbf{v} = \mathbf{0}$. Since

$$B\mathbf{0} = \mathbf{0},$$

the zero n-vector belongs to W. If $\mathbf{v}$ and $\mathbf{w}$ are in W and if c is a scalar, then

$$B(\mathbf{v} + \mathbf{w}) = B\mathbf{v} + B\mathbf{w} = \mathbf{0} + \mathbf{0} = \mathbf{0}$$

$$B(c\mathbf{v}) = c(B\mathbf{v}) = c\mathbf{0} = \mathbf{0}$$

so $\mathbf{v} + \mathbf{w}$ and $c\mathbf{v}$ also belong to W. Thus W is a subspace of R^n.

In the last three examples we showed that the solutions of a *homogeneous* linear constant-coefficient o.d.e. or system of o.d.e.'s and the solutions of an algebraic system $B\mathbf{v} = \mathbf{0}$ all form subspaces of the appropriate vector spaces. By contrast, the solutions of a *non*homogeneous linear o.d.e. $Lx = E(t)$, the solutions of a *non*-

homogeneous system of o.d.e.'s $Dx = Ax + E(t)$, and the solutions of an algebraic system $Bv = k$ with $k \neq 0$ do not form subspaces, since in each of these cases the zero-vector is *not* a solution. The following examples of subsets contain the zero-vector yet still fail to be subspaces.

Example 4.2.5

Let W be the set of all 2-vectors with nonnegative entries. We leave it to you to check (Exercise 11a) that 0 belongs to W and that W is closed under addition. On the other hand,

$$\mathbf{v} = \begin{bmatrix} 1 \\ 2 \end{bmatrix}$$

lies in W, but multiplication by the scalar -1 gives

$$(-1)\mathbf{v} = \begin{bmatrix} -1 \\ -2 \end{bmatrix},$$

which does not lie in W. Since W is not closed under multiplication by scalars, it is not a subspace of R^2.

Example 4.2.6

Let W be the set of all those 2-vectors

$$\begin{bmatrix} x \\ y \end{bmatrix}$$

for which $xy \leq 0$. Then W contains 0 and is closed under multiplication by scalars (Exercise 11b). On the other hand, the vectors

$$\mathbf{v} = \begin{bmatrix} 2 \\ -1 \end{bmatrix} \quad \text{and} \quad \mathbf{w} = \begin{bmatrix} -1 \\ 2 \end{bmatrix}$$

both belong to W, while their sum

$$\mathbf{v} + \mathbf{w} = \begin{bmatrix} 1 \\ 1 \end{bmatrix}$$

does not. Since W is not closed under addition, it is not a subspace of R^2.

Recall from Sections 2.3 and 3.3 that any linear combination of solutions to a homogeneous linear o.d.e. or system of o.d.e.'s will also be a solution. This behavior is typical for subspaces W of any vector space V. If $\mathbf{v}_1, \ldots, \mathbf{v}_k$ all belong to W and if $c_1, \ldots, c_k$ are any scalars, then property (S3) tells us that $c_1\mathbf{v}_1, \ldots, c_k\mathbf{v}_k$ all belong to W. Now property (S2) assures us that the sum $c_1\mathbf{v}_1 + \ldots + c_k\mathbf{v}_k$ also belongs to W. In keeping with our earlier terminology, we will refer to any vector of the form $c_1\mathbf{v}_1 + \ldots + c_k\mathbf{v}_k$ as a **linear combination** of $\mathbf{v}_1, \ldots, \mathbf{v}_k$. We have just seen the following:

Fact: *If $\mathbf{v}_1, \ldots, \mathbf{v}_k$ are vectors that lie in a subspace W of V, then any linear combination of $\mathbf{v}_1, \ldots, \mathbf{v}_k$ also lies in W.*

This fact tells us that if we wish to form a subspace containing given vectors, then we have to include, at the very least, all linear combinations of these vectors. Let's use $\mathbf{Comb}(\mathbf{v}_1, \ldots, \mathbf{v}_k)$ to denote the collection of all linear combinations of $\mathbf{v}_1, \ldots, \mathbf{v}_k$. Then $\mathbf{0}$ belongs to $\mathbf{Comb}(\mathbf{v}_1, \ldots, \mathbf{v}_k)$, since

$$\mathbf{0} = 0\mathbf{v}_1 + \ldots + 0\mathbf{v}_k.$$

If $\mathbf{v}$ and $\mathbf{w}$ lie in $\mathbf{Comb}(\mathbf{v}_1, \ldots, \mathbf{v}_k)$, then there are scalars $a_1, \ldots, a_k$ and $b_1, \ldots, b_k$ so that

$$\mathbf{v} = a_1\mathbf{v}_1 + \ldots + a_k\mathbf{v}_k$$
$$\mathbf{w} = b_1\mathbf{v}_1 + \ldots + b_k\mathbf{v}_k.$$

Now

$$\mathbf{v} + \mathbf{w} = (a_1 + b_1)\mathbf{v}_1 + \ldots + (a_k + b_k)\mathbf{v}_k$$

and if c is any scalar, then

$$c\mathbf{v} = (ca_1)\mathbf{v}_1 + \ldots + (ca_k)\mathbf{v}_k.$$

Thus $\mathbf{v} + \mathbf{w}$ and $c\mathbf{v}$ are in $\mathbf{Comb}(\mathbf{v}_1, \ldots, \mathbf{v}_k)$. Since $\mathbf{Comb}(\mathbf{v}_1, \ldots, \mathbf{v}_k)$ contains $\mathbf{0}$ and is closed under addition and multiplication by scalars, it is a subspace of V.

Fact: *If $\mathbf{v}_1, \ldots, \mathbf{v}_k$ are vectors in the vector space V, then the set $\mathbf{Comb}(\mathbf{v}_1, \ldots, \mathbf{v}_k)$ consisting of all linear combinations of $\mathbf{v}_1, \ldots, \mathbf{v}_k$ is a subspace of V.*

Note that $\mathbf{Comb}(\mathbf{v}_1, \ldots, \mathbf{v}_k)$ contains each of the vectors $\mathbf{v}_i$ (Exercise 19). Since any other subspace of V that contains $\mathbf{v}_1, \ldots, \mathbf{v}_k$ also contains all of their linear combinations, we see that $\textbf{Comb}(\textbf{v}_1, \ldots, \textbf{v}_k)$ *is the smallest subspace of V that contains all of the vectors* $\textbf{v}_1, \ldots, \textbf{v}_k$.

Example 4.2.7

If

$$\textbf{v}_1 = \begin{bmatrix} 1 \\ 0 \\ 0 \end{bmatrix} \quad \text{and} \quad \textbf{v}_2 = \begin{bmatrix} 0 \\ 1 \\ 0 \end{bmatrix},$$

then $\textbf{Comb}(\textbf{v}_1, \textbf{v}_2)$ consists of all 3-vectors of the form

$$c_1\textbf{v}_1 + c_2\textbf{v}_2 = \begin{bmatrix} c_1 \\ c_2 \\ 0 \end{bmatrix}.$$

This is the same as the collection of all those 3-vectors whose third entry is zero.

Example 4.2.8

The linear combinations of $x_1(t) = 1$, $x_2(t) = e^t$, and $x_3(t) = e^{-t}$ are the functions of the form

$$c_1x_1(t) + c_2x_2(t) + c_3x_3(t) = c_1 + c_2e^t + c_3e^{-t}.$$

The subspace $\textbf{Comb}(1, e^t, e^{-t})$ of C_∞ is the same as the set of all solutions of the third-order o.d.e. $(D^3 - D)x = 0$.

Example 4.2.9

If

$$\textbf{e}_1 = \begin{bmatrix} 1 \\ 0 \\ 0 \\ \cdot \\ \cdot \\ \cdot \\ 0 \end{bmatrix}, \quad \textbf{e}_2 = \begin{bmatrix} 0 \\ 1 \\ 0 \\ \cdot \\ \cdot \\ \cdot \\ 0 \end{bmatrix}, \ldots, \textbf{e}_n = \begin{bmatrix} 0 \\ 0 \\ 0 \\ \cdot \\ \cdot \\ \cdot \\ 1 \end{bmatrix},$$

then the subspace $\textbf{Comb}(e_1, \ldots, \textbf{e}_n)$ of R^n consists of all vectors of the form

$$c_1\textbf{e}_1 + c_2\textbf{e}_2 + \ldots + c_n\textbf{e}_n = \begin{bmatrix} c_1 \\ c_2 \\ \cdot \\ \cdot \\ \cdot \\ c_n \end{bmatrix}.$$

Since every n-vector is included in this collection,

$$\textbf{Comb}(e_1, \ldots, \textbf{e}_n) = R^n.$$

Let's summarize.

SUBSPACES

A subset W of a vector space V is called a **subspace** of V provided

(S1) $\textbf{0}$ is in W.

(S2)   If $\textbf{w}_1$ and $\textbf{w}_2$ are in W, then so is $\textbf{w}_1 + \textbf{w}_2$.

(S3)   If $\textbf{w}$ is in W and if c is any scalar, then $c\textbf{w}$ is in W.

A subspace W of a vector space V is itself a vector space.
If $\textbf{v}_1, \ldots, \textbf{v}_k$ are vectors in a subspace W, then any **linear combination** $c_1\textbf{v}_1 + \ldots + c_k\textbf{v}_k$ is also in W.
If $\textbf{v}_1, \ldots, \textbf{v}_k$ are vectors in a vector space V, then the collection $\textbf{Comb}(\textbf{v}_1, \ldots, \textbf{v}_k)$ of all linear combinations of $\textbf{v}_1, \ldots, \textbf{v}_k$ is a subspace of V.

Note

Visualizing the linear combinations of one, two, and three 3-vectors

If $V = R^3$, we can use our picture of V as consisting of arrows in 3-space to help visualize subspaces of the form $\textbf{Comb}(\textbf{v})$, $\textbf{Comb}(\textbf{v}, \textbf{w})$, and $\textbf{Comb}(\textbf{v}, \textbf{w}, \textbf{u})$.

For a single vector $\textbf{v}$, the subspace $\textbf{Comb}(\textbf{v})$ consists of all scalar multiples of $\textbf{v}$. When $\textbf{v} \neq \textbf{0}$, this can be interpreted geometrically as the line through the origin determined by $\textbf{v}$ (see Figure 4.5).

If $\textbf{v}$ and $\textbf{w}$ are both nonzero vectors, then their scalar multiples $\textbf{Comb}(\textbf{v})$ and $\textbf{Comb}(\textbf{w})$ yield lines through the origin determined by $\textbf{v}$ and $\textbf{w}$, respectively. If $\textbf{w}$ is a multiple of $\textbf{v}$, then

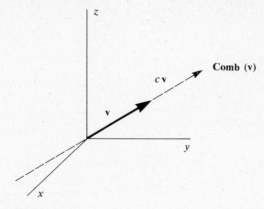

FIGURE 4.5

these lines are the same; in this case, every linear combination $c_1\mathbf{v} + c_2\mathbf{w}$ is a multiple of $\mathbf{v}$, so **Comb**$(\mathbf{v}, \mathbf{w})$ = **Comb**$(\mathbf{v})$. If $\mathbf{w}$ is not a multiple of $\mathbf{v}$, then the lines **Comb**$(\mathbf{v})$ and **Comb**$(\mathbf{w})$ are different; in this case, the parallelogram rule for adding vectors shows that the elements $c_1\mathbf{v} + c_2\mathbf{w}$ of **Comb**$(\mathbf{v}, \mathbf{w})$ are the same as the vectors in the plane through the origin determined by $\mathbf{v}$ and $\mathbf{w}$ (see Figure 4.6).

The case of three vectors $\mathbf{v}$, $\mathbf{w}$, and $\mathbf{u}$ is slightly more involved. If all three vectors are nonzero scalar multiples of a single nonzero vector, then the lines **Comb**$(\mathbf{v})$, **Comb**$(\mathbf{w})$, and **Comb**$(\mathbf{u})$ coincide and equal **Comb**$(\mathbf{v}, \mathbf{w}, \mathbf{u})$ (Exercise 20a). If two of the vectors—say $\mathbf{v}$ and $\mathbf{w}$—determine a plane **Comb**$(\mathbf{v}, \mathbf{w})$ through the origin, and if the third vector $\mathbf{u}$ is a linear combination of $\mathbf{v}$ and $\mathbf{w}$, then $\mathbf{u}$ belongs to **Comb**$(\mathbf{v}, \mathbf{w})$ and **Comb**$(\mathbf{v}, \mathbf{w}, \mathbf{u})$ = **Comb**$(\mathbf{v}, \mathbf{w})$ (Exercise 20b). Finally, if $\mathbf{v}$ and $\mathbf{w}$ determine a plane, and if $\mathbf{u}$ is not a linear combination of $\mathbf{v}$ and $\mathbf{w}$, then $\mathbf{u}$ points out of the plane **Comb**$(\mathbf{v}, \mathbf{w})$; we shall see (Exercise 14 of Section 4.5) that in this case **Comb**$(\mathbf{v}, \mathbf{w}, \mathbf{u})$ is all of R^3.

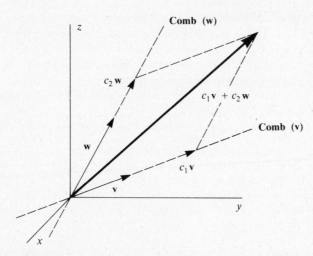

FIGURE 4.6

EXERCISES

1. In each of the following, determine whether the set of all vectors of the given form is a subspace of R^3.

    a. $\begin{bmatrix} x \\ x \\ y \end{bmatrix}$

    b. $\begin{bmatrix} x \\ y \\ x + y \end{bmatrix}$

    c. $\begin{bmatrix} x \\ y \\ xy \end{bmatrix}$

    d. $\begin{bmatrix} x \\ y \\ 1 \end{bmatrix}$

    e. $\begin{bmatrix} x \\ x^2 \\ 0 \end{bmatrix}$

    f. $\begin{bmatrix} x + y \\ y - z \\ 3z \end{bmatrix}$

2. In each of the following, determine whether the set of all those vectors $\begin{bmatrix} x \\ y \\ z \end{bmatrix}$ that satisfy the given conditions is a subspace of R^3.

    a. $z = -x$

    b. $y = 1$

    c. $x^2 = y^2$

    d. $\begin{cases} x + y = 0 \\ 3y - z = 0 \end{cases}$

    e. $\begin{cases} 2x + y = 0 \\ y + z = 1 \end{cases}$

    f. $\begin{cases} xy = 0 \\ z = 0 \end{cases}$

3. In each of the following, determine whether the set of all those functions $x(t)$ in C_∞ that satisfy the given condition is a subspace.

    a. $(D^3 - D)x = 0$

    b. $(D^3 - D)x = e^t$

    c. $\lim_{t \to \infty} x(t) = 0$

    d. $\lim_{t \to \infty} x(t) \geq 0$

    e. $\int_{-1}^{1} x(t)\, dt = 0$

    f. $x(-t) = x(t)$ for all t.

4. Let P be the set of all polynomial functions in C_∞—that is, the set of all functions $x(t)$ that can be written in the form $x(t) = a_0 + a_1 t + \ldots + a_n t^n$ for some integer n and some choice of real constants $a_0, \ldots, a_n$. Show that P is a subspace of C_∞.

5. Let m be a fixed integer, and let P_m be the set of all functions $x(t)$ in C_∞ of the form $x(t) = a_0 + a_1 t + \ldots + a_m t^m$—that is, the set of all polynomial functions of degree at most m, together with the zero function. Show that P_m is a subspace of C_∞.

6. Let λ be a real (respectively complex) eigenvalue of the $n \times n$ matrix A. Let W be the set of all eigenvectors of A with respect to λ, together with the zero-vector. Show that W is a subspace of R^n (respectively C^n).

7. In each of the following, determine whether the set of all those vector-valued functions $\begin{bmatrix} x(t) \\ y(t) \end{bmatrix}$ in C_∞^2 that satisfy the given conditions is a subspace.

    a. $y(t) = -x(t)$

    b. $y(t) = x(t)^2$

    c. $\begin{cases} x'(t) = x(t) - y(t) + 1 \\ y'(t) = x(t) \end{cases}$

    d. $\begin{cases} x'(t) = x(t) - y(t) \\ y'(t) = x(t) \end{cases}$

    e. $y(t) = t^2 x(t)$

    f. $y(t) = x(t) + t$

8. Determine whether the following subsets of the space $R_{2 \times 2}$ of 2×2 matrices (see Exercise 3, Section 4.1) are subspaces.

a. The matrices of the form $\begin{bmatrix} a & b \\ 0 & a \end{bmatrix}$

b. The matrices of the form $\begin{bmatrix} a & a^2 \\ 0 & a \end{bmatrix}$

c. The matrices of the form $\begin{bmatrix} a & b \\ b & c \end{bmatrix}$

d. The 2×2 matrices A for which $A \begin{bmatrix} 1 \\ 2 \end{bmatrix} = \begin{bmatrix} 0 \\ 0 \end{bmatrix}$

9. Determine whether the following are subspaces of the space S of sequences $\{a_n\}$ of real numbers (see Exercise 9, Section 4.1).
   a. The set of all sequences that converge to 0
   b. The set of all sequences that converge to 1
   c. The set of all convergent sequences

10. Determine whether the following subsets of C^2 are subspaces.

    a. The set of all vectors of the form $\begin{bmatrix} x \\ ix \end{bmatrix}$

    b. The set of all vectors of the form $\begin{bmatrix} x \\ i + x \end{bmatrix}$

    c. The set of all those 2-vectors that have only real entries

11. a. Show that the set W of 2-vectors considered in Example 4.2.5 contains $\mathbf{0}$ and is closed under addition.
    b. Show that the set W of 2-vectors considered in Example 4.2.6 contains $\mathbf{0}$ and is closed under multiplication by scalars.

12. Determine whether the following vectors belong to **Comb** $\left(\begin{bmatrix} 1 \\ 2 \\ 0 \end{bmatrix}, \begin{bmatrix} -1 \\ 3 \\ 1 \end{bmatrix} \right)$.

    a. $\begin{bmatrix} 2 \\ -1 \\ -1 \end{bmatrix}$ b. $\begin{bmatrix} 1 \\ 3 \\ 3 \end{bmatrix}$ c. $\begin{bmatrix} 0 \\ 5 \\ 1 \end{bmatrix}$

13. Show that **Comb** $\left(\begin{bmatrix} 1 \\ 0 \end{bmatrix}, \begin{bmatrix} 1 \\ 2 \end{bmatrix} \right) = R^2$.

14. Let W be the subspace of R^3 consisting of all 3-vectors of the form $\begin{bmatrix} x \\ x \\ y \end{bmatrix}$.

    a. Show that **Comb** $\left(\begin{bmatrix} 1 \\ 1 \\ 0 \end{bmatrix}, \begin{bmatrix} 0 \\ 0 \\ 1 \end{bmatrix} \right) = W$.

    b. Show that **Comb** $\left(\begin{bmatrix} 1 \\ 1 \\ 1 \end{bmatrix}, \begin{bmatrix} 2 \\ 2 \\ 1 \end{bmatrix}, \begin{bmatrix} 0 \\ 0 \\ 1 \end{bmatrix} \right) = W$.

15. In each of the following, $\mathbf{Comb}(h_1, h_2, h_3)$ coincides with the set of all solutions of a homogeneous constant-coefficient linear o.d.e. $Lx = 0$. Find L. (*Hint*: Recall undetermined coefficients.)
    a. $h_1 = 1, h_2 = e^t, h_3 = e^{2t}$
    b. $h_1 = \sin t, h_2 = \cos t, h_3 = 1$
    c. $h_1 = e^t, h_2 = e^t + 2e^{-t}, h_3 = e^{-t}$
    d. $h_1 = e^t + e^{-t}, h_2 = e^t - e^{-t}, h_3 = e^t + 2e^{-t}$
    e. $h_1 = 1, h_2 = t, h_3 = t^2$.

16. Determine whether the following functions belong to $\mathbf{Comb}(\sin t, \cos t)$.
    a. $2 \sin t - \cos t$ b. $\sin 2t$
    c. $\sin^2 t$ d. $\sin(t + 2)$

More abstract problems:

17. Suppose that W is a nonempty subset of a vector space V (i.e., W contains at least one vector) and that W is closed under addition and multiplication by scalars. Show that W is a subspace of V.

18. Let W and U be subspaces of a vector space V.
    a. The **intersection** of W and U is the set $W \cap U$ consisting of all those vectors that are simultaneously contained in W and U. Show that $W \cap U$ is a subspace of V.
    b. The **sum** of W and U is the set $W + U$ consisting of all vectors of the form $\mathbf{w} + \mathbf{u}$ with $\mathbf{w}$ in W and $\mathbf{u}$ in U. Show that $W + U$ is a subspace of V.
    c. The **union** of W and U is the set $W \cup U$ consisting of all vectors that belong either to W or to U. Give an example of two subspaces W and U of R^2 such that $W \cup U$ is *not* a subspace of R^2.

19. Show that $\mathbf{Comb}(\mathbf{v}_1, \ldots, \mathbf{v}_k)$ contains each of the vectors $\mathbf{v}_i$.

20. Let $\mathbf{v}, \mathbf{w},$ and $\mathbf{u}$ be nonzero vectors in R^3.
    a. Show that if the lines through the origin determined by $\mathbf{v}, \mathbf{w},$ and $\mathbf{u}$ are the same, then $\mathbf{Comb}(\mathbf{v}, \mathbf{w}, \mathbf{u}) = \mathbf{Comb}(\mathbf{v})$.
    b. Show that if $\mathbf{v}$ and $\mathbf{w}$ determine a plane through the origin and if $\mathbf{u}$ is a linear combination of $\mathbf{v}$ and $\mathbf{w}$, then $\mathbf{Comb}(\mathbf{v}, \mathbf{w}, \mathbf{u}) = \mathbf{Comb}(\mathbf{v}, \mathbf{w})$.

4.3 SPANNING SETS

In the previous section we saw how, starting from a collection $\mathbf{v}_1, \ldots, \mathbf{v}_k$ of vectors in V, we could form the subspace $\mathbf{Comb}(\mathbf{v}_1, \ldots, \mathbf{v}_k)$ of all linear combinations of $\mathbf{v}_1, \ldots, \mathbf{v}_k$. We turn now to the converse problem. Given a subspace W of V, are there vectors $\mathbf{v}_1, \ldots, \mathbf{v}_k$ such that W coincides with $\mathbf{Comb}(\mathbf{v}_1, \ldots, \mathbf{v}_k)$? If there are, then these vectors are said to **span** W.

Definition: *Let W be a subspace of V. A collection $\mathbf{v}_1, \ldots, \mathbf{v}_k$ of vectors is a spanning set for W if $W = \mathbf{Comb}(\mathbf{v}_1, \ldots, \mathbf{v}_k)$.*

Examples 4.2.7 to 4.2.9 provide us with some examples of spanning sets. The subspace W of R^3 consisting of those 3-vectors whose third entry is zero coincides with **Comb**($\mathbf{v}_1$, $\mathbf{v}_2$) where

$$\mathbf{v}_1 = \begin{bmatrix} 1 \\ 0 \\ 0 \end{bmatrix} \quad \text{and} \quad \mathbf{v}_2 = \begin{bmatrix} 0 \\ 1 \\ 0 \end{bmatrix},$$

so $\mathbf{v}_1$ and $\mathbf{v}_2$ form a spanning set for W. The solutions of $(D^3 - D^2)x = 0$ coincide with **Comb**(1, e^t, e^{-t}), so 1, e^t, and e^{-t} span the solution space. If

$$\mathbf{e}_1 = \begin{bmatrix} 1 \\ 0 \\ \cdot \\ \cdot \\ \cdot \\ 0 \end{bmatrix}, \quad \mathbf{e}_2 = \begin{bmatrix} 0 \\ 1 \\ \cdot \\ \cdot \\ \cdot \\ 0 \end{bmatrix}, \ldots, \mathbf{e}_n = \begin{bmatrix} 0 \\ 0 \\ \cdot \\ \cdot \\ \cdot \\ 1 \end{bmatrix},$$

then $R^n = $ **Comb**($\mathbf{e}_1, \ldots, \mathbf{e}_n$), so $\mathbf{e}_1, \ldots, \mathbf{e}_n$ form a spanning set for R^n.

It would be useful, when looking at further examples, to have a test for determining whether given vectors $\mathbf{v}_1, \ldots, \mathbf{v}_k$ span a given subspace. Recall that if $\mathbf{v}_1, \ldots, \mathbf{v}_k$ all lie in W, then so does any linear combination of $\mathbf{v}_1, \ldots, \mathbf{v}_k$; in this case, **Comb**($\mathbf{v}_1, \ldots, \mathbf{v}_k$) is a subset of W. If, in addition, every element in W is a linear combination of $\mathbf{v}_1, \ldots, \mathbf{v}_k$, then each element of W lies in **Comb**($\mathbf{v}_1, \ldots, \mathbf{v}_k$). Thus these two conditions guarantee that W coincides with **Comb**($\mathbf{v}_1, \ldots, \mathbf{v}_k$) so that $\mathbf{v}_1, \ldots, \mathbf{v}_k$ span W.

Fact: *Let W be a subspace of V and let $\mathbf{v}_1, \ldots, \mathbf{v}_k$ be vectors in V. If*

1.   *each of the vectors $\mathbf{v}_1, \ldots, \mathbf{v}_k$ belongs to W, and*
2.   *every element $\mathbf{w}$ of W can be written as a linear combination of $\mathbf{v}_1, \ldots, \mathbf{v}_k$,*

$$\mathbf{w} = c_1\mathbf{v}_1 + \ldots + c_k\mathbf{v}_k$$

for some choice of scalars $c_1, \ldots, c_k$,

then $\mathbf{v}_1, \ldots, \mathbf{v}_k$ span W.

Example 4.3.1

Find a spanning set for the subspace W of R^4 consisting of the solutions to $B\mathbf{v} = \mathbf{0}$, where

$$B = \begin{bmatrix} 1 & 1 & -1 & 0 \\ 1 & 2 & 1 & 1 \\ 3 & 4 & -1 & 1 \end{bmatrix}.$$

To solve the equation, we first reduce the augmented matrix:

$$[B|\mathbf{0}] = \begin{bmatrix} 1 & 1 & -1 & 0 & | & 0 \\ 1 & 2 & 1 & 1 & | & 0 \\ 3 & 4 & -1 & 1 & | & 0 \end{bmatrix} \rightarrow \begin{bmatrix} 1 & 0 & -3 & -1 & | & 0 \\ 0 & 1 & 2 & 1 & | & 0 \\ 0 & 0 & 0 & 0 & | & 0 \end{bmatrix}.$$

We then read off that the solutions are the vectors of the form

$$\mathbf{v} = \begin{bmatrix} 3a + b \\ -2a - b \\ a \\ b \end{bmatrix} = a \begin{bmatrix} 3 \\ -2 \\ 1 \\ 0 \end{bmatrix} + b \begin{bmatrix} 1 \\ -1 \\ 0 \\ 1 \end{bmatrix}.$$

[handwritten margin notes: $a - 3c - d = 0$; $b + 2c + d = 0$; $a - 3c = -b - 2c$; $a + b = c$]

The vectors

$$\mathbf{v}_1 = \begin{bmatrix} 3 \\ -2 \\ 1 \\ 0 \end{bmatrix} \quad \text{and} \quad \mathbf{v}_2 = \begin{bmatrix} 1 \\ -1 \\ 0 \\ 1 \end{bmatrix}$$

are solutions of $B\mathbf{v} = \mathbf{0}$, and every other solution is a linear combination of $\mathbf{v}_1$ and $\mathbf{v}_2$ (with $c_1 = a$ and $c_2 = b$). Thus $\mathbf{v}_1$ and $\mathbf{v}_2$ span W.

Example 4.3.2
    Find a spanning set for the subspace of C_∞ consisting of the solutions to $(D^4 + D^2)x = 0$.
    The characteristic polynomial is $r^4 + r^2 = r^2(r^2 + 1)$, so the general solution of the o.d.e. is

$$x = c_1 + c_2 t + c_3 \cos t + c_4 \sin t.$$

The functions

$$1, \quad t, \quad \cos t, \quad \sin t$$

are solutions with the property that every other solution is a linear combination of these. Hence these four solutions span the solution space.

Example 4.3.3
    Find a spanning set for the subspace of C_∞^3 consisting of the solutions to $D\mathbf{x} =$

$A\mathbf{x}$, where

$$A = \begin{bmatrix} 1 & 0 & 4 \\ 0 & 3 & 0 \\ 1 & 0 & 1 \end{bmatrix}.$$

We found in Example 3.7.1 that the general solution to $D\mathbf{x} = A\mathbf{x}$ is

$$\mathbf{x} = c_1 \begin{bmatrix} -2e^{-t} \\ 0 \\ e^{-t} \end{bmatrix} + c_2 \begin{bmatrix} 0 \\ e^{3t} \\ 0 \end{bmatrix} + c_3 \begin{bmatrix} 2e^{3t} \\ 0 \\ e^{3t} \end{bmatrix}.$$

The specific solutions

$$\begin{bmatrix} -2e^{-t} \\ 0 \\ e^{-t} \end{bmatrix}, \quad \begin{bmatrix} 0 \\ e^{3t} \\ 0 \end{bmatrix}, \quad \begin{bmatrix} 2e^{3t} \\ 0 \\ e^{3t} \end{bmatrix}$$

span the solution space.

Example 4.3.4

Show that the vectors

$$\mathbf{v}_1 = \begin{bmatrix} 1 \\ 1 \end{bmatrix}, \quad \mathbf{v}_2 = \begin{bmatrix} 1 \\ -1 \end{bmatrix}$$

span $W = R^2$.

Each of these vectors belongs to W, so we need only check the second condition of our test. Given an arbitrary vector

$$\mathbf{w} = \begin{bmatrix} x \\ y \end{bmatrix}$$

in W, we must determine whether we can find scalars c_1 and c_2 so that

$$\mathbf{w} = c_1\mathbf{v}_1 + c_2\mathbf{v}_2.$$

If we write this out and equate corresponding entries, we obtain two equations for c_1 and c_2:

$$c_1 + c_2 = x$$

$$c_1 - c_2 = y.$$

We reduce the augmented matrix

$$\begin{bmatrix} 1 & 1 & | & x \\ 1 & -1 & | & y \end{bmatrix} \rightarrow \begin{bmatrix} 1 & 0 & | & \frac{1}{2}(x + y) \\ 0 & 1 & | & \frac{1}{2}(x - y) \end{bmatrix},$$

and read off the solution

$$c_1 = \tfrac{1}{2}(x + y), \quad c_2 = \tfrac{1}{2}(x - y).$$

Thus it is possible to express each vector $\mathbf{w}$ in W as a linear combination of $\mathbf{v}_1$ and $\mathbf{v}_2$:

$$\begin{bmatrix} x \\ y \end{bmatrix} = \frac{1}{2}(x + y)\begin{bmatrix} 1 \\ 1 \end{bmatrix} + \frac{1}{2}(x - y)\begin{bmatrix} 1 \\ -1 \end{bmatrix},$$

and hence $\mathbf{v}_1$ and $\mathbf{v}_2$ span W.

Note that this spanning set is different from the spanning set

$$\mathbf{e}_1 = \begin{bmatrix} 1 \\ 0 \end{bmatrix}, \quad \mathbf{e}_2 = \begin{bmatrix} 0 \\ 1 \end{bmatrix}$$

obtained from Example 4.2.9 by taking $n = 2$.

Example 4.3.5

Show that the vectors

$$\mathbf{w}_1 = \begin{bmatrix} 1 \\ 2 \\ 0 \end{bmatrix}, \quad \mathbf{w}_2 = \begin{bmatrix} 2 \\ 1 \\ 0 \end{bmatrix}, \quad \mathbf{w}_3 = \begin{bmatrix} 1 \\ 1 \\ 0 \end{bmatrix}$$

span the subspace W of R^3 consisting of those 3-vectors whose third entry is zero.

We note first that all three vectors have third entry zero, so all three belong to W. Thus we need only determine whether, given an arbitrary vector in W

$$\mathbf{w} = \begin{bmatrix} x \\ y \\ 0 \end{bmatrix},$$

we can find scalars c_1, c_2, c_3 so that

$$\mathbf{w} = c_1\mathbf{w}_1 + c_2\mathbf{w}_2 + c_3\mathbf{w}_3.$$

If we write out this equation and equate corresponding entries, we obtain equations for c_1, c_2, and c_3:

$$c_1 + 2c_2 + c_3 = x$$

$$2c_1 + c_2 + c_3 = y$$

$$0c_1 + 0c_2 + 0c_3 = 0.$$

To solve these equations, we reduce the augmented matrix:

$$\begin{bmatrix} 1 & 2 & 1 & x \\ 2 & 1 & 1 & y \\ 0 & 0 & 0 & 0 \end{bmatrix} \rightarrow \begin{bmatrix} 1 & 0 & \frac{1}{3} & \frac{1}{3}(2y - x) \\ 0 & 1 & \frac{1}{3} & \frac{1}{3}(2x - y) \\ 0 & 0 & 0 & 0 \end{bmatrix}.$$

Any choice of c_3, say $c_3 = t$, leads to a solution

$$c_1 = \tfrac{1}{3}(2y - x) - \tfrac{1}{3}t, \qquad c_2 = \tfrac{1}{3}(2x - y) - \tfrac{1}{3}t, \qquad c_3 = t.$$

Thus it is possible to express each vector $\mathbf{w}$ in W as a linear combination of $\mathbf{w}_1$, $\mathbf{w}_2$, $\mathbf{w}_3$,

$$\begin{bmatrix} x \\ y \\ 0 \end{bmatrix} = \frac{1}{3}(2y - x - t)\begin{bmatrix} 1 \\ 2 \\ 0 \end{bmatrix} + \frac{1}{3}(2x - y - t)\begin{bmatrix} 2 \\ 1 \\ 0 \end{bmatrix} + t\begin{bmatrix} 1 \\ 1 \\ 0 \end{bmatrix},$$

and hence $\mathbf{w}_1$, $\mathbf{w}_2$, and $\mathbf{w}_3$ span W.

Note that this spanning set for W is different from the spanning set $\mathbf{v}_1$, $\mathbf{v}_2$ in Example 4.2.7.

It is important to keep in mind that we need to check two things in order to decide whether a given set of vectors $\mathbf{v}_1, \ldots, \mathbf{v}_k$ spans a given subspace W. First we must decide whether each of the vectors is itself in W. Then we must decide whether each vector in W is a linear combination of $\mathbf{v}_1, \ldots, \mathbf{v}_k$. *If either of these conditions fails to hold, then the vectors do not span W* (Exercise 11).

Example 4.3.6

Show that the vectors

$$\mathbf{v}_1 = \begin{bmatrix} 1 \\ 0 \\ 0 \end{bmatrix}, \qquad \mathbf{v}_2 = \begin{bmatrix} 0 \\ 1 \\ 0 \end{bmatrix}, \qquad \mathbf{v}_3 = \begin{bmatrix} 1 \\ 1 \\ 1 \end{bmatrix}$$

do *not* span the subspace W of R^3 consisting of the vectors of the form

$$\mathbf{w} = \begin{bmatrix} x \\ y \\ 0 \end{bmatrix}.$$

One of the vectors, namely $\mathbf{v}_3$, does not belong to W. Thus $\mathbf{v}_1$, $\mathbf{v}_2$, $\mathbf{v}_3$ do not span W. Note that each vector $\mathbf{w}$ in W can be expressed as a linear combination of $\mathbf{v}_1$, $\mathbf{v}_2$, $\mathbf{v}_3$.

$$\begin{bmatrix} x \\ y \\ 0 \end{bmatrix} = x\begin{bmatrix} 1 \\ 0 \\ 0 \end{bmatrix} + y\begin{bmatrix} 0 \\ 1 \\ 0 \end{bmatrix} + 0\begin{bmatrix} 1 \\ 1 \\ 1 \end{bmatrix} = x\mathbf{v}_1 + y\mathbf{v}_2 + 0\mathbf{v}_3.$$

Thus $\mathbf{Comb}(\mathbf{v}_1, \mathbf{v}_2, \mathbf{v}_3)$ contains W. However, since $\mathbf{v}_3$ belongs to $\mathbf{Comb}(\mathbf{v}_1, \mathbf{v}_2\ \mathbf{v}_3)$ but not to W, $\mathbf{Comb}(\mathbf{v}_1, \mathbf{v}_2, \mathbf{v}_3)$ is strictly larger than W.

Example 4.3.7
Show that the vectors

$$\mathbf{v}_1 = \begin{bmatrix} 1 \\ 0 \\ 1 \end{bmatrix}, \quad \mathbf{v}_2 = \begin{bmatrix} 1 \\ 1 \\ 0 \end{bmatrix}, \quad \mathbf{v}_3 = \begin{bmatrix} 3 \\ 2 \\ 1 \end{bmatrix}$$

do *not* span $W = R^3$.

Here, each of $\mathbf{v}_1$, $\mathbf{v}_2$, and $\mathbf{v}_3$ belongs to W. However, if we choose an arbitrary element of W,

$$\mathbf{w} = \begin{bmatrix} x \\ y \\ z \end{bmatrix},$$

and attempt to write it as a linear combination of $\mathbf{v}_1$, $\mathbf{v}_2$, $\mathbf{v}_3$,

$$\mathbf{w} = c_1\mathbf{v}_1 + c_2\mathbf{v}_2 + c_3\mathbf{v}_3,$$

we are led to the equations

$$c_1 + c_2 + 3c_3 = x$$
$$c_2 + 2c_3 = y$$
$$c_1 \quad + c_3 = z.$$

To solve these equations for the unknowns c_1, c_2, c_3, we reduce the augmented matrix:

$$\begin{bmatrix} 1 & 1 & 3 & | & x \\ 0 & 1 & 2 & | & y \\ 1 & 0 & 1 & | & z \end{bmatrix} \rightarrow \begin{bmatrix} 1 & 0 & 1 & | & x - y \\ 0 & 1 & 2 & | & y \\ 0 & 0 & 0 & | & -x + y + z \end{bmatrix}.$$

We see from the last row of the reduced matrix that we can solve these equations only if the entries of $\mathbf{w}$ satisfy

$$-x + y + z = 0.$$

The vector

$$\mathbf{w} = \begin{bmatrix} 1 \\ 1 \\ 1 \end{bmatrix}$$

is an example of a vector that belongs to W but that cannot be written as a linear combination of $\mathbf{v}_1, \mathbf{v}_2, \mathbf{v}_3$ (since $-x + y + z = 1 \neq 0$). Thus $\mathbf{v}_1, \mathbf{v}_2, \mathbf{v}_3$ do not span W.

Note that in this example W is strictly larger than $\mathbf{Comb}(\mathbf{v}_1, \mathbf{v}_2, \mathbf{v}_3)$.

The subspaces considered in this section have each had finite spanning sets. In the note following the summary we discuss an example that shows this is not always the case.

SPANNING SETS

A collection $\mathbf{v}_1, \ldots, \mathbf{v}_k$ of vectors in V **spans** the subspace W of V (or forms a **spanning set** for W) provided W coincides with $\mathbf{Comb}(\mathbf{v}_1, \ldots, \mathbf{v}_k)$.

To determine whether a given collection $\mathbf{v}_1, \ldots, \mathbf{v}_k$ of vectors spans a given subspace W, we check to see if

1. each of the vectors $\mathbf{v}_1, \ldots, \mathbf{v}_k$ belongs to W, and
2. every element $\mathbf{w}$ in W is expressible as a linear combination of $\mathbf{v}_1, \ldots, \mathbf{v}_k$.

If both conditions hold, then $\mathbf{v}_1, \ldots, \mathbf{v}_k$ span W. If either fails, then they don't span W.

Note
On the existence of finite spanning sets: an example

One of the important subspaces of C_∞ is the collection P of polynomial functions—that is, the collection of functions $x(t)$ that can be written in the form

$$x(t) = a_0 + a_1 t + \ldots + a_n t^n$$

for some integer n and some choice of real constants $a_0, a_1, \ldots, a_n$. (Some of you have already dealt with this space in Exercise 8, Section 4.1, or in Exercise 4, Section 4.2.) If $x_1(t), \ldots, x_k(t)$ are polynomial functions, then the degree of any linear combination $c_1 x_1(t) + \ldots + c_k x_k(t)$ of these functions will be at most the largest of the degrees of $x_1(t), \ldots, x_k(t)$. In particular, the linear combinations of $x_1(t), \ldots, x_k(t)$ cannot include all of P. Thus P *does not have a finite spanning set.* On the other hand, every element in P can be written as a linear combination of the elements in the infinite set $1, t, \ldots, t^n, \ldots$. P is an example of a space with an *infinite* spanning set.

EXERCISES

1. In each of the following, find a spanning set for the subspace of R^3 consisting of all vectors of the given form.

   a. $\begin{bmatrix} x \\ x \\ 0 \end{bmatrix}$
   b. $\begin{bmatrix} x \\ y \\ x + y \end{bmatrix}$
   c. $\begin{bmatrix} x + y \\ y + z \\ x + z \end{bmatrix}$

   d. $\begin{bmatrix} x + y \\ x - y \\ 2x \end{bmatrix}$
   e. $\begin{bmatrix} x + y \\ x - z \\ y + z \end{bmatrix}$
   f. $\begin{bmatrix} x - y \\ y - x \\ x - y \end{bmatrix}$

2. Find a spanning set for the solution space of $B\mathbf{x} = \mathbf{0}$.

   a. $B = \begin{bmatrix} 1 & 2 & 3 \\ -1 & -2 & -3 \\ 3 & 6 & 9 \end{bmatrix}$
   b. $B = \begin{bmatrix} 1 & 3 & -1 \\ 5 & 2 & 1 \\ 4 & -1 & 2 \end{bmatrix}$

   c. $B = \begin{bmatrix} 1 & 0 & 1 & -1 \\ 2 & 1 & 0 & 1 \\ 1 & 1 & -1 & 2 \\ 4 & 2 & 0 & 2 \end{bmatrix}$
   d. $B = \begin{bmatrix} 0 & 1 & 3 & -2 \\ 1 & 0 & 1 & 1 \\ 0 & 0 & 1 & 2 \\ 2 & 1 & 6 & 2 \end{bmatrix}$

3. Find a spanning set for the solution space of $L\mathbf{x} = 0$.
   a. $L = D^5$ b. $L = D^4 - 8D^2 + 16$ c. $L = D^4 - 1$

4. Find a spanning set for the solution space of $D\mathbf{x} = A\mathbf{x}$.

   a. $A = \begin{bmatrix} -3 & 1 \\ -2 & 0 \end{bmatrix}$ (see Section 3.7, Exercise 6)

   b. $A = \begin{bmatrix} 1 & -1 \\ 1 & 1 \end{bmatrix}$ (see Section 3.8, Exercise 9)

   c. $A = \begin{bmatrix} 0 & 1 \\ -4 & 4 \end{bmatrix}$ (see Section 3.9, Exercise 9)

5. Check to see if the given vectors span R^3.

a. $\begin{bmatrix} 1 \\ -1 \\ 0 \end{bmatrix}, \begin{bmatrix} 0 \\ 1 \\ 2 \end{bmatrix}$ b. $\begin{bmatrix} 1 \\ 1 \\ 0 \end{bmatrix}, \begin{bmatrix} 0 \\ 1 \\ 1 \end{bmatrix}, \begin{bmatrix} 1 \\ 0 \\ 1 \end{bmatrix}$ c. $\begin{bmatrix} 1 \\ 1 \\ 0 \end{bmatrix}, \begin{bmatrix} 0 \\ 1 \\ 1 \end{bmatrix}, \begin{bmatrix} 1 \\ 0 \\ -1 \end{bmatrix}$

d. $\begin{bmatrix} 1 \\ 0 \\ 2 \end{bmatrix}, \begin{bmatrix} 2 \\ 0 \\ 1 \end{bmatrix}, \begin{bmatrix} 0 \\ 1 \\ 0 \end{bmatrix}, \begin{bmatrix} 0 \\ 0 \\ 2 \end{bmatrix}$ e. $\begin{bmatrix} 1 \\ 2 \\ 0 \end{bmatrix}, \begin{bmatrix} 0 \\ 1 \\ 1 \end{bmatrix}, \begin{bmatrix} 1 \\ 3 \\ 1 \end{bmatrix}, \begin{bmatrix} 1 \\ 0 \\ -2 \end{bmatrix}$

6. Check to see if the given vectors span the subspace of R^4 consisting of all vectors of the

form $\begin{bmatrix} x \\ y \\ x + y \\ x - y \end{bmatrix}$.

a. $\begin{bmatrix} 1 \\ 1 \\ 2 \\ 0 \end{bmatrix}, \begin{bmatrix} 1 \\ -1 \\ 0 \\ 2 \end{bmatrix}$ b. $\begin{bmatrix} 1 \\ 1 \\ 2 \\ 0 \end{bmatrix}, \begin{bmatrix} 2 \\ 2 \\ 4 \\ 0 \end{bmatrix}$

c. $\begin{bmatrix} 0 \\ 1 \\ 1 \\ -1 \end{bmatrix}, \begin{bmatrix} 1 \\ 0 \\ 0 \\ 0 \end{bmatrix}, \begin{bmatrix} 1 \\ 0 \\ 1 \\ 1 \end{bmatrix}$ d. $\begin{bmatrix} 1 \\ 1 \\ 2 \\ 0 \end{bmatrix}, \begin{bmatrix} 2 \\ 1 \\ 3 \\ 1 \end{bmatrix}, \begin{bmatrix} 0 \\ 1 \\ 1 \\ -1 \end{bmatrix}$

7. Check to see if the given vectors span the space consisting of those vectors $\begin{bmatrix} x \\ y \\ z \\ u \end{bmatrix}$ for

which $x - y + z = 0$ and $u - x + y + z = 0$.

a. $\begin{bmatrix} 1 \\ 0 \\ -1 \\ 2 \end{bmatrix}, \begin{bmatrix} 1 \\ 1 \\ 0 \\ 0 \end{bmatrix}$ b. $\begin{bmatrix} 1 \\ 0 \\ -1 \\ 2 \end{bmatrix}, \begin{bmatrix} 1 \\ 0 \\ 0 \\ 0 \end{bmatrix}, \begin{bmatrix} 0 \\ 1 \\ 0 \\ 0 \end{bmatrix}$

c. $\begin{bmatrix} 1 \\ 0 \\ -1 \\ 2 \end{bmatrix}, \begin{bmatrix} -1 \\ 0 \\ 1 \\ -2 \end{bmatrix}$ d. $\begin{bmatrix} 1 \\ 0 \\ -1 \\ 2 \end{bmatrix}, \begin{bmatrix} 1 \\ 2 \\ 1 \\ -2 \end{bmatrix}, \begin{bmatrix} 0 \\ 1 \\ 1 \\ -2 \end{bmatrix}$

8. Check to see if the given vectors span C^3.

a. $\begin{bmatrix} i \\ 0 \\ 1 \end{bmatrix}, \begin{bmatrix} 1 \\ i \\ 0 \end{bmatrix}$ b. $\begin{bmatrix} 1 \\ 0 \\ 0 \end{bmatrix}, \begin{bmatrix} 0 \\ 1 \\ 0 \end{bmatrix}, \begin{bmatrix} 0 \\ 0 \\ 1 \end{bmatrix}$

c. $\begin{bmatrix} 1 \\ i \\ 0 \end{bmatrix}, \begin{bmatrix} 0 \\ -i \\ 1 \end{bmatrix}, \begin{bmatrix} 1 \\ 0 \\ 1 \end{bmatrix}$ d. $\begin{bmatrix} 1 \\ i \\ 0 \end{bmatrix}, \begin{bmatrix} 1 \\ -i \\ 0 \end{bmatrix}, \begin{bmatrix} 0 \\ 0 \\ i \end{bmatrix}, \begin{bmatrix} 0 \\ 0 \\ 1 \end{bmatrix}$

9. Check to see if the given polynomials span the subspace P_2 of C_∞ consisting of the functions $x(t)$ that can be written in the form $x(t) = a_0 + a_1 t + a_2 t^2$.

a. $1, t, t^2$ b. $1, 1 + t, 1 + t + t^2$
c. $1 + t, 1 + t^2, t - t^2$ d. $1 + t, 1 + t^2, t + t^2, 1 + t + t^2$

10. Check to see if the given matrices span the space $R_{2 \times 2}$ of 2×2 real matrices.

a. $\begin{bmatrix} 1 & -1 \\ 0 & 0 \end{bmatrix}, \begin{bmatrix} 0 & 0 \\ 1 & -1 \end{bmatrix}, \begin{bmatrix} 1 & 0 \\ 0 & 1 \end{bmatrix}$

b. $\begin{bmatrix} 1 & 0 \\ 0 & 1 \end{bmatrix}, \begin{bmatrix} 0 & 1 \\ 1 & 0 \end{bmatrix}, \begin{bmatrix} 0 & 0 \\ 1 & 0 \end{bmatrix}, \begin{bmatrix} 0 & 0 \\ 0 & 1 \end{bmatrix}$

c. $\begin{bmatrix} -1 & 1 \\ 0 & 1 \end{bmatrix}, \begin{bmatrix} 1 & 0 \\ 1 & 0 \end{bmatrix}, \begin{bmatrix} 0 & 1 \\ 1 & 1 \end{bmatrix}, \begin{bmatrix} 1 & 0 \\ 0 & 0 \end{bmatrix}$

d. $\begin{bmatrix} 1 & 0 \\ 0 & 1 \end{bmatrix}, \begin{bmatrix} 0 & 1 \\ 0 & 0 \end{bmatrix}, \begin{bmatrix} 1 & 1 \\ 0 & 0 \end{bmatrix}, \begin{bmatrix} 0 & 0 \\ 1 & 0 \end{bmatrix}, \begin{bmatrix} 1 & 1 \\ 1 & 1 \end{bmatrix}$

More abstract problems:

11. We saw above that if $\mathbf{v}_1, \ldots, \mathbf{v}_k$ lie in a subspace W, and if every element $\mathbf{w}$ of W can be written as a linear combination of these vectors, then they span W. Show conversely, that if $\mathbf{v}_1, \ldots, \mathbf{v}_k$ span W, then they lie in W, and every element of W is a linear combination of these vectors.

12. Suppose that $\mathbf{v}_1, \ldots, \mathbf{v}_k$ span W and that $\mathbf{v}_j$ is a linear combination of the other vectors $\mathbf{v}_1, \ldots, \mathbf{v}_{j-1}, \mathbf{v}_{j+1}, \ldots, \mathbf{v}_k$. Show that $\mathbf{v}_1, \ldots, \mathbf{v}_{j-1}, \mathbf{v}_{j+1}, \ldots, \mathbf{v}_k$ span W.

13. Let W and U be subspaces of a vector space V. Suppose that $\mathbf{w}_1, \ldots, \mathbf{w}_k$ span W and $\mathbf{u}_1, \ldots, \mathbf{u}_j$ span U. Show that $\mathbf{w}_1, \ldots, \mathbf{w}_k, \mathbf{u}_1, \ldots, \mathbf{u}_j$ span the subspace $W + U$ consisting of the vectors of the form $\mathbf{w} + \mathbf{u}$ with $\mathbf{w}$ in W and $\mathbf{u}$ in U.

4.4 LINEAR INDEPENDENCE AND BASES

If $\mathbf{v}_1, \ldots, \mathbf{v}_k$ span a subspace W, then every vector $\mathbf{w}$ in W can be written as a linear combination of $\mathbf{v}_1, \ldots, \mathbf{v}_k$

$$\mathbf{w} = c_1 \mathbf{v}_1 + \ldots + c_k \mathbf{v}_k.$$

Sometimes, as in Example 4.3.4, the coefficients $c_1, \ldots, c_k$ in this expression are uniquely determined by $\mathbf{w}$. At other times, as in Example 4.3.5, there are multiple choices of $c_1, \ldots, c_k$, all of which yield the same vector $\mathbf{w}$. In this section we will see that the difference between these two cases is determined by considering the special case $\mathbf{w} = \mathbf{0}$.

Given any vectors $\mathbf{v}_1, \ldots, \mathbf{v}_k$, we can always find one choice of scalars so that

$$c_1 \mathbf{v}_1 + \ldots + c_k \mathbf{v}_k = \mathbf{0},$$

since the trivial choice of values

$$c_1 = c_2 = \ldots = c_k = 0$$

always works. If this is the *only* choice that works, then we say that the vectors $v_1, \ldots, v_k$ are **linearly independent**. On the other hand, if there is another choice of scalars that works—that is, if there exist scalars $c_1, \ldots, c_k$, with at least one $c_i \neq 0$, so that $c_1 v_1 + \ldots + c_k v_k = \mathbf{0}$—then we say that $v_1, \ldots, v_k$ are **linearly dependent**.

We have already seen these notions in the context of certain special vector spaces. When V is C_∞, these definitions coincide with those in Section 2.6. When V is R^n, they coincide with those in Section 3.4. We've already seen many examples of linear dependence and independence in these spaces. Let's consider some examples in C_∞^3.

Example 4.4.1

Show that

$$\mathbf{x}_1(t) = \begin{bmatrix} e^t \\ 0 \\ e^t \end{bmatrix}, \qquad \mathbf{x}_2(t) = \begin{bmatrix} te^t \\ e^t \\ 0 \end{bmatrix}, \qquad \mathbf{x}_3(t) = \begin{bmatrix} e^{-t} \\ 0 \\ 0 \end{bmatrix}$$

are linearly independent vectors in C_∞^3.

Suppose

$$c_1 \mathbf{x}_1(t) + c_2 \mathbf{x}_2(t) + c_3 \mathbf{x}_3(t) = \mathbf{0}(t).$$

Then we have

$$
\begin{aligned}
c_1 e^t + c_2 t e^t + c_3 e^{-t} &= 0 \\
c_2 e^t &= 0 \\
c_1 e^t &= 0
\end{aligned}
$$

for all t. The last two equations force

$$c_2 = 0 \quad \text{and} \quad c_1 = 0.$$

Substituting these values into the first equation forces

$$c_3 = 0.$$

Since $c_1 = c_2 = c_3 = 0$ is the only possible choice, the vectors are linearly independent.

Example 4.4.2
Show that

$$\mathbf{x}_1(t) = \begin{bmatrix} e^t \\ 0 \\ e^t \end{bmatrix}, \qquad \mathbf{x}_2(t) = \begin{bmatrix} te^t \\ e^t \\ 0 \end{bmatrix}, \qquad \mathbf{x}_3(t) = \begin{bmatrix} (2t - 3)e^t \\ 2e^t \\ -3e^t \end{bmatrix}$$

are linearly dependent vectors in C_∞^3.

If we look for scalars so that

$$c_1\mathbf{x}_1(t) + c_2\mathbf{x}_2(t) + c_3\mathbf{x}_3(t) = \mathbf{0}(t),$$

we are led to the equations

$$c_1 e^t + c_2 te^t + c_3(2t - 3)e^t = 0$$
$$c_2 e^t + 2c_3 e^t = 0$$
$$c_1 e^t - 3c_3 e^t = 0.$$

To solve these equations, we reduce the augmented matrix:

$$\begin{bmatrix} e^t & te^t & (2t - 3)e^t & | & 0 \\ 0 & e^t & 2e^t & | & 0 \\ e^t & 0 & -3e^t & | & 0 \end{bmatrix} \rightarrow \begin{bmatrix} 1 & 0 & -3 & | & 0 \\ 0 & 1 & 2 & | & 0 \\ 0 & 0 & 0 & | & 0 \end{bmatrix}.$$

We see from the reduced matrix that we can find a solution with at least one $c_i \neq 0$. For example, if we choose $c_3 = 1$, $c_1 = 3$, and $c_2 = -2$, then

$$c_1\mathbf{x}_1(t) + c_2\mathbf{x}_2(t) + c_3\mathbf{x}_3(t) = 3\begin{bmatrix} e^t \\ 0 \\ e^t \end{bmatrix} - 2\begin{bmatrix} te^t \\ e^t \\ 0 \end{bmatrix} + \begin{bmatrix} (2t - 3)e^t \\ 2e^t \\ -3e^t \end{bmatrix} = \begin{bmatrix} 0 \\ 0 \\ 0 \end{bmatrix} = \mathbf{0}(t).$$

Thus $\mathbf{x}_1(t)$, $\mathbf{x}_2(t)$, $\mathbf{x}_3(t)$ are linearly dependent.

Let's now investigate the different ways in which a vector $\mathbf{w}$ in $\mathbf{Comb}(\mathbf{v}_1, \ldots, \mathbf{v}_k)$ can be written as a linear combination of $\mathbf{v}_1, \ldots, \mathbf{v}_k$. Since $\mathbf{w}$ is in $\mathbf{Comb}(\mathbf{v}_1, \ldots, \mathbf{v}_k)$, we are guaranteed at least one such expression

$$\mathbf{w} = a_1\mathbf{v}_1 + \ldots + a_k\mathbf{v}_k.$$

Let's try to find another

$$\mathbf{w} = b_1\mathbf{v}_1 + \ldots + b_k\mathbf{v}_k.$$

If we subtract the first equation from the second, we obtain

$$\mathbf{0} = (b_1 - a_1)\mathbf{v}_1 + \ldots + (b_k - a_k)\mathbf{v}_k.$$

Now if $\mathbf{v}_1, \ldots, \mathbf{v}_k$ are linearly independent, then this last equation forces

$$b_1 - a_1 = b_2 - a_2 = \ldots = b_k - a_k = 0$$

so that

$$b_1 = a_1, \qquad b_2 = a_2, \ldots, b_k = a_k;$$

in this case, any two expressions are identical. On the other hand, if $\mathbf{v}_1, \ldots, \mathbf{v}_k$ are linearly dependent, then there exist scalars $c_1, \ldots, c_k$, with at least one $c_i \neq 0$, so that

$$c_1\mathbf{v}_1 + \ldots + c_k\mathbf{v}_k = \mathbf{0};$$

in this case we can choose $b_1, \ldots, b_k$ so that

$$b_1 - a_1 = c_1, \qquad b_2 - a_2 = c_2, \ldots, b_k - a_k = c_k$$

and thereby obtain an expression with at least one b_i that differs from the corresponding a_i. This shows the following:

Fact: *If the vectors* $\mathbf{v}_1, \ldots, \mathbf{v}_k$ *are linearly independent, then every vector* $\mathbf{w}$ *in* **Comb**$(\mathbf{v}_1, \ldots, \mathbf{v}_k)$ *has a unique expression as a linear combination of* $\mathbf{v}_1, \ldots, \mathbf{v}_k$. *If the vectors are linearly dependent, then every such vector* $\mathbf{w}$ *can be expressed in more than one way as a linear combination of* $\mathbf{v}_1, \ldots, \mathbf{v}_k$.

This observation explains the difference between Examples 4.3.4 and 4.3.5. On the one hand, the vectors

$$\mathbf{v}_1 = \begin{bmatrix} 1 \\ 1 \end{bmatrix}, \qquad \mathbf{v}_2 = \begin{bmatrix} 1 \\ -1 \end{bmatrix}$$

are linearly independent (check this), so each vector in $\mathbf{Comb}(\mathbf{v}_1, \mathbf{v}_2) = R^2$ has a unique expression as a linear combination of $\mathbf{v}_1$ and $\mathbf{v}_2$. We found this expression in Example 4.3.4:

$$\begin{bmatrix} x \\ y \end{bmatrix} = \frac{1}{2}(x + y)\begin{bmatrix} 1 \\ 1 \end{bmatrix} + \frac{1}{2}(x - y)\begin{bmatrix} 1 \\ -1 \end{bmatrix}.$$

On the other hand, the vectors

$$\mathbf{w}_1 = \begin{bmatrix} 1 \\ 2 \\ 0 \end{bmatrix}, \qquad \mathbf{w}_2 = \begin{bmatrix} 2 \\ 1 \\ 0 \end{bmatrix}, \qquad \mathbf{w}_3 = \begin{bmatrix} 1 \\ 1 \\ 0 \end{bmatrix}$$

are linearly dependent, since

$$\mathbf{w}_1 + \mathbf{w}_2 - 3\mathbf{w}_3 = \mathbf{0}.$$

We saw in Example 4.3.5 that an arbitrary vector $\mathbf{w}$ in $\mathbf{Comb}(\mathbf{w}_1, \mathbf{w}_2, \mathbf{w}_3)$ could be written as follows:

$$\begin{bmatrix} x \\ y \\ 0 \end{bmatrix} = \frac{1}{3}(2x - x - t)\begin{bmatrix} 1 \\ 2 \\ 0 \end{bmatrix} + \frac{1}{3}(2x - y - t)\begin{bmatrix} 2 \\ 1 \\ 0 \end{bmatrix} + t\begin{bmatrix} 1 \\ 1 \\ 0 \end{bmatrix}.$$

Each value of t yields a different expression for $\mathbf{w}$.

If we are given a subspace W and if we find vectors $\mathbf{v}_1, \ldots, \mathbf{v}_k$ that span W and at the same time are linearly independent, then we have the desirable situation that every vector $\mathbf{w}$ in W can be written in exactly one way as a linear combination of $\mathbf{v}_1, \ldots, \mathbf{v}_k$. We call such a set of vectors a basis for W.

Definition: *If $\mathbf{v}_1, \ldots, \mathbf{v}_k$ are linearly independent vectors that span the subspace W, then we say that the collection $\mathbf{v}_1, \ldots, \mathbf{v}_k$ is a **basis** for W.*

Example 4.4.3

We saw in Example 4.2.9 that the n-vectors

$$\mathbf{e}_1 = \begin{bmatrix} 1 \\ 0 \\ \cdot \\ \cdot \\ \cdot \\ 0 \end{bmatrix}, \qquad \mathbf{e}_2 = \begin{bmatrix} 0 \\ 1 \\ \cdot \\ \cdot \\ \cdot \\ 0 \end{bmatrix}, \ldots, \mathbf{e}_n = \begin{bmatrix} 0 \\ 0 \\ \cdot \\ \cdot \\ \cdot \\ 1 \end{bmatrix}$$

span R^n. To check for independence of these vectors, we look for scalars $c_1, \ldots, c_n$ so that

$$c_1\mathbf{e}_1 + c_2\mathbf{e}_2 + \ldots + c_n\mathbf{e}_n = \mathbf{0}.$$

Combining the left side into a single vector gives

$$\begin{bmatrix} c_1 \\ \cdot \\ \cdot \\ \cdot \\ c_n \end{bmatrix} = \mathbf{0}$$

so

$$c_1 = \ldots = c_n = 0.$$

Thus $\mathbf{e}_1, \ldots, \mathbf{e}_n$ are independent. Putting these observations together, we see that $\mathbf{e}_1, \ldots, \mathbf{e}_n$ form a basis for R^n.

This particular basis is called the **standard basis** of R^n. When $n = 3$, the vectors $\mathbf{e}_1$, $\mathbf{e}_2$, and $\mathbf{e}_3$ can be visualized as unit vectors along the three coordinate axes. In physics, these vectors are often denoted by $\mathbf{i}$, $\mathbf{j}$, and $\mathbf{k}$.

Example 4.4.4

The vectors

$$\mathbf{v}_1 = \begin{bmatrix} 1 \\ 0 \\ 0 \end{bmatrix}, \qquad \mathbf{v}_2 = \begin{bmatrix} 0 \\ 1 \\ 0 \end{bmatrix}$$

are linearly independent (why?) and span the subspace W of R^3 consisting of those 3-vectors whose third entry is zero (see Example 4.2.7). Thus they form a basis of W.

The vectors

$$\mathbf{w}_1 = \begin{bmatrix} 1 \\ 2 \\ 0 \end{bmatrix}, \qquad \mathbf{w}_2 = \begin{bmatrix} 2 \\ 1 \\ 0 \end{bmatrix}, \qquad \mathbf{w}_3 = \begin{bmatrix} 1 \\ 1 \\ 0 \end{bmatrix}$$

also span W. However, these vectors are linearly dependent, so they do *not* form a basis of W.

Note that an arbitrary vector $\mathbf{w}$ in W can be written as a linear combination of $\mathbf{w}_1$ and $\mathbf{w}_2$ (without using $\mathbf{w}_3$)

$$\mathbf{w} = \begin{bmatrix} x \\ y \\ 0 \end{bmatrix} = \frac{1}{3}(2y - x)\begin{bmatrix} 1 \\ 2 \\ 0 \end{bmatrix} + \frac{1}{3}(2x - y)\begin{bmatrix} 2 \\ 1 \\ 0 \end{bmatrix}$$

so $\mathbf{w}_1$ and $\mathbf{w}_2$ span W. If we look for scalars c_1 and c_2 so that

$$c_1\mathbf{w}_1 + c_2\mathbf{w}_2 = \mathbf{0},$$

we are led to the equations

$$c_1 + 2c_2 = 0$$
$$2c_1 + c_2 = 0.$$

The only solution of these equations is

$$c_1 = c_2 = 0$$

(check this), so $\mathbf{w}_1$ and $\mathbf{w}_2$ are linearly independent. Hence $\mathbf{w}_1$, $\mathbf{w}_2$ form another basis for W.

We have dealt with bases several times before. Indeed, our solution methods for several problems discussed in Chapters 2 and 3 involve the determination of bases. To solve an nth-order linear homogeneous o.d.e. $Lx = 0$, we find n independent solutions $h_1(t), \ldots, h_n(t)$ and read off the general solution $x = c_1h_1(t) + \ldots + c_nh_n(t)$. The functions $h_1(t), \ldots, h_n(t)$ form a basis for the space of solutions to $Lx = 0$. The general solution of an nth-order system of o.d.e.'s $D\mathbf{x} = A\mathbf{x}$ is of the form $\mathbf{x} = c_1\mathbf{h}_1(t) + \ldots + c_n\mathbf{h}_n(t)$, where $\mathbf{h}_1(t), \ldots, \mathbf{h}_n(t)$ have independent initial vectors. The independence of the initial vectors guarantees that the vector valued functions are themselves independent (Exercise 24), so that $\mathbf{h}_1(t), \ldots, \mathbf{h}_n(t)$ form a basis for the solution space of $D\mathbf{x} = A\mathbf{x}$. Finally, recall that if λ is an eigenvalue of the $n \times n$ matrix A, then we can find independent eigenvectors corresponding to λ in the usual way: we reduce the matrix $[A - \lambda I \mid \mathbf{0}]$ and then list the solutions obtained by setting each noncorner variable in turn equal to 1 while setting all other noncorner variables equal to 0. The eigenvectors so obtained form a basis for the space of solutions of $(A - \lambda I)\mathbf{v} = \mathbf{0}$ (see the Fact preceding Example 3.7.4).

The usefulness of a basis for a subspace W lies in our observation that each vector in W has a *unique* expression as a linear combination of the basis vectors. The coefficients in this expression serve as a set of coordinates for identifying the different elements of W.

Definition: *Given a subspace W and a basis* $\mathbf{v}_1, \ldots, \mathbf{v}_k$ *for W, the* **coordinates** *of any element* $\mathbf{w}$ *of W with respect to this basis are the coefficients* $c_1, \ldots, c_k$ *that appear when we write* $\mathbf{w}$ *as a linear combination of* $\mathbf{v}_1, \ldots, \mathbf{v}_k$:

$$\mathbf{w} = c_1\mathbf{v}_1 + \ldots + c_k\mathbf{v}_k.$$

Example 4.4.5

The expression for an arbitrary element of $W = R^n$

$$\mathbf{x} = \begin{bmatrix} x_1 \\ x_2 \\ \cdot \\ \cdot \\ \cdot \\ x_n \end{bmatrix}$$

in terms of the standard basis

$$\mathbf{e}_1 = \begin{bmatrix} 1 \\ 0 \\ \cdot \\ \cdot \\ \cdot \\ 0 \end{bmatrix}, \qquad \mathbf{e}_2 = \begin{bmatrix} 0 \\ 1 \\ \cdot \\ \cdot \\ \cdot \\ 0 \end{bmatrix}, \ldots, \mathbf{e}_n = \begin{bmatrix} 0 \\ 0 \\ \cdot \\ \cdot \\ \cdot \\ 1 \end{bmatrix}$$

is easily seen to be

$$\mathbf{x} = x_1\mathbf{e}_1 + x_2\mathbf{e}_2 + \ldots + x_n\mathbf{e}_n.$$

Thus *the coordinates of the n-vector* $\mathbf{x}$ *with respect to the standard basis of* R^n *are its entries*

$$x_1, \qquad x_2, \ldots, x_n.$$

Example 4.4.6

The functions

$$x_1(t) = 1, \qquad x_2(t) = t, \qquad x_3(t) = e^t$$

form a basis for the solution space of $(D^3 - D^2)x = 0$ (check this). Find the coordinates with respect to this basis of the solution $x(t)$ that satisfies the initial condition

$$x(0) = 1, \qquad x'(0) = -1, \qquad x''(0) = 2.$$

The coordinates are the constants c_1, c_2, c_3 that appear when we write $x(t)$ in the form

$$x(t) = c_1 x_1(t) + c_2 x_2(t) + c_3 x_3(t)$$
$$= c_1 \qquad + c_2 t \qquad + c_3 e^t.$$

Differentiating twice gives

$$x'(t) = \qquad c_2 \qquad + c_3 e^t$$
$$x''(t) = \qquad\qquad\quad c_3 e^t.$$

Substitution of our initial condition yields the equations

$$1 = c_1 \quad + c_3$$
$$-1 = \quad c_2 + c_3$$
$$2 = \qquad\quad c_3,$$

which we solve to find the required coordinates

$$c_1 = -1, \qquad c_2 = -3, \qquad c_3 = 2.$$

Example 4.4.7

In Example 4.4.4 we described two bases for the subspace W of 3-vectors with third entry zero,

$$\mathbf{v}_1 = \begin{bmatrix} 1 \\ 0 \\ 0 \end{bmatrix}, \quad \mathbf{v}_2 = \begin{bmatrix} 0 \\ 1 \\ 0 \end{bmatrix}$$

and

$$\mathbf{w}_1 = \begin{bmatrix} 1 \\ 2 \\ 0 \end{bmatrix}, \quad \mathbf{w}_2 = \begin{bmatrix} 2 \\ 1 \\ 0 \end{bmatrix}.$$

Find the coordinates of

$$\mathbf{w} = \begin{bmatrix} 3 \\ 2 \\ 0 \end{bmatrix}$$

with respect to each of these bases.

To find the coordinates with respect to $\mathbf{v}_1, \mathbf{v}_2$, we write $\mathbf{w}$ as a linear combination of these vectors:

$$\mathbf{w} = \begin{bmatrix} 3 \\ 2 \\ 0 \end{bmatrix} = 3 \begin{bmatrix} 1 \\ 0 \\ 0 \end{bmatrix} + 2 \begin{bmatrix} 0 \\ 1 \\ 0 \end{bmatrix}.$$

The coordinates of $\mathbf{w}$ with respect to this basis are the first two entries of $\mathbf{w}$:

$$3, \ 2.$$

We noted in Example 4.4.4 that

$$\begin{bmatrix} x \\ y \\ 0 \end{bmatrix} = \frac{1}{3} (2y - x)\mathbf{w}_1 + \frac{1}{3} (2x - y)\mathbf{w}_2.$$

Taking $x = 3$ and $y = 2$, we see that

$$\mathbf{w} = \begin{bmatrix} 3 \\ 2 \\ 0 \end{bmatrix} = \frac{1}{3} \mathbf{w}_1 + \frac{4}{3} \mathbf{w}_2.$$

The coordinates of $\mathbf{w}$ with respect to $\mathbf{w}_1, \mathbf{w}_2$ are

$$\frac{1}{3}, \ \frac{4}{3}.$$

The preceding example illustrates an important fact. *Different bases for the same subspace define different coordinate systems.*

Let's summarize.

LINEAR INDEPENDENCE AND BASES

If we can find scalars, at least one of which is not zero, so that

$$c_1 \mathbf{v}_1 + \ldots + c_k \mathbf{v}_k = \mathbf{0},$$

then we say that $\mathbf{v}_1, \ldots, \mathbf{v}_k$ are **linearly dependent**. If the only scalars for which this equation holds are

$$c_1 = c_2 = \ldots = c_k = 0,$$

then $\mathbf{v}_1, \ldots, \mathbf{v}_k$ are **linearly independent**.

If $\mathbf{v}_1, \ldots, \mathbf{v}_k$ are linearly independent vectors that span the subspace W, then we say that $\mathbf{v}_1, \ldots, \mathbf{v}_k$ form a **basis** for W.

If $\mathbf{v}_1, \ldots, \mathbf{v}_k$ form a basis for W, then any vector $\mathbf{w}$ in W can be expressed in exactly one way as a linear combination of $\mathbf{v}_1, \ldots, \mathbf{v}_k$:

$$\mathbf{w} = c_1 \mathbf{v}_1 + \ldots + c_k \mathbf{v}_k.$$

The scalars $c_1, \ldots, c_k$ that appear in this expression are called the **coordinates** of $\mathbf{w}$ with respect to the basis $\mathbf{v}_1, \ldots, \mathbf{v}_k$.

EXERCISES

In Exercises 1 to 6, check to see if the sets of vectors previously considered in the indicated exercise of Section 4.3 are
(i)   linearly independent, and
(ii) a basis for the subspace considered in the indicated exercise.

1. Exercise 5   2. Exercise 6
3. Exercise 7   4. Exercise 8
5. Exercise 9   6. Exercise 10

In Exercises 7 to 10, find a basis for the space considered in the indicated exercise of Section 4.3.

7. Exercise 1   8. Exercise 2
9. Exercise 3   10. Exercise 4
11. Find a basis for C^n.
12. a. Show that $x_1(t) = 1$, $x_2(t) = t$, $x_3(t) = t^2$ form a basis for P_2.
    b. Find a similar basis for P_m.

13. a. Show that

$$M_{11} = \begin{bmatrix} 1 & 0 \\ 0 & 0 \end{bmatrix}, M_{12} = \begin{bmatrix} 0 & 1 \\ 0 & 0 \end{bmatrix}, M_{21} = \begin{bmatrix} 0 & 0 \\ 1 & 0 \end{bmatrix}, M_{22} = \begin{bmatrix} 0 & 0 \\ 0 & 1 \end{bmatrix}$$

   form a basis for $R_{2 \times 2}$.
   b. Find a similar basis for $R_{m \times n}$.
   c. Find a basis for $C_{m \times n}$.

14. a. Recall from Exercise 5 of Section 4.1 that the complex numbers can be viewed as a *real* vector space. Show that 1 and i form a basis for this space.
   b. The complex numbers can also be viewed as a *complex* vector space (with respect to the usual addition and multiplication). Do 1 and i form a basis for this space?

15. Find the coordinates of $\begin{bmatrix} 1 \\ 3 \\ -1 \end{bmatrix}$ with respect to each of the following bases of R^3.

   a. $\begin{bmatrix} 1 \\ 0 \\ 0 \end{bmatrix}, \begin{bmatrix} 0 \\ 2 \\ 0 \end{bmatrix}, \begin{bmatrix} 0 \\ 0 \\ -1 \end{bmatrix}$ b. $\begin{bmatrix} 1 \\ 0 \\ 1 \end{bmatrix}, \begin{bmatrix} 1 \\ 1 \\ 0 \end{bmatrix}, \begin{bmatrix} 0 \\ 0 \\ 1 \end{bmatrix}$

   c. $\begin{bmatrix} 0 \\ 0 \\ 1 \end{bmatrix}, \begin{bmatrix} 0 \\ 1 \\ 0 \end{bmatrix}, \begin{bmatrix} 1 \\ 0 \\ 0 \end{bmatrix}$ d. $\begin{bmatrix} 1 \\ 0 \\ 0 \end{bmatrix}, \begin{bmatrix} 1 \\ 1 \\ 0 \end{bmatrix}, \begin{bmatrix} 1 \\ 1 \\ 1 \end{bmatrix}$

16. Find the coordinates of the given vectors in R^3 with respect to the basis $\begin{bmatrix} 1 \\ 1 \\ 1 \end{bmatrix}, \begin{bmatrix} 0 \\ 1 \\ 1 \end{bmatrix}, \begin{bmatrix} 0 \\ 0 \\ 1 \end{bmatrix}$.

   a. $\begin{bmatrix} 1 \\ -1 \\ 2 \end{bmatrix}$ b. $\begin{bmatrix} 3 \\ 2 \\ 1 \end{bmatrix}$ c. $\begin{bmatrix} 1 \\ 0 \\ 0 \end{bmatrix}$

17. Find the coordinates of each of the following vectors in C^2 with respect to the basis $\begin{bmatrix} 1 \\ i \end{bmatrix}, \begin{bmatrix} 1 \\ -i \end{bmatrix}$.

   a. $\begin{bmatrix} 0 \\ i \end{bmatrix}$ b. $\begin{bmatrix} 1 \\ 0 \end{bmatrix}$ c. $\begin{bmatrix} i \\ 0 \end{bmatrix}$ d. $\begin{bmatrix} 1 \\ 1 \end{bmatrix}$

18. Given the vector space V and a basis v_1, v_2, v_3, v_4, find the vector in V whose coordinates are $-1, 0, 2, 3$.

   a. $V = R^4$; $v_1 = \begin{bmatrix} 1 \\ 0 \\ 0 \\ 0 \end{bmatrix}, v_2 = \begin{bmatrix} 0 \\ 1 \\ 0 \\ 0 \end{bmatrix}, v_3 = \begin{bmatrix} 0 \\ 0 \\ 1 \\ 0 \end{bmatrix}, v_4 = \begin{bmatrix} 0 \\ 0 \\ 0 \\ 1 \end{bmatrix}$

b. $V = R^4$; $\mathbf{v}_1 = \begin{bmatrix} 1 \\ 1 \\ 0 \\ 0 \end{bmatrix}$, $\mathbf{v}_2 = \begin{bmatrix} 1 \\ -1 \\ 0 \\ 0 \end{bmatrix}$, $\mathbf{v}_3 = \begin{bmatrix} 1 \\ 0 \\ 1 \\ 0 \end{bmatrix}$, $\mathbf{v}_4 = \begin{bmatrix} 1 \\ 0 \\ 0 \\ 1 \end{bmatrix}$

c. $V = R^4$; $\mathbf{v}_1 = \begin{bmatrix} 1 \\ 0 \\ 0 \\ 1 \end{bmatrix}$, $\mathbf{v}_2 = \begin{bmatrix} 1 \\ 0 \\ 1 \\ 0 \end{bmatrix}$, $\mathbf{v}_3 = \begin{bmatrix} 1 \\ -1 \\ 0 \\ 0 \end{bmatrix}$, $\mathbf{v}_4 = \begin{bmatrix} 1 \\ 1 \\ 0 \\ 0 \end{bmatrix}$

d. $V = R_{2 \times 2}$; $\mathbf{v}_1 = \begin{bmatrix} 1 & 0 \\ 0 & 0 \end{bmatrix}$, $\mathbf{v}_2 = \begin{bmatrix} 0 & 1 \\ 0 & 0 \end{bmatrix}$, $\mathbf{v}_3 = \begin{bmatrix} 0 & 0 \\ 1 & 0 \end{bmatrix}$, $\mathbf{v}_4 = \begin{bmatrix} 0 & 0 \\ 0 & 1 \end{bmatrix}$

e. $V = P_3$; $\mathbf{v}_1 = 1 + t$, $\mathbf{v}_2 = 1 + t^2$, $\mathbf{v}_3 = 1 + t^3$, $\mathbf{v}_4 = t^3$

More abstract problems:

19. Let $\mathbf{v}_1, \ldots, \mathbf{v}_k$ be vectors in a vector space V.
  a. Show that if one of these vectors is a linear combination of the others,

$$\mathbf{v}_j = a_1\mathbf{v}_1 + \ldots + a_{j-1}\mathbf{v}_{j-1} + a_{j+1}\mathbf{v}_{j+1} + \ldots + a_k\mathbf{v}_k,$$

  then $\mathbf{v}_1, \ldots, \mathbf{v}_k$ are linearly dependent.
  b. Show that if $\mathbf{v}_1, \ldots, \mathbf{v}_k$ are linearly dependent, then one of these vectors can be written as a linear combination of the others.

20. Suppose $\mathbf{v}_1, \ldots, \mathbf{v}_n$ span a subspace W. Use the preceding exercise, together with Exercise 12 of Section 4.3, to show that there is a subset of $\mathbf{v}_1, \ldots, \mathbf{v}_n$ that forms a basis for W. (*Hint:* If $\mathbf{v}_1, \ldots, \mathbf{v}_n$ are linearly independent, they form a basis. Otherwise, one of these vectors is a linear combination of the others. Discard this vector.)

21. Suppose $\mathbf{v}_1, \ldots, \mathbf{v}_j$ are linearly independent vectors that lie in a subspace W. Prove that either $\mathbf{v}_1, \ldots, \mathbf{v}_j$ form a basis for W or there is a vector $\mathbf{v}_{j+1}$ in W so that $\mathbf{v}_1, \ldots, \mathbf{v}_j$, $\mathbf{v}_{j+1}$ are linearly independent. (*Hint:* use Exercise 20b.)

22. Let V be a vector space and let $\mathcal{B} = \{\mathbf{v}_1, \ldots, \mathbf{v}_n\}$ be a basis for V. Then each vector $\mathbf{v}$ in V can be expressed uniquely in the form $\mathbf{v} = c_1\mathbf{v}_1 + \ldots + c_n\mathbf{v}_n$. Define the **coordinate vector** of $\mathbf{v}$ with respect to $\mathcal{B}$ to be the n-vector

$$_{\mathcal{B}}[\mathbf{v}] = \begin{bmatrix} c_1 \\ \cdot \\ \cdot \\ c_n \end{bmatrix}.$$

  Thus the coordinate vector of $\mathbf{v}$ is obtained by placing the coordinates of $\mathbf{v}$ into a column. Show that if $\mathbf{v}$ and $\mathbf{w}$ are vectors in V and if c is a scalar, then
  a. $_{\mathcal{B}}[\mathbf{v} + \mathbf{w}] = {}_{\mathcal{B}}[\mathbf{v}] + {}_{\mathcal{B}}[\mathbf{w}]$ b. $_{\mathcal{B}}[c\mathbf{v}] = c_{\mathcal{B}}[\mathbf{v}]$

23. Let V be a vector space, and let $\mathbf{v}_1, \mathbf{v}_2, \ldots, \mathbf{v}_n$ be a basis for V. Let a be any scalar. Show that $\mathbf{v}_1 + a\mathbf{v}_2, \mathbf{v}_2, \ldots, \mathbf{v}_n$ is also a basis for V.

24. Suppose $\mathbf{h}_1(t), \ldots, \mathbf{h}_n(t)$ are elements of C_∞^n that have independent initial vectors $\mathbf{h}_1(t_0), \ldots, \mathbf{h}_n(t_0)$. Show that $\mathbf{h}_1(t), \ldots, \mathbf{h}_n(t)$ are independent.

4.5 DIMENSION

The general solution of an nth-order homogeneous linear o.d.e. or system of o.d.e.'s is always written as a linear combination of specific solutions that form a basis for the space of solutions. We saw in Sections 2.6 and 3.4 that any n independent solutions form such a basis. Thus, once we have the correct number of independent solutions, we are automatically guaranteed that these solutions also span the solution space. In this section we will see that this is typical of vector spaces that have a finite basis. A key part of our discussion is the verification of an observation you may already have made. While a space may have many different bases, the sizes of these bases must always be the same.

We begin by noting a connection between the number of elements in a basis and the possible sizes of independent sets of vectors. Suppose $\mathbf{v}_1, \ldots, \mathbf{v}_n$ form a basis for a subspace W and that we are given vectors $\mathbf{w}_1, \ldots, \mathbf{w}_m$ in W, where $m > n$. Since $\mathbf{v}_1, \ldots, \mathbf{v}_n$ span W, each $\mathbf{w}_i$ can be written as a linear combination of the $\mathbf{v}_j$'s:

$$\mathbf{w}_1 = a_{11}\mathbf{v}_1 + \ldots + a_{n1}\mathbf{v}_n$$

$$.$$
$$.$$
$$.$$

$$\mathbf{w}_m = a_{1m}\mathbf{v}_1 + \ldots + a_{nm}\mathbf{v}_n.$$

To see if the vectors $\mathbf{w}_1, \ldots, \mathbf{w}_m$ are independent, we must solve the equation

$$c_1\mathbf{w}_1 + \ldots + c_m\mathbf{w}_m = \mathbf{0}$$

for the unknowns $c_1, \ldots, c_m$. By substituting the expressions for the $\mathbf{w}_i$'s into this equation and rearranging the terms, we obtain the equation

$$(c_1a_{11} + \ldots + c_ma_{1m})\mathbf{v}_1 + \ldots + (c_1a_{n1} + \ldots + c_ma_{nm})\mathbf{v}_n = \mathbf{0}.$$

Since $\mathbf{v}_1, \ldots, \mathbf{v}_n$ are independent, this equation holds if and only if the coefficient of each $\mathbf{v}_i$ is 0:

$$c_1a_{11} + \ldots + c_ma_{1m} = 0$$

$$.$$
$$.$$
$$.$$

$$c_1a_{n1} + \ldots + c_ma_{nm} = 0.$$

This system of equations for $c_1, \ldots, c_m$ can be solved by reducing the augmented matrix

$$\begin{bmatrix} a_{11} & \cdots & a_{1m} & 0 \\ \cdot & & \cdot & \cdot \\ \cdot & & \cdot & \cdot \\ \cdot & & \cdot & \cdot \\ a_{n1} & \cdots & a_{nm} & 0 \end{bmatrix}.$$

The number of corner entries in the reduced matrix will be at most the same as the number of rows n. Since $m > n$, at least one of the first m columns will *not* have a corner entry. By taking the corresponding noncorner variable to be 1, we will obtain a solution for $c_1, \ldots, c_m$ with at least one $c_i \neq 0$. Thus $\mathbf{w}_1, \ldots, \mathbf{w}_m$ are dependent. We have shown the following:

Fact: *If W has a basis consisting of n vectors, then any set consisting of more than n vectors in W must be linearly dependent.*

Of course, this means that in a space with a finite basis, *an independent set of vectors can have at most the same number of vectors as the basis.* In particular, if we find two bases for W,

$$\mathbf{u}_1, \ldots, \mathbf{u}_k \quad \text{and} \quad \mathbf{z}_1, \ldots, \mathbf{z}_j.$$

then, on the one hand, the number of vectors in the independent set $\mathbf{z}_1, \ldots, \mathbf{z}_j$ is at most the same as the number of vectors in the basis $\mathbf{u}_1, \ldots, \mathbf{u}_k$—that is, $j \leq k$. On the other hand, the number of vectors in the independent set $\mathbf{u}_1, \ldots, \mathbf{u}_k$ is at most the same as the number of vectors in the basis $\mathbf{z}_1, \ldots, \mathbf{z}_j$—that is, $k \leq j$. Thus $j = k$, and we have another important result.

Fact: *Any two (finite) bases for a subspace must have the same number of vectors.*

In the case that a space W has a finite basis, we say the space is **finite dimensional**, and we refer to the number of vectors in this (or any other) basis as the **dimension** of W. We also refer to the space $\{\mathbf{0}\}$ as finite-dimensional, and we say that it has dimension 0.

Example 4.5.1

We saw in Example 4.4.5 that R^n has a basis (the standard basis) consisting of n vectors. Thus R^n has dimension n.

Example 4.5.2

Suppose L is an nth-order homogeneous linear o.d.e. We saw in Section 2.3 that the existence and uniqueness theorem guarantees that there is a basis for the solution space of $Lx = 0$ consisting of n functions. Thus this solution space has dimension n.

Example 4.5.3

Suppose A is an $n \times n$ matrix with real entries. The existence and uniqueness theorem for linear systems of o.d.e.'s guarantees that there is a basis for the solution space of $Dx = Ax$ consisting of n vector valued functions (see Section 3.3). Thus this solution space has dimension n.

Example 4.5.4

Suppose that A is an $n \times n$ real matrix and that λ is an eigenvalue of A. We observed in Section 3.7 that if we reduce the augmented matrix $[A - \lambda I \mid \mathbf{0}]$ to obtain a matrix whose corresponding system of equations has k_λ noncorner variables, then the space of solutions to $(A - \lambda I)\mathbf{v} = \mathbf{0}$ has a basis consisting of k_λ vectors. Thus this space has dimension k_λ.

Example 4.5.5

Find the dimension of the space W consisting of all 4-vectors of the form

$$\begin{bmatrix} x \\ y \\ x + y \\ z \end{bmatrix}.$$

Each vector in W can be written as

$$\begin{bmatrix} x \\ y \\ x + y \\ z \end{bmatrix} = x \begin{bmatrix} 1 \\ 0 \\ 1 \\ 0 \end{bmatrix} + y \begin{bmatrix} 0 \\ 1 \\ 1 \\ 0 \end{bmatrix} + z \begin{bmatrix} 0 \\ 0 \\ 0 \\ 1 \end{bmatrix},$$

so the vectors

$$\mathbf{v}_1 = \begin{bmatrix} 1 \\ 0 \\ 1 \\ 0 \end{bmatrix}, \qquad \mathbf{v}_2 = \begin{bmatrix} 0 \\ 1 \\ 1 \\ 0 \end{bmatrix}, \qquad \mathbf{v}_3 = \begin{bmatrix} 0 \\ 0 \\ 0 \\ 1 \end{bmatrix}$$

span W. If

$$c_1\mathbf{v}_1 + c_2\mathbf{v}_2 + c_3\mathbf{v}_3 = \mathbf{0},$$

then

$$
\begin{aligned}
c_1 \qquad\qquad &= 0 \\
c_2 \qquad &= 0 \\
c_1 + c_2 \qquad &= 0 \\
c_3 &= 0.
\end{aligned}
$$

Since the only solution is

$$c_1 = c_2 = c_3 = 0,$$

the vectors $\mathbf{v}_1$, $\mathbf{v}_2$, $\mathbf{v}_3$ are independent. Thus these vectors form a basis for W, and therefore W has dimension 3.

Once we know the dimension of a finite dimensional space W, the determination of whether a given set of vectors $\mathbf{w}_1, \ldots, \mathbf{w}_m$ is a basis for W can be shortened. First we must make sure that the vectors are actually in W (since otherwise they can't span W). Then we compare the number of vectors m with the dimension. If these numbers are not equal, then $\mathbf{w}_1, \ldots, \mathbf{w}_m$ cannot be a basis (since any two bases have the same number of vectors). If the number of vectors is the same as the dimension, we check further to see if they are independent (since vectors that are not independent cannot form a basis).

Suppose now that all three conditions have checked out—that is, $\mathbf{w}_1, \ldots, \mathbf{w}_m$ are linearly independent vectors in W and m is the same as the dimension of W. Let $\mathbf{w}$ be any vector in W. Since the set $\mathbf{w}_1, \ldots, \mathbf{w}_m, \mathbf{w}$ has more vectors than the dimension of W—that is, more vectors than a basis for W—these vectors must be linearly dependent. Thus there are scalars, not all 0, so that

$$c_1\mathbf{w}_1 + \ldots + c_m\mathbf{w}_m + c\mathbf{w} = \mathbf{0}.$$

If $c = 0$, then $c_1\mathbf{w}_1 + \ldots + c_m\mathbf{w}_m = \mathbf{0}$, and the independence of $\mathbf{w}_1, \ldots, \mathbf{w}_m$ forces $c_1 = \ldots = c_m = 0$; this is impossible since the scalars are not all 0. Thus $c \neq 0$, and we can solve for $\mathbf{w}$:

$$\mathbf{w} = -(c_1/c)\mathbf{w}_1 - \ldots - (c_m/c)\mathbf{w}_m.$$

Since any vector $\mathbf{w}$ in W is a linear combination of $\mathbf{w}_1, \ldots, \mathbf{w}_m$, these vectors span W. Thus, once we have the correct number of independent vectors in W, we are automatically guaranteed that these vectors also span W. Hence these vectors form a basis for W.

Fact: *If W has dimension m and if* $\mathbf{w}_1, \ldots, \mathbf{w}_m$ *are independent vectors in W, then* $\mathbf{w}_1, \ldots, \mathbf{w}_m$ *form a basis for W.*

Example 4.5.6

Show that

$$\mathbf{w}_1 = \begin{bmatrix} 1 \\ 1 \\ 1 \\ 1 \end{bmatrix}, \qquad \mathbf{w}_2 = \begin{bmatrix} 1 \\ 1 \\ 1 \\ 0 \end{bmatrix}, \qquad \mathbf{w}_3 = \begin{bmatrix} 1 \\ 1 \\ 0 \\ 0 \end{bmatrix}, \qquad \mathbf{w}_4 = \begin{bmatrix} 1 \\ 0 \\ 0 \\ 0 \end{bmatrix}$$

form a basis for R^4.

These vectors are clearly in R^4. There are four of them, and the dimension of R^4 is also four. Thus it suffices to check that the vectors are independent. If

$$c_1\mathbf{w}_1 + c_2\mathbf{w}_2 + c_3\mathbf{w}_3 + c_4\mathbf{w}_4 = \mathbf{0},$$

then

$$c_1 + c_2 + c_3 + c_4 = 0$$
$$c_1 + c_2 + c_3 \qquad = 0$$
$$c_1 + c_2 \qquad\qquad = 0$$
$$c_1 \qquad\qquad\qquad = 0.$$

If we read these equations from the bottom up, we see immediately that

$$c_1 = c_2 = c_3 = c_4 = 0.$$

Thus $\mathbf{w}_1, \mathbf{w}_2, \mathbf{w}_3, \mathbf{w}_4$ are independent. Since we have the right number of independent vectors, these vectors form a basis for R^4.

Example 4.5.7

Show that

$$\mathbf{w}_1 = \begin{bmatrix} 1 \\ 1 \\ 2 \\ 0 \end{bmatrix}, \qquad \mathbf{w}_2 = \begin{bmatrix} 1 \\ -1 \\ 0 \\ 2 \end{bmatrix}, \qquad \mathbf{w}_3 = \begin{bmatrix} 2 \\ 0 \\ 2 \\ 3 \end{bmatrix}$$

form a basis for the space W consisting of all 4-vectors of the form

$$\begin{bmatrix} x \\ y \\ x + y \\ z \end{bmatrix}.$$

To see that $\mathbf{w}_1$ is of the correct form, take $x = y = 1$ and $z = 0$. Similarly, the choices $x = 1$, $y = -1$, and $z = 2$ yield $\mathbf{w}_2$, while the choices $x = 2$, $y = 0$ and $z = 3$ yield $\mathbf{w}_3$. Thus $\mathbf{w}_1$, $\mathbf{w}_2$, and $\mathbf{w}_3$ lie in W. Since W has dimension 3 (see Example 4.5.5), we have the correct number of vectors for a basis. If

$$c_1\mathbf{w}_1 + c_2\mathbf{w}_2 + c_3\mathbf{w}_3 = \mathbf{0},$$

then

$$\begin{aligned}
c_1 + c_2 + 2c_3 &= 0 \\
c_1 - c_2 &= 0 \\
2c_1 + 2c_3 &= 0 \\
 2c_2 + 3c_3 &= 0.
\end{aligned}$$

We reduce the augmented matrix

$$\begin{bmatrix} 1 & 1 & 2 & 0 \\ 1 & -1 & 0 & 0 \\ 2 & 0 & 2 & 0 \\ 0 & 2 & 3 & 0 \end{bmatrix} \rightarrow \begin{bmatrix} 1 & 0 & 0 & 0 \\ 0 & 1 & 0 & 0 \\ 0 & 0 & 1 & 0 \\ 0 & 0 & 0 & 0 \end{bmatrix}$$

and read off the unique solution

$$c_1 = c_2 = c_3 = 0.$$

Thus $\mathbf{w}_1$, $\mathbf{w}_2$, $\mathbf{w}_3$ are independent. Since we have the right number of independent vectors, these vectors form a basis for W.

While each of the spaces we have considered in this section has had a finite basis, there are spaces that don't. One such example is C_∞. To see this, note that if C_∞ had a finite basis consisting of n elements, then any $n + 1$ elements of C_∞ would be dependent. Since the $n + 1$ functions $1, t, \ldots, t^n$ are independent, there can't be such a basis.

Let's summarize.

DIMENSION

If a subspace has a finite basis, then any two bases have the same number of vectors. We refer to this number as the **dimension** of the subspace.

If W has dimension n, then any $n + 1$ vectors in W are linearly dependent.

To see if vectors $\mathbf{w}_1, \ldots, \mathbf{w}_m$ form a basis for a subspace W that we know has dimension n, we check that

1. $\mathbf{w}_1, \ldots, \mathbf{w}_m$ are in W,
2. $m = n$, and
3. $\mathbf{w}_1, \ldots, \mathbf{w}_m$ are independent.

If all three conditions hold, then the vectors form a basis. If any one fails to hold, then they do not.

EXERCISES

In Exercises 1 to 10, find the dimension of the indicated spaces. (Note that each of these spaces has appeared in the exercises of Sections 4.3 and 4.4.)

1. The subspace of R^2 consisting of all vectors of the given form

   a. $\begin{bmatrix} x \\ x \\ 0 \end{bmatrix}$
   b. $\begin{bmatrix} x \\ y \\ x + y \end{bmatrix}$
   c. $\begin{bmatrix} x + y \\ y + z \\ x + z \end{bmatrix}$

   d. $\begin{bmatrix} x + y \\ x - y \\ 2x \end{bmatrix}$
   e. $\begin{bmatrix} x + y \\ x - z \\ y + z \end{bmatrix}$
   f. $\begin{bmatrix} x - y \\ y - x \\ x - y \end{bmatrix}$

2. The subspace of R^4 consisting of

   a. all vectors of the form $\begin{bmatrix} x \\ y \\ x + y \\ x - y \end{bmatrix}$

   b. those vectors $\begin{bmatrix} x \\ y \\ z \\ u \end{bmatrix}$ for which $x - y + z = 0$ and $w - x + y + z = 0$

3. The solution space of $B\mathbf{x} = \mathbf{0}$, where

   a. $B = \begin{bmatrix} 1 & 2 & 3 \\ -1 & -2 & -3 \\ 3 & 6 & 9 \end{bmatrix}$
   b. $B = \begin{bmatrix} 1 & 3 & -1 \\ 5 & 2 & 1 \\ 4 & -1 & 2 \end{bmatrix}$

c. $B = \begin{bmatrix} 1 & 0 & 1 & -1 \\ 2 & 1 & 0 & 1 \\ 1 & 1 & -1 & 2 \\ 4 & 2 & 0 & 2 \end{bmatrix}$ d $B = \begin{bmatrix} 0 & 1 & 3 & -2 \\ 1 & 0 & 1 & 1 \\ 0 & 0 & 1 & 2 \\ 2 & 1 & 6 & 2 \end{bmatrix}$

4. The solution space if $Lx = 0$, where
    a. $L = D^5$ b. $L = D^4 - 8D^2 + 16$ c. $L = D^4 - 1$

5. The solution space of $Dx = Ax$, where

    a. $A = \begin{bmatrix} -3 & 1 \\ -2 & 0 \end{bmatrix}$ b. $A = \begin{bmatrix} 1 & -1 \\ 1 & 1 \end{bmatrix}$ c. $A = \begin{bmatrix} 0 & 1 \\ -4 & 4 \end{bmatrix}$

6. C^n 7. P_m 8. $R_{m \times n}$ 9. $C_{m \times n}$

10. a. The space of complex numbers, as a real vector space.
    b. The space of complex numbers, as a complex vector space.

Show that each of the spaces in Exercises 11 to 13 is *not* finite-dimensional.

11. P 12. C_∞^n

13. The space S of sequences $\{a_n\}$ of real numbers

More abstract problems:

14. Let $\mathbf{u}$, $\mathbf{v}$, and $\mathbf{w}$ be vectors in R^3. Suppose that the lines $\mathbf{Comb(v)}$ and $\mathbf{Comb(w)}$ determined by $\mathbf{v}$ and $\mathbf{w}$ are unequal, and that $\mathbf{u}$ does not lie in the plane $\mathbf{Comb(v, w)}$. Show that $\mathbf{Comb(v, w, u)} = R^3$.

15. Prove that if W has dimension m and if $\mathbf{w}_1, \ldots, \mathbf{w}_m$ are vectors that span W, then $\mathbf{w}_1, \ldots, \mathbf{w}_m$ form a basis for W. (*Hint*: Use Exercise 20 of Section 4.4.)

16. Let $\mathbf{v}_1, \ldots, \mathbf{v}_k$ be independent vectors in an n-dimensional vector space, where $n > k$. Show that it is possible to find vectors $\mathbf{v}_{k+1}, \ldots, \mathbf{v}_n$ so that $\mathbf{v}_1, \ldots, \mathbf{v}_n$ form a basis for V. (*Hint*: Use Exercise 21 of Section 4.4 repeatedly, until you get a basis.)

17. Suppose that any set consisting of more than n vectors in W is linearly dependent. Prove that W is finite-dimensional and that its dimension is at most n. (*Hint*: If $W = \{\mathbf{0}\}$, then it has dimension 0, and we are done. Otherwise, let $\mathbf{v}_1$ be a nonzero vector in V and use Exercise 21 of Section 4.4 repeatedly, until you get a basis.)

18. Suppose that V is an n-dimensional vector space and that W is a subspace of V. Use Exercise 17 to prove
    a. W is finite-dimensional.
    b. The dimension of W is at most n.
    c. The dimension of W is equal to n if and only if $W = V$.

4.6 LINEAR TRANSFORMATIONS

Our treatment of linear o.d.e.'s in Chapter 2 relied heavily on the interpretation of a linear differential operator L as a mapping that assigns to each function x another function Lx. The key properties of this mapping are the principle of superposition

$$L(x + y) = L(x) + L(y)$$

and the principle of proportionality

$$L(cx) = cL(x).$$

These principles tell us that L preserves addition and multiplication by scalars. In this section we consider mappings from one vector space to another that share these properties.

Definition: *Let T be a mapping that assigns to each vector $\mathbf{v}$ in the vector space V another vector $T(\mathbf{v})$ in the vector space W. We say that T is a **linear transformation** from V to W provided*

$$T(\mathbf{v} + \mathbf{u}) = T(\mathbf{v}) + T(\mathbf{u})$$

and

$$T(c\mathbf{v}) = cT(\mathbf{v})$$

for all $\mathbf{v}$ and $\mathbf{u}$ in V and all scalars c.

Let's look at some examples, beginning with linear differential operators.

Example 4.6.1

If L is a linear differential operator with constant coefficients, then L assigns to each vector x in $V = C_\infty$ another vector $L(x)$ in $W = C_\infty$. The principles of superposition and proportionality tell us that L is a linear transformation from C_∞ to C_∞.

Example 4.6.2

Show that the map T from $V = R^4$ to $W = R^3$ given by the rule

$$T\begin{bmatrix} x \\ y \\ z \\ u \end{bmatrix} = \begin{bmatrix} x + y \\ 2u \\ x + y \end{bmatrix}$$

is a linear transformation.

If we apply T to the sum of arbitrary vectors in $V = R^4$, we get

$$T\left(\begin{bmatrix} x_1 \\ y_1 \\ z_1 \\ u_1 \end{bmatrix} + \begin{bmatrix} x_2 \\ y_2 \\ z_2 \\ u_2 \end{bmatrix}\right) = T\begin{bmatrix} x_1 + x_2 \\ y_1 + y_2 \\ z_1 + z_2 \\ u_1 + u_2 \end{bmatrix} = \begin{bmatrix} (x_1 + x_2) + (y_1 + y_2) \\ 2(u_1 + u_2) \\ (x_1 + x_2) + (y_1 + y_2) \end{bmatrix}$$

$$= \begin{bmatrix} (x_1 + y_1) + (x_2 + y_2) \\ (2u_1 + 2u_2) \\ (x_1 + y_1) + (x_2 + y_2) \end{bmatrix} = \begin{bmatrix} x_1 + y_1 \\ 2u_1 \\ x_1 + y_1 \end{bmatrix} + \begin{bmatrix} x_2 + y_2 \\ 2u_2 \\ x_2 + y_2 \end{bmatrix}$$

$$= T\begin{bmatrix} x_1 \\ y_1 \\ z_1 \\ u_1 \end{bmatrix} + T\begin{bmatrix} x_2 \\ y_2 \\ z_2 \\ u_2 \end{bmatrix}.$$

Similarly,

$$T\left(c\begin{bmatrix} x \\ y \\ z \\ u \end{bmatrix}\right) = T\begin{bmatrix} cx \\ cy \\ cz \\ cu \end{bmatrix} = \begin{bmatrix} cx + cy \\ 2cu \\ cx + cy \end{bmatrix} = c\begin{bmatrix} x + y \\ 2u \\ x + y \end{bmatrix} = cT\begin{bmatrix} x \\ y \\ z \\ u \end{bmatrix}.$$

Thus T is a linear transformation from R^4 to R^3.

Example 4.6.3

Suppose B is an $m \times n$ matrix with real entries. If $\mathbf{v}$ is an n-vector, then $B\mathbf{v}$ is an m-vector. Thus the rule

$$T_B(\mathbf{v}) = B\mathbf{v}$$

defines a mapping from $V = R^n$ to $W = R^m$. If $\mathbf{v}$ and $\mathbf{u}$ are in $V = R^n$ and if c is a scalar, then

$$T_B(\mathbf{v} + \mathbf{u}) = B(\mathbf{v} + \mathbf{u}) = B\mathbf{v} + B\mathbf{u} = T_B(\mathbf{v}) + T_B(\mathbf{u})$$

$$T_B(c\mathbf{v}) = B(c\mathbf{v}) = c(B\mathbf{v}) = cT_B(\mathbf{v}).$$

Thus T_B is a linear transformation from R^n to R^m.

Example 4.6.4

Suppose A is an $n \times n$ real matrix. Let $D - A$ be the mapping defined on

$V = C_\infty^n$ by the rule

$$(D - A)\mathbf{x}(t) = D\mathbf{x}(t) - A\mathbf{x}(t).$$

Then $(D - A)\mathbf{x}(t)$ is in $W = C_\infty^n$ (check this!). If $\mathbf{x}$ and $\mathbf{y}$ are in $V = C_\infty^n$ and if c is a scalar, then

$$(D - A)(\mathbf{x} + \mathbf{y}) = D(\mathbf{x} + \mathbf{y}) - A(\mathbf{x} + \mathbf{y}) = D\mathbf{x} + D\mathbf{y} - A\mathbf{x} - A\mathbf{y}$$
$$= D\mathbf{x} - A\mathbf{x} + D\mathbf{y} - A\mathbf{y} = (D - A)\mathbf{x} + (D - A)\mathbf{y}$$
$$(D - A)(c\mathbf{x}) = D(c\mathbf{x}) - A(c\mathbf{x}) = cD\mathbf{x} - cA\mathbf{x}$$
$$= c(D\mathbf{x} - A\mathbf{x}) = c(D - A)\mathbf{x}.$$

Thus $D - A$ is a linear transformation from C_∞^n to C_∞^n.

Associated to each linear transformation T from V to W is an important subset of V, the set of all those vectors in V that are mapped by T to the zero vector of W.

Definition: *If T is a linear transformation from V to W, then the set of all those vectors $\mathbf{v}$ in V such that $T(\mathbf{v}) = \mathbf{0}$ is called the **kernel** of T and is denoted **Ker**(T).*

For any vector $\mathbf{v}$ in V,

$$T(\mathbf{0}) = T(0\mathbf{v}) = 0T(\mathbf{v}) = \mathbf{0},$$

so $\mathbf{0}$ is in **Ker**(T). If $\mathbf{v}$ and $\mathbf{u}$ are in **Ker**(T), then $T(\mathbf{v}) = \mathbf{0}$ and $T(\mathbf{u}) = \mathbf{0}$, so

$$T(\mathbf{v} + \mathbf{u}) = T(\mathbf{v}) + T(\mathbf{u}) = \mathbf{0} + \mathbf{0} = \mathbf{0}$$
$$T(c\mathbf{v}) = cT(\mathbf{v}) = c\mathbf{0} = \mathbf{0}.$$

Thus $\mathbf{v} + \mathbf{u}$ and $c\mathbf{v}$ are in **Ker**(T). We have shown the following:

Fact: *The kernel of a linear transformation from V to W is a subspace of V.*

Example 4.6.5
If L is a linear differential operator with constant coefficients, then **Ker**(L) is the subspace of C_∞ consisting of all solutions of the homogeneous o.d.e. $Lx = 0$.

Example 4.6.6

If T is the linear transformation from R^4 to R^3 given by

$$T\begin{bmatrix} x \\ y \\ z \\ u \end{bmatrix} = \begin{bmatrix} x + y \\ 2u \\ x + y \end{bmatrix}$$

(see Example 4.6.2), then **Ker**(T) is the subspace of R^4 consisting of all 4-vectors

$$\begin{bmatrix} x \\ y \\ z \\ u \end{bmatrix}$$

for which

$$\begin{bmatrix} x + y \\ 2u \\ x + y \end{bmatrix} = \mathbf{0}.$$

This is the same as the set of all 4-vectors of the form

$$\begin{bmatrix} x \\ -x \\ z \\ 0 \end{bmatrix}.$$

Example 4.6.7

If B is an $m \times n$ real matrix, and if T_B is the linear transformation from R^n to R^m given by $T_B(\mathbf{v}) = B\mathbf{v}$ (see Example 4.6.3), then **Ker**(T_B) is the space of all those vectors $\mathbf{v}$ in R^n so that $T_B(\mathbf{v}) = \mathbf{0}$. This is the same as the solution space of the equation $B\mathbf{v} = \mathbf{0}$.

Example 4.6.8

The kernel of the linear transformation $D - A$ discussed in Example 4.6.4 is the set of all those vector valued functions $\mathbf{x}$ in C_∞^n for which $D\mathbf{x} - A\mathbf{x} = \mathbf{0}$. This is the same as the solution space of the homogeneous system $D\mathbf{x} = A\mathbf{x}$.

Another important set of vectors associated to a linear transformation is the set of all those vectors that are of the form $T(\mathbf{v})$.

Definition: *If T is a linear transformation from V to W, then the set of all those vectors* $\mathbf{w}$ *in W that can be written in the form* $\mathbf{w} = T(\mathbf{v})$ *for some* $\mathbf{v}$ *in V is called the* **range** *of T and is denoted* **Range(*T*)**.

We have already observed that $T(\mathbf{0}) = \mathbf{0}$, so that $\mathbf{0}$ is in **Range(*T*)**. If $\mathbf{w}$ and $\mathbf{y}$ are in **Range(*T*)**, then $\mathbf{w} = T(\mathbf{v})$ and $\mathbf{y} = T(\mathbf{x})$ for some $\mathbf{v}$ and $\mathbf{x}$ in V. It follows that

$$\mathbf{w} + \mathbf{y} = T(\mathbf{v}) + T(\mathbf{x}) = T(\mathbf{v} + \mathbf{x})$$

and, for any scalar c,

$$c\mathbf{w} = cT(\mathbf{v}) = T(c\mathbf{v}).$$

Thus $\mathbf{w} + \mathbf{y}$ and $c\mathbf{w}$ are in **Range(*T*)** and we have shown the following:

Fact: *The range of a linear transformation from V to W is a subspace of W.*

Example 4.6.9
If L is a linear differential operator, viewed as a linear transformation from C_∞ to C_∞, then **Range(*L*)** is the set of all functions $E(t)$ in C_∞ for which the equation $Lx = E(t)$ has a solution $x = x(t)$ in C_∞. We saw in Chapter 2 that this o.d.e. has a solution for all such $E(t)$. Thus **Range(*T*)** $= C_\infty$.

Example 4.6.10
If T is the transformation from R^4 to R^3 given by

$$T\begin{bmatrix} x \\ y \\ z \\ u \end{bmatrix} = \begin{bmatrix} x + y \\ 2u \\ x + y \end{bmatrix},$$

then **Range(*T*)** is the subspace of R^3 consisting of all 3-vectors of the form

$$\begin{bmatrix} x + y \\ 2u \\ x + y \end{bmatrix}.$$

Example 4.6.11

If B is an $m \times n$ real matrix and if T_B is the linear transformation from R^n to R^m given by $T_B(\mathbf{v}) = B\mathbf{v}$, then **Range**$(T_B)$ is the space consisting of all n-vectors $\mathbf{w}$ that can be written in the form $\mathbf{w} = B\mathbf{v}$.

In the special case that $m = n$, Cramer's determinant test tells us something about **Range**(T_B). If $\det B \neq 0$, then the equation $B\mathbf{v} = \mathbf{w}$ *always* has a solution for $\mathbf{v}$, so **Range**$(T_B) = R^n$. If $\det B = 0$, then there are some vectors $\mathbf{w}$ for which the equation $B\mathbf{v} = \mathbf{w}$ does not have solutions, so **Range**$(T_B) \neq R^n$.

Example 4.6.12

We saw in Chapter 3 that if $\mathbf{E}(t)$ is in C_∞^n, then a system of the form $D\mathbf{x} = A\mathbf{x} + \mathbf{E}(t)$ could always be solved for $\mathbf{x}$. Since this system can be rewritten as $(D - A)\mathbf{x} = \mathbf{E}(t)$, we see that the linear transformation $D - A$ has range equal to C_∞^n.

If T is a linear transformation from V to W, and if $\mathbf{w}$ is a fixed vector in **Range**(T), then the equation $T(\mathbf{v}) = \mathbf{w}$ has as least one solution $\mathbf{v}_p$ for $\mathbf{v}$:

$$T(\mathbf{v}_p) = \mathbf{w}.$$

If $\mathbf{u}$ is in **Ker**(T), then

$$T(\mathbf{u} + \mathbf{v}_p) = T(\mathbf{u}) + T(\mathbf{v}_p) = \mathbf{0} + \mathbf{w} = \mathbf{w},$$

so $\mathbf{u} + \mathbf{v}_p$ is another solution of $T(\mathbf{v}) = \mathbf{w}$. On the other hand, if $\mathbf{v}_1$ is another solution, then setting $\mathbf{u} = \mathbf{v}_1 - \mathbf{v}_p$ we see that

$$\mathbf{v}_1 = \mathbf{u} + \mathbf{v}_p$$

and

$$T(\mathbf{u}) = T(\mathbf{v}_1 - \mathbf{v}_p) = T(\mathbf{v}_1) - T(\mathbf{v}_p) = \mathbf{w} - \mathbf{w} = \mathbf{0},$$

so $\mathbf{u}$ is in **Ker**(T). Thus *every solution of $T(\mathbf{v}) = \mathbf{w}$ is of the form $\mathbf{u} + \mathbf{v}_p$ with $\mathbf{u}$ in* **Ker**(T). In the case that **Ker**(T) has a finite basis $\mathbf{u}_1, \ldots, \mathbf{u}_n$, the elements $\mathbf{u}$ in **Ker**(T) can be written in the form $\mathbf{u} = c_1\mathbf{u}_1 + \ldots + c_n\mathbf{u}_n$. Combining these observations, we have the following:

Fact: *Let T be a linear transformation from V to W, and let $\mathbf{w}$ be a fixed vector in* **Range**(T). *If $\mathbf{v}_p$ is a particular solution of the equation $T(\mathbf{v}) = \mathbf{w}$, and if $\mathbf{u}_1 \ldots, \mathbf{u}_n$*

form a basis for **Ker**(*T*), *then the solutions of T*(**v**) = **w** *are precisely the vectors of the form*

$$\mathbf{v} = c_1\mathbf{u}_1 + \ldots + c_n\mathbf{u}_n + \mathbf{v}_p.$$

This fact was central to our methods for solving linear differential equations and systems. While we didn't make use of it when dealing with algebraic systems of the form $B\mathbf{v} = \mathbf{w}$, we encourage you to look back at the examples and exercises in Section 3.6 to verify that the answers follow this pattern.

Let's summarize.

LINEAR TRANSFORMATIONS

A mapping *T* that assigns to each vector **v** in the vector space *V* another vector *T*(**v**) in the space *W* is called a **linear transformation** from *V* to *W* provided that

$$T(\mathbf{v} + \mathbf{u}) = T(\mathbf{v}) + T(\mathbf{u})$$

$$T(c\mathbf{v}) = cT(\mathbf{v})$$

for all **v** and **u** in *V* and all scalars *c*.

If *T* is a linear transformation from *V* to *W*, then the subspace of *V* consisting of all those vectors **v** for which *T*(**v**) = **0** is called the **kernel** of *T* and is denoted **Ker**(*T*). The subspace of *W* consisting of all those vectors in *W* that are of the form *T*(**v**) for some **v** in *V* is called the **range** of *T* and is denoted **Range**(*T*).

EXERCISES

In Exercises 1 to 26 you are given a mapping *T* from one vector space *V* to another vector space *W*.

a. Determine whether *T* is a linear transformation.

b. If it is linear, describe its kernel.

c. If it is linear, describe its range.

1. $V = W$ any vector space, $T(\mathbf{v}) = \mathbf{v}$ for each **v** in *V*

2. *V* and *W* any vector spaces, $T(\mathbf{v}) = \mathbf{0}$ for each **v** in *V*

3. $V = W = R^3, T\begin{bmatrix} x \\ y \\ z \end{bmatrix} = \begin{bmatrix} x + y \\ y + z \\ z + x \end{bmatrix}$ 4. $V = W = R^3, T\begin{bmatrix} x \\ y \\ z \end{bmatrix} = \begin{bmatrix} xy \\ y \\ y \end{bmatrix}$

5. $V = W = R^3$, $T \begin{bmatrix} x \\ y \\ z \end{bmatrix} = \begin{bmatrix} x + 1 \\ y \\ 0 \end{bmatrix}$

6. $V = W = R^3$, $T \begin{bmatrix} x \\ y \\ z \end{bmatrix} = \begin{bmatrix} y \\ 2y \\ 3y \end{bmatrix}$

7. $V = R^3$, $W = R^2$, $T \begin{bmatrix} x \\ y \\ z \end{bmatrix} = \begin{bmatrix} 3x \\ 5y \end{bmatrix}$

8. $V = R^3$, $W = R^2$, $T \begin{bmatrix} x \\ y \\ z \end{bmatrix} = \begin{bmatrix} x^2 \\ y^2 \end{bmatrix}$

9. $V = R^3$, $W = R^2$, $T \begin{bmatrix} x \\ y \\ z \end{bmatrix} = \begin{bmatrix} e^x \\ e^y \end{bmatrix}$

10. $V = R^3$, $W = R^2$, $T \begin{bmatrix} x \\ y \\ z \end{bmatrix} = \begin{bmatrix} x + y \\ x + y \end{bmatrix}$

11. $V = R^3$, $W = R^4$, $T \begin{bmatrix} x \\ y \\ z \end{bmatrix} = \begin{bmatrix} x \\ y \\ z \\ x + y \end{bmatrix}$

12. $V = R^3$, $W = R^4$, $T \begin{bmatrix} x \\ y \\ z \end{bmatrix} = \begin{bmatrix} z \\ x \\ x \\ x \end{bmatrix}$

13. $V = W = C_\infty$, $T(x) = (D - 1)x$
14. $V = W = C_\infty$, $T(x) = (D - 1)x + e^t$
15. $V = W = C_\infty$, $T(x) = (D^3 - D^2)x$
16. $V = W = P$, $T(x) = (D - 1)x$ ker: $x' - x = 0$ $x(t) = ce^t$
17. $V = W = P$, $T(x) = (D^3 - D^2)x$

18. $V = P_2$, $W = R^3$, $T(a_0 + a_1 t + a_2 t^2) = \begin{bmatrix} a_0 \\ a_1 \\ a_2 \end{bmatrix}$

19. $V = P_2$, $W = R^3$, $T(a_0 + a_1 t + a_2 t^2) = \begin{bmatrix} a_0 \\ a_0 \\ a_2 \end{bmatrix}$

20. $V = W = C_\infty$, $T[x(t)] = \int_0^t x(y)\, dy$

21. $V = W = C_\infty$, $T[x(t)] = x(t - 1)$

22. $V = R_{2 \times 2}$, $W = R^2$, $T \begin{bmatrix} a & b \\ c & d \end{bmatrix} = \begin{bmatrix} a \\ d \end{bmatrix}$

23. $V = R_{2 \times 2}$, $W = R^2$, $T \begin{bmatrix} a & b \\ c & d \end{bmatrix} = \begin{bmatrix} a & b \\ c & d \end{bmatrix} \begin{bmatrix} 1 \\ -1 \end{bmatrix}$

24. $V = W = C^2$, $T \begin{bmatrix} x \\ y \end{bmatrix} = \begin{bmatrix} x - 3iy \\ x \end{bmatrix}$

25. $V = W = C^2$, $T \begin{bmatrix} x \\ y \end{bmatrix} = \begin{bmatrix} x - i \\ y + i \end{bmatrix}$

26. $V = W = C^2$, $T \begin{bmatrix} x \\ y \end{bmatrix} = \begin{bmatrix} ix \\ 3y \end{bmatrix}$

More abstract problems:

27. Let B be a fixed $m \times n$ complex matrix. Show that the rule $T_B(\mathbf{v}) = B\mathbf{v}$ defines a linear transformation from C^n to C^m.

28. Let T be a linear transformation from V to W. Show that if $\mathbf{v}_1, \ldots, \mathbf{v}_n$ are any vectors in V and if $c_1, \ldots, c_n$ are any scalars, then

$$T(c_1\mathbf{v}_1 + \ldots + c_n\mathbf{v}_n) = c_1T(\mathbf{v}_1) + \ldots + c_nT(\mathbf{v}_n).$$

29. Let T and S be linear transformations from V to W. Suppose that $\mathbf{v}_1, \ldots, \mathbf{v}_n$ form a basis for V and that $T(\mathbf{v}_i) = S(\mathbf{v}_i)$ for $i = 1, 2, \ldots, n$. Prove that $T(\mathbf{v}) = S(\mathbf{v})$ for all $\mathbf{v}$ in V. (*Hint:* $\mathbf{v} = c_1\mathbf{v}_1 + \ldots + c_n\mathbf{v}_n$.)

30. Let T be a linear transformation from V to W. Suppose $\mathbf{v}_1, \ldots, \mathbf{v}_n$ is a basis for V. Show that the vectors $T(\mathbf{v}_1), \ldots, T(\mathbf{v}_n)$ span the range of T.

31. Let T be a linear transformation from V to W, where V is a finite-dimensional vector space. Let $\mathbf{v}_1, \ldots, \mathbf{v}_n$ be a basis for V and let $\mathbf{w}_1, \ldots, \mathbf{w}_n$ be any fixed vectors in W. If $\mathbf{v}$ is in V, then $\mathbf{v}$ has a unique expression of the form $\mathbf{v} = c_1\mathbf{v}_1 + \ldots + c_n\mathbf{v}_n$. Define $T(\mathbf{v})$ to be the vector $T(\mathbf{v}) = c_1\mathbf{w}_1 + \ldots + c_n\mathbf{w}_n$. Show that T is a linear transformation.

32. Let V be a finite-dimensional vector space with basis $\mathcal{B} = \{\mathbf{v}_1, \ldots, \mathbf{v}_n\}$. Define $T(\mathbf{v})$ to be the coordinate vector of $\mathbf{v}$, $T(\mathbf{v}) = {}_{\mathcal{B}}[\mathbf{v}]$ (see Exercise 22, Section 4.4).
    a. Show that T is a linear transformation from V to R^n.
    b. Show that $\mathbf{Ker}(T) = \{\mathbf{0}\}$.
    c. Show that $\mathbf{Range}(T) = R^n$.

33. Let S and T be linear transformations from V to W and let a be a scalar. Define $T + S$ and aT by the rules

$$[T + S](\mathbf{v}) = T(\mathbf{v}) + S(\mathbf{v}) \qquad \text{and} \qquad [aT](\mathbf{v}) = a[T(\mathbf{v})]$$

for each $\mathbf{v}$ in V. Show that $T + S$ and aT are linear transformations from V to W.

4.7 SOME SPECIAL TYPES OF LINEAR TRANSFORMATIONS (Optional)

In this section we discuss some important classes of linear transformations. These classes are defined in relation to the equation $T(\mathbf{v}) = \mathbf{w}$, viewed as an equation for the unknown vector $\mathbf{v}$. The first class consists of the transformations for which solutions to the equation, when they exist, are uniquely determined by $\mathbf{w}$. We then consider transformations with the property that solutions of $T(\mathbf{v}) = \mathbf{w}$ always exist. Finally, we discuss transformations that have both these properties.

Our formal definition of the first class of transformations is stated in terms of vectors of the form $T(\mathbf{v})$. Of course, these are precisely the vectors $\mathbf{w}$ in W for which solutions of $T(\mathbf{v}) = \mathbf{w}$ exist. To say such solutions are uniquely determined by $\mathbf{w}$ is equivalent to saying that $T(\mathbf{v}_1) = \mathbf{w} = T(\mathbf{v}_2)$ implies $\mathbf{v}_1 = \mathbf{v}_2$.

Definition: *A linear transformation T from V to W is said to be **one-to-one** if the only way for the equation $T(\mathbf{v}_1) = T(\mathbf{v}_2)$ to hold is for $\mathbf{v}_1$ to equal $\mathbf{v}_2$.*

Example 4.7.1

Show that the linear transformation T from R^3 to R^3 defined by the rule

$$
T \begin{bmatrix} x \\ y \\ z \end{bmatrix} = \begin{bmatrix} x + y \\ y \\ x + z \end{bmatrix}
$$

is one-to-one.

If T has the same effect on two vectors in R^3, say

$$
T \begin{bmatrix} x_1 \\ y_1 \\ z_1 \end{bmatrix} = T \begin{bmatrix} x_2 \\ y_2 \\ z_2 \end{bmatrix},
$$

then

$$
\begin{bmatrix} x_1 + y_1 \\ y_1 \\ x_1 + z_1 \end{bmatrix} = \begin{bmatrix} x_2 + y_2 \\ y_2 \\ x_2 + z_2 \end{bmatrix}.
$$

Comparison of the second entries yields $y_1 = y_2$. From the first entries we now see that $x_1 = x_2$. Comparison of the third entries then gives $z_1 = z_2$. Thus

$$
\begin{bmatrix} x_1 \\ y_1 \\ z_1 \end{bmatrix} = \begin{bmatrix} x_2 \\ y_2 \\ z_2 \end{bmatrix},
$$

and therefore T is one-to-one.

Example 4.7.2

The linear transformation T from R^4 to R^3 given by the rule

$$
T \begin{bmatrix} x \\ y \\ z \\ u \end{bmatrix} = \begin{bmatrix} x + y \\ u \\ x + y \end{bmatrix}
$$

(see Example 4.6.2) is not one-to-one since, for example,

$$T\begin{bmatrix} 1 \\ 0 \\ 0 \\ 0 \end{bmatrix} = \begin{bmatrix} 1 \\ 0 \\ 1 \end{bmatrix} = T\begin{bmatrix} 0 \\ 1 \\ 0 \\ 0 \end{bmatrix}$$

even though

$$\begin{bmatrix} 1 \\ 0 \\ 0 \\ 0 \end{bmatrix} \neq \begin{bmatrix} 0 \\ 1 \\ 0 \\ 0 \end{bmatrix}.$$

Recall from our discussion at the end of Section 4.6 that if $T(\mathbf{v}_p) = \mathbf{w}$, then the solutions of $T(\mathbf{v}) = \mathbf{w}$ are precisely the vectors of the form $\mathbf{v} = \mathbf{v}_p + \mathbf{u}$ with $\mathbf{u}$ in $\mathbf{Ker}(T)$. Thus the solutions are unique if and only if there are no nonzero vectors $\mathbf{u}$ in $\mathbf{Ker}(T)$. This provides us with another description of one-to-one transformations.

Fact: *Let T be a linear transformation from V to W. Then T is one-to-one if and only if* $\mathbf{Ker}(T) = \{\mathbf{0}\}$.

This fact provides a simple way to check whether a given transformation is one-to-one.

Example 4.7.3
Show that the linear transformation T from R^3 to R^4 given by

$$T\begin{bmatrix} x \\ y \\ z \end{bmatrix} = \begin{bmatrix} x \\ 3x + y \\ z \\ x - z \end{bmatrix}$$

is one-to-one.

The kernel of T consists of all those 3-vectors

$$\begin{bmatrix} x \\ y \\ z \end{bmatrix}$$

for which

$$\begin{bmatrix} x \\ 3x + y \\ z \\ x - z \end{bmatrix} = \begin{bmatrix} 0 \\ 0 \\ 0 \\ 0 \end{bmatrix}.$$

Comparison of the first entries and of the third entries gives

$$x = 0 \quad \text{and} \quad z = 0.$$

From the second entries we now see that

$$y = 0.$$

Thus

$$\mathbf{Ker}(T) = \{\mathbf{0}\},$$

from which it follows that T is one-to-one.

Example 4.7.4

Let A be an $n \times n$ matrix. Show that the linear transformation T_A from R^n to R^n given by $T_A(\mathbf{v}) = A\mathbf{v}$ is one-to-one if and only if det $A \neq 0$.

If det $A \neq 0$, then Cramer's determinant test tells us that the equation $A\mathbf{v} = \mathbf{0}$ has a unique solution, $\mathbf{v} = \mathbf{0}$. Since the solutions of $A\mathbf{v} = \mathbf{0}$ are the elements of $\mathbf{Ker}(T_A)$, the kernel is $\{\mathbf{0}\}$. Thus T_A is one-to-one in this case.

If det $A = \mathbf{0}$, then Cramer's determinant test tells us that $A\mathbf{v} = \mathbf{0}$ has infinitely many solutions for $\mathbf{v}$. Thus $\mathbf{Ker}(T_A) \neq \{\mathbf{0}\}$ and T_A is not one-to-one.

Let's now turn to our second class of transformations, those with the property that the equation $T(\mathbf{v}) = \mathbf{w}$ can always be solved for $\mathbf{v}$.

Definition: *A linear transformation T from V to W is said to map V **onto** W provided that given any $\mathbf{w}$ in W, there is always a vector $\mathbf{v}$ in V so that $T(\mathbf{v}) = \mathbf{w}$.*

Let's see which of the transformations in Examples 4.7.1 to 4.7.4 have this property.

Example 4.7.5

Check to see whether the linear transformation T defined by the rule

$$T\begin{bmatrix} x \\ y \\ z \end{bmatrix} = \begin{bmatrix} x + y \\ y \\ x + z \end{bmatrix}$$

maps R^3 onto R^3.

Given an arbitrary vector $\mathbf{w}$ in W, say

$$\mathbf{w} = \begin{bmatrix} a \\ b \\ c \end{bmatrix},$$

we must determine whether it is possible to find a vector in R^3

$$\mathbf{v} = \begin{bmatrix} x \\ y \\ z \end{bmatrix}$$

so that $T(\mathbf{v}) = \mathbf{w}$. Using the definition of T, and comparing corresponding entries, we are led to the equations

$$
\begin{aligned}
x + y \quad\;\; &= a \\
y \quad\;\; &= b \\
x \quad\;\; + z &= c.
\end{aligned}
$$

Since we were given $\mathbf{w}$ and we are trying to solve for $\mathbf{v}$, we treat these as a system of equations for the unknowns x, y, and z. Reduction of the augmented matrix gives

$$\begin{bmatrix} 1 & 1 & 0 & a \\ 0 & 1 & 0 & b \\ 1 & 0 & 1 & c \end{bmatrix} \rightarrow \begin{bmatrix} 1 & 0 & 0 & a - b \\ 0 & 1 & 0 & b \\ 0 & 0 & 1 & -a + b + c \end{bmatrix},$$

from which we determine a solution

$$x = a - b, \qquad y = b, \qquad z = -a + b + c.$$

The vector

$$\mathbf{v} = \begin{bmatrix} a - b \\ b \\ -a + b + c \end{bmatrix}$$

satisfies $T(\mathbf{v}) = \mathbf{w}$. Therefore T maps R^3 onto R^3.

Example 4.7.6

Check to see whether the linear transformation T defined by the rule

$$T \begin{bmatrix} x \\ y \\ z \\ u \end{bmatrix} = \begin{bmatrix} x + y \\ u \\ x + y \end{bmatrix}$$

maps R^4 onto R^3.

Note that for any $\mathbf{v}$ in R^4, $T(\mathbf{v})$ has equal first and third entries. If we take $\mathbf{w}$ in R^3 to be any vector that does not have this property, for example if we take

$$\mathbf{w} = \begin{bmatrix} 1 \\ 0 \\ 0 \end{bmatrix},$$

then the equation $T(\mathbf{v}) = \mathbf{w}$ cannot be solved for $\mathbf{v}$. Thus T does not map V onto W.

Example 4.7.7

Check to see whether the transformation defined by the rule

$$T \begin{bmatrix} x \\ y \\ z \end{bmatrix} = \begin{bmatrix} x \\ 3x + y \\ z \\ x - z \end{bmatrix}$$

maps R^3 onto R^4.

Given an arbitrary vector in R^4

$$\mathbf{w} = \begin{bmatrix} a \\ b \\ c \\ d \end{bmatrix},$$

we must determine whether it is possible to find a vector in R^3

$$\mathbf{v} = \begin{bmatrix} x \\ y \\ z \end{bmatrix}$$

so that $T(\mathbf{v}) = \mathbf{w}$. This leads to the equations

$$
\begin{aligned}
x &\qquad\qquad = a \\
3x + y &\qquad\quad = b \\
&\quad\; z = c \\
x &\quad - z = d.
\end{aligned}
$$

Reduction of the augmented matrix gives

$$
\begin{bmatrix} 1 & 0 & 0 & a \\ 3 & 1 & 0 & b \\ 0 & 0 & 1 & c \\ 1 & 0 & -1 & d \end{bmatrix} \rightarrow \begin{bmatrix} 1 & 0 & 0 & a \\ 0 & 1 & 0 & -3a + b \\ 0 & 0 & 1 & c \\ 0 & 0 & 0 & -a + c + d \end{bmatrix}.
$$

These equations can only be solved if

$$-a + c + d = 0.$$

In particular, if we take

$$\mathbf{w} = \begin{bmatrix} 0 \\ 0 \\ 0 \\ 1 \end{bmatrix}$$

(so that $-a + c + d = 1 \neq 0$), we cannot find a vector $\mathbf{v}$ with $T(\mathbf{v}) = \mathbf{w}$. Thus T does not map R^3 onto R^4.

Example 4.7.8

Let A be an $n \times n$ matrix. Show that the transformation given by $T_A(\mathbf{v}) = A\mathbf{v}$ maps R^n onto R^n if and only if det $A \neq 0$.

Cramer's determinant test tells us that the equation $A\mathbf{v} = \mathbf{w}$ can be solved for all choices of $\mathbf{w}$ if and only if det $A \neq 0$. Since $A\mathbf{v} = T_A(\mathbf{v})$, this says that T_A maps R^n onto R^n if and only if det $A \neq 0$.

The linear transformation considered in Examples 4.7.1 and 4.7.5 maps R^3 onto R^3 and is also one-to-one. If A is an $n \times n$ matrix with $\det(A) \neq 0$, then the linear transformation T_A maps R^n onto R^n and is one-to-one (see Examples 4.7.4 and 4.7.8). These transformations are examples of the third class we single out.

Definition: *A one-to-one linear transformation T that maps V onto W is said to be an **isomorphism** from V onto W.*

Suppose T is an isomorphism from V onto W. Since T is a mapping from V to W, it associates to each vector $\mathbf{v}$ in V a vector $\mathbf{w} = T(\mathbf{v})$ in W. Since T maps V onto W, *every* vector $\mathbf{w}$ in W arises in this way. Since T is one-to-one, then starting from the vector $\mathbf{w}$ in W there is *exactly one* vector $\mathbf{v}$ in V so that $T(\mathbf{v}) = \mathbf{w}$. Thus T provides us with a way of matching up the vectors in V with the vectors in W. Because T is a linear transformation, the vector space operations are preserved by this matching. We can think of V and W as "the same space" but with the vectors renamed. Given any vector $\mathbf{v}$ in V, the isomorphism tells us its new name $T(\mathbf{v})$ in W.

It is a remarkable fact that the question of whether or not there is an isomorphism from one finite-dimensional vector space onto another can be decided simply by comparing their dimensions (Exercises 23 and 24).

Fact: *Let V and W be finite-dimensional vector spaces. There is an isomorphism from V onto W if and only if the dimensions of V and W are equal.*

SOME SPECIAL TYPES OF LINEAR TRANSFORMATIONS

Let T be a linear transformation from V to W. We say that T is **one-to-one** provided that $T(\mathbf{v}_1) = T(\mathbf{v}_2)$ implies $\mathbf{v}_1 = \mathbf{v}_2$. If every vector $\mathbf{w}$ in W can be written in the form $\mathbf{w} = T(\mathbf{v})$, then we say that T maps V **onto** W. If T is a one-to-one map from V onto W, then we call T an **isomorphism** from V onto W.

A linear transformation T is one-to-one if and only if $\mathbf{Ker}(T) = \{\mathbf{0}\}$.

EXERCISES

In Exercises 1 to 19 you are given a linear transformation T from one vector space V to another vector space W. (Note that each of these maps appeared among the exercises of Section 4.6.)
a. Is T one-to-one?
b. Does T map V onto W?
c. Is T an isomorphism from V onto W?

1. $V = W$ any vector space, $T(\mathbf{v}) = \mathbf{v}$ for each $\mathbf{v}$ in V.
2. V and W any nonzero vector spaces, $T(\mathbf{v}) = \mathbf{0}$ for each $\mathbf{v}$ in V.

3. $V = W = R^3$, $T \begin{bmatrix} x \\ y \\ z \end{bmatrix} = \begin{bmatrix} x + y \\ y + z \\ z + x \end{bmatrix}$ 4. $V = W = R^3$, $T \begin{bmatrix} x \\ y \\ z \end{bmatrix} = \begin{bmatrix} y \\ 2y \\ 3y \end{bmatrix}$

5. $V = R^3$, $W = R^2$, $T \begin{bmatrix} x \\ y \\ z \end{bmatrix} = \begin{bmatrix} 3x \\ 5y \end{bmatrix}$ 6. $V = R^3$, $W = R^2$, $T \begin{bmatrix} x \\ y \\ z \end{bmatrix} = \begin{bmatrix} x + y \\ x + y \end{bmatrix}$

7. $V = R^3$, $W = R^4$, $T \begin{bmatrix} x \\ y \\ z \end{bmatrix} = \begin{bmatrix} x \\ y \\ z \\ x + y \end{bmatrix}$ 8. $V = R^3$, $W = R^4$, $T \begin{bmatrix} x \\ y \\ z \end{bmatrix} = \begin{bmatrix} x \\ x \\ x \\ x \end{bmatrix}$

9. $V = P_2$, $W = R^3$, $T(a_0 + a_1 t + a_2 t^2) = \begin{bmatrix} a_0 \\ a_1 \\ a_2 \end{bmatrix}$

10. $V = P_2$, $W = R^3$, $T(a_0 + a_1 t + a_2 t^2) = \begin{bmatrix} a_0 \\ a_0 \\ a_2 \end{bmatrix}$

11. $V = R_{2 \times 2}$, $W = R^2$, $T \begin{bmatrix} a & b \\ c & d \end{bmatrix} = \begin{bmatrix} a \\ d \end{bmatrix}$

12. $V = R_{2 \times 2}$, $W = R^2$, $T \begin{bmatrix} a & b \\ c & d \end{bmatrix} = \begin{bmatrix} a & b \\ c & d \end{bmatrix} \begin{bmatrix} 1 \\ -1 \end{bmatrix}$

13. $V = W = C^2$, $T \begin{bmatrix} x \\ y \end{bmatrix} = \begin{bmatrix} x - 3iy \\ x \end{bmatrix}$

14. $V = W = C^2$, $T \begin{bmatrix} x \\ y \end{bmatrix} = \begin{bmatrix} ix \\ 3y \end{bmatrix}$

15. $V = W = C_\infty$, $T(x) = (D - 1)x$
16. $V = W = C_\infty$, $T(x) = (D^3 - D^2)x$
17. $V = W = P$, $T(x) = (D - 1)x$
18. $V = W = P$, $T(x) = (D^3 - D^2)x$
19. $V = W = C_\infty$, $T[x(t)] = \int_0^t x(y) \, dy$

Some more abstract problems:

20   a. Let L be a linear differential operator with constant coefficients. Does L map C_∞ onto C_∞? Explain.

b. Let A be an $n \times n$ real matrix. Does $D - A$ map C_∞^n onto C_∞^n? Explain.

21. Let T be a one-to-one linear transformation from V to W. Show that if $v_1, \ldots, v_n$ are linearly independent vectors in V, then $T(v_1), \ldots, T(v_n)$ are linearly independent vectors in W.

22. Let T be a linear transformation from V onto W. Show that if $v_1, \ldots, v_n$ span V, then $T(v_1), \ldots, T(v_n)$ span W.

23. Let T be an isomorphism from V onto W.

a. Show that if $v_1 \ldots, v_n$ form a basis for V, then $T(v_1), \ldots, T(v_n)$ form a basis for W. (*Hint*: See Exercises 21 and 22.)

b. Conclude that if V is finite-dimensional, then W is finite-dimensional, and the dimension of W is the same as that of V.

24. Let V and W be finite-dimensional vector spaces of the same dimension n. Let $\mathbf{v}_1, \ldots, \mathbf{v}_n$ and $\mathbf{w}_1, \ldots, \mathbf{w}_n$ be bases for V and W, respectively. Show that the mapping

$$T(c_1\mathbf{v}_1 + \ldots + c_n\mathbf{v}_n) = c_1\mathbf{w}_1 + \ldots + c_n\mathbf{w}_n$$

described in Exercise 31 of Section 4.6 is an isomorphism from V onto W.

25. Let T be a linear transformation from V to W. Suppose that $\mathbf{v}_1, \ldots, \mathbf{v}_n$ form a basis for V and that $T(\mathbf{v}_1), \ldots, T(\mathbf{v}_n)$ form a basis for W. Show that T is an isomorphism from V onto W.

26. Let V be a finite-dimensional vector space with basis $\mathscr{B} = \{\mathbf{v}_1, \ldots, \mathbf{v}_n\}$, and let T be the linear transformation from V to R^n given by $T(\mathbf{v}) = {}_{\mathscr{B}}[\mathbf{v}]$ (see Exercise 22 of Section 4.4 and Exercise 32 of Section 4.6). Show that T is an isomorphism from V onto R^n.

4.8 INNER PRODUCTS AND ORTHOGONALITY (Optional)

Those of you who have studied multi-dimensional calculus will recall that the dot product of vectors is one of the prime tools in the formulation and solution of a variety of geometric problems in 3-space. In this section we will study generalizations of the dot product to more general vector spaces.

Definition: *An **inner product** on a real vector space V is a rule that assigns to each pair of vectors $\mathbf{v}$, $\mathbf{w}$ a real number, denoted $<\mathbf{v}, \mathbf{w}>$, in such a way that the following laws hold for all vectors $\mathbf{v}$, $\mathbf{w}$, and $\mathbf{u}$, and all scalars c.*

(P1) $<\mathbf{v}, \mathbf{w}> = <\mathbf{w}, \mathbf{v}>$

(P2) $<\mathbf{v} + \mathbf{u}, \mathbf{w}> = <\mathbf{v}, \mathbf{w}> + <\mathbf{u}, \mathbf{w}>$

(P3) $<c\mathbf{v}, \mathbf{w}> = c <\mathbf{v}, \mathbf{w}>$

(P4) $<\mathbf{v}, \mathbf{v}> \geq 0$ and $<\mathbf{v}, \mathbf{v}> = 0$ *if and only if* $\mathbf{v} = \mathbf{0}$

A vector space with a specified inner product $< , >$ on it is often referred to as an **inner product space**.

Example 4.8.1
The **standard inner product** on R^n is defined for two n-vectors

$$\mathbf{v} = \begin{bmatrix} v_1 \\ v_2 \\ \cdot \\ \cdot \\ \cdot \\ v_n \end{bmatrix} \quad \text{and} \quad \mathbf{w} = \begin{bmatrix} w_1 \\ w_2 \\ \cdot \\ \cdot \\ \cdot \\ w_n \end{bmatrix}$$

by the formula

$$<\mathbf{v}, \mathbf{w}> = v_1 w_1 + v_2 w_2 + \ldots + v_n w_n.$$

The verifications of conditions (P1) to (P3) are straightforward calculations, which we leave to you. To see that this inner product satisfies (P4), we calculate

$$<\mathbf{v}, \mathbf{v}> = v_1^2 + v_2^2 + \ldots + v_n^2.$$

Since this is a sum of squares, it can never be negative. Moreover, it is zero if and only if every v_i is zero—that is, if and only if $\mathbf{v} = \mathbf{0}$.

Example 4.8.2

The collection $C[a, b]$ of functions defined and continuous for $a \leq t \leq b$ forms a vector space under the usual addition of functions and multiplication by scalars (Exercise 10, Section 4.1). An inner product can be defined on this space by the formula

$$<x(t), y(t)> = \int_a^b x(t)y(t) \, dt.$$

Again, conditions (P1) to (P3) are easy to verify, and we leave this to you. As to condition (P4), we note that

$$<x(t), x(t)> = \int_a^b [x(t)]^2 \, dt.$$

Since the integrand is non-negative, this integral is never negative. It equals zero only if the integrand (and hence $x(t)$) is identically zero for $a \leq t \leq b$.

We note that other inner products can be defined on the vector spaces in Examples 4.8.1 and 4.8.2 (see, for example, Exercise 16). We shall, however, generally restrict our attention to these two, which are the most direct generalizations of the dot product in 3-space.

If we think of the 2-vector

$$\mathbf{v} = \begin{bmatrix} x \\ y \end{bmatrix}$$

as an arrow in the plane pointing from the origin $(0, 0)$ to the point (x, y) (see Figure 4.1), then the length of the vector is the distance between these points, $\sqrt{x^2 + y^2}$.

In terms of the standard product on R^2, this is the square root of the inner product of **v** with itself. This last expression generalizes to any inner product space. We define the **norm** of the vector **v** to be

$$\|\mathbf{v}\| = \sqrt{<\mathbf{v}, \mathbf{v}>}.$$

For example, the norm of the vector

$$\mathbf{v} = \begin{bmatrix} 3 \\ 2 \\ -1 \\ -2 \end{bmatrix}$$

using the standard inner product on R^4 is

$$\|\mathbf{v}\| = \sqrt{<\mathbf{v}, \mathbf{v}>} = \sqrt{3^2 + 2^2 + (-1)^2 + (-2)^2} = \sqrt{18} = 3\sqrt{2}.$$

The norm of $x(t) = t + 1$, $0 \le t \le 1$, with respect to the inner product on $C[0, 1]$ described in Example 4.8.2 is

$$\|x(t)\| = \sqrt{<x(t), x(t)>} = \left(\int_0^1 (t + 1)^2 \, dt \right)^{1/2} = \left(\frac{1}{3}(t + 1)^3 \Big]_0^1 \right)^{1/2} = \sqrt{\frac{7}{3}}.$$

Note that property (P4) guarantees that the formula defining $\|\mathbf{v}\|$ always makes sense and yields a strictly positive number unless $\mathbf{v} = \mathbf{0}$.

If two 2-vectors, thought of as arrows in the plane, are scalar multiples of each other, then they are parallel (having the same or opposite directions depending on the sign of the scalar). This notion extends to any vector space in a perfectly straightforward way. We say two vectors **v** and **w** in V are **parallel** if there is a scalar c so that $\mathbf{v} = c\mathbf{w}$. For example,

$$\mathbf{v} = \begin{bmatrix} -3 \\ 3 \\ 3 \\ -3 \end{bmatrix} \quad \text{and} \quad \mathbf{w} = \begin{bmatrix} 1 \\ -1 \\ -1 \\ 1 \end{bmatrix}$$

are parallel since $\mathbf{v} = -3\mathbf{w}$.

While inner products do not play a role in testing whether given vectors are parallel, they are used to check for perpendicularity. In R^2 or R^3, two vectors are perpendicular if and only if their standard inner product is zero. In the general setting, we use the word "orthogonal" instead of "perpendicular." Thus we say that two vectors $\mathbf{v}$ and $\mathbf{w}$ in an inner product space are **orthogonal** provided $<\mathbf{v}, \mathbf{w}> = 0$. For example, the vectors

$$\mathbf{v} = \begin{bmatrix} 2 \\ -1 \\ -1 \\ 1 \end{bmatrix} \quad \text{and} \quad \mathbf{w} = \begin{bmatrix} 1 \\ 1 \\ 0 \\ -1 \end{bmatrix}$$

are orthogonal with respect to the standard inner product on R^4, since

$$<\mathbf{v}, \mathbf{w}> = (2)(1) + (-1)(1) + (-1)(0) + (1)(-1) = 0.$$

If we use the inner product on $C[-\pi, \pi]$ described in Example 4.8.2, then

$$<\sin t, \cos t> = \int_{-\pi}^{\pi} \sin t \cos t \, dt = \frac{1}{2} \sin^2 t \Big]_{-\pi}^{\pi} = 0.$$

Thus the functions $\sin t$ and $\cos t$, $-\pi \le t \le \pi$, are orthogonal vectors in this space.

In mechanics, it is often useful to resolve a force into components along certain mutually perpendicular directions. Similarly, it often proves convenient in any inner product space to express a vector $\mathbf{v}$ as the sum of two vectors

$$\mathbf{v} = \mathbf{v}_p + \mathbf{v}_o$$

with $\mathbf{v}_p$ parallel to another given vector $\mathbf{w} \ne \mathbf{0}$ and with $\mathbf{v}_o$ orthogonal to $\mathbf{w}$. Since we want $\mathbf{v}_p$ to be parallel to $\mathbf{w}$,

$$\mathbf{v}_p = c\mathbf{w}.$$

If we substitute this into our expression for $\mathbf{v}$ and solve for $\mathbf{v}_o$, we get

$$\mathbf{v}_o = \mathbf{v} - \mathbf{v}_p = \mathbf{v} - c\mathbf{w}.$$

Since we want $\mathbf{v}_o$ to be orthogonal to $\mathbf{w}$,

$$0 = <\mathbf{v}_o, \mathbf{w}> = <\mathbf{v} - c\mathbf{w}, \mathbf{w}>$$
$$= <\mathbf{v}, \mathbf{w}> - c<\mathbf{w}, \mathbf{w}>.$$

This can be solved for c:

$$c = \frac{<\mathbf{v}, \mathbf{w}>}{<\mathbf{w}, \mathbf{w}>} = \frac{<\mathbf{v}, \mathbf{w}>}{\|\mathbf{w}\|^2}.$$

Substituting back into our expressions for $\mathbf{v}_p$ and $\mathbf{v}_o$, we obtain the following.

Fact: *Given two vectors* $\mathbf{v}$ *and* $\mathbf{w} \neq \mathbf{0}$, *the vector*

$$\mathbf{v}_p = \frac{<\mathbf{v}, \mathbf{w}>}{\|\mathbf{w}\|^2}\, \mathbf{w}$$

is parallel to $\mathbf{w}$, *the vector*

$$\mathbf{v}_o = \mathbf{v} - \frac{<\mathbf{v}, \mathbf{w}>}{\|\mathbf{w}\|^2}\, \mathbf{w}$$

is orthogonal to $\mathbf{w}$, *and*

$$\mathbf{v} = \mathbf{v}_p + \mathbf{v}_o.$$

We refer to the vector $\mathbf{v}_p$ as the **component of v parallel to w**. The vector $\mathbf{v}_o$ is the **component of v orthogonal to w**.

Example 4.8.3
In R^4 with the standard inner product, let

$$\mathbf{v} = \begin{bmatrix} 1 \\ 0 \\ 0 \\ 1 \end{bmatrix} \quad \text{and} \quad \mathbf{w} = \begin{bmatrix} 1 \\ 1 \\ 1 \\ 0 \end{bmatrix}.$$

Then

$$<\mathbf{v}, \mathbf{w}> = 1 \quad \text{and} \quad \|\mathbf{w}\|^2 = 3.$$

The component of $\mathbf{v}$ parallel to $\mathbf{w}$ is

$$\mathbf{v}_p = \frac{<\mathbf{v}, \mathbf{w}>}{\|\mathbf{w}\|^2}\, \mathbf{w} = \frac{1}{3}\, \mathbf{w} = \begin{bmatrix} \frac{1}{3} \\ \frac{1}{3} \\ \frac{1}{3} \\ 0 \end{bmatrix}.$$

The component of **v** orthogonal to **w** is

$$\mathbf{v}_o = \mathbf{v} - \mathbf{v}_p = \begin{bmatrix} 1 \\ 0 \\ 0 \\ 1 \end{bmatrix} - \begin{bmatrix} \frac{1}{3} \\ \frac{1}{3} \\ \frac{1}{3} \\ 0 \end{bmatrix} = \begin{bmatrix} \frac{2}{3} \\ -\frac{1}{3} \\ -\frac{1}{3} \\ 1 \end{bmatrix}.$$

On the other hand, the component of **w** parallel to **v** is

$$\mathbf{w}_p = \frac{<\mathbf{w}, \mathbf{v}>}{\|\mathbf{v}\|^2} \mathbf{v} = \frac{1}{2} \mathbf{v} = \begin{bmatrix} \frac{1}{2} \\ 0 \\ 0 \\ \frac{1}{2} \end{bmatrix}.$$

The component of **w** orthogonal to **v** is

$$\mathbf{w}_o = \mathbf{w} - \mathbf{w}_p = \begin{bmatrix} 1 \\ 1 \\ 1 \\ 0 \end{bmatrix} - \begin{bmatrix} \frac{1}{2} \\ 0 \\ 0 \\ \frac{1}{2} \end{bmatrix} = \begin{bmatrix} \frac{1}{2} \\ 1 \\ 1 \\ -\frac{1}{2} \end{bmatrix}.$$

Example 4.8.4

Decompose the function $x(t) = t$, $0 \leq t \leq \pi$, into components parallel and orthogonal to $y(t) = \sin t$, $0 \leq t \leq \pi$, using the inner product on $C[0, \pi]$ described in Example 4.8.2.

We first calculate

$$<x(t), y(t)> = \int_0^\pi t \sin t \, dt = (-t \cos t + \sin t) \Big]_0^\pi = \pi$$

and

$$\|y(t)\|^2 = \int_0^\pi \sin^2 t \, dt = \left(\frac{1}{2} t - \frac{1}{4} \cos 2t \right) \Big]_0^\pi = \frac{1}{2} \pi.$$

The component of $x(t) = t$ parallel to $y(t) = \sin t$ is

$$x_p(t) = \frac{<x(t), y(t)>}{\|y(t)\|^2} y(t) = \frac{\pi}{\frac{1}{2}\pi} \sin t = 2 \sin t$$

while the component orthogonal to $y(t) = \sin t$ is

$$x_o(t) = x(t) - x_p(t) = t - 2 \sin t.$$

If $\mathbf{w}$ and $\mathbf{v}$ form a basis for a two-dimensional subspace W of an inner product space, then $\mathbf{w}$ and the component of $\mathbf{v}$ orthogonal to $\mathbf{w}$ form another basis consisting of orthogonal vectors. To see this, note first that the component of $\mathbf{v}$ orthogonal to $\mathbf{w}$

$$\mathbf{v}_o = \mathbf{v} - \frac{<\mathbf{v}, \mathbf{w}>}{\|\mathbf{w}\|^2} \mathbf{w}$$

is a linear combination of $\mathbf{v}$ and $\mathbf{w}$, so it lies in W. To show that $\mathbf{w}$ and $\mathbf{v}_o$ form a basis for W, we need only check that they are linearly independent (why?). If

$$\mathbf{0} = c_1\mathbf{w} + c_2\mathbf{v}_o,$$

then

$$
\begin{aligned}
0 = <\mathbf{0}, \mathbf{w}> &= <c_1\mathbf{w} + c_2\mathbf{v}_o, \mathbf{w}> \\
&= c_1<\mathbf{w}, \mathbf{w}> + c_2<\mathbf{v}_o, \mathbf{w}> \\
&= c_1<\mathbf{w}, \mathbf{w}>
\end{aligned}
$$

and

$$
\begin{aligned}
0 = <\mathbf{0}, \mathbf{v}_o> &= <c_1\mathbf{w} + c_2\mathbf{v}_o, \mathbf{v}_o> \\
&= c_1<\mathbf{w}, \mathbf{v}_o> + c_2<\mathbf{v}_o, \mathbf{v}_o> \\
&= c_2<\mathbf{v}_o, \mathbf{v}_o>.
\end{aligned}
$$

Since $\mathbf{w}$ and $\mathbf{v}_o$ are not zero, it follows that

$$c_1 = c_2 = 0.$$

Thus $\mathbf{w}$ and $\mathbf{v}_o$ are independent and hence form a basis for W.

The procedure we have just described can be extended so that starting with a basis for a k-dimensional subspace W of an inner product space, we can find another basis $\mathbf{w}_1, \ldots, \mathbf{w}_k$ for W consisting of vectors with the property that

$$<\mathbf{w}_i, \mathbf{w}_j> = 0 \qquad \text{if } i \neq j.$$

Such a basis is referred to as an **orthogonal basis** for W.

Fact: Gram-Schmidt Orthogonalization Process. *Suppose* $\mathbf{v}_1, \ldots, \mathbf{v}_k$ *form a basis for a subspace W of an inner product space. Define the vectors* $\mathbf{w}_1, \ldots, \mathbf{w}_k$ *as follows:*

$$\mathbf{w}_1 = \mathbf{v}_1$$

$$\mathbf{w}_2 = \mathbf{v}_2 - \frac{<\mathbf{v}_2, \mathbf{w}_1>}{\|\mathbf{w}_1\|^2} \mathbf{w}_1$$

$$\mathbf{w}_3 = \mathbf{v}_3 - \frac{<\mathbf{v}_3, \mathbf{w}_1>}{\|\mathbf{w}_1\|^2} \mathbf{w}_1 - \frac{<\mathbf{v}_3, \mathbf{w}_2>}{\|\mathbf{w}_2\|^2} \mathbf{w}_2$$

.

.

.

$$\mathbf{w}_k = \mathbf{v}_k - \frac{<\mathbf{v}_k, \mathbf{w}_1>}{\|\mathbf{w}_1\|^2} \mathbf{w}_1 - \frac{<\mathbf{v}_k, \mathbf{w}_2>}{\|\mathbf{w}_2\|^2} \mathbf{w}_2 - \cdots - \frac{<\mathbf{v}_k, \mathbf{w}_{k-1}>}{\|\mathbf{w}_{k-1}\|^2} \mathbf{w}_{k-1}.$$

Then $\mathbf{w}_1, \ldots, \mathbf{w}_k$ *form an orthogonal basis for W.*

If we go further and **normalize** the basis $\mathbf{w}_1, \ldots, \mathbf{w}_k$ by replacing each of the vectors $\mathbf{w}_i$ by

$$\mathbf{u}_i = \frac{1}{\|\mathbf{w}_i\|} \mathbf{w}_i,$$

then we obtain an orthogonal basis for W consisting of vectors with norm 1 (see Exercise 20b). Such a basis is called an **orthonormal basis**, and can be characterized as a basis $\mathbf{u}_1, \ldots, \mathbf{u}_k$ for which

$$<\mathbf{u}_i, \mathbf{u}_j> = \begin{cases} 0 & \text{if } i \neq j \\ 1 & \text{if } i = j \end{cases}.$$

Example 4.8.5

Apply the Gram-Schmidt process to the basis of $W = R^3$ given by

$$\mathbf{v}_1 = \begin{bmatrix} 1 \\ 1 \\ 1 \end{bmatrix}, \quad \mathbf{v}_2 = \begin{bmatrix} 0 \\ 1 \\ 1 \end{bmatrix}, \quad \mathbf{v}_3 = \begin{bmatrix} 2 \\ 0 \\ 1 \end{bmatrix}$$

to obtain an orthogonal basis for R^3 with respect to the standard inner product. Normalize the resulting basis to obtain an orthonormal basis.

The Gram-Schmidt process goes as follows.

$$\mathbf{w}_1 = \mathbf{v}_1 = \begin{bmatrix} 1 \\ 1 \\ 1 \end{bmatrix}$$

$$\mathbf{w}_2 = \mathbf{v}_2 - \frac{\langle \mathbf{v}_2, \mathbf{w}_1 \rangle}{\|\mathbf{w}_1\|^2} \mathbf{w}_1 = \begin{bmatrix} 0 \\ 1 \\ 1 \end{bmatrix} - \frac{2}{3} \begin{bmatrix} 1 \\ 1 \\ 1 \end{bmatrix} = \begin{bmatrix} -\frac{2}{3} \\ \frac{1}{3} \\ \frac{1}{3} \end{bmatrix}$$

$$\mathbf{w}_3 = \mathbf{v}_3 - \frac{\langle \mathbf{v}_3, \mathbf{w}_1 \rangle}{\|\mathbf{w}_1\|^2} \mathbf{w}_1 - \frac{\langle \mathbf{v}_3, \mathbf{w}_2 \rangle}{\|\mathbf{w}_2\|^2} \mathbf{w}_2$$

$$= \begin{bmatrix} 2 \\ 0 \\ 1 \end{bmatrix} - \frac{3}{3} \begin{bmatrix} 1 \\ 1 \\ 1 \end{bmatrix} - \frac{-1}{\frac{2}{3}} \begin{bmatrix} -\frac{2}{3} \\ \frac{1}{3} \\ \frac{1}{3} \end{bmatrix} = \begin{bmatrix} 0 \\ -\frac{1}{2} \\ \frac{1}{2} \end{bmatrix}.$$

Thus

$$\mathbf{w}_1 = \begin{bmatrix} 1 \\ 1 \\ 1 \end{bmatrix}, \qquad \mathbf{w}_2 = \begin{bmatrix} -\frac{2}{3} \\ \frac{1}{3} \\ \frac{1}{3} \end{bmatrix}, \qquad \mathbf{w}_3 = \begin{bmatrix} 0 \\ -\frac{1}{2} \\ \frac{1}{2} \end{bmatrix}$$

form an orthogonal basis for R^3.

To obtain an orthonormal basis, we normalize the orthogonal basis by multiplying each $\mathbf{w}_i$ by the reciprocal of its norm:

$$\mathbf{u}_1 = \frac{1}{\|\mathbf{w}_1\|} \mathbf{w}_1 = \begin{bmatrix} 1/\sqrt{3} \\ 1/\sqrt{3} \\ 1/\sqrt{3} \end{bmatrix}, \qquad \mathbf{u}_2 = \frac{1}{\|\mathbf{w}_2\|} \mathbf{w}_2 = \begin{bmatrix} -2/\sqrt{6} \\ 1/\sqrt{6} \\ 1/\sqrt{6} \end{bmatrix},$$

$$\mathbf{u}_3 = \frac{1}{\|\mathbf{w}_3\|} \mathbf{w}_3 = \begin{bmatrix} 0 \\ -1/\sqrt{2} \\ 1/\sqrt{2} \end{bmatrix}.$$

One of the advantages of an orthonormal basis $\mathbf{u}_1, \ldots, \mathbf{u}_k$ of a subspace W is the ease with which we can calculate the coefficients of any given vector $\mathbf{w}$ in W. If

$$\mathbf{w} = c_1 \mathbf{u}_1 + \ldots + c_k \mathbf{u}_k,$$

then

$$\langle \mathbf{w}, \mathbf{u}_i \rangle = \langle c_1 \mathbf{u}_1 + \ldots + c_k \mathbf{u}_k, \mathbf{u}_i \rangle$$

$$= c_1 \langle \mathbf{u}_1, \mathbf{u}_i \rangle + \ldots + \langle \mathbf{u}_k, \mathbf{u}_i \rangle$$

$$= c_i \langle \mathbf{u}_i, \mathbf{u}_i \rangle = c_i.$$

Thus we have the following.

Fact: *If* $\mathbf{u}_1, \ldots, \mathbf{u}_k$ *form an orthonormal basis for W and if* $\mathbf{w}$ *is any vector in W, then*

$$\mathbf{w} = <\mathbf{w}, \mathbf{u}_1> \mathbf{u}_1 + <\mathbf{w}, \mathbf{u}_2> \mathbf{u}_2 + \ldots + <\mathbf{w}, \mathbf{u}_k> \mathbf{u}_k.$$

For example, the vector

$$\mathbf{w} = \begin{bmatrix} 1 \\ 1 \\ 2 \end{bmatrix}$$

in R^3 can be expressed in terms of the orthonormal basis $\mathbf{u}_1, \mathbf{u}_2, \mathbf{u}_3$ found in Example 4.8.5 as

$$\mathbf{w} = <\mathbf{w}, \mathbf{u}_1> \mathbf{u}_1 + <\mathbf{w}, \mathbf{u}_2> \mathbf{u}_2 + <\mathbf{w}, \mathbf{u}_3> \mathbf{u}_3$$

$$= \frac{4}{\sqrt{3}} \begin{bmatrix} 1/\sqrt{3} \\ 1/\sqrt{3} \\ 1/\sqrt{3} \end{bmatrix} + \frac{1}{\sqrt{6}} \begin{bmatrix} -2/\sqrt{6} \\ 1/\sqrt{6} \\ 1/\sqrt{6} \end{bmatrix} + \frac{1}{\sqrt{2}} \begin{bmatrix} 0 \\ -1/\sqrt{2} \\ 1/\sqrt{2} \end{bmatrix}.$$

INNER PRODUCTS AND ORTHOGONALITY

An **inner product** $< , >$ on the real vector space V is a real-valued function satisfying

(P1) $<\mathbf{v}, \mathbf{w}> = <\mathbf{w}, \mathbf{v}>$

(P2) $<\mathbf{v} + \mathbf{u}, \mathbf{w}> = <\mathbf{v}, \mathbf{w}> + <\mathbf{u}, \mathbf{w}>$

(P3) $<c\mathbf{v}, \mathbf{w}> = c <\mathbf{v}, \mathbf{w}>$

(P4) $<\mathbf{v}, \mathbf{v}> \geq 0$ and $<\mathbf{v}, \mathbf{v}> = 0$ if and only if $\mathbf{v} = \mathbf{0}$.

The **norm** of a vector $\mathbf{v}$ with respect to the inner product $< , >$ is

$$\|\mathbf{v}\| = \sqrt{<\mathbf{v}, \mathbf{v}>}.$$

Two vectors $\mathbf{v}$ and $\mathbf{w}$ in V are **parallel** if $\mathbf{w} = c\mathbf{v}$. They are **orthogonal** if $<\mathbf{v}, \mathbf{w}> = 0$.

If we are given two vectors $\mathbf{v}$ and $\mathbf{w} \neq \mathbf{0}$, then we can always express $\mathbf{v}$ as a sum

$$\mathbf{v} = \mathbf{v}_p + \mathbf{v}_o$$

of a **component parallel to w**

$$\mathbf{v}_p = \frac{<\mathbf{v}, \mathbf{w}>}{\|\mathbf{w}\|^2} \mathbf{w}$$

and a **component orthogonal to w**

$$\mathbf{v}_o = \mathbf{v} - \mathbf{v}_p.$$

A basis for a subspace W is said to be an **orthogonal basis** if the inner product of distinct vectors in the basis is always zero. An orthogonal basis consisting of vectors of norm 1 is said to be an **orthonormal basis**.

Given any basis $\mathbf{v}_1, \ldots, \mathbf{v}_k$ for W, the **Gram-Schmidt process** yields an orthogonal basis $\mathbf{w}_1, \ldots, \mathbf{w}_k$ defined by

$$\mathbf{w}_1 = \mathbf{v}_1$$

$$\mathbf{w}_2 = \mathbf{v}_2 - \frac{<\mathbf{v}_2, \mathbf{w}_1>}{\|\mathbf{w}_1\|^2} \mathbf{w}_1$$

$$\mathbf{w}_3 = \mathbf{v}_3 - \frac{<\mathbf{v}_3, \mathbf{w}_1>}{\|\mathbf{w}_1\|^2} \mathbf{w}_1 - \frac{<\mathbf{v}_3, \mathbf{w}_2>}{\|\mathbf{w}_2\|^2} \mathbf{w}_2$$

$$\vdots$$

$$\mathbf{w}_k = \mathbf{v}_k - \frac{<\mathbf{v}_k, \mathbf{w}_1>}{\|\mathbf{w}_1\|^2} \mathbf{w}_1 - \frac{<\mathbf{v}_k, \mathbf{w}_2>}{\|\mathbf{w}_2\|^2} \mathbf{w}_2 - \ldots - \frac{<\mathbf{v}_k, \mathbf{w}_{k-1}>}{\|\mathbf{w}_{k-1}\|^2} \mathbf{w}_{k-1}.$$

We can **normalize** this basis by multiplying each vector by the reciprocal of its norm. This yields an orthonormal basis

$$\mathbf{u}_1 = \frac{1}{\|\mathbf{w}_1\|} \mathbf{w}_1, \ldots, \mathbf{u}_k = \frac{1}{\|\mathbf{w}_k\|} \mathbf{w}_k.$$

If $\mathbf{u}_1, \ldots, \mathbf{u}_k$ form an orthonormal basis for W, then any vector $\mathbf{w}$ in W can be written as

$$\mathbf{w} = <\mathbf{w}, \mathbf{u}_1> \mathbf{u}_1 + \ldots + <\mathbf{w}, \mathbf{u}_k> \mathbf{u}_k.$$

EXERCISES

In Exercises 1 to 3 and 7 to 10, which deal with R^3 or R^4, take $<,>$ to be the standard inner product. In Exercises 4 to 6 and 11 to 14, which deal with spaces of the form $C[a, b]$, take $<,>$ to be the inner product described in Example 4.8.2.

1. For each of the following pairs of vectors
    (i)   calculate $<\mathbf{v}, \mathbf{w}>$ and
    (ii) decide whether $\mathbf{v}$ and $\mathbf{w}$ are orthogonal.

    a. $\mathbf{v} = \begin{bmatrix} 1 \\ 2 \\ -1 \end{bmatrix}, \mathbf{w} = \begin{bmatrix} 1 \\ 3 \\ 1 \end{bmatrix}$ b. $\mathbf{v} = \begin{bmatrix} 1 \\ 2 \\ -1 \end{bmatrix}, \mathbf{w} = \begin{bmatrix} -2 \\ 1 \\ 0 \end{bmatrix}$

    c. $\mathbf{v} = \begin{bmatrix} 1 \\ 2 \\ -1 \end{bmatrix}, \mathbf{w} = \begin{bmatrix} 0 \\ 1 \\ 2 \end{bmatrix}$ d. $\mathbf{v} = \begin{bmatrix} 1 \\ 2 \\ -1 \end{bmatrix}, \mathbf{w} = \begin{bmatrix} -2 \\ 1 \\ 2 \end{bmatrix}$

    e. $\mathbf{v} = \begin{bmatrix} 1 \\ 2 \\ 0 \\ 1 \end{bmatrix}, \mathbf{w} = \begin{bmatrix} -1 \\ 0 \\ 2 \\ 1 \end{bmatrix}$ f. $\mathbf{v} = \begin{bmatrix} 1 \\ 2 \\ 0 \\ 1 \end{bmatrix}, \mathbf{w} = \begin{bmatrix} 2 \\ 1 \\ 0 \\ -2 \end{bmatrix}$

2. For each of the following vectors
    (i)   calculate $\|\mathbf{v}\|$ and
    (ii) find a vector of norm 1 parallel to $\mathbf{v}$.

    a. $\mathbf{v} = \begin{bmatrix} 1 \\ 2 \\ -2 \end{bmatrix}$ b. $\mathbf{v} = \begin{bmatrix} 1 \\ 3 \\ 1 \end{bmatrix}$ c. $\mathbf{v} = \begin{bmatrix} 0 \\ -4 \\ 3 \end{bmatrix}$

    d. $\mathbf{v} = \begin{bmatrix} -2 \\ 1 \\ 1 \end{bmatrix}$ e. $\mathbf{v} = \begin{bmatrix} 2 \\ 2 \\ 0 \\ 1 \end{bmatrix}$ f. $\mathbf{v} = \begin{bmatrix} 2 \\ 2 \\ 2 \\ -2 \end{bmatrix}$

3. For each of the following pairs of vectors,
    (i)   find the component of $\mathbf{v}$ parallel to $\mathbf{w}$ and
    (ii) find the component of $\mathbf{v}$ orthogonal to $\mathbf{w}$.

    a. $\mathbf{v} = \begin{bmatrix} 1 \\ 3 \\ 1 \end{bmatrix}, \mathbf{w} = \begin{bmatrix} 1 \\ 2 \\ -2 \end{bmatrix}$ b. $\mathbf{v} = \begin{bmatrix} 1 \\ 2 \\ -2 \end{bmatrix}, \mathbf{w} = \begin{bmatrix} 1 \\ 3 \\ 1 \end{bmatrix}$

    c. $\mathbf{v} = \begin{bmatrix} 1 \\ 2 \\ -1 \end{bmatrix}, \mathbf{w} = \begin{bmatrix} 0 \\ -4 \\ 3 \end{bmatrix}$ d. $\mathbf{v} = \begin{bmatrix} 2 \\ 2 \\ 0 \\ 1 \end{bmatrix}, \mathbf{w} = \begin{bmatrix} 2 \\ 2 \\ 2 \\ -2 \end{bmatrix}$

4. For each of the following pairs of functions, viewed as elements of $C[-1, 1]$,
    (i)   calculate $<x(t), y(t)>$ and
    (ii) decide whether $x(t)$ and $y(t)$ are orthogonal.
    a. $x(t) = 1, y(t) = t$ b. $x(t) = t, y(t) = t^3$
    c. $x(t) = t, y(t) = t - 1$ d. $x(t) = 1, y(t) = \cos \pi t$

5. For each of the following functions, viewed as elements of $C[-1, 1]$,
   (i) calculate $\|x(t)\|$ and
   (ii) find a function of norm 1 parallel to $x(t)$.
      a. $x(t) = 1$ b. $x(t) = t - 1$ c. $x(t) = t^3$ d. $x(t) = \cos \pi t$

6. For each of the following pairs of functions, viewed as elements of $C[-1, 1]$,
   (i) find the component of $x(t)$ parallel to $y(t)$ and
   (ii) find the component of $x(t)$ orthogonal to $y(t)$.
      a. $x(t) = t, y(t) = t - 1$ b. $x(t) = 1, y(t) = t$
      c. $x(t) = 1, y(t) = t^2$ d. $x(t) = t, y(t) = \sin \pi t$

In Exercises 7 to 13, you are given an inner product space V, a basis $v_1, \ldots, v_k$ for a subspace W of V, and a vector w in W.

a. Use the Gram-Schmidt orthogonalization process to obtain an orthogonal basis for W.

b. Normalize the basis obtained in (a) to get an orthonormal basis for W.

c. Express w as a linear combination of the orthonormal basis found in (b).

7. $V = R^3$; $v_1 = \begin{bmatrix} 0 \\ 3 \\ 0 \end{bmatrix}, v_2 = \begin{bmatrix} 1 \\ 2 \\ -2 \end{bmatrix}; \quad w = \begin{bmatrix} 3 \\ 0 \\ -6 \end{bmatrix}$

8. $V = R^3$; $v_1 = \begin{bmatrix} 1 \\ 2 \\ 0 \end{bmatrix}, v_2 = \begin{bmatrix} 0 \\ 1 \\ 0 \end{bmatrix}; \quad w = \begin{bmatrix} 1 \\ 1 \\ 0 \end{bmatrix}$

9. $V = R^3$; $v_1 = \begin{bmatrix} 1 \\ 1 \\ 0 \end{bmatrix}, v_2 = \begin{bmatrix} 1 \\ 2 \\ 0 \end{bmatrix}, v_3 = \begin{bmatrix} 0 \\ 1 \\ 1 \end{bmatrix}; \quad w = \begin{bmatrix} 1 \\ 0 \\ 1 \end{bmatrix}$

10. $V = R^4$; $v_1 = \begin{bmatrix} 1 \\ 1 \\ 0 \\ 0 \end{bmatrix}, v_2 = \begin{bmatrix} 1 \\ 0 \\ 1 \\ 0 \end{bmatrix}, v_3 = \begin{bmatrix} 0 \\ 1 \\ 0 \\ 1 \end{bmatrix}, v_4 = \begin{bmatrix} 0 \\ 0 \\ 0 \\ 1 \end{bmatrix}; \quad w = \begin{bmatrix} 0 \\ 0 \\ 1 \\ 2 \end{bmatrix}$

11. $V = C[-1, 1]$; $v_1 = 1, v_2 = t, v_3 = t^2$; $w = 1 - t^2$

12. $V = C[0, 1]$; $v_1 = 1, v_2 = t, v_3 = t^2$; $w = 1 - t^2$

13. $V = C[-1, 1]$; $v_1 = 1, v_2 = \cos \pi t, v_3 = \sin \pi t$; $w = \sin \pi t$

14. $V = C[0, 1]$; $v_1 = 1, v_2 = \cos \pi t, v_3 = \sin \pi t$; $w = \sin \pi t$

More abstract exercises:

15. Check that properties (P1) to (P3) hold
    a. for the inner product defined in Example 4.8.1 and
    b. for the inner product defined in Example 4.8.2.

16. Let V be a finite-dimensional vector space with basis $v_1, \ldots, v_n$. Then any vectors v and w in V can be written uniquely as linear combinations of the basis vectors:

$$v = a_1 v_1 + \ldots + a_n v_n, \qquad w = b_1 v_1 + \ldots + b_n v_n.$$

Define

$$<v, w> = a_1 b_1 + \ldots + a_n b_n.$$

a. Show that this rule defines an inner product on V.

b. Show that $v_1, \ldots, v_n$ form an orthonormal basis with respect to this inner product.

17. Let $\mathbf{v} = \begin{bmatrix} 1 \\ 1 \\ 1 \\ 0 \end{bmatrix}$ and $\mathbf{w} = \begin{bmatrix} 1 \\ 1 \\ 1 \\ 2 \end{bmatrix}$. Calculate $<\mathbf{v}, \mathbf{w}>$ and $\|\mathbf{v}\|$ where

   a. $< , >$ is the standard inner product on R^4, and
   b. $< , >$ is the inner product on R^4 determined as described in the preceding exercise by the basis

$$\mathbf{v}_1 = \begin{bmatrix} 1 \\ 1 \\ 0 \\ 0 \end{bmatrix}, \quad \mathbf{v}_2 = \begin{bmatrix} 0 \\ 0 \\ 1 \\ 0 \end{bmatrix}, \quad \mathbf{v}_3 = \begin{bmatrix} 1 \\ 0 \\ 1 \\ 0 \end{bmatrix}, \quad \mathbf{v}_4 = \begin{bmatrix} 0 \\ 1 \\ 0 \\ 2 \end{bmatrix}.$$

18. Show that $<\mathbf{v}, \mathbf{0}> = <\mathbf{0}, \mathbf{v}> = 0$ for all vectors $\mathbf{v}$ in an inner product space.

19. Show that if $\mathbf{v}$, $\mathbf{w}$, and $\mathbf{u}$ are vectors in an inner product space, and if c is a scalar, then $<\mathbf{v}, c\mathbf{w} + \mathbf{u}> = c <\mathbf{v}, \mathbf{w}> + <\mathbf{v}, \mathbf{u}>$.

20. Let $\mathbf{v}$ be a vector in an inner product space V.
   a. Show that $\|c\mathbf{v}\| = |c| \|\mathbf{v}\|$ for any scalar c.
   b. Show that $(1/\|\mathbf{v}\|)\mathbf{v}$ has norm 1.

*21. Let $V = C[-\ell, \ell]$, let $< , >$ be the inner product on V described in Example 4.8.2, and let W be the subspace of V spanned by the functions

$$\frac{1}{2}, \quad \cos\frac{\pi t}{\ell}, \quad \sin\frac{\pi t}{\ell}, \quad \cos\frac{2\pi t}{\ell}, \quad \sin\frac{2\pi t}{\ell}, \ldots, \cos\frac{n\pi t}{\ell}, \quad \sin\frac{n\pi t}{\ell}.$$

   a. Show that these functions form an orthogonal basis for W.
   b. Show that each of these functions has norm ℓ.
   c. Show that if

$$f(t) = \frac{a_0}{2} + a_1 \cos\frac{\pi t}{\ell} + b_1 \sin\frac{\pi t}{\ell} + \ldots + a_n \cos\frac{n\pi t}{\ell} + b_n \sin\frac{n\pi t}{\ell},$$

$$-\ell \leq t \leq \ell, \text{ then}$$

$$a_k = \frac{1}{\ell} \int_{-\ell}^{\ell} f(t) \cos\frac{k\pi t}{\ell} \, dt, \quad k = 0, 1, 2, \ldots$$

$$b_k = \frac{1}{\ell} \int_{-\ell}^{\ell} f(t) \sin\frac{k\pi t}{\ell} \, dt, \quad k = 1, 2, \ldots$$

REVIEW PROBLEMS

1. Let $F[a, b]$ be the set of all real-valued functions $x(t)$ that are defined on the interval $a \leq t \leq b$. We add functions in $F[a, b]$ and multiply them by scalars as usual:

$$(x + y)(t) = x(t) + y(t), \qquad (cx)(t) = c[x(t)].$$

   Check that $F[a, b]$ is a real vector space.

2. Let $C_n[a, b]$ be the set of functions in $F[a, b]$ (see Problem 1) that are continuous and have continuous derivatives of order 1 to n on the interval $a \le t \le b$. Check that $C_n[a, b]$ is a subspace of $F[a, b]$.

3. Let $L = a_n D^n + \ldots + a_0$ be an nth-order linear differential operator that is normal on the interval $a \le t \le b$.
   a. Check that L is a linear transformation from $C_n[a, b]$ to $C[a, b]$ (see Problem 2).
   b. What is $\mathbf{Ker}(L)$? What is the dimension of $\mathbf{Ker}(L)$?
   c. What is $\mathbf{Range}(L)$? (*Hint*: Use the existence and uniqueness theorem.)

4. Determine whether the set V of all real 2-vectors is a real vector space with respect to the given unusual operations.

   a. $\begin{bmatrix} x \\ y \end{bmatrix} \dotplus \begin{bmatrix} u \\ v \end{bmatrix} = \begin{bmatrix} x - u \\ y - v \end{bmatrix}$, $c \begin{bmatrix} x \\ y \end{bmatrix} = \begin{bmatrix} cx \\ cy \end{bmatrix}$

   b. $\begin{bmatrix} x \\ y \end{bmatrix} + \begin{bmatrix} u \\ v \end{bmatrix} = \begin{bmatrix} x + u \\ y + v \end{bmatrix}$, $c \cdot \begin{bmatrix} x \\ y \end{bmatrix} = \begin{bmatrix} c^2 x \\ c^2 y \end{bmatrix}$

   c. $\begin{bmatrix} x \\ y \end{bmatrix} \dotplus \begin{bmatrix} u \\ v \end{bmatrix} = \begin{bmatrix} x + u + 1 \\ y + v + 1 \end{bmatrix}$, $c \cdot \begin{bmatrix} x \\ y \end{bmatrix} = \begin{bmatrix} cx + c - 1 \\ cy + c - 1 \end{bmatrix}$

5. Determine whether the set of all those vectors $\begin{bmatrix} x \\ y \end{bmatrix}$ that satisfy the given condition is a subspace of R^2.
   a. $y = -x$ b. $y = 3x - 2$
   c. $y = \pi x$ d. $xy = 0$

6. Determine whether the set of all vectors of the given form is a subspace of R^3.

   a. $\begin{bmatrix} x \\ y \\ x^3 \end{bmatrix}$ b. $\begin{bmatrix} x \\ y \\ 3x - 4y \end{bmatrix}$

   c. $\begin{bmatrix} x + y \\ x + y \\ x - y \end{bmatrix}$ d. $\begin{bmatrix} x + x \\ 3x + 1 \\ z \end{bmatrix}$

7. Determine whether the set of all functions $x(t)$ in C_∞ that satisfy the given condition is a subspace of C_∞.
   a. $x(-1) + x(1) = 0$ b. $x(-t) + x(t) = 0$ for all t
   c. $x(-1) + x(1) \ge 0$ d. $x(-1) + x'(1) = 0$

8. Determine whether the set of all vector valued functions $\mathbf{x}(t)$ of the given form is a subspace of C_∞^2.

   a. $\mathbf{x}(t) = \begin{bmatrix} x(t) \\ x(t)^2 \end{bmatrix}$ b. $\mathbf{x}(t) = \begin{bmatrix} x(t) \\ x''(t) \end{bmatrix}$

   c. $\mathbf{x}(t) = \begin{bmatrix} c_1 e^t - e^{2t} \\ c_1 e^t + c_2 e^t \end{bmatrix}$ for some choice of constants c_1 and c_2

   d. $\mathbf{x}(t) = \begin{bmatrix} c_1 e^t - c_2 e^t \\ c_1 e^t + c_2 e^t \end{bmatrix}$ for some choice of constants c_1 and c_2

9. Determine whether the following vectors are in $\mathbf{Comb}\left(\begin{bmatrix} 1 \\ 0 \\ 1 \end{bmatrix}, \begin{bmatrix} 2 \\ 3 \\ 0 \end{bmatrix}\right)$.

a. $\begin{bmatrix} 3 \\ 6 \\ 1 \end{bmatrix}$ b. $\begin{bmatrix} 0 \\ 3 \\ 0 \end{bmatrix}$ c. $\begin{bmatrix} 1 \\ 3 \\ -1 \end{bmatrix}$ d. $\begin{bmatrix} 4 \\ 3 \\ 3 \end{bmatrix}$

10. In each of the following,
    (i) decide whether the vectors are linearly independent,
    (ii) decide whether the vectors span R^3, and
    (iii) decide whether the vectors form a basis for R^3.

a. $\begin{bmatrix} 1 \\ 1 \\ 0 \end{bmatrix}, \begin{bmatrix} 1 \\ 2 \\ 0 \end{bmatrix}$ b. $\begin{bmatrix} -1 \\ 2 \\ 0 \end{bmatrix}, \begin{bmatrix} 2 \\ -4 \\ 0 \end{bmatrix}$

c. $\begin{bmatrix} 1 \\ 2 \\ 3 \end{bmatrix}, \begin{bmatrix} 3 \\ 2 \\ 1 \end{bmatrix}, \begin{bmatrix} 2 \\ 3 \\ 1 \end{bmatrix}$ d. $\begin{bmatrix} 1 \\ 0 \\ 1 \end{bmatrix}, \begin{bmatrix} 2 \\ 2 \\ 0 \end{bmatrix}, \begin{bmatrix} 0 \\ -2 \\ 2 \end{bmatrix}$

e. $\begin{bmatrix} 1 \\ 0 \\ 1 \end{bmatrix}, \begin{bmatrix} 0 \\ 1 \\ 1 \end{bmatrix}, \begin{bmatrix} 1 \\ 1 \\ 0 \end{bmatrix}, \begin{bmatrix} 1 \\ 1 \\ 1 \end{bmatrix}$ f. $\begin{bmatrix} 1 \\ 0 \\ 2 \end{bmatrix}, \begin{bmatrix} 0 \\ 2 \\ 1 \end{bmatrix}, \begin{bmatrix} 1 \\ 1 \\ 1 \end{bmatrix}, \begin{bmatrix} 2 \\ 5 \\ 5 \end{bmatrix}$

11. In each of the following,
    (i) decide whether the vectors are linearly independent,
    (ii) decide whether the vectors span the space W consisting of all 4-vectors of the form
    $$\begin{bmatrix} x \\ y \\ x - y \\ y - x \end{bmatrix},$$
    (iii) decide whether the vectors form a basis for W.

a. $\begin{bmatrix} 1 \\ 1 \\ 0 \\ 0 \end{bmatrix}, \begin{bmatrix} 1 \\ 0 \\ 1 \\ -1 \end{bmatrix}$ b. $\begin{bmatrix} 1 \\ 1 \\ 0 \\ 0 \end{bmatrix}, \begin{bmatrix} -1 \\ -1 \\ 0 \\ 0 \end{bmatrix}$

c. $\begin{bmatrix} 1 \\ 0 \\ 1 \\ -1 \end{bmatrix}, \begin{bmatrix} 0 \\ 1 \\ -1 \\ 1 \end{bmatrix}, \begin{bmatrix} 2 \\ 3 \\ -1 \\ 1 \end{bmatrix}$ d. $\begin{bmatrix} 1 \\ 0 \\ 1 \\ 0 \end{bmatrix}, \begin{bmatrix} 0 \\ 1 \\ -1 \\ 0 \end{bmatrix}, \begin{bmatrix} 0 \\ 0 \\ 0 \\ 1 \end{bmatrix}$

12. Find a basis for the subspace of R^4 consisting of all vectors of the given form. What is the dimension of this subspace?

a. $\begin{bmatrix} x + y \\ y - x \\ z \\ z \end{bmatrix}$ b. $\begin{bmatrix} x + y \\ 2x + 3y \\ z \\ 2z \end{bmatrix}$ c. $\begin{bmatrix} x + y \\ 2x + y \\ x - 2y \\ x + z \end{bmatrix}$ d. $\begin{bmatrix} x - y \\ 2x - 2y \\ -x + y \\ x - y \end{bmatrix}$

13. Find a basis for the solution space of $B\mathbf{v} = \mathbf{0}$. What is the dimension of this space?

a. $B = \begin{bmatrix} 1 & 2 \\ 2 & 4 \\ 3 & 6 \end{bmatrix}$ b. $B = \begin{bmatrix} 1 & 2 & 3 \\ 3 & 2 & 1 \\ 2 & 0 & -2 \end{bmatrix}$ c. $B = \begin{bmatrix} 1 & 3 & -2 & 0 \\ 2 & 1 & 2 & 1 \\ 4 & 7 & -2 & 1 \end{bmatrix}$

14. For each of the following bases of R^3,

(i) find the coordinates of $\begin{bmatrix} 3 \\ 2 \\ 0 \end{bmatrix}$ with respect to the basis, and

(ii) find the vector whose coordinates are $1, 0, -1$.

a. $\begin{bmatrix} 1 \\ 0 \\ 0 \end{bmatrix}, \begin{bmatrix} 0 \\ 1 \\ 0 \end{bmatrix}, \begin{bmatrix} 0 \\ 0 \\ 1 \end{bmatrix}$ b. $\begin{bmatrix} 0 \\ 0 \\ 1 \end{bmatrix}, \begin{bmatrix} 1 \\ 0 \\ 0 \end{bmatrix}, \begin{bmatrix} 0 \\ 1 \\ 0 \end{bmatrix}$

c. $\begin{bmatrix} 1 \\ 2 \\ 0 \end{bmatrix}, \begin{bmatrix} 1 \\ 0 \\ 2 \end{bmatrix}, \begin{bmatrix} 2 \\ 1 \\ 0 \end{bmatrix}$ d. $\begin{bmatrix} 1 \\ 1 \\ 1 \end{bmatrix}, \begin{bmatrix} 1 \\ 0 \\ 1 \end{bmatrix}, \begin{bmatrix} 0 \\ 1 \\ 1 \end{bmatrix}$

15. What is the dimension of the solution space of $Lx = 0$?

a. $L = D^4 - D^2$ b. $L = D^8 - 2D^4 + 1$
c. $L = D^9 - 3D^5 + D - 2$ (*Hint:* Think.)

16. What is the dimension of the solution space of $D\mathbf{x} = A\mathbf{x}$?

a. $A = \begin{bmatrix} 1 & 2 \\ 0 & 3 \end{bmatrix}$ b. $A = \begin{bmatrix} 0 & 1 & 0 \\ -1 & 0 & 1 \\ 0 & 0 & 1 \end{bmatrix}$

c. $A = \begin{bmatrix} 1 & 1 & 1 & 1 \\ 1 & 1 & 1 & 2 \\ 1 & 1 & 2 & 3 \\ 1 & 2 & 3 & 4 \end{bmatrix}$ (*Hint:* Think.)

17. Determine whether T is a linear transformation from R^2 to R^2.

a. $T(\mathbf{v}) = 3\mathbf{v}$ b. $T(\mathbf{v}) = 3\mathbf{v} - \begin{bmatrix} 3 \\ 3 \end{bmatrix}$ c. $T\begin{bmatrix} x \\ y \end{bmatrix} = \begin{bmatrix} x + y \\ xy \end{bmatrix}$

d. $T\begin{bmatrix} x \\ y \end{bmatrix} = \begin{bmatrix} x + y \\ x - y \end{bmatrix}$ e. $T\begin{bmatrix} x \\ y \end{bmatrix} = \begin{bmatrix} 2x \\ x/2 \end{bmatrix}$ f. $T\begin{bmatrix} x \\ y \end{bmatrix} = \begin{bmatrix} 2x \\ x^2 \end{bmatrix}$

18. Determine whether T is a linear transformation from C_∞ to C_∞.
a. $T(x) = x'' - 2x' + x$ b. $T(x) = x'' - 2x' + e^t$
c. $T(x) = x'' - 2x' + x^2$ d. $T(x) = x'' - e^t x$

19. Determine whether T defines a linear transformation from C_∞^2 to C_∞^2.

a. $T(\mathbf{x}) = \begin{bmatrix} 1 & 2 \\ 0 & 3 \end{bmatrix}^2 \mathbf{x}$ b. $T(\mathbf{x}) = \begin{bmatrix} 1 & 2 \\ 0 & 3 \end{bmatrix} \mathbf{x} - e^t \mathbf{x}$

c. $T \begin{bmatrix} x(t) \\ y(t) \end{bmatrix} = \begin{bmatrix} x'(t) + 2y(t) \\ 3y'(t) \end{bmatrix}$ d. $T \begin{bmatrix} x(t) \\ y(t) \end{bmatrix} = \begin{bmatrix} x'(t) + 2y(t) \\ e^t + 3y(t) \end{bmatrix}$

20. Each of the following rules defines a linear transformation T from R^n to R^m for some choice of n and m.
    (i) Find the dimension of **Ker**(T).
    (ii) Find the dimension of **Range**(T).

a. $T \begin{bmatrix} x \\ y \end{bmatrix} = \begin{bmatrix} y \\ x \end{bmatrix}$ b. $T \begin{bmatrix} x \\ y \end{bmatrix} = \begin{bmatrix} x \\ 2x \end{bmatrix}$

c. $T \begin{bmatrix} x \\ y \end{bmatrix} + \begin{bmatrix} x + y \\ x + y \\ x + y \end{bmatrix}$ d. $T \begin{bmatrix} x \\ y \end{bmatrix} = \begin{bmatrix} x + y \\ 0 \\ x - y \end{bmatrix}$

e. $T \begin{bmatrix} x \\ y \\ z \end{bmatrix} = \begin{bmatrix} y \\ z \end{bmatrix}$ f. $T \begin{bmatrix} x \\ y \\ z \end{bmatrix} = \begin{bmatrix} x - y \\ y - x \end{bmatrix}$

g. $T \begin{bmatrix} x \\ y \\ z \end{bmatrix} = \begin{bmatrix} x + y \\ x - y \\ x \end{bmatrix}$ h. $T \begin{bmatrix} x \\ y \\ z \end{bmatrix} = \begin{bmatrix} x + y \\ x + z \\ y - z \end{bmatrix}$

21. Each of the following rules defines a linear transformation T from $V = C_\infty$ to $W = C_\infty$ or from $V = C_\infty^2$ to $W = C_\infty^2$.

    (i) Determine the dimension of **Ker**(T).
    (ii) Describe **Range**(T).
    a. $Tx = D^2x - 2Dx$ b. $Tx = x' - tx$
    c. $Tx = D^7x - 3D^5x + 2D^4x + D^2x - 7x$

    d. $Tx = \begin{bmatrix} 0 & 1 \\ -1 & 0 \end{bmatrix} x$ e. $Tx = Dx$

    f. $Tx = \left(D - \begin{bmatrix} 0 & 1 \\ -1 & 0 \end{bmatrix} \right) x$

22. For each of the linear transformations in Exercise 20,
    (i)   decide whether T is one-to-one,
    (ii) decide whether T maps R^n onto R^m, and
    (iii) decide whether T is an isomorphism from R^n onto R^m.

23. For each of the linear transformations in Exercise 21,
    (i)   decide whether T is one-to-one,
    (ii) decide whether T maps V onto W, and
    (iii) decide whether T is an isomorphism from V onto W.

24. Let $V = R^4$ with the standard inner product.
    a. Find an orthonormal basis for the subspace W spanned by the vectors

$$\mathbf{v}_1 = \begin{bmatrix} 1 \\ 1 \\ 0 \\ 0 \end{bmatrix}, \quad \mathbf{v}_2 = \begin{bmatrix} 0 \\ 1 \\ 1 \\ 0 \end{bmatrix}, \quad \mathbf{v}_3 = \begin{bmatrix} 1 \\ 1 \\ 1 \\ 1 \end{bmatrix}.$$

b. Express each of the following vectors as a linear combination of the basis found in (a):

$$\mathbf{w} = \begin{bmatrix} 2 \\ 1 \\ 0 \\ 1 \end{bmatrix}, \qquad \mathbf{u} = \begin{bmatrix} 2 \\ 3 \\ 2 \\ 1 \end{bmatrix}.$$

25. Let $\mathbf{v}_1, \ldots, \mathbf{v}_k$ be fixed vectors in an inner product space V. For each $\mathbf{v}$ in V define

$$T(\mathbf{v}) = \begin{bmatrix} <\mathbf{v}, \mathbf{v}_1> \\ \cdot \\ \cdot \\ \cdot \\ <\mathbf{v}, \mathbf{v}_k> \end{bmatrix}.$$

Show that T is a linear transformation from V to R^n.

26. Let $\mathbf{w}$ be a fixed vector in an inner product space V and define $\mathbf{w}^\perp$ to be the set consisting of all those vectors $\mathbf{v}$ in V such that $<\mathbf{v}, \mathbf{w}> = 0$.
    a. Show that $\mathbf{w}^\perp$ is a subspace of V.
    b. Show that $\mathbf{w}^\perp \cap \mathbf{Comb}(\mathbf{w}) = \{\mathbf{0}\}$.

Matrix Representation
of Linear Transformations

5.1 THE MATRIX OF A LINEAR TRANSFORMATION

We saw in Chapter 3 how to solve an nth-order linear homogeneous system of o.d.e.'s $D\mathbf{x} = A\mathbf{x}$ by finding n independent generalized eigenvectors for A. In this chapter we will deepen our understanding of this process through an investigation of the linear transformation T_A that maps n-vectors to n-vectors by the rule $T_A(\mathbf{v}) = A\mathbf{v}$. We will learn in this section how to associate to any linear transformation T between finite-dimensional vector spaces a matrix whose relation to T is analogous to the relation between A and T_A.

We begin by noting that if $\mathcal{B} = \{\mathbf{v}_1, \ldots, \mathbf{v}_n\}$ is a basis for V, then *any linear transformation T from V to W is completely determined by its values on the basis vectors, $T(\mathbf{v}_1), \ldots, T(\mathbf{v}_n)$.* To see this, suppose $\mathbf{v}$ is any element in V. We know $\mathbf{v}$ can be written uniquely as a linear combination of the vectors in $\mathcal{B}$:

$$\mathbf{v} = b_1\mathbf{v}_1 + \ldots + b_n\mathbf{v}_n.$$

Since T is linear, transforming this equation gives us

$$T(\mathbf{v}) = T(b_1\mathbf{v}_1 + \ldots + b_n\mathbf{v}_n)$$
$$= b_1T(\mathbf{v}_1) + \ldots + b_nT(\mathbf{v}_n).$$

This shows how $T(\mathbf{v})$ can be determined from $T(\mathbf{v}_1), \ldots, T(\mathbf{v}_n)$.

Let's now take a closer look at the values $T(\mathbf{v}_1), \ldots, T(\mathbf{v}_n)$. If W has a basis $\mathcal{C} = \{\mathbf{w}_1, \ldots, \mathbf{w}_m\}$, then each of the vectors $T(\mathbf{v}_j)$ has a unique expression as a combination of the vectors in $\mathcal{C}$:

$$T(\mathbf{v}_1) = a_{11}\mathbf{w}_1 + a_{21}\mathbf{w}_2 + \ldots + a_{m1}\mathbf{w}_m$$
$$T(\mathbf{v}_2) = a_{12}\mathbf{w}_1 + a_{22}\mathbf{w}_2 + \ldots + a_{m2}\mathbf{w}_m$$
$$\cdot$$
$$\cdot$$
$$\cdot$$
$$T(\mathbf{v}_n) = a_{1n}\mathbf{w}_1 + a_{2n}\mathbf{w}_2 + \ldots + a_{mn}\mathbf{w}_m.$$

The scalars in these equations can be tabulated in a matrix known as **the matrix of T with respect to the bases** $\mathcal{B}$ **and** $\mathcal{C}$ and denoted $_\mathcal{C}[T]_\mathcal{B}$. For technical reasons, we write the coefficients in the expression for $T(\mathbf{v}_j)$ as the jth *column* of the matrix:

$$_\mathcal{C}[T]_\mathcal{B} = \begin{bmatrix} a_{11} & a_{12} & \cdots & a_{1n} \\ a_{21} & a_{22} & \cdots & a_{2n} \\ \cdot & \cdot & & \cdot \\ \cdot & \cdot & & \cdot \\ \cdot & \cdot & & \cdot \\ a_{m1} & a_{m2} & \cdots & a_{mn} \end{bmatrix}.$$

Note that this matrix depends on T and on the two bases $\mathcal{B}$ of V and $\mathcal{C}$ of W; this dependence is reflected in our notation for the matrix. Notice also that the order of numbering of the vectors in $\mathcal{B}$ and $\mathcal{C}$ is significant in determining the matrix.

Example 5.1.1

The mapping T from R^3 to R^2 defined by

$$T\begin{bmatrix} x \\ y \\ z \end{bmatrix} = \begin{bmatrix} 2x + y \\ x - y \end{bmatrix}$$

is a linear transformation. Find the matrix of T with respect to the standard bases

$$\mathcal{B} = \left\{ \begin{bmatrix} 1 \\ 0 \\ 0 \end{bmatrix}, \begin{bmatrix} 0 \\ 1 \\ 0 \end{bmatrix}, \begin{bmatrix} 0 \\ 0 \\ 1 \end{bmatrix} \right\}$$

and

$$\mathcal{C} = \left\{ \begin{bmatrix} 1 \\ 0 \end{bmatrix}, \begin{bmatrix} 0 \\ 1 \end{bmatrix} \right\}.$$

We apply T to each of the vectors in $\mathcal{B}$ in turn and write each result in terms of the vectors in $\mathcal{C}$:

$$T\begin{bmatrix} 1 \\ 0 \\ 0 \end{bmatrix} = \begin{bmatrix} 2 \\ 1 \end{bmatrix} = 2\begin{bmatrix} 1 \\ 0 \end{bmatrix} + 1\begin{bmatrix} 0 \\ 1 \end{bmatrix}$$

$$T\begin{bmatrix} 0 \\ 1 \\ 0 \end{bmatrix} = \begin{bmatrix} 1 \\ -1 \end{bmatrix} = 1\begin{bmatrix} 1 \\ 0 \end{bmatrix} + (-1)\begin{bmatrix} 0 \\ 1 \end{bmatrix}$$

$$T\begin{bmatrix} 0 \\ 0 \\ 1 \end{bmatrix} = \begin{bmatrix} 0 \\ 0 \end{bmatrix} = 0\begin{bmatrix} 1 \\ 0 \end{bmatrix} + 0\begin{bmatrix} 0 \\ 1 \end{bmatrix}.$$

The matrix of T is formed by putting the coefficients in these expressions into columns:

$$_{\mathscr{C}}[T]_{\mathscr{B}} = \begin{bmatrix} 2 & 1 & 0 \\ 1 & -1 & 0 \end{bmatrix}.$$

Example 5.1.2

Find the matrix of the transformation T from the previous example with respect to the bases

$$\mathscr{B}' = \left\{ \begin{bmatrix} 0 \\ 0 \\ 1 \end{bmatrix}, \begin{bmatrix} 1 \\ 0 \\ 0 \end{bmatrix}, \begin{bmatrix} 0 \\ 1 \\ 0 \end{bmatrix} \right\} \quad \text{and} \quad \mathscr{C}' = \left\{ \begin{bmatrix} 0 \\ 1 \end{bmatrix}, \begin{bmatrix} 1 \\ 0 \end{bmatrix} \right\}.$$

Again, we apply T to each of the vectors in $\mathscr{B}'$ in turn and express each result in terms of $\mathscr{C}'$:

$$T \begin{bmatrix} 0 \\ 0 \\ 1 \end{bmatrix} = \begin{bmatrix} 0 \\ 0 \end{bmatrix} = 0 \begin{bmatrix} 0 \\ 1 \end{bmatrix} + 0 \begin{bmatrix} 1 \\ 0 \end{bmatrix}$$

$$T \begin{bmatrix} 1 \\ 0 \\ 0 \end{bmatrix} = \begin{bmatrix} 2 \\ 1 \end{bmatrix} = 1 \begin{bmatrix} 0 \\ 1 \end{bmatrix} + 2 \begin{bmatrix} 1 \\ 0 \end{bmatrix}$$

$$T \begin{bmatrix} 0 \\ 1 \\ 0 \end{bmatrix} = \begin{bmatrix} 1 \\ -1 \end{bmatrix} = -1 \begin{bmatrix} 0 \\ 1 \end{bmatrix} + 1 \begin{bmatrix} 1 \\ 0 \end{bmatrix}$$

Thus the matrix of T with respect to bases $\mathscr{B}'$ and $\mathscr{C}'$ is

$$_{\mathscr{C}'}[T]_{\mathscr{B}'} = \begin{bmatrix} 0 & 1 & -1 \\ 0 & 2 & 1 \end{bmatrix}.$$

Notice that the bases $\mathscr{B}'$ and $\mathscr{C}'$ were obtained by reordering the vectors in the bases $\mathscr{B}$ and $\mathscr{C}$ from Example 5.1.1. The resulting matrices are different.

Example 5.1.3

Find the matrix of the transformation T from the previous examples with respect to the bases

$$\mathscr{B}'' = \left\{ \begin{bmatrix} 1 \\ 1 \\ 1 \end{bmatrix}, \begin{bmatrix} 1 \\ 1 \\ 0 \end{bmatrix}, \begin{bmatrix} 1 \\ 0 \\ 0 \end{bmatrix} \right\} \quad \text{and} \quad \mathscr{C}'' = \left\{ \begin{bmatrix} 1 \\ 1 \end{bmatrix}, \begin{bmatrix} 1 \\ -1 \end{bmatrix} \right\}.$$

We apply T to each of the 3-vectors of $\mathcal{B}''$ in turn, obtaining three 2-vectors

$$T\begin{bmatrix} 1 \\ 1 \\ 1 \end{bmatrix} = \begin{bmatrix} 3 \\ 0 \end{bmatrix}, \quad T\begin{bmatrix} 1 \\ 1 \\ 0 \end{bmatrix} = \begin{bmatrix} 3 \\ 0 \end{bmatrix}, \quad T\begin{bmatrix} 1 \\ 0 \\ 0 \end{bmatrix} = \begin{bmatrix} 2 \\ 1 \end{bmatrix}.$$

We determined in Example 4.3.4 how to express an arbitrary 2-vector in terms of $\mathcal{C}''$:

$$\begin{bmatrix} x \\ y \end{bmatrix} = \frac{1}{2}(x - y)\begin{bmatrix} 1 \\ 1 \end{bmatrix} + \frac{1}{2}(x - y)\begin{bmatrix} 1 \\ -1 \end{bmatrix}.$$

Using this expression, we see that

$$T\begin{bmatrix} 1 \\ 1 \\ 1 \end{bmatrix} = \begin{bmatrix} 3 \\ 0 \end{bmatrix} = \frac{3}{2}\begin{bmatrix} 1 \\ 1 \end{bmatrix} + \frac{3}{2}\begin{bmatrix} 1 \\ -1 \end{bmatrix}$$

$$T\begin{bmatrix} 1 \\ 1 \\ 0 \end{bmatrix} = \begin{bmatrix} 3 \\ 0 \end{bmatrix} = \frac{3}{2}\begin{bmatrix} 1 \\ 1 \end{bmatrix} + \frac{3}{2}\begin{bmatrix} 1 \\ -1 \end{bmatrix}$$

$$T\begin{bmatrix} 1 \\ 0 \\ 0 \end{bmatrix} = \begin{bmatrix} 2 \\ 1 \end{bmatrix} = \frac{3}{2}\begin{bmatrix} 1 \\ 1 \end{bmatrix} + \frac{1}{2}\begin{bmatrix} 1 \\ -1 \end{bmatrix}.$$

Thus

$$\mathcal{C}''[T]_{\mathcal{B}''} = \begin{bmatrix} \dfrac{3}{2} & \dfrac{3}{2} & \dfrac{3}{2} \\ \dfrac{3}{2} & \dfrac{3}{2} & \dfrac{1}{2} \end{bmatrix}.$$

Example 5.1.4
Find the matrix of the transformation T_A from R^n to R^m defined by multiplication by the $m \times n$ matrix

$$A = \begin{bmatrix} a_{11} & \cdots & a_{1n} \\ \cdot & & \cdot \\ \cdot & & \cdot \\ \cdot & & \cdot \\ a_{m1} & \cdots & a_{mn} \end{bmatrix}$$

with respect to the standard bases $\mathcal{B}$ and $\mathcal{C}$ of R^n and R^m, respectively.

In this case,

$$
\mathbf{v}_j = \begin{bmatrix} 0 \\ \cdot \\ \cdot \\ \cdot \\ 1 \\ \cdot \\ \cdot \\ \cdot \\ 0 \end{bmatrix} \leftarrow (j\text{th entry is the only nonzero entry})
$$

so that

$$
T_A(\mathbf{v}_j) = A\mathbf{v}_j = \begin{bmatrix} a_{11} & \cdots & a_{1n} \\ \cdot & & \cdot \\ \cdot & & \cdot \\ \cdot & & \cdot \\ a_{m1} & \cdots & a_{mn} \end{bmatrix} \begin{bmatrix} 0 \\ \cdot \\ \cdot \\ \cdot \\ 1 \\ \cdot \\ \cdot \\ \cdot \\ 0 \end{bmatrix} \leftarrow j\text{th entry}
$$

$$
= \begin{bmatrix} a_{1j} \\ \cdot \\ \cdot \\ \cdot \\ a_{mj} \end{bmatrix} = a_{1j} \begin{bmatrix} 1 \\ \cdot \\ \cdot \\ \cdot \\ 0 \end{bmatrix} + \ldots + a_{mj} \begin{bmatrix} 0 \\ \cdot \\ \cdot \\ \cdot \\ 1 \end{bmatrix}.
$$

The scalars on the right form the jth column of $_\mathscr{C}[T_A]_\mathscr{B}$, so

$$
_\mathscr{C}[T_A]_\mathscr{B} = \begin{bmatrix} a_{11} & \cdots & a_{1n} \\ \cdot & & \cdot \\ \cdot & & \cdot \\ \cdot & & \cdot \\ a_{m1} & \cdots & a_{mn} \end{bmatrix}.
$$

In other words, *if* A is an $m \times n$ real matrix, then the matrix of T_A with respect to the standard bases for R^n and R^m is A.

In the special case considered in Example 5.1.4, $T = T_A$ is easily recovered from its matrix A; to find $T(\mathbf{v})$ we simply multiply the n-vector $\mathbf{v}$ by A. In the general case of any transformation T and any bases $\mathscr{B}$ and $\mathscr{C}$, it is also possible to recover T from its matrix $_\mathscr{C}[T]_\mathscr{B}$ by multiplying the matrix times an appropriate n-vector.

To see how this process works, recall our earlier observations. Any vector $\mathbf{v}$ in V has a unique expression in terms of $\mathscr{B} = \{\mathbf{v}_1, \ldots, \mathbf{v}_n\}$.

(1) $$\mathbf{v} = b_1\mathbf{v}_1 + \ldots + b_n\mathbf{v}_n.$$

This yields an expression for $T(\mathbf{v})$ in terms of the $T(\mathbf{v}_j)$'s:

$$T(\mathbf{v}) = b_1T(\mathbf{v}_1) + \ldots + b_nT(\mathbf{v}_n).$$

The matrix

$$\mathscr{C}[T]_{\mathscr{B}} = \begin{bmatrix} a_{11} & \cdots & a_{1n} \\ & \cdot & \\ \cdot & \cdot & \cdot \\ & \cdot & \\ a_{m1} & \cdots & a_{mn} \end{bmatrix}$$

tells us how to express the $T(\mathbf{v}_j)$'s in terms of $\mathscr{C} = \{\mathbf{w}_1, \ldots, \mathbf{w}_m\}$,

$$T(\mathbf{v}_j) = a_{1j}\mathbf{w}_1 + \ldots + a_{mj}\mathbf{w}_m.$$

Substitution into the expression for $T(\mathbf{v})$ gives

(2) $\quad T(\mathbf{v}) = b_1[a_{11}\mathbf{w}_1 + \ldots + a_{m1}\mathbf{w}_m] + \ldots + b_n[a_{1n}\mathbf{w}_1 + \ldots + a_{mn}\mathbf{w}_m]$

$\quad\quad = (b_1a_{11} + \ldots + b_na_{1n})\mathbf{w}_1 + \ldots + (b_1a_{m1} + \ldots + b_na_{mn})\mathbf{w}_m.$

The expressions in parentheses are precisely the entries of a matrix product:

$$\begin{bmatrix} (b_1a_{11} + \ldots + b_na_{1n}) \\ \cdot \\ \cdot \\ \cdot \\ (b_1a_{m1} + \ldots + b_na_{mn}) \end{bmatrix} = \begin{bmatrix} a_{11} & \cdots & a_{1n} \\ & \cdot & \\ \cdot & \cdot & \cdot \\ & \cdot & \\ a_{m1} & \cdots & a_{mn} \end{bmatrix}\begin{bmatrix} b_1 \\ b_2 \\ \cdot \\ \cdot \\ \cdot \\ b_n \end{bmatrix}.$$

The factor on the left in the product is $\mathscr{C}[T]_{\mathscr{B}}$, while the factor on the right is the n-vector whose entries are the coordinates in the expression (1) for $\mathbf{v}$ in terms of $\mathscr{B}$. We refer to this n-vector as the **coordinate vector** of $\mathbf{v}$ with respect to $\mathscr{B}$ and write

$$\mathscr{B}[\mathbf{v}] = \begin{bmatrix} b_1 \\ \cdot \\ \cdot \\ \cdot \\ b_n \end{bmatrix}.$$

The product gives us the coordinates in the expression (2) for $T(\mathbf{v})$ in terms of $\mathscr{C}$:

$$
\mathscr{C}[T(\mathbf{v})] = \begin{bmatrix} (b_1 a_{11} + \ldots + b_n a_{1n}) \\ \cdot \\ \cdot \\ \cdot \\ (b_1 a_{m1} + \ldots + b_n a_{mn}) \end{bmatrix}.
$$

Thus we have the following:

Fact: *Let V and W be vector spaces with bases $\mathscr{B}$ and $\mathscr{C}$, respectively. If T is a linear transformation from V to W, then for any $\mathbf{v}$ in V,*

$$
\mathscr{C}[T(\mathbf{v})] = \mathscr{C}[T]_{\mathscr{B}}\ \mathscr{B}[\mathbf{v}],
$$

where $\mathscr{B}[\mathbf{v}]$ and $\mathscr{C}[T(\mathbf{v})]$ are the coordinate vectors of $\mathbf{v}$ with respect to $\mathscr{B}$ and of $T(\mathbf{v})$ with respect to $\mathscr{C}$.

This formula codifies the method for recovering a linear transformation from its matrix. Given $A = \mathscr{C}[T]_{\mathscr{B}}$ and any vector $\mathbf{v}$ in V, we can calculate $T(\mathbf{v})$ as follows:

1.   Write $\mathbf{v}$ as a linear combination of the vectors in $\mathscr{B}$

$$
\mathbf{v} = b_1 \mathbf{v}_1 + \ldots + b_n \mathbf{v}_n
$$

and form the coordinate vector of $\mathbf{v}$ with respect to $\mathscr{B}$

$$
\mathscr{B}[\mathbf{v}] = \begin{bmatrix} b_1 \\ \cdot \\ \cdot \\ \cdot \\ b_n \end{bmatrix}.
$$

2.   Multiply $\mathscr{C}[T]_{\mathscr{B}} = A$ times $\mathscr{B}[\mathbf{v}]$ to obtain the coordinate vector $\mathscr{C}[T(\mathbf{v})]$:

$$
\mathscr{C}[T(\mathbf{v})] = \mathscr{C}[T]_{\mathscr{B}}\ \mathscr{B}[\mathbf{v}] = \begin{bmatrix} a_{11} & \cdots & a_{1n} \\ \cdot & & \cdot \\ \cdot & & \cdot \\ \cdot & & \cdot \\ a_{m1} & \cdots & a_{mn} \end{bmatrix} \begin{bmatrix} b_1 \\ b_2 \\ \cdot \\ \cdot \\ \cdot \\ b_n \end{bmatrix} = \begin{bmatrix} c_1 \\ \cdot \\ \cdot \\ \cdot \\ c_m \end{bmatrix}.
$$

3.   The value $T(\mathbf{v})$ is the corresponding linear combination of the vectors in $\mathscr{C}$:

$$T(\mathbf{v}) = c_1\mathbf{w}_1 + \ldots + c_m\mathbf{w}_m.$$

Example 5.1.5

The matrix of a linear transformation T from R^2 to R^3 with respect to the bases

$$\mathscr{B} = \left\{ \begin{bmatrix} 1 \\ 0 \end{bmatrix}, \begin{bmatrix} 0 \\ 1 \end{bmatrix} \right\} \quad \text{and} \quad \mathscr{C} = \left\{ \begin{bmatrix} 1 \\ 1 \\ 1 \end{bmatrix}, \begin{bmatrix} 1 \\ 0 \\ 1 \end{bmatrix}, \begin{bmatrix} 0 \\ 1 \\ 0 \end{bmatrix} \right\}$$

is

$$_\mathscr{C}[T]_\mathscr{B} = \begin{bmatrix} 1 & 1 \\ 2 & 2 \\ -1 & 0 \end{bmatrix}.$$

Find the effect of T on a typical vector in R^2

$$\mathbf{v} = \begin{bmatrix} x \\ y \end{bmatrix}.$$

We first write $\mathbf{v}$ in terms of the vectors in $\mathscr{B}$

$$\mathbf{v} = \begin{bmatrix} x \\ y \end{bmatrix} = x \begin{bmatrix} 1 \\ 0 \end{bmatrix} + y \begin{bmatrix} 0 \\ 1 \end{bmatrix}$$

and read off the coordinate vector

$$_\mathscr{B}[\mathbf{v}] = \begin{bmatrix} x \\ y \end{bmatrix}.$$

(Since $\mathscr{B}$ is the standard basis, this coordinate vector is the same as $\mathbf{v}$ itself.) Next, we multiply the matrix of T times the coordinate vector of $\mathbf{v}$ to find the coordinate vector of $T(\mathbf{v})$:

$$_\mathscr{C}[T(\mathbf{v})] = {}_\mathscr{C}[T]_\mathscr{B} \, _\mathscr{B}[\mathbf{v}]$$

$$= \begin{bmatrix} 1 & 1 \\ 2 & 2 \\ -1 & 0 \end{bmatrix} \begin{bmatrix} x \\ y \end{bmatrix} = \begin{bmatrix} x + y \\ 2x + 2y \\ -x \end{bmatrix}.$$

The value $T(\mathbf{v})$ is the corresponding linear combination of the vectors in $\mathscr{C}$:

$$T(\mathbf{v}) = (x + y)\begin{bmatrix} 1 \\ 1 \\ 1 \end{bmatrix} + (2x + 2y)\begin{bmatrix} 1 \\ 0 \\ 1 \end{bmatrix} - x\begin{bmatrix} 0 \\ 1 \\ 0 \end{bmatrix} = \begin{bmatrix} 3x + 3y \\ y \\ 3x + 3y \end{bmatrix}.$$

Example 5.1.6

The matrix of a linear transformation T from R^3 to R^2 with respect to the bases

$$\mathscr{B} = \left\{ \begin{bmatrix} 0 \\ 0 \\ 1 \end{bmatrix}, \begin{bmatrix} 1 \\ 0 \\ 0 \end{bmatrix}, \begin{bmatrix} 1 \\ 1 \\ 0 \end{bmatrix} \right\} \quad \text{and} \quad \mathscr{C} = \left\{ \begin{bmatrix} 1 \\ 1 \end{bmatrix} \begin{bmatrix} 0 \\ 1 \end{bmatrix} \right\}$$

is

$$_\mathscr{C}[T]_\mathscr{B} = \begin{bmatrix} 1 & 2 & 0 \\ -1 & 0 & 1 \end{bmatrix}.$$

Find the effect of T on a typical vector in R^3

$$\mathbf{v} = \begin{bmatrix} x \\ y \\ z \end{bmatrix}.$$

We must first write $\mathbf{v}$ in terms of $\mathscr{B}$

$$\begin{bmatrix} x \\ y \\ z \end{bmatrix} = b_1 \begin{bmatrix} 0 \\ 0 \\ 1 \end{bmatrix} + b_2 \begin{bmatrix} 1 \\ 0 \\ 0 \end{bmatrix} + b_3 \begin{bmatrix} 1 \\ 1 \\ 0 \end{bmatrix}.$$

This leads to the equations

$$b_2 + b_3 = x$$
$$b_3 = y$$
$$b_1 \qquad = z$$

which we solve to find

$$b_1 = z, \qquad b_2 = x - y, \qquad b_3 = y.$$

Thus the coordinate vector of **v** is

$$_\mathscr{B}[\mathbf{v}] = \begin{bmatrix} z \\ x - y \\ y \end{bmatrix}.$$

Next, we multiply the matrix of T times the coordinate vector of **v** to find the coordinate vector of $T(\mathbf{v})$:

$$_\mathscr{C}[T(\mathbf{v})] = {}_\mathscr{C}[T]_\mathscr{B} \; _\mathscr{B}[\mathbf{v}]$$

$$= \begin{bmatrix} 1 & 2 & 0 \\ -1 & 0 & 1 \end{bmatrix} \begin{bmatrix} z \\ x - y \\ y \end{bmatrix} = \begin{bmatrix} 2x - 2y + z \\ y - z \end{bmatrix}.$$

The value $T(\mathbf{v})$ is the corresponding linear combination of the vectors in $\mathscr{C}$:

$$T(\mathbf{v}) = (2x - 2y + z) \begin{bmatrix} 1 \\ 1 \end{bmatrix} + y - z \begin{bmatrix} 0 \\ 1 \end{bmatrix} = \begin{bmatrix} 2x - 2y + z \\ 2x - y \end{bmatrix}.$$

Let's summarize.

THE MATRIX OF A LINEAR TRANSFORMATION

Let V and W be vector spaces with respective bases $\mathscr{B} = \{\mathbf{v}_1, \ldots, \mathbf{v}_n\}$ and $\mathscr{C} = \{\mathbf{w}_1, \ldots, \mathbf{w}_m\}$. Let T be a linear transformation from V to W.

To find the **matrix of T with respect to $\mathscr{B}$ and $\mathscr{C}$:**

1. Calculate the value of T on each of the vectors $\mathbf{v}_j$ in $\mathscr{B}$.
2. Express each result in terms of the vectors in $\mathscr{C}$

$$T(\mathbf{v}_j) = a_{1j}\mathbf{w}_1 + \ldots + a_{mj}\mathbf{w}_m$$

and read off the **coordinate vector** of $T(\mathbf{v}_j)$ with respect to $\mathscr{C}$

$$_\mathscr{C}[T(\mathbf{v}_j)] = \begin{bmatrix} a_{1j} \\ \cdot \\ \cdot \\ \cdot \\ a_{mj} \end{bmatrix}.$$

3. Form the matrix whose columns are these coordinate vectors

$$_\mathscr{C}[T]_\mathscr{B} = \left[_\mathscr{C}[T(\mathbf{v}_1)] \; _\mathscr{C}[T(\mathbf{v}_2)] \; \ldots \; _\mathscr{C}[T(\mathbf{v}_n)]] \right]$$

$$= \begin{bmatrix} a_{11} & \cdots & a_{1n} \\ \cdot & & \cdot \\ \cdot & & \cdot \\ \cdot & & \cdot \\ a_{m1} & \cdots & a_{mn} \end{bmatrix}.$$

To calculate $T(\mathbf{v})$ for any vector $\mathbf{v}$ in V, given $_\mathscr{C}[T]_\mathscr{B}$:

1. Express $\mathbf{v}$ in terms of the vectors in $\mathscr{B}$

$$\mathbf{v} = b_1\mathbf{v}_1 + \ldots + b_n\mathbf{v}_n$$

and read off the coordinate vector of $\mathbf{v}$ with respect to $\mathscr{B}$

$$_\mathscr{B}[\mathbf{v}] = \begin{bmatrix} b_1 \\ \cdot \\ \cdot \\ \cdot \\ b_n \end{bmatrix}.$$

2. Multiply $_\mathscr{C}[T]_\mathscr{B}$ times $_\mathscr{B}[\mathbf{v}]$ to obtain $_\mathscr{C}[T(\mathbf{v})]$:

$$_\mathscr{C}[T(\mathbf{v})] = _\mathscr{C}[T]_\mathscr{B} \, _\mathscr{B}[\mathbf{v}] = \begin{bmatrix} c_1 \\ \cdot \\ \cdot \\ \cdot \\ c_m \end{bmatrix}.$$

3. The value $T(\mathbf{v})$ is the corresponding linear combination of the vectors in $\mathscr{C}$,

$$T(\mathbf{v}) = c_1\mathbf{w}_1 + \ldots + c_m\mathbf{w}_m.$$

Notes
1. *On remembering the formula for $_\mathscr{C}[T(\mathbf{v})]$.*
   The formula

$$_\mathscr{C}[T(\mathbf{v})] = _\mathscr{C}[T]_\mathscr{B} \, _\mathscr{B}[\mathbf{v}]$$

played an important role in our discussion. One way to remember this formula is to view the left side as obtained from the right by "cancelling" adjacent $\mathscr{B}$'s and brackets.

2. On the coordinates of a given vector with respect to two different bases

Let $\mathcal{B}$ and $\mathcal{C}$ be two different bases for the same vector space V. Then the coordinate vectors $_{\mathcal{B}}[\mathbf{v}]$ and $_{\mathcal{C}}[\mathbf{v}]$ will in general be different. However, we can use the formula

$$_{\mathcal{C}}[T(\mathbf{v})] = {}_{\mathcal{C}}[T]_{\mathcal{B}} \, {}_{\mathcal{B}}[\mathbf{v}]$$

to determine the relation between these vectors.

Let $T = I_V$ be the linear transformation from V to $W = V$ given by the rule

$$I_V(\mathbf{v}) = \mathbf{v}$$

(see Exercise 1 of Section 4.6). Then

$$_{\mathcal{C}}[\mathbf{v}] = {}_{\mathcal{C}}[I_V(\mathbf{v})] = {}_{\mathcal{C}}[I_V]_{\mathcal{B}} \, {}_{\mathcal{B}}[\mathbf{v}].$$

The matrix $_{\mathcal{C}}[I_V]_{\mathcal{B}}$ is sometimes referred to as the **transition matrix** from basis $\mathcal{B}$ to basis $\mathcal{C}$. The coordinate vector with respect to $\mathcal{C}$ is obtained by multiplying this transition matrix times the coordinate vector with respect to $\mathcal{B}$.

EXERCISES

In Exercises 1 to 10, you are given vector spaces V and W, a linear transformation T from V to W, a matrix A, bases $\mathcal{B}$ and $\mathcal{B}'$ for V, and bases $\mathcal{C}$ and $\mathcal{C}'$ for W.

a. Find $_{\mathcal{C}}[T]_{\mathcal{B}}$.

b. Find $_{\mathcal{C}'}[T]_{\mathcal{B}'}$.

c. Find the effect on a typical vector $\mathbf{v}$ in V of the linear transformation S satisfying $_{\mathcal{C}}[S]_{\mathcal{B}} = A$.

d. Find the effect on a typical vector $\mathbf{v}$ in V of the linear transformation R satisfying $_{\mathcal{C}'}[R]_{\mathcal{B}'} = A$.

1. $V = R^3$, $W = R^2$, $T\begin{bmatrix} x \\ y \\ z \end{bmatrix} = \begin{bmatrix} 2z \\ x + y - z \end{bmatrix}$, $A = \begin{bmatrix} 1 & 0 & -1 \\ 0 & 2 & 3 \end{bmatrix}$,

$\mathcal{B} = \left\{ \begin{bmatrix} 1 \\ 0 \\ 0 \end{bmatrix}, \begin{bmatrix} 0 \\ 1 \\ 0 \end{bmatrix}, \begin{bmatrix} 0 \\ 0 \\ 1 \end{bmatrix} \right\}$, $\mathcal{B}' = \left\{ \begin{bmatrix} 1 \\ 1 \\ 1 \end{bmatrix}, \begin{bmatrix} 0 \\ 1 \\ 0 \end{bmatrix}, \begin{bmatrix} 1 \\ 0 \\ 0 \end{bmatrix} \right\}$,

$\mathcal{C} = \left\{ \begin{bmatrix} 1 \\ 0 \end{bmatrix}, \begin{bmatrix} 0 \\ 1 \end{bmatrix} \right\}$, $\mathcal{C}' = \left\{ \begin{bmatrix} 1 \\ 1 \end{bmatrix}, \begin{bmatrix} 1 \\ -1 \end{bmatrix} \right\}$

2. $V = R^2$, $W = R^3$, $T\begin{bmatrix} x \\ y \end{bmatrix} = \begin{bmatrix} x \\ x + y \\ x - y \end{bmatrix}$, $A = \begin{bmatrix} 1 & 0 \\ 0 & 2 \\ -1 & 3 \end{bmatrix}$,

$\mathcal{B} = \left\{ \begin{bmatrix} 1 \\ 0 \end{bmatrix}, \begin{bmatrix} 0 \\ 1 \end{bmatrix} \right\}$, $\mathcal{B}' = \left\{ \begin{bmatrix} 1 \\ 1 \end{bmatrix}, \begin{bmatrix} 1 \\ -1 \end{bmatrix} \right\}$,

$$\mathscr{C} = \left\{ \begin{bmatrix} 1 \\ 0 \\ 0 \end{bmatrix}, \begin{bmatrix} 0 \\ 1 \\ 0 \end{bmatrix}, \begin{bmatrix} 0 \\ 0 \\ 1 \end{bmatrix} \right\}, \quad \mathscr{C}' = \left\{ \begin{bmatrix} 1 \\ 1 \\ 1 \end{bmatrix}, \begin{bmatrix} 1 \\ 1 \\ 0 \end{bmatrix}, \begin{bmatrix} 1 \\ 0 \\ 0 \end{bmatrix} \right\}$$

3. $V = W = R^3$, $T \begin{bmatrix} x \\ y \\ z \end{bmatrix} = \begin{bmatrix} 2x - y \\ z \\ 3y \end{bmatrix}$, $A = \begin{bmatrix} 1 & 0 & 1 \\ 0 & 2 & 2 \\ -1 & 0 & 0 \end{bmatrix}$

$$\mathscr{B} = \mathscr{C} = \left\{ \begin{bmatrix} 1 \\ 0 \\ 0 \end{bmatrix}, \begin{bmatrix} 0 \\ 1 \\ 0 \end{bmatrix}, \begin{bmatrix} 0 \\ 0 \\ 1 \end{bmatrix} \right\}, \quad \mathscr{B}' = \mathscr{C}' = \left\{ \begin{bmatrix} 1 \\ 1 \\ 0 \end{bmatrix}, \begin{bmatrix} 0 \\ 0 \\ 1 \end{bmatrix}, \begin{bmatrix} 0 \\ 1 \\ 1 \end{bmatrix} \right\}$$

4. $V = W = R^3$, T, and A as in Exercise 3,

$$\mathscr{B} = \mathscr{C}' = \left\{ \begin{bmatrix} 1 \\ 0 \\ 0 \end{bmatrix}, \begin{bmatrix} 0 \\ 1 \\ 0 \end{bmatrix}, \begin{bmatrix} 0 \\ 0 \\ 1 \end{bmatrix} \right\}, \quad \mathscr{B}' = \mathscr{C} = \left\{ \begin{bmatrix} 1 \\ 1 \\ 0 \end{bmatrix}, \begin{bmatrix} 0 \\ 0 \\ 1 \end{bmatrix}, \begin{bmatrix} 0 \\ 1 \\ 1 \end{bmatrix} \right\}$$

5. $V = W = R^3$, $T \begin{bmatrix} x \\ y \\ z \end{bmatrix} = I_V \begin{bmatrix} x \\ y \\ z \end{bmatrix} = \begin{bmatrix} x \\ y \\ z \end{bmatrix}$, $A = I = \begin{bmatrix} 1 & 0 & 0 \\ 0 & 1 & 0 \\ 0 & 0 & 1 \end{bmatrix}$,

$\mathscr{B}$, $\mathscr{B}'$, $\mathscr{C}$, and $\mathscr{C}'$ as in Exercise 4.

6. $V = P_2$, $W = R^3$, $T(a_0 + a_1 t + a_2 t^2) = \begin{bmatrix} a_0 \\ a_1 \\ a_2 \end{bmatrix}$, $A = \begin{bmatrix} 1 & 0 & 1 \\ 0 & 2 & 2 \\ -1 & 0 & 0 \end{bmatrix}$

$\mathscr{B} = \{1, t, t^2\}$, $\mathscr{B}' = \{1 + t, 1 + t^2, t^2\}$

$$\mathscr{C} = \left\{ \begin{bmatrix} 1 \\ 0 \\ 0 \end{bmatrix}, \begin{bmatrix} 0 \\ 1 \\ 0 \end{bmatrix}, \begin{bmatrix} 0 \\ 0 \\ 1 \end{bmatrix} \right\}, \quad \mathscr{C}' = \left\{ \begin{bmatrix} 1 \\ 1 \\ 0 \end{bmatrix}, \begin{bmatrix} 1 \\ 2 \\ 0 \end{bmatrix}, \begin{bmatrix} 0 \\ 0 \\ 1 \end{bmatrix} \right\}$$

7. $V = W = P_2$, $T(a_0 + a_1 t + a_2 t^2) = a_2 + a_1 t + a_0 t^2$, $A = \begin{bmatrix} 1 & 0 & 1 \\ 0 & 2 & 2 \\ -1 & 0 & 0 \end{bmatrix}$

$\mathscr{B} = \mathscr{C} = \{1, t, t^2\}$, $\mathscr{B}' = \mathscr{C}' = \{1 + t, 1 + t^2, t^2\}$

8. $V = W = R_{2 \times 2}$, $T \begin{bmatrix} a & b \\ c & d \end{bmatrix} = \begin{bmatrix} a & 0 \\ 0 & d \end{bmatrix}$, $A = \begin{bmatrix} 1 & 0 & 1 & 0 \\ 0 & 1 & 0 & 1 \\ 1 & 0 & 0 & 1 \\ 0 & 0 & 0 & 2 \end{bmatrix}$

$$\mathscr{B} = \mathscr{C} = \left\{ \begin{bmatrix} 1 & 0 \\ 0 & 0 \end{bmatrix}, \begin{bmatrix} 0 & 1 \\ 0 & 0 \end{bmatrix}, \begin{bmatrix} 0 & 0 \\ 1 & 0 \end{bmatrix}, \begin{bmatrix} 0 & 0 \\ 0 & 1 \end{bmatrix} \right\}$$

$$\mathscr{B}' = \mathscr{C}' = \left\{ \begin{bmatrix} 1 & 1 \\ 0 & 0 \end{bmatrix}, \begin{bmatrix} 0 & 1 \\ 0 & 0 \end{bmatrix}, \begin{bmatrix} 0 & 0 \\ 1 & 1 \end{bmatrix}, \begin{bmatrix} 0 & 0 \\ 0 & 1 \end{bmatrix} \right\}$$

9. $V = W = C^2$, $T \begin{bmatrix} x \\ y \end{bmatrix} = \begin{bmatrix} x + y \\ x - y \end{bmatrix}$, $A = \begin{bmatrix} i & 0 \\ 1 & 2 \end{bmatrix}$

$$\mathscr{B} = \mathscr{C} = \left\{ \begin{bmatrix} 1 \\ 0 \end{bmatrix}, \begin{bmatrix} 0 \\ 1 \end{bmatrix} \right\}, \quad \mathscr{B}' = \mathscr{C}' = \left\{ \begin{bmatrix} i \\ 1 \end{bmatrix}, \begin{bmatrix} -1 \\ 1 \end{bmatrix} \right\}$$

10. $V = W$ any finite-dimensional vector space, $T(\mathbf{v}) = I_V(\mathbf{v}) = \mathbf{v}$,

$$A = I = \begin{bmatrix} 1 & 0 & \dots & 0 \\ 0 & 1 & \dots & 0 \\ & \cdot & & \cdot \\ & \cdot & & \cdot \\ & \cdot & & \cdot \\ 0 & 0 & \dots & 1 \end{bmatrix}, \quad \mathcal{B} = \mathcal{B}' = \mathcal{C} = \{\mathbf{v}_1, \dots, \mathbf{v}_n\} \text{ any basis,}$$

$\mathcal{C}' = \{\mathbf{v}_n, \mathbf{v}_{n-1}, \dots, \mathbf{v}_1\}$.

More abstract exercises:

In Exercises 11 to 15, V and W are vector spaces with respective bases $\mathcal{B} = \{\mathbf{v}_1, \dots, \mathbf{v}_n\}$ and $\mathcal{C} = \{\mathbf{w}_1, \dots, \mathbf{w}_m\}$.

11.   a. Let T be the zero transformation from V to W, so that $T(\mathbf{v}) = \mathbf{0}$ for each $\mathbf{v}$ in V. Show that $_\mathcal{C}[T]_\mathcal{B}$ is the $m \times n$ zero matrix.
    b. Show, conversely, that if T is a linear transformation for which $_\mathcal{C}[T]_\mathcal{B}$ is the zero matrix, then T is the zero transformation.

12.   Suppose that T and S are linear transformations from V to W and that a is a scalar. Then $T + S$ and aT are also linear transformations from V to W (see Exercise 33, Section 4.6). Show that
    a. $_\mathcal{C}[T + S]_\mathcal{B} = {}_\mathcal{C}[T]_\mathcal{B} + {}_\mathcal{C}[S]_\mathcal{B}$
    b. $_\mathcal{C}[aT]_\mathcal{B} = a\,_\mathcal{C}[T]_\mathcal{B}$

13.   Suppose that $V = W$ and $\mathcal{B} = \mathcal{C}$. Let A be an $n \times n$ matrix, and let $T = T_A$.
    a. Suppose that for each j, $\mathbf{v}_j$ is an eigenvector of A with respect to some eigenvalue λ_j. Show that

$$_\mathcal{B}[T]_\mathcal{B} = \begin{bmatrix} \lambda_1 & & & \\ & \cdot & & 0's \\ & & \cdot & \\ 0's & & & \cdot \\ & & & \lambda_n \end{bmatrix}.$$

    b. Show, conversely, that if $_\mathcal{B}[T]_\mathcal{B}$ has this form, then for each j, $\mathbf{v}_j$ is an eigenvalue with respect to λ_j.

14.   Suppose T is a linear transformation from V to W, where V and W are finite-dimensional vector spaces with bases $\mathcal{B}$ and $\mathcal{C}$, respectively. Suppose further that A is a matrix with the property that for *every* vector $\mathbf{v}$ in V,

$$_\mathcal{C}[T(\mathbf{v})]_\mathcal{B} = A \,_\mathcal{B}[\mathbf{v}].$$

Show that $A = {}_\mathcal{C}[T]_\mathcal{B}$. (*Hint:* See Exercise 31 in Section 3.2.)

5.2 MULTIPLICATION OF LINEAR TRANSFORMATIONS AND MULTIPLICATION OF MATRICES

In this section we define multiplication of linear transformations. This operation is related to multiplication of matrices. Indeed, if the vector spaces involved are finite-dimensional, then the matrix of a product of two linear transformations is the same as the product of their matrices (with respect to appropriate bases).

The product AB of two matrices is only defined if the number of rows of B is the same as the number of columns of A. Similarly, the product ST of two linear transformations is defined only if the space to which T maps i. the same as the space from which S maps. If T is a linear transformation from V to W and if S is a linear transformation from W to U, then the **product** ST is the mapping from V to U defined by the rule

$$[ST](\mathbf{v}) = S[T(\mathbf{v})].$$

Thus, to find the effect of ST on $\mathbf{v}$, we first apply T to $\mathbf{v}$ to get the vector $T(\mathbf{v})$ in W. We then apply S to $T(\mathbf{v})$ to get the vector $S[T(\mathbf{v})]$ in U.

If S and T are linear transformations for which ST is defined, and if $\mathbf{v}_1$ and $\mathbf{v}_2$ are in V, then

$$
\begin{aligned}
[ST](\mathbf{v}_1 + \mathbf{v}_2) &= S[T(\mathbf{v}_1 + \mathbf{v}_2)] && \text{(definition of } ST) \\
&= S[T(\mathbf{v}_1) + T(\mathbf{v}_2)] && (T \text{ is a linear transformation)} \\
&= S[T(\mathbf{v}_1)] + S[T(\mathbf{v}_2)] && (S \text{ is a linear transformation)} \\
&= [ST](\mathbf{v}_1) + [ST](\mathbf{v}_2) && \text{(definition of } ST).
\end{aligned}
$$

If c is a scalar, then

$$
\begin{aligned}
[ST](c\mathbf{v}) &= S[T(c\mathbf{v})] && \text{(definition of } ST) \\
&= S(c[T(\mathbf{v})]) && (T \text{ is a linear transformation)} \\
&= c(S[T(\mathbf{v})]) && (S \text{ is a linear transformation)} \\
&= c([ST](\mathbf{v})) && \text{(definition of } ST).
\end{aligned}
$$

Thus, *the product of two linear transformations is a linear transformation.*

Example 5.2.1

Let T be the linear transformation from $V = R^3$ to $W = R^2$ given by

$$
T\begin{bmatrix} x \\ y \\ z \end{bmatrix} = \begin{bmatrix} 2x + y \\ x - y \end{bmatrix}
$$

and let S be the linear transformation from $W = R^2$ to $U = R^4$ given by

$$
S\begin{bmatrix} x \\ y \end{bmatrix} = \begin{bmatrix} x \\ x \\ y \\ x + y \end{bmatrix}.
$$

Then ST is the linear transformation from R^3 to R^4 given by

$$(ST) \begin{bmatrix} x \\ y \\ z \end{bmatrix} = S\left(T \begin{bmatrix} x \\ y \\ z \end{bmatrix} \right) = S \begin{bmatrix} 2x + y \\ x - y \end{bmatrix} = \begin{bmatrix} 2x + y \\ 2x + y \\ x - y \\ (2x + y) + (x - y) \end{bmatrix} = \begin{bmatrix} 2x + y \\ 2x + y \\ x - y \\ 3x \end{bmatrix}.$$

Note that if S and T are the linear transformations of Example 5.2.1, then ST is defined but TS is not. The next example shows that even when both ST and TS are defined, they may not be equal.

Example 5.2.2

Let T and S be the linear transformations from R^2 to R^2 given by

$$T \begin{bmatrix} x \\ y \end{bmatrix} = \begin{bmatrix} x + y \\ x - y \end{bmatrix} \quad \text{and} \quad S \begin{bmatrix} x \\ y \end{bmatrix} = \begin{bmatrix} y \\ x \end{bmatrix}.$$

Then ST and TS are linear transformations from R^2 to R^2 given by

$$(ST) \begin{bmatrix} x \\ y \end{bmatrix} = S\left(T \begin{bmatrix} x \\ y \end{bmatrix} \right) = S \begin{bmatrix} x + y \\ x - y \end{bmatrix} = \begin{bmatrix} x - y \\ x + y \end{bmatrix}$$

and

$$(TS) \begin{bmatrix} x \\ y \end{bmatrix} = T\left(S \begin{bmatrix} x \\ y \end{bmatrix} \right) = T \begin{bmatrix} y \\ x \end{bmatrix} = \begin{bmatrix} y + x \\ y - x \end{bmatrix}.$$

Note that ST and TS are not equal. For example,

$$(ST) \begin{bmatrix} 1 \\ 0 \end{bmatrix} = \begin{bmatrix} 1 \\ 1 \end{bmatrix}$$

while

$$(TS) \begin{bmatrix} 1 \\ 0 \end{bmatrix} = \begin{bmatrix} 1 \\ -1 \end{bmatrix}.$$

We mentioned, at the beginning of the section, the relationship between multiplication of linear transformations and multiplication of matrices. Formally, this is given by the following.

Fact: *Let V, W, and U be finite-dimensional vector spaces with bases $\mathcal{B}$, $\mathcal{C}$, and $\mathcal{A}$, respectively. Let T be a linear transformation from V to W, and let S be a linear transformation from W to U. Then*

$$_{\mathcal{A}}[ST]_{\mathcal{B}} = {}_{\mathcal{A}}[S]_{\mathcal{C}} \; {}_{\mathcal{C}}[T]_{\mathcal{B}}.$$

Informally, this says that *the matrix of the product is the product of the matrices.* Note, however, that the bases with respect to which we take these matrices have to match up properly. One method for keeping track of the bases is to view the left side as obtained from the right by "cancelling" adjacent $\mathcal{C}$'s and brackets. (Compare Note 1 in Section 5.1.)

Before we verify this fact in general, let's check that it holds in a specific example.

Example 5.2.3
We saw in Example 5.2.1 that if T is the linear transformation from $V = R^3$ to $W = R^2$ given by

$$T\begin{bmatrix} x \\ y \\ z \end{bmatrix} = \begin{bmatrix} 2x + y \\ x - y \end{bmatrix}$$

and if S is the linear transformation from $W = R^2$ to $U = R^4$ given by

$$S\begin{bmatrix} x \\ y \end{bmatrix} = \begin{bmatrix} x \\ x \\ y \\ x + y \end{bmatrix},$$

then ST is the linear transformation from V to U given by

$$(ST)\begin{bmatrix} x \\ y \\ z \end{bmatrix} = \begin{bmatrix} 2x + y \\ 2x + y \\ x - y \\ 3x \end{bmatrix}.$$

Let's calculate the matrices of these transformations with respect to the bases

$$\mathcal{B} = \left\{ \begin{bmatrix} 1 \\ 0 \\ 0 \end{bmatrix}, \begin{bmatrix} 0 \\ 1 \\ 0 \end{bmatrix}, \begin{bmatrix} 0 \\ 0 \\ 1 \end{bmatrix} \right\}$$

$$\mathcal{C} = \left\{ \begin{bmatrix} 1 \\ 0 \end{bmatrix}, \begin{bmatrix} 0 \\ 1 \end{bmatrix} \right\}$$

$$\mathcal{A} = \left\{ \begin{bmatrix} 0 \\ 1 \\ 0 \\ 0 \end{bmatrix}, \begin{bmatrix} 1 \\ 0 \\ 0 \\ 0 \end{bmatrix}, \begin{bmatrix} 0 \\ 0 \\ 1 \\ 1 \end{bmatrix}, \begin{bmatrix} 0 \\ 0 \\ 0 \\ -1 \end{bmatrix} \right\}$$

of V, W, and U, respectively.

We calculated $_\mathcal{C}[T]_\mathcal{B}$ in Example 5.1.1:

$$_\mathcal{C}[T]_\mathcal{B} = \begin{bmatrix} 2 & 1 & 0 \\ 1 & -1 & 0 \end{bmatrix}.$$

To find $_\mathcal{A}[S]_\mathcal{C}$, we apply S to the vectors in $\mathcal{C}$ and express the results in terms of $\mathcal{A}$:

$$S\begin{bmatrix} 1 \\ 0 \end{bmatrix} = \begin{bmatrix} 1 \\ 1 \\ 0 \\ 1 \end{bmatrix} = 1\begin{bmatrix} 0 \\ 1 \\ 0 \\ 0 \end{bmatrix} + 1\begin{bmatrix} 1 \\ 0 \\ 0 \\ 0 \end{bmatrix} + 0\begin{bmatrix} 0 \\ 0 \\ 1 \\ 1 \end{bmatrix} + (-1)\begin{bmatrix} 0 \\ 0 \\ 0 \\ -1 \end{bmatrix}$$

$$S\begin{bmatrix} 0 \\ 1 \end{bmatrix} = \begin{bmatrix} 0 \\ 0 \\ 1 \\ 1 \end{bmatrix} = 0\begin{bmatrix} 0 \\ 1 \\ 0 \\ 0 \end{bmatrix} + 0\begin{bmatrix} 0 \\ 0 \\ 0 \\ 0 \end{bmatrix} + 1\begin{bmatrix} 0 \\ 0 \\ 1 \\ 1 \end{bmatrix} + 0\begin{bmatrix} 0 \\ 0 \\ 0 \\ -1 \end{bmatrix}.$$

We place the scalars in these expressions into the columns of a matrix to obtain

$$_\mathcal{A}[S]_\mathcal{C} = \begin{bmatrix} 1 & 0 \\ 1 & 0 \\ 0 & 1 \\ -1 & 0 \end{bmatrix}.$$

The product of these matrices is

$$_\mathcal{A}[S]_\mathcal{C} \, _\mathcal{C}[T]_\mathcal{B} = \begin{bmatrix} 1 & 0 \\ 1 & 0 \\ 0 & 1 \\ -1 & 0 \end{bmatrix}\begin{bmatrix} 2 & 1 & 0 \\ 1 & -1 & 0 \end{bmatrix} = \begin{bmatrix} 2 & 1 & 0 \\ 2 & 1 & 0 \\ 1 & -1 & 0 \\ -2 & -1 & 0 \end{bmatrix}.$$

To calculate $_\mathcal{A}[ST]_\mathcal{B}$ directly, we apply ST to the vectors in $\mathcal{B}$ and express the results in terms of $\mathcal{A}$:

$$(ST)\begin{bmatrix}1\\0\\0\end{bmatrix} = \begin{bmatrix}2\\2\\1\\3\end{bmatrix} = 2\begin{bmatrix}0\\1\\0\\0\end{bmatrix} + 2\begin{bmatrix}1\\0\\0\\0\end{bmatrix} + 1\begin{bmatrix}0\\0\\1\\1\end{bmatrix} + (-2)\begin{bmatrix}0\\0\\0\\-1\end{bmatrix}$$

$$(ST)\begin{bmatrix}0\\1\\0\end{bmatrix} = \begin{bmatrix}1\\1\\-1\\0\end{bmatrix} = 1\begin{bmatrix}0\\1\\0\\0\end{bmatrix} + 1\begin{bmatrix}1\\0\\0\\0\end{bmatrix} + (-1)\begin{bmatrix}0\\0\\1\\1\end{bmatrix} + (-1)\begin{bmatrix}0\\0\\0\\-1\end{bmatrix}$$

$$(ST)\begin{bmatrix}0\\0\\1\end{bmatrix} = \begin{bmatrix}0\\0\\0\\0\end{bmatrix} = 0\begin{bmatrix}0\\1\\0\\0\end{bmatrix} + 0\begin{bmatrix}1\\0\\0\\0\end{bmatrix} + 0\begin{bmatrix}0\\0\\1\\1\end{bmatrix} + 0\begin{bmatrix}0\\0\\0\\-1\end{bmatrix}.$$

Thus the matrix of the product is

$$\mathscr{C}[ST]_{\mathscr{B}} = \begin{bmatrix}2 & 1 & 0\\2 & 1 & 0\\1 & -1 & 0\\-2 & -1 & 0\end{bmatrix}$$

$$= {}_{\mathscr{A}}[S]_{\mathscr{C}}\ {}_{\mathscr{C}}[T]_{\mathscr{B}}.$$

Let's now verify the general pattern. Suppose V, W, and U are finite-dimensional vector spaces with bases $\mathscr{B}$, $\mathscr{C}$, and $\mathscr{A}$, respectively. If T is any linear transformation from V to W, then the effect of T on a typical vector $\mathbf{v}$ in V can be expressed in terms of $\mathscr{C}[T]_{\mathscr{B}}$ as follows:

(1) $$\mathscr{C}[T(\mathbf{v})] = \mathscr{C}[T]_{\mathscr{B}}\ {}_{\mathscr{B}}[\mathbf{v}].$$

Similarly, the effect of any linear transformation S from W to U on a typical vector $\mathbf{w}$ in W is given by

(2) $$\mathscr{A}[S(\mathbf{w})] = \mathscr{A}[S]_{\mathscr{C}}\ {}_{\mathscr{C}}[\mathbf{w}].$$

We can use these equations to calculate the effect of the product transformation ST on $\mathbf{v}$:

$$\mathscr{A}[(ST)(\mathbf{v})] = \mathscr{A}[S(T(\mathbf{v}))] \qquad \text{(definition of } ST\text{)}$$

$$= \mathscr{A}[S]_{\mathscr{C}}\ {}_{\mathscr{C}}[T(\mathbf{v})] \qquad \text{(equation (2) with } \mathbf{w} = T(\mathbf{v})\text{)}$$

$$= \mathscr{A}[S]_{\mathscr{C}}\ (\mathscr{C}[T]_{\mathscr{B}}\ {}_{\mathscr{B}}[\mathbf{v}]) \qquad \text{(equation (1))}$$

$$= (\mathscr{A}[S]_{\mathscr{C}}\ {}_{\mathscr{C}}[T]_{\mathscr{B}})\ {}_{\mathscr{B}}[\mathbf{v}] \qquad \text{(property of matrix multiplication)}$$

On the other hand, for each **v** in V,

$$_{\mathcal{A}}[(ST)(\mathbf{v})] = {}_{\mathcal{A}}[ST]_{\mathcal{B}} \, {}_{\mathcal{B}}[\mathbf{v}].$$

Thus, for each **v** in V,

$$_{\mathcal{A}}[ST]_{\mathcal{B}} \, {}_{\mathcal{B}}[\mathbf{v}] = ({}_{\mathcal{A}}[S]_{\mathcal{C}} \, {}_{\mathcal{C}}[T]_{\mathcal{B}}) \, {}_{\mathcal{B}}[\mathbf{v}].$$

It follows (Exercise 31, Section 3.2) that

$$_{\mathcal{A}}[ST]_{\mathcal{B}} = {}_{\mathcal{A}}[S]_{\mathcal{C}} \, {}_{\mathcal{C}}[T]_{\mathcal{B}}.$$

Let's summarize.

**MULTIPLICATION OF LINEAR TRANSFORMATIONS
AND MULTIPLICATION OF MATRICES**

Let T be a linear transformation from V to W and let S be a linear transformation from W to U. Define the **product** ST by the rule

$$[ST](\mathbf{v}) = S[T(\mathbf{v})].$$

Then ST is a linear transformation from V to U.
  If $\mathcal{B}$, $\mathcal{C}$, and $\mathcal{A}$ are finite bases for V, W, and U, respectively, then

$$_{\mathcal{A}}[ST]_{\mathcal{B}} = {}_{\mathcal{A}}[S]_{\mathcal{C}} \, {}_{\mathcal{C}}[T]_{\mathcal{B}}.$$

That is, *the matrix of the product of two transformations is the product of their matrices.*

EXERCISES

In Exercises 1 to 8, you are given vector spaces V, W, and U with respective bases $\mathcal{B}$, $\mathcal{C}$, and $\mathcal{A}$, a linear transformation T from V to W, and a linear transformation S from W to U.

a. Find the effect of ST on a typical vector **v** in V.

b. Decide whether TS is defined, and if it is, whether $TS = ST$.

c. Calculate $_{\mathcal{A}}[S]_{\mathcal{C}}$, $_{\mathcal{C}}[T]_{\mathcal{B}}$ and $_{\mathcal{A}}[S]_{\mathcal{C}} \, {}_{\mathcal{C}}[T]_{\mathcal{B}}$.

d. Calculate $_{\mathcal{A}}[ST]_{\mathcal{B}}$ directly, and compare it with the product in part c.

1. $V = R^2$, $W = U = R^3$, $\mathscr{B} = \left\{ \begin{bmatrix} 1 \\ 2 \end{bmatrix}, \begin{bmatrix} 2 \\ 1 \end{bmatrix} \right\}$, $\mathscr{C} = \mathscr{A} = \left\{ \begin{bmatrix} 1 \\ 0 \\ 0 \end{bmatrix}, \begin{bmatrix} 0 \\ 1 \\ 0 \end{bmatrix}, \begin{bmatrix} 0 \\ 0 \\ 1 \end{bmatrix} \right\}$,

   $T \begin{bmatrix} x \\ y \end{bmatrix} = \begin{bmatrix} x + y \\ 2y \\ x - y \end{bmatrix}$, $S \begin{bmatrix} x \\ y \\ z \end{bmatrix} = \begin{bmatrix} 2x - y \\ x + z \\ y \end{bmatrix}$

2. $V = R^2$, $W = R^3$, $U = R^2$, $\mathscr{B} = \mathscr{A} = \left\{ \begin{bmatrix} 1 \\ 0 \end{bmatrix}, \begin{bmatrix} 0 \\ 1 \end{bmatrix} \right\}$,

   $\mathscr{C} = \left\{ \begin{bmatrix} 1 \\ 1 \\ 1 \end{bmatrix}, \begin{bmatrix} 0 \\ 1 \\ 0 \end{bmatrix}, \begin{bmatrix} 1 \\ 0 \\ 0 \end{bmatrix} \right\}$, $T \begin{bmatrix} x \\ y \end{bmatrix} = \begin{bmatrix} x + y \\ 2y \\ x - y \end{bmatrix}$, $S \begin{bmatrix} x \\ y \\ z \end{bmatrix} = \begin{bmatrix} 3y + z \\ z \end{bmatrix}$

3. $V = W = R^3$, $\mathscr{B} = \mathscr{C} = \mathscr{A} = \left\{ \begin{bmatrix} 1 \\ 0 \\ 0 \end{bmatrix}, \begin{bmatrix} 0 \\ 1 \\ 1 \end{bmatrix}, \begin{bmatrix} 0 \\ 0 \\ 1 \end{bmatrix} \right\}$,

   $T \begin{bmatrix} x \\ y \\ z \end{bmatrix} = \begin{bmatrix} 3x \\ y - z \\ 0 \end{bmatrix}$, $S \begin{bmatrix} x \\ y \\ z \end{bmatrix} = \begin{bmatrix} z \\ z \\ x \end{bmatrix}$

4. $V = W = R^3$, $\mathscr{B} = \left\{ \begin{bmatrix} 1 \\ 0 \\ 0 \end{bmatrix}, \begin{bmatrix} 0 \\ 1 \\ 0 \end{bmatrix}, \begin{bmatrix} 0 \\ 0 \\ 1 \end{bmatrix} \right\}$, $\mathscr{C} = \left\{ \begin{bmatrix} 1 \\ 1 \\ 0 \end{bmatrix}, \begin{bmatrix} 0 \\ 1 \\ 1 \end{bmatrix}, \begin{bmatrix} 0 \\ 1 \\ 0 \end{bmatrix} \right\}$,

   $\mathscr{A} = \left\{ \begin{bmatrix} 1 \\ 0 \\ 1 \end{bmatrix}, \begin{bmatrix} 0 \\ 1 \\ 0 \end{bmatrix}, \begin{bmatrix} 0 \\ 0 \\ 1 \end{bmatrix} \right\}$, $T \begin{bmatrix} x \\ y \\ z \end{bmatrix} = \begin{bmatrix} y \\ z \\ 3x \end{bmatrix}$, $S \begin{bmatrix} x \\ y \\ z \end{bmatrix} = \begin{bmatrix} x + y \\ x + y \\ x + y \end{bmatrix}$

5. $V = W = P_3$, $\mathscr{B} = \mathscr{C} = \mathscr{A} = \{1, t, t^2\}$,
   $T(a_0 + a_1t + a_2t^2) = a_0 + a_1t$, $S(a_0 + at + a_2t^2) = (a_0 - a_2) + (a_1 - a_2)t + a_2t^2$

6. $V = W = P_3$, $\mathscr{A} = \mathscr{B} = \{1, t, t^2\}$, $\mathscr{C} = \{1 + t^2, t + t^2, t^2\}$,
   $T(a_0 + a_1t + a_2t^2) = a_2$, $S(a_0 + a_1t + a_2t^2) = a_0$

7. $V = W = C^2$, $\mathscr{B} = \mathscr{C} = \left\{ \begin{bmatrix} 1 \\ 0 \end{bmatrix}, \begin{bmatrix} 0 \\ 1 \end{bmatrix} \right\}$, $\mathscr{A} = \left\{ \begin{bmatrix} i \\ 0 \end{bmatrix}, \begin{bmatrix} i \\ 1 \end{bmatrix} \right\}$,

   $T \begin{bmatrix} x \\ y \end{bmatrix} = \begin{bmatrix} y \\ x \end{bmatrix}$, $S \begin{bmatrix} x \\ y \end{bmatrix} = \begin{bmatrix} 2x + y \\ 0 \end{bmatrix}$

8. $V = W = R_{2 \times 2}$, $\mathscr{B} = \mathscr{C} = \mathscr{A} = \left\{ \begin{bmatrix} 1 & 0 \\ 0 & 0 \end{bmatrix}, \begin{bmatrix} 0 & 1 \\ 0 & 0 \end{bmatrix}, \begin{bmatrix} 0 & 0 \\ 1 & 0 \end{bmatrix}, \begin{bmatrix} 0 & 0 \\ 0 & 1 \end{bmatrix} \right\}$,

   $T \begin{bmatrix} a & b \\ c & d \end{bmatrix} = \begin{bmatrix} a & c \\ b & d \end{bmatrix}$, $S \begin{bmatrix} a & b \\ c & d \end{bmatrix} = \begin{bmatrix} a & 0 \\ 0 & d \end{bmatrix}$

In Exercises 9 to 11, you are given linear transformations T and S from C_∞ to C_∞.
a. Find $(ST)[x(t)]$.
b. Find $(TS)[x(t)]$.
c. Are ST and TS equal?

9. $T[x(t)] = Dx(t),\ S[x(t)] = \int_0^t x(s)\ ds$

10. $T[x(t)] = (D - 2)x(t),\ S[x(t)] = (D - 1)x(t)$

11. $T[x(t)] = (D - t)x(t),\ S[x(t)] = (D - 1)x(t)$

More abstract exercises:

12. Suppose that T is a linear transformation from V to W, that S is a linear transformation from W to U, and that R is a linear transformation from U to X. Show that $R(ST) = (RS)T$.

13. For any vector space Y, let I_Y be the linear transformation given by $I_Y(\mathbf{y}) = \mathbf{y}$ for each $\mathbf{y}$ in Y. Show that if T is a linear transformation from V to W, then $I_W T = T$ and $T I_V = T$.

14. Let T be a linear transformation from V to W and let S be a linear transformation from W to U.
     a. Show that if T maps V onto W and S maps W onto U, then ST maps V onto U.
     b. Show that if T and S are both one-to-one, then so is ST.
     c. Show that if T and S are both isomorphisms, then so is ST.

15. Find an example of two nonzero linear transformations from R^2 to R^2 whose product is zero.

5.3 CHANGE OF BASES

In the examples of Section 5.1 we calculated the matrix of a single linear transformation with respect to different pairs of bases and obtained different matrices. In this section, we determine the relationship between these matrices.

Our discussion relies heavily on the use of a mapping that some of you have already seen in the exercises and notes of earlier sections. If U is any vector space, then the **identity** on U is the mapping I_U from U to itself defined by the rule

$$I_U(\mathbf{u}) = \mathbf{u}$$

for each $\mathbf{u}$ in U. This mapping certainly preserves addition and scalar multiplication, so it is a linear transformation from U to U.

If T is any linear transformation from V to W, then so are $(I_W T)$ and $(I_W T)I_V$. For each $\mathbf{v}$ in V,

$$[(I_W T)I_V](\mathbf{v}) = (I_W T)[I_V(\mathbf{v})] \qquad \text{(definition of } (I_W T)I_V)$$

$$= (I_W T)[\mathbf{v}] \qquad \text{(definition of } I_V)$$

$$= I_W[T(\mathbf{v})] \qquad \text{(definition of } I_W T)$$

$$= T(\mathbf{v}) \qquad \text{(definition of } I_W).$$

Thus

$$(I_W T)I_V = T.$$

Suppose now that $\mathcal{B}$ and $\mathcal{B}'$ are bases for V and that $\mathcal{C}$ and $\mathcal{C}'$ are bases for W. We would like to describe $_{\mathcal{C}'}[T]_{\mathcal{B}'}$ in terms of $_{\mathcal{C}}[T]_{\mathcal{B}}$. To do so, we use the fact that $T = (I_W T)I_V$ and apply our result that the matrix of a product is the product of the matrices twice:

$$
\begin{aligned}
_{\mathcal{C}'}[T]_{\mathcal{B}'} &= {}_{\mathcal{C}'}[(I_W T)I_V]_{\mathcal{B}'} \\
&= {}_{\mathcal{C}'}[(I_W T)]_{\mathcal{B}}\, _{\mathcal{B}}[I_V]_{\mathcal{B}'} \\
&= {}_{\mathcal{C}'}[I_W]_{\mathcal{C}}\, _{\mathcal{C}}[T]_{\mathcal{B}}\, _{\mathcal{B}}[I_V]_{\mathcal{B}'}.
\end{aligned}
$$

Thus we have the following description of the effect of a change of bases.

Fact: *Let T be a linear transformation from V to W. Suppose that $\mathcal{B}$ and $\mathcal{B}'$ are bases of V, and that $\mathcal{C}$ and $\mathcal{C}'$ are bases of W. Then*

$$
_{\mathcal{C}'}[T]_{\mathcal{B}'} = {}_{\mathcal{C}'}[I_W]_{\mathcal{C}}\, _{\mathcal{C}}[T]_{\mathcal{B}}\, _{\mathcal{B}}[I_V]_{\mathcal{B}'}.
$$

Once again, it is important to note that the bases in this formula have to match up properly. Here, as before, we can view the left side as obtained from the right by cancelling adjacent $\mathcal{C}$'s and adjacent $\mathcal{B}$'s.

Example 5.3.1

In Examples 5.1.1 and 5.1.2 we calculated the matrix of a linear transformation T with respect to two pairs of bases of $V = R^3$ and $W = R^2$:

$$
\mathcal{B} = \left\{ \begin{bmatrix} 1 \\ 0 \\ 0 \end{bmatrix}, \begin{bmatrix} 0 \\ 1 \\ 0 \end{bmatrix}, \begin{bmatrix} 0 \\ 0 \\ 1 \end{bmatrix} \right\} \quad \text{and} \quad \mathcal{C} = \left\{ \begin{bmatrix} 1 \\ 0 \end{bmatrix}, \begin{bmatrix} 0 \\ 1 \end{bmatrix} \right\}
$$

and

$$
\mathcal{B}' = \left\{ \begin{bmatrix} 0 \\ 0 \\ 1 \end{bmatrix}, \begin{bmatrix} 1 \\ 0 \\ 0 \end{bmatrix}, \begin{bmatrix} 0 \\ 1 \\ 0 \end{bmatrix} \right\} \quad \text{and} \quad \mathcal{C}' = \left\{ \begin{bmatrix} 0 \\ 1 \end{bmatrix}, \begin{bmatrix} 1 \\ 0 \end{bmatrix} \right\}.
$$

These matrices are

$$
_{\mathcal{C}}[T]_{\mathcal{B}} = \begin{bmatrix} 2 & 1 & 0 \\ 1 & -1 & 0 \end{bmatrix} \quad \text{and} \quad _{\mathcal{C}'}[T]_{\mathcal{B}'} = \begin{bmatrix} 0 & 1 & -1 \\ 0 & 2 & 1 \end{bmatrix}.
$$

Let's verify the change-of-basis formula in this case.

To calculate $_{\mathscr{C}'}[I_W]_{\mathscr{C}}$, we apply I_W to the vectors in $\mathscr{C}$ and express the results in terms of $\mathscr{C}'$:

$$I_W \begin{bmatrix} 1 \\ 0 \end{bmatrix} = \begin{bmatrix} 1 \\ 0 \end{bmatrix} = 0 \begin{bmatrix} 0 \\ 1 \end{bmatrix} + 1 \begin{bmatrix} 1 \\ 0 \end{bmatrix}$$

$$I_W \begin{bmatrix} 0 \\ 1 \end{bmatrix} = \begin{bmatrix} 0 \\ 1 \end{bmatrix} = 1 \begin{bmatrix} 0 \\ 1 \end{bmatrix} + 0 \begin{bmatrix} 1 \\ 0 \end{bmatrix}.$$

Thus

$$_{\mathscr{C}'}[I_W]_{\mathscr{C}} = \begin{bmatrix} 0 & 1 \\ 1 & 0 \end{bmatrix}.$$

To find $_{\mathscr{B}}[I_V]_{\mathscr{B}'}$, we calculate

$$I_V \begin{bmatrix} 0 \\ 0 \\ 1 \end{bmatrix} = \begin{bmatrix} 0 \\ 0 \\ 1 \end{bmatrix} = 0 \begin{bmatrix} 1 \\ 0 \\ 0 \end{bmatrix} + 0 \begin{bmatrix} 0 \\ 1 \\ 0 \end{bmatrix} + 1 \begin{bmatrix} 0 \\ 0 \\ 1 \end{bmatrix}$$

$$I_V \begin{bmatrix} 1 \\ 0 \\ 0 \end{bmatrix} = \begin{bmatrix} 1 \\ 0 \\ 0 \end{bmatrix} = 1 \begin{bmatrix} 1 \\ 0 \\ 0 \end{bmatrix} + 0 \begin{bmatrix} 0 \\ 1 \\ 0 \end{bmatrix} + 0 \begin{bmatrix} 0 \\ 0 \\ 1 \end{bmatrix}$$

$$I_V \begin{bmatrix} 0 \\ 1 \\ 0 \end{bmatrix} = \begin{bmatrix} 0 \\ 1 \\ 0 \end{bmatrix} = 0 \begin{bmatrix} 1 \\ 0 \\ 0 \end{bmatrix} + 1 \begin{bmatrix} 0 \\ 1 \\ 0 \end{bmatrix} + 0 \begin{bmatrix} 0 \\ 0 \\ 1 \end{bmatrix}.$$

Thus

$$_{\mathscr{B}}[I_V]_{\mathscr{B}'} = \begin{bmatrix} 0 & 1 & 0 \\ 0 & 0 & 1 \\ 1 & 0 & 0 \end{bmatrix}.$$

Now

$$_{\mathscr{C}'}[I_W]_{\mathscr{C}} \, _{\mathscr{C}}[T]_{\mathscr{B}} \, _{\mathscr{B}}[I_V]_{\mathscr{B}'} = \begin{bmatrix} 0 & 1 \\ 1 & 0 \end{bmatrix} \begin{bmatrix} 2 & 1 & 0 \\ 1 & -1 & 0 \end{bmatrix} \begin{bmatrix} 0 & 1 & 0 \\ 0 & 0 & 1 \\ 1 & 0 & 0 \end{bmatrix}$$

$$= \begin{bmatrix} 1 & -1 & 0 \\ 2 & 1 & 0 \end{bmatrix} \begin{bmatrix} 0 & 1 & 0 \\ 0 & 0 & 1 \\ 1 & 0 & 0 \end{bmatrix} = \begin{bmatrix} 0 & 1 & -1 \\ 0 & 2 & 1 \end{bmatrix}$$

$$= \, _{\mathscr{C}'}[T]_{\mathscr{B}'}.$$

We calculated the matrix of T with respect to yet another pair of bases, $\mathcal{B}''$ and $\mathcal{C}''$, in Example 5.1.3. We leave it to you (Exercise 11) to check the relationship between $_{\mathcal{C}''}[T]_{\mathcal{B}''}$ and $_{\mathcal{C}}[T]_{\mathcal{B}}$.

Example 5.3.2

Let $\mathcal{B}$, $\mathcal{B}'$, $\mathcal{C}$, $\mathcal{C}'$ be as in Example 5.3.1. Let S be the linear transformation from $V = R^3$ to $W = R^2$ determined by

$$_{\mathcal{C}}[S]_{\mathcal{B}} = \begin{bmatrix} 1 & 0 & 3 \\ -1 & 2 & 0 \end{bmatrix}.$$

Find $_{\mathcal{C}'}[S]_{\mathcal{B}'}$.

The change-of-basis formula tells us that

$$_{\mathcal{C}'}[S]_{\mathcal{B}'} = {}_{\mathcal{C}'}[I_W]_{\mathcal{C}} \; _{\mathcal{C}}[S]_{\mathcal{B}} \; _{\mathcal{B}}[I_V]_{\mathcal{B}'}.$$

We found in Example 5.3.1 that

$$_{\mathcal{C}'}[I_W]_{\mathcal{C}} = \begin{bmatrix} 0 & 1 \\ 1 & 0 \end{bmatrix}$$

and

$$_{\mathcal{B}}[I_V]_{\mathcal{B}'} = \begin{bmatrix} 0 & 1 & 0 \\ 0 & 0 & 1 \\ 1 & 0 & 0 \end{bmatrix}.$$

Therefore

$$_{\mathcal{C}'}[S]_{\mathcal{B}'} = \begin{bmatrix} 0 & 1 \\ 1 & 0 \end{bmatrix} \begin{bmatrix} 1 & 0 & 3 \\ -1 & 2 & 0 \end{bmatrix} \begin{bmatrix} 0 & 1 & 0 \\ 0 & 0 & 1 \\ 1 & 0 & 0 \end{bmatrix}.$$

$$= \begin{bmatrix} -1 & 2 & 0 \\ 1 & 0 & 3 \end{bmatrix} \begin{bmatrix} 0 & 1 & 0 \\ 0 & 0 & 1 \\ 1 & 0 & 0 \end{bmatrix} = \begin{bmatrix} 0 & -1 & 2 \\ 3 & 1 & 0 \end{bmatrix}.$$

Our primary interest is in a special case of the above change-of-basis formula. The linear transformations we are most interested in are of the form $T = T_A$, where A is an $n \times n$ matrix. In particular, they are transformations from a vector space $V = R^n$ (or C^n) to itself. The matrices we want to relate are matrices of the form

$_{\mathscr{B}'}[T]_{\mathscr{B}'}$ and $_{\mathscr{B}}[T]_{\mathscr{B}}$. To do this, we apply our formula in the special case that $V = W$, $\mathscr{B} = \mathscr{C}$, and $\mathscr{B}' = \mathscr{C}'$.

Fact: *If T is a linear transformation from V to V and if $\mathscr{B}$ and $\mathscr{B}'$ are bases of V, then*

$$_{\mathscr{B}'}[T]_{\mathscr{B}'} = \ _{\mathscr{B}'}[I_V]_{\mathscr{B}} \ _{\mathscr{B}}[T]_{\mathscr{B}} \ _{\mathscr{B}}[I_V]_{\mathscr{B}'}.$$

Let's check this version of the change-of-basis formula in a specific example.

Example 5.3.3

Let $T = T_A$, where

$$A = \begin{bmatrix} 1 & -1 & 0 \\ 0 & 2 & 0 \\ 1 & 0 & 1 \end{bmatrix},$$

let $\mathscr{B}$ be the standard basis of R^3, and let

$$\mathscr{B}' = \left\{ \begin{bmatrix} 1 \\ 0 \\ 1 \end{bmatrix}, \begin{bmatrix} 1 \\ 1 \\ 0 \end{bmatrix}, \begin{bmatrix} 0 \\ 0 \\ 1 \end{bmatrix} \right\}.$$

We know from Example 5.1.4 that the matrix of $T = T_A$ with respect to the bases $\mathscr{B}$ and $\mathscr{C} = \mathscr{B}$ is A itself:

$$_{\mathscr{B}}[T]_{\mathscr{B}} = A.$$

To find $_{\mathscr{B}'}[T]_{\mathscr{B}'}$ directly, we calculate

$$T\begin{bmatrix} 1 \\ 0 \\ 1 \end{bmatrix} = A\begin{bmatrix} 1 \\ 0 \\ 1 \end{bmatrix} = \begin{bmatrix} 1 \\ 0 \\ 2 \end{bmatrix} = 1\begin{bmatrix} 1 \\ 0 \\ 1 \end{bmatrix} + 0\begin{bmatrix} 1 \\ 1 \\ 0 \end{bmatrix} + (-1)\begin{bmatrix} 0 \\ 0 \\ -1 \end{bmatrix}$$

$$T\begin{bmatrix} 1 \\ 1 \\ 0 \end{bmatrix} = A\begin{bmatrix} 1 \\ 1 \\ 0 \end{bmatrix} = \begin{bmatrix} 0 \\ 2 \\ 1 \end{bmatrix} = (-2)\begin{bmatrix} 1 \\ 0 \\ 1 \end{bmatrix} + 2\begin{bmatrix} 1 \\ 1 \\ 0 \end{bmatrix} + (-3)\begin{bmatrix} 0 \\ 0 \\ -1 \end{bmatrix}$$

$$I\begin{bmatrix} 0 \\ 0 \\ -1 \end{bmatrix} = A\begin{bmatrix} 0 \\ 0 \\ -1 \end{bmatrix} = \begin{bmatrix} 0 \\ 0 \\ -1 \end{bmatrix} = 0\begin{bmatrix} 1 \\ 0 \\ 1 \end{bmatrix} + 0\begin{bmatrix} 1 \\ 1 \\ 0 \end{bmatrix} + 1\begin{bmatrix} 0 \\ 0 \\ -1 \end{bmatrix}.$$

Thus

$$
_{\mathscr{B}'}[T]_{\mathscr{B}'} = \begin{bmatrix} 1 & -2 & 0 \\ 0 & 2 & 0 \\ -1 & -3 & 1 \end{bmatrix}.
$$

To find $_{\mathscr{B}'}[I_V]_{\mathscr{B}}$, we calculate

$$
I_V \begin{bmatrix} 1 \\ 0 \\ 0 \end{bmatrix} = \begin{bmatrix} 1 \\ 0 \\ 0 \end{bmatrix} = 1 \begin{bmatrix} 1 \\ 0 \\ 0 \end{bmatrix} + 0 \begin{bmatrix} 1 \\ 1 \\ 0 \end{bmatrix} + 1 \begin{bmatrix} 0 \\ 0 \\ -1 \end{bmatrix}
$$

$$
I_V \begin{bmatrix} 0 \\ 1 \\ 0 \end{bmatrix} = \begin{bmatrix} 0 \\ 1 \\ 0 \end{bmatrix} = (-1) \begin{bmatrix} 1 \\ 0 \\ 0 \end{bmatrix} + 1 \begin{bmatrix} 1 \\ 1 \\ 0 \end{bmatrix} + (-1) \begin{bmatrix} 0 \\ 0 \\ -1 \end{bmatrix}
$$

$$
I_V \begin{bmatrix} 0 \\ 0 \\ 1 \end{bmatrix} = \begin{bmatrix} 0 \\ 0 \\ 1 \end{bmatrix} = 0 \begin{bmatrix} 1 \\ 0 \\ 0 \end{bmatrix} + 0 \begin{bmatrix} 1 \\ 1 \\ 0 \end{bmatrix} + (-1) \begin{bmatrix} 0 \\ 0 \\ -1 \end{bmatrix}.
$$

Thus

$$
_{\mathscr{B}'}[I_V]_{\mathscr{B}} = \begin{bmatrix} 1 & -1 & 0 \\ 0 & 1 & 0 \\ 1 & -1 & -1 \end{bmatrix}.
$$

On the other hand,

$$
I_V \begin{bmatrix} 1 \\ 0 \\ 1 \end{bmatrix} = \begin{bmatrix} 1 \\ 0 \\ 1 \end{bmatrix} = 1 \begin{bmatrix} 1 \\ 0 \\ 0 \end{bmatrix} + 0 \begin{bmatrix} 0 \\ 1 \\ 0 \end{bmatrix} + 1 \begin{bmatrix} 0 \\ 0 \\ 1 \end{bmatrix}
$$

$$
I_V \begin{bmatrix} 1 \\ 1 \\ 0 \end{bmatrix} = \begin{bmatrix} 1 \\ 1 \\ 0 \end{bmatrix} = 1 \begin{bmatrix} 1 \\ 0 \\ 0 \end{bmatrix} + 1 \begin{bmatrix} 0 \\ 1 \\ 0 \end{bmatrix} + 0 \begin{bmatrix} 0 \\ 0 \\ 1 \end{bmatrix}
$$

$$
I_V \begin{bmatrix} 0 \\ 0 \\ -1 \end{bmatrix} = \begin{bmatrix} 0 \\ 0 \\ -1 \end{bmatrix} = 0 \begin{bmatrix} 1 \\ 0 \\ 0 \end{bmatrix} + 0 \begin{bmatrix} 0 \\ 1 \\ 0 \end{bmatrix} + (-1) \begin{bmatrix} 0 \\ 0 \\ 1 \end{bmatrix}
$$

so that

$$
_{\mathscr{B}}[I_V]_{\mathscr{B}'} = \begin{bmatrix} 1 & 1 & 0 \\ 0 & 1 & 0 \\ 1 & 0 & -1 \end{bmatrix}.
$$

Now

$$_{\mathscr{B}'}[I_V]_{\mathscr{B}} \ _{\mathscr{B}}[T]_{\mathscr{B}} \ _{\mathscr{B}}[I_V]_{\mathscr{B}'} = \ _{\mathscr{B}'}[I_V]_{\mathscr{B}} \ A \ _{\mathscr{B}}[I_V]_{\mathscr{B}'}$$

$$= \begin{bmatrix} 1 & -1 & 0 \\ 0 & 1 & 0 \\ 1 & -1 & -1 \end{bmatrix} \begin{bmatrix} 1 & -1 & 0 \\ 0 & 2 & 0 \\ 1 & 0 & 1 \end{bmatrix} \begin{bmatrix} 1 & 1 & 0 \\ 0 & 1 & 0 \\ 1 & 0 & -1 \end{bmatrix}$$

$$= \begin{bmatrix} 1 & -3 & 0 \\ 0 & 2 & 0 \\ 0 & -3 & -1 \end{bmatrix} \begin{bmatrix} 1 & 1 & 0 \\ 0 & 1 & 0 \\ 1 & 0 & -1 \end{bmatrix} = \begin{bmatrix} 1 & -2 & 0 \\ 0 & 2 & 0 \\ -1 & -3 & 1 \end{bmatrix}$$

$$= \ _{\mathscr{B}'}[T]_{\mathscr{B}'}.$$

We note, in closing, that each of our examples involved the calculation of a matrix of the form $_{\mathscr{B}}[I_V]_{\mathscr{B}'}$ where V was R^n and $\mathscr{B}$ was the standard basis. In each case, the matrix turned out to be the matrix whose columns were the vectors in $\mathscr{B}'$. We leave it to you (Exercise 12) to check that this is typical.

Fact: *If $V = R^n$ and if $\mathscr{B}$ is the standard basis, then $_{\mathscr{B}}[I_V]_{\mathscr{B}'}$ is the matrix whose columns are the vectors in $\mathscr{B}'$.*

In the following section, we see how to use this information to calculate the matrix $_{\mathscr{B}'}[I_V]_{\mathscr{B}}$, as well.

CHANGE OF BASES

Let T be a linear transformation from V to W, let $\mathscr{B}$ and $\mathscr{B}'$ be bases of V, and let $\mathscr{C}$ and $\mathscr{C}'$ be bases of W. Then

$$_{\mathscr{C}'}[T]_{\mathscr{B}'} = \ _{\mathscr{C}'}[I_W]_{\mathscr{C}} \ _{\mathscr{C}}[T]_{\mathscr{B}} \ _{\mathscr{B}}[I_V]_{\mathscr{B}'}.$$

In the special case where $V = W$, $\mathscr{B} = \mathscr{C}$, and $\mathscr{B}' = \mathscr{C}'$, this gives

$$_{\mathscr{B}'}[T]_{\mathscr{B}'} = \ _{\mathscr{B}'}[I_V]_{\mathscr{B}} \ _{\mathscr{B}}[T]_{\mathscr{B}} \ _{\mathscr{B}}[I_V]_{\mathscr{B}'}.$$

EXERCISES

In Exercises 1 to 10, let V, W, T, $\mathscr{B}$, $\mathscr{B}'$, $\mathscr{C}$, $\mathscr{C}'$, and $A = \ _{\mathscr{C}}[S]_{\mathscr{B}}$ be as in the indicated exercise of Section 5.1.

(a) Find $_{\mathscr{C}'}[I_W]_{\mathscr{C}}$.   (b) Find $_{\mathscr{B}}[I_V]_{\mathscr{B}'}$.

(c) Use the change-of-basis formula to calculate $_{\mathscr{C}'}[T]_{\mathscr{B}'}$ from $_{\mathscr{C}}[T]_{\mathscr{B}}$ and your answers to (a) and (b). Verify that your answer agrees with the matrix $_{\mathscr{C}'}[T]_{\mathscr{B}'}$ calculated directly in Section 5.1.

(d) Use the change-of-basis formula to calculate $_{\mathscr{C}'}[S]_{\mathscr{B}'}$.

1. Exercise 1   2. Exercise 2   3. Exercise 3

4. Exercise 4   5. Exercise 5   6. Exercise 6

7. Exercise 7   8. Exercise 8   9. Exercise 9

10. Exercise 10

11. Verify that in Example 5.3.1

$$_{\mathscr{C}''}[T]_{\mathscr{B}''} = {_{\mathscr{C}''}}[I_W]_{\mathscr{C}} \; {_{\mathscr{C}}}[T]_{\mathscr{B}} \; {_{\mathscr{B}}}[I_V]_{\mathscr{B}''}.$$

More abstract exercises:

12. Show that if $V = R^n$ and $\mathscr{B}$ is the standard basis for V, then $_{\mathscr{B}}[I_V]_{\mathscr{B}'}$ is the matrix whose columns are the vectors in $\mathscr{B}'$.

13. Show that if V is an n-dimensional vector space with bases $\mathscr{B}$ and $\mathscr{B}'$, then $_{\mathscr{B}'}[I_V]_{\mathscr{B}} \; {_{\mathscr{B}}}[I_V]_{\mathscr{B}'} = {_{\mathscr{B}}}[I_V]_{\mathscr{B}'} \; {_{\mathscr{B}'}}[I_V]_{\mathscr{B}} = I$, the $n \times n$ identity matrix.

5.4 INVERTIBLE MATRICES

When T is a linear transformation from a finite-dimensional vector space V to itself, the change-of-basis formula relating $_{\mathscr{B}'}[T]_{\mathscr{B}'}$ and $_{\mathscr{B}}[T]_{\mathscr{B}}$ involves the two matrices

$$P = {_{\mathscr{B}}}[I_V]_{\mathscr{B}'} \quad \text{and} \quad Q = {_{\mathscr{B}'}}[I_V]_{\mathscr{B}}.$$

If we multiply these matrices and use the fact that a product of two matrices is the matrix of the product, we see that

$$PQ = {_{\mathscr{B}}}[I_V]_{\mathscr{B}'} \; {_{\mathscr{B}'}}[I_V]_{\mathscr{B}} = {_{\mathscr{B}}}[I_V I_V]_{\mathscr{B}} = {_{\mathscr{B}}}[I_V]_{\mathscr{B}}.$$

Since the identity transformation satisfies $I_V(\mathbf{v}_j) = \mathbf{v}_j$ for each vector $\mathbf{v}_j$ in $\mathscr{B}$, the last matrix above is just the $n \times n$ identity matrix. Thus we have the condition

$$PQ = I.$$

In this section, we consider $n \times n$ matrices satisfying this condition. Among our goals is a systematic method for finding Q when given P.

Definition: *Given the $n \times n$ matrix P, if there exists an $n \times n$ matrix Q such that $PQ = I$, then we say that P is **invertible** and we refer to Q as an **inverse** of P.*

Cramer's determinant test can be used to obtain a test for invertibility. On the one hand, if

$$PQ = I,$$

then for any n-vector $\mathbf{r}$ we have

$$P(Q\mathbf{r}) = (PQ)\mathbf{r} = I\mathbf{r} = \mathbf{r}.$$

Thus the equation

$$P\mathbf{x} = \mathbf{r}$$

always has a solution

$$\mathbf{x} = Q\mathbf{r}.$$

It follows from Cramer's determinant test that

$$\det P \neq 0.$$

On the other hand, suppose that P is an $n \times n$ matrix with

$$\det P \neq 0.$$

Then, for any n-vector $\mathbf{r}$, the equation

$$P\mathbf{x} = \mathbf{r}$$

has a solution. In particular, if we take $\mathbf{r}$ in turn to be each of the columns of the identity matrix

$$\mathbf{e}_1 = \begin{bmatrix} 1 \\ 0 \\ \cdot \\ \cdot \\ \cdot \\ 0 \end{bmatrix}, \quad \mathbf{e}_2 = \begin{bmatrix} 0 \\ 1 \\ \cdot \\ \cdot \\ \cdot \\ 0 \end{bmatrix}, \ldots, \quad \mathbf{e}_n = \begin{bmatrix} 0 \\ 0 \\ \cdot \\ \cdot \\ \cdot \\ 1 \end{bmatrix},$$

we can find vectors $\mathbf{q}_1, \ldots, \mathbf{q}_n$ satisfying

$$P\mathbf{q}_j = \mathbf{e}_j.$$

If we form the matrix whose columns are $\mathbf{q}_1, \ldots, \mathbf{q}_n$, then we can calculate the product column by column to get

$$PQ = P[\mathbf{q}_1 \ldots \mathbf{q}_n] = [P\mathbf{q}_1 \ldots P\mathbf{q}_n] = [\mathbf{e}_1 \ldots \mathbf{e}_n] = I.$$

Thus P is invertible in this case.

We have obtained the following test for invertibility.

Fact: *The $n \times n$ matrix P is invertible if and only if* $\det P \neq 0$.

Example 5.4.1

Is the matrix

$$P = \begin{bmatrix} 1 & 0 & 3 \\ 3 & 1 & 2 \\ 2 & 3 & 1 \end{bmatrix}$$

invertible?

The determinant of P can be calculated by expansion along the first row:

$$\det P = 1 \det \begin{bmatrix} 1 & 2 \\ 3 & 1 \end{bmatrix} - 0 + 3 \det \begin{bmatrix} 3 & 1 \\ 3 & 3 \end{bmatrix} = -5 + 3(7) = 16.$$

Since this is not zero, P is invertible.

Example 5.4.2

Is the matrix

$$P = \begin{bmatrix} 1 & 2 & 1 \\ 2 & 1 & 2 \\ 3 & 3 & 3 \end{bmatrix}$$

invertible?

Expansion by minors along the last row gives

$$\det P = 3 \det \begin{bmatrix} 2 & 1 \\ 1 & 2 \end{bmatrix} - 3 \det \begin{bmatrix} 1 & 1 \\ 2 & 2 \end{bmatrix} + 3 \det \begin{bmatrix} 1 & 2 \\ 2 & 1 \end{bmatrix}$$

$$= 3 \cdot 3 - 3 \cdot 0 + 3(-3) = 0.$$

Thus P is not invertible.

A closer look at the determinant test for invertibility yields further informa-tion about this situation. If P is invertible—that is, if there is a matrix Q satisfying $PQ = I$—then the jth column $\mathbf{q}_j$ of Q has to be a solution of the equation $P\mathbf{x} = \mathbf{e}_j$, where $\mathbf{e}_j$ is the jth column of the identity matrix. Since det $P \neq 0$, Cramer's deter-minant test tells us that the solution of this equation is *unique*. Thus Q is uniquely determined by P. In other words, *an invertible matrix has only one inverse*. We can stop referring to Q as *an* inverse of P; rather, it is *the* inverse of P and is denoted $Q = P^{-1}$.

Let's turn now to the question of finding the inverse of a given invertible matrix P. To find the jth column $\mathbf{q}_j$ of P^{-1}, we must solve the equation $P\mathbf{x} = \mathbf{e}_j$. One way to do this is to reduce the augmented matrix $[P|\mathbf{e}_j]$. The uniqueness of the solution insures that the reduced matrix will represent an algebraic system without noncorner variables. Hence each of the first n columns of the reduced matrix will have a corner entry, and thus the reduced matrix will have the form

$$\begin{bmatrix} 1 & 0 & \cdots & 0 & q_{1j} \\ 0 & 1 & & 0 & q_{2j} \\ \cdot & \cdot & & \cdot & \cdot \\ \cdot & \cdot & & \cdot & \cdot \\ \cdot & \cdot & & \cdot & \cdot \\ 0 & 0 & \cdots & 1 & q_{nj} \end{bmatrix}.$$

The column to the right of the vertical line will be the desired solution of $P\mathbf{x} = \mathbf{e}_j$. That is, $[P|\mathbf{e}_j]$ will reduce to $[I|\mathbf{q}_j]$ where $\mathbf{q}_j$ is the jth column of P^{-1}. If we want to find all the columns of P^{-1}, then instead of reducing n different $n \times (n + 1)$ matrices

$$[P|\mathbf{e}_1] \to [I|\mathbf{q}_1], \ [P|\mathbf{e}_2] \to [I|\mathbf{q}_2], \ \ldots, \ [P|\mathbf{e}_n] \to [I|\mathbf{q}_n],$$

we can simply reduce the single $n \times 2n$ matrix

$$[P|I] = [P|\mathbf{e}_1, \mathbf{e}_2 \ldots \mathbf{e}_n] \to [I|\mathbf{q}_1 \ \mathbf{q}_2 \ldots \ \mathbf{q}_n].$$

The last n columns of the reduced matrix will be the columns of P^{-1}. Thus, if P is invertible, then $[P|I]$ will reduce to $[I|Q]$ where $Q = P^{-1}$.

Suppose, on the other hand, that P is an $n \times n$ matrix and that the $n \times 2n$ matrix $[P|I] = [P|\mathbf{e}_1 \ \ldots \ \mathbf{e}_n]$ reduces to a matrix of the form $[I|Q] = [I|\mathbf{q}_1 \ \ldots \ \mathbf{q}_n]$. Then the columns $\mathbf{q}_j$ of Q satisfy $P\mathbf{q}_j = \mathbf{e}_j$, so that $PQ = I$. Thus P is invertible and $Q = P^{-1}$.

This provides us with another test for the invertibility of a given matrix P. The advantage of this test is that in the case when P is invertible, we don't have to do any further work to find P^{-1}.

Fact: *The $n \times n$ matrix P is invertible if and only if the $n \times 2n$ matrix $[P|I]$ reduces to a matrix of the form $[I|Q]$. In this case, $Q = P^{-1}$.*

Example 5.4.3

Find the inverse (if any) of the 4 × 4 matrix

$$P = \begin{bmatrix} 1 & 2 & 3 & 4 \\ 0 & 1 & 1 & 2 \\ 0 & 0 & 2 & 1 \\ 1 & 1 & 1 & 1 \end{bmatrix}.$$

We reduce the 4 × 8 matrix $[P|I]$:

$$[P|I] = \begin{bmatrix} 1 & 2 & 3 & 4 & | & 1 & 0 & 0 & 0 \\ 0 & 1 & 1 & 2 & | & 0 & 1 & 0 & 0 \\ 0 & 0 & 2 & 1 & | & 0 & 0 & 1 & 0 \\ 1 & 1 & 1 & 1 & | & 0 & 0 & 0 & 1 \end{bmatrix}$$

$$\rightarrow \begin{bmatrix} 1 & 0 & 0 & 0 & | & 2 & -3 & -1 & -1 \\ 0 & 1 & 0 & 0 & | & -3 & 4 & 1 & 3 \\ 0 & 0 & 1 & 0 & | & -1 & 1 & 1 & 1 \\ 0 & 0 & 0 & 1 & | & 2 & -2 & -1 & 2 \end{bmatrix}.$$

The columns to the right of the vertical line form the inverse

$$P^{-1} = \begin{bmatrix} 2 & -3 & -1 & -1 \\ -3 & 4 & 1 & 3 \\ -1 & 1 & 1 & 1 \\ 2 & -2 & -1 & 2 \end{bmatrix}.$$

You should check by multiplying that $PP^{-1} = I$.

Example 5.4.4

Find the inverse (if any) of the 3 × 3 matrix

$$P = \begin{bmatrix} 1 & 1 & 2 \\ 2 & 1 & -1 \\ 0 & -1 & -5 \end{bmatrix}.$$

We reduce the 3 × 6 matrix $[P \mid I]$:

$$[P|I] = \begin{bmatrix} 1 & 1 & 2 & | & 1 & 0 & 0 \\ 2 & 1 & -1 & | & 0 & 1 & 0 \\ 0 & -1 & -5 & | & 0 & 0 & 1 \end{bmatrix} \rightarrow \begin{bmatrix} 1 & 0 & -3 & | & -1 & 1 & 0 \\ 0 & 1 & 5 & | & 2 & -1 & 0 \\ 0 & 0 & 0 & | & 2 & -1 & 1 \end{bmatrix}.$$

Since this is not of the form $[I \mid Q]$, the matrix P is *not* invertible.

We began our discussion in this section by noting that if V is a finite-dimensional vector space with bases $\mathscr{B}$ and $\mathscr{B}'$, then $_\mathscr{B}[I_V]_{\mathscr{B}'}\,_{\mathscr{B}'}[I_V]_\mathscr{B} = I$. In our new terminology, this says that $P = {_\mathscr{B}}[I_V]_{\mathscr{B}'}$ is invertible and that $P^{-1} = {_{\mathscr{B}'}}[I_V]_\mathscr{B}$. Our method for calculating inverses of matrices provides an alternative for calculating the change-of-basis matrix P^{-1}. This alternative is particularly attractive when P is easy to calculate, for example when one of the bases is the standard basis.

Example 5.4.5

Given the bases for $V = R^3$,

$$\mathscr{B} = \left\{ \begin{bmatrix} 1 \\ 0 \\ 0 \end{bmatrix}, \begin{bmatrix} 0 \\ 1 \\ 0 \end{bmatrix}, \begin{bmatrix} 0 \\ 0 \\ 1 \end{bmatrix} \right\}, \qquad \mathscr{B}' = \left\{ \begin{bmatrix} 1 \\ 1 \\ 1 \end{bmatrix}, \begin{bmatrix} 1 \\ 1 \\ 0 \end{bmatrix}, \begin{bmatrix} 1 \\ 0 \\ 0 \end{bmatrix} \right\},$$

find the change-of-basis matrices $_\mathscr{B}[I_V]_{\mathscr{B}'}$ and $_{\mathscr{B}'}[I_V]_\mathscr{B}$.

We know that since $\mathscr{B}$ is the standard basis, $P = {_\mathscr{B}}[I_V]_{\mathscr{B}'}$ is the matrix whose columns are the vectors in $\mathscr{B}'$:

$$_\mathscr{B}[I_V]_{\mathscr{B}'} = P = \begin{bmatrix} 1 & 1 & 1 \\ 1 & 1 & 0 \\ 1 & 0 & 0 \end{bmatrix}.$$

We can calculate $_{\mathscr{B}'}[I_V]_\mathscr{B} = P^{-1}$ by reducing $[P|I]$:

$$[P|I] = \begin{bmatrix} 1 & 1 & 1 & | & 1 & 0 & 0 \\ 1 & 1 & 0 & | & 0 & 1 & 0 \\ 1 & 0 & 0 & | & 0 & 0 & 1 \end{bmatrix} \rightarrow \begin{bmatrix} 1 & 0 & 0 & | & 0 & 0 & 1 \\ 0 & 1 & 0 & | & 0 & 1 & -1 \\ 0 & 0 & 1 & | & 1 & -1 & 0 \end{bmatrix} = [I|P^{-1}].$$

Therefore

$$_{\mathscr{B}'}[I_V]_\mathscr{B} = P^{-1} = \begin{bmatrix} 0 & 0 & 1 \\ 0 & 1 & -1 \\ 1 & -1 & 0 \end{bmatrix}.$$

You should verify this by calculating $_{\mathscr{B}'}[I_V]_\mathscr{B}$ directly.

Let's now take another look at the change-of-basis formula for a linear transformation T from a finite-dimensional vector space V to itself. if $\mathscr{B}$ and $\mathscr{B}'$ are bases for V, then

$$_{\mathscr{B}'}[T]_{\mathscr{B}'} = {_{\mathscr{B}'}}[I_V]_\mathscr{B}\,_\mathscr{B}[T]_\mathscr{B}\,_\mathscr{B}[I_V]_{\mathscr{B}'}.$$

We saw above that $P = {}_{\mathcal{B}}[I_V]_{\mathcal{B}'}$ is invertible and ${}_{\mathcal{B}'}[I_V]_{\mathcal{B}} = P^{-1}$. Thus the matrices $A = {}_{\mathcal{B}}[T]_{\mathcal{B}}$ and $A' = {}_{\mathcal{B}'}[T]_{\mathcal{B}'}$ satisfy

$$A' = P^{-1}AP.$$

Matrices that are related in this way are said to be similar.

Definition: *Given the $n \times n$ matrices A and A', if there exists an $n \times n$ invertible matrix P such that $A' = P^{-1}AP$, then we say that A' is **similar** to A, with **similarity matrix P.***

Using this terminology, our change-of-basis formula tells us that *if T is a linear transformation from V to V, and if B and B' are bases for V, then $A' = {}_{\mathcal{B}'}[T]_{\mathcal{B}'}$ is similar to $A = {}_{\mathcal{B}}[T]_{\mathcal{B}}$ with similarity matrix $P = {}_{\mathcal{B}}[I_V]_{\mathcal{B}'}$.*

INVERTIBLE MATRICES

Suppose P is an $n \times n$ matrix. If there is some $n \times n$ matrix Q such that $PQ = I$, then we say P is **invertible** and Q is an **inverse** of P. Each invertible matrix has only one inverse, which we denote P^{-1}.

An $n \times n$ matrix P is invertible if and only if det $P \neq 0$.

The $n \times n$ matrix P is invertible if and only if the $n \times 2n$ matrix $[P|I]$ reduces to a matrix of the form $[I|Q]$. In this case, $Q = P^{-1}$.

Let A and A' be $n \times n$ matrices. If there is an $n \times n$ invertible matrix P so that $A' = P^{-1}AP$, then we say that A' is **similar** to A, with **similarity matrix** P.

Suppose that T is a linear transformation from V to V and that $\mathcal{B}$ and $\mathcal{B}'$ are bases for V. Then $P = {}_{\mathcal{B}}[I]_{\mathcal{B}'}$ is invertible, and $A' = {}_{\mathcal{B}'}[T]_{\mathcal{B}'}$ is similar to $A = {}_{\mathcal{B}}[T]_{\mathcal{B}}$ with similarity matrix P.

Notes
1.   *On the product $P^{-1}P$*
If P is an $n \times n$ invertible matrix, then P^{-1} is defined to be the $n \times n$ matrix satisfying $PP^{-1} = I$. In fact, the product in the opposite order also gives the identity:

$$P^{-1}P = I.$$

To see this, note that

$$P(P^{-1}P - I) = P(P^{-1}P) - PI = (PP^{-1})P - P = IP - P = P - P = 0.$$

Thus the columns of $(P^{-1}P - I)$ are solutions of $P\mathbf{x} = \mathbf{0}$. But $\det P \neq 0$, so this equation has only the solution $\mathbf{x} = 0$. Thus the columns of $(P^{-1}P - I)$ are all zero. Since $(P^{-1}P - I) = 0$, we have the desired result $P^{-1}P = I$.

2. Invertible matrices are change-of-basis matrices

The condition defining invertibility was motivated by the properties of change-of-basis matrices $_{\mathcal{B}}[I_V]_{\mathcal{B}'}$. In fact, every invertible matrix P has this form. To see this, we note that the columns of P form a basis for R^n (Exercise 25). If we take V to be R^n, $\mathcal{B}$ to be the standard basis for V, and $\mathcal{B}'$ to be the basis consisting of the columns of P, then $_{\mathcal{B}}[I_V]_{\mathcal{B}'} = P$.

3. Similar matrices are the matrices of a linear transformation with respect to different bases

We saw above that if T is a linear transformation from V to V, and if $\mathcal{B}$ and $\mathcal{B}'$ are bases for V, then $A' = {}_{\mathcal{B}'}[T]_{\mathcal{B}'}$ is similar to $A = {}_{\mathcal{B}}[T]_{\mathcal{B}}$. In this note we will see, conversely, that all similar matrices arise in this way.

Suppose $A' = P^{-1}AP$. Let $V = R^n$, let $\mathcal{B}$ be the standard basis for R^n, and let $\mathcal{B}'$ be the basis for R^n consisting of the columns of P (see Note 2). Then

$$P = {}_{\mathcal{B}}[I_V]_{\mathcal{B}'} \quad \text{and} \quad P^{-1} = {}_{\mathcal{B}'}[I_V]_{\mathcal{B}}.$$

Setting $T = T_A$, we see that

$$_{\mathcal{B}}[T]_{\mathcal{B}} = A$$

and

$$\begin{aligned}
{}_{\mathcal{B}'}[T]_{\mathcal{B}'} &= {}_{\mathcal{B}'}[I_V]_{\mathcal{B}} \, {}_{\mathcal{B}}[T]_{\mathcal{B}} \, {}_{\mathcal{B}}[I_V]_{\mathcal{B}'} \\
&= P^{-1}AP \\
&= A'.
\end{aligned}$$

EXERCISES

In Exercises 1 to 6, determine whether the given matrix is invertible.

1. $\begin{bmatrix} 1 & -1 \\ 1 & 1 \end{bmatrix}$

2. $\begin{bmatrix} -1 & 1 \\ 1 & -1 \end{bmatrix}$

3. $\begin{bmatrix} 1 & -1 & 3 \\ 1 & 1 & -3 \\ 3 & -1 & 3 \end{bmatrix}$

4. $\begin{bmatrix} 1 & -1 & 3 \\ 1 & 1 & -3 \\ 2 & 0 & -1 \end{bmatrix}$

5. $\begin{bmatrix} 1 & 0 & 1 \\ 0 & -1 & 2 \\ -2 & -1 & 0 \end{bmatrix}$

6. $\begin{bmatrix} 1 & -1 & 1 & 1 \\ 3 & 2 & 0 & 0 \\ 1 & 0 & 0 & -1 \\ 0 & 1 & 2 & 0 \end{bmatrix}$

In Exercises 7 to 15, find the inverse (if any) of the given matrix.

7. $\begin{bmatrix} 0 & 2 \\ 3 & 0 \end{bmatrix}$

8. $\begin{bmatrix} 1 & 2 \\ 2 & 1 \end{bmatrix}$

9. $\begin{bmatrix} 1 & 2 \\ -2 & -4 \end{bmatrix}$

10. $\begin{bmatrix} 1 & 0 & 1 \\ 1 & 1 & 0 \\ 1 & 0 & -1 \end{bmatrix}$

11. $\begin{bmatrix} 1 & 1 & 1 \\ 1 & 2 & -1 \\ 0 & 1 & -2 \end{bmatrix}$

12. $\begin{bmatrix} 1 & 1 & 1 \\ 1 & 2 & -1 \\ 1 & 4 & 1 \end{bmatrix}$

13.
$$\begin{bmatrix} 1 & 0 & 1 & 0 \\ 0 & -1 & -1 & 0 \\ 0 & 1 & 1 & -1 \\ 1 & 1 & 0 & -1 \end{bmatrix}$$
14.
$$\begin{bmatrix} 1 & 0 & 1 & 1 \\ 0 & 1 & 0 & 0 \\ 0 & 0 & -4 & -2 \\ 1 & 1 & 1 & 1 \end{bmatrix}$$
15.
$$\begin{bmatrix} 0 & 0 & 0 & 1 & 0 \\ 1 & 3 & 2 & 0 & 3 \\ 0 & 1 & 0 & 1 & 0 \\ 0 & 0 & 1 & 6 & 1 \\ 0 & -5 & 0 & 0 & 2 \end{bmatrix}$$

In Exercises 16 to 20, you are given a basis $\mathcal{B}'$ for $V = R^3$. Take A to be the 3×3 matrix

$$A = \begin{bmatrix} 1 & 0 & 4 \\ 0 & 3 & 0 \\ 1 & 0 & 1 \end{bmatrix}$$

and take $\mathcal{B}$ to be the standard basis for R^3 (so that $_\mathcal{B}[T_A]_\mathcal{B} = A$). Find

a. $_\mathcal{B}[I_V]_{\mathcal{B}'}$ b. $_{\mathcal{B}'}[I_V]_\mathcal{B}$ c. $_{\mathcal{B}'}[T_A]_{\mathcal{B}'}$

16. $\mathcal{B}' = \left\{ \begin{bmatrix} 0 \\ 0 \\ 1 \end{bmatrix}, \begin{bmatrix} 1 \\ 0 \\ 0 \end{bmatrix}, \begin{bmatrix} 0 \\ 1 \\ 0 \end{bmatrix} \right\}$

17. $\mathcal{B}' = \left\{ \begin{bmatrix} -2 \\ 0 \\ 1 \end{bmatrix}, \begin{bmatrix} 0 \\ 1 \\ 0 \end{bmatrix}, \begin{bmatrix} 2 \\ 0 \\ 1 \end{bmatrix} \right\}$

18. $\mathcal{B}' = \left\{ \begin{bmatrix} 1 \\ 0 \\ 0 \end{bmatrix}, \begin{bmatrix} 0 \\ 1 \\ 1 \end{bmatrix}, \begin{bmatrix} 1 \\ 0 \\ 1 \end{bmatrix} \right\}$

19. $\mathcal{B}' = \left\{ \begin{bmatrix} 1 \\ 0 \\ -1 \end{bmatrix}, \begin{bmatrix} 1 \\ 1 \\ 0 \end{bmatrix}, \begin{bmatrix} 0 \\ -1 \\ 1 \end{bmatrix} \right\}$

20. $\mathcal{B}' = \left\{ \begin{bmatrix} 1 \\ 1 \\ -1 \end{bmatrix}, \begin{bmatrix} 3 \\ 0 \\ -2 \end{bmatrix}, \begin{bmatrix} -2 \\ 1 \\ 2 \end{bmatrix} \right\}$

More abstract exercises:

21. a. Use the fact that $I \cdot I = I$ to conclude that I is invertible and $I^{-1} = I$.
    b. Find at least three other 2×2 invertible matrices satisfying $P^{-1} = P$. (*Hint:* Try matrices of the form $\begin{bmatrix} a & 0 \\ 0 & d \end{bmatrix}$.)

22. Suppose P is an invertible $n \times n$ matrix. Show that P^{-1} is also invertible and that $(P^{-1})^{-1} = P$. (*Hint:* See Note 1.)

23. Suppose P and R are invertible $n \times n$ matrices. Show that PR is also invertible and $(PR)^{-1} = R^{-1}P^{-1}$.

24. Let P be an $n \times n$ matrix.
    a. Show that P is invertible if and only if the transformation T_P maps R^n onto R^n.
    b. Show that P is invertible if and only if T_P is one-to-one.
    c. Conclude that P is invertible if and only if T_P is an isomorphism from R^n onto R^n.

25. Show that the columns of an $n \times n$ invertible matrix P form a basis for R^n. (*Hint:* First note from Exercise 24c that T_P is an isomorphism from R^n onto R^n. Then use Exercise 23a of Section 4.7 with $v_1 = e_1, \ldots, v_n = e_n$.)

26. Let P be an $n \times n$ invertible matrix and let A and A' be $n \times n$ matrices that satisfy $A' = P^{-1}AP$. We noted in Execise 21 of Section 3.9 that the determinant of a product of $n \times n$ matrices is equal to the product of the determinants. Use this to show
   a. $(\det P)(\det P^{-1}) = 1$
   b. $\det A' = \det A$
   c. $\det (A' - \lambda I) = \det (A - \lambda I)$ (*Hint:* first check that $(A' - \lambda I) = P(A - \lambda I)P^{-1}$.)
27. What goes wrong if we apply the reasoning of Note 2 to a noninvertible matrix P?

5.5 DIAGONALIZABLE MATRICES AND BLOCK DIAGONAL FORM

In this section we will look back at the method we used in Chapter 3 to solve nth-order homogeneous systems $D\mathbf{x} = A\mathbf{x}$. We will see that the core of this method is really the determination of a basis $\mathcal{B}'$ for C^n so that the matrix $A' = {}_{\mathcal{B}'}[T_A]_{\mathcal{B}'}$ has a particularly simple form.

The systems that were easiest to solve were those for which we could find n independent eigenvectors $\mathbf{v}_1, \ldots, \mathbf{v}_n$ corresponding to the eigenvalues $\lambda_1, \ldots, \lambda_n$ of A. Note that we allow the possibility that several of the eigenvalues λ_j may be equal (repeated roots) and that some may be complex. We insist, however, that the eigenvectors $\mathbf{v}_j$ be independent. Since we have n independent vectors in the n-dimensional space C^n, we know from Section 4.5 that these vectors form a basis for C^n:

$$\mathcal{B}' = \{\mathbf{v}_1, \ldots, \mathbf{v}_n\}.$$

Since these basis vectors are eigenvectors of A,

$$T_A(\mathbf{v}_j) = A\mathbf{v}_j = \lambda_j \mathbf{v}_j.$$

Thus the matrix $A' = {}_{\mathcal{B}'}[T_A]_{\mathcal{B}'}$ has the simple form

$$A' = {}_{\mathcal{B}'}[T_A]_{\mathcal{B}'} = \begin{bmatrix} \lambda_1 & 0 & \cdots & 0 \\ 0 & \lambda_2 & \cdots & 0 \\ & & \ddots & \\ & & & \\ 0 & 0 & \cdots & \lambda_n \end{bmatrix}.$$

Of course, the matrix of T_A with respect to the standard basis $\mathcal{B}$ is

$$_{\mathcal{B}}[T_A]_{\mathcal{B}} = A.$$

Furthermore, since $\mathcal{B}$ is the standard basis, the change-of-basis matrix $P = {}_{\mathcal{B}}[I_V]_{\mathcal{B}'}$ is just the matrix whose columns are the vectors in $\mathcal{B}'$,

$$P = [\mathbf{v}_1 \ldots \mathbf{v}_n].$$

The change-of-basis formula tells us that

$$A' = P^{-1}AP.$$

In the language introduced at the end of the last section, A' is similar to A, with similarity matrix $P = [\mathbf{v}_1 \ldots \mathbf{v}_n]$.

The situation we have just encountered can be described in purely matrix terms. An $n \times n$ matrix A' that has only zeroes off the diagonal

$$A' = \begin{bmatrix} \lambda_1 & 0 & \cdots & 0 \\ 0 & \lambda_2 & \cdots & 0 \\ & & \cdot & \\ & & \cdot & \\ & & \cdot & \\ 0 & 0 & \cdots & \lambda_n \end{bmatrix}$$

is called a **diagonal matrix**. If we are given an $n \times n$ matrix A, and if we can find an $n \times n$ diagonal matrix A' that is similar to A, then we say that A is **diagonalizable**. The process of finding A' and the similarity matrix P is called **diagonalizing** A. The preceding paragraph tells us that *if A has n independent eigenvectors, then A is diagonalizable*. It even tells us how to diagonalize A.

Example 5.5.1

Diagonalize the matrix

$$A = \begin{bmatrix} 1 & 0 & 4 \\ 0 & 3 & 0 \\ 1 & 0 & 1 \end{bmatrix}$$

considered in Examples 3.5.4 and 3.5.5.

We saw in Example 3.5.4 that A has eigenvalues

$$\lambda_1 = -1, \qquad \lambda_2 = 3, \qquad \lambda_3 = 3,$$

with corresponding independent eigenvectors

$$\mathbf{v}_1 = \begin{bmatrix} -2 \\ 0 \\ 1 \end{bmatrix}, \qquad \mathbf{v}_2 = \begin{bmatrix} 0 \\ 1 \\ 0 \end{bmatrix}, \qquad \mathbf{v}_3 = \begin{bmatrix} 2 \\ 0 \\ 1 \end{bmatrix}.$$

It follows that we should have

$$A' = \begin{bmatrix} -1 & 0 & 0 \\ 0 & 3 & 0 \\ 0 & 0 & 3 \end{bmatrix}$$

and that P can be formed using the $\mathbf{v}_j$'s as columns

$$P = \begin{bmatrix} -2 & 0 & 2 \\ 0 & 1 & 0 \\ 1 & 0 & 1 \end{bmatrix}.$$

To verify that $A' = P^{-1}AP$, we first calculate P^{-1}. We reduce $[P|I]$ to find $[I|P^{-1}]$.

$$[P \mid I] = \begin{bmatrix} -2 & 0 & 2 & 1 & 0 & 0 \\ 0 & 1 & 0 & 0 & 1 & 0 \\ 1 & 0 & 1 & 0 & 0 & 1 \end{bmatrix} \rightarrow \begin{bmatrix} 1 & 0 & 0 & -\frac{1}{4} & 0 & \frac{1}{4} \\ 0 & 1 & 0 & 0 & 1 & 0 \\ 0 & 0 & 1 & \frac{1}{2} & 0 & \frac{1}{2} \end{bmatrix}.$$

Therefore

$$P^{-1}AP = \begin{bmatrix} -\frac{1}{4} & 0 & \frac{1}{4} \\ 0 & 1 & 0 \\ \frac{1}{2} & 0 & \frac{1}{2} \end{bmatrix} \begin{bmatrix} 1 & 0 & 4 \\ 0 & 3 & 0 \\ 1 & 0 & 1 \end{bmatrix} \begin{bmatrix} -2 & 0 & 2 \\ 0 & 1 & 0 \\ 1 & 0 & 1 \end{bmatrix}$$

$$= \begin{bmatrix} \frac{1}{4} & 0 & -\frac{1}{2} \\ 0 & 3 & 0 \\ \frac{3}{4} & 0 & \frac{3}{2} \end{bmatrix} \begin{bmatrix} -2 & 0 & 2 \\ 0 & 1 & 0 \\ 1 & 0 & 1 \end{bmatrix} = \begin{bmatrix} -1 & 0 & 0 \\ 0 & 3 & 0 \\ 0 & 0 & 3 \end{bmatrix} = A'.$$

Unfortunately, not every matrix is diagonalizable—that is, there are $n \times n$ matrices A for which we cannot find n independent eigenvectors. We saw in Chapter 3 that in this case the solutions to the differential system $D\mathbf{x} = A\mathbf{x}$ are obtained from generalized eigenvectors of A.

Recall that if λ is a root of the characteristic polynomial of A with multiplicity m, then the generalized eigenvectors of A are the nonzero solutions of $(A - \lambda I)^m\mathbf{v} = \mathbf{0}$. For our present purposes, we prefer to view these generalized eigenvectors as the nonzero vectors in the subspace of C^n consisting of all solutions of $(A - \lambda I)^m \mathbf{v} = \mathbf{0}$. We refer to this subspace as the **generalized eigenspace** of A corresponding to λ, and we denote it $G(\lambda)$.

If we take a vector $\mathbf{w}$ in $G(\lambda)$, so that $(A - \lambda I)^m \mathbf{w} = \mathbf{0}$, then the vector $T_A(\mathbf{w}) = A\mathbf{w}$ can be written

$$T_A(\mathbf{w}) = (A - \lambda I)\mathbf{w} + \lambda \mathbf{w}.$$

The second vector in this sum, $\lambda \mathbf{w}$, is a multiple of a vector in $G(\lambda)$, so it also lies in $G(\lambda)$. If we multiply the first vector in the sum by $(A - \lambda I)^m$, we get

$$(A - \lambda I)^m (A - \lambda I)\mathbf{w} = (A - \lambda I)^{m+1}\mathbf{w} = (A - \lambda I)(A - \lambda I)^m \mathbf{w}$$
$$= (A - \lambda I)\mathbf{0} = \mathbf{0}.$$

Thus the vector $(A - \lambda I)\mathbf{w}$ is also in $G(\lambda)$. Since $T_A(\mathbf{w})$ is the sum of two vectors in $G(\lambda)$, it too lies in $G(\lambda)$. Thus we have the following:

Fact: *For any vector w in $G(\lambda)$, $T_A(\mathbf{w})$ is also in $G(\lambda)$.*

Suppose now that A has distinct eigenvalues $\lambda_1, \ldots, \lambda_k$. Then we can find a basis for each of the generalized eigenspaces $G(\lambda_i)$. Let's say that

$$\mathbf{v}_1, \ldots, \mathbf{v}_a \text{ form a basis for } G(\lambda_1)$$
$$\mathbf{v}_{a+1}, \ldots, \mathbf{v}_b \text{ form a basis for } G(\lambda_2)$$
$$\cdot$$
$$\cdot$$
$$\cdot$$
$$\mathbf{v}_{g+1}, \ldots, \mathbf{v}_h \text{ form a basis for } G(\lambda_k).$$

The success of our method in Section 3.10 stemmed from the fact that by combining these vectors we obtain a basis for C^n:

$$\mathcal{B}' = \{\mathbf{v}_1, \ldots, \mathbf{v}_a, \mathbf{v}_{a+1}, \ldots, \mathbf{v}_b, \ldots, \mathbf{v}_{g+1}, \ldots, \mathbf{v}_h\}.$$

To find the matrix of T_A with respect to $\mathcal{B}'$, we must calculate the effect of T_A on each vector $\mathbf{v}_j$ in $\mathcal{B}'$. Let's consider the $\mathbf{v}_j$'s in groups, depending on which generalized eigenspace $G(\lambda_i)$ they lie in. The basis vectors that lie in a single $G(\lambda_i)$ will be numbered consecutively, $\mathbf{v}_{r+1}, \ldots, \mathbf{v}_s$, for some choice of r and s. If $\mathbf{v}_j$ is one of these vectors—that is, if $r + 1 \leq j \leq s$—then $\mathbf{v}_j$ is in $G(\lambda_i)$, so $T_A(\mathbf{v}_j)$ is also in $G(\lambda_i)$. Then $T_A(\mathbf{v}_j)$ is a linear combination of those vectors in $\mathcal{B}'$ that lie in $G(\lambda_i)$:

$$T_A(\mathbf{v}_j) = 0\mathbf{v}_1 + \ldots + 0\mathbf{v}_a + \ldots$$
$$+ c_{r+1,j}\mathbf{v}_{r+1} + \ldots + c_{s,j}\mathbf{v}_s + \ldots + 0\mathbf{v}_{g+1} + \ldots + 0\mathbf{v}_n.$$

Thus, when we form the matrix $_{\mathcal{B}'}[T_A]_{\mathcal{B}'}$, the columns numbered $j = r + 1$ to $j = s$ have zero entries in all rows except those numbered $r + 1$ to s. This new matrix can be thought of as made up of blocks, as described in the following.

Fact: *Suppose A is an $n \times n$ matrix whose characteristic polynomial has distinct roots $\lambda_1, \ldots, \lambda_k$ of multiplicities $m_1, \ldots, m_k$, respectively. Suppose further that $\mathcal{B}'$ is a basis for C^n consisting of generalized eigenvectors for A, with those corresponding to λ_1 preceding those corresponding to λ_2, and so on. Then the matrix of T_A with respect to $\mathcal{B}'$ has the **block diagonal form***

$$_{\mathcal{B}'}[T_A]_{\mathcal{B}'} = \begin{bmatrix} B_1 & & & \\ & B_2 & 0's & \\ & & \cdot & \\ & & & \cdot \\ 0's & & & \cdot \\ & & & B_k \end{bmatrix}$$

where each B_i is a square matrix (whose size is the number of vectors in $\mathcal{B}'$ that are generalized eigenvectors corresponding to λ_j). Furthermore, each B_i satisfies

$$(B_i - \lambda_i I)^{m_i} = 0.$$

The last observation follows (see Exercise 19) from the block form of $_{\mathcal{B}'}[T_A]_{\mathcal{B}'}$ and the fact that every vector in $G(\lambda_i)$ is a solution of $(A - \lambda_i I)^{m_i} \mathbf{v} = \mathbf{0}$.

Example 5.5.2

Find a matrix A', in block diagonal form, similar to the matrix

$$A = \begin{bmatrix} 3 & 2 & 2 & 2 \\ 0 & 2 & 0 & 0 \\ 0 & 0 & 3 & 0 \\ 0 & 1 & 1 & 3 \end{bmatrix}.$$

The characteristic equation of A is det $(A - \lambda I) = (2 - \lambda)(3 - \lambda)^3$, so there are two distinct roots $\lambda_1 = 2$ and $\lambda_2 = 3$ with multiplicities $m_1 = 1$ and $m_2 = 3$, respectively. The generalized eigenspace $G(\lambda_1)$ corresponding to $\lambda_1 = 2$ consists of the solutions of $(A - 2I)\mathbf{v} = \mathbf{0}$. We reduce the augmented matrix

$$[A - 2I | \mathbf{0}] = \begin{bmatrix} 1 & 2 & 2 & 2 & | & 0 \\ 0 & 0 & 0 & 0 & | & 0 \\ 0 & 0 & 1 & 0 & | & 0 \\ 0 & 1 & 0 & 1 & | & 0 \end{bmatrix} \rightarrow \begin{bmatrix} 1 & 0 & 0 & 0 & | & 0 \\ 0 & 1 & 0 & 1 & | & 0 \\ 0 & 0 & 1 & 0 & | & 0 \\ 0 & 0 & 0 & 0 & | & 0 \end{bmatrix}$$

and read off a basis for $G(\lambda_1)$,

$$\mathbf{v}_1 = \begin{bmatrix} 0 \\ -1 \\ 0 \\ 1 \end{bmatrix}.$$

The generalized eigenspace $G(\lambda_2)$ corresponding to $\lambda_2 = 3$ consists of the solutions of $(A - 3I)^3\mathbf{v} = \mathbf{0}$. We reduce this augmented matrix

$$[(A - 3I)^3 \mid \mathbf{0}] = \begin{bmatrix} 0 & 0 & 0 & 0 & | & 0 \\ 0 & -1 & 0 & 0 & | & 0 \\ 0 & 0 & 0 & 0 & | & 0 \\ 0 & 1 & 0 & 0 & | & 0 \end{bmatrix} \rightarrow \begin{bmatrix} 0 & 1 & 0 & 0 & | & 0 \\ 0 & 0 & 0 & 0 & | & 0 \\ 0 & 0 & 0 & 0 & | & 0 \\ 0 & 0 & 0 & 0 & | & 0 \end{bmatrix}$$

and read off a basis for $G(\lambda_2)$,

$$\mathbf{v}_2 = \begin{bmatrix} 1 \\ 0 \\ 0 \\ 0 \end{bmatrix}, \quad \mathbf{v}_3 = \begin{bmatrix} 0 \\ 0 \\ 1 \\ 0 \end{bmatrix}, \quad \mathbf{v}_4 = \begin{bmatrix} 0 \\ 0 \\ 0 \\ 1 \end{bmatrix}.$$

By combining these vectors, we obtain a basis $\mathcal{B}' = \{\mathbf{v}_1, \mathbf{v}_2, \mathbf{v}_3, \mathbf{v}_4\}$ for C^4. Since $\mathbf{v}_1$ is an eigenvector of A corresponding to $\lambda_1 = 2$,

$$T_A(\mathbf{v}_1) = A\mathbf{v}_1 = 2\mathbf{v}_1.$$

We can easily calculate that

$$T_A(\mathbf{v}_2) = A\mathbf{v}_2 = \begin{bmatrix} 3 & 2 & 2 & 2 \\ 0 & 2 & 0 & 0 \\ 0 & 0 & 3 & 0 \\ 0 & 1 & 1 & 3 \end{bmatrix}\begin{bmatrix} 1 \\ 0 \\ 0 \\ 0 \end{bmatrix} = \begin{bmatrix} 3 \\ 0 \\ 0 \\ 0 \end{bmatrix} = 3\mathbf{v}_2$$

$$T_A(\mathbf{v}_3) = A\mathbf{v}_3 = \begin{bmatrix} 3 & 2 & 2 & 2 \\ 0 & 2 & 0 & 0 \\ 0 & 0 & 3 & 0 \\ 0 & 1 & 1 & 3 \end{bmatrix}\begin{bmatrix} 0 \\ 0 \\ 1 \\ 0 \end{bmatrix} = \begin{bmatrix} 2 \\ 0 \\ 3 \\ 1 \end{bmatrix} = 2\mathbf{v}_2 + 3\mathbf{v}_3 + \mathbf{v}_4$$

$$T_A(\mathbf{v}_4) = A\mathbf{v}_4 = \begin{bmatrix} 3 & 2 & 2 & 2 \\ 0 & 2 & 0 & 0 \\ 0 & 0 & 3 & 0 \\ 0 & 1 & 1 & 3 \end{bmatrix}\begin{bmatrix} 0 \\ 0 \\ 0 \\ 1 \end{bmatrix} = \begin{bmatrix} 2 \\ 0 \\ 0 \\ 3 \end{bmatrix} = 2\mathbf{v}_2 + \quad\quad + 3\mathbf{v}_4.$$

Thus

$$A' = {}_{\mathscr{B}'}[T_A]_{\mathscr{B}'} = \begin{bmatrix} 2 & \vdots & 0 & 0 & 0 \\ \hdotsfor{5} \\ 0 & \vdots & 3 & 2 & 2 \\ 0 & \vdots & 0 & 3 & 0 \\ 0 & \vdots & 0 & 1 & 3 \end{bmatrix}.$$

The dotted lines above indicate the two blocks

$$B_1 = [2] \quad \text{and} \quad B_2 = \begin{bmatrix} 3 & 2 & 2 \\ 0 & 3 & 0 \\ 0 & 1 & 3 \end{bmatrix}.$$

You should check that

$$[B_1 - 2I] = 0 \quad \text{and} \quad [B_2 - 3I]^3 = 0.$$

You may also wish to check the similarity relation $A' = P^{-1}AP$, with

$$P = [\mathbf{v}_1\ \mathbf{v}_2\ \mathbf{v}_3\ \mathbf{v}_4] = \begin{bmatrix} 0 & 1 & 0 & 0 \\ -1 & 0 & 0 & 0 \\ 0 & 0 & 1 & 0 \\ 1 & 0 & 0 & 1 \end{bmatrix}.$$

We note in closing that whenever λ was an eigenvalue with multiplicity m, we were able to find m independent generalized eigenvectors corresponding to λ. A theorem of linear algebra (which we already mentioned in Section 3.10) guarantees that this is always possible:

Theorem: *The dimension of the generalized eigenspace $G(\lambda)$ corresponding to the eigenvalue λ of A equals the multiplicity m of λ as a root of the characteristic polynomial.*

A consequence of this theorem is the fact that *the block B_i in block diagonal form is $m_i \times m_i$, where m_i is the multiplicity of λ_i.*

Let's summarize.

DIAGONALIZABLE MATRICES AND BLOCK DIAGONAL FORM

Suppose that A is an $n \times n$ matrix and that $\mathbf{v}_1, \ldots, \mathbf{v}_n$ are independent eigenvectors of A corresponding to the respective eigenvalues $\lambda_1, \ldots, \lambda_n$ (some of which may be equal). Then A is **diagonalizable.** That is, there is a matrix $P = [\mathbf{v}_1 \ldots \mathbf{v}_n]$ so that

$$P^{-1}AP = \begin{bmatrix} \lambda_1 & 0 & \cdots & 0 \\ 0 & \lambda_2 & \cdots & 0 \\ & & \ddots & \\ & & & \ddots & \\ 0 & 0 & \cdots & \lambda_n \end{bmatrix}.$$

Suppose A is an $n \times n$ matrix whose characteristic polynomial has distinct roots $\lambda_1, \ldots, \lambda_k$ of multiplicities $m_1, \ldots, m_k$, respectively. Then there exists a basis $\mathcal{B}'$ for C^n consisting of m_1 generalized eigenvectors for A corresponding to λ_1, followed by m_2 generalized eigenvectors for A corresponding to λ_2, and so on. The matrix of T_A with respect to $\mathcal{B}'$ has the **block diagonal form**

$$_{\mathcal{B}'}[T_A]_{\mathcal{B}'} = \begin{bmatrix} B_1 & & & \\ & B_2 & & 0's \\ & & \ddots & \\ 0's & & & \ddots \\ & & & & B_k \end{bmatrix}$$

where each B_i is an $m_i \times m_i$ matrix satisfying $(B_i - \lambda_i I)^{m_i} = 0$. In particular, every matrix is similar to one in block diagonal form.

Note

Diagonal Matrices and Uncoupled Systems of O.D.E.'s

Recall from the discussion in Example 3.5.6 that when A was diagonalizable, we could think of diagonalization as finding new variables that measure the state of the system and whose action is independent. The language of linear transformations gives us a way to formulate this perspective.

Let us rewrite the system $D\mathbf{x} = A\mathbf{x}$ in the form $D\mathbf{x} = T_A(\mathbf{x})$. Here, we think of a vector $\mathbf{x}$ in C^n as the "state" of a system, and T_A is the transformation that assigns to each "state" vector $\mathbf{x}$ a "velocity" vector $\mathbf{v} = T_A(\mathbf{x})$. The choice of a basis $\mathcal{B}$ for C^n amounts to choosing a system of coordinates $_{\mathcal{B}}[\mathbf{x}]$, which we regard as variables $x_1, \ldots, x_n$ that measure the state of the system. The standard basis gives one such choice of variables—for example, the charge $x_1 = Q$ and the currents $x_2 = I_2, x_3 = I_3$ in the circuit described in Examples 3.1.3 and 3.5.6. A different basis $\mathcal{B}'$ gives a different choice of variables $y_1, \ldots, y_n$ to measure the state $\mathbf{x}$. When the matrix $_{\mathcal{B}'}[T_A]_{\mathcal{B}'}$ is diagonal, the differential equation describing the evolution of each of these variables y_i does not involve any of the other variables. We say in this case that the equations are "uncoupled."

An uncoupled system of linear o.d.e.'s is easy to solve: each equation has the form

$$Dy_i = \lambda_i y_i,$$

and its solution is an exponential function

$$y_i(t) = c_i e^{\lambda_i t}.$$

When the equations for the x_i's are coupled but the matrix is diagonalizable, this means it is possible to choose new variables y_i for which the o.d.e.'s are uncoupled. Thus, each $y_i(t)$ is an exponential function as above. We then must use the change-of-basis matrix $P = {}_{\mathcal{B}}[I_V]_{\mathcal{B}'}$ to express the variables $x_1, \ldots, x_n$ in terms of the known solutions $y_1, \ldots, y_n$. The relation is

$$
\begin{bmatrix} x_1(t) \\ \cdot \\ \cdot \\ \cdot \\ x_n(t) \end{bmatrix} = {}_{\mathcal{B}}[\mathbf{x}] = P \; {}_{\mathcal{B}'}[\mathbf{x}] = P \begin{bmatrix} y_1(t) \\ \cdot \\ \cdot \\ \cdot \\ y_n(t) \end{bmatrix} = P \begin{bmatrix} c_1 e^{\lambda_1 t} \\ \cdot \\ \cdot \\ \cdot \\ c_n e^{\lambda_n t} \end{bmatrix}.
$$

You should check that, since the columns $\mathbf{v}_1, \ldots, \mathbf{v}_n$ of P are the eigenvectors of A, this last expression is the same as the solution we found in Section 3.5,

$$
\begin{bmatrix} x_1(t) \\ \cdot \\ \cdot \\ \cdot \\ x_n(t) \end{bmatrix} = c_1 e^{\lambda_1 t} \mathbf{v}_1 + \ldots + c_n e^{\lambda_n t} \mathbf{v}_n.
$$

EXERCISES

In Exercises 1 to 8 you are given a matrix A that was previously considered in the indicated exercise.
a.   Find a diagonal matrix A' that is similar to A.
b.   Find an invertible matrix P such that $A' = P^{-1}AP$.
Note that the matrices in Exercises 7 and 8 have complex eigenvalues.

1. $A = \begin{bmatrix} 0 & 2 \\ -1 & 3 \end{bmatrix}$ (Exercise 14, Section 3.6.)

2. $A = \begin{bmatrix} 1 & 1 \\ 1 & 1 \end{bmatrix}$ (Exercise 15, Section 3.6.)

3. $A = \begin{bmatrix} 1 & 1 & 0 \\ 1 & 1 & 0 \\ 0 & 0 & -1 \end{bmatrix}$ (Exercise 9, Section 3.7.)

4. $A = \begin{bmatrix} 1 & 1 & 0 \\ 1 & 1 & 0 \\ 1 & 1 & 0 \end{bmatrix}$ (Exercise 10, Section 3.7.)

5. $A = \begin{bmatrix} 0 & -2 & 2 \\ 1 & 3 & -2 \\ 2 & 4 & -3 \end{bmatrix}$ (Exercise 19, Section 3.6.)

6. $A = \begin{bmatrix} 1 & 0 & 1 & 0 \\ 0 & 2 & 0 & 2 \\ 0 & 0 & 1 & 1 \\ 0 & 0 & 0 & 4 \end{bmatrix}$ (Exercise 12, Section 3.7.)

7. $A = \begin{bmatrix} 0 & 1 \\ -1 & 0 \end{bmatrix}$ (Exercise 1, Section 3.8.)

8. $A = \begin{bmatrix} 1 & 0 & -1 \\ 0 & 2 & 0 \\ 1 & 0 & 1 \end{bmatrix}$ (Exercise 3, Section 3.8.)

In Exercises 9 to 18 you are given a matrix A that was previously considered in the indicated exercise.
a. Find a matrix A' in block diagonal form that is similar to A.
b. Find an invertible matrix P such that $A' = P^{-1}AP$.
Note that the matrices in Exercises 16 to 18 have complex eigenvalues.

9. $A = \begin{bmatrix} 0 & 1 & 0 \\ 0 & 0 & 1 \\ 1 & -3 & 3 \end{bmatrix}$ (Exercise 1, Section 3.10.)

10. $A = \begin{bmatrix} 1 & 0 & 0 \\ 1 & 0 & 1 \\ 1 & -1 & 2 \end{bmatrix}$ (Exercise 2, Section 3.10.)

11. $A = \begin{bmatrix} 1 & 1 & 0 & 0 \\ 0 & 1 & 1 & 0 \\ 0 & 0 & 1 & -1 \\ 0 & 0 & 0 & 2 \end{bmatrix}$ (Exercise 4, Section 3.10.)

12. $A = \begin{bmatrix} 2 & 0 & -1 & 1 \\ 0 & 0 & -1 & 0 \\ 0 & -1 & 0 & 0 \\ -1 & -1 & 0 & 0 \end{bmatrix}$ (Exercise 3, Section 3.10.)

13. $A = \begin{bmatrix} 2 & 1 & 0 & 0 \\ 0 & 1 & 1 & 0 \\ -1 & -1 & 0 & 0 \\ 0 & 0 & 0 & 1 \end{bmatrix}$ (Exercise 6, Section 3.10.)

14. $A = \begin{bmatrix} 2 & 0 & 0 & 0 \\ 0 & 2 & 0 & 0 \\ 1 & 0 & 1 & 1 \\ 1 & 0 & -1 & 3 \end{bmatrix}$ (Exercise 7, Section 3.10.)

15. $A = \begin{bmatrix} 0 & 0 & 0 & 0 & 1 \\ 1 & -1 & 0 & -1 & 1 \\ 1 & 0 & -1 & -1 & 0 \\ 0 & 0 & 0 & 0 & 1 \\ 1 & 0 & 0 & -1 & 0 \end{bmatrix}$ (Exercise 9, Section 3.10.)

16. $A = \begin{bmatrix} 0 & 2 & 0 & 0 \\ -2 & 0 & 1 & 0 \\ 0 & 0 & 0 & 2 \\ 0 & 0 & -2 & 0 \end{bmatrix}$ (Exercise 5, Section 3.10.)

17. $A = \begin{bmatrix} 1 & 1 & 0 & 0 \\ -1 & 1 & 1 & 0 \\ 0 & 0 & 1 & -1 \\ 0 & 0 & 1 & 1 \end{bmatrix}$ (Exercise 8, Section 3.10.)

18. $A = \begin{bmatrix} 2 & 1 & 0 & 0 & 0 & 0 \\ 0 & 2 & 1 & 0 & 0 & 0 \\ 0 & 0 & 2 & 1 & 0 & 0 \\ 0 & 0 & 0 & 2 & 0 & 0 \\ 0 & 0 & 0 & 0 & 0 & 1 \\ 0 & 0 & 0 & 0 & -2 & -2 \end{bmatrix}$ (Exercise 10, Section 3.10.)

*19. Let A be an $n \times n$ matrix whose characteristic polynomial has distinct roots $\lambda_1, \ldots, \lambda_k$ of multiplicities $m_1, \ldots, m_k$, respectively. Let $\mathcal{B}'$ be a basis of C^n obtained by listing first the vectors in a basis $\mathcal{B}_1'$ for $G(\lambda_1)$, then the vectors in a basis $\mathcal{B}_2'$ for $G(\lambda_2)$, and so on. Then the matrix $_{\mathcal{B}'}[T_A]_{\mathcal{B}'}$ has block diagonal form

$$_{\mathcal{B}'}[T_A]_{\mathcal{B}'} = \begin{bmatrix} B_1 & & & \\ & B_2 & & 0\text{'s} \\ & & \ddots & \\ 0\text{'s} & & & B_k \end{bmatrix}.$$

a. Show that for each vector $\mathbf{w}$ in $G(\lambda_i)$,

$$(A - \lambda_i I)\mathbf{w} = (B_i - \lambda_i I)_{\mathcal{B}_i}[\mathbf{w}].$$

b. Use (a) together with Exercise 11 of Section 5.1 to show that

$$(B_i - \lambda_i I)^{m_i} = 0.$$

5.6 JORDAN FORM (Optional)

We saw in Section 5.5 that given any $n \times n$ matrix A, we can always find a matrix A' in diagonal block form that is similar to A. In this section we discuss a deeper theorem in linear algebra that guarantees the existence of a special block diagonal form in which the blocks themselves can be broken up into smaller blocks.

The smaller blocks in this special form will be $m \times m$ matrices with λ's on the diagonal, 1's on the subdiagonal, and zeroes elsewhere:

$$J = \begin{bmatrix} \lambda & & & \\ 1 & \cdot & & 0\text{'s} \\ & \cdot & \cdot & \\ & & \cdot & \cdot \\ 0\text{'s} & & & \cdot & \cdot \\ & & & 1 & \lambda \end{bmatrix}.$$

Matrices of this form are referred to as **Jordan blocks.** We also refer to the 1×1 matrix $J = [\lambda]$ as a Jordan block.

The theorem (which we won't try to prove) is the following.

Theorem: Jordan Form.   *Suppose A is an $n \times n$ matrix whose characteristic polynomial has distinct roots $\lambda_1, \ldots, \lambda_k$ with multiplicities $m_1, \ldots, m_k$, respectively. Then there is a matrix A', similar to A, so that A' has the block diagonal form*

$$A' = \begin{bmatrix} B_1 & & & \\ & B_2 & & 0\text{'s} \\ & & \cdot & \\ & & & \cdot \\ 0\text{'s} & & & \cdot \\ & & & & B_k \end{bmatrix}$$

where each B_i in turn is of the form

$$B_i = \begin{bmatrix} J_{i1} & & \\ & \cdot & 0\text{'s} \\ & & \cdot \\ 0\text{'s} & & \cdot \\ & & & J_{il_i} \end{bmatrix}$$

with $J_{i1}, \ldots, J_{il_i}$, Jordan blocks having λ_i on the diagonal.

A matrix A' that has the form specified in the theorem is said to be in **Jordan form**. The matrix

$$A' = \begin{bmatrix} 2 & 0 & 0 & \vdots & 0 & 0 & 0 & 0 \\ 1 & 2 & 0 & \vdots & 0 & 0 & 0 & 0 \\ 0 & 1 & 2 & \vdots & 0 & 0 & 0 & 0 \\ \cdots & & & & & & & \\ 0 & 0 & 0 & \vdots & 2 & \vdots & 0 & 0 & 0 \\ 0 & 0 & 0 & & 0 & \vdots & 1 & 0 & \vdots & 0 \\ 0 & 0 & 0 & & 0 & \vdots & 1 & 1 & \vdots & 0 \\ 0 & 0 & 0 & & 0 & & 0 & 0 & \vdots & 3 \end{bmatrix}$$

is an example of a matrix in Jordan form. The dotted lines indicate the Jordan blocks of A'.

If the matrix A' in Jordan form is similar to A, then $A' = {}_{\mathcal{B}'}[T_A]_{\mathcal{B}'}$ for a particularly nice basis $\mathcal{B}' = \{\mathbf{v}_1, \ldots, \mathbf{v}_n\}$ for C^n. To see what this basis looks like, let's suppose that the first Jordan block of A' is $r \times r$ and has λ's on the diagonal:

$$A' = \begin{bmatrix} \lambda & & & & & \vdots & 0 \\ 1 & \cdot & & 0\text{'s} & & \vdots & 0 \\ & \cdot & \cdot & & & \vdots & \\ & & \cdot & \cdot & & \vdots & 0 \\ 0\text{'s} & & & \cdot & \cdot & \vdots & \\ & & & & 1 & \lambda & \vdots \\ \cdots & \cdots & \cdots & \cdots & \cdots & \cdots & \cdots \\ & & 0 & & & \vdots & \end{bmatrix}.$$

Using the definition of $_{\mathscr{B}'}[T_A]_{\mathscr{B}'}$, we see that

$$A\mathbf{v}_1 = \lambda\mathbf{v}_1 + \mathbf{v}_2$$

$$A\mathbf{v}_2 = \lambda\mathbf{v}_2 + \mathbf{v}_3$$

$$\cdot$$
$$\cdot$$
$$\cdot$$

$$A\mathbf{v}_{r-1} = \lambda\mathbf{v}_{r-1} + \mathbf{v}_r$$

$$A\mathbf{v}_r = \lambda\mathbf{v}_r.$$

Therefore

$$(A - \lambda I)\mathbf{v}_1 = \mathbf{v}_2$$

$$(A - \lambda I)\mathbf{v}_2 = \mathbf{v}_3$$

$$\cdot$$
$$\cdot$$
$$\cdot$$

$$(A - \lambda I)\mathbf{v}_{r-1} = \mathbf{v}_r$$

$$(A - \lambda I)\mathbf{v}_r = \mathbf{0}.$$

We refer to any nonzero vectors $\mathbf{v}_1, \ldots, \mathbf{v}_r$ satisfying these equations as a **string of generalized eigenvectors** corresponding to λ. (We leave it to you to check—Exercise 17—that nonzero vectors satisfying these equations are independent generalized eigenvectors.)

If the second Jordan block of A' is $s \times s$ and has μ's on the diagonal (with μ possibly equal to λ), then the next s vectors in $\mathscr{B}'$, $\mathbf{u}_1 = \mathbf{v}_{r+1}$, $\mathbf{u}_2 = \mathbf{v}_{r+2}, \ldots,$ $\mathbf{u}_s = \mathbf{v}_{r+s}$ satisfy

$$(A - \mu I)\mathbf{u}_1 = \mathbf{u}_2$$

.

.

.

$$(A - \mu I)\mathbf{u}_{s-1} = \mathbf{u}_s$$

$$(A - \mu I)\mathbf{u}_s = 0.$$

Thus $\mathbf{u}_1, \ldots, \mathbf{u}_s$ form a string of generalized eigenvectors corresponding to μ.

Continuing in this way, we see that the basis $\mathscr{B}'$ consists of strings of generalized eigenvectors. Thus the existence of a matrix A' in Jordan form, similar to A, is equivalent to the following.

Fact: *Corresponding to each $n \times n$ matrix A, there is a basis for C^n that consists of strings of generalized eigenvectors corresponding to the various eigenvalues $\lambda_1, \ldots, \lambda_k$ of A.*

To find the basis $\mathscr{B}'$, we must find strings of generalized eigenvectors. This is simplified by the observation that if $\mathbf{v}_1, \ldots, \mathbf{v}_r$ form a string corresponding to λ, then

$$(A - \lambda I)\mathbf{v}_1 = \mathbf{v}_2$$
$$(A - \lambda I)^2 \mathbf{v}_1 = (A - \lambda I)\mathbf{v}_2 = \mathbf{v}_3$$

.

.

.

$$(A - \lambda I)^{r-1}\mathbf{v}_1 = \mathbf{v}_r$$
$$(A - \lambda I)^r \mathbf{v}_1 = (A - \lambda I)\mathbf{v}_r = \mathbf{0}.$$

In particular, $\mathbf{v}_1$ *is a solution of* $(A - \lambda I)^r \mathbf{v} = \mathbf{0}$, *but,* $\mathbf{v}_1$ *is not a solution of* $(A - \lambda I)^{r-1}\mathbf{v} = \mathbf{0}$. Once we find $\mathbf{v}_1$, the other vectors are determined:

$$\mathbf{v}_2 = (A - \lambda I)\mathbf{v}_1, \qquad \mathbf{v}_3 = (A - \lambda I)\mathbf{v}_2, \ldots, \mathbf{v}_r = (A - \lambda I)\mathbf{v}_{r-1}.$$

Example 5.6.1

Find a matrix A' in Jordan form similar to the matrix

$$A = \begin{bmatrix} 3 & 2 & 2 & 2 \\ 0 & 2 & 0 & 0 \\ 0 & 0 & 3 & 0 \\ 0 & 1 & 1 & 3 \end{bmatrix}$$

considered in Example 5.5.2.

We know from Example 5.5.2 that $\lambda_1 = 2$ is a root of det $(A - \lambda I)$ with multiplicity $m_1 = 1$ and that each eigenvector of A is a multiple of

$$\mathbf{v}_1 = \begin{bmatrix} 0 \\ -1 \\ 0 \\ 1 \end{bmatrix}.$$

The eigenvalue $\lambda_2 = 3$ is a root of multiplicity $m_2 = 3$. Let's see if we can find a string $\mathbf{u}_1, \mathbf{u}_2, \mathbf{u}_3$ of generalized eigenvectors corresponding to 3. The first of these vectors $\mathbf{u}_1$ has to be a solution of $(A - \lambda I)^3\mathbf{u} = \mathbf{0}$, but we want it *not* to be a solution of $(A - \lambda I)^2\mathbf{u} = \mathbf{0}$. We can determine the solutions of these two systems by reducing their augmented matrices:

$$[(A - 3I)^3 \mid \mathbf{0}] = \begin{bmatrix} 0 & 0 & 0 & 0 & \mid 0 \\ 0 & -1 & 0 & 0 & \mid 0 \\ 0 & 0 & 0 & 0 & \mid 0 \\ 0 & 1 & 0 & 0 & \mid 0 \end{bmatrix} \rightarrow \begin{bmatrix} 0 & 0 & 0 & 0 & \mid 0 \\ 0 & 1 & 0 & 0 & \mid 0 \\ 0 & 0 & 0 & 0 & \mid 0 \\ 0 & 0 & 0 & 0 & \mid 0 \end{bmatrix}$$

$$[(A - 3I)^2 \mid \mathbf{0}] = \begin{bmatrix} 0 & 0 & 2 & 0 & \mid 0 \\ 0 & 1 & 0 & 0 & \mid 0 \\ 0 & 0 & 0 & 0 & \mid 0 \\ 0 & -1 & 0 & 0 & \mid 0 \end{bmatrix} \rightarrow \begin{bmatrix} 0 & 1 & 0 & 0 & \mid 0 \\ 0 & 0 & 1 & 0 & \mid 0 \\ 0 & 0 & 0 & 0 & \mid 0 \\ 0 & 0 & 0 & 0 & \mid 0 \end{bmatrix}.$$

Since we want $\mathbf{u}_1$ to be a solution of $(A - 3I)^3\mathbf{u} = \mathbf{0}$, its second entry must be 0. Since we want it *not* to be a solution of $(A - 3I)^2\mathbf{u} = \mathbf{0}$, we should take its third entry to be nonzero. Let's take

$$\mathbf{u}_1 = \begin{bmatrix} 0 \\ 0 \\ 1 \\ 0 \end{bmatrix}.$$

Now that we have chosen $\mathbf{u}_1$, we calculate

$$\mathbf{u}_2 = (A - 3I)\mathbf{u}_1 = \begin{bmatrix} 0 & 2 & 2 & 2 \\ 0 & -1 & 0 & 0 \\ 0 & 0 & 0 & 0 \\ 0 & 1 & 1 & 0 \end{bmatrix}\begin{bmatrix} 0 \\ 0 \\ 1 \\ 0 \end{bmatrix} = \begin{bmatrix} 2 \\ 0 \\ 0 \\ 1 \end{bmatrix}$$

$$\mathbf{u}_3 = (A - 3I)\mathbf{u}_2 = \begin{bmatrix} 0 & 2 & 2 & 2 \\ 0 & -1 & 0 & 0 \\ 0 & 0 & 0 & 0 \\ 0 & 1 & 1 & 0 \end{bmatrix}\begin{bmatrix} 2 \\ 0 \\ 0 \\ 1 \end{bmatrix} = \begin{bmatrix} 2 \\ 0 \\ 0 \\ 0 \end{bmatrix}.$$

By combining the vectors associated to $\lambda_1 = 2$ and $\lambda_2 = 3$, we obtain a basis $\mathcal{B}' = \{\mathbf{v}_1, \mathbf{u}_1, \mathbf{u}_2, \mathbf{u}_3\}$ for C^4. These vectors were chosen so that

$$T_A(\mathbf{v}_1) = A\mathbf{v}_1 = 2\mathbf{v}_1$$

$$T_A(\mathbf{u}_1) = A\mathbf{u}_1 = 3\mathbf{u}_1 + (A - 3I)\mathbf{u}_1 = 3\mathbf{u}_1 + \mathbf{u}_2$$

$$T_A(\mathbf{u}_2) = A\mathbf{u}_2 = 3\mathbf{u}_2 + (A - 3I)\mathbf{u}_2 = \qquad 3\mathbf{u}_2 + \mathbf{u}_3$$

$$T_A(\mathbf{u}_3) = A\mathbf{u}_3 = 3\mathbf{u}_3 + (A - 3I)\mathbf{u}_3 = \qquad\qquad 3\mathbf{u}_3.$$

Thus

$$A' = {}_{\mathcal{B}'}[T_A]_{\mathcal{B}'} = \begin{bmatrix} 2 & \vdots & 0 & 0 & 0 \\ \hdashline 0 & \vdots & 3 & 0 & 0 \\ 0 & \vdots & 1 & 3 & 0 \\ 0 & \vdots & 0 & 1 & 3 \end{bmatrix}.$$

We note that the matrix

$$P = [\mathbf{v}_1 \ \mathbf{u}_1 \ \mathbf{u}_2 \ \mathbf{u}_3] = \begin{bmatrix} 0 & 0 & 2 & 2 \\ -1 & 0 & 0 & 0 \\ 0 & 1 & 0 & 0 \\ 1 & 0 & 1 & 0 \end{bmatrix}$$

has the property that $A' = P^{-1}AP$.

In the preceding example we were lucky to find a single string of generalized eigenvectors containing as many vectors as the multiplicity of the eigenvalue. This will not always be possible.

Example 5.6.2

Find a matrix A' in Jordan form similar to

$$A = \begin{bmatrix} -5 & 2 & -1 & 2 \\ 0 & -4 & 0 & 0 \\ 0 & 1 & -5 & 1 \\ 1 & 2 & -1 & -2 \end{bmatrix}.$$

The characteristic polynomial $\det (A - \lambda I) = (-4 - \lambda)^4$ has only one root $\lambda = -4$ of multiplicity $m = 4$. We calculate

$$(A + 4I) = \begin{bmatrix} -1 & 2 & -1 & 2 \\ 0 & 0 & 0 & 0 \\ 0 & 1 & -1 & 1 \\ -1 & 2 & -1 & 2 \end{bmatrix}, \qquad (A + 4I)^2 = \begin{bmatrix} -1 & 1 & 0 & 1 \\ 0 & 0 & 0 & 0 \\ -1 & 1 & 0 & 1 \\ -1 & 1 & 0 & 1 \end{bmatrix}$$

and

$$(A + 4I)^3 = (A + 4I)^4 = 0.$$

Since $(A + 4I)^3 = (A + 4I)^4$, the solutions of $(A + 4I)^4 \mathbf{v} = \mathbf{0}$ and $(A + 4I)^3 \mathbf{v} = \mathbf{0}$ are the same. Thus, we cannot find a string of generalized eigenvectors consisting of $r = 4$ vectors. On the other hand, every 4-vector is a solution of $(A + 4I)^3 \mathbf{v} = \mathbf{0}$, while the solutions of $(A + 4I)^2 \mathbf{v} = \mathbf{0}$ consist of the 4-vectors whose first entries are equal to the sum of their second and fourth entries (check this). In particular,

$$\mathbf{v}_1 = \begin{bmatrix} 1 \\ 0 \\ 0 \\ 0 \end{bmatrix}$$

is a solution of $(A + 4I)^3 \mathbf{v} = \mathbf{0}$ that does not solve $(A + 4I)^2 \mathbf{v} = \mathbf{0}$. The vectors

$$\mathbf{v}_1 = \begin{bmatrix} 1 \\ 0 \\ 0 \\ 0 \end{bmatrix}, \qquad \mathbf{v}_2 = (A + 4I)\mathbf{v}_1 = \begin{bmatrix} -1 \\ 0 \\ 0 \\ -1 \end{bmatrix}, \qquad \mathbf{v}_3 = (A + 4I)\mathbf{v}_2 = \begin{bmatrix} -1 \\ 0 \\ -1 \\ -1 \end{bmatrix}$$

form a string of generalized eigenvectors.

To complete our basis for C^4, we must find a string of generalized eigenvectors of length one (that is, an eigenvector) independent of $\mathbf{v}_1$, $\mathbf{v}_2$, $\mathbf{v}_3$. The eigenvectors of A satisfy $(A + 4I)\mathbf{v} = \mathbf{0}$, which we solve by reducing the augmented matrix:

$$[A + 4I | \mathbf{0}] = \begin{bmatrix} -1 & 2 & -1 & 2 & | & 0 \\ 0 & 0 & 0 & 0 & | & 0 \\ 0 & 1 & -1 & 1 & | & 0 \\ -1 & 2 & -1 & 2 & | & 0 \end{bmatrix} \rightarrow \begin{bmatrix} 1 & 0 & -1 & 0 & | & 0 \\ 0 & 1 & -1 & 1 & | & 0 \\ 0 & 0 & 0 & 0 & | & 0 \\ 0 & 0 & 0 & 0 & | & 0 \end{bmatrix}.$$

We want a solution that is independent of $\mathbf{v}_1$, $\mathbf{v}_2$, and $\mathbf{v}_3$. We leave it to you to check that the solution

$$\mathbf{v}_4 = \begin{bmatrix} 0 \\ -1 \\ 0 \\ 1 \end{bmatrix}$$

works.

The vectors in the basis $\mathcal{B}' = \{\mathbf{v}_1, \mathbf{v}_2, \mathbf{v}_3, \mathbf{v}_4\}$ were chosen so that

$$T_A(\mathbf{v}_1) = A\mathbf{v}_1 = -4\mathbf{v}_1 + (A + 4I)\mathbf{v}_1 = -4\mathbf{v}_1 + \mathbf{v}_2$$

$$T_A(\mathbf{v}_2) = A\mathbf{v}_2 = -4\mathbf{v}_2 + (A + 4I)\mathbf{v}_2 = -4\mathbf{v}_2 + \mathbf{v}_3$$

$$T_A(\mathbf{v}_3) = A\mathbf{v}_3 = -4\mathbf{v}_3 + (A + 4I)\mathbf{v}_3 = -4\mathbf{v}_3$$

$$T_A(\mathbf{v}_4) = A\mathbf{v}_4 = -4\mathbf{v}_4.$$

Thus

$$A' = {}_{\mathcal{B}'}[T_A]_{\mathcal{B}'} = \left[\begin{array}{ccc:c} -4 & 0 & 0 & 0 \\ 1 & -4 & 0 & 0 \\ 0 & 1 & -4 & 0 \\ \hdashline 0 & 0 & 0 & -4 \end{array}\right].$$

The matrix

$$P = [\mathbf{v}_1 \ \mathbf{v}_2 \ \mathbf{v}_3 \ \mathbf{v}_4] = \begin{bmatrix} 1 & -1 & -1 & 0 \\ 0 & 0 & 0 & -1 \\ 0 & 0 & -1 & 0 \\ 0 & -1 & -1 & 1 \end{bmatrix}$$

has the property that $P^{-1}AP = A'$.

From the point of view of homogeneous systems of o.d.e.'s $D\mathbf{x} = A\mathbf{x}$, strings of generalized eigenvectors have the advantage that the associated solutions of $D\mathbf{x} = A\mathbf{x}$ are particularly easy to calculate. Recall that if $(A - \lambda I)^m \mathbf{v} = \mathbf{0}$, then the solution of $D\mathbf{x} = A\mathbf{x}$ associated to $\mathbf{v}$ is

$$\mathbf{h}(t) = e^{\lambda t}\left[\mathbf{v} + t(A - \lambda I)\mathbf{v} + \frac{1}{2}t^2(A - \lambda I)^2\mathbf{v} + \dots \right.$$

$$\left. + \frac{1}{(m-1)!}t^{m-1}(A - \lambda I)^{m-1}\mathbf{v}\right].$$

If $\mathbf{v}_1, \ldots, \mathbf{v}_r$ form a string of generalized eigenvectors, then the associated solutions are (see Exercise 15, Section 3.10)

$$\mathbf{h}_1(t) = e^{\lambda t}\left[\mathbf{v}_1 + t\mathbf{v}_2 + \frac{1}{2}t^2\mathbf{v}_3 + \ldots + \frac{1}{(r-1)!}t^{r-1}\mathbf{v}_r\right]$$

$$\mathbf{h}_2(t) = e^{\lambda t}\left[\mathbf{v}_2 + t\mathbf{v}_3 + \ldots + \frac{1}{(r-2)!}t^{r-2}\mathbf{v}_r\right]$$

$$\cdot$$
$$\cdot$$
$$\cdot$$

$$\mathbf{h}_r(t) = e^{\lambda t}\mathbf{v}_r.$$

For example, if A is as in Example 5.6.2, then the solutions of $D\mathbf{x} = A\mathbf{x}$ associated to the vectors $\mathbf{v}_1, \mathbf{v}_2, \mathbf{v}_3, \mathbf{v}_4$ are

$$\mathbf{h}_1(t) = e^{-4t}\left(\begin{bmatrix}1\\0\\0\\0\end{bmatrix} + t\begin{bmatrix}-1\\0\\0\\-1\end{bmatrix} + \frac{1}{2}t^2\begin{bmatrix}-1\\0\\-1\\-1\end{bmatrix}\right) = e^{-4t}\begin{bmatrix}1 - t - \frac{1}{2}t^2\\0\\-\frac{1}{2}t^2\\-t - \frac{1}{2}t^2\end{bmatrix},$$

$$\mathbf{h}_2(t) = e^{-4t}\left(\begin{bmatrix}-1\\0\\0\\-1\end{bmatrix} + t\begin{bmatrix}-1\\0\\-1\\-1\end{bmatrix}\right) = e^{-4t}\begin{bmatrix}-1 - t\\0\\-t\\-1 - t\end{bmatrix},$$

$$\mathbf{h}_3(t) = e^{-4t}\begin{bmatrix}-1\\0\\-1\\-1\end{bmatrix},$$

$$\mathbf{h}_4(t) = e^{-4t}\begin{bmatrix}0\\-1\\0\\1\end{bmatrix},$$

and these four solutions generate the general solution of $D\mathbf{x} = A\mathbf{x}$. We note, however, that the ease of calculating these solutions is offset by the extra work required to find a basis yielding Jordan form (as opposed to a basis consisting of randomly chosen generalized eigenvectors and yielding only block diagonal form).

JORDAN FORM

If $\mathbf{v}_1$ is a solution of $(A - \lambda I)^r\mathbf{v} = \mathbf{0}$, but $\mathbf{v}_1$ is *not* a solution of $(A - \lambda I)^{r-1}\mathbf{v} = \mathbf{0}$, then the vectors

$$\mathbf{v}_1, \qquad \mathbf{v}_2 = (A - \lambda I)\mathbf{v}_1, \ldots, \mathbf{v}_r = (A - \lambda I)\mathbf{v}_{r-1}$$

form a **string of generalized eigenvectors** corresponding to λ.

If A is any $n \times n$ matrix, then there is a basis $\mathcal{B}'$ for C^n consisting of strings of generalized eigenvectors corresponding to the various eigenvalues $\lambda_1, \ldots, \lambda_k$ of A. The matrix $A' = {}_{\mathcal{B}'}[T_A]_{\mathcal{B}'}$ has the block diagonal form

$$A' = \begin{bmatrix} B_1 & & & \\ & B_2 & & 0\text{'s} \\ & & \cdot & \\ & & & \cdot \\ 0\text{'s} & & & \cdot \\ & & & & B_k \end{bmatrix}$$

where each B_i is in turn of the form

$$B_i = \begin{bmatrix} J_{i1} & & \\ & \cdot & 0\text{'s} \\ & & \cdot \\ 0\text{'s} & & \cdot \\ & & & J_{il_i} \end{bmatrix}$$

with

$$J_{ij} = \begin{bmatrix} \lambda_i & & & & \\ 1 & \cdot & & 0\text{'s} & \\ & \cdot & \cdot & & \\ & & \cdot & \cdot & \\ 0\text{'s} & & & \cdot & \\ & & & 1 & \lambda_i \end{bmatrix}.$$

The matrix A' is said to be in **Jordan form**, and the matrices J_{ij} are called the **Jordan blocks** of A'.

EXERCISES

In Exercises 1 to 16, you are given a matrix A that was previously considered in the indicated exercise.
a. Find a matrix A' in Jordan form that is similar to A.
b. Find an invertible matrix P such that $A' = P^{-1}AP$.
c. Find the general solution of $D\mathbf{x} = A\mathbf{x}$.
Note that the matrices in Exercises 13 to 16 have complex eigenvalues.

1. $A = \begin{bmatrix} 1 & -1 \\ 1 & 3 \end{bmatrix}$ (Exercise 3, Section 3.9)

2. $A = \begin{bmatrix} -3 & 1 \\ -1 & -1 \end{bmatrix}$ (Exercise 4, Section 3.9)

3. $A = \begin{bmatrix} 1 & -1 & -1 \\ 0 & -1 & -1 \\ 0 & 0 & 1 \end{bmatrix}$ (Exercise 5, Section 3.9)

4. $A = \begin{bmatrix} 2 & -1 & -4 \\ 0 & 2 & -4 \\ 0 & 1 & -2 \end{bmatrix}$ (Exercise 6, Section 3.9)

5. $A = \begin{bmatrix} -2 & 0 & 0 & 0 \\ 0 & -2 & 2 & -1 \\ 0 & 0 & -4 & 9 \\ 0 & 0 & -4 & 8 \end{bmatrix}$ (Exercise 8, Section 3.9)

6. $A = \begin{bmatrix} 0 & 1 & 0 \\ 0 & 0 & 1 \\ 1 & -3 & 3 \end{bmatrix}$ (Exercise 1, Section 3.10, and Exercise 9, Section 5.5)

7. $A = \begin{bmatrix} 1 & 0 & 0 \\ 1 & 0 & 1 \\ 1 & -1 & 2 \end{bmatrix}$ (Exercise 2, Section 3.10, and Exercise 10, Section 5.5)

8. $A = \begin{bmatrix} 1 & 1 & 0 & 0 \\ 0 & 1 & 1 & 0 \\ 0 & 0 & 1 & -1 \\ 0 & 0 & 0 & 2 \end{bmatrix}$ (Exercise 4, Section 3.10, and Exercise 11, Section 5.5)

9. $A = \begin{bmatrix} 2 & 0 & -1 & 1 \\ 0 & 0 & -1 & 0 \\ 0 & -1 & 0 & 0 \\ -1 & -1 & 0 & 0 \end{bmatrix}$ (Exercise 3, Section 3.10, and Exercise 12, Section 5.5)

10. $A = \begin{bmatrix} 2 & 1 & 0 & 0 \\ 0 & 1 & 1 & 0 \\ -1 & -1 & 0 & 0 \\ 0 & 0 & 0 & 1 \end{bmatrix}$ (Exercise 6, Section 3.10, and Exercise 13, Section 5.5)

11. $A = \begin{bmatrix} 2 & 0 & 0 & 0 \\ 0 & 2 & 0 & 0 \\ 1 & 0 & 1 & 0 \\ 1 & 0 & -1 & 3 \end{bmatrix}$ (Exercise 7, Section 3.10, and Exercise 14, Section 5.5)

12. $A = \begin{bmatrix} 0 & 0 & 0 & 0 & 1 \\ 1 & -1 & 0 & -1 & 1 \\ 1 & 0 & -1 & -1 & 0 \\ 0 & 0 & 0 & 0 & 1 \\ 1 & 0 & 0 & -1 & 0 \end{bmatrix}$ (Exercise 9, Section 3.10, and Exercise 15, Section 5.5)

13. $A = \begin{bmatrix} 0 & 2 & 0 & 0 \\ -2 & 0 & 1 & 0 \\ 0 & 0 & 0 & 2 \\ 0 & 0 & -2 & 0 \end{bmatrix}$ (Exercise 5, Section 3.10, and Exercise 16, Section 5.5)

14. $A = \begin{bmatrix} 1 & 1 & 0 & 0 \\ -1 & 1 & 1 & 0 \\ 0 & 0 & 1 & -1 \\ 0 & 0 & 1 & 1 \end{bmatrix}$ (Exercise 8, Section 3.10, and Exercise 17, Section 5.5)

15. $A = \begin{bmatrix} 2 & 1 & 0 & 0 & 0 & 0 \\ 0 & 2 & 1 & 0 & 0 & 0 \\ 0 & 0 & 2 & 1 & 0 & 0 \\ 0 & 0 & 0 & 2 & 0 & 0 \\ 0 & 0 & 0 & 0 & 0 & 1 \\ 0 & 0 & 0 & 0 & -2 & -2 \end{bmatrix}$ (Exercise 10, Section 3.10, and Exercise 18, Section 5.5)

16. $A = \begin{bmatrix} 2 & 0 & 1 & 0 \\ 0 & 1 & 0 & 1 \\ 0 & 0 & 2 & 1 \\ 0 & -1 & 0 & 1 \end{bmatrix}$ (Exercise 7, Section 3.9)

More abstract exercises:

17. Suppose that A is an $n \times n$ matrix and that the nonzero vectors $v_1, \ldots, v_r$ satisfy

$$(A - \lambda I)v_1 = v_2, \ldots, (A - \lambda I)v_{r-1} = v_r, \qquad (A - \lambda I)v_r = 0.$$

   a. Show that $(A - \lambda I)^r v_i = 0$ for $i = 1, 2, \ldots, r$.
   b. Show that $v_1, \ldots, v_r$ are linearly independent (compare Exercise 15, Section 3.10).

18. a. An $n \times n$ matrix B is *lower triangular* if every entry above the diagonal is zero:

$$B = \begin{bmatrix} b_{11} & & 0's \\ & \cdot & \cdot \\ \cdot & & \cdot \\ b_{n1} & \cdots & b_{nn} \end{bmatrix}.$$

Show that the determinant of a lower triangular matrix is the product of its diagonal entries:

$$\det B = b_{11} b_{22} \ldots b_{nn}.$$

   (*Hint:* Expand repeatedly by minors along the last column.)
   b. Using (a), show that if A' is an $n \times n$ matrix in Jordan form, then its characteristic polynomial is

$$P(\lambda) = (d_{11} - \lambda)(d_{22} - \lambda) \ldots (d_{nn} - \lambda)$$

where the d_{ii}'s are the diagonal entries of A'.

19. Suppose A' is a matrix in Jordan form that is similar to A.
   a. Use Exercise 26 of Section 5.4 and Exercise 18 to show that

$$\det (A - \lambda I) = (d_{11} - \lambda)(d_{22} - \lambda) \ldots (d_{nn} - \lambda)$$

where the d_{ii}'s are the diagonal entries of A'.

FIVE / MATRIX REPRESENTATION OF LINEAR TRANSFORMATIONS

b. Conclude that each eigenvalue of A appears on the diagonal of A' as often as its multiplicity as a root of det $(A - \lambda I)$.

REVIEW PROBLEMS

In Exercises 1 to 5, you are given a linear transformation T from the vector space V to itself and two bases $\mathcal{B}$ and $\mathcal{B}'$ for V.

a. Find $_{\mathcal{B}}[T]_{\mathcal{B}}$.

b. Use the change-of-basis formula to find $_{\mathcal{B}'}[T]_{\mathcal{B}'}$.

c. Find a matrix in block diagonal form similar to the matrices in (a) and (b).

d. Find a matrix in Jordan normal form similar to the matrices in (a) and (b).

1. $V = R^3$

$$\mathcal{B} = \left\{ \begin{bmatrix} 1 \\ 0 \\ 0 \end{bmatrix}, \begin{bmatrix} 0 \\ 1 \\ 0 \end{bmatrix}, \begin{bmatrix} 0 \\ 0 \\ 1 \end{bmatrix} \right\}$$

$$\mathcal{B}' = \left\{ \begin{bmatrix} 1 \\ 1 \\ 1 \end{bmatrix}, \begin{bmatrix} 1 \\ 1 \\ 0 \end{bmatrix}, \begin{bmatrix} 1 \\ 0 \\ 0 \end{bmatrix} \right\}$$

$$T\left(\begin{bmatrix} x \\ y \\ z \end{bmatrix} \right) = \begin{bmatrix} x - 2y + z \\ x - 5y - z \\ 2y \end{bmatrix}$$

2. $V = R^3$

$$\mathcal{B} = \left\{ \begin{bmatrix} 1 \\ 0 \\ 0 \end{bmatrix}, \begin{bmatrix} 0 \\ 1 \\ 0 \end{bmatrix}, \begin{bmatrix} 0 \\ 0 \\ 1 \end{bmatrix} \right\}$$

$$\mathcal{B}' = \left\{ \begin{bmatrix} 1 \\ 1 \\ 1 \end{bmatrix}, \begin{bmatrix} 1 \\ 1 \\ 0 \end{bmatrix}, \begin{bmatrix} 1 \\ 0 \\ 0 \end{bmatrix} \right\}$$

$$T = T_A, \text{ where } A = \begin{bmatrix} 2 & 1 & 1 \\ 0 & 2 & 1 \\ 0 & 0 & 2 \end{bmatrix}$$

3. $V = P_3$

$\mathcal{B} = \{1, t, t^2, t^3\}$

$\mathcal{B}' = \{1, t + 1, t^2 - t, t^3 + t^2 + t\}$

$T(x) = Dx$

4. $V = P_3$

$\mathcal{B} = \{1, t, t^2, t^3\}$

$\mathcal{B}' = \{1, t + 1, t^2 - t, t^3 + t^2 + t\}$

$T(x(t)) = x(t - 1)$

5. $V = R_{2 \times 2}$

$$\mathcal{B} = \left\{ \begin{bmatrix} 1 & 0 \\ 0 & 0 \end{bmatrix}, \begin{bmatrix} 0 & 1 \\ 0 & 0 \end{bmatrix}, \begin{bmatrix} 0 & 0 \\ 1 & 0 \end{bmatrix}, \begin{bmatrix} 0 & 0 \\ 0 & 1 \end{bmatrix} \right\}$$

$$\mathcal{B}' = \left\{ \begin{bmatrix} 1 & 0 \\ 0 & 1 \end{bmatrix}, \begin{bmatrix} 1 & 0 \\ 0 & -1 \end{bmatrix}, \begin{bmatrix} 0 & 1 \\ 1 & 0 \end{bmatrix}, \begin{bmatrix} 0 & 1 \\ -1 & 0 \end{bmatrix} \right\}$$

$$T(M) = Q^{-1}MQ, \quad \text{where} \quad Q = \begin{bmatrix} 0 & 1 \\ 1 & 0 \end{bmatrix}$$

In Exercises 6 to 12, you are given a matrix A.

a. Find a matrix B that is in block diagonal form and similar to A.

b. Find a matrix J that is in Jordan form and similar to A.

c. Find the general solution of $D\mathbf{x} = A\mathbf{x}$.

6. $A = \begin{bmatrix} 0 & -1 & -1 \\ 4 & -4 & -4 \\ 0 & 0 & 0 \end{bmatrix}$.

7. $A = \begin{bmatrix} 2 & -2 & 2 \\ 0 & 1 & 1 \\ 0 & -1 & 3 \end{bmatrix}$.

8. $A = \begin{bmatrix} 1 & -2 & 3 \\ 0 & 1 & 1 \\ -1 & -1 & 4 \end{bmatrix}$.

9. $A = \begin{bmatrix} 2 & 0 & 1 \\ -1 & 1 & -1 \\ 0 & 1 & 2 \end{bmatrix}$.

10. $A = \begin{bmatrix} 1 & 0 & 0 & 0 & 0 \\ 1 & 1 & 0 & 0 & 0 \\ 2 & 1 & 1 & 0 & -1 \\ 2 & 2 & 0 & -1 & 0 \\ 0 & 0 & 0 & 0 & -1 \end{bmatrix}$.

11. $A = \begin{bmatrix} 1 & 0 & 0 & 0 & 0 & 0 \\ 1 & 1 & 0 & 0 & 0 & 0 \\ 2 & 0 & 1 & 0 & 0 & 2 \\ 0 & 0 & 1 & 1 & 0 & 1 \\ -1 & -2 & 0 & 0 & -1 & 0 \\ -2 & 0 & 0 & 0 & 0 & -1 \end{bmatrix}$.

12. $A = \begin{bmatrix} 1 & 0 & 0 & 0 & 0 & 0 \\ 1 & 1 & 0 & 0 & 0 & 0 \\ 0 & 1 & 1 & 0 & 0 & 1 \\ 0 & 0 & 0 & 1 & 0 & 0 \\ -1 & 0 & 0 & 1 & 1 & 0 \\ 0 & 0 & 0 & 0 & 0 & 1 \end{bmatrix}$.

The Laplace Transform

6.1 OLD MODELS FROM A NEW VIEWPOINT

With the exception of the graphing technique in Section 1.4, our methods so far have aimed at finding a list of *all* solutions of an o.d.e. or system. To find the specific solution satisfying given initial conditions, we have had to find the general solution first and then solve for the coefficients to fit our initial data. While the general solution gives us a better picture of the possible patterns of behavior for a model, concrete problems often require only a specific solution. The technique we consider in this chapter solves initial value problems directly, without requiring the general solution. An added bonus from this method is its effectiveness in handling forcing terms defined "in pieces."

In this section we will reexamine several models from earlier chapters to see how initial value problems and forcing terms defined in pieces arise. The reader will note that every problem we pose could be solved by methods from Chapters 1 to 3. However, by the end of the chapter it should be clear that the Laplace transform method often leads to specific solutions much more simply than our old methods.

Example 6.1.1   Controlled Immigration

In Section 1.1 we considered several specific models for populations with given growth and immigration rates. In general, a natural growth rate $g(t)$ together with an immigration rate $E(t)$ is modeled by the first-order o.d.e.

(N) $$Dx = g(t)x + E(t).$$

Suppose the population of Mania has a natural growth rate of 5% per year. From January 1965 to January 1975, government policy allowed immigration into Mania at the rate of $(1/1000)e^{t/20}$ million per year (t measured in years since 1965). Then an abrupt change in policy led to the sealing of all borders and an end to immigration. Since January 1965, when the population was five million, the Maniac government has kept the census figures secret. We would like to estimate the population of Mania in January 1985.

411

To use our old methods, we break the problem into two parts. First, we consider equation (N) with $g(t) = 1/20$ and $E(t) = (1/1000)e^{t/20}$:

(N$_1$)
$$\left(D - \frac{1}{20}\right)x = \frac{1}{1000}\,e^{t/20}.$$

We find the general solution of (N$_1$) and then determine the specific solution satisfying the initial condition $x(0) = 5$. You can check that the solution of this initial value problem is

$$x(t) = 5e^{t/20} + \frac{t}{1000}\,e^{t/20}.$$

This describes the population from January 1965 to January 1975. In particular, the population in January 1975 was

$$x(10) = 5e^{1/2} + \frac{1}{100}\,e^{1/2} \approx 8.26 \text{ million.}$$

Now, to study the population since January 1975, we consider equation (N) with $g(t) = 1/20$ and $E(t) = 0$:

(N$_2$)
$$\left(D - \frac{1}{20}\right)x = 0.$$

For a prediction of the 1985 population, we must find the specific solution of (N$_2$) satisfying the initial condition $x(10) = 8.26$. The solution of this initial value problem is

$$x(t) = 8.26\,e^{(t-10)/20}.$$

The population in January 1985 is

$$x(20) = 8.26\,e^{1/2} \approx 13.62.$$

The Laplace transform gives the same prediction (see Exercise 34, Section 6.5) by attacking directly the o.d.e. that models the behavior of the Maniac population:

(N$_3$)
$$\left(D - \frac{1}{20}\right)x = \begin{cases} \dfrac{1}{1000}\,e^{t/20}, & t < 10 \\[2mm] 0, & t > 10 \end{cases}$$

with initial condition

$$x(0) = 5.$$

The right hand side of (N₃) is a typical example of a function defined in pieces; its graph is sketched in Figure 6.1.

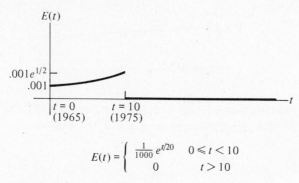

$$E(t) = \begin{cases} \frac{1}{1000} e^{t/20} & 0 \leq t < 10 \\ 0 & t > 10 \end{cases}$$

FIGURE 6.1

Example 6.1.2 Damped Forced Springs

Recall from Section 2.1 that a mass m attached to a horizontal spring with constant k, encountering damping b and driven by the external force $E(t)$, is modeled by the equation

$$(mD^2 + bD + k)x = E(t).$$

Specific physical problems are naturally formulated as initial value problems. For example, if the mass is pulled out p units from equilibrium and let go, the initial data are

$$x(0) = p, \qquad x'(0) = 0.$$

If the mass starts from the equilibrium position but is nudged so that the spring is contracting at a rate q, the initial data are

$$x(0) = 0, \qquad x'(0) = -q.$$

Certain physical solutions lead to forcing terms defined in pieces. If the mass is constrained to move along a shaft that is spun (see Figure 6.2), then the forcing term comes from centrifugal force. Assume the shaft is at rest until time t_1, spins

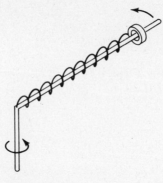

FIGURE 6.2

steadily faster between times t_1 and t_2 (when the force reaches F), and continues spinning at a constant angular velocity from time t_2 on. Then

$$E(t) = \begin{cases} 0, & t < t_1 \\ F(t - t_1)/(t_2 - t_1), & t_1 < t < t_2 \\ F, & t > t_2 \end{cases}$$

(see Figure 6.3). Note that our old methods would require us to solve three successive initial value problems to predict the long-term behavior of this model.

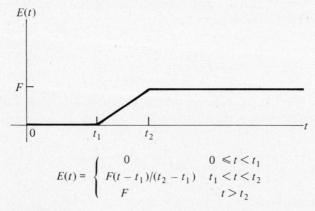

$$E(t) = \begin{cases} 0 & 0 \leqslant t < t_1 \\ F(t - t_1)/(t_2 - t_1) & t_1 < t < t_2 \\ F & t > t_2 \end{cases}$$

FIGURE 6.3

Example 6.1.3 An *LRC* Circuit

The circuit sketched in Figure 6.4 has one loop involving an inductance L, a resistance R, a capacitance C, and a time-dependent voltage $V(t)$. The charge Q on

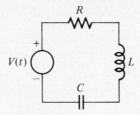

FIGURE 6.4

the capacitor is modeled (see Section 3.1 and Exercise 3 in Section 2.1) by the o.d.e.

$$\left(LD^2 + RD + \frac{1}{C}\right)Q = V(t).$$

An initial charge q_0 and initial current i_0 are formulated as the initial conditions

$$Q(0) = q_0, \qquad Q'(0) = i_0.$$

If the circuit is initially unforced, but is plugged into an alternating voltage source $V(t) = E \sin \beta t$ at time t_1, then $V(t)$ is defined in pieces by

$$V(t) = \begin{cases} 0, & t < t_1 \\ E \sin \beta t, & t > t_1. \end{cases}$$

The graph of this function is sketched in Figure 6.5.

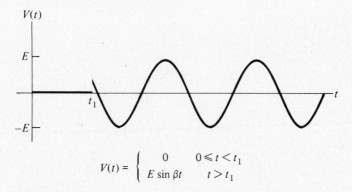

$$V(t) = \begin{cases} 0 & 0 \leqslant t < t_1 \\ E \sin \beta t & t > t_1 \end{cases}$$

FIGURE 6.5

While alternating current or voltage is often modeled by sinusoidal functions, a different model is provided by the **square wave**, sketched in Figure 6.6. The square

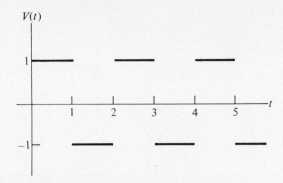

FIGURE 6.6 SQUARE WAVE.

wave is defined in many pieces by

$$V(t) = \begin{cases} +1 & \text{if } n < t < n + 1, n \text{ even} \\ -1 & \text{if } n < t < n + 1, n \text{ odd.} \end{cases}$$

Our final examples deal with systems of o.d.e.'s.

Example 6.1.4   A Multiloop Circuit with a Switch

The circuit in Figure 6.7 is that of Example 3.1.3, except that a switch is provided to allow us to change the voltage source from $V_1(t)$ (position "a") to $V_2(t)$ (position "b"). To model a circuit that has the switch at "a" until time t_1 and at "b" from then on, we can use the equations of Example 3.1.3,

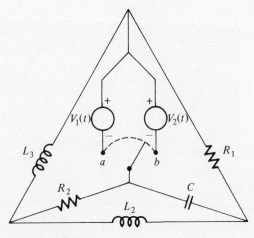

FIGURE 6.7

$$DQ = -\frac{1}{R_1 C} Q - I_2 \qquad\qquad + \frac{V(t)}{R_1}$$

$$DI_2 = \frac{1}{L_2 C} Q - \frac{R_2}{L_2} I_2 + \frac{R_2}{L_2} I_3$$

$$DI_3 = \qquad\qquad \frac{R_2}{L_3} I_2 - \frac{R_2}{L_3} I_3 - \frac{V(t)}{L_3}$$

with $V(t)$ defined in pieces by

$$V(t) = \begin{cases} V_1(t), & t < t_1 \\ V_2(t), & t > t_1. \end{cases}$$

To apply the methods of Chapter 3 to this circuit, we would have to solve two initial value problems, each involving a third-order system. The Laplace transform method will let us handle only one initial value problem involving a single system.

Example 6.1.5   Forced Coupled Springs

The physical setup in Figure 6.8 is that of Example 2.1.4, but with an external force $E(t)$ pulling the mass m_2. The force analysis in Example 2.1.4 is easily modified to take account of this extra force and leads to two second-order o.d.e.'s,

$$m_1 D^2 x_1 = -(k_1 + k_2)x_1 + k_2 x_2$$
$$m_2 D^2 x_2 = \qquad k_2 x_1 - k_2 x_2 + E(t).$$

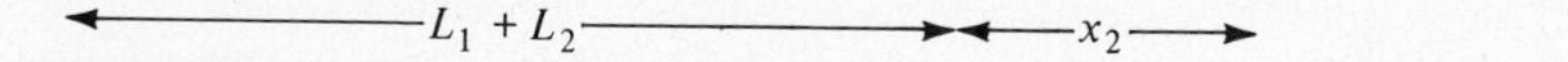

FIGURE 6.8

Recall that in Section 2.1 we worked hard to change this system into a single fourth-order o.d.e. for x_1. An adaptation of the methods of Chapter 3 would require that we introduce two new variables ($v_1 = Dx_1$ and $v_2 = Dx_2$) and study an equivalent system of four first-order o.d.e.'s (for x_1, v_1, x_2, and v_2). The Laplace transform requires no preliminary changes; it can be applied directly to a system of higher-order o.d.e.'s.

EXERCISES

Your final answer to each problem should consist of an o.d.e. (or system of o.d.e.'s) together with a set of initial conditions at $t = 0$ for the variable(s) in the problem.

1. *A Savings Account:* Suppose the savings account of Exercise 19, Section 1.1, was opened with an initial deposit of $1000. After two years the income from the other investment increased to $500 per year. Set up an initial value problem modeling the growth of the account.

2. *Ice Water:* Suppose that when $t = 0$ the water in a tank is at 70°F while the surroundings are at 40°F. Assume that the water loses heat according to Newton's law of cooling (Exercise 21, Section 1.1) with constant of proportionality $\gamma = 1/10$. Suppose the temperature of the surroundings decreases at 2° per hour until it reaches 30°F and remains at 30°F thereafter. Write an initial value problem to model the temperature of the water.

3. *A Tipped Spring:* Suppose a 16-pound weight is attached to a frictionless horizontal spring system with constant $k = 2$. At $t = 0$ the spring is compressed $1/2$ foot and released. After 2 seconds, the system is tipped so it hangs vertically. Write an initial value problem for the amount $x = x(t)$ by which the spring is stretched. Be sure to take into account that once the system is tipped, gravity acts against the spring.

4. *A Vertical Spring:* Suppose a 16-pound weight is hung from a vertical spring with natural length $L = 3$ feet and constant $k = 4$. The only forces acting on the weight are gravity, the restoring force of the spring, and friction with damping constant $b = 3$. Denote the amount by which the spring is stretched by $x = x(t)$.
   a. Find the equilibrium position of the spring. (*Hint:* In this position, the restoring force exactly balances the gravitational force.)
   b. Suppose that at $t = 0$ the mass is raised $1/2$ foot above the equilibrium position and released. Set up an initial value problem modeling the behavior of the weight.
   c. Suppose that at $t = 0$ the mass is $1/2$ foot below the equilibrium position and moving upward at $1/2$ foot per second. Set up an initial value problem modeling the behavior of the weight.

5. *Another LRC Circuit:* The voltage source in the circuit of Example 6.1.3 is turned on at $t = 0$, at which time there is no current or charge. The voltage increases at 1 volt per second until it reaches 10 volts; it stays at 10 volts thereafter. Find an o.d.e. for the charge Q on the capacitor and formulate an initial value problem modeling this situation.

6. *An LRC Circuit with a Switch:* The switch in the circuit of Figure 6.9 allows us to change the voltage source from $V_1(t)$ to $V_2(t)$. At $t = 0$ there is no charge or current, and $V_1(t)$ is connected. After 2 seconds, the voltage source is changed to $V_2(t)$ and, 3 seconds later, back to $V_1(t)$. Find an o.d.e. for the charge Q on the capacitor and formulate an initial value problem to model the behavior of the circuit.

7. *A Moving Spring:* Suppose that at $t = 0$ the weight suspended from the elevator of Exercise 5, Section 2.1, is at rest, even with the ground floor. Suppose further that the spring has natural length $L = 10$ feet and spring constant $k = 2$ and that the damping constant is $b = 2$. Set up an initial value problem to model the height of the weight if at $t = 0$ the bottom of the elevator is (a) 16 feet above ground level or (b) 20 feet above ground level.

8. A fisherman attaches a baited hook to the bottom of the floating box of Exercise 1, Section 2.1, pushes the box down so its bottom is $1/4$ ft from the surface of the water, and releases it. After 3 seconds, a fish bites and pulls straight down with a force of 2 pounds. Find an initial value problem modeling the motion of the box, ignoring friction.

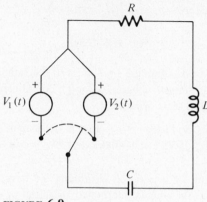

FIGURE 6.9

Exercises 9 and 10 involve systems of o.d.e.'s.

9. *A Two-Loop Circuit:* Suppose that at $t = 0$ the currents through the resistors of the circuit in Example 3.1.2 are $I_1(0) = I_2(0) = 0$. Find $Q(0)$ if $R_1 = 1$ ohm, $R_2 = 3$ ohms, $L = 1$ henry, $C = 1$ farad, and $V(t) = 10$ volts, and formulate the initial value problem modeling this circuit.

10. *Radioactive Decay:* Suppose the radioactive substances A and B decay into substances B and C at the rates of $k_A = 50\%$ and $k_B = 10\%$ per year, as in Exercise 2, Section 3.1. At $t = 0$, there are 500 grams of each of the three substances. Each year, 250 grams of substance A are steadily added to the mix. During the first year, 100 grams of substance C are extracted. From then on, 125 grams per year of substance C are extracted. Write an initial value problem modeling the amounts of substances A, B, and C.

6.2 DEFINITIONS AND BASIC CALCULATIONS

In this section we introduce the Laplace transform, which will enable us in later sections to solve initial value problems directly without requiring the general solution. Philosophically, the Laplace transform technique is analogous to the method of undetermined coefficients considered in Section 2.9. There, we solved a nonhomogeneous equation by applying an appropriate operator (an annihilator of the forcing term) to both sides. This changed the original problem into another (a higher-order homogeneous equation), which we solved using algebra (factoring the new characteristic polynomial). We then used the solution of the new problem to find the solution of the original one. Similarly, the Laplace transform is an operator that, when applied to both sides of a (linear, constant-coefficient) differential equation or system, with initial conditions, yields an algebra problem. We will use the solution of this algebra problem to find the desired solution of the initial value o.d.e. problem.

Definition: *The **Laplace transform** assigns to the function $f(t)$ a new function, $F(s)$, defined by the formula*

$$F(s) = \int_0^\infty e^{-st} f(t) \, dt.$$

F(s) is called the Laplace transform of f(t), and we write

$$F(s) = \mathcal{L}[f(t)].$$

At first sight, the formula defining $F(s) = \mathcal{L}[f(t)]$ is rather imposing. It involves two variables, an infinite limit of integration, and a complicated integrand. But let's take a closer look at each feature separately.

i.   The formula defines a *function F(s)*—that is, it assigns a number $F(s)$ to each numerical value of s. This means that *s acts like a constant inside the integral sign*.

ii. Since the upper limit of integration is infinite, we have an *improper integral*. Recall from calculus that this is interpreted as a limit of proper integrals:

$$\int_0^\infty e^{-st}f(t)\,dt = \lim_{h\to\infty}\left[\int_0^h e^{-st}f(t)\,dt\right].$$

iii. Even though the integrand may seem complicated at first, the formulas for the transforms of the functions we most often deal with turn out to be quite reasonable. What's more, we only have to derive each of them once.

To give some feeling for the Laplace transform, let's calculate some specific examples.

Example 6.2.1
Find $\mathcal{L}[e^{\lambda t}]$.
The definition tells us that

$$\mathcal{L}[e^{\lambda t}] = \int_0^\infty e^{-st}e^{\lambda t}\,dt = \lim_{h\to\infty}\int_0^h e^{-(s-\lambda)t}\,dt.$$

We obtain different values for the integral in the last expression, depending on whether or not $s - \lambda = 0$:

$$\mathcal{L}[e^{\lambda t}] = \lim_{h\to\infty}\begin{cases} h & \text{if } s - \lambda = 0 \\ -\dfrac{e^{-(s-\lambda)h}}{s-\lambda} + \dfrac{1}{s-\lambda} & \text{if } s - \lambda \neq 0. \end{cases}$$

This limit makes sense only for $s - \lambda > 0$. Thus, the function $F(s)$ is defined only for $s > \lambda$, and

$$F(s) = \mathcal{L}[e^{\lambda t}] = \frac{1}{s-\lambda} \qquad \text{for } s > \lambda.$$

Note that when $\lambda = 0$ this reads

$$\mathscr{L}[1] = \frac{1}{s} \qquad \text{for } s > 0.$$

Example 6.2.2

Find $\mathscr{L}[\cos \beta t]$, $\beta \neq 0$.

The definition says

$$\mathscr{L}[\cos \beta t] = \int_0^\infty e^{-st} \cos \beta t \, dt = \lim_{h \to \infty} \int_0^h e^{-st} \cos \beta t \, dt.$$

To evaluate the integral, we use integration by parts twice (first with $u = e^{-st}$ and $dv = \cos \beta t \, dt$, then with $U = e^{-st}$ and $dV = \sin \beta t \, dt$):

$$\int_0^h e^{-st} \cos \beta t \, dt = \frac{e^{-st} \sin \beta t}{\beta} \Bigg]_0^h + \frac{s}{\beta} \int_0^h e^{-st} \sin \beta t \, dt$$

$$= \frac{e^{-st} \sin \beta t}{\beta} \Bigg]_0^h - \frac{se^{-st} \cos \beta t}{\beta^2} \Bigg]_0^h - \frac{s^2}{\beta^2} \int_0^h e^{-st} \cos \beta t \, dt.$$

We solve this equation for the integral:

$$\int_0^h e^{-st} \cos \beta t \, dt = \frac{\beta e^{-st} \sin \beta t}{s^2 + \beta^2} \Bigg]_0^h - \frac{se^{-st} \cos \beta t}{s^2 + \beta^2} \Bigg]_0^h$$

$$= \frac{e^{-sh}[\beta \sin(\beta h) - s \cos(\beta h)]}{s^2 + \beta^2} + \frac{s}{s^2 + \beta^2}$$

Taking limits as $h \to \infty$, we get

$$\mathscr{L}[\cos \beta t] = \frac{s}{s^2 + \beta^2} \qquad \text{for } s > 0.$$

Example 6.2.3

Find $\mathscr{L}[\sin \beta t]$, $\beta \neq 0$.

We could do this problem by mimicking the previous example. But instead, note that the first integration by parts in that example gave the equation

$$\int_0^h e^{-st} \cos \beta t \, dt = \frac{e^{-st} \sin \beta t}{\beta} \Bigg]_0^h + \frac{s}{\beta} \int_0^h e^{-st} \sin \beta t \, dt,$$

which we can solve for the integral on the right-hand side (after carrying out the indicated evaluation of the first term):

$$\int_0^h e^{-st} \sin \beta t \, dt = \frac{-\beta}{s} \frac{e^{-sh} \sin \beta h}{\beta} + \frac{\beta}{s} \int_0^h e^{-st} \cos \beta t \, dt.$$

As $h \to \infty$, the left side of this equation approaches $\mathscr{L}[\sin \beta t]$; if $s > 0$, then the first term on the right goes to 0, while the second tends to β/s times $\mathscr{L}[\cos \beta t]$. Hence

$$\mathscr{L}[\sin \beta t] = \frac{\beta}{s} \mathscr{L}[\cos \beta t] = \frac{\beta}{s} \frac{s}{s^2 + \beta^2} = \frac{\beta}{s^2 + \beta^2}, \qquad s > 0.$$

Example 6.2.4

Find $\mathscr{L}[t^n]$ when n is a positive integer.

By definition,

$$\mathscr{L}[t^n] = \int_0^\infty e^{-st} t^n \, dt = \lim_{h \to \infty} \int_0^h e^{-st} t^n \, dt.$$

Integration by parts (with $u = t^n$ and $dv = e^{-st} \, dt$) gives

$$\int_0^h e^{-st} t^n \, dt = \frac{-h^n e^{-sh}}{s} + \frac{n}{s} \int_0^h e^{-st} t^{n-1} \, dt.$$

As $h \to \infty$, the left side approaches $\mathscr{L}[t^n]$. If $s > 0$, then by L'Hôpital's rule the first term on the right approaches 0, while the second approaches n/s times $\mathscr{L}[t^{n-1}]$. Thus

$$\mathscr{L}[t^n] = \frac{n}{s} \mathscr{L}[t^{n-1}] \qquad \text{for } s > 0.$$

Repeated use of this reduction formula ultimately gives us a formula for $\mathscr{L}[t^n]$ in terms of $\mathscr{L}[t^0] = \mathscr{L}[1]$, which we already know from Example 6.2.1:

$$\mathscr{L}[t^n] = \frac{n}{s} \mathscr{L}[t^{n-1}] = \frac{n(n-1)}{s^2} \mathscr{L}[t^{n-2}] = \ldots = \frac{n!}{s^n} \mathscr{L}[1]$$

$$= \frac{n!}{s^{n+1}} \qquad \text{for } s > 0.$$

Example 6.2.5

Find $\mathcal{L}[5e^{2t} - t^3]$.

The definition says

$$\mathcal{L}[5e^{2t} - t^3] = \int_0^\infty e^{-st}(5e^{2t} - t^3) \, dt.$$

We could use integration by parts to evaluate this integral, but familiar properties of the integral allow us to break it up instead:

$$\mathcal{L}[5e^{2t} - t^3] = 5\int_0^\infty e^{-st}e^{2t} \, dt - \int_0^\infty e^{-st}t^3 \, dt.$$

The integrals on the right side are themselves Laplace transforms—$\mathcal{L}[e^{2t}]$ and $\mathcal{L}[t^3]$, respectively. We can use the results of Examples 6.2.1 and 6.2.4. to find these, so that

$$\mathcal{L}[5e^{2t} - t^3] = 5\,\mathcal{L}[e^{2t}] - \mathcal{L}[t^3] = \frac{5}{s-2} - \frac{6}{s^4}.$$

Note that $\mathcal{L}[e^{2t}]$ is defined only for $s > 2$, while $\mathcal{L}[t^3]$ is defined for $s > 0$. Thus, our new formula will be valid for all values of s that are higher than both 0 and 2— that is, for $s > 2$.

The observation contained in the last example applies to the transforms of linear combinations of any known functions. We state it formally:

Fact: *The Laplace transform is linear. That is, for any two functions $f_1(t)$ and $f_2(t)$ and constants c_1 and c_2,*

$$\mathcal{L}[c_1 f_1(t) + c_2 f_2(t)] = c_1\,\mathcal{L}[f_1(t)] + c_2\,\mathcal{L}[f_2(t)].$$

Example 6.2.5 illustrates how linearity, together with knowledge of the transforms of a few basic functions, enables us to calculate the transforms of their linear combinations. Let's look at another such example.

Example 6.2.6

Find $\mathcal{L}[3 - e^{-3t} + 5 \sin 2t]$.

By linearity,

$$\mathcal{L}[3 - e^{-3t} + 5 \sin 2t] = 3 \, \mathcal{L}[1] - \mathcal{L}[e^{-3t}] + 5 \, \mathcal{L}[\sin 2t].$$

Using the results of Examples 6.2.1 and 6.2.3, we get

$$\mathcal{L}[3 - e^{-t} + 5 \sin 2t] = \frac{3}{s} - \frac{1}{s + 3} + \frac{10}{s^2 + 4} \qquad \text{for } s > 0.$$

The Laplace transform will allow us to deal easily with o.d.e.'s whose forcing terms are defined "in pieces." We next calculate a specific example; a general, efficient method for such examples will be worked out in Section 6.5.

Example 6.2.7

Find $\mathcal{L}[u_a(t)]$ where $a > 0$ and

$$u_a(t) = \begin{cases} 0 & \text{if } 0 \le t \le a \\ 1 & \text{if } a < t. \end{cases}$$

The definition of $\mathcal{L}$ says

$$\mathcal{L}[u_a(t)] = \int_0^\infty e^{-st} u_a(t) \, dt.$$

We calculate this integral in pieces, one for each part of the formula defining $u_a(t)$:

$$\mathcal{L}[u_a(t)] = \int_0^a e^{-st} u_a(t) \, dt + \int_a^\infty e^{-st} u_a(t) \, dt$$

$$= \int_0^a e^{-st} 0 \, dt \quad + \int_a^\infty e^{-st} 1 \, dt$$

$$= 0 + \lim_{h \to \infty} \int_a^h e^{-st} \, dt$$

$$= \lim_{h \to \infty} \begin{cases} h - a & \text{if } s = 0 \\ \dfrac{-e^{-sh}}{s} + \dfrac{e^{-sa}}{s} & \text{if } s \ne 0 \end{cases}$$

$$= \frac{e^{-sa}}{s} \qquad \text{for } s > 0.$$

The last step of the Laplace transform method for solving initial value problems will require us to retrieve a function from its Laplace transform. To conclude this section, let's look back at the formulas we have derived from this new perspective.

Definition: *If $F(s) = \mathcal{L}[f(t)]$, then we say that $f(t)$ is an **inverse Laplace transform** of $F(s)$ and write*

$$f(t) = \mathcal{L}^{-1}[F(s)].$$

The relation between $\mathcal{L}$ and $\mathcal{L}^{-1}$ is like the relation between differentiation and integration. Just as a table of integrals starts from a "backward" reading of differentiation formulas, so an inverse transform table begins with transform formulas read backward. We list here some Laplace transform formulas (obtained from the results of Examples 6.2.1 to 6.2.4), together with the corresponding inverse Laplace transform formulas.

$$\mathcal{L}[e^{\lambda t}] = \frac{1}{s - \lambda} \qquad\qquad \mathcal{L}^{-1}\left[\frac{1}{s - \lambda}\right] = e^{\lambda t}$$

$$\mathcal{L}[1] = \frac{1}{s} \qquad\qquad \mathcal{L}^{-1}\left[\frac{1}{s}\right] = 1$$

$$\mathcal{L}[\cos \beta t] = \frac{s}{s^2 + \beta^2} \qquad\qquad \mathcal{L}^{-1}\left[\frac{s}{s^2 + \beta^2}\right] = \cos \beta t$$

$$\mathcal{L}\left[\frac{1}{\beta} \sin \beta t\right] = \frac{1}{s^2 + \beta^2} \qquad\qquad \mathcal{L}^{-1}\left[\frac{1}{s^2 + \beta^2}\right] = \frac{1}{\beta} \sin \beta t$$

$$\mathcal{L}\left[\frac{t^{n-1}}{(n - 1)!}\right] = \frac{1}{s^n} \qquad\qquad \mathcal{L}^{-1}\left[\frac{1}{s^n}\right] = \frac{t^{n-1}}{(n - 1)!}$$

Recall also the linearity of $\mathcal{L}$: if $\mathcal{L}[f_1(t)] = F_1(s)$ and $\mathcal{L}[f_2(t)] = F_2(s)$, then $\mathcal{L}[c_1 f_1(t) + c_2 f_2(t)] = c_1 F_1(s) + c_2 F_2(s)$. Read backward, this gives

$$\mathcal{L}^{-1}[c_1 F_1(s) + c_2 F_2(s)] = c_1 f_1(t) + c_2 f_2(t) = c_1 \mathcal{L}^{-1}[F_1(s)] + c_2 \mathcal{L}^{-1}[F_2(s)].$$

Thus, $\mathcal{L}^{-1}$, like $\mathcal{L}$, is *linear!* Just as previously for $\mathcal{L}$, if we know the inverse transforms of a few basic functions, linearity lets us find the inverse transforms of their linear combinations.

Example 6.2.8

Find $\mathcal{L}^{-1}\left[\dfrac{5}{s+1} - \dfrac{6}{s^2+4} + \dfrac{1}{s^4}\right]$.

Using linearity and the list of inverse transforms, we have

$$\mathcal{L}^{-1}\left[\frac{5}{s+1} - \frac{6}{s^2+4} + \frac{1}{s^4}\right]$$

$$= 5\,\mathcal{L}^{-1}\left[\frac{1}{s+1}\right] - 6\,\mathcal{L}^{-1}\left[\frac{1}{s^2+4}\right] + \mathcal{L}^{-1}\left[\frac{1}{s^4}\right]$$

$$= 5e^{-t} - 3\sin 2t + \frac{1}{6}t^3.$$

The manipulation rules that we obtain in the following sections will greatly increase our repertoire of transforms. However, the effective use of the Laplace transform depends on knowing well the basic formulas listed in the following summary.

BASIC FORMULAS FOR THE LAPLACE TRANSFORM

Definitions

$$\mathcal{L}[f(t)] = \int_0^\infty e^{-st}f(t)\,dt$$

$$\mathcal{L}^{-1}[F(s)] = f(t) \text{ provided } F(s) = \mathcal{L}[f(t)]$$

Linearity

$$\mathcal{L}[c_1 f_1(t) + c_2 f_2(t)] = c_1\,\mathcal{L}[f_1(t)] + c_2\,\mathcal{L}[f_2(t)]$$

$$\mathcal{L}^{-1}[c_1 F_1(s) + c_2 F_2(s)] = c_1\,\mathcal{L}^{-1}[F_1(s)] + c_2\,\mathcal{L}^{-1}[F_2(s)]$$

Basic Transforms and Inverse Transforms

$$\mathcal{L}[e^{\lambda t}] = \frac{1}{s-\lambda} \qquad\qquad \mathcal{L}^{-1}\left[\frac{1}{s-\lambda}\right] = e^{\lambda t}$$

$$\mathcal{L}[1] = \frac{1}{s} \qquad\qquad \mathcal{L}^{-1}\left[\frac{1}{s}\right] = 1$$

$$\mathscr{L}[t^n] = \frac{n!}{s^{n+1}} \qquad\qquad \mathscr{L}^{-1}\left[\frac{1}{s^n}\right] = \frac{t^{n-1}}{(n-1)!}$$

$$\mathscr{L}[\cos \beta t] = \frac{s}{s^2 + \beta^2} \qquad\qquad \mathscr{L}^{-1}\left[\frac{s}{s^2 + \beta^2}\right] = \cos \beta t$$

$$\mathscr{L}[\sin \beta t] = \frac{\beta}{s^2 + \beta^2} \qquad\qquad \mathscr{L}^{-1}\left[\frac{1}{s^2 + \beta^2}\right] = \frac{1}{\beta} \sin \beta t$$

The Laplace transform is named after the French mathematician Marquis Pierré-Simon de Laplace, in whose masterwork on probability ("Théorie Analytique des Probabilités," 1812) the integral expression defining $\mathscr{L}[f(t)]$ appears.

Notes

1. On the domain of $\mathscr{L}$

What is the domain of $\mathscr{L}$? That is, for what functions $f(t)$ does the improper integral in the definition of $\mathscr{L}[f(t)]$ converge for at least some values of s? An exhaustive answer to this question would involve technicalities beyond the scope of this book. However, we can formulate fairly general conditions on $f(t)$ that do insure that $\mathscr{L}[f(t)]$ is defined.

There are two kinds of difficulties that can occur in trying to define the improper integral

$$\int_0^\infty e^{-st}f(t)\, dt = \lim_{h\to\infty} \int_0^h e^{-st}f(t)\, dt.$$

The first is that for some h, the proper definite integral

$$\int_0^h e^{-st}f(t)\, dt$$

may be undefined. Recall from calculus that a function $f(t)$ is **piecewise continuous** on $0 \le t \le h$ provided it is continuous there, except possibly at a finite number of points $t_1, \ldots, t_k$ at which the two one-sided limits

$$\lim_{t\to t_i^-} f(t) \qquad \text{and} \qquad \lim_{t\to t_i^+} f(t)$$

exist but are unequal. (The functions defined in pieces that came up in Sections 6.1 and 6.2 are all examples of such functions.) If $f(t)$ is piecewise continuous on $0 \le t \le h$, then $\int_0^h f(t)\, dt$ exists. Thus, the first problem does not arise if $f(t)$ is piecewise continuous on every interval $0 \le t \le h$.

The second possible problem is that, even though the proper integrals exist, they may not tend to a finite limit as $h \to \infty$. One way to avoid this problem is to require that the integrand approach 0 rapidly as $t \to \infty$. To make this precise, we say a function $f(t)$ is of **exponential order** if there is a constant c so that

$$\lim_{t\to\infty} e^{-ct}f(t) = 0.$$

When this limit condition holds for a given value of c, then one can show that for $s > c$, the proper integrals defining the Laplace transform do converge. We can therefore formulate the following general criterion:

Fact: *If $f(t)$ is piecewise continuous on each interval $0 \le t \le h$ and of exponential order, then $\mathscr{L}[f(t)] = \int_0^\infty e^{-st}f(t)\,dt$ will be defined for all $s > c$.*

Although all the examples we consider in the text of this chapter are piecewise continuous and of exponential order (see Exercises 25 and 26), there are functions that are not included in this class but that still have Laplace transforms (see Exercise 24).

2. The linearity of $\mathscr{L}$—a technicality

Recall from calculus that two functions are equal provided (i) they have the same domain and (ii) they assign the same value to each element of the domain. Now, if we let $f_1(t) = e^t + 1$ and $f_2(t) = e^t$, then

$$\mathscr{L}[f_1(t)] = \frac{1}{s-1} + \frac{1}{s}, \qquad s > 1$$

and

$$\mathscr{L}[f_2(t)] = \frac{1}{s-1}, \qquad s > 1.$$

so that

$$\mathscr{L}[f_1(t)] - \mathscr{L}[f_2(t)] = \frac{1}{s}, \qquad s > 1.$$

On the other hand, $f_1(t) - f_2(t) = 1$, so that

$$\mathscr{L}[f_1(t) - f_2(t)] = \mathscr{L}[1] = \frac{1}{s}, \qquad s > 0.$$

Even though $\mathscr{L}[f_1(t)] - \mathscr{L}[f_2(t)]$ and $\mathscr{L}[f_1(t) - f_2(t)]$ assign the same value to every $s > 1$, they have different domains and so are not equal in the above sense.

When we say the Laplace transform is linear, we are using a different notion of equality for functions. We consider two functions $F(s)$ and $G(s)$ to be equal provided they agree eventually, that is, provided $F(s) = G(s)$ for all s larger than some fixed value s_0.

3. The definition of $\mathscr{L}^{-1}$—another technicality

Our definition tells us $\mathscr{L}^{-1}[F(s)] = f(t)$ provided $\mathscr{L}[f(t)] = F(s)$. Unfortunately, there do exist different functions $f(t)$ and $g(t)$ with the same Laplace transform. Thus it is possible to have $f(t) = \mathscr{L}^{-1}[F(s)]$ and $g(t) = \mathscr{L}^{-1}[F(s)]$ even though $f(t) \ne g(t)$. This is why we called $f(t)$ *an* inverse transform, rather than *the* inverse transform, of $F(s)$.

One way this difficulty can arise is if $f(t)$ and $g(t)$ agree except at a number of isolated points. Fortunately, for the functions we will be dealing with this is the *only* way the problem can arise:

Theorem   (Lerch):   *If $f(t)$ and $g(t)$ are piecewise continuous and of exponential order, and if $\mathcal{L}[f(t)] = \mathcal{L}[g(t)]$, then $f(t) = g(t)$ for all $t > 0$, except possibly at points where one or both of $f(t)$ and $g(t)$ are discontinuous.*

Note in particular that if $f(t)$ and $g(t)$ are both continuous and $\mathcal{L}[f(t)] = \mathcal{L}[g(t)]$, then $f(t) = g(t)$ for all $t > 0$. Thus a given function $F(s)$ can have at most one *continuous* inverse transform defined on $(0, \infty)$.

EXERCISES

In Exercises 1 to 7, calculate $F(s) = \mathcal{L}[f(t)]$ directly from the definition and indicate the values of s for which the integral defining $F(s)$ converges.

1. $f(t) = e^{4t}$
2. $f(t) = e^{-t}$
3. $f(t) = t^2$
4. $f(t) = te^{3t}$
5. $f(t) = \sin 2t$
6. $f(t) = e^t \sin 2t$
7. $f(t) = \begin{cases} 1 & \text{if } t < 3 \\ 0 & \text{if } t > 3 \end{cases}$

In Exercises 8 to 14, calculate $F(s) = \mathcal{L}[f(t)]$ using the linearity of $\mathcal{L}$ together with the basic formulas summarized at the end of this section.

8. $f(t) = e^{-t}$
9. $f(t) = t^4$
10. $f(t) = \sin 2t$
11. $f(t) = t^2 - 7 + \cos 2t$
12. $f(t) = e^{2t+3}$
13. $f(t) = (t + 1)(t + 2)$
14. $f(t) = \sin (t + \pi/6)$ (*Hint:* Use trig identities.)

In Exercises 15 to 22, calculate $f(t) = \mathcal{L}^{-1}[F(s)]$ using the linearity of $\mathcal{L}^{-1}$ together with the basic formulas summarized at the end of this section.

15. $F(s) = \dfrac{1}{s - 2}$
16. $F(s) = \dfrac{2}{s - 1}$
17. $F(s) = \dfrac{1}{2s - 1}$
18. $F(s) = \dfrac{1}{s + 2}$
19. $F(s) = \dfrac{2}{s^5}$
20. $F(s) = \dfrac{s}{s^2 + 4}$
21. $F(s) = \dfrac{1}{s^2 + 3}$
22. $F(s) = \dfrac{3}{s^2 + 1} - \dfrac{20}{s^4} + \dfrac{3}{s}$

More advanced problems:

23. Is the Laplace transform of a product, $\mathcal{L}[f(t)g(t)]$, the same as the product of the transforms, $\mathcal{L}[f(t)] \, \mathcal{L}[g(t)]$? (*Hint:* Try some examples.)

24. Work through the following determination of $\mathcal{L}[t^{-1/2}]$:
   a. Show that for $s > 0$ the substitution $y = \sqrt{st}$ gives

$$\mathcal{L}[t^{-1/2}] = \int_0^\infty t^{-1/2} e^{-st} \, dt = \frac{2}{\sqrt{s}} \int_0^\infty e^{-y^2} \, dy.$$

   b. Rewrite the square of this integral in the form

$$(\mathcal{L}[t^{-1/2}])^2 = \frac{4}{s} \int_0^\infty e^{-x^2} \, dx \int_0^\infty e^{-y^2} \, dy = \frac{4}{s} \int_0^\infty \int_0^\infty e^{-x^2-y^2} \, dx \, dy.$$

   c. Convert this double integral to an integral in polar coordinates and evaluate to obtain

$$(\mathcal{L}[t^{-1/2}])^2 = \frac{4}{s} \int_0^{\pi/2} \int_0^\infty e^{-r^2} r \, dr \, d\theta = \frac{\pi}{s}.$$

   d. Conclude that $\mathcal{L}[t^{-1/2}] = \sqrt{\pi/s}$ for $s > 0$. Note that $t^{-1/2}$ is not piecewise continuous on any interval of the form $[0, h]$, since $\lim_{t \to 0^+} t^{-1/2}$ does not exist.

Exercises 25 and 26 together with the results of Chapter 2 guarantee that all solutions of any homogeneous constant-coefficient linear o.d.e. are of exponential order.

25. a. Work through the following demonstration that if n is a nonnegative integer, then $f(t) = t^n e^{\alpha t} \cos \beta t$ is of exponential order:
   i.   Show that if $c = \alpha + 1$, then $|e^{-ct}f(t)| \le t^n e^{-t}$ for $t > 0$.
   ii. Show that $\lim_{t \to \infty} (t^n \, e^{-t}) = 0$. (*Hint:* Use L'Hôpital's rule to handle the case $n > 0$.)
   iii. Conclude that $\lim_{t \to \infty} e^{-ct}f(t) = 0$.

   b. Repeat part (a) for $f(t) = t^n e^{\alpha t} \sin \beta t$.

26. Show that if $f(t)$ and $g(t)$ are of exponential order, then so is any linear combination $h(t) = af(t) + bg(t)$. (*Hint:* If $\lim_{t \to \infty} e^{-\alpha t}f(t) = 0 = \lim_{t \to \infty} e^{-\beta t}g(t)$, then $\lim_{t \to \infty} e^{-ct}h(t) = 0$, where c is the larger of α and β.)

27. Which of the following functions are (a) of exponential order and (b) piecewise continuous on every interval of the form $[0, h]$?

   i. $1/t$
   ii. e^{t^2}
   iii. e^{-t^2}
   iv. $\arctan t$
   v. $\ln t$
   vi. $t^{-1/2}$

28. Show that the integral defining $\mathcal{L}[e^{t^2}]$ does not converge for any s. (*Hint:* $t^2 - st > 0$ for $t > s$.)

29. Show that $f(t)$ is of exponential order if and only if there exist constants a and b and a value t_0 so that

$$|f(t)| \le ae^{bt} \qquad \text{for all } t \ge t_0.$$

6.3 THE LAPLACE TRANSFORM AND INITIAL VALUE PROBLEMS

In this section we see how the Laplace transform is used to solve initial value problems.

The first step in solving an o.d.e. with constant coefficients

$$a_n D^n x + a_{n-1} D^{n-1} x + \ldots + a_0 x = g(t)$$

using the Laplace transform is to apply $\mathscr{L}$ to both sides. Since $\mathscr{L}$ is linear we get

$$a_n \, \mathscr{L}[D^n x] + a_{n-1} \, \mathscr{L}[D^{n-1} x] + \ldots + a_0 \, \mathscr{L}[x] = \mathscr{L}[g(t)].$$

We already know how to find the right-hand term, $\mathscr{L}[g(t)]$, for many choices of $g(t)$. However, we need a way of dealing with the terms $\mathscr{L}[D^k x]$ appearing on the left side. We will try to express all terms of this form in terms of the transform $\mathscr{L}[x]$ of our unknown.

Let's start by considering $\mathscr{L}[Dx] = \mathscr{L}[x'(t)]$. By definition,

$$\mathscr{L}[x'(t)] = \int_0^\infty e^{-st} x'(t) \, dt = \lim_{h \to \infty} \int_0^h e^{-st} x'(t) \, dt.$$

Integration by parts (with $u = e^{-st}$ and $dv = x'(t) \, dt$) gives

$$\mathscr{L}[x'(t)] = \lim_{h \to \infty} \left\{ e^{-st} x(t) \Big]_0^h + s \int_0^h e^{-st} x(t) \, dt \right\}$$

$$= \lim_{h \to \infty} \{ e^{-sh} x(h) \} - x(0) + s \, \mathscr{L}[x(t)].$$

For the functions we are interested in (see Note 1, Section 6.2),

$$\lim_{h \to \infty} e^{-sh} x(h) = 0$$

as long as s is sufficiently large. Thus we have the following.

Fact: First Differentiation Formula ($k = 1$). $\mathscr{L}[Dx] = s \, \mathscr{L}[x] - x(0).$

We consider how this formula is used in an extremely simple problem.

Example 6.3.1

Solve the initial value problem

$$Dx = t, \qquad x(0) = 2.$$

If we apply $\mathcal{L}$ to both sides of the o.d.e., we have

$$\mathcal{L}[Dx] = \mathcal{L}[t].$$

Using the formula obtained in Example 6.2.4 for $\mathcal{L}[t]$ and the first differentiation formula for $\mathcal{L}[Dx]$, we rewrite this equation as

$$s\,\mathcal{L}[x] - x(0) = \frac{1}{s^2}.$$

Since we require $x(0)$ to be 2,

$$s\,\mathcal{L}[x] - 2 = \frac{1}{s^2}.$$

We solve for $\mathcal{L}[x]$:

$$\mathcal{L}[x] = \frac{1}{s^3} + \frac{2}{s}.$$

Now we take inverse transforms to find x:

$$x = \mathcal{L}^{-1}\left[\frac{1}{s^3} + \frac{2}{s}\right] = \mathcal{L}^{-1}\left[\frac{1}{s^3}\right] + 2\,\mathcal{L}^{-1}\left[\frac{1}{s}\right]$$

$$= \frac{1}{2}t^2 + 2.$$

To deal with o.d.e.'s of order higher than 1, we must find formulas for terms of the form $\mathcal{L}[D^k x]$ with $k > 1$. For $k = 2$ we note that $D^2 x = Dx'(t)$, so that

$$\mathcal{L}[D^2 x] = \mathcal{L}[Dx'(t)].$$

But now we can use the first differentiation formula for $k = 1$ to rewrite $\mathcal{L}[Dx'(t)]$ in terms of $\mathcal{L}[x'(t)]$:

$$\mathcal{L}[D^2 x] = s\,\mathcal{L}[x'(t)] - x'(0).$$

Of course, we already know how to express $\mathcal{L}[x'(t)]$ in terms of $\mathcal{L}[x]$; substituting into the preceding equation, we find

$$\mathcal{L}[D^2x] = s\,\{s\,\mathcal{L}[x] - x(0)\} - x'(0)$$
$$= s^2\,\mathcal{L}[x] - sx(0) - x'(0).$$

Repeated application of this process yields the following general formula:

Fact: First Differentiation Formula

$$\mathcal{L}[D^kx] = s^k\,\mathcal{L}[x] - s^{k-1}x(0) - s^{k-2}x'(0) - \ldots - x^{(k-1)}(0).$$

Example 6.3.2

Solve the initial value problem

$$D^3x - D^2x = 0, \qquad x(0) = x'(0) = x''(0) = 3.$$

If we apply $\mathcal{L}$ to both sides of our o.d.e., we get

$$\mathcal{L}[D^3x] - \mathcal{L}[D^2x] = \mathcal{L}[0] = 0.$$

By the first differentiation formula,

$$\mathcal{L}[D^3x] = s^3\,\mathcal{L}[x] - s^2x(0) - sx'(0) - x''(0)$$
$$= s^3\,\mathcal{L}[x] - 3s^2 - 3s - 3$$

and

$$\mathcal{L}[D^2x] = s^2\,\mathcal{L}[x] - sx(0) - x'(0)$$
$$= s^2\,\mathcal{L}[x] - 3s - 3.$$

Substitution into the transformed o.d.e. gives

$$(s^3 - s^2)\,\mathcal{L}[x] - 3s^2 = 0,$$

which we solve for $\mathcal{L}[x]$:

$$\mathcal{L}[x] = \frac{3s^2}{s^3 - s^2} = \frac{3}{s - 1}.$$

Then

$$x = \mathcal{L}^{-1}\left[\frac{3}{s-1}\right] = 3\,\mathcal{L}^{-1}\left[\frac{1}{s-1}\right]$$

$$= 3e^t.$$

These examples illustrate the three steps in solving an initial value problem by Laplace transforms. First, we *transform the o.d.e., incorporating the initial data by means of the first differentiation formula.* Then, we *solve algebraically for $\mathcal{L}[x]$ in terms of s.* Finally, we *obtain x as the inverse transform of $\mathcal{L}[x]$.*

In the examples so far, the third step was unusually easy. Most of the time we will need to rewrite $\mathcal{L}[x]$ in order to recognize its inverse transform. This is done by means of partial fraction decomposition of quotients of polynomials. A reminder of how such decompositions look is given in Note 3. We illustrate their use with three examples.

Example 6.3.3

Solve the initial value problem

$$Dx - x = 2\sin t, \qquad x(0) = 0.$$

We transform both sides of the o.d.e.:

$$\mathcal{L}[Dx] - \mathcal{L}[x] = \mathcal{L}[2\sin t]$$

$$(s\,\mathcal{L}[x] - x(0)) - \mathcal{L}[x] = \frac{2}{s^2 + 1}.$$

Now, we solve for $\mathcal{L}[x]$:

$$\mathcal{L}[x] = \frac{2}{(s-1)(s^2+1)}$$

so that

$$x = \mathcal{L}^{-1}\left[\frac{2}{(s-1)(s^2+1)}\right].$$

To find this inverse transform, we look for the partial fraction decomposition of $\mathscr{L}[x]$, which is of the form

$$\frac{2}{(s - 1)(s^2 + 1)} = \frac{A}{s - 1} + \frac{Bs + C}{s^2 + 1}.$$

To find the values of the constants A, B, and C, we multiply both sides by $(s - 1)(s^2 + 1)$:

$$\begin{aligned} 2 &= A(s^2 + 1) + (Bs + C)(s - 1) \\ &= (A + B)s^2 + (-B + C)s + (A - C) \end{aligned}$$

and equate the coefficients of s^2, s, and 1 on either side of the equation:

$$\begin{aligned} A + B \qquad &= 0 \\ -B + C &= 0 \\ A \qquad - C &= 2. \end{aligned}$$

The solution of this algebraic system is

$$A = 1, \qquad B = C = -1,$$

so that

$$\frac{2}{(s - 1)(s^2 + 1)} = \frac{1}{s - 1} + \frac{-s - 1}{s^2 + 1}.$$

This allows us to use the inverse transform formulas from the last section to find x:

$$\begin{aligned} x &= \mathscr{L}^{-1}\left[\frac{2}{(s - 1)(s^2 + 1)}\right] = \mathscr{L}^{-1}\left[\frac{1}{s - 1} + \frac{-s - 1}{s^2 + 1}\right] \\ &= \mathscr{L}^{-1}\left[\frac{1}{s - 1}\right] - \mathscr{L}^{-1}\left[\frac{s}{s^2 + 1}\right] - \mathscr{L}^{-1}\left[\frac{1}{s^2 + 1}\right] \\ &= e^t - \cos t - \sin t. \end{aligned}$$

Example 6.3.4

Solve the initial value problem

$$D^2x - x = 0, \qquad x(0) = 3, x'(0) = 1.$$

We transform both sides of the o.d.e.:

$$\mathcal{L}[D^2 x] - \mathcal{L}[x] = 0$$

$$(s^2 \mathcal{L}[x] - sx(0) - x'(0)) - \mathcal{L}[x] = 0$$

$$(s^2 \mathcal{L}[x] - 3s - 1) - \mathcal{L}[x] = 0.$$

Solving for $\mathcal{L}[x]$, we have

$$\mathcal{L}[x] = \frac{3s + 1}{s^2 - 1} = \frac{3s + 1}{(s - 1)(s + 1)}.$$

Thus, we know that

$$x = \mathcal{L}^{-1}\left[\frac{3s + 1}{(s - 1)(s + 1)}\right].$$

To find this inverse transform, we look for the partial fraction decomposition of $\mathcal{L}[x]$, which has the form

$$\frac{3s + 1}{(s - 1)(s + 1)} = \frac{A}{s - 1} + \frac{B}{s + 1}.$$

We multiply by $(s - 1)(s + 1)$:

$$3s + 1 = A(s + 1) + B(s - 1) = (A + B)s + (A - B)$$

and equate coefficients:

$$A + B = 3$$

$$A - B = 1.$$

This gives

$$A = 2, \quad B = 1.$$

Thus

$$\frac{3s + 1}{(s - 1)(s + 1)} = \frac{2}{s - 1} + \frac{1}{s + 1},$$

and we can find x:

$$x = \mathcal{L}^{-1}\left[\frac{2}{s-1} + \frac{1}{s+1}\right] = 2e^t + e^{-t}.$$

Example 6.3.5

Solve the initial value problem

$$D^2x - 2Dx = 4, \qquad x(0) = -1, x'(0) = 2.$$

We transform the o.d.e.:

$$\mathcal{L}[D^2x - 2Dx] = \frac{4}{s}$$

$$(s^2 \mathcal{L}[x] + s - 2) - 2(s \mathcal{L}[x] + 1) = \frac{4}{s}$$

Solving for $\mathcal{L}[x]$ we have

$$(s^2 - 2s) \mathcal{L}[x] = \frac{4}{s} - s + 4 = \frac{-s^2 + 4s + 4}{s}$$

$$\mathcal{L}[x] = \frac{-s^2 + 4s + 4}{s^2(s-2)}.$$

The partial fraction decomposition of the right side is

$$\frac{-s^2 + 4s + 4}{s^2(s-2)} = \frac{-3}{s} + \frac{-2}{s^2} + \frac{2}{s-2}$$

so that

$$x = \mathcal{L}^{-1}\left[\frac{-3}{s} - \frac{2}{s^2} + \frac{2}{s-2}\right] = -3 - 2t + 2e^{2t}.$$

Let's summarize the method.

THE LAPLACE TRANSFORM AND INITIAL VALUE PROBLEMS

To solve an o.d.e. with constant coefficients, subject to initial conditions at $t = 0$:

1. Transform both sides of the o.d.e., incorporating the initial data by means of the **first differentiation formula:**

$$\mathcal{L}[D^k x] = s^k \, \mathcal{L}[x] - s^{k-1}x(0) - s^{k-2}x'(0) - \ldots - x^{(k-1)}(0).$$

2. Solve algebraically for $\mathcal{L}[x]$ in terms of s.
3. Obtain x as the inverse Laplace transform of $\mathcal{L}[x]$. In this last step we often use the linearity of $\mathcal{L}^{-1}$ and partial fractions.

Notes

1. The first differentiation formula—a technicality

A careful look at our argument for the first differentiation formula when $k = 1$ shows that we assumed $x(t)$ and $x'(t)$ were reasonably well behaved (e.g., we assumed $\lim_{h \to \infty} e^{-sh}x(h) = 0$ for large values of s). Since the general formula is obtained by applying this case to successive derivatives of x, the formula is valid only if these derivatives are also well behaved. A more careful statement of the formula would read as follows:

Theorem: *Suppose $x(t)$, $x'(t)$, . . ., $x^{(k-1)}(t)$ are continuous and of exponential order, and suppose $x^{(k)}(t)$ is piecewise continuous and of exponential order. Then $\mathcal{L}[D^k x] = s^k \, \mathcal{L}[x] - s^{k-1}x(0) - \ldots - x^{(k-1)}(0)$.*

We will be using the method described in the summary to solve nth-order linear constant-coefficient o.d.e.'s whose forcing terms are piecewise continuous and of exponential order. The solutions we obtain for such o.d.e.'s will satisfy the hypotheses of the preceding theorem (with $k = n$).

2. Partial fractions

Each polynomial $q(s)$ with real coefficients can, at least in theory, be factored as a number (the leading coefficient) times a product of irreducible polynomials of two kinds: (i) **linear factors** $s - a$, where a is a real root of the polynomial, and (ii) **irreducible quadratic factors** $s^2 + bs + c$ (with $b^2 < 4c$), corresponding to pairs of complex roots. If $p(s)$ is a polynomial whose degree is strictly less than the degree of $q(s)$, then the rational expression $p(s)/q(s)$ can be written as a sum according to the following rules:

i. If $(s - a)^m$ is the highest power of $(s - a)$ that divides $q(s)$, then the sum should include terms of the form

$$\frac{A_1}{s - a} + \frac{A_2}{(s - a)^2} + \ldots + \frac{A_m}{(s - a)^m}.$$

ii. If $(s^2 + bs + c)^m$ is the highest power of the irreducible quadratic $s^2 + bs + c$ that divides $q(s)$, then the sum should include terms of the form

$$\frac{B_1 s + C_1}{(s^2 + bs + c)} + \frac{B_2 s + C_2}{(s^2 + bs + c)^2} + \ldots + \frac{B_m s + C_m}{(s^2 + bs + c)^m}.$$

In general, we obtain the **partial fraction decomposition** of a rational expression $p(s)/q(s)$, once we know how to factor $q(s)$, by first using long division to rewrite the original quotient as a polynomial plus a new quotient whose numerator has degree less than the denominator $q(s)$ and then writing this new quotient as a sum of terms of the form (i) and (ii), just described, corresponding to all the factors of $q(s)$.

3.   *The first differentiation formula and inverse transforms*

Sometimes we can use a restatement of the first differentiation formula in place of partial fractions to calculate inverse transforms. To obtain this restatement, first note that the function

$$g(t) = \int_0^t f(t) \, dt$$

satisfies

$$Dg(t) = f(t) \quad \text{and} \quad g(0) = 0.$$

Taking Laplace transforms, we get

$$s \, \mathcal{L}[g(t)] - 0 = \mathcal{L}[f(t)]$$

$$\mathcal{L}[g(t)] = \frac{1}{s} \mathcal{L}[f(t)].$$

Setting $F(s) = \mathcal{L}[f(t)]$, we see that

$$\mathcal{L}\left[\int_0^t f(t) \, dt\right] = \frac{1}{s} F(s)$$

or, reformulating this in terms of $\mathcal{L}^{-1}$,

$$\mathcal{L}^{-1}\left[\frac{1}{s} F(s)\right] = \int_0^t \mathcal{L}^{-1}[F(s)] \, dt.$$

For example, we see that

$$\mathcal{L}^{-1}\left[\frac{1}{s(s^2 + \beta^2)}\right] = \int_0^t \mathcal{L}^{-1}\left[\frac{1}{s^2 + \beta^2}\right] dt = \frac{1}{\beta} \int_0^t \sin \beta t \, dt$$

$$= -\frac{1}{\beta^2} \cos \beta t + \frac{1}{\beta^2}.$$

Repeated application of the formula for $\mathcal{L}^{-1}[F(s)/s]$ gives us the following.

Fact: First Differentiation Formula—Inverse Transform Version.

$$\mathcal{L}^{-1}\left[\frac{1}{s^k} F(s)\right] = \underbrace{\int_0^t \int_0^t \cdots \int_0^t}_{k \text{ integrals}} \mathcal{L}^{-1}[F(s)] \, dt \ldots dt$$

For example,

$$\mathcal{L}^{-1}\left[\frac{1}{s^3(s-2)}\right] = \int_0^t \int_0^t \int_0^t \mathcal{L}^{-1}\left[\frac{1}{s-2}\right] dt\, dt\, dt$$

$$= \int_0^t \int_0^t \int_0^t e^{2t}\, dt\, dt\, dt = \int_0^t \int_0^t \left(\frac{e^{2t}}{2} - \frac{1}{2}\right) dt\, dt$$

$$= \int_0^t \left(\frac{e^{2t}}{4} - \frac{t}{2} - \frac{1}{4}\right) dt = \frac{e^{2t}}{8} - \frac{t^2}{4} - \frac{t}{4} - \frac{1}{8}.$$

Of course, this formula should not be used unless $\mathcal{L}^{-1}[F(s)]$ is both easy to find and easy to integrate.

4. Initial conditions at nonzero time

The Laplace transform of an o.d.e. depends on the initial conditions that hold at $t = 0$. To solve an o.d.e.

(N) $(a_n D^n + \ldots + a_0)x = E(t)$

with conditions imposed at time $t = u$, $u \neq 0$, we can restate the problem in terms of the time $\tau = t - u$ elapsed since imposition of the given conditions. The substitution $t = \tau + u$ changes the o.d.e. (N) into

(N') $(a_n D^n + \ldots + a_0)y = E(\tau + u)$.

If $y = y(\tau)$ solves (N') with certain conditions at $\tau = 0$, then $x = y(t - u)$ solves (N) with the same conditions imposed at $t = u$. This observation should be used to do Exercises 26 to 29.

EXERCISES

In Exercises 1 to 6, use the first differentiation formula to find an expression for $\mathcal{L}[x]$, where x is the solution of the given initial value problem.

1. $(D - 1)x = 0;$ $x(0) = -3$

2. $(D - 1)x = e^{3t};$ $x(0) = 3$

3. $(D^2 - 1)x = e^{2t};$ $x(0) = x'(0) = 0$

4. $(D^2 - 1)x = e^{2t};$ $x(0) = 0, x'(0) = 1$

5. $(D^2 + 1)x = \cos 3t;$ $x(0) = x'(0) = 0$

6. $(D^2 + 1)x = \cos 3t;$ $x(0) = 0, x'(0) = 3$

In Exercises 7 to 9, use the fact that $x = f(t)$ is the solution of the given initial value problem to find $\mathcal{L}[f(t)]$.

7. $f(t) = \cos \beta t;$ $(D^2 + \beta^2)x = 0, x(0) = 1, x'(0) = 0$

8. $f(t) = t \cos \beta t;$ $(D^4 + 2\beta^2 D^2 + \beta^4)x = 0, x(0) = 0, x'(0) = 1,$
                                       $x''(0) = 0, x'''(0) = -3\beta^2$

9. $f(t) = (t + 2)e^{-t};$ $(D^2 + 2D + 1)x = 0, x(0) = 2, x'(0) = -1$

Find the inverse transforms of the functions in Exercises 10 to 15.

10. $\dfrac{1}{(s + 1)(s - 3)}$ 11. $\dfrac{1}{s(s + 1)(s - 3)}$ 12. $\dfrac{s + 4}{s^2 + 4s + 3}$

13. $\dfrac{1}{s^4 - 1}$ 14. $\dfrac{1}{(s^2 + 1)(s^2 + 4)}$ 15. $\dfrac{s + 1}{(s^2 + 4)s}$

Use the Laplace transform to solve the initial value problems in Exercises 16 to 25.

16. $x'' - 2x' - 3x = 0$; $x(0) = 2,\ x'(0) = -2$

17. $x'' - 2x' - 3x = 1$; $x(0) = 2,\ x'(0) = -2$

18. $(D^2 + 4)x = \sin t$; $x(0) = x'(0) = 0$

19. $(D^2 + 4)x = e^t$; $x(0) = x'(0) = 0$

20. $(D^2 + 4)x = t$; $x(0) = -1,\ x'(0) = 0$

21. $(D^2 + 2D - 3)x = 5e^{2t}$; $x(0) = 2,\ x'(0) = 3$

22. $(D^3 - 4D)x = 0$; $x(0) = 4,\ x'(0) = x''(0) = 8$

23. $(D^3 - 4D)x = e^t$; $x(0) = 5,\ x'(0) = x''(0) = 9$

24. $(D^4 - 1)x = 0$; $x(0) = 1,\ x'(0) = x''(0) = x'''(0) = 0$

25. $(D^4 - 1)x = 1$; $x(0) = 1,\ x'(0) = x''(0) = x'''(0) = 0$

Use the substitution $\tau = t - u$ (discussed in Note 4) and the Laplace transform to solve the initial value problems in Exercises 26 to 29.

26. $(D^2 - 1)x = 3t$; $x(1) = x'(1) = 0$

27. $(D^2 - 1)x = 3t$; $x(1) = x'(1) = 2$

28. $(D^2 - 1)x = e^{3t}$; $x(1) = x'(1) = 0$

29. $(D^2 - 1)x = \sin t$; $x(\pi) = x'(\pi) = 0$

In Exercises 30 to 34, which refer back to our models, solve the initial value problem of the indicated exercise.

30. Exercise 22, Section 1.1 (compare Exercise 16, Section 1.2)

31. Exercise 21(b), Section 1.3

32. Exercise 4(b), Section 6.1

33. Exercise 4(c), Section 6.1

34. Exercise 3, Section 2.1, with $L = 1,\ C = 1/9,\ R = 0,\ V(t) = \sin 2t$, and subject to $Q(0) = 1,\ I(0) = -1$

More abstract problems:

35. Suppose $f(t)$ is a polynomial of degree n. Show that

$$\mathscr{L}[f((t)] = \frac{f(0)}{s} + \frac{f'(0)}{s^2} + \ldots + \frac{f^{(n)}(0)}{s^{n+1}}.$$

[*Hint:* Use the first differentiation formula, or else express the coefficients of $f(t)$ in terms of the initial values $f^{(i)}(0)$.]

36. a. Suppose $x = h(t)$ is a solution of the homogeneous o.d.e. $P(D)x = 0$, where $P(D) = a_2D^2 + a_1D + a_0$. Show that $\mathcal{L}[h(t)]$ has the form $g(s)/P(s)$, where $g(s) = b_1s + b_0$ for some constants b_0, b_1.
     b. What is the analogous form for $\mathcal{L}[h(t)]$ if $x = h(t)$ solves $P(D)x = 0$ with $P(D) = a_nD^n + \ldots + a_1D + a_0$?

37. a. Suppose $q(s) = (s - m_1)(s - m_2) \ldots (s - m_k)$ with $m_1, \ldots, m_k$ distinct. The partial fraction decomposition of $p(s)/q(s)$ is

$$\frac{p(s)}{q(s)} = \frac{A_1}{s - m_1} + \frac{A_2}{s - m_2} + \ldots + \frac{A_k}{s - m_k}.$$

Multiply both sides of this equation by $(s - m_1)$ and substitute $s = m_1$ to obtain an expression for A_1. Interpret this in terms of p and the factorization of q.
     b. Show that your expression in part (a) is the same as $A_1 = p(m_1)/q'(m_1)$.

38. a. Suppose $q(s) = (s - m_1)^p(s - m_2) \ldots (s - m_k)$ with $m_1, \ldots, m_k$ distinct and $p \geq 1$. The partial fraction decomposition of $p(s)/q(s)$ is

$$\frac{p(s)}{q(s)} = \frac{B_1}{s - m_1} + \frac{B_2}{(s - m_1)^2} + \ldots + \frac{B_p}{(s - m_1)^p} + \frac{A_2}{s - m_2} + \ldots + \frac{A_k}{s - m_k}.$$

Multiply both sides of this equation by $(s - m_1)^p$ to obtain an equation that we will call (*).
     b. Substitute $s = m_1$ in (*) to obtain an expression for B_p.
     c. Differentiate both sides of (*) r times, $r < p$, and substitute $s = m_1$ to obtain an expression for B_{p-r}.
     d. Do the expressions for B_{p-r}, $r = 0, \ldots, p - 1$ obtained in parts (b) and (c) remain valid when $m_i = m_j$ for some $i \neq j$ with $i, j > 1$?

6.4 FURTHER PROPERTIES OF THE LAPLACE TRANSFORM AND INVERSE TRANSFORM

We saw in the last section how the Laplace transform is used to solve initial value problems. In this section we will develop some rules of manipulation that will greatly extend our repertoire of Laplace transforms and make its use more efficient.

We begin by considering the effect of the Laplace transform on a function of the form $e^{\alpha t}f(t)$, assuming we know $F(s) = \mathcal{L}[f(t)]$. The definition says

$$\mathcal{L}[e^{\alpha t}f(t)] = \int_0^\infty e^{-st}e^{\alpha t}f(t) \, dt = \int_0^\infty e^{-(s-\alpha)t}f(t) \, dt.$$

This last integral is just the value at $s - \alpha$ of the Laplace transform of $f(t)$—that is, it is $F(s - \alpha)$. In other words, we have the following.

Fact: First Shift Formula. *If $\mathcal{L}[f(t)] = F(s)$, then*

$$\mathcal{L}[e^{\alpha t}f(t)] = F(s - \alpha).$$

Example 6.4.1

Find $\mathcal{L}[t^3 e^{2t}]$.

The first shift formula tells us that $\mathcal{L}[t^3 e^{2t}]$ is obtained from $\mathcal{L}[t^3]$ by replacing s by $s - 2$. Since

$$\mathcal{L}[t^3] = F(s) = \frac{6}{s^4}$$

this means

$$\mathcal{L}[t^3 e^{2t}] = F(s - 2) = \frac{6}{(s - 2)^4}.$$

Example 6.4.2

Find $\mathcal{L}[e^{-t}\cos 3t]$.

$\mathcal{L}[e^{-t}\cos 3t]$ is obtained from

$$\mathcal{L}[\cos 3t] = F(s) = \frac{s}{s^2 + 9}$$

by replacing s with $s - (-1) = s + 1$. Thus

$$\mathcal{L}[e^{-t}\cos 3t] = F(s + 1) = \frac{s + 1}{(s + 1)^2 + 9} = \frac{s + 1}{s^2 + 2s + 10}.$$

Of course, every transform formula can be rewritten as an inverse transform formula. We can rewrite the first shift formula as

$$\mathcal{L}^{-1}[F(s - \alpha)] = e^{\alpha t}\,\mathcal{L}^{-1}[F(s)].$$

If we replace s with $s + \alpha$, we obtain a formula that is easier to work with.

Fact: First Shift Formula—Inverse Version

$$\mathcal{L}^{-1}[F(s)] = e^{\alpha t}\, \mathcal{L}^{-1}[F(s + \alpha)]$$

This formula says that we can substitute $s + \alpha$ for s *inside* the transform, provided we put the exponential factor $e^{\alpha t}$ *outside*.

Example 6.4.3

Find $\mathcal{L}^{-1}\left[\dfrac{3}{(s - 2)^5}\right]$.

We know how to inverse-transform powers of s; in the present case, we note that the substitution of $s + 2$ for s would turn our problem into one of this type. The shift formula tells us how to carry out this substitution:

$$\mathcal{L}^{-1}\left[\frac{3}{(s - 2)^5}\right] = e^{2t}\, \mathcal{L}^{-1}\left[\frac{3}{(s + 2 - 2)^5}\right] = e^{2t}\, \mathcal{L}^{-1}\left[\frac{3}{s^5}\right]$$

$$= \frac{1}{8}\, e^{2t}t^4.$$

Example 6.4.4

Find $\mathcal{L}^{-1}\left[\dfrac{2}{(s + 4)^3}\right]$.

The inverse version of the first shift formula says

$$\mathcal{L}^{-1}\left[\frac{2}{(s + 4)^3}\right] = e^{-4t}\, \mathcal{L}^{-1}\left[\frac{2}{(s - 4 + 4)^3}\right] = e^{-4t}\, \mathcal{L}^{-1}\left[\frac{2}{s^3}\right]$$

$$= e^{-4t}t^2.$$

Example 6.4.5

Find $\mathcal{L}^{-1}\left[\dfrac{s}{(s - 1)^2 + 4}\right]$.

If the denominator had the form $s^2 + 4$, we could handle this using trigonometric functions. Therefore we try the substitution of $s + 1$ for s, which changes $(s - 1)^2$ into s^2. Note that we must also perform this substitution in the numerator.

$$\mathcal{L}^{-1}\left[\frac{s}{(s-1)^2+4}\right] = e^t\,\mathcal{L}^{-1}\left[\frac{s+1}{(s+1-1)^2+4}\right]$$

$$= e^t\,\mathcal{L}^{-1}\left[\frac{s}{s^2+4} + \frac{1}{s^2+4}\right]$$

$$= e^t\cos 2t + \frac{1}{2}\,e^t\sin 2t.$$

The last example was presented to us in a form suitable for shifting, but in practice it is more likely to occur in the form $s/(s^2 - 2s + 5)$. We could recover the more useful form by completing the square in the denominator. In general, to deal with terms of the form $(Bs + C)/(s^2 - bs + c)$ we first check to see whether the denominator can be factored. If it can, we use partial fractions; if it can't, we complete the square and shift.

The following example shows how this can come up in the context of solving an initial value problem.

Example 6.4.6

Solve the initial value problem

$$(D^2 + 2D + 2)x = 25te^t, \qquad x(0) = x'(0) = 0.$$

We transform both sides of the o.d.e. (using the first differentiation formula on the left and the first shift formula on the right):

$$s^2\,\mathcal{L}[x] + 2s\,\mathcal{L}[x] + 2\,\mathcal{L}[x] = \frac{25}{(s-1)^2}.$$

We solve for $\mathcal{L}[x]$:

$$\mathcal{L}[x] = \frac{25}{(s-1)^2(s^2+2s+2)}.$$

Expansion in partial fractions yields

$$\mathcal{L}[x] = \frac{-4}{s-1} + \frac{5}{(s-1)^2} + \frac{4s+7}{s^2+2s+2}$$

so that

$$x = -4\,\mathcal{L}^{-1}\left[\frac{1}{s-1}\right] + 5\,\mathcal{L}^{-1}\left[\frac{1}{(s-1)^2}\right] + \mathcal{L}^{-1}\left[\frac{4s+7}{s^2+2s+2}\right].$$

The first two terms in our expression for x are easily handled by the shift:

$$-4 \mathcal{L}^{-1}\left[\frac{1}{s-1}\right] + 5 \mathcal{L}^{-1}\left[\frac{1}{(s-1)^2}\right] = -4e^t \mathcal{L}^{-1}\left[\frac{1}{s}\right] + 5e^t \mathcal{L}^{-1}\left[\frac{1}{s^2}\right]$$

$$= -4e^t + 5te^t.$$

The third term requires us to complete the square in the denominator before shifting:

$$\mathcal{L}^{-1}\left[\frac{4s+7}{s^2+2s+2}\right] = \mathcal{L}^{-1}\left[\frac{4s+7}{(s+1)^2+1}\right]$$

$$= e^{-t} \mathcal{L}^{-1}\left[\frac{4(s-1)+7}{s^2+1}\right] = e^{-t} \mathcal{L}^{-1}\left[\frac{4s+3}{s^2+1}\right]$$

$$= 4e^{-t} \mathcal{L}^{-1}\left[\frac{s}{s^2+1}\right] + 3e^{-t} \mathcal{L}^{-1}\left[\frac{1}{s^2+1}\right]$$

$$= 4e^{-t} \cos t + 3e^{-t} \sin t.$$

Putting it all together, the solution of the initial value problem is

$$x = -4e^t + 5te^t + 4e^{-t} \cos t + 3e^{-t} \sin t.$$

Another important formula comes from differentiating the Laplace transform. By definition,

$$\frac{d}{ds} \mathcal{L}[f(t)] = \frac{d}{ds} \int_0^\infty e^{-st} f(t) \, dt.$$

For the functions we are interested in [$f(t)$ piecewise continuous and of exponential order], the differentiation can be carried out inside the integral sign. Thus,

$$\frac{d}{ds} \mathcal{L}[f(t)] = \int_0^\infty \left(\frac{\partial}{\partial s} e^{-st} f(t)\right) dt = -\int_0^\infty e^{-st} t f(t) \, dt$$

$$= -\mathcal{L}[tf(t)]$$

or

$$\mathcal{L}[tf(t)] = -\frac{d}{ds} \mathcal{L}[f(t)].$$

Repeated application of the last formula gives a more general version.

Fact: Second Differentiation Formula

$$\mathcal{L}[t^n f(t)] = (-1)^n \frac{d^n}{ds^n} \mathcal{L}[f(t)].$$

Example 6.4.7

Find $\mathcal{L}[t^2 \sin 3t]$.

The second differentiation formula says

$$\mathcal{L}[t^2 \sin 3t] = (-1)^2 \frac{d^2}{ds^2} \mathcal{L}[\sin 3t]$$

$$= \frac{d^2}{ds^2} \left(\frac{3}{s^2 + 9} \right) = \frac{d}{ds} \left(\frac{-6s}{(s^2 + 9)^2} \right)$$

$$= \frac{18(s^2 - 3)}{(s^2 + 9)^3}.$$

Example 6.4.8

Find $\mathcal{L}[te^{2t} \cos 3t]$.

We first use the second differentiation formula to find $\mathcal{L}[t \cos 3t]$:

$$\mathcal{L}[t \cos 3t] = -\frac{d}{ds} \mathcal{L}[\cos 3t] = -\frac{d}{ds} \left(\frac{s}{s^2 + 9} \right)$$

$$= \frac{s^2 - 9}{(s^2 + 9)^2}.$$

We can now use the first shift formula to get

$$\mathcal{L}[te^{2t} \cos 3t] = \frac{(s - 2)^2 - 9}{((s - 2)^2 + 9)^2} = \frac{s^2 - 4s - 5}{(s^2 - 4s + 13)^2}.$$

We note in passing that it is also possible to state the second differentiation formula in terms of the inverse transform (see Note 1). However, most problems to which this version can be applied can also be handled by the convolution methods we shall see in Section 6.7.

Let's summarize our new manipulation rules.

FURTHER PROPERTIES OF LAPLACE TRANSFORMS

First Shift Formula

Transform version:

$$\mathcal{L}[e^{\alpha t}f(t)] = F(s - \alpha), \text{ where } F(s) = \mathcal{L}[f(t)].$$

Inverse transform version:

$$\mathcal{L}^{-1}[F(s)] = e^{\alpha t}\,\mathcal{L}^{-1}[F(s + \alpha)].$$

Second Differentiation Formula

$$\mathcal{L}[t^n f(t)] = (-1)^n \frac{d^n}{ds^n}\,\mathcal{L}[f(t)].$$

Note

The second differentiation formula—inverse version

A straightforward restatement of the second differentiation formula in terms of $\mathcal{L}^{-1}$ is the following.

Fact: Second Differentiation Formula—Inverse Version

$$\mathcal{L}^{-1}\left[\frac{d^n}{ds^n}\,F(s)\right] = (-1)^n t^n\,\mathcal{L}^{-1}[F(s)].$$

This formula could be used, for example, to calculate $\mathcal{L}^{-1}\left[\dfrac{s}{(s^2 + \beta^2)^2}\right]$, provided we notice that

$$\frac{s}{(s^2 + \beta^2)^2} = -\frac{1}{2}\frac{d}{ds}\left(\frac{1}{s^2 + \beta^2}\right).$$

It follows that

$$\mathcal{L}^{-1}\left[\frac{s}{(s^2 + \beta^2)^2}\right] = -\frac{1}{2}\,\mathcal{L}^{-1}\left[\frac{d}{ds}\left(\frac{1}{(s^2 + \beta^2)^2}\right)\right],$$

which we calculate by the second differentiation formula (inverse version):

$$\mathscr{L}^{-1}\left[\frac{s}{(s^2 + \beta^2)^2}\right] = -\frac{1}{2}(-1)t\,\mathscr{L}^{-1}\left[\frac{1}{s^2 + \beta^2}\right] = \frac{t}{2\beta}\sin\beta t.$$

EXERCISES

Calculate the Laplace transforms of the functions in Exercises 1 to 13.

1. te^{3t} 2. t^2e^{3t} 3. $3t\sin 2t$ 4. $(t + 2)e^{-t}$

5. $t^2\sin 3t$ 6. $t^2\cos 3t$ 7. t^2e^{mt} 8. t^ne^{mt}

9. $e^{3t}\sin 2t$ 10. $e^{-t}\cos 3t$ 11. $te^{2t}\sin 3t$ 12. $te^{2t}\cos 3t$

13. $t\sin^2 2t$ (*Hint:* Use trig identities.)

Calculate the inverse Laplace transforms of the functions in Exercises 14 to 24.

14. $\dfrac{1}{(s - 2)^6}$ 15. $\dfrac{1}{(s + 2)^4}$ 16. $\dfrac{s - 1}{(s - 2)^2(s - 3)}$

17. $\dfrac{1}{s^2 + 6s + 9}$ 18. $\dfrac{s + 1}{s^2 + 6s + 9}$ 19. $\dfrac{1}{s^3 + 6s^2 + 9s}$

20. $\dfrac{1}{s^2 + 4s + 3}$ 21. $\dfrac{1}{s^2 + 4s + 5}$ 22. $\dfrac{1}{s^2 + 3s + 3}$

23. $\dfrac{s + 1}{s^2 + 3s + 3}$ 24. $\dfrac{1}{s^3 - 1}$

In Exercises 25 to 32, use Laplace transforms to solve the initial value problems.

25. $(D - 1)x = t^3e^t$; $x(0) = 0$

26. $(D^2 + 2D - 3)x = e^t$; $x(0) = 1, x'(0) = 0$

27. $(D^2 - 2D + 1)x = t^3e^t$; $x(0) = -1, x'(0) = 2$

28. $(D^2 + 2D + 2)x = 0$; $x(0) = x'(0) = 1$

29. $(D^2 - 1)x = e^t\cos 2t$; $x(0) = x'(0) = 0$

30. $(D^2 - D)x = 4te^{2t}$; $x(0) = x'(0) = 0$

31. $(D^3 - 1)x = 0$; $x(0) = x'(0) = 0, x''(0) = 1$

32. $(D^4 - 2D^2 + 1)x = 0$; $x(0) = x'(0) = x''(0) = 0, x'''(0) = -1$

In Exercises 33 to 37, which refer back to our models, solve the initial value problem of the indicated exercise.

33. Exercise 3, Section 2.1, with $V(t) = 0, R = 2, C = L = 1$, and subject to $Q(0) = 0$ and $I(0) = 1$

34. Exercise 3, Section 2.1, with $V(t) = 10, L = R = 1, C = 2$, and subject to $Q(0) = I(0) = 0$

35. Exercise 25(b), Section 2.8

36. Exercise 7(a), Section 6.1

37. Exercise 7(b), Section 6.1

More advanced problems:

38. Use the result of Exercise 24, Section 6.2, and the differentiation and shift formulas to find
     a. $\mathcal{L}[t^{1/2}]$ b. $\mathcal{L}[t^{3/2}]$ c. $\mathcal{L}[t^{-1/2} e^{3t}]$

39. a. Use the inverse version of the *first* differentiation formula (Note 3, Section 6.3)
     to obtain an expression for $\mathcal{L}^{-1}\left[\dfrac{1}{(s^2 + 1)^2}\right] = \mathcal{L}^{-1}\left[\dfrac{1}{s}\dfrac{s}{(s^2 + 1)^2}\right]$ in terms of
     $\mathcal{L}^{-1}\left[\dfrac{s}{(s^2 + 1)^2}\right]$.

     b. Substitute the value of $\mathcal{L}^{-1}\left[\dfrac{s}{(s^2 + 1)^2}\right]$ found in the note at the end of this section
     and calculate $\mathcal{L}^{-1}\left[\dfrac{1}{(s^2 + 1)^2}\right]$.

     c. Use the inverse version of the *second* differentiation formula to calculate
     $$\mathcal{L}^{-1}\left[\frac{s}{(s^2 + 1)^3}\right] = \mathcal{L}^{-1}\left[-\frac{1}{4}\frac{d}{ds}\left(\frac{1}{(s^2 + 1)^2}\right)\right].$$

     d. Calculate $\mathcal{L}^{-1}\left[\dfrac{1}{(s^2 + 1)^3}\right] = \mathcal{L}^{-1}\left[\dfrac{1}{s}\dfrac{s}{(s^2 + 1)^3}\right]$.

40. a. Use the differentiation formulas to express $\mathcal{L}[tx'(t)]$ in terms of $\mathcal{L}[x(t)]$.
     b. Transform the initial value problem

     (H) $(D^2 - 2tD + 2)x = 0;$ $x(0) = 0, x'(0) = 1$

     to obtain a first-order o.d.e. for $\mathcal{L}[x]$ as a function of s.
     c. Solve the equation obtained in (b) for $\mathcal{L}[x]$; your answer will contain a constant of integration.
     d. Find a value for the constant in (c) so that $\mathcal{L}[x]$ is the transform of a polynomial; check that this polynomial solves (H).

41. Follow the procedure described in Exercise 40 to find a polynomial solution of the initial value problem

     $$(D^2 - 2tD + 4)x = 0;   x(0) = 1, x'(0) = 0.$$

     (*Hint:* $\int s^3 e^{s^2/4} ds = \int s^2 s e^{s^2/4} ds$ can be calculated using integration by parts.)

42. a. Use the differentiation formulas to express $\mathcal{L}[t^2 x''(t)]$ in terms of $\mathcal{L}[x(t)]$.
     b. Transform the **Bessel equation**

     (H) $t^2 x'' + t x' + (t^2 - n^2)x = 0.$

     Note that the transformed equation is independent of the initial conditions and is no simpler than the original equation.

6.5 FUNCTIONS DEFINED IN PIECES

We saw in Section 6.1 how circuits with switches, and mechanical systems with suddenly changing forces, are most simply and accurately described using **functions defined in pieces**—that is, functions whose values are defined by several formulas, each applied over a different piece of the domain. In this section we develop a quick scheme for computing the transforms of such functions.

Our first step in this scheme is to develop a better notation for functions defined in pieces. This is accomplished through the use of the **unit step function:**

$$u_a(t) = \begin{cases} 0 & \text{if } t < a \\ 1 & \text{if } t \geq a. \end{cases}$$

We sketch the graph of $u_a(t)$ in Figure 6.10. (Recall that we calculated the transform of $u_a(t)$ for $a > 0$ in Example 6.2.7.)

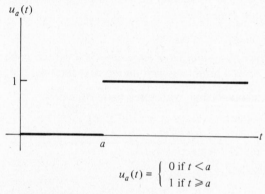

$$u_a(t) = \begin{cases} 0 \text{ if } t < a \\ 1 \text{ if } t \geq a \end{cases}$$

FIGURE 6.10

The unit step function $u_a(t)$ has the effect of a mathematical "on" switch at $t = a$. If we multiply a function $f(t)$ by $u_a(t)$, the product will be zero until $t = a$ and will switch to $f(t)$ thereafter:

$$u_a(t)f(t) = \begin{cases} 0 & \text{if } t < a \\ f(t) & \text{if } t \geq a. \end{cases}$$

Thus, for example, the function

$$g(t) = \begin{cases} 0 & \text{if } t < 2 \\ e^{-t} & \text{if } t \geq 2 \end{cases}$$

can be written as

$$g(t) = u_2(t)e^{-t}.$$

Of course, the functions in which we are interested often switch from one nonzero formula to another. For example, we might have

$$g(t) = \begin{cases} t & \text{if } t < 3 \\ e^{-t} & \text{if } t \geq 3. \end{cases}$$

We can think of this as a function that starts out equal to t. At time $t = 3$, a switch does two things: it turns *off* the first formula and turns *on* the second. To turn *off* the formula $g(t) = t$ at $t = 3$, we *subtract* $u_3(t)t$, while to turn *on* $g(t) = e^{-t}$ at $t = 3$, we *add* $u_3(t)e^{-t}$. Thus

$$g(t) = t + u_3(t)(-t + e^{-t}).$$

In general, to express a function defined in pieces by means of step function notation, we need only remember

i. $u_a(t)$ switches on at $t = a$, and
ii. at every interface, we *subtract* a term to switch *off* the previous formula and simultaneously *add* a term to switch *on* the next formula.

We illustrate this procedure with some more examples.

Example 6.5.1

Rewrite $g(t) = |2t - 1|$ in step function notation.
The definition of absolute value gives

$$g(t) = \begin{cases} -(2t - 1) & \text{if } 2t - 1 < 0 \\ 2t - 1 & \text{if } 2t - 1 \geq 0 \end{cases}$$

$$= \begin{cases} 1 - 2t & \text{if } t < 1/2 \\ 2t - 1 & \text{if } t \geq 1/2. \end{cases}$$

The initial formula is $g(t) = 1 - 2t$. At time $t = 1/2$ we use $u_{1/2}(t)$ to switch *off* $1 - 2t$ and to switch *on* $2t - 1$:

$$g(t) = 1 - 2t + u_{1/2}(t)\{-(1 - 2t) + (2t - 1)\}$$
$$= 1 - 2t + u_{1/2}(t)(4t - 2).$$

Example 6.5.2

Rewrite the following function in step function notation:

$$g(t) = \begin{cases} 2 & \text{if } t < 1 \\ 3t & \text{if } 1 \le t < 2 \\ 5 & \text{if } 2 \le t. \end{cases}$$

Here, the initial formula is the constant 2, and there are two switching times, $t = 1$ and $t = 2$. At $t = 1$, we use $u_1(t)$ to switch off $g(t) = 2$ and to switch on $g(t) = 3t$. At $t = 2$, we use $u_2(t)$ to switch off the *previous* formula, $g(t) = 3t$, and to switch on $g(t) = 5$. Thus

$$g(t) = 2 + u_1(t)(-2 + 3t) + u_2(t)(-3t + 5).$$

Step function notation lets us rewrite any function defined in pieces as a sum of terms of the form $u_a(t)f(t)$ and reduces the calculation of the transform of such a function to finding the transform of this new kind of term. Keep in mind that $\mathscr{L}$ is an integral over the domain $0 \le t$. Any formula that affects only negative values of t has no effect on the transform. Thus we need only consider transforming terms of the form $u_a(t)f(t)$, with $a \ge 0$.

The definition of $\mathscr{L}$ gives

$$\mathscr{L}[u_a(t)f(t)] = \int_0^\infty e^{-st}u_a(t)f(t) \, dt.$$

We break up this integral at $t = a$ and substitute the value of $u_a(t)$ on each piece of the domain separately:

$$\mathscr{L}[u_a(t)f(t)] = \int_0^a e^{-st} \, 0 \, f(t) \, dt + \int_a^\infty e^{-st} \, 1 \, f(t) \, dt$$

$$= \int_a^\infty e^{-st} f(t) \, dt.$$

This differs from the Laplace transform of $f(t)$ in that the lower limit of integration is not zero. However, if we set

$$\tau = t - a,$$

then $d\tau = dt$, $\tau = 0$ when $t = a$, and $\tau \to \infty$ as $t \to \infty$. Hence, by rewriting our integral in terms of τ, we have

$$\mathcal{L}[u_a(t)f(t)] = \int_{\tau=0}^{\infty} e^{-s(\tau+a)}f(\tau + a)\, d\tau$$

$$= e^{-as} \int_0^{\infty} e^{-s\tau}f(\tau + a)\, d\tau.$$

The last integral is just the transform of $f(t + a)$. Thus we have the following.

Fact: Second Shift Formula.   *For $a \ge 0$,*

$$\mathcal{L}[u_a(t)f(t)] = e^{-as}\, \mathcal{L}[f(t + a)].$$

We apply this formula to calculate the transforms of some of the functions we considered earlier in this section.

Example 6.5.3
Find $\mathcal{L}[u_2(t)e^{-t}]$.
The second shift formula says

$$\mathcal{L}[u_2(t)e^{-t}] = e^{-2s}\, \mathcal{L}[e^{-(t+2)}] = e^{-2s}\, \mathcal{L}[e^{-2}\, e^{-t}]$$

$$= e^{-2s}e^{-2}\, \mathcal{L}[e^{-t}]$$

$$= e^{-2(s+1)}\, \frac{1}{s + 1}.$$

Example 6.5.4
Find $\mathcal{L}[|2t - 1|]$.
We saw in Example 6.5.1 that

$$|2t - 1| = 1 - 2t + u_{1/2}(t)(4t - 2).$$

Thus

$$\mathcal{L}[|2t - 1|] = \mathcal{L}[1] - \mathcal{L}[2t] + \mathcal{L}[u_{1/2}(t)(4t - 2)]$$

$$= \frac{1}{s} - \frac{2}{s^2} + e^{-s/2}\, \mathcal{L}\left[4\left(t + \frac{1}{2}\right) - 2\right]$$

$$= \frac{1}{s} - \frac{2}{s^2} + e^{-s/2} \mathcal{L}[4t]$$

$$= \frac{1}{s} - \frac{2}{s^2} + \frac{4e^{-s/2}}{s^2}.$$

Example 6.5.5

Find $\mathcal{L}[g(t)]$, where

$$g(t) = \begin{cases} 2 & \text{if } t < 1 \\ 3t & \text{if } 1 \le t < 2 \\ 5 & \text{if } 2 \le t. \end{cases}$$

Using the result of Example 6.5.2, we have

$$\mathcal{L}[g(t)] = \mathcal{L}[2 + u_1(t)(-2 + 3t) + u_2(t)(-3t + 5)]$$
$$= \mathcal{L}[2] + \mathcal{L}[u_1(t)(-2 + 3t)] + \mathcal{L}[u_2(t)(-3t + 5)]$$
$$= \frac{2}{s} + e^{-s} \mathcal{L}[-2 + 3(t + 1)] + e^{-2s} \mathcal{L}[-3(t + 2) + 5]$$
$$= \frac{2}{s} + e^{-s} \mathcal{L}[1 + 3t] + e^{-2s} \mathcal{L}[-1 - 3t]$$
$$= \frac{2}{s} + \left(e^{-s} - e^{-2s}\right)\left(\frac{1}{s} + \frac{3}{s^2}\right).$$

If we set $f(t - a) = h(t)$, then the second shift formula says

$$\mathcal{L}[u_a(t)f(t - a)] = \mathcal{L}[u_a(t)h(t)] = e^{-as} \mathcal{L}[h(t + a)]$$
$$= e^{-as} \mathcal{L}[f(t)].$$

This translates easily into the following inverse transform statement:

Fact: Second Shift Formula—Inverse Version. *If $\mathcal{L}^{-1}[F(s)] = f(t)$, then*

$$\mathcal{L}^{-1}[e^{-as} F(s)] = u_a(t)f(t - a).$$

Example 6.5.6

Find $\mathcal{L}^{-1}\left[\dfrac{e^{-3s}}{s^2 + 4}\right]$.

We first find

$$f(t) = \mathcal{L}^{-1}\left[\frac{1}{s^2 + 4}\right] = \frac{1}{2}\sin 2t.$$

Then the inverse transform version of the second shift formula says

$$\mathcal{L}^{-1}\left[\frac{e^{-3s}}{s^2 + 4}\right] = u_3(t)f(t - 3) = u_3(t)\frac{\sin 2(t - 3)}{2}$$

$$= \begin{cases} 0 & \text{if } 0 \le t < 3 \\ \dfrac{1}{2}\sin(2t - 6) & \text{if } 3 \le t. \end{cases}$$

Example 6.5.7

Solve the initial value problem

$$D^2x - x = \begin{cases} t & \text{if } t < 1 \\ 0 & \text{if } t \ge 1 \end{cases} \qquad x(0) = x'(0) = 0.$$

We first rewrite the forcing term using $u_1(t)$:

$$D^2x - x = t + u_1(t)(-t).$$

Now we transform both sides and solve for $\mathcal{L}[x]$:

$$s^2\,\mathcal{L}[x] - \mathcal{L}[x] = \frac{1}{s^2} + e^{-s}\mathcal{L}[-(t + 1)]$$

$$= \frac{1}{s^2} - e^{-s}\left(\frac{1}{s^2} + \frac{1}{s}\right)$$

$$\mathcal{L}[x] = \frac{1}{s^2(s^2 - 1)} - e^{-s}\left(\frac{1}{s^2(s^2 - 1)} + \frac{1}{s(s^2 - 1)}\right).$$

Thus

$$x = \mathcal{L}^{-1}\left[\frac{1}{s^2(s^2 - 1)}\right] + \mathcal{L}^{-1}\left[e^{-s}\left(-\frac{1}{s^2(s^2 - 1)} - \frac{1}{s(s^2 - 1)}\right)\right].$$

We find and substitute the partial fraction decompositions for $1/s^2(s^2 - 1)$ and $1/s(s^2 - 1)$ to get

$$x = \mathcal{L}^{-1}\left[-\frac{1}{s^2} + \frac{1/2}{s-1} - \frac{1/2}{s+1} \right] + \mathcal{L}^{-1}\left[e^{-s}\left(\frac{1}{s} + \frac{1}{s^2} - \frac{1}{s-1} \right) \right].$$

The first part of our expression is easy to find:

$$\mathcal{L}^{-1}\left[-\frac{1}{s^2} + \frac{1/2}{s-1} - \frac{1/2}{s+1} \right] = -t + \frac{1}{2}e^t - \frac{1}{2}e^{-t}.$$

To find the second part, we begin by calculating

$$f(t) = \mathcal{L}^{-1}\left[\frac{1}{s} + \frac{1}{s^2} - \frac{1}{s-1} \right] = 1 + t - e^t$$

and then use the inverse version of the second shift formula:

$$\mathcal{L}^{-1}\left[e^{-s}\left(\frac{1}{s} + \frac{1}{s^2} - \frac{1}{s-1} \right) \right] = u_1(t)f(t-1)$$

$$= u_1(t)(t - e^{t-1}).$$

Thus

$$x = -t + \frac{1}{2}e^t - \frac{1}{2}e^{-t} + u_1(t)(t - e^{t-1}),$$

or, written in pieces,

$$x = \begin{cases} -t + \dfrac{1}{2}e^t - \dfrac{1}{2}e^{-t} & \text{if } t < 1 \\[3mm] \left(\dfrac{1}{2} - e^{-1} \right)e^t - \dfrac{1}{2}e^{-t} & \text{if } t \geq 1. \end{cases}$$

Note that even though the forcing function is discontinuous at $t = 1$, our solution is continuous.

The ease with which the Laplace transform lets us handle functions defined in pieces is one of its main advantages.

FUNCTIONS DEFINED IN PIECES

Any function $g(t)$ that is defined in pieces can be written as a sum of terms of the form $u_a(t)f(t)$, where $u_a(t)$ is the **unit step function**

$$u_a(t) = \begin{cases} 0 & \text{if } t < a \\ 1 & \text{if } t \geq a. \end{cases}$$

To obtain this expression for $g(t)$, one notes that

i. $u_a(t)$ switches "on" at $t = a$, and
ii. at every interface of $g(t)$, we want to switch "off" the immediately preceding formula and switch "on" the next one.

The Laplace transforms of functions defined in pieces are governed by the

Second Shift Formula

Transform version: for $a > 0$,

$$\mathcal{L}[u_a(t)f(t)] = e^{-as}\,\mathcal{L}[f(t + a)].$$

Inverse transform version: for $a > 0$,

$$\mathcal{L}^{-1}[e^{-as}F(s)] = u_a(t)f(t - a), \text{ where } f(t) = \mathcal{L}^{-1}[F(s)].$$

The unit step function $u_a(t)$ is sometimes called the "Heaviside function" after Oliver Heaviside, who pioneered such operational methods in his treatise "Electromagnetic Theory" (1899); another pioneer in this development was George Boole, after whom Boolean algebra is named.

Notes
1. Value at an interface
As we defined it, the unit step function $u_a(t)$ takes on the value 1 at $t = a$. Thus, strictly speaking, the first statement of our summary is true only if, at each interface, the value of $g(t)$ is given by the *later* formula. In general, our procedure for expressing $g(t)$ as a sum of terms of the form $u_a(t)f(t)$ yields a function $h(t)$ that equals $g(t)$ *except possibly* at the interfaces. For example, the function

$$g(t) = \begin{cases} t & \text{if } t \leq 2 \\ 1 & \text{if } t > 2 \end{cases}$$

agrees with

$$h(t) = t + u_2(t)(-t + 1)$$

for all values of t except $t = 2$, where $g(2) = 2$ while $h(2) = 1$. Keep in mind, however, that we are interested in step function notation only as a tool for finding Laplace transforms. Two functions that agree except at a few isolated points have the same integral and hence the same Laplace transform. Thus, we can simply ignore the difference between $g(t)$ and $h(t)$ when taking transforms.

2. Periodic functions

A function $f(t)$ is **periodic**, of **period** $p > 0$, provided

$$f(t + p) = f(t)$$

for all t. We have learned to transform two kinds of periodic functions: $\sin \beta t$ and $\cos \beta t$, each with period $2\pi/\beta$. In practice (especially in circuit theory), it is often useful to deal with models involving other kinds of periodic functions. Two common examples are the **square wave**

$$f(t) = \begin{cases} 1 & \text{if} \quad 2n \quad \leq t < 2n + 1, \quad n \text{ an integer} \\ -1 & \text{if } 2n + 1 \quad \leq t < 2n + 2, \quad n \text{ an integer} \end{cases}$$

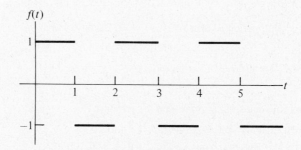

FIGURE 6.11 SQUARE WAVE.

of period $p = 2$ (see Figure 6.11), and the **saw-tooth function** (also called the fractional part of t),

$$f(t) = t - n \quad \text{when } n \leq t < n + 1,$$

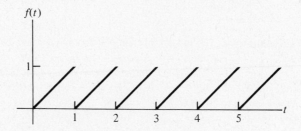

FIGURE 6.12 SAW-TOOTH FUNCTION.

of period $p = 1$ (see Figure 6.12).

It seems clear that one ought to be able to determine the transform of a periodic function just from the knowledge of one cycle. In fact, such a formula can be obtained by noting that if $f(t)$ has period p, then

$$f(t) - u_p(t)f(t) = \begin{cases} f(t) & \text{if } 0 \leq t < p \\ 0 & \text{if } p \leq t. \end{cases}$$

Transforming this expression, we get

$$\mathcal{L}[f(t) - u_p(t)f(t)] = \int_0^\infty e^{-st}\{f(t) - u_p(t)f(t)\}\, dt$$

$$= \int_0^p e^{-st}f(t)\, dt.$$

We can use the second shift formula to rewrite the left side:

$$\mathcal{L}[f(t)] - e^{-ps}\,\mathcal{L}[f(t + p)] = \int_0^p e^{-st}f(t)\, dt.$$

But if f is periodic with period p, then $f(t + p) = f(t)$ and we have

$$\mathcal{L}[f(t)] - e^{-ps}\,\mathcal{L}[f(t)] = \int_0^p e^{-st}f(t)\, dt.$$

We can solve this equation for $\mathcal{L}[f(t)]$ to find the formula

$$\mathcal{L}[f(t)] = \frac{1}{1 - e^{-ps}} \int_0^p e^{-st}f(t)\, dt$$

for f periodic of period $p > 0$.

You should use this formula to work out the transforms of the square wave and the sawtooth function (see Exercise 36).

EXERCISES

In Exercises 1 to 12, (a) express $f(t)$ in step function notation and (b) find $\mathcal{L}[f(t)]$.

1. $f(t) = \begin{cases} 0 & t < 3 \\ t - 3 & t \geq 3 \end{cases}$
2. $f(t) = \begin{cases} t - 3 & t < 3 \\ 0 & t \geq 3 \end{cases}$

3. $f(t) = \begin{cases} 0 & t < \pi \\ \sin t & t \geq \pi \end{cases}$
4. $f(t) = \begin{cases} 0 & t < 2 \\ e^{-t} & t \geq 2 \end{cases}$

5. $f(t) = \begin{cases} 0 & t < 1 \\ e^t & 1 \leq t < 2 \\ 0 & 2 \leq t \end{cases}$
6. $f(t) = \begin{cases} t^2 & t < 1 \\ t^2 - 1 & 1 \leq t < 2 \\ t^2 - 2 & 2 \leq t \end{cases}$

7. $f(t) = |t - 3|$
8. $f(t) = |(t - 2)(t - 1)|$

9. $f(t)$ as in Figure 6.13(a)

10. $f(t)$ as in Figure 6.13(b)

11. $f(t)$ as in Figure 6.13(c)

12. $f(t)$ as in Figure 6.13(d)

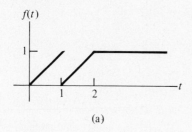

(a)

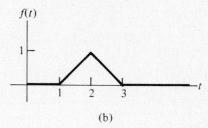

(b)

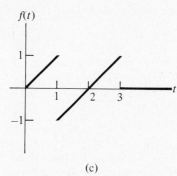

(c)

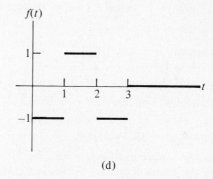

(d)

FIGURE 6.13

Find the inverse Laplace transforms of the functions in Exercises 13 to 20.

13. $\dfrac{e^{-4s}}{s + 4}$

14. $\dfrac{e^{-4s}}{s^2 + 4}$

15. $\dfrac{e^{-4s}}{(s + 4)^2}$

16. $\dfrac{se^{-\pi s}}{s^2 + 4}$

17. $\dfrac{s + e^{-\pi s}}{s^2 + 1}$

18. $\dfrac{e^{-(s+1)}}{s + 1}$

19. $\dfrac{e^{-s}}{s(s + 1)}$

20. $\dfrac{e^{-2s}}{s(s + 1)}$

Solve the initial value problems in Exercises 21 to 29.

21. $(D - 1)x = \begin{cases} 0 & t < 3 \\ t - 3 & 3 \le t \end{cases}$ $x(0) = -5$

22. $x'' + 2x' + x = \begin{cases} 0 & t < 1 \\ 1 & t \ge 1 \end{cases}$ $x(0) = 0,\ x'(0) = -3$

23. $(D^2 - 4)x = \begin{cases} 0 & t < 2 \\ 2 & t \ge 2 \end{cases}$ $x(0) = 1,\ x'(0) = 0$

24. $(D^2 - 4)x = \begin{cases} 2 & t < 2 \\ 0 & t \ge 2 \end{cases}$ $x(0) = 1,\ x'(0) = 0$

25. $(D^2 + 1)x = \begin{cases} 0 & t < 2 \\ e^{-t} & t \ge 2 \end{cases}$ $x(0) = x'(0) = 0$

26. $(D^2 + 1)x = \begin{cases} e^{-t} & t < 2 \\ 0 & t \geq 2 \end{cases}$ $x(0) = x'(0) = 0$

27. $(D^2 + 1)x = \begin{cases} \sin 2t & t < \pi \\ -\sin 2t & t \geq \pi \end{cases}$ $x(0) = x'(0) = 0$

28. $(D^3 - D)x = \begin{cases} 1 & t < 2 \\ 0 & t \geq 2 \end{cases}$ $x(0) = x'(0) = x''(0) = 0$

29. $(D + 1)^3 x = \begin{cases} e^{-t} & t < 3 \\ 0 & t \geq 3 \end{cases}$ $x(0) = x'(0) = x''(0) = 0$

In Exercises 30 to 33, which refer to our models, (a) solve the initial value problem of the indicated exercise in Section 6.1 and (b) find the value of the solution at each of the indicated times.

30. Exercise 1; $t = 1/2$, $t = 3$
31. Exercise 3; $t = 1$, $t = 4$
32. Exercise 5, with $L = 1$, $R = 0$, $C = 1/4$; $t = 5$, $t = 20$
33. Exercise 6, with $L = 1$, $R = 3$, $C = 1/2$, and

    i. $V_1(t) = 0$, $V_2(t) = 10$; $t = 1$, $t = 3$, $t = 6$
    ii. $V_1(t) = 10$, $V_2(t) = 0$; $t = 1$, $t = 3$, $t = 6$

34. a. Use Laplace transforms to solve the initial value problem (N_3) of Example 6.1.1.
    b. Use the result from (a) to estimate the Maniac population in January 1985.

More advanced problems:

35. Use the result of Exercise 24, Section 6.2, to find $\mathcal{L}[f(t)]$ where

$$f(t) = \begin{cases} 0, & t < k \\ (t - k)^{-1/2}, & t \geq k. \end{cases}$$

36. Use the formula for the Laplace transform of a periodic function (Note 2) to find the transform of

    a. the square wave (see Note 2)
    b. the saw-tooth function (see Note 2)
    c. $|\sin t|$

37. Show that if $k \geq 0$, then

$$\mathcal{L}[f(t + k)] = e^{ks}\,\mathcal{L}[f(t)] - e^{ks}\int_0^k e^{-st} f(t)\, dt.$$

[Hint: Find $\mathcal{L}[f(t) - u_k(t)g(t - k)]$ where $g(t) = f(t + k)$.]

38. Suppose $f(t + w) = -f(t)$ for all t.
    a. Show that $f(t)$ is periodic of period $2w$.
    b. Find a formula for $\mathcal{L}[f(t)]$ in terms of the behavior of $f(t)$ for $0 \leq t \leq w$. [Hint: Consider $f(t) - u_w(t)f(t)$.]

c. The half-wave rectification of $f(t)$ is the function $h(t)$ that is periodic of period $2w$ and satisfies

$$h(t) = \begin{cases} f(t) & 0 \le t < w \\ 0 & w \le t < 2w. \end{cases}$$

Find $\mathscr{L}[h(t)]$.

6.6 CONVOLUTION

In this section we develop a direct way to inverse-transform products. The method uses an operation on functions, the convolution, which has many uses beyond the specific application we consider here.

Definition: *Given two functions $f(t)$ and $g(t)$, we define a new function, called the* **convolution** *of f and g and denoted $f * g$, by the rule*

$$(f * g)(t) = \int_0^t f(t - u)g(u)\, du.$$

This formula, like the definition of $\mathscr{L}[f(t)]$, becomes easier to understand if we pause to consider some features of the defining integral.

i.   The formula assigns a numerical value, $(f * g)(t)$, to each specific value of t, so that as far as the integration is concerned, t *acts like a constant.*

ii.   The limits of integration refer to u: we integrate from $u = 0$ to $u = t$.

iii.   Although the second function enters the integrand in a straightforward way, the first has an argument depending on both t and u. In practice, we try to rewrite $f(t - u)$ in terms of functions of t and u alone before integrating.

We calculate a simple example to get a feeling for how this operation works.

Example 6.6.1
Find $(f * g)(t)$ when $f(t) = e^{2t}$ and $g(t) = e^{3t}$.
The definition says

$$(f * g)(t) = \int_0^t f(t - u)g(u)\, du = \int_0^t e^{2(t-u)}\, e^{3u}\, du$$

$$= e^{2t} \int_0^t e^u\, du = e^{2t}(e^t - 1) = e^{3t} - e^{2t}.$$

The convolution is written so as to resemble a product. In fact, it shares a number of properties with products:

the distributive law

$$f * (c_1 g_1 + c_2 g_2) = c_1(f * g_1) + c_2(f * g_2)$$

the associative law

$$f * (g * h) = (f * g) * h$$

the commutative law

$$f * g = g * f$$

(see Exercises 31 and 32). However, one must be careful not to carry the analogy with ordinary products too far. For example, it is tempting to assume that the convolution of a function $g(t)$ with the constant $f(t) = 1$ leaves $g(t)$ unchanged. But this is *not* so; in fact, for any function $g(t)$,

$$(1 * g)(t) = \int_0^t g(u) \, du.$$

In Note 3 we discuss a physical motivation for the formula giving the convolution. For the moment, though, we concentrate on its usefulness in calculating inverse Laplace transforms. This is a result of the fact that the transform turns convolutions into products.

Fact: Convolution Formula.   *If $f(t)$ and $g(t)$ both have Laplace transforms, then*

$$\mathcal{L}[(f * g)(t)] = \mathcal{L}[f(t)] \, \mathcal{L}[g(t)].$$

A proof of this fact is sketched in a note at the end of this section. We make use of this formula primarily in its inverse-transform version:

Fact: Convolution Formula—Inverse Transform Version.   *If $F(s)$ and $G(s)$ have inverse Laplace transforms, then*

$$\mathcal{L}^{-1}[F(s)G(s)] = \mathcal{L}^{-1}[F(s)] * \mathcal{L}^{-1}[G(s)].$$

For practice, we calculate three specific inverse transforms using this formula.

Example 6.6.2

Find $\mathcal{L}^{-1}\left[\dfrac{1}{(s-2)(s-3)}\right]$.

Of course, we could do this problem by partial fractions. Alternatively, we can use the convolution formula and the result of Example 6.6.1:

$$\mathcal{L}^{-1}\left[\frac{1}{(s-2)(s-3)}\right] = \mathcal{L}^{-1}\left[\frac{1}{(s-2)}\right] * \mathcal{L}^{-1}\left[\frac{1}{(s-3)}\right]$$

$$= e^{2t} * e^{3t} = e^{3t} - e^{2t}.$$

Example 6.6.3

Find $\mathcal{L}^{-1}\left[\dfrac{s}{(s^2+1)^2}\right]$.

Note that partial fractions can't help us here, since the function $F(s)$ is already written in its partial fraction decomposition. On the other hand, if we regard it as the product of $s/(s^2+1)$ and $1/(s^2+1)$, whose inverse transforms we know, then

$$\mathcal{L}^{-1}\left[\frac{s}{(s^2+1)^2}\right] = \mathcal{L}^{-1}\left[\frac{s}{s^2+1} \cdot \frac{1}{s^2+1}\right] = \mathcal{L}^{-1}\left[\frac{s}{s^2+1}\right] * \mathcal{L}^{-1}\left[\frac{1}{s^2+1}\right]$$

$$= \cos t * \sin t = \int_0^t \cos(t-u)\sin u\, du.$$

We calculate and simplify this integral with the help of several trigonometric identities:

$$\mathcal{L}^{-1}\left[\frac{s}{(s^2+1)^2}\right] = \int_0^t (\cos t \cos u + \sin t \sin u)\sin u\, du$$

$$= (\cos t)\int_0^t \cos u \sin u\, du + (\sin t)\int_0^t \sin^2 u\, du$$

$$= (\cos t)\int_0^t \sin u\, d(\sin u) + (\sin t)\int_0^t \left(\frac{1}{2} - \frac{1}{2}\cos 2u\right) du$$

$$= (\cos t)\left(\frac{1}{2}\sin^2 u\right)\Big]_0^t + (\sin t)\left(\frac{u}{2} - \frac{1}{4}\sin 2u\right)\Big]_0^t$$

$$= \frac{1}{2}\cos t \sin^2 t + \frac{t}{2}\sin t - \frac{1}{4}\sin t \sin 2t$$

$$= \frac{1}{2}\cos t \sin^2 t + \frac{t}{2}\sin t - \frac{1}{2}\cos t \sin^2 t$$

$$= \frac{t}{2}\sin t.$$

(You might compare this with the calculation in the note to Section 6.4, which depended on cleverness in recognizing $F(s)$ as a derivative.)

Example 6.6.4

Find $\mathcal{L}^{-1}\left[\dfrac{1}{(s^2 + 1)^2}\right]$.

Again, we regard $F(s)$ as a product and calculate its inverse transform as a convolution, with the aid of trigonometric identities:

$$\mathcal{L}^{-1}\left[\frac{1}{(s^2 + 1)^2}\right] = \mathcal{L}^{-1}\left[\frac{1}{s^2 + 1}\right] * \mathcal{L}^{-1}\left[\frac{1}{s^2 + 1}\right]$$

$$= \sin t * \sin t = \int_0^t \sin(t - u) \sin u \, du$$

$$= \int_0^t (\sin t \cos u - \cos t \sin u) \sin u \, du$$

$$= (\sin t) \int_0^t \cos u \sin u \, du - (\cos t) \int_0^t \sin^2 u \, du$$

$$= (\sin t)\left(\frac{1}{2}\sin^2 t\right) - (\cos t)\left(\frac{t}{2} - \frac{1}{4}\sin 2t\right)$$

$$= \frac{1}{2}\sin t \sin^2 t - \frac{t}{2}\cos t + \frac{1}{2}\sin t \cos^2 t$$

$$= \frac{1}{2}\sin t - \frac{t}{2}\cos t.$$

(An alternate calculation of this problem is outlined in Exercise 39, Section 6.4.)

Inverse Laplace transforms like the two in Examples 6.6.3 and 6.6.4 arise naturally in certain initial value problems.

Example 6.6.5

Find the solution of the initial value problem

$$(D^2 + 1)x = \cos t, \qquad x(0) = x'(0) = 0.$$

When we transform both sides of the o.d.e., incorporating the initial data via the first differentiation formula, we obtain

$$(s^2 + 1)\, \mathcal{L}[x] = \mathcal{L}[\cos t] = \frac{s}{s^2 + 1}.$$

Solving for $\mathcal{L}[x]$, we get

$$\mathcal{L}[x] = \frac{s}{(s^2 + 1)^2}.$$

Inverse-transforming to get x, we are led to Example 6.6.3:

$$x = \mathcal{L}^{-1}\left[\frac{s}{(s^2 + 1)^2}\right] = \cos t * \sin t = \frac{t}{2} \sin t.$$

In cases where partial fractions and the convolution are both applicable for finding inverse transforms, the convolution is usually harder. Nevertheless, it can be a useful auxiliary device, as illustrated in the following example.

Example 6.6.6

Solve the initial value problem

$$(D^4 - 1)x = \begin{cases} 2 \text{ if } t < 3 \\ 0 \text{ if } t \ge 3 \end{cases} \qquad x(0) = x'(0) = x''(0) = 0,\ x'''(0) = 2.$$

In step function notation, the o.d.e. reads

$$(D^4 - 1)x = 2 - 2u_3(t),$$

so that its transform is

$$s^4\, \mathcal{L}[x] - 2 - \mathcal{L}[x] = \frac{2}{s} - \frac{2}{s}\, e^{-3s}.$$

Solving for $\mathcal{L}[x]$, we find

$$\mathcal{L}[x] = \frac{2}{s^4 - 1} + \frac{2}{s(s^4 - 1)} - \frac{2e^{-3s}}{s(s^4 - 1)}.$$

The first term can be decomposed into partial fractions as

$$\frac{2}{s^4 - 1} = \frac{1}{2} \frac{1}{s - 1} - \frac{1}{2} \frac{1}{s + 1} - \frac{1}{s^2 + 1}$$

so its inverse transform is

$$\mathcal{L}^{-1} \left[\frac{2}{s^4 - 1} \right] = \frac{1}{2} e^t - \frac{1}{2} e^{-t} - \sin t.$$

The second term could also be decomposed into partial fractions. However, note that it is the same as the first term multiplied by $1/s$. We can thus use the convolution formula to obtain its inverse transform from the previous one:

$$\mathcal{L}^{-1} \left[\frac{2}{s(s^4 - 1)} \right] = \mathcal{L}^{-1} \left[\frac{1}{s} \right] * \mathcal{L}^{-1} \left[\frac{2}{s^4 - 1} \right] = 1 * \left(\frac{1}{2} e^t - \frac{1}{2} e^{-t} - \sin t \right)$$

$$= \int_0^t \left(\frac{1}{2} e^u - \frac{1}{2} e^{-u} - \sin u \right) du$$

$$= \frac{1}{2} (e^t - 1) + \frac{1}{2} (e^{-t} - 1) + (\cos t - 1)$$

$$= \frac{1}{2} e^t + \frac{1}{2} e^{-t} + \cos t - 2.$$

Finally, the third term is an exponential times the second. We obtain its inverse transform from that of the second term by means of the second shift formula:

$$\mathcal{L}^{-1} \left[\frac{2e^{-3s}}{s(s^4 - 1)} \right] = u_3(t) \left\{ \frac{1}{2} e^{t-3} + \frac{1}{2} e^{-(t-3)} + \cos(t - 3) - 2 \right\}.$$

Combining all three terms, we have the solution of our initial value problem:

$$x = e^t + \cos t - \sin t - 2 - u_3(t) \left\{ \frac{1}{2} e^{t-3} + \frac{1}{2} e^{-(t-3)} + \cos (t - 3) - 2 \right\}.$$

Let's summarize.

CONVOLUTION

Definition

$$(f * g)(t) = \int_0^t f(t - u) g(u) \, du$$

Algebraic Properties

distributive: $f * (c_1g_1 + c_2g_2) = c_1(f * g_1) + c_2(f * g_2)$

associative: $f * (g * h) = (f * g) * h$

commutative: $f * g = g * f$

Convolution Formula

$$\mathcal{L}[(f * g)(t)] = \mathcal{L}[f(t)] \, \mathcal{L}[g(t)]$$
$$\mathcal{L}^{-1}[F(s)G(s)] = \mathcal{L}^{-1}[F(s)] * \mathcal{L}^{-1}[G(s)]$$

Notes
1. *Proof of the convolution formula*
Substituting the definition of $f * g$ into the definition of the Laplace transform, we get

$$\mathcal{L}[(f * g)(t)] = \int_0^\infty e^{-st}(f * g)(t) \, dt = \int_0^\infty e^{-st} \left(\int_0^t f(t - u)g(u) \, du \right) dt$$
$$= \int_0^\infty \int_0^t e^{-st}f(t - u)g(u) \, du \, dt.$$

The region of integration,

$$0 \le u \le t, \qquad 0 \le t < \infty,$$

is sketched in Figure 6.14. It can be rewritten as

$$u \le t < \infty, \qquad 0 \le u < \infty.$$

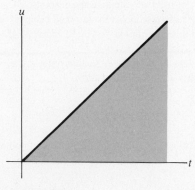

FIGURE 6.14

Thus, reversing the order of integration leads to

$$\mathcal{L}[(f * g)(t)] = \int_0^\infty \int_u^\infty e^{-st} f(t - u) g(u) \, dt \, du.$$

Now, setting $w = t - u$ and $v = u$, and noting the change in the limits ($u \le t < \infty$ becomes $0 \le w < \infty$), we obtain

$$\mathcal{L}[(f * g)(t)] = \int_0^\infty \int_0^\infty e^{-s(w+v)} f(w) g(v) \, dw \, dv,$$

which can be written as a product:

$$\mathcal{L}[(f * g)(t)] = \int_0^\infty e^{-sw} f(w) \, dw \int_0^\infty e^{-sv} g(v) \, dv$$

$$= \mathcal{L}[f(t)] \, \mathcal{L}[g(t)].$$

2. A special case of the convolution formula

We have noted already that convolution of a function $g(t)$ with the constant function $f(t) = 1$ leads to the integral of g:

$$(1 * g)(t) = \int_0^t g(u) \, du.$$

Since the constant $f(t) = 1$ is the inverse transform of $F(s) = 1/s$, by setting $G(s) = \mathcal{L}[g(t)]$ we obtain the following integral formula:

$$\mathcal{L}^{-1}\left[\frac{1}{s} G(s)\right] = \mathcal{L}^{-1}\left[\frac{1}{s}\right] * \mathcal{L}^{-1}[G(s)] = \int_0^t \mathcal{L}^{-1}[G(s)] \, du.$$

You might recognize this as the inverse transform version of the first differentiation formula (Note 3, Section 6.3).

3. Convolution and the response of physical systems

The convolution can be used to calculate the response of an electrical or mechanical device to external forcing from knowledge of its unforced behavior and of the forcing function. The heuristic derivation of this calculation serves to motivate the form of the convolution integral.

Many simple devices are modeled by a second-order o.d.e. with constant coefficients. For simplicity, we will work with the standard form,

(N) $(D^2 + a_1 D + a_0)x = E(t).$

Assuming we have at our disposal the solutions of the related homogeneous o.d.e.

(H) $(D^2 + a_1 D + a_0)x = 0$

we want to combine this information and the forcing function $E(t)$ into a formula for the solution $x(t)$ of (N) starting from absolute rest:

$$x(0) = 0, \qquad x'(0) = 0.$$

If we can decompose the forcing $E(t)$ into a sum of simple forces, then the principle of superposition (Section 2.2) lets us write the cumulative response as the sum of the individual responses to the constituent forces. For our analysis, we think of $E(t)$ as a sum of "bursts" of force, each exerted over a short period of time. A single such burst, lasting for time Δu and terminating at time $t = u$, is modeled by the function defined in pieces

$$E(t) = \begin{cases} 0, & t < u - \Delta u \\ E(t), & u - \Delta u < t < u \\ 0, & t > u. \end{cases}$$

Over a short time, $E(t)$ will not vary too much, so we assume Δu is small and think of $E(t)$ as constant over this period:

$$E(t) = E_u, \qquad u - \Delta u < t < u.$$

For this single burst, the device is unforced before $t = u - \Delta u$, and $x = 0$ and $x' = 0$ at $t = 0$, so this is still true at $t = u - \Delta u$. Over the short period $u - \Delta u < t < u$ the quantities x and x' will not vary greatly from zero. Thus we can approximately calculate the effect of $E(t)$ over this time period by ignoring the x and x' terms in the o.d.e.; that is,

$$x'' \approx E_u \qquad \text{for } u - \Delta u < t < u.$$

Integration yields

$$x'(u) \approx E_u\, \Delta u \text{ and } x(u) \approx E_u \cdot \frac{(\Delta u)^2}{2}.$$

When Δu is small, $(\Delta u)^2$ is extremely small, so our solution $x(t)$ approximately satisfies the conditions

$$x' \approx E_u\, \Delta u \text{ and } x \approx 0 \text{ at } t = u.$$

The device is again unforced after $t = u$, so that $x(t)$ for $t > u$ will coincide with the solution of (H) satisfying the conditions just stated. If these conditions were imposed at $t = 0$, we would simply write $x(t) = E_u\, \Delta u\, h(t)$, where $h(t)$ is the solution of (H) satisfying

$$h(0) = 0, \qquad h'(0) = 1.$$

To move these conditions to $t = u$, we replace t in $h(t)$ with the time since u, or $t - u$ (see Note 4, Section 6.3) and obtain an approximate expression for the response to a single burst:

$$x(t) \approx E_u\, \Delta u\, h(t - u), \qquad t > u.$$

The cumulative response at time t to many bursts is the sum of the constituent responses for $u < t$:

$$x(t) \approx \sum_{0 < u < t} E(u) \, h(t - u) \, \Delta u.$$

The accuracy of our approximations should improve as the durations of the individual bursts become shorter, so we expect the expression to become exact in the limit as $\Delta u \to 0$. But this limit is just the definition of the definite integral,

$$x(t) = \int_0^t E(u) \, h(t - u) \, du,$$

which defines the convolution of E with h. Thus we arrive at the following.

Fact: *The solution of the constant-coefficient o.d.e. in standard form*

$$(D^2 + a_1 D + a_0)x = E(t)$$

starting from absolute rest ($x(0) = x'(0) = 0$) is given by

$$x(t) = (h * E)(t)$$

where $h(t)$ is the solution of the related homogeneous o.d.e. satisfying $h(0) = 0$, $h'(0) = 1$.

We stress that this discussion was not a proof. A proof using Laplace transforms is outlined in Exercise 33. A straightforward extension to higher-order constant-coefficient o.d.e.'s is discussed in Exercise 34, while the more involved second-order version for variable coefficients is covered in Exercise 35.

EXERCISES

Evaluate the convolutions in Exercises 1 to 7.

1. $t * t^3$ 2. $1 * 1$ 3. $t * e^{-4t}$ 4. $t^2 * e^{-4t}$

5. $t * \sin 2t$ 6. $e^t * \cos t$ 7. $\cos t * \sin 2t$

Verify the formulas in Exercises 8 to 10.

8. $(\sin \alpha t) * (\cos \alpha t) = \dfrac{t}{2} \sin \alpha t$

9. $(\sin \alpha t) * (\sin \alpha t) = \dfrac{1}{2\alpha} \sin \alpha t - \dfrac{t}{2} \cos \alpha t$

10. $(\cos \alpha t) * (\cos \alpha t) = \dfrac{1}{2\alpha} \sin \alpha t + \dfrac{t}{2} \cos \alpha t$

Use the convolution formula to find the inverse Laplace transforms of the functions in Exercises 11 to 18. You may find the formulas in Exercises 8 to 10 useful in doing some of these problems.

11. $\dfrac{3}{s(s^2 + 4)}$

12. $\dfrac{3}{s(s + 4)}$

13. $\dfrac{3}{s^2(s + 4)}$

14. $\dfrac{1}{(s^2 + 4)^2}$

15. $\dfrac{s}{(s^2 + 4)^2}$

16. $\dfrac{s^2}{(s^2 + 4)^2}$

17. $\dfrac{s}{(s - 1)(s^2 + 1)}$ (*Hint:* Cf. Exercise 6)   18. $\dfrac{s}{(s^2 + 1)(s^2 + 4)}$ (*Hint:* Cf. Exercise 7)

Solve the initial value problems in Exercises 19 to 25.

19. $(D^2 + 1)^2x = 0$; $x(0) = x'(0) = x''(0) = 0, x'''(0) = 1$

20. $(D^2 + 1)^2x = 0$; $x(0) = x'(0) = 0, x''(0) = 1, x'''(0) = 0$

21. $(D^2 + 1)^2x = 0$; $x(0) = 0, x'(0) = 1, x''(0) = x'''(0) = 0$

22. $(D^2 + 1)^2x = 0$; $x(0) = 1, x'(0) = x''(0) = x'''(0) = 0$ (*Hint:* Partial fractions may help.)

23. $(D^2 + 1)^2x = \begin{cases} 0, & t < 1 \\ 3, & t \geq 1 \end{cases}$ $x(0) = x'(0) = x''(0) = x'''(0) = 0$

24. $(D^2 + 2D + 2)^2x = 0$; $x(0) = x'(0) = x''(0) = 0, x'''(0) = -3$

25. $(D^2 + 2D + 2)^2x = 0$; $x(0) = x'(0) = 0, x''(0) = -3, x'''(0) = 0$

26. Suppose that when $t = 0$, the current in the circuit of Example 6.1.3 is $i_0 = 1$, while the charge is $q_0 = 0$. Find the charge at times $t = \pi/2$ and $t = 2\pi$ if

   a. $L = 1, R = 0, C = 1,$ $V(t) = \sin t$

   b. $L = 1, R = 0, C = 1,$ $V(t) = \begin{cases} 0, & 0 \leq t < \pi \\ -\sin t, & \pi \leq t \end{cases}$

   c. $L = 1, R = 0, C = \frac{1}{4},$ $V(t) = \sin 2t$
   d. $L = 1, R = 2, C = \frac{1}{2},$ $V(t) = e^{-t} \sin t$

Some more advanced problems:

27. Use Laplace transforms to show that $(\cos t) * \left(\dfrac{1}{\alpha} \sin \alpha t \right) = (\sin t) * (\cos \alpha t)$.

28. a. Use the result of Exercise 24, Section 6.2, and Laplace transforms to show that $(t^{-1/2}) * (t^{-1/2}) = \pi$.
   b. Find $(t^{1/2}) * (t^{1/2})$.

29. Calculate $t * u_a(t)$.

30. a. Evaluate $(\sin t) * (t \sin t)$.
   b. Evaluate $(\sin t) * (t \cos t)$.
   c. Use (a), (b), and the convolution formula to find

$$\mathscr{L}^{-1}\left[\frac{s}{(s^2 + 1)^3} \right] \quad \text{and} \quad \mathscr{L}^{-1}\left[\frac{1}{(s^2 + 1)^3} \right].$$

d. Solve the initial value problem

$$(D^2 + 1)^3 x = 0; \qquad x(0) = x'(0) = x''(0) = x'''(0) = 0, x^{(iv)}(0) = x^{(v)}(0) = 1.$$

More abstract problems:

31. Use a change of variable in the definition to show that $f * g = g * f$. (*Caution:* Be careful about the limits of integration!)

32. Verify the algebraic properties of the convolution listed in the summary in case the functions f, g, g_1, g_2, and h are continuous and have Laplace transforms.

33. a. Show that if $x = h(t)$ is the solution of the constant-coefficient initial value problem

    (H) $(D^2 + a_1 D + a_0)x = 0; \qquad x(0) = 0, x'(0) = 1$

    then

    $$\mathcal{L}[h(t)] = \frac{1}{s^2 + a_1 s + a_0}.$$

    b. Show that if $x = x(t)$ is the solution of the initial value problem

    (N) $(D^2 + a_1 D + a_0)x = E(t); \qquad x(0) = x'(0) = 0$

    then

    $$\mathcal{L}[x(t)] = \mathcal{L}[h(t)]\mathcal{L}[E(t)].$$

    c. Apply the convolution formula to the equation in (b) to show that

    $$x(t) = h(t) * E(t).$$

34. Following the outline of Exercise 33, show that the solution of the constant-coefficient initial value problem

    (N) $(D^n + a_{n-1}D^{n-1} + \ldots + a_1 D + a_0)x = E(t);$
    $x(0) = x'(0) = \ldots = x^{(n-1)}(0) = 0$

    is $x(t) = h(t) * E(t)$, where $x = h(t)$ is the solution of the initial value problem

    (H) $(D^n + a_{n-1}D^{n-1} + \ldots + a_1 D + a_0)x = 0;$
    $x(0) = x'(0) = \ldots = x^{(n-2)}(0) = 0, x^{(n-1)}(0) = 1.$

35. Suppose $h_1(t)$ and $h_2(t)$ are linearly independent solutions of the homogeneous o.d.e.

    (H) $D^2 x + a_1(t)Dx + a_0(t)x = 0.$

    a. Use Cramer's rule to show that the specific solution of (H) subject to the conditions at $t = u$

    $$x(u) = 0 \qquad \text{and} \qquad x'(u) = 1$$

is given by the expression

$$H(t, u) = \frac{-h_2(u)}{w(u)} h_1(t) + \frac{h_1(u)}{w(u)} h_2(t)$$

where $w(u)$ is the Wronskian

$$w(u) = \det \begin{bmatrix} h_1(u) & h_2(u) \\ h_1'(u) & h_2'(u) \end{bmatrix}.$$

The function of two variables $H(t, u)$ is called the **Green's function** for (H).

b. Use variation of parameters to show that the general solution of the nonhomogeneous o.d.e.

(N) $$D^2x + a_1(t)Dx + a_0(t)x = E(t)$$

is given by the expression

$$x(t) = \left[-\int_T^t \frac{h_2(u)}{w(u)} E(u) \, du + c_1 \right] h_1(t) + \left[\int_T^t \frac{h_1(u)}{w(u)} E(u) \, du + c_2 \right] h_2(t).$$

c. Use (a) and (b) to show that the solution of the initial value problem

(N) $D^2x + a_1(t)Dx + a_0(t)x = E(t);$ $x(T) = x'(T) = 0$

is given by the expression

$$x(t) = \int_T^t H(t, u)E(u) \, du.$$

d. Find the Green's functions for the homogeneous o.d.e.'s related to the o.d.e.'s given in Exercises 12 and 13 of Section 2.10.

6.7 REVIEW: LAPLACE TRANSFORM SOLUTION OF INITIAL VALUE PROBLEMS

This section is devoted to working through some initial value problems that illustrate the role of the methods from the previous sections in the overall solution process. We also make a few comments about some general patterns that come up.

Example 6.7.1

Solve the initial value problem

$$(D^2 + 3D + 2)x = 2 \sin 2t; \qquad x(0) = x'(0) = 0$$

that models the *LRC* circuit of Example 6.1.3 with inductance $L = 1$, resistance $R = 3$, and capacitance $C = 1/2$, when an alternating voltage $V(t) = 2 \sin 2t$ is plugged in at time $t = 0$.

The transform of the o.d.e. is

$$s^2 \mathcal{L}[x] + 3s \mathcal{L}[x] + 2 \mathcal{L}[x] = \frac{4}{s^2 + 4}$$

so that

$$\mathcal{L}[x] = \frac{1}{(s^2 + 3s + 2)} \frac{4}{(s^2 + 4)}.$$

The partial fraction decomposition of this expression has the form

$$\frac{4}{(s + 1)(s + 2)(s^2 + 4)} = \frac{A}{s + 1} + \frac{B}{s + 2} + \frac{Cs + E}{s^2 + 4}.$$

The resulting equations for the coefficients are

$$
\begin{aligned}
A + B + C &= 0 \\
2A + B + 3C + E &= 0 \\
4A + 4B + 2C + 3E &= 0 \\
8A + 4B + 2E &= 4
\end{aligned}
$$

with solution

$$A = \frac{4}{5}, \qquad B = \frac{-1}{2}, \qquad C = \frac{-3}{10}, \qquad E = \frac{-2}{10}.$$

Thus

$$
\begin{aligned}
x &= \mathcal{L}^{-1}\left[\frac{4}{(s + 1)(s + 2)(s^2 + 4)} \right] \\
&= \frac{4}{5} \mathcal{L}^{-1}\left[\frac{1}{s + 1} \right] - \frac{1}{2} \mathcal{L}^{-1}\left[\frac{1}{s + 2} \right] - \frac{1}{10} \mathcal{L}^{-1}\left[\frac{3s + 2}{s^2 + 4} \right] \\
&= \frac{4}{5} e^{-t} - \frac{1}{2} e^{-2t} - \frac{3}{10} \cos 2t - \frac{1}{10} \sin 2t.
\end{aligned}
$$

In this example, the transform of the left side of the o.d.e. was $\mathcal{L}[x]$ times $(s^2 + 3s + 2)$, which was the characteristic polynomial of the o.d.e. In general, the initial value problem

$$P(D)x = E(t); \qquad x(0) = x'(0) = \ldots = x^{(n-1)}(0) = 0$$

will transform (using the first differentiation formula) to

$$P(s)\,\mathcal{L}[x] = \mathcal{L}[E(t)]$$

so that

$$\mathcal{L}[x] = \frac{1}{P(s)}\,\mathcal{L}[E(t)].$$

When nonzero initial conditions are imposed, we obtain extra terms, as illustrated in the next example.

Example 6.7.2

Solve the initial value problem

$$(D^2 + 8D + 15)x = 15; \qquad x(0) = 3,\, x'(0) = 0,$$

which models a 1-gram mass attached to a horizontal spring with constant $k = 15$, subject to damping $b = 8$ and a constant force $E(t) = 15$. The initial conditions indicate that the mass is pulled 3 cm from the equilibrium position and released.

The transform of the o.d.e. is

$$(s^2\,\mathcal{L}[x] - 3s) + 8(s\,\mathcal{L}[x] - 3) + 15\,\mathcal{L}[x] = \frac{15}{s}$$

so

$$\mathcal{L}[x] = \frac{15}{s(s^2 + 8s + 15)} + \frac{3s + 24}{s^2 + 8s + 15}.$$

We combine the two terms and look for a single partial fraction decomposition:

$$\frac{15 + 3s^2 + 24s}{s(s^2 + 8s + 15)} = \frac{A}{s} + \frac{B}{s + 3} + \frac{C}{s + 5}.$$

The equations for the coefficients

$$A + B + C = 3$$
$$8A + 5B + 3C = 24$$
$$15A \qquad\qquad = 15$$

have the solution

$$A = 1, \qquad B = 5, \qquad C = -3.$$

Thus,

$$x = \mathcal{L}^{-1}\left[\frac{1}{s} + \frac{5}{s+3} - \frac{3}{s+5}\right] = 1 + 5e^{-3t} - 3e^{-5t}.$$

The expression for $\mathcal{L}[x]$ in this last example is the sum of two terms. The first term has the same form as we obtained in the first example: the transform of the forcing term, divided by the characteristic polynomial. This is the transform of the particular solution of the nonhomogeneous o.d.e. that satisfies initial conditions of absolute rest. The second term, on the other hand, is a polynomial depending on the nonzero initial conditions (and not the forcing term) divided by the characteristic polynomial. This term would be unchanged if we replaced the forcing term with 0 and looked for the solution of the related homogeneous o.d.e. satisfying our given initial condition.

Again, this is a general pattern. The solution of any initial value problem involving the constant-coefficient o.d.e.

(N) $P(D)x = E(t)$

with given initial data at $t = 0$ will lead to an expression of the form

$$\mathcal{L}[x] = \frac{1}{P(s)}\,\mathcal{L}[E(t)] + \frac{Q(s)}{P(s)}$$

where $Q(s)$ is determined solely by $P(D)$ and the initial data. We recognize this as a special case of the general form "general (N) = particular (N) + general (H)": the solution of our initial value problem breaks into the particular solution of (N) satisfying initial conditions of absolute rest, plus the solution of the related homogeneous equation (H) fitting the given initial data.

Example 6.7.3

Solve the initial value problem

$$(3D^2 + 6D + 4)x = \frac{11}{2}\,e^{-t}\sin 2t; \qquad x(0) = 0,\ x'(0) = 3,$$

which models an *LRC* circuit with inductance $L = 3$, resistance $R = 6$, capacitance

$C = 1/4$, and an alternating voltage of decreasing amplitude $V(t) = \dfrac{11}{2} e^{-t} \sin 2t$.

The initial conditions indicate a charge $q_0 = 0$ and current $i_0 = 3$ at time $t = 0$.

To transform the right-hand side of the o.d.e., we need to apply the first shift formula. First, we find

$$\mathcal{L}[\sin 2t] = \frac{2}{s^2 + 4} = F(s)$$

and then we apply the shift formula:

$$\mathcal{L}\left[\frac{11}{2} e^{-t} \sin 2t\right] = \frac{11}{2} F(s + 1) = \frac{11}{(s + 1)^2 + 4}.$$

We transform the o.d.e. and solve for $\mathcal{L}[x]$:

$$(3s^2\, \mathcal{L}[x] - 3) + 6s\, \mathcal{L}[x] + 4\, \mathcal{L}[x] = \frac{11}{(s + 1)^2 + 4}$$

$$\mathcal{L}[x] = \frac{3}{3s^2 + 6s + 4} + \frac{11}{(3s^2 + 6s + 4)([s + 1]^2 + 4)}.$$

Thus

$$x = \mathcal{L}^{-1}\left[\frac{3}{3s^2 + 6s + 4} + \frac{11}{(3s^2 + 6s + 4)([s + 1]^2 + 4)}\right].$$

Upon completion of the square of the first denominator

$$3s^2 + 6s + 4 = 3(s^2 + 2s + 1 - 1) + 4 = 3(s + 1)^2 + 1,$$

we notice that the first shift formula (with $\alpha = -1$) will simplify the entire expression:

$$x = \mathcal{L}^{-1}\left[\frac{3}{3(s + 1)^2 + 1} + \frac{11}{(3[s + 1]^2 + 1)([s + 1]^2 + 4)}\right]$$

$$= e^{-t} \mathcal{L}^{-1}\left[\frac{3}{3s^2 + 1} + \frac{11}{(3s^2 + 1)(s^2 + 4)}\right].$$

Now, we look for a partial fraction decomposition of the second quotient:

$$\frac{11}{(3s^2 + 1)(s^2 + 4)} = \frac{As + B}{3s^2 + 1} + \frac{Cs + E}{s^2 + 4}.$$

The resulting equations

$$
\begin{array}{rcrcrcl}
A & & + 3C & & & = & 0 \\
& B & & & + 3E & = & 0 \\
4A & & + C & & & = & 0 \\
& 4B & & & + E & = & 11
\end{array}
$$

have the solution

$$
A = C = 0, \qquad B = 3, \qquad E = -1.
$$

Thus

$$
\begin{aligned}
x &= e^{-t} \mathcal{L}^{-1} \left[\frac{3}{3s^2 + 1} + \frac{3}{3s^2 + 1} - \frac{1}{s^2 + 4} \right] \\
&= e^{-t} \mathcal{L}^{-1} \left[\frac{6}{3s^2 + 1} - \frac{1}{s^2 + 4} \right] = e^{-t} \mathcal{L}^{-1} \left[\frac{2}{s^2 + \frac{1}{3}} - \frac{1}{s^2 + 4} \right] \\
&= 2\sqrt{3}\, e^{-t} \sin\left(\frac{t}{\sqrt{3}} \right) - \frac{1}{2} e^{-t} \sin 2t.
\end{aligned}
$$

Our final example involves a forcing term defined in pieces.

Example 6.7.4
Solve the initial value problem

$$
(D^2 + 9)x = \begin{cases} 0, & t < 1 \\ 9(t - 1), & 1 < t < 2 \\ 9, & t > 2 \end{cases} \qquad x(0) = x'(0) = 0
$$

which models the centrifuge problem in Example 6.1.2 with $m = 1$, $b = 0$, $k = 9$, $t_1 = 1$, $t_2 = 2$, and $F = 9$.

We begin by rewriting the forcing term in step function notation:

$$
\begin{aligned}
(D^2 + 9)x &= u_1(t)[9(t - 1)] + u_2(t)[-9(t - 1) + 9] \\
&= u_1(t)[9(t - 1)] + u_2(t)[-9(t - 2)].
\end{aligned}
$$

We transform the o.d.e., using the second shift formula on the right side:

$$
(s^2 + 9)\, \mathcal{L}[x] = e^{-s} \frac{9}{s^2} - e^{-2s} \frac{9}{s^2}.
$$

Solving for $\mathscr{L}[x]$, we have

$$\mathscr{L}[x] = e^{-s}\frac{9}{s^2(s^2+9)} - e^{-2s}\frac{9}{s^2(s^2+9)}.$$

We will ultimately need to use the second shift formula to inverse-transform both terms of the preceding expression, so we will need to find

$$f(t) = \mathscr{L}^{-1}\left[\frac{9}{s^2(s^2+9)}\right].$$

The partial fraction decomposition of the quotient is

$$\frac{9}{s^2(s^2+9)} = \frac{1}{s^2} - \frac{1}{s^2+9}$$

so

$$f(t) = t - \frac{1}{3}\sin 3t.$$

Now we use the second shift formula to find x:

$$x = \mathscr{L}^{-1}\left[e^{-s}\frac{9}{s^2(s^2+9)}\right] - \mathscr{L}^{-1}\left[e^{-2s}\frac{9}{s^2(s^2+9)}\right]$$

$$= u_1(t)f(t-1) - u_2(t)f(t-2)$$

$$= u_1(t)[(t-1) - \frac{1}{3}\sin 3(t-1)] - u_2(t)[(t-2) - \frac{1}{3}\sin 3(t-2)].$$

Written in pieces, this solution reads

$$x(t) = \begin{cases} 0, & \text{for } t < 1 \\ t - 1 - \dfrac{1}{3}\sin 3(t-1), & \text{for } 1 < t < 2 \\ 1 - \dfrac{1}{3}\sin 3(t-1) + \dfrac{1}{3}\sin 3(t-2), & \text{for } t > 2. \end{cases}$$

We note from this last example that the second shift formula comes into play only when the forcing function is defined in pieces, and the "switch" times for the solution are the same as those of the forcing.

In general, the effective use of the Laplace transform requires familiarity with the basic transforms and the manipulation rules and comes with practice. We summarize the basic formulas from previous sections and include a few formulas discussed in earlier notes.

TABLE OF LAPLACE TRANSFORM FORMULAS

$$\mathcal{L}[t^n] = \frac{n!}{s^{n+1}}$$

$$\mathcal{L}^{-1}\left[\frac{1}{s^n}\right] = \frac{1}{(n-1)!}\, t^{n-1}$$

$$\mathcal{L}[e^{at}] = \frac{1}{s-a}$$

$$\mathcal{L}^{-1}\left[\frac{1}{s-a}\right] = e^{at}$$

$$\mathcal{L}[\sin at] = \frac{a}{s^2 + a^2}$$

$$\mathcal{L}^{-1}\left[\frac{1}{s^2 + a^2}\right] = \frac{1}{a}\sin at$$

$$\mathcal{L}[\cos at] = \frac{s}{s^2 + a^2}$$

$$\mathcal{L}^{-1}\left[\frac{s}{s^2 + a^2}\right] = \cos at$$

First Differentiation Formula

$$\mathcal{L}[f^{(n)}(t)] = s^n\, \mathcal{L}[f(t)] - s^{n-1}f(0) - s^{n-2}f'(0) - \ldots - f^{(n-1)}(0)$$

$$\mathcal{L}\left[\int_0^t f(u)\, du\right] = \frac{1}{s}\, \mathcal{L}[f(t)]$$

$$\mathcal{L}^{-1}\left[\frac{1}{s}\, F(s)\right] = \int_0^t \mathcal{L}^{-1}[F(s)]\, du$$

In the following formulas, $F(s) = \mathcal{L}[f(t)]$, so $f(t) = \mathcal{L}^{-1}[F(s)]$.

First Shift Formula

$$\mathcal{L}[e^{at}f(t)] = F(s - a)$$

$$\mathcal{L}^{-1}[F(s)] = e^{at}\, \mathcal{L}^{-1}[F(s + a)]$$

Second Differentiation Formula

$$\mathcal{L}[t^n f(t)] = (-1)^n\, \frac{d^n}{ds^n}\, \mathcal{L}[f(t)]$$

$$\mathcal{L}^{-1}\left[\frac{d^n F(s)}{ds^n}\right] = (-1)^n t^n f(t)$$

Second Shift Formula

$$\mathcal{L}[u_a(t)g(t)] = e^{-as}\, \mathcal{L}[g(t + a)]$$

$$\mathcal{L}^{-1}[e^{-as}F(s)] = u_a(t)f(t - a)$$

Convolution

$$\mathcal{L}^{-1}[F(s)G(s)] = \mathcal{L}^{-1}[F(s)] * \mathcal{L}^{-1}[G(s)]$$

where

$$(f * g)(t) = \int_0^t f(t - u)g(u) \, du.$$

Periodic Functions

If $f(t + p) = f(t)$ for all t, then

$$\mathcal{L}[f(t)] = \frac{\int_0^p e^{-st}f(t) \, dt}{1 - e^{-ps}}.$$

EXERCISES (REVIEW PROBLEMS FOR CHAPTER 6)

Solve the initial value problems in Exercises 1 to 19.

1. $(9D^2 - 4)x = 0$; $x(0) = x'(0) = 3$

2. $(D^2 + 3)x = 0$; $x(0) = x'(0) = -4$

3. $(D^2 + 2D + 5)x = 0$; $x(0) = x'(0) = 2$

4. $(2D^2 + 2D + 1)x = 0$; $x(0) = 0, x'(0) = 2$

5. $x'' + 4x' + 4x = te^{-2t}$; $x(0) = 0, x'(0) = 1$

6. $x'' + 4x' + 4x = te^{-2t}$; $x(0) = 1, x'(0) = 0$

7. $x'' - 2x' + 10x = t^2 e^t$; $x(0) = 0, x'(0) = -6$

8. $(D - 1)x = \begin{cases} 1, & t \le 1 \\ t, & 1 \le t < 2 \\ t^2 - 2, & 2 \le t \end{cases}$ $x(0) = 0$

9. $(D^2 + 3D + 2)x = \begin{cases} e^{2t}, & t < 1 \\ e^{3t}, & t \ge 1 \end{cases}$ $x(0) = x'(0) = 0$

10. $(D^2 - 2D + 1)x = \begin{cases} 0, & t < 2 \\ t - 2, & t \ge 2 \end{cases}$ $x(0) = x'(0) = 1$

11. $(D^2 + 1)x = \begin{cases} 0, & t < 1 \\ t, & t \ge 1 \end{cases}$ $x(0) = 0, x'(0) = 1$

12. $(D^2 + 1)x = \begin{cases} t, & t < 1 \\ 0, & t \ge 1 \end{cases}$ $x(0) = 0, x'(0) = 1$

13. $(D^2 - 2D + 1)x = \begin{cases} t, & t < 2 \\ e^t, & t \geq 2 \end{cases}$ $x(0) = x'(0) = 0$

14. $(D^2 + 1)x = \begin{cases} \sin t, & t \leq \pi \\ 0, & t \geq \pi \end{cases}$ $x(0) = x'(0) = 0$

15. $(D^3 - D)x = 0$; $x(0) = 1, x'(0) = x''(0) = 0$

16. $(D^3 - D)x = 0$; $x(0) = 0, x'(0) = 1, x''(0) = 0$

17. $(D^3 - D)x = 1$; $x(0) = x'(0) = x''(0) = 0$

18. $(D^4 - 1)x = 0$; $x(0) = x'(0) = x''(0) = 0, x'''(0) = 1$

19. $(D^4 - 4D^3 + 6D^2 - 4D + 1)x = 60e^t$; $x(0) = x'(0) = x''(0) = 0, x'''(0) = 1$

Exercises 20 to 23 refer back to our models.

20. How long will it take for the water in Exercise 2, Section 6.1, to reach 32°F?

21. a. Find a formula describing the motion of the box in Exercise 8, Section 6.1.
    b. Show that at least one third of the box will be out of the water at all times.

22. Suppose the voltage in the circuit of Example 6.1.3 is the square wave

$$V(t) = \begin{cases} 1 & \text{if } n < t < n + 1, \, n \text{ even} \\ -1 & \text{if } n < t < n + 1, \, n \text{ odd} \end{cases}$$

sketched in Figure 6.6. Suppose also that $L = 1$, $R = 0$, $C = 1/9$, $Q(0) = 3$, and $I(0) = 0$. Find the charge and current after
a. 1/2 sec;   b. 3/2 sec;   c. 5/2 sec.
[*Hint:* The values of $V(t)$ for $t > 5/2$ do not affect the answer.]

6.8 LAPLACE TRANSFORMS FOR SYSTEMS

The method of Laplace transforms readily extends to initial value problems for linear differential systems with constant coefficients. We illustrate with several examples.

Example 6.8.1
Solve the system of o.d.e.'s

$$x' = 3x - 2y$$
$$y' = 4x - y$$

subject to the initial conditions

$$x(0) = 1, \quad y(0) = -1.$$

As in the case of a single equation, we transform both sides of each equation, incorporating the initial data by means of the first differentiation formula. This gives

$$s \, \mathcal{L}[x] - 1 = 3 \, \mathcal{L}[x] - 2 \, \mathcal{L}[y]$$
$$s \, \mathcal{L}[y] + 1 = 4 \, \mathcal{L}[x] - \mathcal{L}[y].$$

If we move all terms involving $\mathcal{L}[x]$ or $\mathcal{L}[y]$ to one side, we recognize this as a linear algebraic system of equations in the unknowns $\mathcal{L}[x]$ and $\mathcal{L}[y]$:

$$(s - 3) \, \mathcal{L}[x] + 2 \, \mathcal{L}[y] = +1$$
$$-4 \, \mathcal{L}[x] + (s + 1) \, \mathcal{L}[y] = -1.$$

This system can be solved for $\mathcal{L}[x]$ and $\mathcal{L}[y]$ in terms of s (for example, by Cramer's rule):

$$\mathcal{L}[x] = \frac{s + 3}{s^2 - 2s + 5}, \qquad \mathcal{L}[y] = \frac{-s + 7}{s^2 - 2s + 5}.$$

We find x and y by applying the inverse transform [this involves completing the square of $s^2 - 2s + 5 = (s - 1)^2 + 4$ and using the first shift formula]:

$$x(t) = \mathcal{L}^{-1}\left[\frac{s + 3}{s^2 - 2s + 5}\right] = e^t \, \mathcal{L}^{-1}\left[\frac{s + 4}{s^2 + 4}\right]$$
$$= e^t \cos 2t + 2e^t \sin 2t$$
$$y(t) = \mathcal{L}^{-1}\left[\frac{-s + 7}{s^2 - 2s + 5}\right] = e^t \, \mathcal{L}^{-1}\left[\frac{-s + 6}{s^2 + 4}\right]$$
$$= -e^t \cos 2t + 3e^t \sin 2t.$$

Example 6.8.2

Solve the system of o.d.e.'s

$$DQ = -Q - I_2 \qquad + V(t)$$
$$DI_2 = Q - I_2 + I_3$$
$$DI_3 \qquad I_2 - I_3 - V(t)$$

where

$$V(t) = \begin{cases} 0, & t < 1 \\ 10, & t > 1 \end{cases}$$

and subject to the initial conditions

$$Q(0) = 10, \qquad I_2(0) = I_3(0) = 0.$$

This models the multiloop circuit of Example 6.1.4 with $R_1 = R_2 = L_1 = L_2 = C = 1$, $V_1(t) = 0$, and $V_2(t) = 10$ (a battery). At $t = 0$ there is a charge of 10 coulombs on the capacitor but no current.

We rewrite $V(t)$ in step function notation, $V(t) = 10u_1(t)$, and transform the system:

$$s \mathcal{L}[Q] - 10 = - \mathcal{L}[Q] - \mathcal{L}[I_2] \qquad\qquad + 10\frac{e^{-s}}{s}$$

$$s \mathcal{L}[I_2] \qquad = \qquad \mathcal{L}[Q] - \mathcal{L}[I_2] + \mathcal{L}[I_3]$$

$$s \mathcal{L}[I_3] \qquad = \qquad\qquad \mathcal{L}[I_2] - \mathcal{L}[I_3] - 10\frac{e^{-s}}{s}.$$

Some rearrangement of the terms leads to

$$(s + 1) \mathcal{L}[Q] + \mathcal{L}[I_2] \qquad\qquad\qquad = 10 + 10\frac{e^{-s}}{s}$$

$$- \mathcal{L}[Q] + (s + 1) \mathcal{L}[I_2] - \mathcal{L}[I_3] = 0$$

$$- \mathcal{L}[I_2] + (s + 1) \mathcal{L}[I_3] = -\frac{10e^{-s}}{s}.$$

We solve this system for $\mathcal{L}[Q]$, $\mathcal{L}[I_2]$, and $\mathcal{L}[I_3]$ to get

$$\mathcal{L}[Q] = \frac{10}{s + 1} - \frac{10}{(s + 1)^3} + \frac{10e^{-s}}{s(s + 1)}$$

$$= \frac{10}{s + 1} - \frac{10}{(s + 1)^3} + 10e^{-s}\left(\frac{1}{s} - \frac{1}{s + 1}\right)$$

$$\mathcal{L}[I_2] = \frac{10}{(s + 1)^2}$$

$$\mathcal{L}[I_3] = \frac{10}{(s + 1)^3} - \frac{10e^{-s}}{s(s + 1)}$$

$$= \frac{10}{(s + 1)^3} - 10e^{-s}\left(\frac{1}{s} - \frac{1}{s + 1}\right).$$

The inverse transforms of these expressions can be found using the shift formulas:

$$Q = 10e^{-t} - 5t^2e^{-t} + 10u_1(t)[1 - e^{-(t-1)}]$$
$$I_2 = 10te^{-t}$$
$$I_3 = 5t^2e^{-t} - 10u_1(t)[1 - e^{-(t-1)}].$$

Finally, we note that the Laplace transform can be applied equally well to systems of o.d.e.'s of order higher than one, as shown by the following example.

Example 6.8.3

Find the solution of the system of o.d.e.'s

$$4D^2x_1 = -8x_1 + 2x_2$$
$$D^2x_2 = 2x_1 - 2x_2 + 24$$

subject to the initial conditions

$$x_1(0) = x_2(0) = x_1'(0) = x_2'(0) = 0.$$

This models the system of Example 6.1.5 with masses $m_1 = 4$ and $m_2 = 1$, spring constants $k_1 = 6$ and $k_2 = 2$, and a constant force $E(t) = 24$, starting from absolute rest.

We transform the system, obtaining

$$4s^2 \, \mathcal{L}[x_1] = -8 \, \mathcal{L}[x_1] + 2 \, \mathcal{L}[x_2]$$
$$s^2 \, \mathcal{L}[x_2] = 2 \, \mathcal{L}[x_1] - 2 \, \mathcal{L}[x_2] + \frac{24}{s}.$$

Rearranging, we obtain

$$4(s^2 + 2) \, \mathcal{L}[x_1] - \qquad 2 \, \mathcal{L}[x_2] = 0$$
$$-2 \, \mathcal{L}[x_1] + (s^2 + 2) \, \mathcal{L}[x_2] = \frac{24}{s}.$$

Solving for $\mathcal{L}[x_1]$ and $\mathcal{L}[x_2]$, we have

$$\mathcal{L}[x_1] = \frac{12}{s(s^2 + 3)(s^2 + 1)}, \qquad \mathcal{L}[x_2] = \frac{24(s^2 + 2)}{s(s^2 + 3)(s^2 + 1)}.$$

Using partial fractions, we obtain

$$x_1 = \mathcal{L}^{-1}\left[\frac{12}{s(s + 3)(s^2 + 1)}\right] = \mathcal{L}^{-1}\left[\frac{4}{s} + \frac{2s}{s^2 + 3} - \frac{6s}{s^2 + 1}\right]$$

$$= 4 + 2\cos\sqrt{3}\,t - 6\cos t$$

$$x_2 = \mathcal{L}^{-1}\left[\frac{24(s^2 + 2)}{s(s^2 + 3)(s^2 + 1)}\right] = \mathcal{L}^{-1}\left[\frac{16}{s} - \frac{4s}{s^2 + 3} - \frac{12s}{s^2 + 1}\right]$$

$$= 16 - 4\cos\sqrt{3}\,t - 12\cos t.$$

Let's summarize.

LAPLACE TRANSFORMS FOR SYSTEMS

To solve an initial value problem for a system of n linear constant-coefficient o.d.e.'s (of any order) in the unknowns $x_1, \ldots, x_n$:

i.   Transform each o.d.e., incorporating the initial data by the first differentiation formula.

ii. Gather all terms involving $\mathcal{L}[x_1], \ldots, \mathcal{L}[x_n]$ on one side and all other terms on the other.

iii. Solve the resulting linear algebraic system for $\mathcal{L}[x_1], \ldots, \mathcal{L}[x_n]$ in terms of s.

iv. Inverse-transform each $\mathcal{L}[x_i]$ to obtain $x_i(t)$.

Note
On spurious solutions
When we apply the Laplace transform to a system of o.d.e.'s with specified initial conditions, we obtain a new system of equations that is satisfied by the transform of a solution to our initial value problem—*provided such a solution exists*. The existence of a solution is insured by the theorem in Section 3.3 whenever the system consists of n linear constant-coefficient o.d.e.'s in standard form

$$Dx_1 = a_{11}x_1 + \ldots + a_{1n}x_n + E_1(t)$$

$$\vdots$$

$$Dx_n = a_{n1}x_1 + \ldots + a_{nn}x_n + E_n(t)$$

with the $E_i(t)$'s continuous, and the initial conditions specify a value for each variable at $t = 0$. However, when this theorem does not apply, application of the Laplace transform method can lead to spurious solutions. For example, the system

(S)
$$Dx + y = 0$$
$$D^2x + y + Dy = 1$$

with initial conditions of absolute rest

$$x(0) = x'(0) = 0, \qquad y(0) = 0$$

can be formally solved by Laplace transforms [see Exercise 16(a)], but the method yields

$$x = -t, \qquad y = 1.$$

This pair of functions solves (S) but fails to satisfy the initial conditions. In fact, no solution of (S) can satisfy the given initial conditions [see Exercise 16(b)].

   This example shows that when we use Laplace transforms to solve an initial value problem for which the system is not in standard form, we should substitute our answer back to check that it actually solves the initial value problem with which we began.

EXERCISES

In Exercises 1 to 10, find the solution of the given system that satisfies the given initial condition.

1. $x' = x + 2y$
    $y' = 2x + y$ $x(0) = 5, y(0) = -5$

2. $x' = x + 2y + e^t$
    $y' = 2x + y$ $x(0) = y(0) = 0$

3. $x' = x + 2y + u_1(t)$
    $y' = 2x + y$ $x(0) = 5, y(0) = -5$

4. $x' = x - 2y$
    $y' = 2x + y$ $x(0) = 5, y(0) = -5$

5. $x'' = -3x - 4y$
    $y' = 2x + 3y$ $x(0) = 0, x'(0) = 1, y(0) = 0$

6. $x'' = 3x + 6y + 6e^{2t}$
    $y' = x + 2y$ $x(0) = x'(0) = y(0) = 0$

7. $x'' = 5x - 3y$
    $y'' = 3x - 5y$ $x(0) = 8, x'(0) = 16, y(0) = y'(0) = 0$

8. $x' = x - y$
    $y' = \quad - y + z$ $x(0) = 0, y(0) = 12, z(0) = 0$
    $z' = \qquad y - z$

9. $\begin{aligned} x' &= x - y \\ y' &= \quad - y + z + 12 \\ z' &= \quad\quad y - z \end{aligned}$ $x(0) = y(0) = z(0) = 0$

10. $\begin{aligned} x' &= \quad x + 5y \\ y' &= -x - \quad y \\ z' &= \quad x \quad\quad + 2z + e^{2t} \end{aligned}$ $x(0) = 8, y(0) = z(0) = 0$

Exercises 11 to 15 refer to our models.

11. Find the solution of the system of equations describing the circuit of Example 3.1.2 under the following assumptions.
    a. $V(t) = 0, R_1 = 2, R_2 = 3, L = 6, C = 1/6$, and $I_2(0) = 1, I_1(0) = -3$. [Compare this to Exercise 15(b), Section 3.7.]
    b. $V(t) = 0, R_1 = R_2 = L = C = 1$, and $I_2(0) = I_1(0) = 2$. [Compare this to Exercise 14(b), Section 3.8.]
    c. $V(t) = 0, R_1 = 1, R_2 = 3, C = L = 1$, and $Q(0) = 1, I_2(0) = 2$. [Compare this to Exercise 14(b), Section 3.9.]
    d. $V(t) = 10, R_1 = 1, R_2 = 3, C = L = 1$, and $I_1(0) = I_2(0) = 0$. (See Exercise 9, Section 6.1.)

12. Solve the initial value problem of Exercise 17, Section 3.11, describing Professor Kay's experiment.

13. An undergraduate assistant calculates that the reaction described in Exercise 17, Section 3.11, will never run dry, so Professor Kay decides to risk extracting more of substance C. One year after he began the experiment, he starts extracting 125 grams per year of substance C.
    a. Solve the system describing this experiment (see Exercise 10, Section 6.1).
    b. Will the reaction run dry?

14. Use Laplace transforms to solve the initial value problem solved in Example 3.11.3.

15. Solve the initial value problem of Exercise 19(b), Section 3.11.

16. *Spurious Solutions:* Suppose x and y satisfy the system

    (S) $\begin{aligned} Dx + y \quad\quad &= 0 \\ D^2x + y + Dy &= 1 \end{aligned}$

    and initial conditions

    (I) $x(0) = 0, \quad x'(0) = 0, \quad y(0) = 0.$

    a. Show that the Laplace transforms of x and y must be

    $$\mathcal{L}[x] = -\frac{1}{s^2}, \quad \mathcal{L}[y] = \frac{1}{s},$$

    so that $x = -t, y = 1$. Note that these functions don't satisfy (I).
    b. By differentiating the first equation of (S) and substituting in the second, show that the general solution of (S) is

    $$x = c - t, \quad y = 1,$$

    so that the initial value problem specified by (I) has no solution.

17. *Systems in Matrix Form:* The Laplace transform of a vector valued function

$$\mathbf{x}(t) = \begin{bmatrix} x_1(t) \\ \cdot \\ \cdot \\ \cdot \\ x_n(t) \end{bmatrix}$$

is defined by the rule

$$\mathcal{L}[\mathbf{x}(t)] = \begin{bmatrix} \mathcal{L}[x_1(t)] \\ \cdot \\ \cdot \\ \cdot \\ \mathcal{L}[x_n(t)] \end{bmatrix}.$$

a. Show that $\mathcal{L}[D\mathbf{x}(t)] = s\,\mathcal{L}[\mathbf{x}(t)] - \mathbf{x}(0)$.
b. If A is an $n \times n$ matrix with constant entries, show that $\mathcal{L}[A\mathbf{x}(t)] = A\,\mathcal{L}[\mathbf{x}(t)]$.
c. Apply the Laplace transform to both sides of the constant-coefficient system

$$D\mathbf{x} = A\mathbf{x} + \mathbf{E}(t)$$

and regroup the resulting terms to show that

$$(A - sI)\,\mathcal{L}[\mathbf{x}] = -\mathbf{x}(0) - \mathcal{L}[\mathbf{E}(t)].$$

d. Cramer's rule can now be used to obtain an expression for $\mathcal{L}[x_i(t)]$ as a quotient of determinants, with det $[A - sI]$ in the denominator (valid for those values of s for which det $[A - sI] \neq 0$). How does this denominator compare with the characteristic polynomial of A discussed in Section 3.5?
e. When the system is homogeneous ($\mathbf{E}(t) = \mathbf{0}$), the numerator of the expression for $\mathcal{L}[x_i(t)]$ is a polynomial in s of degree less than n. What does this tell us about the relationship between the functions $x_i(t)$ and the roots of the characteristic polynomial? Compare with Sections 3.7 to 3.10.

REVIEW PROBLEMS: See Section 6.7.

Linear Equations with Variable Coefficients: Power Series

7.1 TEMPERATURE MODELS: O.D.E.'S FROM P.D.E.'S

Our attention in Chapters 2, 3, and 6 was focused primarily on linear o.d.e.'s with constant coefficients. In this chapter we will develop an approach to solving certain o.d.e.'s with variable coefficients. While the solutions of these equations may not have finite expressions in terms of elementary functions (polynomials, exponentials, trigonometric functions, and so on), we will be able to find expressions for them in terms of power series. You may recall from calculus that the first several terms of a power series expression can be used to obtain a good approximation to the value of a function at a specified point. (Indeed, if our goal is to obtain a decimal expression for such a value, a series expression can be more useful than a closed form expression involving exponentials and trigonometric functions.) In addition, series expressions can often be used to obtain powerful information about the behavior of solutions of o.d.e.'s.

Although linear o.d.e.'s with variable coefficients can be obtained by varying our earlier models (see Exercises 1 to 5), in practice they usually arise when we look for special solutions to certain partial differential equations (p.d.e.'s). The way such problems lead to o.d.e.'s is illustrated in this section by several versions of the problem of finding a steady-state (i.e., time-independent) temperature distribution in a body.

It turns out that the steady-state temperature $u(x, y)$ in a two-dimensional plate whose flat surfaces are insulated must satisfy the p.d.e.

$$(L_2) \qquad \frac{\partial^2 u}{\partial x^2} + \frac{\partial^2 u}{\partial y^2} = 0$$

while the steady-state temperature $u(x, y, z)$ in a three-dimensional solid satisfies

$$(L_3) \qquad \frac{\partial^2 u}{\partial x^2} + \frac{\partial^2 u}{\partial y^2} + \frac{\partial^2 u}{\partial z^2} = 0.$$

These p.d.e.'s are known as the two- and three-dimensional **Laplace equations.**

As a first step toward solving these p.d.e.'s, we look for solutions that can be written as products of functions of one variable. Our hope is that we can replace the p.d.e. by o.d.e.'s for these functions. In the following example we illustrate how the search for such special solutions to the two-dimensional Laplace equation can lead to familiar o.d.e.'s. The solutions to these o.d.e.'s can be combined to find solutions of the Laplace equation that match given boundary conditions (see, for example, Exercise 13(d), Section 7.5, and Exercise 15(d), Section 7.8).

Example 7.1.1   Temperature in a Rectangular Plate

Suppose the temperature in a thin rectangular plate, insulated above and below, is controlled by heating (or cooling) elements along its edges (see Figure 7.1). We wish to predict the steady-state temperature, $u(x,y)$, that satisfies the two-dimensional Laplace equation (L_2) inside the rectangle.

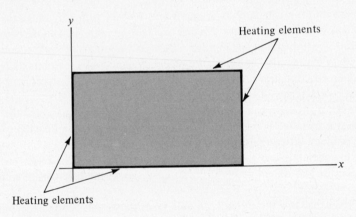

FIGURE 7.1

We look for solutions of (L_2) in the special form

(S) $$u = X(x)Y(y).$$

Substituting (S) into (L_2), we obtain

$$\frac{d^2X}{dx^2} Y + X \frac{d^2Y}{dy^2} = 0$$

or

$$\frac{1}{X} \frac{d^2X}{dx^2} = -\frac{1}{Y} \frac{d^2Y}{dy^2}.$$

Note that the left side of this equation is a function of x alone, while the right is independent of x. The only way this can happen is if both sides have a common *constant* value,

$$\frac{1}{X}\frac{d^2X}{dx^2} = -\frac{1}{Y}\frac{d^2Y}{dy^2} = \lambda.$$

When we write these equations out separately, we obtain two ordinary differential equations,

$$\frac{1}{X}\frac{d^2X}{dx^2} = \lambda \quad \text{and} \quad -\frac{1}{Y}\frac{d^2Y}{dy^2} = \lambda.$$

These can be rewritten in the more familiar forms

(H_x) $\qquad\qquad\qquad\qquad\qquad [D^2 - \lambda]X = 0$

(H_y) $\qquad\qquad\qquad\qquad\qquad [D^2 + \lambda]Y = 0$

where D stands for d/dx in (H_x) and for d/dy in (H_y).

Note that the solutions of (H_x) and (H_y) depend on the specific values of λ; these in turn are determined by the temperature along the edges of the plate (see, for example, Exercise 6). Once we solve (H_x) and (H_y), we substitute into (S) to obtain solutions to (L_2).

In the next examples we see how o.d.e.'s with variable coefficients arise when we try this procedure on the Laplace equation in polar, cylindrical, or spherical coordinates.

Example 7.1.2   Temperature in a Circular Plate

To investigate the steady-state temperature distribution in a circular plate of radius 1 (insulated above and below) when we know the temperature on the boundary, we are led to the two-dimensional Laplace equation (L_2) in the interior of the disc $x^2 + y^2 < 1$. For such problems, it is most convenient to switch to polar coordinates. Rewritten in terms of r and θ, (L_2) becomes (Exercise 9)

(H) $\qquad\qquad\qquad\qquad \dfrac{\partial^2 u}{\partial r^2} + \dfrac{1}{r}\dfrac{\partial u}{\partial r} + \dfrac{1}{r^2}\dfrac{\partial^2 u}{\partial \theta^2} = 0.$

We look for solutions of (H) in the form

(S) $$u = R(r)\Theta(\theta).$$

Substituting (S) into (H), we obtain

$$\frac{d^2R}{dr^2}\Theta + \frac{1}{r}\frac{dR}{dr}\Theta + \frac{1}{r^2}R\frac{d^2\Theta}{d\theta^2} = 0.$$

Subtracting the last term from both sides, multiplying by r^2 and dividing by $u = R\Theta$, we come to

$$\frac{r^2}{R}\frac{d^2R}{dr^2} + \frac{r}{R}\frac{dR}{dr} = -\frac{1}{\Theta}\frac{d^2\Theta}{d\theta^2}.$$

Since the left side of this equation is a function of r alone, while the right side depends only on θ, the common value must be a constant, λ. This leads to two o.d.e.'s,

(H$_\theta$) $$-\frac{1}{\Theta}D^2\Theta = \lambda$$

(H$_r$) $$\frac{r^2}{R}D^2R + \frac{r}{R}DR = \lambda$$

where D stands for $\dfrac{d}{d\theta}$ in (H$_\theta$) and for $\dfrac{d}{dr}$ in (H$_r$).

The first of these equations can be rewritten as

(H$_\theta$) $$(D^2 + \lambda)\Theta = 0.$$

Since Θ is a function of the angular coordinate θ, it must satisfy the geometric condition $\Theta(\theta + 2\pi) = \Theta(\theta)$ for all θ. If this is to hold for all solutions of (H$_\theta$), then λ must be a perfect square (Exercise 7):

$$\lambda = n^2, \qquad n \geq 0 \text{ an integer.}$$

Of course, the value of λ in (H$_r$) must agree with that in (H$_\theta$). Thus, we are led to the o.d.e. for R(r),

(H$_r$) $$\frac{r^2}{R}D^2R + \frac{r}{R}DR = n^2.$$

Multiplying by R and rearranging terms gives

$$[r^2D^2 + rD - n^2]R = 0.$$

This is a special case of the **Cauchy-Euler equation**

(CE)
$$p_0 t^2 \frac{d^2 x}{dt^2} + q_0 t \frac{dx}{dt} + r_0 x = 0$$

(with $p_0 = q_0 = 1$ and $r_0 = -n^2$). We will learn to solve (CE) in Section 7.5.

Example 7.1.3   Temperature in a Plug

A metal plug in the shape of a (solid) cylinder is insulated along the surface; the temperature is controlled by two rings on the top and bottom edges (see Figure 7.2). We wish to predict the steady-state temperature in the plug, $u(x,y,z)$, which must satisfy the three-dimensional Laplace equation (L_3).

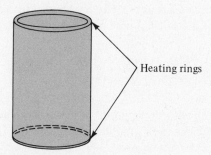

Heating rings

FIGURE 7.2

We start by looking for solutions of (L_3) in the form

(S_1)
$$u = U(x,y)Z(z).$$

Substituting (S_1) into (L_3), we get

$$\frac{\partial^2 U}{\partial x^2} Z + \frac{\partial^2 U}{\partial y^2} Z + U \frac{d^2 Z}{dz^2} = 0$$

or

$$\frac{1}{U} \left[\frac{\partial^2 U}{\partial x^2} + \frac{\partial^2 U}{\partial y^2} \right] = -\frac{1}{Z} \frac{d^2 Z}{dz^2}.$$

Since the right side is a function of z alone, while the left side is independent of z, each is constant and we obtain two o.d.e.'s

(H$_z$)
$$-\frac{1}{Z}\frac{d^2Z}{dz^2} = \lambda$$

(H$_{xy}$)
$$\frac{1}{U}\left[\frac{\partial^2 U}{\partial x^2} + \frac{\partial^2 U}{\partial y^2}\right] = \lambda.$$

The first equation is again the familiar o.d.e.

$$[D^2 + \lambda]Z = 0$$

(where $D = d/dz$) whose solutions depend on λ. It turns out that physical considerations lead to the requirement that λ is nonpositive, so

$$\lambda = -\gamma^2 \qquad \text{for some } \gamma \geq 0.$$

The shape of the plug leads us to try to rewrite the function U in terms of polar coordinates r and θ. The p.d.e. (H$_{xy}$) now takes the form (Exercise 9)

(H$_{r\theta}$)
$$\frac{1}{U}\left[\frac{\partial^2 U}{\partial r^2} + \frac{1}{r}\frac{\partial U}{\partial r} + \frac{1}{r^2}\frac{\partial^2 U}{\partial \theta^2}\right] = -\gamma^2.$$

When $\gamma = 0$, this is equivalent to the p.d.e. of the previous example. When $\gamma > 0$, we can still look for solutions to (H$_{r\theta}$) in the separated form

(S$_2$)
$$U = R(r)\Theta(\theta).$$

Substituting (S$_2$) into (H$_{r\theta}$), we obtain

$$\frac{1}{R}\frac{d^2R}{dr^2} + \frac{1}{rR}\frac{dR}{dr} + \frac{1}{r^2\Theta}\frac{d^2\Theta}{d\theta^2} = -\gamma^2$$

or, multiplying by r^2 and rearranging terms,

$$\frac{r^2}{R}\frac{d^2R}{dr^2} + \frac{r}{R}\frac{dR}{dr} + (\gamma r)^2 = -\frac{1}{\Theta}\frac{d\Theta}{d\theta^2}.$$

Once again, the two sides of this equation depend on different variables, so we set each equal to a common constant, κ:

(H$_\theta$)
$$-\frac{1}{\Theta}\frac{d^2\Theta}{d\theta^2} = \kappa$$

(H$_r$)
$$\frac{r^2}{R}\frac{d^2R}{dr^2} + \frac{r}{R}\frac{dR}{dr} + (\gamma r)^2 = \kappa.$$

As in Example 7.1.2, the geometric condition $\Theta(\theta + 2\pi) = \Theta(\theta)$ forces κ to be a perfect square:

$$\kappa = n^2, \qquad n \geq 0 \text{ an integer.}$$

The substitution $t = \gamma r$ then puts (H$_r$) in the form of the **Bessel equation**

(B$_\mu$)
$$[t^2D^2 + tD + (t^2 - \mu^2)]x = 0$$

with $D = d/dt$, $x = R$, and $\mu = n$. We will discuss this equation in Sections 7.7 and 7.8.

Example 7.1.4   Temperature in a Planet

Suppose the temperature at the surface of a spherical planet is found to be a function of the latitude alone (see Figure 7.3). We would like to predict the internal temperature distribution u, which we assume to be steady-state.

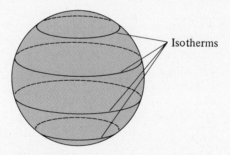

Isotherms

FIGURE 7.3

Again, u must satisfy the three-dimensional Laplace equation (L$_3$), but here we expect it to be most convenient to express u in terms of spherical coordinates ρ, ϕ, and θ (see Figure 7.4). We have assumed that the temperature on the surface depends on the latitude ϕ and not on the longitude θ, so we expect the same to be true inside the planet. When $u = u(\rho,\phi)$ is independent of θ, the three-dimensional Laplace equation becomes (Exercise 10)

(H)
$$\rho\frac{\partial^2}{\partial\rho^2}(\rho u) + \frac{1}{\sin\phi}\frac{\partial}{\partial\phi}\left(\sin\phi\frac{\partial u}{\partial\phi}\right) = 0.$$

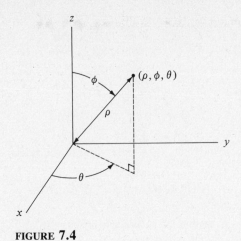

FIGURE 7.4

We look for a solution of (H) in the separated form

(S) $$u = R(\rho)\Phi(\phi).$$

Substituting (S) into (H), we obtain

$$\rho\Phi \frac{d^2}{d\rho^2}(\rho R) + \frac{R}{\sin\phi}\frac{d}{d\phi}\left(\sin\phi \frac{d\Phi}{d\phi}\right) = 0$$

or

$$\frac{\rho}{R}\frac{d^2}{d\rho^2}(\rho R) = -\frac{1}{\Phi\sin\phi}\frac{d}{d\phi}\left(\sin\phi \frac{d\Phi}{d\phi}\right).$$

As before, the two sides of this equation depend on distinct variables, so the common value is a constant:

(H$_\rho$) $$\frac{\rho}{R}\frac{d^2}{d\rho^2}(\rho R) = \lambda$$

(H$_\phi$) $$-\frac{1}{\Phi\sin\phi}\frac{d}{d\phi}\left(\sin\phi \frac{d\Phi}{d\phi}\right) = \lambda.$$

The first o.d.e. (H$_\rho$) can be written in the form

$$[\rho^2 D^2 + 2\rho D - \lambda]R = 0,$$

which is another case of the Cauchy-Euler equation described in Example 7.1.2 (this time with $p_0 = 1$, $q_0 = 2$, and $r_0 = -\lambda$). The second o.d.e. (H_ϕ) does not resemble any of our earlier examples. It turns out that if we want solutions of (H_ϕ) to be twice differentiable at the poles, $\phi = 0$ and $\phi = \pi$, then we are forced to choose $\lambda = n(n + 1)$ where n is a nonnegative integer. If in addition we make the substitution $t = \cos \phi$, we obtain (Exercise 8)

$$\frac{1}{\Phi} \frac{d}{dt} \left[(t^2 - 1) \frac{d\Phi}{dt} \right] = n(n + 1).$$

After some manipulation, this takes the form of the **Legendre equation:**

$$[(t^2 - 1)D^2 + 2tD - n(n + 1)]\Phi = 0.$$

EXERCISES

1. *A Circuit with Varying Resistance:* The resistance of metals varies with temperature. Suppose the resistor of Exercise 3, Section 2.1 (or Example 3.1.1) is heated so that its resistance at time t is $R(t) = 10 + t$. Find an o.d.e. for the charge Q on the capacitor if $L = C = 1$ and $V(t) = 10$.

2. *A Varying Spring Constant:* When a spring is heated, the spring constant may change. Write an o.d.e. modeling the spring system of Example 2.1.2, with $m = b = 10$, under the assumption that the spring "constant" at time t is $k(t) = 10 - t$.

3. *Varying Friction:* As oil is heated, it becomes more slippery. Write an o.d.e. modeling the spring system of Example 2.1.2, with $m = k = 10$, under the assumption that the damping coefficient at time t is $b(t) = 10 - t$.

4. *Varying Mass:* Suppose the mass attached to the spring of Example 2.1.2 is a 10-gram block of ice that melts at the rate of 1 gram per second. Write an o.d.e. modeling the motion of the system if $k = 10$ and $b = 11$. [*Note:* When mass varies, Newton's second law takes the form $F = d(mv)/dt$. Use this, *not* $F = ma$, as the basis for this model.]

5. *A Submarine:* Recall the Principle of Archimedes (Exercise 1, Section 2.1), which states that water buoys up a submerged object by a force equal to the weight of the water it displaces. Suppose a submarine weighing 50 tons (100,000 lb) is resting on the ocean floor and that the buoyant force exactly counters the force of gravity at this stage. Now, water is pumped out of its tanks at 100 lb/sec.
    a. Write a formula for the mass $m(t)$ of the submarine after t seconds.
    b. Write a formula for the net force $F(t)$ acting on the submarine after t seconds.
    c. Use Newton's second law in the form for varying mass [$F = d(mv)/dt$, not $F = ma$] to derive a second-order o.d.e. modeling the position of the submarine, ignoring friction.

6. *The Rectangular Plate:* Suppose the temperatures of the bottom edge ($y = 0$) and the top edge ($y = h$) of the plate in Example 7.1.1 are held at $0°$; that is,

    (B) $u(x, 0) = u(x, h) = 0$ for all x.

a. Show that if $u = X(x)Y(y)$ satisfies (B), but is not identically zero, then

(B₁) $Y(0) = Y(h) = 0.$

b. Show that if $\lambda < 0$, then the only solution of (H_y) that satisfies (B₁) is the trivial solution $Y = 0$.

c. Show that if $\lambda = 0$, then (H_y) has no nontrivial solutions that satisfy (B₁).

d. Show that if $\lambda > 0$, then (H_y) has nontrivial solutions that satisfy (B₁) only if $\sin h\sqrt{\lambda} = 0$, or $\lambda = n^2\pi^2/h^2$ where n is a nonnegative integer.

e. Show that if $\lambda = n^2\pi^2/h^2$, then the nontrivial solutions of (H_y) that satisfy (B₁) are constant multiples of

$$y_n = \sin (n\pi y/h),$$

while (H_x) has two linearly independent solutions,

$$x_n = e^{n\pi x/h} \quad \text{and} \quad x_n^* = e^{-n\pi x/h}.$$

f. Corresponding to the solutions of (H_y) and (H_x) described in (e), obtain solutions to (L₂),

$$u_n = e^{-n\pi x/h} \sin(n\pi y/h) \quad \text{and} \quad u_n^* = e^{-n\pi x/h} \cos(n\pi y/h)$$

that satisfy (B). Note that any linear combination of these (infinitely many) functions also solves (L₂) and satisfies (B).

Exercises 7 to 10 fill some of the gaps in the discussion of Examples 7.1.2 to 7.1.4.

7.   a. Show that if $\lambda < 0$, then the only solution of

(Hθ) $$\frac{d^2\Theta}{d\theta^2} + \lambda\Theta = 0$$

that satisfies the geometric condition

(G) $\Theta(\theta + 2\pi) = \Theta(\theta)$ for all θ

is $\Theta = 0$.

b. Show that if $\lambda = 0$, then the only solutions of (Hθ) that satisfy (G) are constant multiples of $\Theta = 1$.

c. Show that if $\lambda > 0$, then (Hθ) has nontrivial solutions that satisfy (G) only if $\lambda = n^2$ where n is a nonnegative integer.

8.   a. Show that the substitution $t = \cos \phi$ changes the equation

(Hφ) $$-\frac{1}{\Phi \sin \phi} \frac{d}{d\phi} \left[\sin \phi \frac{d\Phi}{d\phi} \right] = n(n + 1)$$

into

$$\frac{1}{\Phi} \frac{d}{dt} \left[(t^2 - 1) \frac{d\Phi}{dt} \right] = n(n + 1).$$

b. Show that this last equation can be rewritten in the form

$$[(t^2 - 1)D^2 + 2tD - n(n + 1)]\Phi = 0.$$

9. *Conversion to Polar Coordinates*
   a. Use the chain rule for partial derivatives to show that if $v = v(x, y)$, where $x = r \cos \theta$ and $y = r \sin \theta$, then

$$\frac{\partial v}{\partial r} = \cos \theta \frac{\partial v}{\partial x} + \sin \theta \frac{\partial v}{\partial y}$$

$$\frac{\partial v}{\partial \theta} = -r \sin \theta \frac{\partial v}{\partial x} + r \cos \theta \frac{\partial v}{\partial y}.$$

b. Solve the preceding equations for $\partial v / \partial x$ and $\partial v / \partial y$ in terms of $\partial v / \partial r$ and $\partial v / \partial \theta$:

$$\frac{\partial v}{\partial x} = \cos \theta \frac{\partial v}{\partial r} - \frac{1}{r} \sin \theta \frac{\partial v}{\partial \theta}$$

$$\frac{\partial v}{\partial y} = \sin \theta \frac{\partial v}{\partial r} + \frac{1}{r} \cos \theta \frac{\partial v}{\partial \theta}.$$

c. By applying the equations in (b), first to $v = u(x, y)$ and then to $v = \partial u / \partial x$ and $v = \partial u / \partial y$, show that

$$\frac{\partial^2 u}{\partial x^2} + \frac{\partial^2 u}{\partial y^2} = \frac{\partial^2 u}{\partial r^2} + \frac{1}{r} \frac{\partial u}{\partial r} + \frac{1}{r^2} \frac{\partial^2 u}{\partial \theta^2}.$$

d. Conclude that the change of variables $x = r \cos \theta$ and $y = r \sin \theta$ changes

$$\frac{\partial^2 u}{\partial x^2} + \frac{\partial^2 u}{\partial y^2} = \beta$$

into

$$\frac{\partial^2 u}{\partial r^2} + \frac{1}{r} \frac{\partial u}{\partial r} + \frac{1}{r^2} \frac{\partial^2 u}{\partial \theta^2} = \beta.$$

*10. *Conversion to Spherical Coordinates*
   a. Suppose $v = v(x, y, z)$, where $x = \rho \sin \phi \cos \theta$, $y = \rho \sin \phi \sin \theta$, and $z = \rho \cos \phi$. Use the chain rule to express $\partial v / \partial \rho$, $\partial v / \partial \phi$, and $\partial v / \partial \theta$ in terms of $\partial v / \partial x$, $\partial v / \partial y$, and $\partial v / \partial z$.
   b. Solve the equations in (a) for $\partial v / \partial x$, $\partial v / \partial y$, and $\partial v / \partial z$ in terms of $\partial v / \partial \rho$, $\partial v / \partial \phi$, and $\partial v / \partial \theta$.

c. By applying the equation in (b) to $v = u(x, y, z)$, $v = \partial u / \partial x$, $v = \partial u / \partial y$, and $v = \partial u / \partial z$, show that

$$\frac{\partial^2 u}{\partial x^2} + \frac{\partial^2 u}{\partial y^2} + \frac{\partial^2 u}{\partial z^2} = \frac{1}{\rho} \frac{\partial^2}{\partial \rho^2} (\rho u) + \frac{1}{\rho^2 \sin \phi} \frac{\partial}{\partial \phi} \left(\sin \phi \frac{\partial u}{\partial \phi} \right) + \frac{1}{\rho^2 \sin^2 \phi} \frac{\partial^2 u}{\partial \theta^2}.$$

d. Conclude that if u is independent of θ, then the change of variables $x = \rho \sin \phi \cos \theta$, $y = \rho \sin \phi \sin \theta$, and $z = \rho \cos \phi$ changes the three-dimensional Laplace equation (L_3) into

$$\rho \frac{\partial^2}{\partial \rho^2} (\rho u) + \frac{1}{\sin \phi} \frac{\partial}{\partial \phi} \left(\sin \phi \frac{\partial u}{\partial \phi} \right) = 0.$$

7.2 REVIEW OF POWER SERIES

In this section we recall some features of power series that we will find useful in solving o.d.e.'s. Our review is brief; you can find a more elaborate treatment in most good calculus texts.

A **series** is a formal sum

$$\sum_{k=0}^{\infty} u_k = u_0 + u_1 + \ldots + u_k + \ldots .$$

To evaluate this series, we consider the **partial sums**

$$S_n = \sum_{k=0}^{n} u_k = u_0 + u_1 + \ldots + u_n.$$

If the partial sums approach a limit

$$S = \lim_{n \to \infty} S_n,$$

we call S the **sum** of the series,

$$\sum_{k=0}^{\infty} u_k = S,$$

and say the series **converges** to S. If the partial sums do not approach a limit, we say the series **diverges**.

There are many tests for determining whether a series converges. The most useful one for our purposes is the ratio test.

Fact: Ratio Test. *Given the series* $\sum\limits_{k=0}^{\infty} u_k$, *suppose that*

$$\rho = \lim_{n \to \infty} \left| \frac{u_{n+1}}{u_n} \right|.$$

Then:

  i. *If* $\rho < 1$, *the series converges.*
 ii. *If* $\rho > 1$, *the series diverges.*
iii. *If* $\rho = 1$, *the test fails. The series may converge or diverge.*

We will be dealing with power series. A **power series about** $t = t_0$ is a series of the form

$$\sum_{k=0}^{\infty} b_k(t - t_0)^k = b_0 + b_1(t - t_0) + \ldots + b_k(t - t_0)^k + \ldots$$

where the b_k's and t_0 are constants, while t is a variable. Note that the partial sums of a power series are polynomials. The basic convergence properties of power series are given by the following statement.

Fact: *Associated to the power series* $\sum\limits_{k=0}^{\infty} b_k(t - t_0)^k$ *is a "number" R, with* $0 \le R \le \infty$, *so that*

  i. *if* $0 < R < \infty$, *the series converges whenever* $|t - t_0| < R$ *and diverges whenever* $|t - t_0| > R$;
 ii. *if* $R = \infty$, *the series converges for all t; and*
iii. *if* $R = 0$, *the series converges only when* $t = t_0$.

The "number" R is called the **radius of convergence** of the power series. In cases (i) and (ii), when $R > 0$, we refer to the values of t with $|t - t_0| < R$ (that is, $t_0 - R < t < t_0 + R$) as the **interval of convergence** of the power series. Note that in case (i) nothing was said about the endpoints ($t = t_0 \pm R$) of this interval; the series may converge or diverge at either point.

The radius of convergence of a power series can usually be calculated by means of the ratio test.

Example 7.2.1

Find the interval of convergence of $\displaystyle\sum_{k=0}^{\infty} t^k$.

We apply the ratio test. Since

$$\rho = \lim_{n\to\infty} \left| \frac{u_{n+1}}{u_n} \right| = \lim_{n\to\infty} \left| \frac{t^{n+1}}{t^n} \right| = \lim_{n\to\infty} |t| = |t|,$$

the series converges for $|t| < 1$ and diverges for $|t| > 1$. The interval of convergence is $|t| < 1$.

Example 7.2.2

Find the interval of convergence of $\displaystyle\sum_{k=0}^{\infty} (t^k/k!)$.

By the ratio test, since

$$\rho = \lim_{n\to\infty} \left| \frac{u_{n+1}}{u_n} \right| = \lim_{n\to\infty} \left| \frac{t^{n+1}/(n+1)!}{t^n/n!} \right|$$

$$= \lim_{n\to\infty} \left| \frac{t^{n+1} n!}{(n+1)! \, t^n} \right| = \lim_{n\to\infty} \left| \frac{t}{n+1} \right| = 0 < 1,$$

the series always converges. The interval of convergence is the whole real line, $|t| < \infty$.

If a power series has $R > 0$, then it defines a function on its interval of convergence. Conversely, given a function $f(t)$, it may be possible to find a power series about $t = t_0$ so that

$$f(t) = \sum_{k=0}^{\infty} b_k(t - t_0)^k \qquad \text{for } |t - t_0| < R$$

where $R > 0$. If such a series exists, we say that $f(t)$ is **analytic** at $t = t_0$.

As examples, we list some basic functions that are analytic at $t = 0$, together with expressions for these functions as series. (See Examples 7.2.4 and 7.2.6 and Exercise 31.)

Fact: *The following functions are analytic at t = 0.*

$$b_0 + b_1 t + \ldots + b_m t^m = b_0 + b_1 t + \ldots + b_m t^m + 0 t^{m+1} + \ldots, \quad |t| < \infty$$

$$e^t = \sum_{k=0}^{\infty} \frac{t^k}{k!}, \quad |t| < \infty$$

$$\sin t = \sum_{k=0}^{\infty} \frac{(-1)^k t^{2k+1}}{(2k+1)!}, \quad |t| < \infty$$

$$\cos t = \sum_{k=0}^{\infty} \frac{(-1)^k t^{2k}}{(2k)!}, \quad |t| < \infty$$

$$\frac{1}{1-t} = \sum_{k=0}^{\infty} t^k, \quad |t| < 1$$

In this chapter, we will be looking for analytic solutions of an o.d.e. by substituting a formal power series $\sum_{k=0}^{\infty} b_k(t - t_0)^k$ into the equation and attempting to determine the constants b_k. In order to carry out this process we will need to know when a power series is identically equal to zero, how to do arithmetic with power series, and how to differentiate power series. Equality of a series with zero is governed by the following (see Exercise 32):

Fact: *If* $\displaystyle\sum_{k=0}^{\infty} b_k(t - t_0)^k = 0$ *for all t satisfying* $|t - t_0| < R$, *where R > 0, then* $b_k = 0$ *for every k.*

The basic arithmetic operations for power series at $t = t_0$ are analogous to those for polynomials. In particular, we multiply a series by a number, or add two series, termwise.

Fact: *If the series expressions*

$$f(t) = \sum_{k=0}^{\infty} b_k(t - t_0)^k \quad and \quad g(t) = \sum_{k=0}^{\infty} c_k(t - t_0)^k$$

are both valid (at least) for $|t - t_0| < R$, *where R > 0, then so are the expressions*

$$\alpha f(t) = \sum_{k=0}^{\infty} (\alpha b_k)(t - t_0)^k$$

$$f(t) + g(t) = \sum_{k=0}^{\infty} (b_k + c_k)(t - t_0)^k.$$

The multiplication and division of series is more complicated (see Notes 1 and 2); the multiplication of a power series by a polynomial is illustrated in the following examples.

Example 7.2.3

Find the terms up to t^4 in the expression for

$$e^{2t} - (2 + t) \cos t$$

as a power series about $t = 0$.

The series expression for e^t described above is valid for all values of t. We obtain the expression for e^{2t} by replacing t with $2t$:

$$e^{2t} = 1 + (2t) + \frac{(2t)^2}{2} + \frac{(2t)^3}{6} + \frac{(2t)^4}{24} + \cdots$$

$$= 1 + 2t + 2t^2 + \frac{4}{3}t^3 + \frac{2}{3}t^4 + \cdots.$$

Next, we multiply the series for $\cos t$ by $(2 + t)$:

$$(2 + t)\cos t = (2 + t)\left[1 - \frac{t^2}{2} + \frac{t^4}{24} - \cdots \right]$$

$$= (2 + t) - \left(t^2 + \frac{t^3}{2} \right) + \left(\frac{t^4}{12} + \frac{t^5}{24} \right) + \cdots.$$

Finally, we subtract the expression for $(2 + t)\cos t$ from the expression for e^{2t}:

$$e^{2t} = \quad 1 + 2t + 2t^2 + \frac{4}{3}t^3 + \frac{2}{3}t^4 + \cdots$$

$$(2 + t)\cos t = \quad 2 + t - t^2 - \frac{1}{2}t^3 + \frac{1}{12}t^4 + \cdots$$

$$\overline{e^{2t} - (2 + t)\cos t = -1 + t + 3t^2 + \frac{11}{6}t^3 + \frac{7}{12}t^4 + \cdots.}$$

Example 7.2.4

Verify that

$$\frac{1}{1 - t} = \sum_{k=0}^{\infty} t^k \qquad \text{for } |t| < 1.$$

We multiply the series on the right by $(1 - t)$; since this series converges only for $|t| < 1$ (see Example 7.2.1), we must restrict our attention to this interval:

$$(1 - t) \sum_{k=0}^{\infty} t^k = \sum_{k=0}^{\infty} t^k - t \sum_{k=0}^{\infty} t^k = \sum_{k=0}^{\infty} t^k - \sum_{k=0}^{\infty} t^{k+1}, \qquad |t| < 1.$$

To perform our subtraction termwise, we must match up the powers of t. We accomplish this by index substitution—substitute $j = k$ in the first series and $j = k + 1$ in the second, remembering that this also affects the limits of summation:

$$(1 - t) \sum_{k=0}^{\infty} t^k = \sum_{j=0}^{\infty} t^j - \sum_{j=1}^{\infty} t^j, \qquad |t| < 1.$$

Now we have to match up the limits of summation. We do this by breaking the first sum in two:

$$(1 - t) \sum_{k=0}^{\infty} t^k = \left(1 + \sum_{j=1}^{\infty} t^j\right) - \sum_{j=1}^{\infty} t^j = 1, \qquad |t| < 1.$$

Thus, dividing both sides by $1 - t$, we obtain the desired equality:

$$\sum_{k=0}^{\infty} t^k = \frac{1}{1 - t}, \qquad |t| < 1.$$

We can differentiate an analytic function by differentiating its power series term by term.

Fact: *If the series expression*

$$f(t) = \sum_{k=0}^{\infty} b_k(t - t_0)^k = b_0 + b_1(t - t_0) + b_2(t - t_0)^2 + \ldots$$

is valid for $|t - t_0| < R$ (where $R > 0$), then so is the expression obtained by termwise differentiation

$$f'(t) = \sum_{k=1}^{\infty} k b_k(t - t_0)^{k-1} = b_1 + 2b_2(t - t_0) + \ldots .$$

Repeated application of this differentiation formula yields the following important relation between the derivatives at $t = t_0$ of an analytic function $f(t)$ and the coefficients of its power series expression about $t = t_0$.

Fact: *If* $f(t) = \sum\limits_{k=0}^{\infty} b_k(t - t_0)^k$ *for* $|t - t_0| < R, R > 0$, *then the derivatives of*
f at $t = t_0$ *are*

$$f^{(k)}(t_0) = k!b_k \qquad for \ k = 0, 1, \dots .$$

Example 7.2.5

Find a series expression for $1/(1 - t)^2$ valid for $|t| < 1$.
We know that

$$\frac{1}{1 - t} = \sum_{k=0}^{\infty} t^k, \qquad |t| < 1.$$

Differentiation gives

$$\frac{1}{(1 - t)^2} = \sum_{k=1}^{\infty} kt^{k-1}, \qquad |t| < 1.$$

The limits of summation can be changed by the index substitution $j = k - 1$ (and $k = j + 1$):

$$\frac{1}{(1 - t)^2} = \sum_{j=0}^{\infty} (j + 1)t^j, \qquad |t| < 1.$$

Example 7.2.6

Verify that

$$e^t = \sum_{k=0}^{\infty} \frac{t^k}{k!} \qquad for \ all \ t.$$

Denote the series on the right side by $f(t)$, recalling (Example 7.2.2) that it converges for all t:

$$f(t) = \sum_{k=0}^{\infty} \frac{t^k}{k!}, \qquad |t| < \infty.$$

Differentiation yields

$$f'(t) = \sum_{k=1}^{\infty} \frac{kt^{k-1}}{k!} = \sum_{k=1}^{\infty} \frac{t^{k-1}}{(k - 1)!}$$

and the index substitution $j = k - 1$ (or $k = j + 1$) gives

$$f'(t) = \sum_{j=0}^{\infty} \frac{t^j}{j!} = f(t), \qquad |t| < \infty.$$

Thus $x = f(t)$ is a solution of

$$(D - 1)x = 0.$$

Then

$$f(t) = ce^t,$$

and since $f(0) = 1$, we have $c = 1$. It follows that $f(t) = e^t$, or

$$\sum_{k=0}^{\infty} \frac{t^k}{k!} = e^t, \qquad |t| < \infty.$$

Our rules for manipulating series enable us to find power series expressions for a wide variety of functions, starting from our expressions for e^t, $\sin t$, $\cos t$, and $1/(1 - t)$. We discuss briefly the problem of finding power series expressions for more complicated functions in the notes following our summary.

POWER SERIES REVIEW

Definitions

A **power series about** $t = t_0$ is a formal expression

$$p(t) = \sum_{k=0}^{\infty} b_k(t - t_0)^k = b_0 + b_1(t - t_0) + \ldots + b_k(t - t_0)^k + \ldots.$$

If for some value $t = t_1$ of t

$$\lim_{n \to \infty}[b_0 + b_1(t_1 - t_0) + \ldots + b_n(t_1 - t_0)^n]$$

exists, then we say the series **converges** at $t = t_1$ and set $p(t_1)$ equal to this limit. If the limit does not exist, we say the series **diverges** at $t = t_1$. Associated with the power series is a **radius of convergence** R, with $0 \leq R \leq \infty$, so that the series converges for all t in the **interval of convergence** $|t - t_0| < R$ and diverges when $|t - t_0| > R$.

A function $f(t)$ is **analytic** at $t = t_0$ if there is a power series about $t = t_0$ with radius of convergence $R > 0$ such that

$$f(t) = \sum_{k=0}^{\infty} b_k(t - t_0)^k, \qquad |t - t_0| < R.$$

Properties of Analytic Functions

Suppose that the series expressions

$$f(t) = \sum_{k=0}^{\infty} b_k(t - t_0)^k \qquad \text{and} \qquad g(t) = \sum_{k=0}^{\infty} c_k(t - t_0)^k$$

are both valid for $|t - t_0| < R, R > 0$.

1. If $f(t) = 0$ for $|t - t_0| < R$, then $b_k = 0$ for $k = 0, 1, 2, \ldots$.

2. $\alpha f(t) = \sum_{k=0}^{\infty} (\alpha b_k)(t - t_0)^k, \qquad |t - t_0| < R.$

3. $f(t) + g(t) = \sum_{k=0}^{\infty} (b_k + c_k)(t - t_0)^k, \qquad |t - t_0| < R.$

4. $f'(t) = \sum_{k=1}^{\infty} k b_k(t - t_0)^{k-1}, \qquad |t - t_0| < R.$

5. $f^{(k)}(t_0) = k! b_k, \qquad k = 0, 1, \ldots$.

Notes
1. *Products of power series*
    If the power series expressions

$$f(t) = \sum_{k=0}^{\infty} b_k(t - t_0)^k \qquad \text{and} \qquad g(t) = \sum_{k=0}^{\infty} c_k(t - t_0)^k$$

are both valid for $|t - t_0| < R$, then so is the expression

$$f(t)g(t) = \sum_{k=0}^{\infty} (b_0 c_k + b_1 c_{k-1} + \ldots + b_k c_0)(t - t_0)^k.$$

That is, power series can be multiplied like polynomials. For example, we can obtain the first terms of a series about $t = 0$ for $e^t \cos t$ by multiplying the respective series for e^t and $\cos t$:

$$e^t = 1 + t + \frac{1}{2}t^2 + \frac{1}{6}t^3 + \frac{1}{24}t^4 + \ldots$$

$$\cos t = 1 \qquad\quad - \frac{1}{2}t^2 \qquad\quad + \frac{1}{24}t^4 + \ldots$$

$$1 \times e^t = 1 + t + \frac{1}{2}t^2 + \frac{1}{6}t^3 + \frac{1}{24}t^4 + \ldots$$

$$-\frac{1}{2}t^2 \times e^t = \qquad\quad - \frac{1}{2}t^2 - \frac{1}{2}t^3 - \frac{1}{4}t^4 + \ldots$$

$$\frac{1}{24}t^4 \times e^t = \qquad\qquad\qquad\qquad\quad \frac{1}{24}t^4 + \ldots$$

$$e^t \cos t = 1 + t \qquad\qquad - \frac{1}{3}t^3 - \frac{1}{6}t^4 + \ldots .$$

2. Quotients of analytic functions

If $f(t)$ and $g(t)$ have expressions as power series about $t = t_0$ that are both valid for $|t - t_0| < R$, and furthermore $g(z) \neq 0$ for every complex number z with $|z - t_0| < R$, then $f(t)/g(t)$ has an expression as a power series about $t = t_0$, also valid for $|t - t_0| < R$. Note, however, that *complex* zeros of the denominator affect the convergence of the quotient series even at *real* values of t. For example, the series for $1/(1 + t^2)$ (see Exercise 29) diverges for real t with $|t| > 1$ even though $1 + t^2$ is never zero for real values of t.

One can find any desired number of terms for a quotient series by means of long division. For example, we can find the first terms of the series for $1/(1 - t)$ as follows:

$$
\begin{array}{r}
1 + t + t^2 + \ldots \\
1 - t \overline{)\ 1 } \\
\underline{1 - t} \\
t \\
\underline{t - t^2} \\
t^2 \\
\underline{t^2 - t^3} \\
t^3
\end{array}
$$

3. Taylor series

Suppose we are given a function $f(t)$ for which we want to find a power series expression about $t = t_0$. If there is such an expression

$$f(t) = \sum_{k=0}^{\infty} b_k (t - t_0)^k$$

valid for $|t - t_0| < R$, where $R > 0$, then

$$f^{(k)}(t_0) = k!b_k$$

so the coefficients must be given by

$$b_k = \frac{f^{(k)}(t_0)}{k!}.$$

The series with these coefficients,

$$\sum_{k=0}^{\infty} \frac{f^{(k)}(t_0)}{k!} (t - t_0)^k,$$

is known as the **Taylor series for f(t) about $t = t_0$.** It is the only series about $t = t_0$ that could possibly equal $f(t)$.

In practice, it may be easy to decide where the Taylor series converges but difficult to decide whether its limit equals $f(t)$. Indeed, there are examples of functions $f(t)$ whose Taylor series converges, but not to $f(t)$.

Examples 7.2.4 to 7.2.6 illustrated some of the methods used to show that a function is equal to a given series. Another method is to study the "remainder term"

$$R_n(t, t_0) = f(t) - \sum_{k=0}^{n} \frac{f^{(k)}(t_0)}{k!} (t - t_0)^k.$$

The function $f(t)$ equals its Taylor series precisely when

$$\lim_{n \to \infty} R_n(t, t_0) = 0.$$

EXERCISES

In Exercises 1 to 9, find the interval of convergence.

1. $\displaystyle\sum_{k=0}^{\infty} \frac{2^k t^k}{(k + 1)!}$

2. $\displaystyle\sum_{k=0}^{\infty} \frac{2^k t^k}{(k + 1)^2}$

3. $\displaystyle\sum_{k=0}^{\infty} \frac{(-1)^k k! t^k}{(k^2 + 1)}$

4. $\displaystyle\sum_{k=0}^{\infty} (-1)^k t^{2k}$

5. $\displaystyle\sum_{k=0}^{\infty} \frac{(k^2 + 1)t^k}{(k + 1)3^k}$

6. $\displaystyle\sum_{k=0}^{\infty} \frac{k^2 t^k}{(2k)!}$

7. $\displaystyle\sum_{k=0}^{\infty} \frac{(t - 2)^k}{(3k + 1)^2}$

8. $\displaystyle\sum_{k=0}^{\infty} \frac{(2t + 1)^k}{3^k}$

9. $\displaystyle\sum_{k=0}^{\infty} \frac{k!(2t - 1)^k}{(2k + 1)!}$

In Exercises 10 to 17, perform the indicated formal operation.

10. $\displaystyle\sum_{k=0}^{\infty} t^k - \sum_{k=0}^{\infty} (k + 1)t^k$

11. $\displaystyle\sum_{k=0}^{\infty} t^k + \sum_{k=0}^{\infty} (-1)^k t^k$

12. $\displaystyle\sum_{k=0}^{\infty} t^k + \sum_{k=0}^{\infty} t^{k+1}$

13. $\displaystyle(1 + t) \sum_{k=0}^{\infty} (-1)^k t^k$

14. $\displaystyle(t^2 - 1) \sum_{k=0}^{\infty} 3kt^k$

15. $\displaystyle\frac{d}{dt}\left(\sum_{k=0}^{\infty} (3k + 1)t^k\right)$

16. $\displaystyle\frac{d^2}{dt^2}\left(\sum_{k=0}^{\infty} \frac{(-1)^k t^{2k+1}}{(2k + 1)!}\right)$

17. $\displaystyle[t^2 D^2 - 3tD] \sum_{k=1}^{\infty} \frac{3t^k}{k^2}$

In Exercises 18 to 22, find the terms up to t^5 of the given product or quotient (see Notes 1 and 2).

18. $\displaystyle\left(\sum_{k=0}^{\infty} t^k\right)\left(\sum_{k=0}^{\infty} (-1)^k t^k\right)$

19. $\displaystyle\left(\sum_{k=0}^{\infty} \frac{t^k}{k!}\right)^2$

20. $\displaystyle\left(\sum_{k=0}^{\infty} \frac{(-1)^k t^{2k}}{(2k)!}\right)\left(\sum_{k=0}^{\infty} \frac{(-1)^k t^{2k+1}}{(2k + 1)!}\right)$

21. $\displaystyle\left(\sum_{k=0}^{\infty} t^k\right) \Big/ (1 + t^2)$

22. $\displaystyle\left(\sum_{k=0}^{\infty} \frac{2^k t^k}{k!}\right) \Big/ \left(\sum_{k=0}^{\infty} \frac{t^k}{k!}\right)$

Starting with the basic series expressions for $\sin t$, $\cos t$, e^t, and $1/(1 - t)$, find series expressions for the functions in Exercises 23 to 30.

23. $\displaystyle\sin \frac{t}{2}$

24. $\cos 2t$

25. e^{-t}

26. $\displaystyle\frac{1}{1 + t}$

27. $\displaystyle\frac{1}{(1 + t)^2}$

28. $\displaystyle\frac{1}{(1 + t)^3}$

29. $\displaystyle\frac{1}{1 + t^2}$

30. $\cos^2 t$ (*Hint:* Use a trig identity or see Note 1.)

Some more abstract problems:

31. *Series Expressions for sin t and cos t:*

   a. Show that $\displaystyle\sum_{k=0}^{\infty} (-1)^k t^{2k+1}/(2k + 1)!$ converges for all t.

   b. Verify that $\sin t$ and the series in (a) are equal by showing that they solve the same initial value problem:

$$(D^2 + 1)x = 0, \qquad x(0) = 0, \ x'(0) = 1.$$

   c. Obtain an expression for $\cos t$ by differentiating the expression for $\sin t$.

32. *Equality of a Series with Zero:* Show that if $f(t) = \sum_{k=0}^{\infty} b_k(t - t_0)^k = 0$ for all t, $|t - t_0| < R$ $(R > 0)$, then $b_k = 0$, $k = 0, 1, \ldots$. [*Hint:* Express b_k in terms of $f^{(k)}(t_0)$.]

*33. Consider the function

$$f(t) = \begin{cases} e^{-1/t^2} & \text{for } t \neq 0 \\ 0 & \text{for } t = 0. \end{cases}$$

   a. Show that $D(e^{-1/t^2}) = (2/t^3)e^{-1/t^2}$ for $t \neq 0$.

b. Show that in general $D^n(e^{-1/t^2}) = (P_n(t)/t^{3n})e^{-1/t^2}$ for $t \neq 0$, where $P_n(t)$ is a polynomial of degree at most $2(n - 1)$.

c. Show that $f^{(n)}(0) = 0$ for each n. (*Hint:* Use the fact that for any α, $t^\alpha e^{-1/t^2} \to 0$ as $t \to 0$.)

d. Show that the Taylor series for $f(t)$ expanded about $t = 0$ (see Note 3) converges for all $t \neq 0$, but the sum does not equal $f(t)$ for $t > 0$.

7.3 SOLUTIONS ABOUT ORDINARY POINTS

We will see in this section how to find power series expressions for the solutions to linear o.d.e.'s

(N) $$[a_n(t)D^n + a_{n-1}(t)D^{n-1} + \ldots + a_0(t)]x = E(t)$$

in case the coefficients $a_n(t), \ldots, a_0(t)$ and the forcing term $E(t)$ are all polynomials. We will find such series by substituting the expression

$$x(t) = \sum_{k=0}^{\infty} b_k(t - t_0)^k$$

into (N) and solving for the coefficients b_k. This straightforward approach will always work when t_0 is a point at which the leading coefficient does not vanish. Such points [where $a_n(t_0) \neq 0$] are called the **ordinary points** of (N) to distinguish them from the **singular points** of (N), where $a_n(t_0) = 0$.

The following theorem justifies our search for power series solutions of (N) about an ordinary point. Moreover, it guarantees the validity of the answers we will obtain as long as t is closer to t_0 than the nearest complex root of $a_n(t)$.

Theorem: *Suppose that (N) is a linear nth-order o.d.e. with polynomial coefficients and forcing term and that t_0 is an ordinary point of (N). If $a_n(z) \neq 0$ for every complex number z satisfying $|z - t_0| < R$, then any solution of (N) has an expression as a power series about $t = t_0$ that is valid for (at least) $|t - t_0| < R$.*

To see how the procedure works in practice, we start with a familiar o.d.e.

Example 7.3.1

Find a power series expression about $t = 0$ for the general solution of

(H) $$(D^2 - 1)x = 0.$$

Note that (H) has no singular points.

We look for solutions of (H) in the form

(S) $$x(t) = \sum_{k=0}^{\infty} b_k t^k.$$

We differentiate (S) twice

$$x'(t) = \sum_{k=1}^{\infty} k b_k t^{k-1}, \qquad x''(t) = \sum_{k=2}^{\infty} k(k-1) b_k t^{k-2}$$

and substitute into (H):

$$\sum_{k=2}^{\infty} k(k-1) b_k t^{k-2} - \sum_{k=0}^{\infty} b_k t^k = 0.$$

In order to combine these two series into one, we need to match up powers of t. To this end, we perform the index substitutions $j = k - 2$ (or $k = j + 2$) in the first series and $j = k$ in the second; we must remember to adjust the limits of summation as well as the summands. We get

$$\sum_{j=0}^{\infty} (j + 2)(j + 1) b_{j+2} t^j - \sum_{j=0}^{\infty} b_j t^j = 0$$

or

$$\sum_{j=0}^{\infty} [(j + 2)(j + 1) b_{j+2} - b_j] t^j = 0.$$

Since a power series is identically zero only if all its coefficients are zero, we conclude that

$$(j + 2)(j + 1) b_{j+2} - b_j = 0$$

or

(R) $$b_{j+2} = \frac{1}{(j + 2)(j + 1)} b_j \qquad \text{for } j = 0, 1, \ldots .$$

An equation such as (R), which relates a given coefficient to earlier ones, is called a **recurrence relation**.

Although (R) tells us nothing about b_0 or b_1, we can use it to find all of the later coefficients in terms of these two. When $j = 0$ and $j = 1$, respectively, (R) gives

$$b_2 = \frac{1}{2 \cdot 1} b_0, \qquad b_3 = \frac{1}{3 \cdot 2} b_1.$$

Now these expressions can be substituted back into (R) with $j = 2$ and 3, respectively, to find

$$b_4 = \frac{1}{4 \cdot 3} b_2 = \frac{1}{4 \cdot 3 \cdot 2 \cdot 1} b_0, \qquad b_5 = \frac{1}{5 \cdot 4} b_3 = \frac{1}{5 \cdot 4 \cdot 3 \cdot 2} b_1.$$

Continuing in this way, we can express all even-numbered coefficients as multiples of b_0 and all odd-numbered coefficients as multiples of b_1:

$$b_{2m} = \frac{1}{(2m)!} b_0, \qquad b_{2m+1} = \frac{1}{(2m + 1)!} b_1.$$

We substitute these values into (S) and separate the terms involving b_0 from those involving b_1 to obtain

$$x(t) = b_0 \sum_{m=0}^{\infty} \frac{1}{(2m)!} t^{2m} + b_1 \sum_{m=0}^{\infty} \frac{1}{(2m + 1)!} t^{2m+1}.$$

Since the leading coefficient of (H) never vanishes, this series expression is valid for all t.

It should not surprise us that our answer involves two arbitrary constants, b_0 and b_1, since we are solving a second-order o.d.e. These constants have a simple relation to initial conditions:

$$x(0) = b_0, \qquad x'(0) = b_1.$$

You should verify that substituting $b_0 = b_1 = 1$ (respectively $b_0 = -b_1 = 1$) into (R) and (S) gives a series expression for the solution $x = e^t$ (respectively $x = e^{-t}$), which played an important role in our earlier method for solving (H).

This example illustrates the general pattern. We substitute a formal power series for $x(t)$ into the o.d.e., combine terms, and obtain a recurrence relation. We then use the recurrence relation to determine the coefficients of the series.

We next consider a variable-coefficient example.

Example 7.3.2

Solve about $t = 0$

(H) $$[(t^2 - 1)D^2 - 2]x = 0.$$

Note that $t = 0$ is an ordinary point of (H); the only singular points are $t = \pm 1$.
We look for solutions in the form

(S) $$x(t) = \sum_{k=0}^{\infty} b_k t^k.$$

Substitution into (H) yields

$$(t^2 - 1) \sum_{k=2}^{\infty} k(k - 1)b_k t^{k-2} - 2 \sum_{k=0}^{\infty} b_k t^k = 0$$

or

$$\sum_{k=2}^{\infty} k(k - 1)b_k t^k - \sum_{k=2}^{\infty} k(k - 1)b_k t^{k-2} - \sum_{k=0}^{\infty} 2b_k t^k = 0.$$

To match up the powers of t, we substitute $j = k - 2$ (or $k = j + 2$) in the
middle series and $j = k$ in the other two:

$$\sum_{j=2}^{\infty} j(j - 1)b_j t^j - \sum_{j=0}^{\infty} (j + 2)(j + 1)b_{j+2} t^j - \sum_{j=0}^{\infty} 2b_j t^j = 0.$$

Now the limits of summation don't match, so we separate the terms for $j = 0$ and
$j = 1$ from the last two series.

$$\left[\sum_{j=2}^{\infty} j(j - 1)b_j t^j \right] - \left[2b_2 + 6b_3 t + \sum_{j=2}^{\infty} (j + 2)(j + 1)b_{j+2} t^j \right]$$

$$- \left[2b_0 + 2b_1 t + \sum_{j=2}^{\infty} 2b_j t^j \right] = 0.$$

Combining terms, we have

$$-(2b_2 + 2b_0) - (6b_3 + 2b_1)t$$

$$+ \sum_{j=2}^{\infty} [j(j - 1)b_j - (j + 2)(j + 1)b_{j+2} - 2b_j]t^j = 0.$$

This can happen only if the coefficient of each power of t is zero. Thus

$$b_2 = -b_0$$

$$b_3 = -\frac{1}{3} b_1$$

and

(R) $\qquad b_{j+2} = \dfrac{j(j-1)-2}{(j+2)(j+1)} b_j = \dfrac{j-2}{j+2} b_j, \qquad j = 2, 3, \ldots$

As in the last example, the even-numbered coefficients are multiples of b_0:

$$b_2 = -b_0, \quad b_4 = \frac{0}{4} b_2 = 0, \quad b_6 = \frac{4}{6} b_4 = 0, \ldots, b_{2m} = 0, \ldots,$$

while the odd-numbered coefficients are multiples of b_1:

$$b_3 = -\frac{1}{3} b_1, \quad b_5 = \frac{1}{5} b_3 = -\frac{1}{3 \cdot 5} b_1,$$

$$b_7 = \frac{3}{7} b_5 = -\frac{1}{5 \cdot 7} b_1, \quad b_9 = \frac{5}{9} b_7 = -\frac{1}{7 \cdot 9} b_1, \ldots,$$

$$b_{2m+1} = \frac{2m-3}{2m+1} b_{2m-1} = -\frac{1}{(2m+1)(2m-1)} b_1, \ldots.$$

Note that the general formula for b_{2m} works from $m = 2$ on, while the formula for b_{2m+1} works starting from $m = 0$.

Substituting these values back into (S) and separating the terms involving b_0 from those involving b_1, we obtain

$$x(t) = b_0(1 - t^2) + b_1 \sum_{m=0}^{\infty} \left[\frac{-1}{(2m+1)(2m-1)} \right] t^{2m+1}.$$

This expression will be valid provided we stay away from the zeroes, ± 1, of the leading coefficient of (H)—that is, for $|t| < 1$.

So far we have managed to obtain explicit formulas describing all the coefficients of our solutions. Sometimes the recurrence relations are sufficiently complicated that we cannot find a general pattern. Nonetheless, we can use the recurrence relations to calculate explicitly any specified finite number of coefficients. Since this usually allows us to obtain good approximations to the values of our solutions at specified points (see Section 7.4), we content ourselves with finding the first few terms of the solution.

Example 7.3.3

Solve about $t = 0$:

(H)
$$[D^2 - tD + t]x = 0.$$

We look for solutions in the form

$$x(t) = \sum_{k=0}^{\infty} b_k t^k.$$

Substitution into (H) yields

$$\sum_{k=2}^{\infty} k(k - 1)b_k t^{k-2} - \sum_{k=1}^{\infty} kb_k t^k + \sum_{k=0}^{\infty} b_k t^{k+1} = 0.$$

We match up powers by substituting $j = k - 2$ (or $k = j + 2$) in the first series, $j = k$ in the second, and $j = k + 1$ (or $k = j - 1$) in the third:

$$\sum_{j=0}^{\infty} (j + 2)(j + 1)b_{j+2} t^j - \sum_{j=1}^{\infty} jb_j t^j + \sum_{j=1}^{\infty} b_{j-1} t^j = 0.$$

We separate the $j = 0$ term from the first series and combine terms:

$$2b_2 + \sum_{j=1}^{\infty} [(j + 2)(j + 1)b_{j+2} - jb_j + b_{j-1}]t^j = 0.$$

This can happen only if

$$b_2 = 0$$

and

(R)
$$b_{j+2} = \frac{jb_j - b_{j-1}}{(j + 2)(j + 1)}, \qquad j = 1, 2, \ldots.$$

We can use (R) to find the first few coefficients:

$$b_3 = \frac{1}{6}(b_1 - b_0)$$

$$b_4 = \frac{1}{12}(2b_2 - b_1) = -\frac{1}{12}b_1$$

$$b_5 = \frac{1}{20}(3b_3 - b_2) = \frac{1}{40}(b_1 - b_0).$$

Thus, the first few terms of our solution are

$$x(t) = b_0 + b_1 t + 0t^2 + \frac{1}{6}(b_1 - b_0)t^3 - \frac{1}{12}b_1 t^4 + \frac{1}{40}(b_1 - b_0)t^5 + \ldots$$

$$= b_0\left(1 - \frac{1}{6}t^3 - \frac{1}{40}t^5 + \ldots\right) + b_1\left(t + \frac{1}{6}t^3 - \frac{1}{12}t^4 + \frac{1}{40}t^5 + \ldots\right).$$

The expression is valid for all t.

If $t = 0$ is a singular point, or if we want to match initial conditions at an ordinary point $t_0 \neq 0$, then we may wish to express our solutions as power series about $t = t_0 \neq 0$. Since most of us find it easier to work with a series about $t = 0$, the usual procedure is to first make the substitution $T = t - t_0$, as shown in the following example.

Example 7.3.4

Find the solution of

(H) $$[D^2 - (t - 2)]x = 0$$

subject to the conditions at $t = 2$

$$x(2) = 1, \qquad x'(2) = 0.$$

Since the condition is to be satisfied at $t = 2$, we look for a series about $t = 2$:

(S) $$x(t) = \sum_{k=0}^{\infty} b_k(t - 2)^k.$$

Our conditions tell us that

$$b_0 = 1, \qquad b_1 = 0.$$

To simplify our calculations, we make the substitution

$$T = t - 2 \quad (\text{or } t = T + 2),$$

which changes (H) and (S) to

(H')
$$[D^2 - T]x = 0$$

and

(S')
$$x(T + 2) = \sum_{k=0}^{\infty} b_k T^k.$$

Substituting (S') into (H') we obtain

$$\sum_{k=2}^{\infty} k(k - 1)b_k T^{k-2} - \sum_{k=0}^{\infty} b_k T^{k+1} = 0.$$

The index substitutions $j = k - 2$ (or $k = j + 2$) in the first series and $j = k + 1$ (or $k = j - 1$) in the second, together with a separation of the $j = 0$ term in the first series, lead to

$$2b_2 + \sum_{j=1}^{\infty} [(j + 2)(j + 1)b_{j+2} - b_{j-1}]T^j = 0.$$

This implies

$$b_2 = 0$$

and

(R)
$$b_{j+2} = \frac{1}{(j + 2)(j + 1)} b_{j-1}, \qquad j = 1, 2, 3, \ldots .$$

We know specific values for b_0, b_1, and b_2 and have a formula for b_{j+2} in terms of the coefficient *three* steps back, b_{j-1}. Since b_1 and b_2 are both zero, we have

$$0 = b_4 = b_7 = \ldots = b_{3m+1} \ldots$$

$$0 = b_5 = b_8 = \ldots = b_{3m+2} \ldots .$$

The first few terms of the form b_{3m} are

$$b_0 = 1, \quad b_3 = \frac{1}{3 \cdot 2} b_0 = \frac{1}{3 \cdot 2}, \quad b_6 = \frac{1}{6 \cdot 5} b_3 = \frac{1}{6 \cdot 5 \cdot 3 \cdot 2},$$

$$b_9 = \frac{1}{9 \cdot 8} b_6 = \frac{1}{9 \cdot 8 \cdot 6 \cdot 5 \cdot 3 \cdot 2}.$$

The denominator of b_{3m} looks like $(3m)!$, except that the factors $1, 4, 7, \ldots,$ $3m - 2$ are missing. Thus,

$$b_{3m} = \frac{1 \cdot 4 \cdot 7 \cdot \ldots \cdot (3m - 2)}{(3m)!}.$$

This general formula starts to work at $m = 1$.

Now, putting our expressions back into (S′), we get

$$x(T + 2) = 1 + \sum_{m=1}^{\infty} \frac{1 \cdot 4 \cdot 7 \cdot \ldots \cdot (3m - 2)}{(3m)!} T^{3m}$$

or, using $T = t - 2$,

$$x(t) = 1 + \sum_{m=1}^{\infty} \frac{1 \cdot 4 \cdot 7 \cdot \ldots \cdot (3m - 2)}{(3m)!} (t - 2)^{3m}.$$

The expression is valid for all t.

This technique also works nicely for nonhomogeneous equations, as in our final example.

Example 7.3.5

Solve about $t = 0$

(N) $[D^2 + tD + 1]x = t.$

We seek solutions in the form

(S) $x(t) = \sum_{k=0}^{\infty} b_k t^k.$

Substituting (S) into (N) yields

$$\sum_{k=2}^{\infty} k(k - 1)b_k t^{k-2} + \sum_{k=1}^{\infty} kb_k t^k + \sum_{k=0}^{\infty} b_k t^k = t.$$

We bring the forcing term to the left-hand side, make substitutions ($j = k - 2$, $j = k$, $j = k$) to match up powers of t, and separate the $j = 0$ and $j = 1$ terms to obtain

$$(2b_2 + b_0) + (6b_3 + 2b_1 - 1)t + \sum_{j=2}^{\infty} [(j + 2)(j + 1)b_{j+2} + (j + 1)b_j]t^j = 0.$$

This leads to

$$b_2 = -\frac{1}{2} b_0, \qquad b_3 = \frac{1}{6} - \frac{1}{3} b_1,$$

and

(R) $$b_{j+2} = -\frac{1}{j + 2} b_j, \qquad j = 2, 3, \ldots.$$

We use (R) to express the even-numbered coefficients in terms of b_0 and the odd-numbered ones in terms of b_1:

$$b_2 = -\frac{1}{2} b_0, \quad b_4 = -\frac{1}{4} b_2 = \frac{1}{4 \cdot 2} b_0, \quad b_6 = -\frac{1}{6} b_4 = \frac{1}{6 \cdot 4 \cdot 2} b_0, \ldots,$$

$$b_{2m} = \frac{(-1)^m}{(2m) \cdot \ldots \cdot 6 \cdot 4 \cdot 2} b_0, \ldots,$$

$$b_3 = -\frac{1}{3} b_1 + \frac{1}{6}, \quad b_5 = -\frac{1}{5} b_3 = \frac{1}{5 \cdot 3} b_1 - \frac{1}{6 \cdot 5},$$

$$b_7 = -\frac{1}{7} b_5 = \frac{-1}{7 \cdot 5 \cdot 3} b_1 + \frac{1}{7 \cdot 6 \cdot 5}, \ldots,$$

$$b_{2m+1} = \frac{(-1)^m}{(2m + 1) \cdot \ldots \cdot 5 \cdot 3 \cdot 1} b_1$$

$$+ \frac{(-1)^{m+1}}{2[(2m + 1) \cdot \ldots \cdot 5 \cdot 3 \cdot 1]}, \ldots.$$

Substitution in (S) gives us

$$x(t) = b_0\left(1 + \sum_{m=1}^{\infty} \frac{(-1)^m}{(2m) \cdot \ldots \cdot 4 \cdot 2} t^{2m}\right)$$

$$+ b_1\left(t + \sum_{m=1}^{\infty} \frac{(-1)^m}{(2m + 1) \cdot \ldots \cdot 5 \cdot 3 \cdot 1} t^{2m+1}\right)$$

$$+ \frac{1}{2} \sum_{m=1}^{\infty} \frac{(-1)^{m+1}}{(2m + 1) \cdot \ldots \cdot 5 \cdot 3 \cdot 1} t^{2m+1}.$$

The expression is valid for all t.

Although all our examples were second-order o.d.e.'s, the same procedure works about ordinary points for o.d.e.'s of any order.

SOLUTIONS ABOUT ORDINARY POINTS

Suppose the coefficients $a_i(t)$ and forcing term $E(t)$ in the o.d.e.

(N) $$[a_n(t)D^n + \ldots + a_0(t)]x = E(t)$$

are all polynomials. We call $t = t_0$ an **ordinary point** of (N) if $a_n(t_0) \neq 0$ and a **singular point** if $a_n(t_0) = 0$.

If $t = 0$ is an ordinary point of (N), we can find a series expression

(S) $$x(t) = \sum_{k=0}^{\infty} b_k t^k$$

for the solutions of (N) as follows:

1. Substitute (S) into (N).
2. Combine terms so as to rewrite the equation in the form of a power series set identically equal to zero. (This requires index substitutions to match up powers of t, and possibly separating out a few extra terms to match up the limits of summation.)
3. Set each (combined) coefficient of the resulting series equal to zero, to obtain a **recurrence relation** (R) between the general coefficient in (S) and earlier ones.

4. If a pattern is evident, write down an explicit formula for the coefficients of (S); otherwise, we can still use (R) to write down any specified finite number of coefficients.

5. Substitute the coefficients back into (S) to obtain a power series expression for the solutions.

The resulting series expression will be valid for $|t| < R$, provided the leading coefficient $a_n(t)$ of (N) has no (complex) zeroes with $|z| < R$.

If $t = t_0$ is a nonzero ordinary point of (N), we can find a series expression about $t = t_0$

$$\text{(S)} \qquad x(t) = \sum_{k=0}^{\infty} b_k(t - t_0)^k$$

for the solutions of (N). We substitute $T = t - t_0$ (or $t = T + t_0$) and use the preceding procedure to find a solution

$$\text{(S')} \qquad x(T + t_0) = \sum_{k=0}^{\infty} b_k T^k$$

to the modified o.d.e.

$$\text{(N')} \qquad [a_n(T + t_0)D^n + \ldots + a_0(T + t_0)]x = E(T + t_0).$$

We then substitute $T = t - t_0$ back into (S') to get (S).

Note
Concerning our hypotheses

The assumption in this section that the coefficients and forcing term in (N) are polynomials can be weakened. The general theorem, of which we quoted a special case, is the following.

Theorem: *Suppose that*

i. *$a_n(t), \ldots, a_1(t)$, and $E(t)$ are analytic functions whose power series expressions about $t = t_0$ are all valid (at least) for $|t - t_0| < R$, where $R > 0$; and*
ii. *$a_n(z) \neq 0$ for any complex number z satisfying $|z - t_0| < R$.*

Then every solution of $[a_n(t)D^n + \ldots + a_0(t)]x = E(t)$ is analytic at $t = t_0$, with a power series expression about $t = t_0$ valid (at least) for $|t - t_0| < R$.

The procedure described in the summary actually works under these more general hypotheses. We replace each $a_i(t)$ and $E(t)$ with its power series expression and then proceed as before. However, if one (or more) of the coefficients is not a polynomial, then step 1 will require multiplication of series (see Note 1 in Section 7.2 and Exercises 18 and 19). Also, we cannot in general expect a simple recurrence relation for the b_k's; we can hope to actually calculate only the first few terms.

EXERCISES

In Exercises 1 to 6, (a) find the recurrence relation for the coefficients of series solutions about $t = 0$, and (b) write out the terms to t^5 of the general solution.

1. $[D^2 + tD + 4]x = 0$

2. $[D^2 + 4t]x = 0$

3. $[D^2 + D + t]x = 0$

4. $[D^2 + 2tD + t]x = 0$

5. $[D^2 + 3tD - 4]x = t - 2$

6. $[D^3 + tD]x = 0$

In Exercises 7 to 10, (a) find the recurrence relation for the coefficients of series solutions about $t = 0$, and (b) write out the terms to t^4 of the solution matching the given initial condition. Compare the cited exercise in Section 7.1.

7. $[D^2 + (10 + t)D + 1]x = 10;$ $x(0) = x'(0) = 0$ (Exercise 1)

8. $[10D^2 + 10D + (10 - t)]x = 0;$ $x(0) = 1, x'(0) = 0$ (Exercise 2)

9. $[10D^2 + (10 - t)D + 10]x = 0;$ $x(0) = 1, x'(0) = 0$ (Exercise 3)

10. $[(10 - t)D^2 + 10D + 10]x = 0;$ $x(0) = 1, x'(0) = 0$ (Exercise 4)

In Exercises 11 and 12, (a) find the recurrence relation for the coefficients of series solutions about $t = t_0$, and (b) write out the terms to $(t - t_0)^4$ of the general solution.

11. $[tD^2 - 4]x = 0;$ $t_0 = 1$

12. $[tD^2 - 4]x = 0;$ $t_0 = 2$

In Exercises 13 and 14, (a) find the recurrence relation for the coefficients of series solutions about $t = 0$, and (b) find the general form for the coefficients.

13. $[D^2 + tD + 1]x = 0$

14. $[(t^2 + 1)D^2 - 2]x = 0$

Exercises 15 to 17 deal with some important equations from mathematical physics.

15. a. Find the recurrence relation for solutions about $t = 0$ of the **Legendre equation** $[(t^2 - 1)D^2 + 2tD - \mu(\mu + 1)]x = 0$, where μ is a constant.
   b. Show that if μ is a nonnegative integer, then this o.d.e. has a polynomial solution of degree μ.

16. a. Find the recurrence relation for solutions of the **Hermite equation** $[D^2 - 2tD + \mu]x = 0$, where μ is a constant.
   b. Show that if $\mu = 2N$ where N is a nonnegative integer, then this o.d.e. has a polynomial solution of degree N.

17. Find a series expression about $t = 0$ for the general solution of the **Airy equation** $(D^2 + t)x = 0$.

The o.d.e.'s in Exercises 18 to 20 involve analytic functions that are not polynomials (see the note at the end of the section). Find the terms to t^5 of the series solution to the given initial value problem.

18. $[D^2 + \sin t]x = 0;$ $x(0) = 0, x'(0) = 1$

19. $[D^2 + \sin t]x = 0;$ $x(0) = 1, x'(0) = 0$

20. $[D^2 + t]x = e^t;$ $x(0) = 1, x'(0) = 0$

7.4 POWER SERIES ON PROGRAMMABLE CALCULATORS (Optional)

The recurrence relations that arise in finding series solutions to o.d.e.'s are especially well suited to the use of computers and programmable hand calculators. In this section we discuss general algorithms for finding power series on any program-mable device, as well as specific programs that illustrate how these algorithms can be implemented on two popular calculators, the Texas Instruments TI-57 and the Hewlett-Packard HP-25. (We have picked the TI-57 and HP-25 as the least powerful program-mable calculators of their types, illustrating the use of algebraic and reverse Polish notation, respectively.) The principles discussed here can easily be adapted to other devices, including computers. Their implementation shows how specific features of a device affect the practical implications of algorithms.

Our discussion focuses on the o.d.e.

$$[D^2 + (t + 1)D + 1]x = 0.$$

Substitution of the expression

$$x(t) = \sum_{k=0}^{\infty} b_k t^k$$

into the o.d.e. leads to the recurrence relation

$$b_{j+2} = \frac{-b_j - b_{j+1}}{j + 2}, \qquad j = 0, 1, \ldots .$$

(check this). Let's begin with an algorithm for calculating coefficients.

Example 7.4.1
Design an algorithm to calculate the coefficient b_N, $N = 10$, of the series defined by the preceding recurrence relation, when $b_0 = 1$ and $b_1 = 3$.

The heart of our problem is to calculate a given coefficient b_k, assuming we have already determined the coefficients b_j for $j = 0, 1, 2, \ldots, k - 1$. We can rewrite the recurrence relation as a formula for b_k in terms of earlier data by means of the index substitution $k = j + 2$ (or $j = k - 2$):

$$b_k = \frac{-b_{k-2} - b_{k-1}}{k}, \qquad k = 2, 3, \ldots .$$

Calculation of b_k requires three pieces of previous data: the coefficients two steps and one step back, b_{k-2} and b_{k-1}, together with the index k. If these have previously been stored in memory registers A, B, and C,

$$A \leftarrow b_{k-2}, \qquad B \leftarrow b_{k-1}, \qquad C \leftarrow k,$$

then our first step is to calculate b_k and store it for future use:

1. Calculate $\dfrac{-A - B}{C}$ and store it in register D.

If $k \geq 10$, then we have accomplished our goal. Otherwise, we must continue on to calculate b_{k+1}. If we have previously stored 10 in memory register E,

$$E \leftarrow 10,$$

then we decide whether or not to stop on the basis of a comparison of the contents of registers C and E:

2. If $C \geq E$, stop. Otherwise, continue on.

We would now like to calculate b_{k+1} by repeating step 1, but first we must replace the numbers in registers A, B, and C (b_{k-2}, b_{k-1}, and k, respectively) with the coefficients two steps and one step back from b_{k+1} and the new index (b_{k-1}, b_k, and $k + 1$, respectively):

3. Replace   a. A with B
               b. B with D
               c. C with $C + 1$
and return to step 1.

Note that after we replace A with B, the value of b_{k-2} will no longer be stored in memory. Of course, now that we have calculated b_k, we no longer need b_{k-2}.

Now, if we start by storing the values

$$A \leftarrow b_0 = 1, \qquad B \leftarrow b_1 = 3, \qquad C \leftarrow 2, \qquad E \leftarrow N = 10,$$

and set the machine going at step 1, it will

1. Calculate b_2 and store it in D.

2.   Compare 2 to 10. Since $2 < 10$, the machine will continue on.

3.   Store b_1 in A, b_2 in B, and 3 in C.

The machine will then return to steps 1 and 2 to

1.   Calculate b_3 and store it in D.

2.   Compare 3 to 10. Since $3 < 10$, the machine will continue on.

When it finally stops, the value of b_{10} will be in register D.

Although we asked only for the value of b_{10}, it is often desirable to have a record of the coefficients $b_2, \ldots, b_9$ that were calculated in the process of finding b_{10}. On a computer we would instruct the machine to print the results of each calculation. On our calculators a "pause" command between steps 1 and 2 will hold the value of b_k in the display for a predetermined time, to allow the user to copy it. The flowchart in Figure 7.5 summarizes our algorithm, incorporating this additional feature.

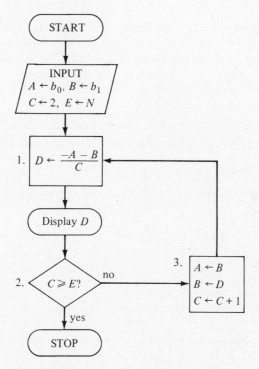

FIGURE **7.5** CALCULATING $b_2, \ldots, b_N$.

We turn now to the implementation of the algorithm on the TI-57 and HP-25.

Example 7.4.2

Write a program for the TI-57 to calculate and display in turn the coefficients b_k for $k = 2, \ldots, N$.

The TI-57 has eight memory registers, numbered 0 to 7, and allows up to 50 program steps, numbered 00 to 49. We will use memories 0 to 3 to store A, B, C, and D from our algorithm. Since the value in E will be compared to other quantities, we store it in register 7, which is specially designated for this purpose.

In Figure 7.6 we list a sequence of key strokes (together with their location codes) that implement the flowchart in Figure 7.5. Commands 00 to 08 carry out the calculation of b_k and store the result. Commands 09 to 11 cause the calculator to pause for about $2\frac{1}{4}$ seconds to allow us to note the value of b_k; a longer pause can be achieved by inserting more pause commands. Commands 12 and 13 ask whether the content of register 0, which holds k, is greater than or equal to the content of register 7, which holds N. If the answer is yes, the calculator performs the very next command,

REGISTER	CONTENT	INITIAL SETTING
0	$A = b_{k-2}$	b_0
1	$B = b_{k-1}$	b_1
2	$C = k$	2
3	$D = b_k$	
7	$E = N$	N

STEP	KEY STROKES	LOCATION CODE
00	RCL 0	33 0
01	+	75
02	RCL 1	33 1
03	=	85
04	÷	45
05	RCL 2	33 2
06	=	85
07	+/−	84
08	STO 3	32 3
09	2nd Pause	36
10	2nd Pause	36
11	2nd Pause	36
12	RCL 2	33 2
13	2nd $x \geq t$	76
14	R/S	81
15	RCL 3	33 3
16	2nd EXC 1	38 1
17	STO 0	32 0
18	1	01
19	SUM 2	34 2
20	RST	71

FIGURE 7.6   TI-57 PROGRAM FOR $b_2, \ldots, b_N$.

number 14, which stops the program. If the answer is no, it performs step 15, which enters b_k in the display. Step 16 interchanges the current display, b_k, and the content of register 1, b_{k-1}. Step 17 stores the new display, b_{k-1}, in register 0, erasing the previous entry, b_{k-2}. Steps 18 and 19 increase the content of register 2 by 1. Step 20 starts the calculation all over again with step 00.

To program the calculator, we press LRN and key in the strokes as shown. We should check that the program has been keyed in correctly. To do this, press LRN, RST, and LRN again; then proceed through the program step by step, using the SST key and checking the display against the location codes.

To run the program, we press LRN and then RST to reset the calculator at command 00. We enter the appropriate values in registers 0, 1, 2, and 7, press R/S, and watch the display. For example, to calculate b_{10} ($N = 10$) using $b_0 = 1$ and $b_1 = 3$, we first key in

$$
\begin{array}{ll}
1 \text{ STO } 0 & (A = b_0 = 1) \\
3 \text{ STO } 1 & (B = b_1 = 3) \\
2 \text{ STO } 2 & (C = 2) \\
10 \text{ STO } 7 & (E = N = 10).
\end{array}
$$

When we push R/S, the calculator will display

$$
\begin{array}{ll}
-\ 2 & (= b_2) \\
-0.3333333 & (= b_3) \\
0.5833333 & (= b_4) \\
-0.05 & (= b_5) \\
-0.0888889 & (= b_6) \\
0.0198413 & (= b_7) \\
0.008631 & (= b_8) \\
0.0031636 & (= b_9) \\
0.0005467 & (= b_{10}) \\
10 &
\end{array}
$$

The last number is the display when the program stops; it is the current value of register 2.

Example 7.4.3

Write a program for the HP-25 to calculate and display in turn the coefficients b_k for $k = 2, \ldots, N$.

The HP-25 has eight memory registers, numbered 0 to 7, and a "stack" of four temporary memories (including the display). It allows 49 usable program steps, numbered 01 to 49. We again use memories 0 to 3 to store A, B, C, and D. The comparison

REGISTER	CONTENT	INITIAL SETTING
0	$A = b_{k-2}$	b_0
1	$B = b_{k-1}$	b_1
2	$C = k$	2
3	$D = b_k$	
7	$E = N$	N

STEP	KEY STROKES	LOCATION CODE	STACK CONTENTS
01	RCL 0	24 00	$[A,*,*,*]$
02	RCL 1	24 01	$[B,A,*,*]$
03	+	51	$[B + A,*,*,*]$
04	RCL 2	24 02	$[C,B + A,*,*]$
05	÷	71	$[(B + A)/C,*,*,*]$
06	CHS	32	$[-(B + A)/C,*,*,*]$
07	STO 3	23 03	$[D,*,*,*]$
08	f Pause	14 74	
09	f Pause	14 74	
10	f Pause	14 74	
11	RCL 7	24 07	$[E,D,*,*]$
12	RCL 2	24 02	$[C,E,D,*]$
13	f $x \geq y$	14 51	
14	GTO 00	13 00	
15	RCL 1	24 01	$[B,C,E,D]$
16	STO 0	23 00	
17	RCL 3	24 03	$[D,B,C,E]$
18	STO 1	23 01	
19	1	01	$[1,D,B,C]$
20	STO + 2	23 51 02	
21	GTO 01	13 01	

FIGURE 7.7   HP-25 PROGRAM FOR $b_2, \ldots, b_N$.

commands on HP calculators are not tied to specific memory registers, so we are free to use any of memories 4 to 7 for E; however, to facilitate comparison with Example 7.4.2, we stick with memory 7 for E.

In Figure 7.7, we list a sequence of key strokes (together with their location codes and the stack contents whenever they change) that defines a program for the HP-25 parallel to the TI program in Example 7.4.2. Commands 01 to 07 calculate and store b_k. Commands 08 to 10 provide a pause to display b_k. Commands 11 to 13 compare k to N. If $k \geq N$, the machine performs command 14 (going to step 00 automatically stops the program). If $k < N$, the machine performs steps 15 to 20, which update the contents of the memory registers, and step 21, which starts the calculation over again.

We program the calculator by switching to PRGM mode and keying in the strokes. To check the program, we switch back to RUN mode and press f PRGM to reset at step 00, then switch back to PRGM mode and go through with the SST key, comparing the display with the location codes in Figure 7.7. To run the program, we switch to RUN mode, press f PRGM, key in the initial values (as in Example 7.4.2), and then press R/S. Note that the HP-25 is automatically set to display only two decimal places, so the display will round the numbers to two decimal places unless we first set the display at seven places with the command f FIX 7. Then, with initial values as in Example 7.4.2, we will observe the same sequence of displays.

So far, we have considered only the calculation of the *coefficients* of the series for $x(t)$. However, in practice we are usually interested in the numerical value of $x(t)$ for a given choice of t. This value is approximated by the partial sums of the series. One way to calculate these partial sums is to supplement the algorithm of Example 7.4.1 with some extra steps that evaluate t^k, multiply it by b_k, and then sum the result into a memory that stores the partial sum. However, if we are not interested in the coefficients per se, it is more efficient to treat the calculation of the terms $T_k = b_k t^k$ of the series directly.

Example 7.4.4

Design an algorithm to calculate the partial sum

$$S_N = \sum_{j=0}^{N} b_j t^j = b_0 + b_1 t + \ldots + b_N t^N$$

of the series for $x(t)$.

The recurrence relation for b_k can be multiplied on both sides by t^k and rewritten to yield an expression for $T_k = b_k t^k$ in terms of $T_{k-2} = b_{k-2} t^{k-2}$ and $T_{k-1} = b_{k-1} t^{k-1}$:

$$T_k = b_k t^k = \frac{-b_{k-2} t^k - b_{k-1} t^k}{k} = \frac{-T_{k-2} t^2 - T_{k-1} t}{k}.$$

This is a recurrence relation analogous to the one in Example 7.4.1. In order to calculate the terms $T_2, \ldots, T_N$, we will store T_{k-2}, T_{k-1}, k, N, and t in memory. After each calculation of a term T_k, we will store the result in one register and sum it into another register where we keep track of the partial sum to that point. In all, we will use seven memory registers:

$$A \leftarrow T_{k-2}, \quad B \leftarrow T_{k-1}, \quad C \leftarrow k, \quad D \leftarrow T_k, \quad E \leftarrow N,$$

$$F \leftarrow t, \quad G \leftarrow \sum_{j=0}^{k-1} T_j = \sum_{j=0}^{k-1} b_k t^j.$$

The flowchart in Figure 7.8 describes an algorithm that requires us to provide initial settings for memory registers A, B, C, E, and F. Note that the initial setting for B is b_1 rather than T_1. The first step of this algorithm is the calculation of $T_1 = b_1t$ and $b_0 + b_1t$. We could eliminate this step if we were willing to calculate T_1 and $b_0 + b_1t$ separately and store the results in registers B and G as two of the initial settings. The central loop calculates the partial sums $\Sigma_{j=0}^2 T_j, \ldots, \Sigma_{j=0}^N T_j$ in turn. When the process ends, the required partial sum will be in register G.

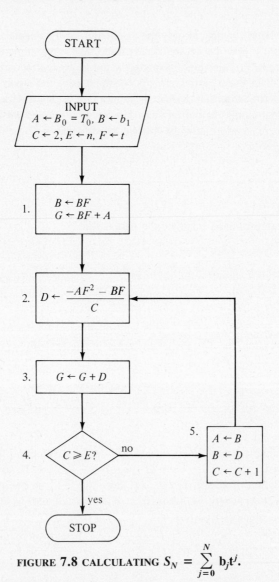

FIGURE 7.8 CALCULATING $S_N = \displaystyle\sum_{j=0}^{N} b_j t^j.$

In practical situations, we are usually interested in the numerical value of $x(t)$ *to some specified accuracy.* Our next example illustrates one approach to deciding whether a given partial sum has achieved a desired degree of accuracy. We compare the partial sum to a fixed number of previously calculated partial sums. If these successive partial sums all agree to within a specified error allowance $\varepsilon > 0$, it is hoped that all later partial sums, and hence the sum of the series $x(t)$, will also agree with these to within ε. Although this hope is not always justified (see Note 1), the test does provide a good educated guess for the value of $x(t)$.

Example 7.4.5

Modify the algorithm in Example 7.4.4 so that the calculation terminates the first time that a partial sum differs from each of the previous two by less than a specified error, $\varepsilon > 0$.

Since the difference between the kth and $(k - 1)$st partial sums is

$$\sum_{j=0}^{k} b_j t^j - \sum_{j=0}^{k-1} b_j t^j = b_k t^k = T_k$$

while the difference between the kth and $(k - 2)$nd partial sums is

$$\sum_{j=0}^{k} b_j t^j - \sum_{j=0}^{k-2} b_j t^j = b_{k-1} t^{k-1} + b_k t^k = T_{k-1} + T_k,$$

we want the calculator to stop as soon as

$$|T_k| < \varepsilon \qquad \text{and} \qquad |T_{k-1} + T_k| < \varepsilon.$$

To incorporate this into our algorithm, we need to have previously stored ε in memory; we can use register E for this, since N is no longer needed:

$$E \leftarrow \varepsilon.$$

We replace the comparison, step 4, in the algorithm of Example 7.4.4 with two comparisons:

4.1   Calculate $|D|$ and compare it to ε. If $|D| \geq \varepsilon$, go to step 5. Otherwise, go to 4.2.

4.2   Calculate $|D + B|$ and compare it to ε. If $|D + B| \geq \varepsilon$, go to step 5. Otherwise, stop.

The resulting algorithm is summarized in Figure 7.9.

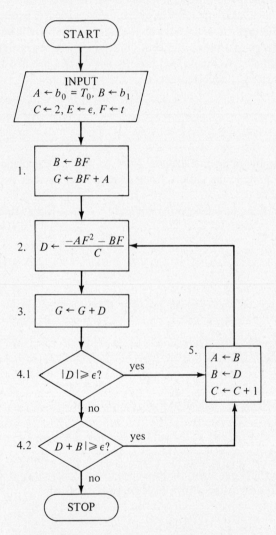

FIGURE 7.9 CALCULATING PARTIAL SUMS TO A DESIRED ACCURACY.

The programs in Figures 7.10 and 7.11 implement the algorithm on the TI-57 and HP-25. When we use these programs, the value of the desired partial sum will be in the display when the calculator stops.

REGISTER	CONTENT	INITIAL SETTING
0	$A = T_{k-2}$ ($T_{k-2}t$, steps 9–36)	$b_0 = T_0$
1	$B = T_{k-1}$	b_1
2	$C = k$	2
3	$D = T_k$	
4	$F = t$	t
5	$G = S_{k-1} = \sum_{j=0}^{k-1} T_j$	
7	$E = \varepsilon$	ε

STEP	KEY STROKES	LOCATION CODE
00	RCL 4	33 4
01	2nd PRD 1	39 1
02	RCL 1	33 1
03	+	75
04	RCL 0	33 0
05	=	85
06	STO 5	32 5
07	2nd Lbl 0	86 0
08	RCL 4	33 4
09	2nd PRD 0	39 0
10	×	55
11	(	43
12	RCL 0	33 0
13	+	75
14	RCL 1	33 1
15	)	44
16	÷	45
17	RCL 2	33 2
18	=	85
19	+/−	84
20	STO 3	32 3
21	SUM 5	34 5
22	2nd $|x|$	40
23	2nd $x \geq t$	76
24	GTO 1	51 1
25	RCL 3	33 3
26	+	75
27	RCL 1	33 1
28	=	85
29	2nd $|x|$	40
30	2nd $x \geq t$	76
31	GTO 1	51 1
32	RCL 5	33 5
33	R/S	81
34	2nd Lbl 1	86 1
35	RCL 3	33 3
36	2nd EXC 1	38 1
37	STO 0	32 0
38	1	01
39	SUM 2	34 2
40	GTO 0	51 0

FIGURE 7.10   TI-57 PROGRAM FOR CALCULATING PARTIAL SUMS.

REGISTER	CONTENT	INITIAL SETTING
0	$A = T_{k-2}$ ($T_{k-2}t$, steps 8–33)	$b_0 = T_0$
1	$B = T_{k-1}$	b_1
2	$C = k$	2
3	$D = T_k$	
4	$F = t$	t
5	$G = S_{k-1} = \sum_{j=0}^{k-1} T_j$	
7	$E = \varepsilon$	ε

STEP	KEY STROKES	LOCATION CODE	STACK CONTENTS
01	RCL 4	24 04	$[F,*,*,]$
02	STO x 1	23 61 01	
03	RCL 1	24 01	$[B,F,*,*]$
04	RCL 0	24 00	$[A,B,F,*]$
05	+	51	$[A + B,F,*,*]$
06	STO 5	23 05	
07	RCL 4	24 04	$[F,*,*,*]$
08	STO x 0	23 61 00	
09	RCL 0	24 00	$[AF,F,*,*]$
10	RCL 1	24 01	$[B,AF,F,*]$
11	+	51	$[B + AF,F,*,*]$
12	×	61	$[(B + AF)F,*,*,*]$
13	RCL 2	24 02	$[C,(B + AF)F,*,*]$
14	÷	71	$[(B + AF)F/C,*,*,*]$
15	CHS	32	$[-(B + AF)F/C,*,*,*]$
16	STO 3	23 03	$[D,*,*,*]$
17	STO + 5	23 51 05	
18	g ABS	15 03	$[|D|,*,*,*]$
19	RCL 7	24 07	$[E, |D|,*,*]$
20	$x \rightleftarrows y$	21	$[|D|,E,*,*]$
21	f $x \geq y$	14 51	
22	GTO 32	13 32	
23	R↓	22	$[E,*,*,*]$
24	RCL 1	24 01	$[B,E,*,*]$
25	RCL 3	24 03	$[D,B,E,*]$
26	+	51	$[D + B,E,*,*]$
27	g ABS	15 03	$[|D + B|,E,*,*]$
28	f $x \geq y$	14 51	
29	GTO 32	13 32	
30	RCL 5	24 05	$[G, |D + B|,E,*]$
31	GTO 00	13 00	
32	RCL 3	24 03	$[D,*,*,*]$
33	RCL 1	24 01	$[B,D,*,*]$
34	STO 0	23 00	
35	R↓	22	$[D,*,*,B]$
36	STO 1	23 01	
37	1	01	$[1,D,*,*]$
38	STO + 2	23 51 02	
39	GTO 07	13 07	

FIGURE 7.11 HP-25 PROGRAM FOR CALCULATING PARTIAL SUMS.

The preceding examples illustrate some general principles that apply to calculating the coefficients or numerical value of any power series that is defined by a recurrence relation. We formulate these in the following summary.

POWER SERIES ON PROGRAMMABLE CALCULATORS

To calculate the coefficients b_k or partial sums S_k of a power series defined by a recurrence relation:

1. Design an **algorithm** for the problem.
   a. Express the **recurrence relation** as a formula for the kth coefficient b_k (or term T_k) in terms of earlier ones.
   b. Allocate **memory** for the data needed in this calculation.
   c. Design a **loop** that will calculate b_k (or S_k) from this data, and either terminate the calculation (at the appropriate step) or update the memories and return to the start with all the data for calculating b_{k+1} (or S_{k+1}) in place.
   d. Check to see whether **additional memory** is needed to retain data not stored in the memories allocated in (b) but needed for repetitions of the loop.
   e. Formulate any necessary **initial steps** that must precede the first repetition of the loop; make sure that at the end of these steps the memories are all set up for this first repetition.
2. Implement the algorithm as a **program** for your specific calculator.
   a. Keep track of the **contents of each memory register** or stack entry as you run through the program.
   b. If desired, include "Pause" statements to **display** intermediate data.
   c. Learn to use **special features,** such as "EXC" keys, subroutines, or a stack, to keep the number of program steps and memories within the limits of the calculator.
   d. When writing the program, note the **location codes** of your commands.
3. **Key the program** into your calculator in "Learn" or "Program" mode. Reset the calculator and **check your entries** step by step, comparing location codes.
4. Clear the memories, key in the initial data, set the program at the first step, and **run.**

Notes

1. Accuracy

Heuristic accuracy tests like the one in Example 7.4.5 are fallible. For example, the series about $t = 0$ for $\sin (t^2)$

$$\sin (t^2) = 0 + 0 + t^2 + 0 + 0 + 0 - \frac{t^6}{6} + \ldots$$

has as its first few partial sums

$$S_0 = 0 = S_1, \qquad S_2 = t^2 = S_3 = S_4 = S_5;$$

a test that compares three (or even four) successive partial sums would conclude that this series sums to t^2. In particular, when $t = \sqrt{\pi}$, the test would suggest that $\sin \pi = \pi$, which is of course very far from the true value, $\sin \pi = 0$.

Nonetheless, we can use a heuristic test to make an educated guess that a specific partial sum $S_N = \sum_{j=0}^{N} b_j t^j$ agrees with the true sum $S = \sum_{j=0}^{\infty} b_j t^j$ within an accuracy specified by $\varepsilon > 0$. To be certain of the accuracy, we must then use more rigorous, abstract estimates to show that the "tail" of the series adds up to less than ε:

$$|S - S_N| = \left| \sum_{j=N+1}^{\infty} b_j t^j \right| < \varepsilon.$$

Even if we have determined that the partial sum S_N is within ε of S, there remains the possibility that roundoff error will contribute significantly to our calculation of S_N. Roundoff error results from the fact that every machine uses a finite approximation to represent numbers. For example, on the TI-57 if we divide 1 by 3, multiply the result by 3, and then subtract 1, we obtain not zero but -1×10^{-10}, because 1/3 is represented on the machine as .3333333333. Such minor errors could in principle accumulate (especially if many calculations are involved) to significantly affect our calculated value of S_N. The analysis of roundoff error is a difficult art, however, and we shall not go into it here.

2. Programming other calculators

The algorithms and programs discussed here can be adapted to most programmable calculators. To decide which of the two versions of programs to use as a model, you need to determine whether your calculator uses algebraic or reverse Polish ("parenthesis-free") logic. If the calculation of $x + y$ is keyed in as x, $+$, y, $=$, then your calculator uses algebraic logic, as in the TI-57. If this calculator is keyed in as x, y, $+$, then your calculator uses reverse Polish notation, as in the HP-25.

EXERCISES

In each of Exercises 1 to 3, we seek information about the series solution $x(t) = \sum_{k=0}^{\infty} b_k t^k$ to the given initial value problem. For each initial value problem,

a.   Find the coefficients $b_0, b_1, \ldots, b_{10}$ (as in Examples 7.4.1 to 7.4.3).

b.   Approximate $x(t)$, for $t = 0.1$ and 1.0, by calculating the partial sum $S_{10}(t) = \sum_{k=0}^{10} b_k t^k$ (as in Example 7.4.4).

c.   Approximate $x(t)$, for $t = 0.1$ and 1.0, by calculating partial sums until a partial sum $S_N(t)$ differs from each of the previous two, $S_{N-1}(t)$ and $S_{N-2}(t)$, by less than $\varepsilon = 0.0001$ (as in Example 7.4.5). What is the value of N when the calculation stops?

d.   Based on your answers (and ignoring roundoff error), estimate the value of $x(t)$ for $t = 0.1$ and $t = 1.0$ to three decimal places.

1. $[D^2 + (t + 1)D + 1]x = 0$ (discussed in the text)
    i. $x(0) = 1, x'(0) = 0$ ii. $x(0) = 0, x'(0) = 1$
    iii. $x(0) = -2, x'(0) = 3$

2. $[D^2 + tD + 1]x = 0$ (compare Exercise 13, Section 7.3)
    i. $x(0) = 1, x'(0) = 0$ ii. $x(0) = 0, x'(0) = 1$
    iii. $x(0) = -2, x'(0) = 3$

3. $[D^2 + 4t]x = 0$ (compare Exercise 2, Section 7.3)
    i. $x(0) = 1, x'(0) = 0$ ii. $x(0) = 0, x'(0) = 1$
    iii. $x(0) = -2, x'(0) = 3$

7.5 THE CAUCHY-EULER EQUATION

In this section we investigate the solutions of the second-order **Cauchy-Euler equation**

(H) $$[p_0 t^2 D^2 + q_0 t D + r_0]x = 0$$

where p_0, q_0, and r_0 are constants. This equation has a singular point at $t = 0$. Although solutions of (H) do not require series, the cases that arise here will help us understand a wider class of equations, which we will solve using series in the next two sections. We will restrict attention to the interval $t > 0$; to find solutions for $t < 0$, we can substitute $T = -t$ and assume $T > 0$.

We begin by noting that if we substitute $x = t^m$ into the left-hand side of (H), we obtain a multiple of t^m:

$$[p_0 t^2 D^2 + q_0 t D + r_0]t^m = (p_0 m(m - 1) + q_0 m + r_0)t^m.$$

Thus, we see immediately that *if m satisfies the equation*

(I) $$p_0 m(m - 1) + q_0 m + r_0 = 0,$$

then $x = t^m$ is a solution of (H). This quadratic equation is called the **indicial equation** of (H).

Since distinct powers of t are linearly independent (see Exercise 14), we see that *if the indicial equation has distinct real roots m_1 and m_2, then the general solution of (H) is*

$$x = c_1 t^{m_1} + c_2 t^{m_2}.$$

Example 7.5.1

Solve

(H)
$$[2t^2D^2 + tD - 1]x = 0, \qquad t > 0.$$

Substituting $x = t^m$ into the left-hand side gives

$$[2t^2D^2 + tD - 1]t^m = (2m(m - 1) + m - 1)t^m,$$

so the indicial equation of (H) is

(I)
$$2m(m - 1) + m - 1 = 0$$

or

$$(2m + 1)(m - 1) = 0.$$

The roots of (I) are

$$m = -\frac{1}{2}, \qquad m = 1.$$

Corresponding to these roots we obtain the solutions $x_1 = t^{-1/2}$ and $x_2 = t$, which generate the general solution of (H):

$$x = c_1 t^{-1/2} + c_2 t.$$

Example 7.5.2

Solve

(H)
$$[t^2D^2 - 2tD + 2]x = 0, \qquad t > 0.$$

Substituting $x = t^m$ into the left-hand side of (H), we get

$$[t^2D^2 - 2tD + 2]t^m = (m(m - 1) - 2m + 2)t^m,$$

and the indicial equation of (H) is

(I)
$$m(m - 1) - 2m + 2 = 0$$

or

$$(m - 1)(m - 2) = 0.$$

We obtain the general solution of (H) from the roots, $m = 1$ and $m = 2$, of (I):

$$x = c_1t + c_2t^2.$$

If the indicial equation has only one root m_1, then $x = t^{m_1}$ is still a solution of (H), but we need another to generate the general solution of (H). In the following example, we use a method similar to variation of parameters to find this second solution.

Example 7.5.3
Solve

(H) $$[t^2D^2 + 5tD + 4]x = 0, \qquad t > 0.$$

Substitution of $x = t^m$ into (H) gives us

$$[t^2D^2 + 5tD + 4]t^m = (m(m - 1) + 5m + 4)t^m = 0,$$

so the indicial equation is

(I) $$m(m - 1) + 5m + 4 = 0$$

or

$$m^2 + 4m + 4 = 0.$$

This equation has only one root, $m = -2$. Corresponding to this root we have one solution of (H),

$$x_1 = t^{-2}.$$

We look for a second solution of the form

(S$_2$) $$x_2 = x_1k(t) = t^{-2}k(t).$$

The first two derivatives of this function are

$$x_2' = t^{-2}k'(t) - 2t^{-3}k(t)$$
$$x_2'' = t^{-2}k''(t) - 4t^{-3}k'(t) + 6t^{-4}k(t).$$

Substitution into (H) gives

$$[k''(t) - 4t^{-1}k'(t) + 6t^{-2}k(t)] + [5t^{-1}k'(t) - 10t^{-2}k(t)] + 4t^{-2}k(t) = 0$$

or

$$k''(t) + t^{-1}k'(t) = 0.$$

We rewrite this in the form

$$\frac{1}{k'(t)} \frac{d\,k'(t)}{dt} = \frac{-1}{t}$$

and integrate both sides to get

$$\ln |k'(t)| = -\ln t + a.$$

Thus

$$k'(t) = \frac{b}{t}$$

and

$$k(t) = b \ln t + c.$$

Taking $b = 1$ and $c = 0$, we substitute back into (S$_2$) to obtain our second solution to (H):

$$x_2 = t^{-2} \ln t.$$

The solutions x_1 and x_2 are linearly independent on $t > 0$ (see Exercise 15), so the general solution of (H) is

$$x = c_1 x_1 + c_2 x_2 = c_1 t^{-2} + c_2 t^{-2} \ln t.$$

The pattern of Example 7.5.3 is a general one: *When the indicial equation has only one root $m = m_1$, the general solution of (H) on $t > 0$ is*

$$x = c_1 t^{m_1} + c_2 t^{m_1} \ln t.$$

We leave it to the reader to verify this (see Exercise 17).

There is one final possibility: The indicial equation may have complex roots.

Example 7.5.4

Solve

(H) $$[t^2D^2 + 3tD + 2]x = 0, \qquad t > 0.$$

Substitution of $x = t^m$ in the left-hand side gives

$$[t^2D^2 + 3tD + 2]t^m = (m(m - 1) + 3m + 2)t^m,$$

so the indicial equation is

(I) $$m(m - 1) + 3m + 2 = 0$$

or

$$m^2 + 2m + 2 = 0.$$

The roots of (I) are

$$m = -1 \pm i.$$

As in Sections 2.8 and 3.8, we reason that (H) should have two complex solutions

$$x_1(t) = t^{-1+i}, \qquad x_2 = t^{-1-i}.$$

To make use of these expressions, we recall the definition of t^m for general m,

$$t^m = e^{m \ln t},$$

and Euler's formula,

$$e^{u+iv} = e^u(\cos v + i \sin v).$$

Using these formulas, we rewrite x_1:

$$x_1(t) = t^{-1+i} = t^{-1}t^i = t^{-1}e^{i \ln t}$$
$$= t^{-1}[\cos (\ln t) + i \sin (\ln t)].$$

Similarly

$$x_2(t) = t^{-1}[\cos (\ln t) - i \sin (\ln t)].$$

The functions

$$\frac{1}{2}[x_1(t) + x_2(t)] = t^{-1} \cos(\ln t)$$

and

$$\frac{1}{2i}[x_1(t) - x_2(t)] = t^{-1} \sin(\ln t)$$

are real-valued solutions of (H), as can be verified by substitution. Since they are linearly independent on $t > 0$ (Exercise 16), the general solution of (H) is

$$x = c_1 t^{-1} \cos(\ln t) + c_2 t^{-1} \sin(\ln t).$$

This example is also typical (see Exercise 18). *When the indicial equation has complex roots $\alpha \pm \beta i$, the general solution of (H) is*

$$x = c_1 t^{\alpha} \cos(\beta \ln t) + c_2 t^{\alpha} \sin(\beta \ln t).$$

Notice that

$$t^{\alpha + \beta i} = t^{\alpha} t^{\beta i} = t^{\alpha} e^{i\beta \ln t}$$
$$= t^{\alpha} \cos(\beta \ln t) + i t^{\alpha} \sin(\beta \ln t),$$

so that the functions generating the general solution are the real and imaginary parts of the complex solution $t^{\alpha + \beta i}$.

We summarize our observations.

THE CAUCHY-EULER EQUATION

To solve the Cauchy-Euler equation

(H) $$[p_0 t^2 D^2 + q_0 tD + r_0]x = 0, \qquad t > 0:$$

1.   Calculate the effect of substituting $x = t^m$ into the left-hand side:

$$[t^2 p_0 D^2 + t q_0 D + r_0]t^m = (p_0 m(m-1) + q_0 m + r_0)t^m.$$

2. Find the roots $m = m_1$ and $m = m_2$ of the **indicial equation**

(I) $$p_0 m(m - 1) + q_0 m + r_0 = 0.$$

3. If the roots are real and distinct, the general solution of (H) for $t > 0$ is

$$x = c_1 t^{m_1} + c_2 t^{m_2}.$$

4. If there is only one root, $m = m_1 = m_2$, the general solution of (H) for $t > 0$ is

$$x = c_1 t^{m_1} + c_2 t^{m_1} \ln t.$$

5. If the roots are complex, $m = \alpha \pm \beta i$, the general solution of (H) for $t > 0$ is generated by the real and imaginary parts of $t^{\alpha + \beta i}$:

$$x = c_1 t^{\alpha} \cos (\beta \ln t) + c_2 t^{\alpha} \sin (\beta \ln t).$$

EXERCISES

In Exercises 1 to 12, find the general solution for $t > 0$.

1. $[t^2 D^2 + tD - 1]x = 0$ 2. $[4t^2 D^2 - 3]x = 0$

3. $[t^2 D^2 - tD + 1]x = 0$ 4. $[t^2 D^2 + tD + 1]x = 0$

5. $[9t^2 D^2 + 15tD + 1]x = 0$ 6. $[t^2 D^2 - 2tD - 18]x = 0$

7. $[t^2 D^2 + 3tD + 5]x = 0$ 8. $[t^2 D^2 - 6tD + 12]x = 0$

9. $[t^2 D^2 + tD - 3]x = 0$ 10. $[t^2 D^2 + 7tD + 9]x = 0$

11. $[t^2 D^2 - 3tD + 5]x = 0$ 12. $[3t^2 D^2 - 3tD + 1]x = 0$

13. *Temperature in a Circular Plate:* Recall, from Example 7.1.2, that if $u = R(r)\Theta(\theta)$ is a solution of the two-dimensional Laplace equation (in polar coordinates), then $\Theta(\theta)$ satisfies

(H_θ) $$\frac{d^2\Theta}{d\theta^2} + n^2\Theta = 0$$

while $R(r)$ satisfies

(H_r) $$r^2 \frac{d^2 R}{dr^2} + r \frac{dR}{dr} - n^2 R = 0.$$

a. Find the general solution of (H_r).

b. Use the fact that the temperature in the plate must be bounded as r approaches 0 to conclude that the physically meaningful solutions of (H_r) are constant multiples of $R_n(r) = r^n$.

c. Corresponding to each nonnegative integer n, obtain two solutions of the two-dimensional Laplace equation by multiplying R_n with generators for the general solution of (H_θ).

d. Find a linear combination of the functions described in (c) (possibly with different values of n) that satisfies the boundary condition

$$u(1, \theta) = \cos \theta - 2 \sin 3\theta, \qquad 0 \le \theta \le 2\pi.$$

More abstract problems:

14. If m_1 and m_2 are distinct real numbers (not necessarily positive integers), show that the functions t^{m_1} and t^{m_2} are linearly independent on the interval $0 < t < \infty$.

15. Show that the functions t^{m_1} and $t^{m_1} \ln t$ are linearly independent on $0 < t < \infty$.

16. If $\beta \ne 0$, show that $t^\alpha \cos (\beta \ln t)$ and $t^\alpha \sin (\beta \ln t)$ are linearly independent on $0 < t < \infty$.

17. Suppose the indicial equation has a double root $m = m_1$.
    a. Show that $q_0 = p_0 - 2p_0m_1$ and $r_0 = p_0m_1^2$.
    b. Show that t^{m_1} and $t^{m_1} \ln t$ are solutions of the Cauchy-Euler equation.

18. Suppose the indicial equation has roots $\alpha \pm \beta i$ with $\beta \ne 0$.
    a. Show that $q_0 = p_0 - 2\alpha p_0$ and $r_0 = p_0(\alpha^2 + \beta^2)$.
    b. Show that $t^\alpha \cos (\beta \ln t)$ and $t^\alpha \sin (\beta \ln t)$ are solutions of the Cauchy-Euler equation.

19. a. Show that upon substituting $T = -t$, the Cauchy-Euler equation

$$p_0 t^2 \frac{d^2x}{dt^2} + q_0 t \frac{dx}{dt} + r_0 x = 0$$

becomes

$$p_0 T^2 \frac{d^2x}{dT^2} + q_0 T \frac{dx}{dT} + r_0 x = 0.$$

b. Show that if $x = t^m$ is a solution of the Cauchy-Euler equation for $t > 0$, then $x = |t|^m$ is a solution for $t < 0$.

20. Use the result of Exercise 19(b) to find the general solution for $t < 0$ of
    a. $t^2x'' - 6x = 0$ b. $t^2x'' + tx' - 4x = 0$
    c. $t^2x'' - 3tx' + 4x = 0$ d. $t^2x'' + tx' + 4x = 0$

21. a. Show that for $t > 0$, the substitution $s = \ln t$ (so that $t = e^s$) transforms the Cauchy-Euler equation into a homogeneous equation with constant coefficients (cf. Exercise 22, Section 2.9).
    b. Relate the solutions of this transformed equation to the three cases treated in the summary.

7.6 REGULAR SINGULAR POINTS: FROBENIUS SERIES

In this and the following two sections, we use series to handle a class of second-order equations with relatively well-behaved singular points. The method, developed by the German mathematicians L. Fuchs (1866) and G. Frobenius (1873) [building on techniques used by Euler (1764)], is known as the **method of Frobenius.**

Recall that $t = t_0$ is a singular point of an o.d.e. with polynomial coefficients if it is a root of the leading coefficient. This means that $t - t_0$ is a factor of the leading coefficient; we will deal with singular points at which the powers of $t - t_0$ dividing the coefficients are subject to certain restrictions:

Definition: *We say that $t = t_0$ is a **regular singular point** of an o.d.e. with polynomial coefficients provided the equation can be written in the form*

(H) $$[(t - t_0)^2 p(t)D^2 + (t - t_0)q(t)D + r(t)]x = 0$$

where $p(t)$, $q(t)$, and $r(t)$ are polynomials and $p(t_0) \neq 0$.

The Cauchy-Euler equation discussed in the previous section,

$$[t^2 p_0 D^2 + t q_0 D + r_0]x = 0,$$

has a regular singular point at $t = 0$. So does **Bessel's equation:**

$$[t^2 D^2 + tD + (t^2 - \mu^2)]x = 0.$$

Another o.d.e. that comes up frequently in mathematical physics is **Laguerre's equation**

$$[tD^2 + (1 - t)D + \mu]x = 0.$$

This o.d.e. has a regular singular point at $t = 0$; to see this, multiply through by t to get

$$[t^2 D^2 + t(1 - t)D + \mu t]x = 0.$$

Another example is the **Legendre equation**

$$[(1 - t^2)D^2 - 2tD + \mu(\mu + 1)]x = 0,$$

with regular singular points at $t = \pm 1$ (multiply through by $1 \mp t$ to check this).

We shall concentrate on the case $t_0 = 0$. As we saw in Section 7.3, other cases can be reduced to this one by substituting $T = t - t_0$. We shall also concentrate on finding solutions for $t > 0$; solutions with $t < 0$ can be found by substituting $T = -t$ with $T > 0$.

We know from the Cauchy-Euler equation that *integer* powers do not suffice for solving o.d.e.'s with regular singular points. Instead, we look for solutions of the form

$$x(t) = t^m \sum_{k=0}^{\infty} b_k t^k = \sum_{k=0}^{\infty} b_k t^{m+k}$$

where $b_0 \neq 0$, and m need not be an integer. We refer to an expression of this form as a **Frobenius series.**

Let's see how substituting a Frobenius series into an o.d.e. can lead to the general solution.

Example 7.6.1
Solve

(H) $$[2t^2D^2 + (t^2 + t)D - 1]x = 0, \qquad t > 0.$$

Note that (H) has a regular singular point at $t = 0$, with $p(t) = 2$, $q(t) = t + 1$, and $r(t) = -1$. We look for solutions of the form

(S) $$x = t^m \sum_{k=0}^{\infty} b_k t^k = \sum_{k=0}^{\infty} b_k t^{k+m}, \; b_0 \neq 0.$$

In order to substitute (S) into (H), we first calculate the effect of $L = 2t^2D^2 + (t^2 + t)D - 1$ on an arbitrary power of t:

$$
\begin{aligned}
L[t^s] &= 2s(s - 1)t^s + st^{s+1} + st^s - t^s \\
&= (2s + 1)(s - 1)t^s + st^{s+1}.
\end{aligned}
$$

Using this, we calculate the effect of L on the Frobenius series (S):

$$
\begin{aligned}
Lx &= \sum_{k=0}^{\infty} b_k L[t^{m+k}] \\
&= \sum_{k=0}^{\infty} b_k[(2m + 2k + 1)(m + k - 1)t^{m+k} + (m + k)t^{m+k+1}] \\
&= \sum_{j=0}^{\infty} b_j(2m + 2j + 1)(m + j - 1)t^{m+j} + \sum_{j=1}^{\infty} b_{j-1}(m + j - 1)t^{m+j} \\
&= b_0(2m + 1)(m - 1)t^m \\
&\quad + \sum_{j=1}^{\infty} [b_j(2m + 2j + 1)(m + j - 1) + b_{j-1}(m + j - 1)]t^{m+j}.
\end{aligned}
$$

If this is to be identically zero, the coefficient of each power of t must vanish. Thus

$$b_0(2m + 1)(m - 1) = 0$$

and

(R) $$b_j(2m + 2j + 1)(m + j - 1) + b_{j-1}(m + j - 1) = 0,$$
$$j = 1, 2, \ldots .$$

Since $b_0 \neq 0$, the first of these equations gives

(I) $$(2m + 1)(m - 1) = 0,$$

which has two roots,

$$m = -1/2 \text{ and } m = 1.$$

If we choose the second root, $m = 1$, then (R) reads

(R_1) $$b_j(2j + 3)(j) + b_{j-1}(j) = 0$$

or

$$b_j = -\frac{1}{2j + 3} b_{j-1}.$$

This recurrence relation lets us find the coefficients in terms of b_0:

$$b_1 = -\frac{1}{5} b_0, \quad b_2 = \frac{(-1)^2}{5 \cdot 7} b_0, \ldots, b_k = \frac{(-1)^k}{5 \cdot 7 \cdot \ldots \cdot (2k + 3)} b_0, \ldots.$$

Setting $b_0 = 1$, this gives us a solution to (H):

(S_1) $$x_1(t) = t\left(1 + \sum_{k=1}^{\infty} \frac{(-1)^k}{5 \cdot 7 \cdot \ldots \cdot (2k + 3)} t^k\right).$$

On the other hand, if we take $m = -1/2$, then (R) reads

(R_2) $$b_j(2j)\left(j - \frac{3}{2}\right) + b_{j-1}\left(j - \frac{3}{2}\right) = 0$$

or

$$b_j = -\frac{1}{2j} b_{j-1}.$$

Thus

$$b_1 = -\frac{1}{2} b_0, \quad b_2 = \frac{(-1)^2}{2 \cdot 2^2} b_0, \quad b_3 = \frac{(-1)^3}{2 \cdot 3 \cdot 2^3} b_0,$$

$$b_4 = \frac{(-1)^4}{2 \cdot 3 \cdot 4 \cdot 2^4} b_0, \ldots, b_k = \frac{(-1)^k}{k! 2^k} b_0, \ldots.$$

Setting $b_0 = 1$, we obtain a second solution of (H):

(S$_2$)
$$x_2(t) = t^{-1/2} \sum_{k=0}^{\infty} \frac{(-1)^k}{k! 2^k} t^k.$$

We note that $x_1(t)$ and $x_2(t)$ define linearly independent solutions of (H) for $t > 0$ (see Exercise 22), so that the general solution to (H) for $t > 0$ is

$$x = c_1 x_1(t) + c_2 x_2(t)$$

$$= c_1 t \left(1 + \sum_{k=1}^{\infty} \frac{(-1)^k}{5 \cdot 7 \cdot \ldots \cdot (2k + 3)} t^k \right) + c_2 t^{-1/2} \left(\sum_{k=0}^{\infty} \frac{(-1)^k}{k! 2^k} t^k \right).$$

This example illustrates a general pattern. If

$$L = t^2 p(t) D^2 + t q(t) D + r(t),$$

then

$$L[t^s] = p(t) s(s - 1) t^s + q(t) s t^s + r(t) t^s.$$

Since $p(t)$, $q(t)$, and $r(t)$ are polynomials, this can be expanded in powers of t (with coefficients depending on s):

$$L[t^s] = f_0(s) t^s + f_1(s) t^{s+1} + \ldots + f_n(s) t^{s+n}.$$

Note that the coefficient $f_0(s)$ of t^s can be described in terms of the constant coefficients of $p(t)$, $q(t)$, and $r(t)$:

$$f_0(s) = p_0 s(s - 1) + q_0 s + r_0.$$

To solve the homogeneous o.d.e.

(H)
$$Lx = [t^2 p(t) D^2 + t q(t) D + r(t)] x = 0$$

we substitute the Frobenius series

$$x = t^m \sum_{k=0}^{\infty} b_k t^k = \sum_{k=0}^{\infty} b_k t^{m+k}, \qquad b_0 \neq 0$$

and obtain

$$0 = Lx = \sum_{k=0}^{\infty} b_k L[t^{m+k}]$$

$$= \sum_{k=0}^{\infty} b_k f_0(m + k)t^{m+k} + \sum_{k=0}^{\infty} b_k f_1(m + k)t^{m+k+1} + \ldots$$

$$+ \sum_{k=0}^{\infty} b_k f_n(m + k)t^{m+k+n}$$

$$= \sum_{j=0}^{\infty} b_j f_0(m + j)t^{m+j} + \sum_{j=1}^{\infty} b_{j-1} f_1(m + j - 1)t^{m+j} + \ldots$$

$$+ \sum_{j=n}^{\infty} b_{j-n} f_n(m + j - n)t^{m+j}.$$

For this last expression to be zero, the combined coefficient of each power of t should be zero. The coefficient of t^m is

$$b_0 f_0(m) = 0,$$

while the coefficients of higher powers give the recurrence relation

(R) $$b_j f_0(m + j) + b_{j-1} f_1(m + j - 1) + \ldots + b_0 f_j(m) = 0,$$
$$j = 1, 2, 3, \ldots$$

where we take the functions $f_k(s)$ to be identically zero if $k > n$. Since $b_0 \neq 0$, the first equation can be rewritten $f_0(m) = 0$, or

(I) $$p_0 m(m - 1) + q_0 m + r_0 = 0.$$

This is called the **indicial equation** of (H) (compare the previous section, in which (H) is the Cauchy-Euler equation). *Once we find the roots, $m = m_1$ and $m = m_2$, of the indicial equation (I), we try to use the recurrence relation (R) to determine the coefficients of the corresponding solution to (H).*

Of course, if $m_1 = m_2$ we cannot expect to obtain two independent solutions in this way. In fact, even when $m_1 \neq m_2$, we may find it impossible to obtain two independent solutions as Frobenius series; if the roots differ by an integer, say

$m_1 = m_2 + j$ ($j > 0$), then $f_0(m_2 + j) = f_0(m_1) = 0$, and we cannot divide (R) by $f_0(m_2 + j)$ to obtain an expression for b_j. We consider two examples with roots differing by an integer. In the first, we will still manage to find two independent Frobenius series solutions, while in the second we will be able to find only one.

Example 7.6.2
Solve

(H) $$Lx = [t^2D^2 + (t^2 + t)D - 1]x = 0, \qquad t > 0.$$

Again we look for Frobenius series solutions

(S) $$x = t^m \sum_{k=0}^{\infty} b_k t^k = \sum_{k=0}^{\infty} b_k t^{m+k}, \qquad b_0 \neq 0.$$

Since

$$L[t^s] = s(s - 1)t^s + (t^2 + t)st^{s-1} - t^s$$
$$= (s + 1)(s - 1)t^s + st^{s+1},$$

substituting (S) into (H) gives

$$0 = Lx = \sum_{k=0}^{\infty} b_k L[t^{m+k}]$$

$$= \sum_{k=0}^{\infty} b_k(m + k + 1)(m + k - 1)t^{m+k} + \sum_{k=0}^{\infty} b_k(m + k)t^{m+k+1}$$

$$= \sum_{j=0}^{\infty} b_j(m + j + 1)(m + j - 1)t^{m+j} + \sum_{j=1}^{\infty} b_{j-1}(m + j - 1)t^{m+j}$$

$$= b_0(m + 1)(m - 1)t^m$$

$$+ \sum_{j=1}^{\infty} [b_j(m + j + 1)(m + j - 1) + b_{j-1}(m + j - 1)]t^{m+j}.$$

We need

(I) $$(m + 1)(m - 1) = 0$$

and

(R) $$b_j(m + j + 1)(m + j - 1) + b_{j-1}(m + j - 1) = 0, \qquad j = 1, 2, 3, \ldots$$

The roots of the indicial equation (I) are

$$m = \pm 1.$$

Substituting $m = 1$ into (R) gives

(R$_1$) $$b_j(j + 2)j + b_{j-1}j = 0$$

or

$$b_j = -\frac{1}{j + 2} b_{j-1}.$$

Thus,

$$b_1 = -\frac{1}{3} b_0, \quad b_2 = \frac{(-1)^2}{4 \cdot 3} b_0, \ldots,$$

$$b_k = \frac{(-1)^k}{(k + 2) \ldots 4 \cdot 3} b_0 = \frac{(-1)^k \cdot 2}{(k + 2)!} b_0, \ldots.$$

If we set $b_0 = 1$, the resulting solution of (H) is

(S$_1$) $$x_1(t) = t\left(1 + \sum_{k=1}^{\infty} \frac{(-1)^k 2}{(k + 2)!} t^k\right).$$

Substituting the other root, $m = -1$, into (R) leads to

(R$_2$) $$b_j j(j - 2) + b_{j-1}(j - 2) = 0.$$

When $j = 1$, this gives

$$b_1 = -b_0.$$

When $j = 2$, (R$_2$) reads $0 = 0$. This means that there is no restriction on b_2; in particular, we can pick

$$b_2 = 0.$$

Now (R$_2$) gives

$$b_3 = 0, \quad b_4 = 0, \ldots, b_k = 0, \ldots.$$

Setting $b_0 = 1$ yields the solution of (H)

(S$_2$) $$x_2(t) = t^{-1}(1 - t).$$

The functions $x_1(t)$ and $x_2(t)$ are linearly independent solutions of (H) for $t > 0$, so the general solution (for $t > 0$) is

$$x = c_1 x_1(t) + c_2 x_2(t) = c_1 t\left(1 + \sum_{k=1}^{\infty} \frac{(-1)^k 2}{(k + 2)!} t^k\right) + c_2 t^{-1}(1 - t).$$

Example 7.6.3
Solve

(H) $$Lx = [t^2 D^2 + (t^3 + t)D - 1]x = 0, \qquad t > 0.$$

You can check that substituting the Frobenius series

(S) $$x = t^m \sum_{k=0}^{\infty} b_k t^k = \sum_{k=0}^{\infty} b_k t^{m+k}, \qquad b_0 \neq 0$$

into (H) leads to

$$0 = Lx = b_0(m + 1)(m - 1)t^m + b_1(m + 2)mt^{m+1} +$$

$$\sum_{j=2}^{\infty} [b_j(m + j + 1)(m + j - 1) + b_{j-2}(m + j - 2)]t^{m+j}.$$

This implies

(I) $$(m + 1)(m - 1) = 0$$
$$b_1(m + 2)m = 0$$

and

(R) $b_j(m + j + 1)(m + j - 1) + b_{j-2}(m + j - 2) = 0, \qquad j = 2, 3, \ldots.$

The roots of the indicial equation (I) are

$$m = \pm 1.$$

Using $m = +1$, we get

$$3b_1 = 0$$

and

(R$_1$) $b_j(j + 2)j + b_{j-2}(j - 1) = 0, \qquad j = 2, 3, \ldots .$

Thus

$$b_2 = \frac{-1}{2^2 \cdot 2}\, b_0, \quad b_4 = \frac{(-1)^2 \cdot 3}{2^4(3 \cdot 2)(2 \cdot 1)}\, b_0, \ldots$$

and

$$b_1 = 0, \quad b_3 = 0, \ldots .$$

The general pattern here is difficult to discover, but it can be verified that if $b_0 = 1$, we have, for $k = 1, 2, \ldots,$

$$b_{2k} = \frac{(-1)^k(2k - 1)!}{2^{3k-1}(k + 1)!k!(k - 1)!}, \qquad b_{2k-1} = 0.$$

Thus, one solution of (H) is

(S$_1$) $x_1(t) = t\left(1 + \displaystyle\sum_{k=1}^{\infty} \frac{(-1)^k(2k - 1)!}{2^{3k-1}(k + 1)!k!(k - 1)!}\, t^{2k}\right).$

On the other hand, substitution of $m = -1$ leads to

$$-b_1 = 0$$

and

(R$_2$) $b_j j(j - 2) + b_{j-2}(j - 3) = 0, \qquad j = 2, 3, \ldots .$

But for $j = 2$, (R$_2$) reads

$$-b_0 = 0.$$

This means in particular that *we cannot find a Frobenius series solution (with $b_0 \neq 0$) corresponding to $m = -1$.* We will return to this example in Section 7.8.

Another potential source of trouble is the case when *the roots of the indicial equation are complex*. Fortunately, the device that worked for the Cauchy-Euler equation saves us here: *The general solution to (H) is generated by the real and imaginary parts of the complex solution that corresponds to one of the complex roots.*

Example 7.6.4

Solve

(H) $$Lx = [t^2D^2 + tD + (1 - t)]x = 0, \qquad t > 0.$$

You can verify that the indicial equation and recurrence relation for (H) are, respectively,

(I) $$m^2 + 1 = 0$$

and

(R) $$b_j[(m + j)^2 + 1] - b_{j-1} = 0, \qquad j = 1, 2, \ldots .$$

The roots of the indicial equation (I) are

$$m = \pm i.$$

Using $m = i$, the recurrence relation becomes

(R$_1$) $$b_j[(j + i)^2 + 1] - b_{j-1} = 0$$

or

$$b_j = \frac{b_{j-1}}{(j + 2i)j}.$$

We rationalize the denominator by multiplying top and bottom by $j - 2i$:

$$b_j = \frac{(j - 2i)}{(j^2 + 4)j} b_{j-1}.$$

If we set $b_0 = 1$, we find

$$b_1 = \frac{1 - 2i}{5 \cdot 1}, \quad b_2 = \frac{2 - 2i}{8 \cdot 2} b_1 = \frac{-(1 + 3i)}{40}, \ldots .$$

We won't attempt a general pattern here but content ourselves with these coefficients. The corresponding complex Frobenius series is

$$x = t^i \left[1 + \left(\frac{1 - 2i}{5} \right)t - \left(\frac{1 + 3i}{40} \right)t^2 + \ldots \right]$$

$$= [\cos (\ln t) + i \sin (\ln t)] \left[1 + \left(\frac{1 - 2i}{5} \right)t - \left(\frac{1 + 3i}{40} \right)t^2 + \ldots \right]$$

$$= \left\{ \cos (\ln t) \left[1 + \frac{t}{5} - \frac{t^2}{40} + \ldots \right] + \sin (\ln t) \left[\frac{2t}{5} + \frac{3t^2}{40} + \ldots \right] \right\}$$

$$+ i \left\{ \cos (\ln t) \left[-\frac{2}{5}t - \frac{3}{40}t^2 + \ldots \right] + \sin (\ln t) \left[1 + \frac{t}{5} - \frac{t^2}{40} + \ldots \right] \right\}.$$

The general solution to (H) is generated by the real and imaginary parts of this function:

$$x = c_1 \left\{ \cos (\ln t) \left[1 + \frac{t}{5} - \frac{t^2}{40} + \ldots \right] + \sin (\ln t) \left[\frac{2t}{5} + \frac{3t^2}{40} + \ldots \right] \right\}$$

$$+ c_2 \left\{ \cos (\ln t) \left[-\frac{2t}{5} - \frac{3t^2}{40} + \ldots \right] + \sin (\ln t) \left[1 + \frac{t}{5} - \frac{t^2}{40} + \ldots \right] \right\}.$$

We now summarize the method, noting that (as in Section 7.3) the solutions it yields will be valid for $0 < t < R$, provided the leading coefficient of (H), $p(t)$, has no complex roots z with $|z| < R$.

REGULAR SINGULAR POINTS: FROBENIUS SERIES

An o.d.e. has a **regular singular point** at $t = 0$ if it can be written in the form

(H) $$Lx = [t^2 p(t)D^2 + tq(t)D + r(t)]x = 0$$

where $p(t)$, $q(t)$, and $r(t)$ are polynomials and $p(0) \neq 0$.

In this situation we look for solutions of (H) in the form of **Frobenius series,**

(S)
$$x = t^m \sum_{k=0}^{\infty} b_k t^k = \sum_{k=0}^{\infty} b_k t^{m+k}, \qquad b_0 \neq 0$$

as follows.

1. Calculate $L[t^s]$ and use the result to find $Lx = \sum_{k=0}^{\infty} b_k L[t^{m+k}]$.

2. Rewrite Lx as a Frobenius series by combining like powers of t (using index substitutions).

3. Set each combined coefficient equal to zero to obtain the **indicial equation**

    (I)
    $$p_0 m(m - 1) + q_0 m + r_0 = 0$$

    and a general **recurrence relation** (R).

4. If the indicial equation (I) has distinct real roots, then substituting these values for m in (R) leads to solutions of (H) that generate the general solution, except possibly when the roots differ by an integer.

5. If the indicial equation (I) has complex roots, then substituting either root into (R) leads to a complex solution of (H) whose real and imaginary parts generate the general solution.

6. If the indicial equation (I) has a double root, we can still find one solution of (H), using (R), but need a second solution to generate the general solution of (H).

The Frobenius series solutions obtained by this method will be valid for $0 < t < R$, provided $p(z) \neq 0$ for every complex number z with $|z| < R$.

Note
Extending the method

Our discussion of Frobenius series in this section focused on o.d.e.'s

(H)
$$Lx = [t^2 p(t) D^2 + t q(t) D + r(t)] x = 0$$

with $p(t)$, $q(t)$, and $r(t)$ polynomials. The method just summarized also applies when $p(t)$, $q(t)$, and $r(t)$ are given by power series

$$p(t) = \sum_{n=0}^{\infty} p_n t^n, \quad q(t) = \sum_{k=0}^{\infty} q_n t^n, \quad r(t) = \sum_{n=0}^{\infty} r_n t^n.$$

The indicial equation maintains the same form,

(I) $$p_0 m(m - 1) + q_0 m + r_0 = 0,$$

but now the recurrence relation, which arises from a multiplication of series, can be extremely complicated. The Frobenius series expressions for the solutions will be valid for $0 < t < R$ provided all three coefficient series converge for $|t| < R$ and $p(z) \neq 0$ for all complex z with $|z| < R$.

EXERCISES

In Exercises 1 to 12, find an expression for the general solution in terms of Frobenius series. In Exercises 1 to 8, you need only find the first four coefficients of each of the two generating solutions. In Exercises 9 to 12, find the general pattern for the coefficients.

1. $[9t^2 D^2 + 9tD + (6t - 1)]x = 0, \qquad 0 < t$

2. $[3t^2 D^2 + t(3t + 4)D + (t - 2)]x = 0, \qquad 0 < t$

3. $[4t^2 D^2 + 4tD + (4t^2 - 1)]x = 0, \qquad 0 < t$

4. $[9t^2 D^2 + 9tD + (9t^2 - 4)]x = 0, \qquad 0 < t$

5. $[4t^2 D^2 + 4tD + (4t^2 - 9)]x = 0, \qquad 0 < t$

6. $[t^2 D^2 + t(1 - t)D + 1]x = 0, \qquad 0 < t$

7. $[t^2(t + 2)D^2 + t(t + 1)D + (t^2 - 1)]x = 0, \qquad 0 < t < 2$

8. $[t^2 D^2 - tD + 5(1 - t)]x = 0, \qquad 0 < t$

9. $[2t^2 D^2 - t(t - 1)D - 1]x = 0, \qquad 0 < t$

10. $[4t^2(t + 1)D^2 + (t - 3)]x = 0, \qquad 0 < t < 1$

11. $[2tD^2 + D + t]x = 0, \qquad 0 < t$

12. $[3t^2(t - 1)D^2 + 4tD - 2]x = 0, \qquad 0 < t < 1$

In Exercises 13 to 16, find a Frobenius series solution. Check that any other Frobenius series solution is a constant multiple of the one you find.

13. $[t^2 D^2 + tD + t^2]x = 0, \qquad 0 < t$

14. $[t^2 D^2 - tD + (t + 1)]x = 0, \qquad 0 < t$

15. $[t^2 D^2 + tD + (t - 1)]x = 0, \qquad 0 < t$

16. $[t^2 D^2 + tD + (t^2 - 4)]x = 0, \qquad 0 < t$

In Exercises 17 to 21, find the general solution by making an appropriate change of variable and solving the resulting equation.

17. $[2t^2 D^2 - t(t - 1)D - 1]x = 0, \qquad t < 0 \qquad$ (compare Exercise 9)

18. $[2tD^2 + D + t]x = 0, \qquad t < 0 \qquad$ (compare Exercise 11)

19. $[2(t - 1)^2D^2 + t(t - 1)D - 1]x = 0,$ $t > 1$

20. $[2(t - 1)^2D^2 + t(t - 1)D - 1]x = 0,$ $t < 1$

21. $[(t^2 + 2t + 1)D^2 + (t^2 + 3t + 2)D - 1]x = 0,$ $t > -1$

22. *Linear Independence of Frobenius Series:* Suppose $f(t) = t^{m_1} \Sigma_{k=0}^{\infty} b_k t^k$ and $g(t) = t^{m_2} \Sigma_{k=0}^{\infty} c_k t^k$, where $b_0 \neq 0$, $c_0 \neq 0$, and $m_1 < m_2$. Show that $f(t)$ and $g(t)$ are linearly independent on any interval of the form $0 < t < a$. [*Hint:* Multiply the equation $Af(t) + Bg(t) = 0$ by t^{-m_1} and see what happens as t approaches 0.]

7.7 A CASE STUDY: BESSEL FUNCTIONS OF THE FIRST KIND

In this section, we will use Frobenius series to find nontrivial solutions for Bessel's equation,

(B$_\mu$) $[t^2D^2 + tD + (t^2 - \mu^2)]x = 0,$ $t > 0,$

where $\mu \geq 0$. The "Bessel functions" that arise in solving (B$_\mu$) play an important role in many problems of mathematical physics.

The indicial equation of (B$_\mu$),

$$m^2 - \mu^2 = 0,$$

has roots $m = \pm\mu$. We saw in the last section that we can always find a nontrivial Frobenius series solution corresponding to the larger root, $m = \mu$. We begin by finding such solutions to Bessel's equation with $\mu = 0$ and $\mu = 1$.

Example 7.7.1

Find a nontrivial Frobenius series solution to Bessel's equation with $\mu = 0$,

(B$_0$) $[t^2D^2 + tD + t^2]x = 0,$ $t > 0.$

Substituting

(S) $x = t^m \sum_{k=0}^{\infty} b_k t^k = \sum_{k=0}^{\infty} b_k t^{m+k},$ $b_0 \neq 0$

into (B$_0$) yields the equation

$$\sum_{k=0}^{\infty} (m + k)(m + k - 1)b_k t^{m+k} + \sum_{k=0}^{\infty} (m + k)b_k t^{m+k} + \sum_{k=0}^{\infty} b_k t^{m+k+2} = 0$$

or

$$m^2 b_0 t^m + (m + 1)^2 b_1 t^{m+1} + \sum_{j=2}^{\infty} [(m + j)^2 b_j + b_{j-2}] t^{m+j} = 0.$$

Thus

$$m^2 b_0 = 0$$

$$(m + 1)^2 b_1 = 0$$

and

$$(m + j)^2 b_j + b_{j-2} = 0, \qquad j = 2, 3, \ldots .$$

Since $b_0 \neq 0$, we must take

$$m = 0;$$

then

$$b_1 = 0$$

and

$$b_j = -\frac{1}{j^2} b_{j-2}, \qquad j = 2, 3, \ldots .$$

Since $b_1 = 0$, every odd-numbered coefficient vanishes. On the other hand, the even-numbered coefficients satisfy

$$b_2 = \frac{-b_0}{2^2}, \quad b_4 = \frac{-b_2}{4^2} = \frac{(-1)^2 b_0}{4^2 \cdot 2^2}, \ldots ,$$

$$b_{2n} = \frac{(-1)^n b_0}{(2n)^2 (2n - 2)^2 \ldots 2^2} = \frac{(-1)^n b_0}{2^{2n} (n!)^2}, \ldots .$$

The solution of (B_0) obtained by taking $b_0 = 1$ is called the **Bessel function of the first kind of order 0** and is denoted $J_0(t)$:

$$J_0(t) = \sum_{n=0}^{\infty} \frac{(-1)^n t^{2n}}{2^{2n} (n!)^2} = \sum_{n=0}^{\infty} \frac{(-1)^n}{(n!)^2} \left(\frac{t}{2}\right)^{2n}.$$

Since 0 is the only root of the indicial equation of (B_0), any Frobenius series solution of (B_0) is a constant multiple of $J_0(t)$.

Example 7.7.2

Find a nontrivial Frobenius series solution of Bessel's equation with $\mu = 1$,

(B$_1$)
$$[t^2 D^2 + tD + (t^2 - 1)]x = 0, \qquad t > 0.$$

Substituting

(S)
$$x = t^m \sum_{k=0}^{\infty} b_k t^k = \sum_{k=0}^{\infty} b_k t^{m+k}, \qquad b_0 \neq 0$$

into (B$_1$) yields the equation

$$\sum_{k=0}^{\infty} (m + k)(m + k - 1)b_k t^{m+k}$$

$$+ \sum_{k=0}^{\infty} (m + k)b_k t^{m+k} + \sum_{k=0}^{\infty} b_k t^{m+k+2} - \sum_{k=0}^{\infty} b_k t^{m+k} = 0$$

or

$$(m^2 - 1)b_0 t^m + (m^2 + 2m)b_1 t^{m+1} + \sum_{j=2}^{\infty} \{[(m + j)^2 - 1]b_j + b_{j-2}\}t^{m+j} = 0.$$

Thus

$$(m^2 - 1)b_0 = 0$$
$$(m^2 + 2m)b_1 = 0$$

and

(R)
$$[(m + j)^2 - 1]b_j + b_{j-2} = 0, \qquad j = 2, 3, \ldots .$$

Since $b_0 \neq 0$, the first two equations give

$$m = \pm 1$$

and

$$b_1 = 0.$$

If we take $m = +1$, the recurrence relation (R) gives

$$b_j = -\frac{1}{(j + 2)j} b_{j-2}, \qquad j = 2, 3, \ldots .$$

Since $b_1 = 0$, the odd-numbered coefficients vanish. The even-numbered coefficients satisfy

$$b_2 = -\frac{b_0}{4 \cdot 2} = -\frac{b_0}{2^2(2 \cdot 1)}, \quad b_4 = -\frac{b_2}{6 \cdot 4} = \frac{b_0}{2^4(3 \cdot 2)(2 \cdot 1)}, \cdots,$$

$$b_{2n} = \frac{(-1)^n b_0}{2^{2n}(n + 1)! n!}, \cdots .$$

The solution obtained when $b_0 = 1/2$ is called the **Bessel function of the first kind of order 1** and is denoted $J_1(t)$:

$$J_1(t) = \frac{1}{2} \sum_{n=0}^{\infty} \frac{(-1)^n t^{2n+1}}{2^{2n}(n + 1)! n!} = \sum_{n=0}^{\infty} \frac{(-1)^n}{(n + 1)! n!} \left(\frac{t}{2}\right)^{2n+1} .$$

If we take $m = -1$, the recurrence relation reads

$$(-2j + j^2)b_j + b_{j-2} = 0, \qquad j = 2, 3, \ldots .$$

Taking $j = 2$, we see that $b_0 = 0$. Thus there are no nontrivial Frobenius series solutions corresponding to $m = -1$; any Frobenius series solution of (B_1) is a constant multiple of $J_1(t)$.

In general, substituting

(S) $$x = t^m \sum_{k=0}^{\infty} b_k t^k = \sum_{k=0}^{\infty} b_k t^{m+k}, \qquad b_0 \neq 0$$

into Bessel's equation

(B_μ) $$[t^2 D^2 + tD + (t^2 - \mu^2)]x = 0$$

leads to the equation

$$(m^2 - \mu^2)b_0 t^m + [(m + 1)^2 - \mu^2]b_1 t^{m+1}$$

$$+ \sum_{j=2}^{\infty} \{[(m + j)^2 - \mu^2]b_j + b_{j-2}\}t^{m+j} = 0.$$

Hence

(I)
$$m^2 = \mu^2$$

(R_1)
$$[(m + 1)^2 - \mu^2]b_1 = 0$$

and

(R_j)
$$[(m + j)^2 - \mu^2]b_j + b_{j-2} = 0, \qquad j = 2, 3, \ldots.$$

If we take $m = \mu$, we get

$$(2\mu + 1)b_1 = 0$$

and

$$b_j = -\frac{1}{(2\mu + j)j} b_{j-2}, \qquad j = 2, 3, \ldots.$$

Thus b_1 and the other odd-numbered coefficients vanish; the even-numbered coefficients satisfy

$$b_{2n} = \frac{(-1)^n b_0}{2^{2n}(n!)[(\mu + n) \ldots (\mu + 2)(\mu + 1)]}.$$

When μ is an integer, the choice

$$b_0 = \frac{1}{2^\mu(\mu!)}$$

leads to a solution of (B_μ) that can be written in a neat form:

$$J_\mu(t) = \sum_{n=0}^{\infty} \frac{(-1)^n t^{2n+\mu}}{2^{2n+\mu} n!(\mu + n)!} = \sum_{n=0}^{\infty} \frac{(-1)^n}{n!(\mu + n)!} \left(\frac{t}{2}\right)^{2n+\mu}.$$

When μ is not an integer, an appropriate choice of b_0 still yields a solution that has a neat form. To describe this choice, we need to extend the notion of factorial to nonintegers. This is done by the **gamma function**, defined for $s > 0$ by the integral formula

$$\Gamma(s) = \int_0^{\infty} e^{-t} t^{s-1} \, dt.$$

This improper integral converges for $s > 0$ and defines a function with the following properties (see Exercise 1):

1. $\Gamma(s + 1) = s\Gamma(s)$ for all $s > 0$
2. $\Gamma(n + 1) = n!$ for all integers $n \geq 0$

Our choice for b_0 when μ is an integer can be written as

$$b_0 = \frac{1}{2^\mu \Gamma(\mu + 1)}.$$

Even when μ is not an integer, we can make this same choice of b_0. Using property 1, we then see that

$$b_{2n} = \frac{(-1)^n}{2^{2n+\mu} n! [(\mu + n) \ldots (\mu + 2)(\mu + 1)]\Gamma(\mu + 1)}$$

$$= \frac{(-1)^n}{2^{2n+\mu} n! \Gamma(\mu + n + 1)}.$$

The resulting solution of (B_μ) is called the **Bessel function of the first kind of order** μ and is denoted $J_\mu(t)$:

$$J_\mu(t) = \sum_{n=0}^{\infty} \frac{(-1)^n}{n! \Gamma(\mu + n + 1)} \left(\frac{t}{2}\right)^{2n+\mu}.$$

We have, then, the following:

Fact: *For any $\mu \geq 0$, $x = J_\mu(t)$ is a nontrivial solution of (B_μ).*

In Examples 7.7.1 and 7.7.2, the Frobenius series solutions of (B_μ) were all constant multiples of $J_\mu(t)$. The following two examples illustrate cases in which the smaller root $m = -\mu$ of the indicial equation leads to a solution of (B_μ) independent of $J_\mu(t)$.

Example 7.7.3

Find the general solution to Bessel's equation with $\mu = 1/3$,

$(B_{1/3})$ $\left[t^2 D^2 + tD + \left(t^2 - \frac{1}{9}\right)\right] x = 0, \quad t > 0.$

We already know a Frobenius series solution corresponding to $m = 1/3$, namely, the Bessel function of the first kind

$$J_{1/3}(t) = \sum_{n=0}^{\infty} \frac{(-1)^n}{n!\Gamma(n + \frac{4}{3})} \left(\frac{t}{2}\right)^{2n + 1/3}.$$

In addition, since the roots $m = \pm 1/3$ of the indicial equation differ by $2/3$, which is not an integer, we know that $(B_{1/3})$ also has a Frobenius series solution corresponding to $m = -1/3$.

If we substitute $m = -1/3$ into equations (R_1) and (R_j) above, we get

$$\left[\left(\frac{2}{3}\right)^2 - \frac{1}{9}\right] b_1 = 0$$

and

$$\left[\left(-\frac{1}{3} + j\right)^2 - \frac{1}{9}\right] b_j + b_{j-2} = 0, \qquad j = 2, 3, \ldots.$$

It follows that the odd-numbered coefficients vanish, while

$$b_{2n} = \frac{(-1)^n b_0}{2^{2n}(n!)\left[\left(-\frac{1}{3} + n\right) \cdots \left(-\frac{1}{3} + 2\right)\left(-\frac{1}{3} + 1\right)\right]}, \qquad n = 1, 2, \ldots.$$

The choice

$$b_0 = \frac{1}{2^{-1/3}\Gamma(-\frac{1}{3} + 1)}$$

gives us the **Bessel function of the first kind of order $-1/3$**:

$$J_{-1/3}(t) = \sum_{n=0}^{\infty} \frac{(-1)^n}{n!\Gamma(-\frac{1}{3} + n + 1)} \left(\frac{t}{2}\right)^{2n - 1/3}.$$

This function is independent of $J_{1/3}(t)$ (Exercise 7), so the general solution of $(B_{1/3})$ is

$$x = c_1 J_{1/3}(t) + c_2 J_{-1/3}(t).$$

Example 7.7.4

Find the general solution of Bessel's equation with $\mu = 1/2$,

$(B_{1/2})$ $$\left[t^2 D^2 + tD + \left(t^2 - \frac{1}{4} \right) \right] x = 0, \qquad t > 0.$$

The Bessel function of the first kind

$$J_{1/2}(t) = \sum_{n=0}^{\infty} \frac{(-1)^n}{n! \Gamma(n + \frac{3}{2})} \left(\frac{t}{2} \right)^{2n+1/2}$$

is a solution corresponding to the larger root $m = 1/2$ of the indicial equation. Since the difference between the roots is 1, which is an integer, we are not guaranteed a solution corresponding to $m = -1/2$. Nonetheless, we can try to find one.

Substitution of $m = -1/2$ into (R_1) and (R_j) gives

$$\left[\left(1 - \frac{1}{2} \right)^2 - \left(\frac{1}{2} \right)^2 \right] b_1 = 0$$

and

$$\left[\left(-\frac{1}{2} + j \right)^2 - \left(\frac{1}{2} \right)^2 \right] b_j + b_{j-2} = 0, \qquad j = 2, 3, \ldots.$$

Since the left side of the first of these equations is 0, there is no restriction on b_1. We pick $b_1 = 0$, so that all odd-numbered coefficients vanish. The even-numbered coefficients satisfy

$$b_{2n} = \frac{(-1)^n b_0}{2^{2n} n! \left(-\frac{1}{2} + n \right) \cdots \left(-\frac{1}{2} + 2 \right) \left(-\frac{1}{2} + 1 \right)}.$$

Following our earlier pattern, we set

$$b_0 = \frac{1}{2^{-1/2} \Gamma(-\frac{1}{2} + 1)}$$

to obtain the **Bessel function of the first kind of order $-1/2$**:

$$J_{-1/2}(t) = \sum_{n=0}^{\infty} \frac{(-1)^n}{n! \Gamma(-\frac{1}{2} + n + 1)} \left(\frac{t}{2} \right)^{2n-1/2}.$$

This is independent of $J_{1/2}(t)$, so the general solution of $(B_{1/2})$ is

$$x = c_1 J_{1/2}(t) + c_2 J_{-1/2}(t).$$

In general, when the difference 2μ between the roots $\pm\mu$ is not an integer, we are guaranteed a solution corresponding to $-\mu$. Substitution of $m = -\mu$ into (R_1) and (R_j) gives

$$(-2\mu + 1)b_1 = 0$$

and

$$(-2\mu + j)jb_j + b_{j-2} = 0, \qquad j = 2, 3, \ldots.$$

It follows that the odd-numbered coefficients are 0, while

$$b_{2n} = \frac{(-1)^n b_0}{2^{2n}(n!)(-\mu + n)(-\mu + n - 1) \ldots (-\mu + 1)}.$$

Thus we are led to the solution

$$x = b_0 t^{-\mu} + \sum_{n=1}^{\infty} \frac{(-1)^n b_0}{2^{2n}(n!)(-\mu + n)(-\mu + n - 1) \ldots (-\mu + 1)} t^{2n-\mu}.$$

Notice that this formula makes sense as long as μ is not an integer (even if 2μ is an integer). Indeed, substitution into (B_μ) will show that it defines a solution of (B_μ) even when μ is half an odd integer.

We would like to follow the pattern already set by choosing $b_0 = 1/2^{-\mu}\Gamma(-\mu + 1)$. However, when $-\mu + 1 < 0$, this requires us to define $\Gamma(s)$ for negative values of s. We start by noting that if $-1 < s < 0$, then $s + 1$ is positive, so $\Gamma(s + 1)$ makes sense. To preserve property 1 of the gamma function, we define

$$\Gamma(s) = \frac{\Gamma(s + 1)}{s}, \qquad -1 < s < 0.$$

Now, if $-2 < s < -1$, then $-1 < s + 1 < 0$, so $\Gamma(s + 1)$ makes sense. In order to preserve property 1, we define

$$\Gamma(s) = \frac{\Gamma(s + 1)}{s} = \frac{\Gamma(s + 2)}{(s + 1)s}, \qquad -2 < s < 0.$$

Continuing this way, we can define $\Gamma(s)$ for all $s < 0$ except -1, -2, . . ., in such a way that property 1 remains true.

If μ is not an integer, the choice

$$b_0 = \frac{1}{2^{-\mu}\Gamma(-\mu + 1)}$$

leads to the **Bessel function of the first kind of order** $-\mu$,

$$J_{-\mu}(t) = \sum_{n=0}^{\infty} \frac{(-1)^n}{n!\Gamma(-\mu + n + 1)} \left(\frac{t}{2}\right)^{2n-\mu}.$$

This function is a solution of (B_μ), independent of $J_\mu(t)$. Thus we have the following.

Fact: *If μ is not an integer, then the general solution of (B_μ) is*

$$x = c_1 J_\mu(t) + c_2 J_{-\mu}(t).$$

When the order μ is an integer, it is not possible to find a second Frobenius series solution that is independent of $J_\mu(t)$. This case requires the techniques of the next section and leads to "Bessel functions of the second kind."

Bessel functions were studied in special cases by Euler, the Bernoullis, Lagrange, Fourier, and Poisson. They were first systematically studied by the German mathematician F. W. Bessel (1824). We develop a few properties of these functions in the exercises that follow our summary.

BESSEL FUNCTIONS OF THE FIRST KIND

The **gamma function** is defined for all real numbers $s \neq 0$, -1, -2, . . . by

$$\Gamma(s) = \begin{cases} \displaystyle\int_0^\infty e^{-t}t^{s-1}\,dt & \text{if } s > 0; \\[2ex] \dfrac{\Gamma(s + j)}{(s + j - 1) \ldots (s + 1)s} & \begin{array}{l}\text{if } -j < s < -j + 1, \\ j \text{ a positive integer.}\end{array} \end{cases}$$

It has the following properties:

1. $\Gamma(s + 1) = s\Gamma(s)$ whenever both sides are defined.
2. $\Gamma(n + 1) = n!$ for any integer $n \geq 0$.

For any real number $\rho \neq -1, -2, \ldots$, the **Bessel function of the first kind of order ρ** is defined for $t > 0$ by

$$J_\rho(t) = \sum_{n=0}^{\infty} \frac{(-1)^n}{n!\Gamma(\rho + n + 1)} \left(\frac{t}{2}\right)^{2n+\rho}.$$

The Bessel equation

(B_μ) $$[t^2 D^2 + tD + (t^2 - \mu^2)]x = 0$$

where $\mu \geq 0$, always has $x = J_\mu(t)$ as a solution. If μ is not an integer, then the general solution of (B_μ) is

$$x = c_1 J_\mu(t) + c_2 J_{-\mu}(t).$$

If μ is an integer, then every Frobenius series solution of (B_μ) is a constant multiple of $J_\mu(t)$.

EXERCISES

Exercises 1 to 4 deal with the gamma function.

1. a. Show that $\Gamma(s) > 0$ for $0 < s \leq 1$.
    b. Use integration by parts and L'Hôpital's rule to show that $\Gamma(s + 1) = s\Gamma(s)$ for all $s > 0$.
    c. Show that $\Gamma(1) = 1$.
    d. Use the results of parts (b) and (c) to show that $\Gamma(n + 1) = n!$ for all integers $n \geq 0$.

2. a. By substituting $x = t^{1/2}$ in the definition of $\Gamma(1/2)$, show that $\Gamma(1/2) = 2 \int_0^\infty e^{-x^2} dx$, so that

$$[\Gamma(1/2)]^2 = 4 \int_0^\infty e^{-x^2} dx \int_0^\infty e^{-y^2} dy = 4 \int_0^\infty \int_0^\infty e^{-x^2-y^2} dx \, dy.$$

   b. By converting the double integral in (a) to polar coordinates, show that

$$[\Gamma(1/2)]^2 = 4 \int_0^{\pi/2} \int_0^\infty e^{-r^2} r \, dr \, d\theta = \pi,$$

   or $\Gamma(1/2) = \sqrt{\pi}$.

3. a. Use the fact that $\Gamma(1/2) = \sqrt{\pi}$ (see Exercise 2) to show that $\Gamma(3/2) = \sqrt{\pi}/2$.
    b. More generally, show that if k is a positive integer, then

$$\Gamma\left(\frac{2k + 1}{2}\right) = \frac{1 \cdot 3 \cdot 5 \ldots (2k - 1)}{2^k} \sqrt{\pi}.$$

4. a. Use the fact that $\Gamma(1/2) = \sqrt{\pi}$ (see Exercise 2) to show that $\Gamma(-1/2) = -2\sqrt{\pi}$.

    b. Find $\Gamma\left(-\dfrac{2k+1}{2}\right)$, where k is a positive integer.

In Exercises 5 to 14 we develop some properties of Bessel functions of the first kind, $J_\rho(t)$, where ρ is a real number but $\rho \neq -1, -2, \ldots$.

5. Use the ratio test to show that the series $\sum_{n=0}^{\infty} (-1)^n (t/2)^{2n}/n!\Gamma(\rho + n + 1)$ converges for all t, so that $J_\rho(t)$ is defined (at least) for $0 < t < \infty$.

6. a. Check that $J_0(0) = 1$, and $J_\rho(0) = 0$ for $\rho > 0$.

    b. Show that if $0 > \rho \neq -1, -2, \ldots$, then $J_\rho(t)$ becomes unbounded as t approaches 0. [*Hint:* What happens to $t^{-\rho}J_\rho(t)$?]

7. Use the results of Exercise 6 to show that if $\mu > 0$ is not an integer, then $J_\mu(t)$ and $J_{-\mu}(t)$ are linearly independent on $0 < t < \infty$.

8. Differentiate the series expression for $J_0(t)$ to show that $J_0'(t) = -J_1(t)$.

9. a. Multiply the series expression for $J_\rho(t)$ by t^ρ and differentiate to show that

$$\frac{d}{dt}\,[t^\rho J_\rho(t)] = t^\rho J_{\rho-1}(t).$$

    [When $\rho = 0$, interpret $J_{\rho-1}(t) = J_{-1}(t)$ to mean $-J_1(t)$.]

    b. Expand the left side of the equation in (a) and cancel $t^{\rho-1}$ to obtain

$$tJ_\rho'(t) + \rho J_\rho(t) = tJ_{\rho-1}(t).$$

10. By an argument similar to the one used in Exercise 9, show that

    a. $\dfrac{d}{dt}\,[t^{-\rho}J_\rho(t)] = -t^{-\rho}J_{\rho+1}(t)$

    b. $tJ_\rho'(t) - \rho J_\rho(t) = -tJ_{\rho+1}(t)$

11. By subtracting the formula in Exercise 10(b) from the one in Exercise 9(b), show that

$$J_{\rho+1}(t) = \frac{2\rho J_\rho(t) - tJ_{\rho-1}(t)}{t}.$$

12. Use the formula in Exercise 11 to express $J_2(t)$, $J_3(t)$, and $J_4(t)$ in terms of $J_0(t)$ and $J_1(t)$.

13. a. Using the result of Exercise 3(b), show that

$$J_{-1/2}(t) = \sqrt{\frac{2}{\pi t}} \sum_{n=0}^{\infty} \frac{(-1)^n t^{2n}}{(2n)!} = \sqrt{\frac{2}{\pi t}} \cos t.$$

    b. Using the formula in Exercise 10(b), show that $J_{1/2}(t) = \sqrt{2/\pi t}\, \sin t$.

    c. Using the formula in Exercise 11, first with $\rho = 1/2$ and then with $\rho = -1/2$, obtain expressions for $J_{3/2}(t)$ and $J_{-3/2}(t)$ in terms of $\sin t$, $\cos t$, powers of t, and constants. [*Note:* By continuing in this way, we can obtain expressions for any Bessel function of the form $J_{n/2}(t)$, where n is an odd integer. It turns out that these are the only Bessel functions that have such (finite) expressions in terms of elementary functions.]

It can be shown that each function $J_\rho(t)$ has infinitely many zeroes. In Exercises 14 to 16, we outline a proof that the zeroes of $J_\rho(t)$ and $J_{\rho+1}(t)$ on $0 < t < \infty$ are distinct and alternate.

14. a. Use the formula in Exercise 10(b) to show that if $J_\rho(T) = J_{\rho+1}(T) = 0$, where $T > 0$, then $J_\rho'(T) = 0$ as well.

   b. Use the uniqueness of solutions of initial value problems to show that if $J_\rho(T) = J_{\rho+1}(T) = 0$, where $T > 0$, then $J_\rho(t) = 0$ for all $t > 0$.

   c. Conclude that $J_\rho(T)$ and $J_{\rho+1}(T)$ cannot have a common zero $T > 0$.

15. Suppose T_1 and T_2 are zeroes of $J_\rho(t)$, where $0 < T_1 < T_2$. Show that $J_{\rho+1}(t)$ has a zero between T_1 and T_2. [*Hint:* Apply Rolle's Theorem to the function $f(t) = t^{-\rho}J_\rho(t)$ on the interval $T_1 \le t \le T_2$, and use the formula in Exercise 10(a).]

16. Suppose T_1 and T_2 are zeroes of $J_{\rho+1}(t)$, where $0 < T_1 < T_2$. Show that $J_\rho(t)$ has a zero between T_1 and T_2. [*Hint:* Apply Rolle's Theorem to $g(t) = t^{\rho+1}J_{\rho+1}(t)$, and use the formula in Exercise 9(a) with ρ replaced by $\rho + 1$.]

17. *The Graphs of $J_0(t)$ and $J_1(t)$:*

   a. Write a program for your calculator or computer that takes the initial data t, μ, and $b_0 = 1/2^\mu \mu!$ and then uses the recurrence relation

$$b_{2k}t^{\mu+2k} = \frac{-t^2}{4(\mu + k)k}\, b_{2k-2}t^{\mu+2k-2}$$

to calculate the partial sums

$$S_{2n} = \sum_{k=0}^{n} b_{2k}t^{\mu+2k}$$

until $|b_{2n}t^{\mu+2n}|$ is less than $\varepsilon = 0.001$. Use your program to obtain approximate values for $J_0(t)$ and $J_1(t)$ with $t = 0.5, 1, 1.5, \ldots, 9$. [*Hint:* It might be useful to write your program in such a way that once $J_\mu(t)$ has been calculated, all registers are set at appropriate values for calculating $J_\mu(t + 0.5)$.]

   b. Using your results from (a), sketch the graphs of $J_0(t)$ and $J_1(t)$ for $0 \le t \le 9$. [*Hint:* It helps to use different scales on the horizontal and vertical axes. The values of $J_0(t)$ and $J_1(t)$ all lie in the interval $-0.5 \le x \le 1.0$ when $t > 0$.] Note that your graphs should oscillate with decreasing magnitude as t increases. This behavior is exhibited on $0 < t < \infty$ by each of the functions $J_\rho(t)$.

7.8 REGULAR SINGULAR POINTS: EXCEPTIONAL CASES

Our discussion in Section 7.6 of the equation

(H) $$[t^2p(t)D^2 + tq(t)D + r(t)]x = 0, \qquad t > 0$$

with a regular singular point at $t = 0$ is incomplete in two cases. When the roots m_1 and m_2 of the indicial equation

(I) $$p_0 m(m - 1) + q_0 m + r_0 = 0$$

are equal ($m_1 - m_2 = 0$), or *sometimes* when they differ by a nonzero integer ($m_1 - m_2 = n > 0$), the method discussed in Section 7.6 yields a Frobenius series solution $x_1(t)$ corresponding to m_1, but not the general solution. In this section, we complete our discussion of these cases by finding a solution to (H) independent of $x_1(t)$.

We treat first the case of a double root $m_1 = m_2$ of (I). Recall that when the indicial equation of the Cauchy-Euler equation had a double root, then one solution $x_1(t)$ was a power of t, while a second was obtained by multiplying $x_1(t)$ by $\ln t$. In the present case, the first solution is multiplied by $\ln t$ and then adjusted by adding a second Frobenius series (see Note 1).

Fact: *If the indicial equation (I) has a double root $m_1 = m_2 = m$, then the general solution of (H) is $x = c_1 x_1(t) + c_2 x_2(t)$ where the first solution is a Frobenius series,*

$$x_1(t) = t^m \sum_{k=0}^{\infty} b_k t^k, \qquad b_0 \neq 0$$

while the second has the form

$$x_2(t) = x_1(t) \ln t + t^m \sum_{k=0}^{\infty} b_k^* t^k.$$

We note that the second solution can be replaced by any function of the form $x_2(t) + cx_1(t)$. By adjusting c, we can obtain a second solution with any value we wish for b_0^*. It is customary in this case to take $b_0^* = 0$.

Example 7.8.1

Find the general solution of

(H) $$Lx = [t^2 D^2 + t(t + 3)D + 1]x = 0.$$

We start by looking for Frobenius series solutions. Since

$$L[t^s] = t^2 s(s - 1)t^{s-2} + t(t + 3)st^{s-1} + t^s = (s + 1)^2 t^s + st^{s+1}$$

we have

$$L\left[\sum_{k=0}^{\infty} b_k t^{m+k} \right] = \sum_{k=0}^{\infty} (m + k + 1)^2 b_k t^{m+k} + \sum_{k=0}^{\infty} (m + k)b_k t^{m+k+1}$$

$$= (m + 1)^2 b_0 t^m + \sum_{j=1}^{\infty} [(m + j + 1)^2 b_j + (m + j - 1)b_{j-1}]t^{m+j}.$$

For this to be identically zero, we need

(I)
$$(m + 1)^2 = 0$$

and

(R)
$$(m + j + 1)^2 b_j + (m + j - 1)b_{j-1} = 0, \qquad j = 1, 2, \ldots$$

The indicial equation (I) has a double root $m = -1$. Substituting this root into (R) gives the recurrence relation

$$b_j = \frac{-(j - 2)}{j^2} b_{j-1}.$$

Setting $b_0 = 1$, we get

$$b_0 = 1, \quad b_1 = \frac{-(1 - 2)}{1} = 1, \quad b_2 = 0, \quad b_3 = 0, \ldots, b_n = 0, \ldots$$

Thus, one solution of (H) is

(S$_1$)
$$x_1(t) = t^{-1}(1 + t) = t^{-1} + 1.$$

We next look for a second solution of the form

(S$_2$)
$$x_2(t) = x_1(t) \ln t + t^{-1} \sum_{k=0}^{\infty} b_k^* t^k$$

$$= (t^{-1} + 1) \ln t + \sum_{k=0}^{\infty} b_k^* t^{-1+k}.$$

We can use our earlier calculation of the effect of L on an arbitrary Frobenius series to find its effect on the second summand of $x_2(t)$:

$$L\left[\sum_{k=0}^{\infty} b_k^* t^{-1+k} \right] = \sum_{j=1}^{\infty} [j^2 b_j^* + (j - 2)b_{j-1}^*]t^{-1+j}.$$

The effect of L on the first summand is

$$
\begin{aligned}
L[(t^{-1} + 1) \ln t] &= t^2 D^2[(t^{-1} + 1) \ln t] \\
&\quad + (t^2 + 3t)D[(t^{-1} + 1) \ln t] + (t^{-1} + 1) \ln t \\
&= t^2(-3t^{-3} - t^{-2} + 2t^{-3} \ln t) \\
&\quad + (t^2 + 3t)(t^{-2} + t^{-1} - t^{-2} \ln t) + (t^{-1} + 1) \ln t \\
&= 3 + t.
\end{aligned}
$$

Combining these, we see that

$$Lx_2(t) = 3 + t + \sum_{j=1}^{\infty} [j^2 b_j^* + (j - 2)b_{j-1}^*]t^{-1+j}$$

$$= (3 + b_1^* - b_0^*) + (1 + 4b_2^*)t + \sum_{j=3}^{\infty} [j^2 b_j^* + (j - 2)b_{j-1}^*]t^{-1+j}.$$

If this is to be identically zero, we must have

$$b_1^* = b_0^* - 3$$

$$b_2^* = -\frac{1}{4}$$

and

(R*) $$b_j^* = -\frac{(j - 2)}{j^2} b_{j-1}^*, \qquad j = 3, 4, \ldots.$$

There is no restriction on b_0^*. If we choose

$$b_0^* = 0,$$

then

$$b_1^* = -3, \quad b_2^* = -\frac{1}{4} = -\frac{1}{2^2}$$

and

$$b_3^* = -\frac{1}{3^2} b_2^* = \frac{1}{3^2 2^2},$$

$$b_4^* = -\frac{2}{4^2} b_3^* = -\frac{2}{4^2 3^2 2^2},$$

$$b_5^* = -\frac{3}{5^2} b_4^* = \frac{3 \cdot 2}{5^2 4^2 3^2 2^2}, \ldots,$$

$$b_k^* = \frac{(-1)^{k+1}}{(k!)^2} (k - 2)! = \frac{(-1)^{k-1}}{k!k(k - 1)}, \ldots.$$

Note that the general formula for b_k^* is valid only from $k = 2$ on. Our second solution of (H) is

$$x_2(t) = (t^{-1} + 1) \ln t - 3 + \sum_{k=2}^{\infty} \frac{(-1)^{k-1}}{k!k(k-1)} t^{-1+k}$$

(S$_2$)

$$= (t^{-1} + 1) \ln t - 3 + \sum_{n=1}^{\infty} \frac{(-1)^n}{(n+1)!(n+1)n} t^n.$$

The general solution of (H) is

$$x = c_1 x_1(t) + c_2 x_2(t)$$

$$= c_1(t^{-1} + 1) + c_2\left[(t^{-1} + 1) \ln t - 3 + \sum_{n=1}^{\infty} \frac{(-1)^n}{(n+1)!(n+1)n} t^n\right].$$

In Example 7.8.1, substitution of $(t^{-1} + 1) \ln t$ into L eliminated the terms involving $\ln t$. It is easy to verify (Exercise 16) that this is a general pattern. Indeed, if $x_1(t)$ is a solution of (H), then

(E) $$L[x_1(t) \ln t] = 2tp(t)x_1'(t) + [q(t) - p(t)]x_1(t).$$

We use this fact to simplify our work in the next example.

Example 7.8.2
Find the general solution of Bessel's equation with $\mu = 0$,

(B$_0$) $$[t^2 D^2 + tD + t^2]x = 0.$$

We have already seen in Example 7.7.1 that the indicial equation of (B$_0$) has a double root $m = 0$ and that a corresponding Frobenius series solution is the Bessel function of the first kind

(S$_1$) $$J_0(t) = \sum_{k=0}^{\infty} \frac{(-1)^k}{(k!)^2}\left(\frac{t}{2}\right)^{2k} = \sum_{k=0}^{\infty} \frac{(-1)^k t^{2k}}{(k!)^2 2^{2k}}.$$

We look for a second solution in the form

(S₂) $$K_0(t) = J_0(t) \ln t + t^0 \sum_{k=0}^{\infty} b_k^* t^k.$$

Substitution of (S₂) into (B₀) gives

$$2tJ_0'(t) + b_1^* t + \sum_{j=2}^{\infty} [j^2 b_j^* + b_{j-2}^*]t^j = 0$$

or

$$b_1^* t + \sum_{j=2}^{\infty} [j^2 b_j^* + b_{j-2}^*]t^j = -2tJ_0'(t)$$

$$= \sum_{k=0}^{\infty} \frac{(-1)^{k+1}4k}{(k!)^2 2^{2k}} t^{2k}.$$

Since the right side of the last equation has no odd powers of t, we conclude that

$$b_1^* = 0$$

and

$$j^2 b_j^* + b_{j-2} = 0, \qquad j = 3, 5, \ldots .$$

It follows that

$$b_j^* = 0 \qquad \text{for } j \text{ odd.}$$

On the other hand, comparing even powers of t, we have (taking $j = 2k$)

$$(2k)^2 b_{2k}^* + b_{2k-2}^* = \frac{(-1)^{k+1}4k}{(k!)^2 2^{2k}}$$

or

$$b_{2k}^* = -\frac{1}{4k^2} b_{2k-2}^* + \frac{(-1)^{k+1}}{4^k (k!)^2 k}, \qquad k = 1, 2, \ldots .$$

Once again, there is no restriction on b_0^*. Taking

$$b_0^* = 0$$

we see that

$$b_2^* = -\frac{1}{4}b_0^* + \frac{(-1)^2}{4(1!)^2} = \frac{1}{4}$$

$$b_4^* = -\frac{1}{4 \cdot 2^2}b_2^* + \frac{(-1)^3}{4^2(2!)^2 2} = -\frac{1}{4^2 2^2}\left[1 + \frac{1}{2}\right]$$

$$b_6^* = -\frac{1}{4 \cdot 3^2}b_4^* + \frac{(-1)^4}{4^3(3!)^2 3} = \frac{1}{4^3(3!)^2}\left[1 + \frac{1}{2} + \frac{1}{3}\right]$$

and in general

$$b_{2k}^* = \frac{(-1)^{k+1}}{4^k(k!)^2}\left[1 + \frac{1}{2} + \frac{1}{3} + \ldots + \frac{1}{k}\right].$$

Thus, our second solution to (B_0) is

$$(S_2) \qquad K_0(t) = J_0(t)\ln t + \sum_{k=1}^{\infty}\frac{(-1)^{k+1}H_k}{(k!)^2}\left(\frac{t}{2}\right)^{2k}$$

where H_k is the kth partial sum of the harmonic series,

$$H_k = 1 + \frac{1}{2} + \frac{1}{3} + \ldots + \frac{1}{k}.$$

The general solution of (B_0) is

$$x = c_1 J_0(t) + c_2 K_0(t).$$

Any function that solves (B_0) and is not a multiple of $J_0(t)$ [such as $K_0(t)$] is called a **Bessel function of the second kind of order 0.** It is customary when working with Bessel functions to use as the second solution a specially chosen linear combination of $J_0(t)$ and $K_0(t)$, which is usually denoted $Y_0(t)$.

Besides the case of a double root, we often need to go beyond Frobenius series to handle the case in which the roots, although distinct, differ by an integer, say $m_1 - m_2 = n > 0$. In this case, we can always find a Frobenius series solution $x_1(t)$ corresponding to the larger root m_1. We can also try to find a second Frobenius series solution corresponding to the smaller root m_2. If this works, as in Examples

7.6.2 and 7.7.4, we are done. If not, we look for a second solution in a form similar to that in the double root case.

Fact: *If the indicial equation (I) has roots m_1 and m_2 that differ by a nonzero integer, $m_1 - m_2 = n > 0$, then the general solution of (H) is $x = c_1 x_1(t) + c_2 x_2(t)$ where the first solution is a Frobenius series corresponding to m_1,*

$$x_1(t) = t^{m_1} \sum_{k=0}^{\infty} b_k t^k, \qquad b_0 \neq 0,$$

while the second is either a Frobenius series corresponding to m_2 or else is such a series plus $x_1(t) \ln t$:

$$x_2(t) = t^{m_2} \sum_{k=0}^{\infty} b_k^* t^k, \qquad b_0^* \neq 0$$

or

$$x_2(t) = x_1(t) \ln t + t^{m_2} \sum_{k=0}^{\infty} b_k^* t^k.$$

When (I) had a double root, we were free to choose any value we wished for b_0^*, including 0. When the roots differ by an integer, $m_1 - m_2 = n$, we are free to choose any value we wish for b_n^*.

Example 7.8.3

Find the first terms of the general solution to

(H) $$Lx = [t^2 D^2 + (t^3 + t)D - 1]x = 0, \qquad t > 0.$$

Recall from Example 7.6.3 that the roots of the indicial equation are $m_1 = 1$ and $m_2 = -1$, that a solution corresponding to $m = 1$ is

(S$_1$) $$x_1(t) = t - \frac{1}{8} t^3 + \frac{1}{64} t^5 + \ldots,$$

and that there is no Frobenius series solution corresponding to $m = -1$. Accordingly, we look for a second solution in the form

(S$_2$) $$x_2(t) = x_1(t) \ln t + t^{-1} \sum_{k=0}^{\infty} b_k^* t^k.$$

The effect of L on an arbitrary Frobenius series was calculated in Example 7.6.3; using that calculation, we see that

$$L\left[t^{-1} \sum_{k=0}^{\infty} b_k^* t^k\right] = -b_1^* - b_0^* t + 3b_3^* t^2 + (8b_4^* + b_2^*)t^3$$

$$+ (15b_5^* + 2b_3^*)t^4 + \ldots .$$

The effect of L on the first summand of $x_2(t)$ can be worked out using the formula (E) preceding Example 7.8.2:

$$L[x_1(t) \ln t] = 2tx_1'(t) + (t^2 + 1 - 1)x_1(t)$$

$$= \left[2t - \frac{3}{4}t^3 + \frac{5}{32}t^5 + \ldots\right] + \left[t^3 - \frac{1}{8}t^5 + \ldots\right]$$

$$= 2t + \frac{1}{4}t^3 + \frac{1}{32}t^5 + \ldots .$$

Combining these, we see that

$$L[x_2(t)] = -b_1^* + (2 - b_0^*)t + 3b_3^* t^2 + \left(8b_4^* + b_2^* + \frac{1}{4}\right)t^3$$

$$+ (15b_5^* + 2b_3^*)t^4 + \ldots .$$

Setting coefficients of t^k equal to zero, we find

$$b_1^* = 0, \quad b_0^* = 2, \quad b_3^* = 0, \quad b_4^* = -\frac{1}{8}b_2^* - \frac{1}{32}, \quad b_5^* = 0, \ldots .$$

There is no restriction on b_2^*. If we choose

$$b_2^* = 0,$$

then

$$b_4^* = -\frac{1}{32}$$

and our second solution is

$$x_2(t) = x_1(t) \ln t + \sum_{k=0}^{\infty} b_k^* t^{k-1}$$

(S$_2$)

$$= \left(t - \frac{1}{8}t^3 + \frac{1}{64}t^5 + \ldots\right) \ln t + \left(2t^{-1} - \frac{1}{32}t^3 + \ldots\right).$$

The general solution of (H) is

$$x = c_1 x_1(t) + c_2 x_2(t)$$

$$= c_1\left(t - \frac{1}{8}t^3 + \frac{1}{64}t^5 + \dots\right) + c_2\left(2t^{-1} - \frac{1}{32}t^3 + \dots\right)$$

$$+ c_2\left(t - \frac{1}{8}t^3 + \frac{1}{64}t^5 + \dots\right)\ln t.$$

Example 7.8.4

Find the general solution to Bessel's equation with $\mu = 1$,

$$(\text{B}_1) \qquad Lx = [t^2 D^2 + tD + (t^2 - 1)]x = 0, \qquad t > 0.$$

We know from Example 7.7.2 that the roots of the indicial equation are $m_1 = 1$ and $m_2 = -1$, that a solution corresponding to m_1 is

$$J_1(t) = \sum_{k=0}^{\infty} \frac{(-1)^k}{k!(k+1)!}\left(\frac{t}{2}\right)^{2k+1} = \sum_{k=0}^{\infty} \frac{(-1)^k t^{2k+1}}{k!(k+1)!2^{2k+1}},$$

and that there is no Frobenius series solution corresponding to -1. We look for a second solution of the form

$$K_1(t) = J_1(t)\ln t + t^{-1}\sum_{k=0}^{\infty} b_k^* t^k.$$

The effect of L on the first summand of $K_1(t)$ is

$$L[J_1(t)\ln t] = 2tJ_1'(t) = \sum_{k=0}^{\infty} \frac{(-1)^k(2k+1)}{k!(k+1)!\,2^{2k}} t^{2k+1},$$

while

$$L\left[\sum_{k=0}^{\infty} b_k^* t^{-1+k}\right] = -b_1^* + \sum_{j=2}^{\infty} [j(j-2)b_j^* + b_{j-2}^*]t^{j-1}.$$

These two series must cancel, or

$$-b_1^* + \sum_{j=2}^{\infty} [j(j-2)b_j^* + b_{j-2}^*]t^{j-1} = -\sum_{k=0}^{\infty} \frac{(-1)^k(2k+1)}{k!(k+1)!2^{2k}} t^{2k+1}.$$

Since the right-hand side contains no even powers of t,

$$-b_1^* = 0$$

and

$$j(j - 2)b_j^* + b_{j-2}^* = 0, \qquad j = 3, 5, 7, \ldots.$$

It follows that

$$b_j^* = 0 \qquad \text{for } j \text{ odd.}$$

Equating coefficients of t^{2n-1} (take $j = 2n$ on the left and $k = n - 1$ on the right), we see that

$$2n(2n - 2)b_{2n}^* + b_{2n-2}^* = \frac{(-1)^n(2n - 1)}{(n - 1)!n!2^{2n-2}}, \qquad n = 1, 2, \ldots.$$

Thus

$$b_0^* = -1$$

and

$$b_{2n}^* = \frac{1}{4n(n - 1)} \left[\frac{(-1)^n(2n - 1)}{4^{n-1}n!(n - 1)!} - b_{2n-2}^* \right], \qquad n = 2, 3, \ldots.$$

These relations place no restrictions on b_2^*. The recurrence relation is difficult to translate into a pattern, but you can verify that the choice

$$b_2^* = 1$$

leads to

$$b_{2n}^* = \frac{(-1)^n}{4^n n!(n - 1)!} [H_n + H_{n-1}], \qquad n = 2, 3, \ldots$$

where, as before, H_n is the nth partial sum of the harmonic series. Thus, our second solution is

$$(\text{S}_2) \qquad K_1(t) = J_1(t) \ln t - t^{-1} + t + \sum_{j=2}^{\infty} \frac{(-1)^j[H_j + H_{j-1}]}{4^j j!(j - 1)!} t^{2j-1}.$$

The general solution of (B_1) is

$$x = c_1 J_1(t) + c_2 K_1(t).$$

Again, any function [such as $K_1(t)$] that solves (B_1) and is not a multiple of $J_1(t)$ is a **Bessel function of the second kind of order 1.** The solutions of (B_1) are usually described in terms of $J_1(t)$ and a specially chosen linear combination of $J_1(t)$ and $K_1(t)$.

These exceptional cases exhaust the possibilities for regular singular points. Some other approaches to finding the second solution in these cases are sketched in the notes following our summary.

REGULAR SINGULAR POINTS: EXCEPTIONAL CASES

Suppose the equation

(H) $L[x] = [t^2 p(t)D^2 + tq(t)D + r(t)]x = 0, \qquad t > 0,$

with a regular singular point at $t = 0$, has an indicial equation

(I) $p_0 m(m - 1) + q_0 m + r_0 = 0$

with roots m_1 and m_2 that differ by an integer, $m_1 - m_2 = n \geq 0$. Then (H) always has a Frobenius series solution

$$x_1(t) = t^{m_1} \sum_{k=0}^{\infty} b_k t^k, \qquad b_0 \neq 0.$$

If every Frobenius series solution of (H) is a constant multiple of $x_1(t)$, then (H) has a second solution of the form

$$x_2(t) = x_1(t) \ln t + t^{m_2} \sum_{k=0}^{\infty} b_k^* t^k.$$

(S_2)

To find $x_2(t)$, we substitute into (H), make a free choice of b_n^*, and determine the remaining coefficients b_k^*. The substitution can be facilitated by the formula

(E) $L[x_1(t) \ln t] = 2tp(t)x_1'(t) + [q(t) - p(t)]x_1(t).$

Notes
1. Repeated roots
In this note we describe an alternate approach to the case in which the indicial equation has a double root $m = m_1 = m_2$; you are invited to work through Examples 7.8.1 and 7.8.2 using this approach. Our discussion also gives a justification for the form of the second solution adopted in the text.

When we apply $L = t^2 p(t)D^2 + tq(t)D + r(t)$ to an arbitrary Frobenius series

$$\text{(S)} \qquad\qquad x = \sum_{k=0}^{\infty} b_k t^{m+k}$$

we obtain another Frobenius series,

$$\text{(LS)} \qquad\qquad L[x] = [f_0(m)b_0]\, t^m + \ldots$$

in which the coefficient of the lowest power of t is a multiple of $f_0(m) = p_0 m(m - 1) + q_0 m + r_0$, the left side of the indicial equation. If we equate the coefficient of each power of t in (LS) *except the lowest* to zero, we obtain a recurrence relation that can be used to find expressions $b_k(m)$ for the coefficients in terms of b_0 and the variable m. If we substitute these coefficients into (S), setting $b_0 = 1$, we obtain a Frobenius series

$$x(m, t) = \sum_{k=0}^{\infty} b_k(m)t^{m+k}, \qquad b_0(m) = 1$$

that represents a function of two variables with the property that, for each fixed value of m,

$$\text{(*)} \qquad\qquad L[x(m, t)] = f_0(m)t^m.$$

In particular, when $m = m_1$ is a zero of f_0,

$$L[x(m_1, t)] = 0$$

so that $x(m_1, t)$ is a solution of (H). This is precisely the solution $x_1(t)$ that we found in Section 7.6:

$$x_1(t) = x(m_1, t).$$

Since the indicial equation has a repeated root, we can write $f_0(m) = p_0(m - m_1)^2$. Now if we differentiate (*) with respect to m, we get

$$\frac{\partial}{\partial m} L[x(m, t)] = \frac{\partial}{\partial m} [f_0(m)t^m] = \frac{\partial}{\partial m} [p_0(m - m_1)^2 t^m]$$

$$= 2p_0(m - m_1)t^m + p_0(m - m_1)^2 t^m \ln t$$

or, using the equality of cross-partials,

$$L\left[\frac{\partial}{\partial m} x(m, t)\right] = 2p_0(m - m_1)t^m + p_0(m - m_1)^2 t^m \ln t.$$

In particular, if we set

$$x_2(m, t) = \frac{\partial}{\partial m} x(m, t)$$

and substitute $m = m_1$, we see that

$$L[x_2(m_1, t)] = 2p_0(m_1 - m_1)t^{m_1} + p_0(m_1 - m_1)^2 t^{m_1} \ln t = 0.$$

Thus, another solution of (H) is

$$x_2(t) = x_2(m_1, t).$$

But

$$x_2(m, t) = \frac{\partial}{\partial m} x(m, t) = \sum_{k=0}^{\infty} \frac{\partial}{\partial m} [b_k(m)t^{m+k}]$$

$$= \sum_{k=0}^{\infty} [b_k'(m)t^{m+k} + b_k(m)t^{m+k} \ln t]$$

$$= \sum_{k=0}^{\infty} b_k'(m)t^{m+k} + x(m, t) \ln t.$$

Setting $b_k^* = b_k'(m_1)$, we have

$$x_2(t) = x_2(m_1, t) = \sum_{k=0}^{\infty} b_k'(m_1)t^{m_1+k} + x(m_1, t) \ln t$$

$$= t^{m_1} \sum_{k=0}^{\infty} b_k^* t^k + x_1(t) \ln t,$$

which is precisely the form of the second solution used in the text.

2. Roots differing by an integer

In the case when the roots of the indicial equation $m = m_1, m_2$ differ by a positive integer $n = m_1 - m_2 > 0$, the approach of the preceding note can be applied in a modified form. Again, we apply L to an arbitrary Frobenius series

(S) $$x = \sum_{k=0}^{\infty} b_k t^{m+k}$$

and equate the coefficient of every power of t except the lowest to zero. By means of a recurrence relation [with $b_0(m) = 1$], this leads to a function of two variables

$$x(m, t) = \sum_{k=0}^{\infty} b_k(m)t^{m+k}$$

such that

(*) $$L[x(m, t)] = f_0(m)t^m.$$

When $m = m_1$, the higher root, we always obtain a solution to (H):

$$x_1(t) = x(m_1, t).$$

However, solving for $b_n(m)$ in the recurrence relation involves division by $(m - m_2)$, so that when $m = m_2$, the lower root, we cannot solve for $b_n(m_2)$.

If both sides of the recurrence relation for $b_n(m)$ vanish when $m = m_2$, then there is no restriction on $b_n(m_2)$; we can pick an *arbitrary value for* $b_n(m_2)$ and continue solving the recurrence relations to find a second Frobenius series solution to (H):

$$x_2(t) = x(m_2; t), \qquad b_n(m_2) \text{ arbitrary.}$$

On the other hand, if the right side of the relation for $b_n(m)$ does not vanish identically when $m = m_2$, then every Frobenius series solution to (H) is a multiple of the first solution, $x_1(t)$. In this case, we solve the recurrence relations using $\bar{b}_0(m) = m - m_2$ in place of $b_0(m) = 1$. The first coefficients $\bar{b}_j(m)$, $j < n$, will then be multiples of $(m - m_2)$, so that when we come to $\bar{b}_n(m)$ it will be possible to cancel the factors $(m - m_2)$ in the expression for $\bar{b}_n(m)$ and continue solving the recurrence relations. This leads to a new Frobenius series

$$y(m, t) = \sum_{k=0}^{\infty} \bar{b}_k(m) t^{m+k},$$

which equals $(m - m_2) x(m, t)$ for $m > m_2$ but is also defined when $m = m_2$. This series represents a solution to (H) when $m = m_2$, but it will be simply a multiple of $x_1(t)$. However, as before we can differentiate $y(m, t)$ with respect to m

$$x_2(m, t) = \frac{\partial}{\partial m} y(m, t)$$

and obtain a function of two variables that, when $m = m_2$, is a solution to (H) independent of $x_1(t)$:

$$x_2(t) = x_2(m_2, t).$$

It is again true that $x_2(m_2, t)$ has the form used in the text, but the technicalities of the proof are much fussier in this case.

EXERCISES

In Exercises 1 to 10, find the general solution. In Exercises 1 to 6, you need only find the first four coefficients of each of the two generating solutions. In Exercises 7 to 10, find the general pattern for the coefficients.

1. $[t^2 D^2 - tD + (t + 1)]x = 0,$ $0 < t$ (see Exercise 14, Section 7.6)
2. $[t^2 D^2 + tD + (t - 1)]x = 0,$ $0 < t$ (see Exercise 15, Section 7.6)
3. $[t^2 D^2 + tD + (t^2 - 4)]x = 0,$ $0 < t$ (see Exercise 16, Section 7.6)

4. $[8t^2D^2 + 2t(t + 4)D + (t - 2)]x = 0, \qquad 0 < t$

5. $[9t^2D^2 + 15tD + (t + 1)]x = 0, \qquad 0 < t$

6. $[t^2(t + 1)D^2 + t(t + 1)D - (4t + 1)]x = 0, \qquad 0 < t < 1$

7. $[t^2D^2 + t(1 - t)D - 1]x = 0, \qquad 0 < t$

8. $[tD^2 + (1 - t)D + 1]x = 0, \qquad 0 < t$

9. $[t^2(t - 2)D^2 - 2t(1 - t)D - 2t]x = 0, \qquad 0 < t < 2$

10. $[t^2D^2 + t(t + 4)D + (t + 2)]x = 0, \qquad 0 < t$

11. Show that if $\mu = 0$ or 1, then the only solutions of the Bessel equation that remain bounded as t approaches 0 are the constant multiples of $J_\mu(t)$. [*Hint:* What happens to $K_\mu(t)$ as t approaches 0?]

12. a. Show that if μ is a nonnegative integer, then the Laguerre equation

    (H) $$[tD^2 + (1 - t)D + \mu]x = 0$$

    has a nonzero solution $x_\mu(t)$ that is a polynomial.

    b. Show that the only solutions of (H) that remain bounded as t approaches 0 are the constant multiples of $x_\mu(t)$.

13. Consider the Legendre equation

    (H) $$[(t^2 - 1)D^2 + 2tD - \mu(\mu + 1)]x = 0$$

    with μ a nonnegative integer.

    a. Show that the general solution of (H) is generated by a polynomial $x_\mu(1 - t)$ and a solution that becomes unbounded as t approaches 1. (*Hint:* Substitute $T = 1 - t$. What form does the solution of the resulting equation take?)

    b. Show that the only solutions of (H) that remain bounded as t approaches 1 are the constant multiples of $x_\mu(1 - t)$.

    c. Show that the only solutions of (H) that remain bounded as t approaches -1 are the constant multiples of a polynomial solution $x_\mu^*(1 + t)$. (*Hint:* Substitute $T = 1 + t$.)

    d. Show that $x_\mu(1 - t)$ and $x_\mu^*(1 + t)$ are constant multiples of each other.

14. *Legendre Polynomials:* For each nonnegative integer n, define a polynomial by

    $$P_n(t) = \frac{1}{2^n n!} D^n(t^2 - 1)^n.$$

    a. Find $P_0(t)$, $P_1(t)$ and $P_2(t)$.

    b. Show that $(t^2 - 1)D(t^2 - 1)^n - 2nt(t^2 - 1)^n = 0$.

    c. Differentiate this equation to show that

    $$(t^2 - 1)D^2(t^2 - 1)^n - 2(n - 1)tD(t^2 - 1)^n - 2n(t^2 - 1)^n = 0.$$

    d. Show that repeated differentiation of this equation leads to

    $$(t^2 - 1)D^{k+2}(t^2 - 1)^n - 2(n - k - 1)tD^{k+1}(t^2 - 1)^n$$
    $$- 2[n + (n - 1) + \ldots + (n - k)]D^k(t^2 - 1)^n = 0.$$

In particular, $k = n$ gives

$$(t^2 - 1)D^{n+2}(t^2 - 1)^n + 2tD^{n+1}(t^2 - 1)^n$$
$$- 2[n + (n - 1) + \ldots + 1]D^n(t^2 - 1)^n = 0.$$

e. Use the fact that $n + (n - 1) + \ldots + 1 = n(n + 1)/2$ to show that $x = P_n(t)$ solves the Legendre equation with $\mu = n$,

(H) $$[(t^2 - 1)D^2 + 2tD - n(n + 1)]x = 0.$$

f. Use the result of Exercise 13 to show that any solution of (H) that remains bounded as t approaches 1 or -1 is a constant multiple of $P_n(t)$.

15. *Temperature in the planet* of Example 7.1.4 is modeled by the equation

(H) $$\rho \frac{\partial^2}{\partial \rho^2}(\rho u) + \frac{1}{\sin \phi}\frac{\partial}{\partial \phi}\left(\sin \phi \frac{\partial u}{\partial \phi}\right) = 0.$$

Recall that if $u = R(\rho)\,\Phi(\phi)$ solves (H), then R must solve the Cauchy-Euler equation

(H_ρ) $$\rho^2 \frac{d^2R}{d\rho^2} + 2\rho \frac{dR}{d\rho} - n(n + 1)R = 0,$$

while Φ must solve the Legendre equation

(H_t) $$(t^2 - 1)\frac{d^2\Phi}{dt^2} + 2t\frac{d\Phi}{dt} - n(n + 1)\Phi = 0$$

where $t = \cos \phi$ and n is a nonnegative integer.

a. Use the fact that the temperature in the planet has to remain bounded near the origin to conclude that the physically meaningful solutions of (H_ρ) are constant multiples of $R_n = \rho^n$.

b. Use the boundedness of the temperature near the poles ($\phi = 0$ and $\phi = \pi$), together with the result of Exercise 14, to conclude that the physically meaningful solutions of (H_t) are constant multiples of $\Phi_n = P_n(t)$.

c. Check that any linear combination of the functions

$$u_n = R_n\Phi_n = \rho^n P_n(t) = \rho^n P_n(\cos \phi)$$

solves (H).

d. Find a linear combination of the functions u_n that matches the boundary condition $u(1,\phi) = \sin^2 \phi$, $0 < \phi < \pi$. [*Hint:* First check that $u(1,\phi) = 1 - t^2 = 2/3 P_0(t) - 2/3\, P_2(t)$.]

16. Show that if $x_1(t)$ is a solution of

(H) $$Lx = [t^2p(t)D^2 + tq(t)D + r(t)]x = 0, \qquad t > 0,$$

then $L[x_1(t) \ln t] = 2tp(t)x_1'(t) + [q(t) - p(t)]x_1(t).$

REVIEW PROBLEMS

In Exercises 1 to 18, use the methods of this chapter to find an expression for the general solution on the given interval. When you cannot find a pattern for the terms of a series, find the first four coefficients.

1. $[(t^2 - 1)D^2 + tD - 1]x = 0, \quad 0 < t < 1$
2. $[t^2D^2 + 4tD + 2]x = 0, \quad 0 < t$
3. $[16t^2D^2 + 16tD + (16t^2 - 1)]x = 0, \quad 0 < t$
4. $[t^2D^2 + t(4 - t)D + 2(1 - t)]x = 0, \quad 0 < t$
5. $[9t^2D^2 - 3tD + 4]x = 0, \quad 0 < t$
6. $[t^2(t - 2)D^2 - 2t(1 - t)D - 6t]x = 0, \quad 0 < t < 2$
7. $[D^2 - 4tD]x = 0, \quad 0 < t$
8. $[t^2(2 + t)D^2 + t(1 + t)D - 1]x = 0, \quad 0 < t < 2$
9. $[3t^2(t + 1)D^2 - 4tD + 2]x = 0, \quad 0 < t < 1$
10. $[D^2 + t]x = 0, \quad 0 < t$
11. $[t^2D^2 + tD - (t + 1)]x = 0, \quad 0 < t$
12. $[8t^2D^2 + 2t(4 - t)D - (t + 2)]x = 0, \quad 0 < t$
13. $[t^2(t + 1)D^2 + 3t(1 - t)D + 1]x = 0, \quad 0 < t < 1$
14. $[2t^2D^2 + 6tD + 12]x = 0, \quad 0 < t$
15. $[(t^2 + 1)D^2 + (t + 1)D + 1]x = 0, \quad 0 < t < 1$
16. $[t^2(t + 1)D^2 + t(t + 1)D + 1]x = 0, \quad 0 < t < 1$
17. $[t^2D^2 + t(4 - t)D + (2 - t)]x = 0, \quad 0 < t$
18. $[3t^2D^2 + t(4 - 3t)D - (t + 2)]x = 0, \quad 0 < t$
19. Find a series expression for the solution of the equation in Exercise 1 that satisfies the initial conditions $x(0) = 1$, $x'(0) = 2$.
20. Find the terms up to t^5 in a power series expression for the solution of $[(t^2 - 1)D^2 - 1]x = t^2 + 3$ satisfying the initial conditions $x(0) = -3$, $x'(0) = -5$.
21. The submarine of Exercise 5, Section 7.1, starts from rest at the bottom of the ocean.
    a. Find a series equation for the distance from the submarine to the ocean bottom at time t.
    b. Use the first three nonzero terms of this series to estimate the position of the submarine after 10 seconds.
22. Find the first five terms for a power series about $t_0 = -1$ expressing the general solution to $[tD^2 - 4]x = 0$, $-2 < t < 0$.
23. Find the first five terms of a power series about $t_0 = 1$ expressing the general solution to $[tD^2 + tD + 1]x = 0$, $0 < t < 2$.
24. Show that if μ is a nonnegative integer, then the **Chebyshev equation** $[(1 - t^2)D^2 - tD + \mu^2]x = 0$ has a polynomial solution of degree at most μ.
25. The **Airy equation** is

(H) $$[D^2 + t]x = 0.$$

a. Show that the substitution $x = t^{1/2}y$ changes (H) into

(H$_1$)
$$\left[t^2D^2 + tD + \left(t^3 - \frac{1}{4} \right) \right] y = 0.$$

b. Show that substituting $t = \left(\frac{3}{2} s \right)^{2/3}$ leads to

$$\frac{dy}{dt} = \left(\frac{3}{2} s \right)^{1/3} \frac{dy}{ds},$$

$$\frac{d^2y}{dt^2} = \left(\frac{3}{2} s \right)^{1/3} \left[\left(\frac{3}{2} s \right)^{1/3} \frac{d^2y}{ds^2} + \frac{1}{2} \left(\frac{3}{2} s \right)^{-2/3} \frac{dy}{ds} \right]$$

and converts (H$_1$) into Bessel's equation with $\mu = 1/3$,

(H$_2$)
$$\left[s^2 \frac{d^2y}{ds^2} + s \frac{dy}{ds} + \left(s^2 - \frac{1}{9} \right) \right] y = 0.$$

c. Use (a) and (b) to show that the general solution to the Airy equation on $0 < t$ can be written in the form

$$x(t) = c_1 \sqrt{t} \, J_{1/3} \left(\frac{2}{3} t^{3/2} \right) + c_2 \sqrt{t} \, J_{-1/3} \left(\frac{2}{3} t^{3/2} \right)$$

where $J_{1/3}$ and $J_{-1/3}$ are the Bessel functions of the first kind of orders $1/3$ and $-1/3$, respectively.

Supplementary Reading List

General References About Differential Equations

Boyce, W. E., and DiPrima, R. C., *Elementary Differential Equations and Boundary Value Problems,* 3rd ed., Wiley, New York, 1977.

Braun, M., *Differential Equations and Their Applications,* 2nd ed., Springer Verlag, New York, 1978.

Hildebrand, F. B., *Advanced Calculus for Applications,* 2nd ed., Prentice-Hall, Englewood Cliffs, NJ, 1976.

Kaplan, W., *Ordinary Differential Equations,* Addison-Wesley, Reading, MA, 1958.

Kreider, D. L., Kuller, R. G., Ostberg, D. R., and Perkins, F. W., *An Introduction to Linear Analysis,* Addison-Wesley, Reading, MA, 1966.

Rainville, E. D., and Bedient, P. E., *Elementary Differential Equations,* 6th ed., Macmillan, New York, 1981.

Simmons, G. F., *Differential Equations with Applications and Historical Notes,* McGraw-Hill, New York, 1972.

Applications

Cannon, R. H., *Dynamics of Physical Systems*, McGraw-Hill, New York, 1967.

Papoulis, A., *Circuits and Systems: A Modern Approach,* Holt, Rinehart and Winston, New York, 1980.

Pielou, E., *Mathematical Ecology,* Wiley, New York, 1977.

Pollard, H., *Applied Mathematics: An Introduction,* Addison-Wesley, Reading, MA, 1972.

Sommerfeld, A., *Partial Differential Equations in Physics*, translated by E. Straus, Academic Press, New York, 1949.

Thomson, W. T., *Theory of Vibrations with Applications,* 2nd ed., Prentice-Hall, Englewood Cliffs, NJ, 1981.

More Theoretical Treatments of Differential Equations

Arnold, V. I., *Ordinary Differential Equations,* translated by R. Silverman, MIT Press, Cambridge, MA, 1973.

Birkhoff, G., and Rota, G. C., *Ordinary Differential Equations*, 3rd ed., Wiley, New York, 1978.

Coddington, E. A., and Levinson, N., *Theory of Ordinary Differential Equations,* McGraw-Hill, New York, 1955.

Hirsch, W. H., and Smale, S., *Differential Equations, Dynamical Systems and Linear Algebra,* Academic Press, New York, 1974.

Hurewicz, W., *Lectures on Ordinary Differential Equations,* MIT Press, Cambridge, MA, 1958.

History

Boyer, C. B., *A History of Mathematics*, Wiley, New York, 1968.
Eves, H., *An Introduction to the History of Mathematics*, 5th ed., Saunders College Publishing, Philadelphia, 1983.
Kline, M., *Mathematical Thought from Ancient to Modern Times*, Oxford University Press, New York, 1972.

Linear Algebra

Halmos, P. R., *Finite-Dimensional Vector Spaces*, Springer Verlag, New York, 1974.
Hoffman, K., and Kunze, R., *Linear Algebra*, 2nd ed., Prentice-Hall, Englewood Cliffs, NJ, 1971.

Answers to Odd-Numbered Exercises

CHAPTER ONE

Section 1.1

1a. 3 b. 4 c. 7 d. 3   3.   No

5.   Yes   7. $k = -2, 3$

9. $k = -7$ 11. $k = 1/2$

13a. $x = \dfrac{t^3}{2} + \dfrac{t^2}{2} + c_1 t + c_2$ b. $x = \dfrac{t^3}{2} + \dfrac{t^2}{2} + 3t + 2$

15a. $x = t^3 + c_1 \dfrac{t^2}{2} + c_2 t + c_3 = t^3 + k_1 t^2 + c_2 t + c_3$

   b. $x = t^3 - 3t^2 + 3t - 1$

17a. $x = te^t - 2e^t + c_1 t + c_2$ b. $x = te^t - 2e^t + t + 2$

19. $\dfrac{dx}{dt} = .08x + 400$ 21. $\dfrac{dx}{dt} = -\gamma(x - y)$

23. $\dfrac{dx}{dt} = \gamma(M - x)$

25. $\dfrac{dp}{dt} = \gamma(ap + b - s)$, where $w = ap + b$

Section 1.2

1. $x = ke^{2t/3}$ 3. $x = -\dfrac{e^{-2t}}{2} + e^{-t} + c$

5. $x = kte^{t^2/2} - 1$ 7. $t = 2 \ln |x^5 + 3x + 2| + c$

9. $x = 3e^{-5t/3}$ 11. $x = 2t/(2 - t)$

13. $x = \tan\left(1 - \dfrac{1}{t}\right)$

15a. $x = t \sqrt[3]{-3 \ln |t|} + c$ b. $t = x \ln |x| + cx$

   c. $kt^3 = \dfrac{x}{x + 3t}$, so $x = \dfrac{3kt^4}{1 - kt^3}$

17a. $x = x_0 e^{-\gamma t}$ b. $\gamma = \dfrac{\ln 2}{T}$ c. $x = x_0 2^{-t/22}$

19. $20 \ln 2 \approx 13.9$ years

Section 1.3

1a. linear, nonhomogeneous b. linear, homogeneous c. linear, homogeneous
 d. nonlinear e. linear, nonhomogeneous f. nonlinear g. nonlinear
 h. linear, nonhomogeneous

3. $x = ce^{-2t/3} + \tfrac{1}{2}$

5. $x = ce^{-3t} + \dfrac{t}{3} - \dfrac{1}{9}$

7. $x = c \cos t + t \cos t$

9. $x = ce^{-3t} - \dfrac{t^2}{3} + \dfrac{2t}{9} - \dfrac{11}{27}$

11. $x = c\sqrt{\dfrac{t+1}{t-1}} - \dfrac{2}{\sqrt{t-1}}$

13. $x = 5e^{2t} - 4$

15. $x = 3\sqrt{t^2 + 1}$

17. $x = \tfrac{3}{2} e^{t^2/2} - 1$

19. $v = \dfrac{32m}{\gamma}(1 - e^{-\gamma t/m})$

21a. $y = 40 - 3t$ b. $x = 70 - 3t$ c. $55°$

29. $k = x(0)$ and $c = x(0) + 1$, so $ce^t - e^t - 1 = ke^t + (e^t - t - 1)$

Section 1.4

1.

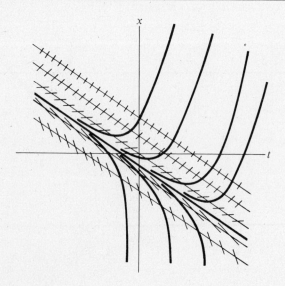

3.

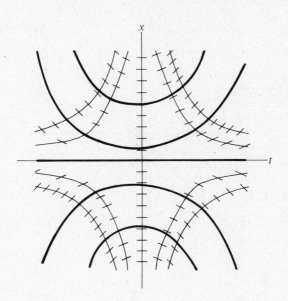

5.

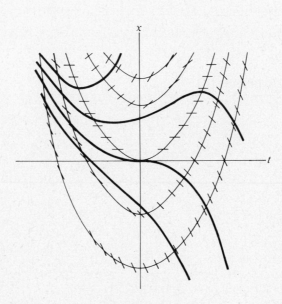

7.

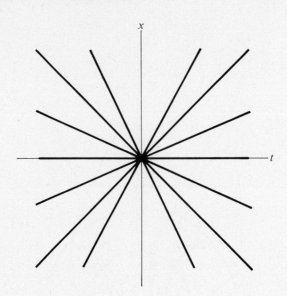

Section 1.5

1. Exact, $\dfrac{t^2}{2} + \dfrac{x^2}{2} = c$, so $x = \pm\sqrt{k - t^2}$

3. Not exact   5. Not exact

7. Not exact   9. Not exact

11. Exact, $x = c(t^2 + 1)/t$

Review Problems

1. (i) $x = k\sqrt{(t + 1)/(t - 1)}$ (ii) $x = \sqrt{(t + 1)/3(t - 1)}$

3. (i) $\dfrac{x}{1 + x} = ke^t$, so $x = \dfrac{ke^t}{1 - ke^t}$ (ii) $x = \dfrac{e^t}{2 - e^t}$

5. (i) $x = k/t$ (ii) $x = 2/t$

7. (i) $3x - 3\ln|x + 1| = \ln t - \dfrac{1}{t} + c, \ x = -1$

(ii) $3x - 3\ln|x + 1| = \ln t - \dfrac{1}{t} + \dfrac{7}{2} - 4\ln 2$

9. (i) $x = \cos t \,(\ln|\sec t + \tan t| + c)$
   (ii) $x = \cos t \,(\ln|\sec t + \tan t| + 1)$

11. (i) $x^2 + 2tx - t^2 = c$, so $x = -t \pm \sqrt{2t^2 + c}$
     (ii) $x = -t + \sqrt{2t^2 + 2}$

13. (i) $x = cte^{-1/t} + t$ (ii) $x = t$

15. (i) $(x - t)e^{x/t} = k$ (ii) $(x - t)e^{x/t} = -e^{1/2}$

17a. $\dfrac{2 \ln (1/2)}{\ln (3/4)} \approx 4.82$ hours   b. $45 - \dfrac{5}{\ln (3/4)} \approx 62.38°$

   c. $30 - \dfrac{25}{2 \ln (3/4)} \approx 73.45°$

CHAPTER TWO

Section 2.1

1a. $\dfrac{d^2x}{dt^2} + 125x = 32$ b. $\dfrac{d^2x}{dt^2} + 2b\dfrac{dx}{dt} + 125x = 32$

3. $L\dfrac{d^2Q}{dt^2} + R\dfrac{dQ}{dt} + \dfrac{1}{C}Q = V(t)$

5a. $\dfrac{d^2x}{dt^2} = -2b\dfrac{dx}{dt} + 2k(y - x - L) - 32$

 b. $\dfrac{d^3x}{dt^3} + 2b\dfrac{d^2x}{dt^2} + 2k\dfrac{dx}{dt} = 4k$

7a. $y = \dfrac{1}{44}\dfrac{d^2x}{dt^2} + \dfrac{62.5x}{22} - \dfrac{16}{22}$ b. $\dfrac{d^2y}{dt^2} = -44y + 7$

 c. $\dfrac{d^4x}{dt^4} + 169\dfrac{d^2x}{dt^2} + 5500x = 1716$

Section 2.2

1. Linear, nonhomogeneous; $(D^2 - tD - 5t)x = -25$

3. Not linear   5. Not linear

7. Linear, nonhomogeneous; $(D^4 + 5t^3D)x = \sqrt{t^2 - 1}$

9. $L[e^t] = 0, L[3e^{2t}] = 3e^{2t}$

11. $L[e^{2t}] = 7e^{2t}, L[t^3] = 6t + 9t^2 - 3t^3,$
     $L[5e^{2t} - 2t^3] = 35e^{2t} - 12t - 18t^2 + 6t^3$

13. $L[t] = 1, L[e^t] = e^t, L[te^t] = (t + 1)e^t$

15. $x = (t - 1)e^t + c_1e^t + c_2$ 17. $x = \tfrac{1}{2}e^{2t} + c_1e^t + c_2$

19. $x = c_1 t^4 + c_2 t^2 + c_3 t + c_4$

21a. $A = \frac{1}{2}, B = -\frac{9}{4}$ b. $x = c_1 e^{2t} + c_2 e^{-t} + \frac{1}{2} t - \frac{9}{4}$

23a. $A = -2$ b. $x = c_1 + c_2 e^{2t} + c_3 e^{-t} - 2t$

25a. $A = -2, B = 0, C = -1$ b. $x = c_1 e^t + c_2 e^{-t} - 2 - t^2$

Section 2.3

1a. $\dfrac{1}{6} e^{3t} - \dfrac{1}{6} e^{-3t}$ b. $\dfrac{1}{2} e^{3t} - \dfrac{1}{2} e^{-3t}$

3a. $x = \sin 2t + \cos 2t$ b. $x = 2 \sin 2t + 4 \cos 2t$

5a. No   b. $x = e^t$ 7a. No   b. $x = te^t - e^{-t}$

9. Yes   11. Yes

13. No   15. No

17a. e^t, e^{-t} b. Yes   19a. $1, t^{-1}$ b. Yes

21a. e^t, e^{2t}, e^{3t} b. Yes   23a. e^{-2t}, e^{2t} b. No

Section 2.4

1. -2 3. 0   5. 20   7. 8

9. $6e^{2t}$ 11. 120   13. 0   15. 120

17. No   19. Yes   21. Yes   23. Yes

25. None   27. Unique   29. $x = \dfrac{8}{7}$ 31. $x = 1$

Section 2.5

1. Yes   3. Yes   5. Yes   7. Yes

9. Yes   11. Yes   13. No   15. Yes

19. $x = c_1 \cos t + c_2 \sin t$

Section 2.6

1. $x = c_1 e^{t\sqrt{5}} + c_2 e^{-t\sqrt{5}}$ 3. $x = c_1 + c_2 e^{5t/2}$

5. $x = c_1 e^{2t/3} + c_2 t e^{2t/3}$

7. $x = c_1 + c_2 t + c_3 t^2 + c_4 e^t + c_5 t e^t$

9. $x = c_1 + c_2 t + c_3 e^{-t} + c_4 t e^{-t} + c_5 t^2 e^{-t} + c_6 e^{2t} + c_7 e^{-5t/3} + c_8 e^{3t/2}$

11. $x = c_1 e^{t\sqrt{2}} + c_2 e^{-t\sqrt{2}} + c_3 e^{-(1+\sqrt{5})t/2} + c_4 e^{-(1-\sqrt{5})t/2}$

13. $x = c_1 e^t + c_2 t e^t + c_3 e^{-t} + c_4 t e^{-t}$ 15. $x = 8 + 2e^{5t}$

17. $x = -e^t + 3t e^t + e^{-2t}$ 19. $Q = t e^{-t}$

21a. $r = \{-b \pm \sqrt{b^2 + 2mk(-3 \pm \sqrt{5})}\}/2m$

  b. $x = (c_1 + c_2 t + c_3 e^{t \sqrt[4]{5}} + c_4 e^{-t \sqrt[4]{5}}) e^{-(1+\sqrt{5})t/2}$

23. $L[e^t] = 4e^t$, $L[e^t \sin t] = e^t(3 \sin t + 4 \cos t)$, $L[e^{-t} \sin t] = -e^{-t} \sin t$

25. $L[e^{\alpha t} \cos \beta t] = L[e^{\alpha t} \sin \beta t] = 0$ 27. $(D - \lambda)^{k-2}$

Section 2.7

1. $x = c_1 \cos \dfrac{t}{3} + c_2 \sin \dfrac{t}{3}$

3. $x = c_1 e^{2t} + c_2 e^{-2t} + c_3 \cos 2t + c_4 \sin 2t$

5. $x = c_1 \cos t + c_2 \sin t + c_3 e^{-t/2} \cos \dfrac{t\sqrt{3}}{2} + c_4 e^{-t/2} \sin \dfrac{t\sqrt{3}}{2}$

7. $x = c_1 e^t + c_2 \cos 3t + c_3 \sin 3t$

9. $x = c_1 e^t + c_2 e^{-t} + c_3 \cos t + c_4 \sin t$

11. $x = (c_1 + c_2 t + c_3 t^2 + c_4 t^3) \cos t + (c_5 + c_6 t + c_7 t^2 + c_8 t^3) \sin t$

13. $x = 2 \cos \dfrac{t}{4} + 36 \sin \dfrac{t}{4}$ 15. $x = 0$

17. $x = \dfrac{5}{2} e^{-t/5} \sin \dfrac{2t}{5}$ 19. $(D - 3)(D - 1)$

21. $(D - 2)^3 (D - 1)^2$ 23. $(D^2 + 4)^2$

25a. $x = \left(c_1 \cos \dfrac{t}{2} + c_2 \sin \dfrac{t}{2} \right) e^{-3t/2}$ b. $x = \left(-2 \cos \dfrac{t}{2} - 4 \sin \dfrac{t}{2} \right) e^{-3t/2}$

27. $Q = Q_0 \cos (t/\sqrt{LC})$ 29. $x = c_1 e^{-t/5} \cos \dfrac{2t}{5} + c_2 e^{-t/5} \sin \dfrac{2t}{5}$

Section 2.8

1. $x = c_1 e^{-t} + c_2 t e^{-t} + 3 + t$

3. $x = c_1 + c_2 t + c_3 e^{-2t} + c_4 t e^{-2t} + \frac{1}{4} t^2$

5. $x = c_1 \cos t + c_2 \sin t + \frac{1}{2} t \sin t$

7. $x = c_1 e^{-3t} \cos t + c_2 e^{-3t} \sin t - 2e^t \cos t + 4e^t \sin t$

9. $x = c_1 e^{t/3} + c_2 e^{-t/3} - \frac{9}{50} \cos t - \frac{1}{10} t \sin t$

11. $x = \frac{3}{8} \sin t - \frac{1}{8} \sin 3t$

13. $x = 9e^{-t/5} \cos \dfrac{2t}{5} - 8e^{-t/5} \sin \dfrac{2t}{5} - 10 + 5t$

15. $p(t) = k_1 + k_2 t + k_3 t^2 + (k_4 + k_5 t)e^{-t/2} \cos t\sqrt{5}$
$\qquad\qquad + (k_6 + k_7 t)e^{-t/2} \sin t\sqrt{5} + k_8 t e^{3t}$

17a. $x = c_1 \cos 5\sqrt{5}\, t + c_2 \sin 5\sqrt{5}\, t + \frac{32}{125}$

  b. $x = c_1 e^{-10t} \cos 5t + c_2 e^{-10t} \sin 5t + \frac{32}{125}$

  c. $x = c_1 e^{-25t} + c_2 e^{-5t} + \frac{32}{125}$

21. $x = c_1 \cos 5\sqrt{5}\, t + c_2 \sin 5\sqrt{5}\, t + c_3 \cos 2\sqrt{11}\, t + c_4 \sin 2\sqrt{11}\, t + \frac{39}{125}$

23. $x = c_1 t^2 + c_2 t + 4 - t \ln t$

25. $x = c_1 t^{-2} + c_2 t^{-2} \ln t + \frac{1}{16} t^2 - \frac{1}{3} t$

Section 2.9

1. $x = c_1 e^t + c_2 e^{2t} + 3(t - 1)e^{2t}$

3. $x = c_1 \cos t + c_2 \sin t - \cos t \ln |\sec t + \tan t|$

5. $x = c_1 e^{t/3} + c_2 t e^{t/3} + \frac{1}{28} t^{7/3} e^{t/3}$

7. $x = (c_1 + c_2 t + c_3 t^2 + \frac{1}{2} \ln t + \frac{3}{4})e^t$

9. $x = (t + 2) \sin t + (1 + \ln |\cos t|) \cos t$

11. $x = \dfrac{-4}{(\ln 2)^3} e^t + \dfrac{3}{(\ln 2)^4} t e^t + t^{-3} e^t$

13. $x = c_1 t^{-1} + c_2 e^t - \dfrac{t}{2} - 1$

15. $x = c_1 t + c_2 t^2 + c_3 t^{-1} + \frac{5}{36} t^{-1} + \frac{1}{6} t^{-1} \ln t$

17. $x = c_1 t^{-2} + c_2 t^{-1} + \frac{1}{42} t^5$ 19. $x = c_1 + c_2 t^{-1} + \frac{4}{15} t^{3/2}$

21. $x = (-c_1 e^{1/t} + c_2)t$ 23. $x = -c_1(t + 1) + c_2 e^t + 4$

25a. $x = c_1 e^{-3t} + c_2 e^{-t} - \frac{1}{4} e^{-t} + \frac{1}{2} t e^{-t}$

  b. $x = c_1 \cos 2t + c_2 \sin 2t + \cos^2 2t - \frac{1}{3} \cos^4 2t + \frac{1}{3} \sin^4 2t$
$\qquad = c_1 \cos 2t + c_2 \sin 2t + \frac{2}{3} - \frac{1}{3} \sin^2 2t$

27a. $(D^2 + 6D + 8)x = 8(h(t) - L - 4)$

   b. $x = c_1 e^{-2t} + c_2 e^{-4t} + 4e^{-2t}\int he^{2t}\, dt - 4e^{-4t}\int he^{4t}\, dt - L - 4$

Section 2.10

3a. $x = \frac{1}{10}\sin 10t$

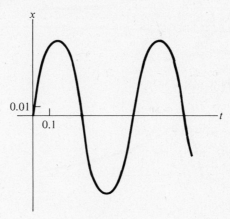

b. $x = \frac{1}{8} e^{-6t}\sin 8t$

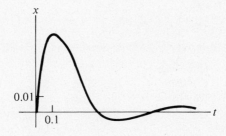

c. $x = \frac{1}{6} e^{-8t}\sin 6t$

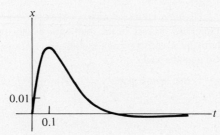

d. $x = te^{-10t}$

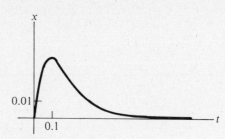

e. $x = \frac{1}{48}(-e^{-50t} + e^{-2t})$

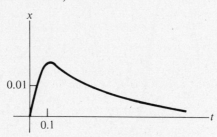

5. Oscillations with extremely large amplitude.

7a. $x = c_1 e^{-t/2} \cos \dfrac{t\sqrt{3}}{2} + c_2 e^{-t/2} \sin \dfrac{t\sqrt{3}}{2} = A e^{-t/2} \cos \left(\dfrac{t\sqrt{3}}{2} - \alpha \right)$

b. $x = c_1 e^{-(5+2\sqrt{6})t} + c_2 e^{-(5-2\sqrt{6})t}$ c. $x = c_1 e^{-t} + c_2 t e^{-t}$

9a. (i) True   (ii) False   (iii) True
 b. (i) False   (ii) False   (iii) True
 c. (i) False   (ii) False   (iii) True

Section 2.11

1. In case a, the shaft turns counterclockwise, then reverses direction, gradually approaching equilibrium. In case b, the shaft turns clockwise, overshooting equilibrium, then turns back and approaches equilibrium. In case c, the shaft turns clockwise toward equilibrium.

3. $s = \frac{1}{6} x^4 - 2x^3 + 36x$; lowest point $x = 3$, $s = 135/2$

5a. $\left(D^2 - \dfrac{g}{L} \right) \theta = 0$ b. No

7a. $F_{\text{tan}} = w \sin \theta + (5 - 5\sqrt{2} \sin \theta) \cos \theta$

c. $w \dfrac{d^2\theta}{dt^2} = 16w\sqrt{2}\sin\theta + 80\sqrt{2}\cos\theta - 160\sin\theta\cos\theta$

d. $w \dfrac{d^2\theta}{dt^2} + 16(5 - w)\theta = 4\pi(5 - w) + 16w$

e. $\theta = -\dfrac{1}{4}\cos 8t + \dfrac{(\pi + 1)}{4}$

Review Problems

1a. $L[1] = L[t] = L\left[\dfrac{1}{t}\right] = 0,\ L[t^3] = 24t$ b. $x = c_1 + c_2 t + c_3\dfrac{1}{t} + t^3$

3. $x = c_1 e^{t/3} + c_2 t e^{t/3} + \dfrac{1}{t} e^{t/3}$

5. $x = (c_1 + c_2 t + c_3 t^2 + t^2\ln t - \tfrac{3}{2}t^2)e^{-t}$

7. $x = c_1 + c_2 t + c_3 t^2 + c_4\cos t + c_5\sin t + \tfrac{1}{2}t^3 - e^{-t}$

9. $x = 2e^{-t/2}\cos\dfrac{3t}{2} - \dfrac{2}{3}e^{-t/2}\sin\dfrac{3t}{2}$ 11. $x = -e^t + 4te^t + 2e^{-t}$

13. $x = c_1\cos(\ln t) + c_2\sin(\ln t) + \dfrac{3}{2t}$

15. $x = c_1\cos t + c_2\sin t - \tfrac{1}{2}t\cos t$ 17. $x = c_1 t + c_2 t\ln t + \dfrac{1}{4t}$

19. $x = c_1 e^{t/2} + c_2 e^{-t/2} + \tfrac{16}{3}e^t - 2te^t + 2te^{t/2}$

CHAPTER THREE

Section 3.1

1. $\dfrac{dp}{dt} = \gamma ap - \gamma s + \gamma b$

$\dfrac{ds}{dt} = \beta p$ where $w = ap + b$.

3a. $\dfrac{dx_1}{dt} = -\dfrac{P}{V}x_1 + \dfrac{P}{V}x_2$ b. $\dfrac{dx_1}{dt} = -\dfrac{P}{V_1}x_1 + \dfrac{P}{V_1}x_2$

$\dfrac{dx_2}{dt} = \dfrac{P}{V}x_1 - \dfrac{P}{V}x_2$ $\dfrac{dx_2}{dt} = \dfrac{P}{V_2}x_1 - \dfrac{P}{V_2}x_2$

5. $\dfrac{dC}{dt} = .11C + .03M$ 7. $\dfrac{dQ}{dt} = I_2$

$\dfrac{dM}{dt} = .04C + .07M$ $\dfrac{dI_2}{dt} = -\dfrac{1}{4}Q - I_2 + e^{-t/2}$

9. $\dfrac{dQ}{dt} = -\ Q + I_3$

   $\dfrac{dI_3}{dt} = -\dfrac{1}{2}Q\ \ \ \ + 2e^{-t/2}$

Section 3.2

1a. $\begin{bmatrix} 0 \\ 2 \end{bmatrix}$ b. $\begin{bmatrix} 0 \\ 9 \end{bmatrix}$ c. $\begin{bmatrix} 7 \\ 7 \end{bmatrix}$

d. $\begin{bmatrix} -1 \\ 8 \end{bmatrix}$ e. $\begin{bmatrix} -1 \\ -2 \end{bmatrix}$ f. $\begin{bmatrix} -2 \\ 26 \end{bmatrix}$

g. $\begin{bmatrix} -3 & 3 \\ -6 & -9 \end{bmatrix}$ h. $\begin{bmatrix} -1 & -5 \\ 2 & 1 \end{bmatrix}$ i. $\begin{bmatrix} 4 \\ 10 \end{bmatrix}$

In Exercises 3 to 9, $\mathbf{x} = \begin{bmatrix} x \\ y \end{bmatrix}$ or $\mathbf{x} = \begin{bmatrix} x \\ y \\ z \end{bmatrix}$.

3. Linear, nonhomogeneous, order 2; $D\mathbf{x} = \begin{bmatrix} -1 & 1 \\ 2 & 0 \end{bmatrix}\mathbf{x} + \begin{bmatrix} t \\ t \end{bmatrix}$

5. Linear, homogeneous, order 2; $D\mathbf{x} = \begin{bmatrix} 5 & -6 \\ 2 & 1 \end{bmatrix}\mathbf{x}$

7. Linear, nonhomogeneous, order 3; $D\mathbf{x} = \begin{bmatrix} 0 & -t & -1 \\ -\dfrac{1}{t} & 0 & -\dfrac{1}{t} \\ 1 & -t & 0 \end{bmatrix}\mathbf{x} + \begin{bmatrix} t \\ 1 \\ 0 \end{bmatrix}$

9. Linear, nonhomogeneous, order 3; $D\mathbf{x} = \begin{bmatrix} 1 & 3 & 0 \\ 2 & 1 & 0 \\ 1 & 3t & 0 \end{bmatrix}\mathbf{x} + \begin{bmatrix} t^2 \\ t \\ -t^2 \end{bmatrix}$

In Exercises 10 to 17, $\mathbf{x} = \begin{bmatrix} x_1 \\ x_2 \end{bmatrix} = \begin{bmatrix} x \\ x' \end{bmatrix}$ or $\mathbf{x} = \begin{bmatrix} x_1 \\ x_2 \\ x_3 \end{bmatrix} = \begin{bmatrix} x \\ x' \\ x'' \end{bmatrix}$.

11a. $D\mathbf{x} = \begin{bmatrix} 0 & 1 \\ 1 & 0 \end{bmatrix}\mathbf{x} + \begin{bmatrix} 0 \\ t \end{bmatrix}$ b. $\mathbf{x} = c_1\begin{bmatrix} e^t \\ e^t \end{bmatrix} + c_2\begin{bmatrix} e^{-t} \\ -e^{-t} \end{bmatrix} + \begin{bmatrix} -t \\ -1 \end{bmatrix}$

13a. $D\mathbf{x} = \begin{bmatrix} 0 & 1 \\ -1 & 0 \end{bmatrix}\mathbf{x} + \begin{bmatrix} 0 \\ 1 \end{bmatrix}$ b. $\mathbf{x} = c_1\begin{bmatrix} \cos t \\ -\sin t \end{bmatrix} + c_2\begin{bmatrix} \sin t \\ \cos t \end{bmatrix} + \begin{bmatrix} 1 \\ 0 \end{bmatrix}$

15a. $D\mathbf{x} = \begin{bmatrix} 0 & 1 & 0 \\ 0 & 0 & 1 \\ -1 & 1 & 1 \end{bmatrix} \mathbf{x} + \begin{bmatrix} 0 \\ 0 \\ 4 \end{bmatrix}$

b. $\mathbf{x} = c_1 \begin{bmatrix} e^t \\ e^t \\ e^t \end{bmatrix} + c_2 \begin{bmatrix} te^t \\ (t+1)e^t \\ (t+2)e^t \end{bmatrix} + c_3 \begin{bmatrix} e^{-t} \\ -e^{-t} \\ e^{-t} \end{bmatrix} + \begin{bmatrix} 4 \\ 0 \\ 0 \end{bmatrix}$

17a. $D\mathbf{x} = \begin{bmatrix} 0 & 1 & 0 \\ 0 & 0 & 1 \\ 0 & 1 & 0 \end{bmatrix} \mathbf{x} + \begin{bmatrix} 0 \\ 0 \\ 1 \end{bmatrix}$

b. $\mathbf{x} = c_1 \begin{bmatrix} 1 \\ 0 \\ 0 \end{bmatrix} + c_2 \begin{bmatrix} e^t \\ e^t \\ e^t \end{bmatrix} + c_3 \begin{bmatrix} e^{-t} \\ -e^{-t} \\ e^{-t} \end{bmatrix} + \begin{bmatrix} -t \\ -1 \\ 0 \end{bmatrix}$

19a. $\begin{bmatrix} \cos t \\ -\sin t \end{bmatrix}, \begin{bmatrix} 0 \\ -1 \end{bmatrix}$ b. $\begin{bmatrix} \cos t + t \\ -\sin t - 1 \end{bmatrix}, \begin{bmatrix} 0 \\ -1 \end{bmatrix}$ c. No, yes

21a. $\begin{bmatrix} e^t \\ 0 \\ 2e^{2t} \end{bmatrix}, \begin{bmatrix} 2e^t \\ 0 \\ 4e^{2t} \end{bmatrix}$ b. $\begin{bmatrix} e^t \\ 0 \\ 2e^{2t} \end{bmatrix}, \begin{bmatrix} 2e^t \\ 0 \\ 4e^{2t} \end{bmatrix}$ c. Yes, yes

23a. $x_1' = \qquad\qquad v_1$ b. $x' = \qquad\qquad v$

$v_1' = \dfrac{-(k_1 + k_2)}{m_1} x_1 + \dfrac{k_2}{m_1} x_2$ $v' = -2kx - 2bv + 2ky - 2kL - 32$

$x_2' = \qquad\qquad v_2$ $y' = \qquad\qquad\qquad 2$

$v_2' = \dfrac{k_2}{m_2} x_1 - \dfrac{k_2}{m_2} x_2$

29. b. No   c. No

Section 3.3

1a. Both are solutions   b. Yes

3a. $\mathbf{h}_1$ is a solution, $\mathbf{h}_2$ is not   b. No

5a. $\mathbf{h}_1$ is a solution, $\mathbf{h}_2$ is not   b. No

7a. All are solutions   b. No   9a. All are solutions   b. Yes

11. Yes   13. No

15. Yes

Section 3.4

1. Independent   3. Dependent

5. Independent   7. Independent

9. Independent

Section 3.5

1a. $\lambda^2 - 3\lambda + 2$ b. $\lambda = 1, 2$

3a. $\lambda^2 - 2\lambda - 2$ b. $\lambda = 1 \pm \sqrt{3}$

5a. $(2 - \lambda)(\lambda - 3)(\lambda - 1)$ b. $\lambda = 1, 2, 3$

When comparing your answers for Exercises 7 to 13 with the answers below, keep in mind that any nonzero linear combination of eigenvectors corresponding to λ will also be an eigenvector corresponding to λ.

7. $\lambda = 2$, $\mathbf{v} = \begin{bmatrix} 1 \\ 1 \end{bmatrix}$; $\lambda = -2$, $\mathbf{w} = \begin{bmatrix} -1 \\ 3 \end{bmatrix}$

9. $\lambda = 1$, $\mathbf{v} = \begin{bmatrix} 1 \\ 0 \\ 0 \end{bmatrix}$; $\lambda = 2$, $\mathbf{w} = \begin{bmatrix} 1 \\ 1 \\ 0 \end{bmatrix}$; $\lambda = -3$, $\mathbf{u} = \begin{bmatrix} -3 \\ 2 \\ 10 \end{bmatrix}$

11. $\lambda = 1$, $\mathbf{v} = \begin{bmatrix} 1 \\ 0 \\ 0 \end{bmatrix}$; $\lambda = -1$, $\mathbf{w} = \begin{bmatrix} 1 \\ 2 \\ 0 \end{bmatrix}$

13. $\mathbf{x} = c_1 \begin{bmatrix} e^t \\ 0 \\ 0 \end{bmatrix} + c_2 \begin{bmatrix} e^{2t} \\ e^{2t} \\ 0 \end{bmatrix} + c_3 \begin{bmatrix} -3e^{-3t} \\ 2e^{-3t} \\ 10e^{-3t} \end{bmatrix}$

15a. $\mathbf{x} = c_1 \begin{bmatrix} e^{-2t} \\ e^{-2t} \end{bmatrix}$ b. $\mathbf{x}(0) = \begin{bmatrix} 1 \\ 0 \end{bmatrix}$

Section 3.6

1. $\begin{bmatrix} 1 & 0 & 0 & 1 \\ 0 & 1 & 0 & 0 \\ 0 & 0 & 1 & 0 \\ 0 & 0 & 0 & 0 \end{bmatrix}$ 3. $\begin{bmatrix} 1 & 0 & 2 & 0 \\ 0 & 1 & 0 & 0 \\ 0 & 0 & 0 & 1 \\ 0 & 0 & 0 & 0 \end{bmatrix}$

5a. $x = y = z = w = u = 0$ b. $\mathbf{x} = \mathbf{0}$

7a. $x_1 = -a + 2, x_2 = 1, x_3 = a$ b. $\mathbf{x} = a\begin{bmatrix} -1 \\ 0 \\ 1 \end{bmatrix} + \begin{bmatrix} 2 \\ 1 \\ 0 \end{bmatrix}$

9a. $x_1 = 1/2, x_2 = 0, x_3 = -1/2, x_4 = 0$ b. $\mathbf{x} = \begin{bmatrix} 1/2 \\ 0 \\ -1/2 \\ 0 \end{bmatrix}$

11. No solutions   13. $\mathbf{u} = \begin{bmatrix} (-7a + 5)/3 \\ -(a + 1)/3 \\ a \end{bmatrix}$

In Exercises 14 to 19, the eigenvectors are the nonzero vectors of the given forms.

15. $\lambda = 0, \mathbf{v} = a\begin{bmatrix} -1 \\ 1 \end{bmatrix}; \lambda = 2, \mathbf{w} = b\begin{bmatrix} 1 \\ 1 \end{bmatrix}$

17. $\lambda = 2, \mathbf{v} = a\begin{bmatrix} -1 \\ 1 \end{bmatrix}$

19. $\lambda = 1, \mathbf{v} = a\begin{bmatrix} 2 \\ 0 \\ 1 \end{bmatrix} + b\begin{bmatrix} -2 \\ 1 \\ 0 \end{bmatrix}; \lambda = -2, \mathbf{u} = c\begin{bmatrix} -1 \\ 1 \\ 2 \end{bmatrix}$

21a. $x = c_1e^t + c_2e^{-t} + c_3\cos t + c_4\sin t; x = \frac{3}{2}e^t - \frac{1}{2}\cos t - \frac{1}{2}\sin t$

  b. $x = c_1e^t + c_2te^t + c_3e^{-t} + c_4te^{-t}; x = \frac{5}{4}e^t - \frac{1}{2}te^t - \frac{1}{4}e^{-t}$

25. $x + y + z = 1$
     $x + y + z = 2$

Section 3.7

1. $\mathbf{x} = c_1\begin{bmatrix} 2e^t \\ e^t \end{bmatrix} + c_2\begin{bmatrix} e^{2t} \\ e^{2t} \end{bmatrix}$

3. $\mathbf{x} = c_1\begin{bmatrix} e^{(1+\sqrt{3})t} \\ \sqrt{3}e^{(1+\sqrt{3})t} \end{bmatrix} + c_2\begin{bmatrix} e^{(1-\sqrt{3})t} \\ -\sqrt{3}e^{(1-\sqrt{3})t} \end{bmatrix}$

5. $\mathbf{x} = c_1\begin{bmatrix} 2e^t \\ 0 \\ e^t \end{bmatrix} + c_2\begin{bmatrix} -2e^t \\ e^t \\ 0 \end{bmatrix} + c_3\begin{bmatrix} -e^{-2t} \\ e^{-2t} \\ 2e^{-2t} \end{bmatrix}$

7a. $x_1 = c_1e^{3t} + c_2e^{-t}$ b. $x_1 = 2e^{3t} - e^{-t}$
     $x_2 = c_1e^{3t} - c_2e^{-t}$ $x_2 = 2e^{3t} + e^{-t}$

9a. $\mathbf{x} = c_1\begin{bmatrix} e^{2t} \\ e^{2t} \\ 0 \end{bmatrix} + c_2\begin{bmatrix} -1 \\ 1 \\ 0 \end{bmatrix} + c_3\begin{bmatrix} 0 \\ 0 \\ e^{-t} \end{bmatrix}$ b. $\mathbf{x} = \begin{bmatrix} 3e^{2t} - 1 \\ 3e^{2t} + 1 \\ 2e^{-t} \end{bmatrix}$

11a. $x_1 = c_1 e^{2t} + c_2 e^t$
$x_2 = 3c_1 e^{2t} + 2c_2 e^t$
$x_3 = c_1 e^{2t} + c_2 e^t + c_3 e^{-t}$

b. $x_1 = e^t$
$x_2 = 2e^t$
$x_3 = e^t + 2e^{-t}$

13a. $x_1 = 4c_2 e^{2t} + 6c_3 e^t$
$x_2 = c_1 e^{2t}$
$x_3 = 3c_2 e^{2t} + 5c_3 e^t + c_4 e^{-2t}$
$x_4 = 4c_2 e^{2t} + 5c_3 e^t$

b. $x_1 = 25e^{2t} - 24e^t$
$x_2 = e^{2t}$
$x_3 = \frac{75}{4} e^{2t} - 20e^t + \frac{9}{4} e^{-2t}$
$x_4 = 25e^{2t} - 20e^t$

15a. $Q = -c_1 e^{-t} - 2c_2 e^{-5t/2}$
$I_2 = 2c_1 e^{-t} + c_2 e^{-5t/2}$

b. $Q = -e^{-t} + 2e^{-5t/2}$
$I_2 = 2e^{-t} - e^{-5t/2}$

17a. $x_1 = c_1 - c_2 e^{-t/25}$
$x_2 = c_1 + c_2 e^{-t/25}$

b. $x_1 = c_1 - 2c_2 e^{-3t/100}$
$x_2 = c_1 + c_2 e^{-3t/100}$

Section 3.8

1. $\mathbf{x} = c_1 \begin{bmatrix} \sin t \\ \cos t \end{bmatrix} + c_2 \begin{bmatrix} -\cos t \\ \sin t \end{bmatrix}$

3. $\mathbf{x} = c_1 \begin{bmatrix} 0 \\ e^{2t} \\ 0 \end{bmatrix} + c_2 \begin{bmatrix} -e^t \sin t \\ 0 \\ e^t \cos t \end{bmatrix} + c_3 \begin{bmatrix} e^t \cos t \\ 0 \\ e^t \sin t \end{bmatrix}$

5. $\mathbf{x} = c_1 \begin{bmatrix} 2e^{-t} \\ 0 \\ e^{-t} \end{bmatrix} + c_2 \begin{bmatrix} e^{-t/2}\left(-\cos \dfrac{t\sqrt{3}}{2} + \sqrt{3} \sin \dfrac{t\sqrt{3}}{2} \right) \\ 2e^{-t/2} \cos \dfrac{t\sqrt{3}}{3} \\ 0 \end{bmatrix}$

$+ c_3 \begin{bmatrix} e^{-t/2}\left(-\sqrt{3} \cos \dfrac{t\sqrt{3}}{2} - \sin \dfrac{t\sqrt{3}}{2} \right) \\ 2e^{-t/2} \sin \dfrac{t\sqrt{3}}{2} \\ 0 \end{bmatrix}$

7. $\mathbf{x} = c_1 \begin{bmatrix} -e^{-t} \sin 2t \\ e^{-t} \sin 2t \\ 0 \\ 2e^{-t} \cos 2t \end{bmatrix} + c_2 \begin{bmatrix} e^{-t} \cos 2t \\ -e^{-t} \cos 2t \\ 0 \\ 2e^{-t} \sin 2t \end{bmatrix} + c_3 \begin{bmatrix} e^{-t}(-\cos 2t + 2 \sin 2t) \\ 0 \\ 5e^{-t} \cos 2t \\ 0 \end{bmatrix}$

$+ c_4 \begin{bmatrix} e^{-t}(-2 \cos 2t - \sin 2t) \\ 0 \\ 5e^{-t} \sin 2t \\ 0 \end{bmatrix}$

9. $\mathbf{x} = c_1 \begin{bmatrix} -e^t \sin t \\ e^t \cos t \end{bmatrix} + c_2 \begin{bmatrix} e^t \cos t \\ e^t \sin t \end{bmatrix}$ b. $\mathbf{x} = \begin{bmatrix} -3e^t \sin t + 2e^t \cos t \\ 3e^t \cos t + 2e^t \sin t \end{bmatrix}$

11a. $x_1 = c_1 e^t(-\cos t\sqrt{2} - \sqrt{2} \sin t\sqrt{2}) + c_2 e^t(\sqrt{2} \cos t\sqrt{2} - \sin t\sqrt{2})$
$x_2 = 3c_1 e^t \cos t\sqrt{2}$ $+ 3c_2 e^t \sin t\sqrt{2}$

b. $x_1 = e^t(2 \cos t\sqrt{2} - \dfrac{3\sqrt{2}}{2} \sin t\sqrt{2})$, $x_2 = e^t(\cos t\sqrt{2} + \dfrac{7\sqrt{2}}{2} \sin t\sqrt{2})$

13.
$Q = c_1 e^{-t/2}\left(-\cos \dfrac{t\sqrt{3}}{2} + \sqrt{3} \sin \dfrac{t\sqrt{3}}{2}\right) + c_2 e^{-t/2}\left(-\sqrt{3} \cos \dfrac{t\sqrt{3}}{2} - \sin \dfrac{t\sqrt{3}}{2}\right)$

$I = 2c_1 e^{-t/2} \cos \dfrac{t\sqrt{3}}{2}$ $+ 2c_2 e^{-t/2} \sin \dfrac{t\sqrt{3}}{2}$

15. $x_1 = c_1 \sin t - c_2 \cos t - 2\sqrt{6}\, c_3 \sin t\sqrt{6} + 2\sqrt{6}\, c_4 \cos t\sqrt{6}$
$v_1 = c_1 \cos t + c_2 \sin t - 12\, c_3 \cos t\sqrt{6} - 12\, c_4 \sin t\sqrt{6}$
$x_2 = 2c_1 \sin t - 2c_2 \cos t + \sqrt{6}\, c_3 \sin t\sqrt{6} - \sqrt{6}\, c_4 \cos t\sqrt{6}$
$v_2 = 2c_1 \cos t + 2c_2 \sin t + 6\, c_3 \cos t\sqrt{6} + 6\, c_4 \sin t\sqrt{6}$

Section 3.9

1a. $\begin{bmatrix} 3 & 2 & 1 \\ 2 & 1 & 1 \\ -1 & 1 & 0 \end{bmatrix}$ b. $\begin{bmatrix} 6 \\ 3 \\ -3 \end{bmatrix}$ c. $\begin{bmatrix} 6 \\ 3 \\ -3 \end{bmatrix}$

d. $\begin{bmatrix} 4 & -1 & 1 \\ 2 & 0 & 1 \\ 2 & -1 & 0 \end{bmatrix}$ e. $\begin{bmatrix} 8 & -3 & -1 \\ 2 & -1 & 0 \\ -3 & 1 & 0 \end{bmatrix}$ f. $\begin{bmatrix} 22 & -8 & -3 \\ 5 & -2 & -1 \\ -8 & 3 & 1 \end{bmatrix}$

g. $\begin{bmatrix} 12 & 7 & 5 \\ 7 & 4 & 3 \\ -5 & -3 & -2 \end{bmatrix}$ h. $\begin{bmatrix} 9 & 12 & -3 \\ 3 & 3 & 0 \\ -4 & -5 & 1 \end{bmatrix}$ i. $\begin{bmatrix} 16 & 0 & 0 \\ 5 & 1 & 0 \\ 0 & 0 & 1 \end{bmatrix}$

3. $\mathbf{x} = c_1 \begin{bmatrix} (1 - t)e^{2t} \\ te^{2t} \end{bmatrix}$ 5. $\mathbf{x} = c_1 \begin{bmatrix} e^{-t} \\ 2e^{-t} \\ e^{-t} \end{bmatrix} + c_2 \begin{bmatrix} e^t \\ 0 \\ 0 \end{bmatrix} + c_3 \begin{bmatrix} -te^t \\ -e^t \\ 2e^t \end{bmatrix}$

7. $\mathbf{x} = c_1 \begin{bmatrix} -e^t \sin t \\ 2e^t \sin t \\ e^t (\sin t + \cos t) \\ 2e^t \cos t \end{bmatrix} + c_2 \begin{bmatrix} e^t \cos t \\ -2e^t \cos t \\ -e^t (\sin t + \cos t) \\ 2e^t \sin t \end{bmatrix} + c_3 \begin{bmatrix} e^{2t} \\ 0 \\ 0 \\ 0 \end{bmatrix} + c_4 \begin{bmatrix} te^{2t} \\ 0 \\ e^{2t} \\ 0 \end{bmatrix}$

9a. $\mathbf{x} = c_1 \begin{bmatrix} (1 - 2t)e^{2t} \\ -4te^{2t} \end{bmatrix} + c_2 \begin{bmatrix} te^{2t} \\ (1 + 2t)e^{2t} \end{bmatrix}$ b. $\mathbf{x} = \begin{bmatrix} (3 - 2t)e^{2t} \\ (4 - 4t)e^{2t} \end{bmatrix}$

11a. $x_1 = c_1(1 - t)e^{3t} - c_2 te^{3t}$, $x_2 = c_1 te^{3t} + c_2(1 + t)e^{3t}$
b. $x_1 = (3 - 7t)e^{3t}$, $x_2 = (4 - 7t)e^{3t}$

13. $Q = c_1(1 + t)e^{-t} + c_2te^{-t}, \ I = -c_1te^{-t} + c_2(1 - t)e^{-t}$

17b. Yes. For example, $\begin{bmatrix} 1 & 1 \\ -1 & -1 \end{bmatrix}$.

Section 3.10

1. $\mathbf{x} = c_1 \begin{bmatrix} (1 - t + \frac{1}{2}t^2)e^t \\ \frac{1}{2}t^2e^t \\ (t + \frac{1}{2}t^2)e^t \end{bmatrix} + c_2 \begin{bmatrix} (t - t^2)e^t \\ (1 - t - t^2)e^t \\ (-3t - t^2)e^t \end{bmatrix} + c_3 \begin{bmatrix} \frac{1}{2}t^2e^t \\ (t + \frac{1}{2}t^2)e^t \\ (1 + 2t + \frac{1}{2}t^2)e^t \end{bmatrix}$

3. $\mathbf{x} = c_1 \begin{bmatrix} 0 \\ e^{-t} \\ e^{-t} \\ e^{-t} \end{bmatrix} + c_2 \begin{bmatrix} (1 + t)e^t \\ 0 \\ 0 \\ -te^t \end{bmatrix} + c_3 \begin{bmatrix} -te^t \\ -e^t \\ e^t \\ te^t \end{bmatrix} + c_4 \begin{bmatrix} te^t \\ 0 \\ 0 \\ (1 - t)e^t \end{bmatrix}$

5. $\mathbf{x} = c_1 \begin{bmatrix} -2t\cos t + \sin 2t \\ 2t\sin 2t \\ 4\sin 2t \\ 4\cos 2t \end{bmatrix} + c_2 \begin{bmatrix} -\cos 2t - 2t\sin 2t \\ -2t\cos 2t \\ -4\cos 2t \\ 4\sin 2t \end{bmatrix}$

$+ c_3 \begin{bmatrix} \sin 2t \\ \cos 2t \\ 0 \\ 0 \end{bmatrix} + c_4 \begin{bmatrix} -\cos 2t \\ \sin 2t \\ 0 \\ 0 \end{bmatrix}$

7. $\mathbf{x} = c_1 \begin{bmatrix} e^{2t} \\ 0 \\ te^{2t} \\ te^{2t} \end{bmatrix} + c_2 \begin{bmatrix} 0 \\ e^{2t} \\ 0 \\ 0 \end{bmatrix} + c_3 \begin{bmatrix} 0 \\ 0 \\ (1 - t)e^{2t} \\ -te^{2t} \end{bmatrix} + c_4 \begin{bmatrix} 0 \\ 0 \\ te^{2t} \\ (1 + t)e^{2t} \end{bmatrix}$

9. $\mathbf{x} = c_1 \begin{bmatrix} 0 \\ e^{-t} \\ 0 \\ 0 \\ 0 \end{bmatrix} + c_2 \begin{bmatrix} 0 \\ 0 \\ e^{-t} \\ 0 \\ 0 \end{bmatrix} + c_3 \begin{bmatrix} (1 + \frac{1}{2}t^2) \\ t \\ 1 \\ \frac{1}{2}t^2 \\ t \end{bmatrix} + c_4 \begin{bmatrix} 1 \\ 0 \\ 0 \\ 1 \\ 0 \end{bmatrix} + c_5 \begin{bmatrix} t \\ 1 \\ 0 \\ t \\ 1 \end{bmatrix}$

11a. $x_1 = c_1e^{-t} + c_2te^{-t}, \ x_2 = c_2e^{-t}, \ x_3 = c_3e^{-t} + c_4te^{-t}, \ x_4 = c_4e^{-t}$

 b. $x_1 = (1 + t)e^{-t}, \ x_2 = .e^{-t}, \ x_3 = -e^{-t}, \ x_4 = 0$

Section 3.11

1. $\mathbf{x} = c_1 \begin{bmatrix} 2e^t \\ e^t \end{bmatrix} + c_2 \begin{bmatrix} e^{2t} \\ e^{2t} \end{bmatrix} + \begin{bmatrix} -e^t \\ -e^t \end{bmatrix}$

3. $\mathbf{x} = c_1 \begin{bmatrix} e^{(1+\sqrt{3})t} \\ \sqrt{3}\,e^{(1+\sqrt{3})t} \end{bmatrix} + c_2 \begin{bmatrix} e^{(1-\sqrt{3})t} \\ -\sqrt{3}\,e^{(1-\sqrt{3})t} \end{bmatrix} + \begin{bmatrix} 0 \\ -1 \end{bmatrix}$

5. $\mathbf{x} = c_1 \begin{bmatrix} e^t \\ e^t \\ e^t \end{bmatrix} + c_2 \begin{bmatrix} 0 \\ 2e^{2t} \\ e^{2t} \end{bmatrix} + c_3 \begin{bmatrix} -e^{3t} \\ 5e^{3t} \\ 3e^{3t} \end{bmatrix} + \begin{bmatrix} te^t \\ te^t \\ te^t \end{bmatrix}$

7. $\mathbf{x} = c_1 \begin{bmatrix} 0 \\ e^{2t} \\ e^{2t} \end{bmatrix} + c_2 \begin{bmatrix} 2e^{2t}\cos t \\ e^{2t}(\cos t + \sin t) \\ 2e^{2t}\cos t \end{bmatrix} + c_3 \begin{bmatrix} 2e^{2t}\sin t \\ e^{2t}(\sin t - \cos t) \\ 2e^{2t}\sin t \end{bmatrix} + \begin{bmatrix} -2e^{2t} \\ -e^{2t} \\ -2e^{2t} \end{bmatrix}$

9. $\mathbf{x} = c_1 \begin{bmatrix} (1-t)e^{-2t} \\ -te^{-2t} \end{bmatrix} + c_2 \begin{bmatrix} te^{-2t} \\ (t+1)e^{-2t} \end{bmatrix} + \begin{bmatrix} (t^2 - \frac{1}{6}t^3)e^{-2t} \\ (\frac{1}{2}t^2 - \frac{1}{6}t^3)e^{-2t} \end{bmatrix}$

11. $\mathbf{x} = c_1 \begin{bmatrix} e^t \\ te^t \\ te^t \end{bmatrix} + c_2 \begin{bmatrix} 0 \\ (1-t)e^t \\ -te^t \end{bmatrix} + c_3 \begin{bmatrix} 0 \\ te^t \\ (1+t)e^t \end{bmatrix} + \begin{bmatrix} 0 \\ 1 \\ 0 \end{bmatrix}$

13a. $x_1 = c_1 + c_2 e^{2t} - \frac{1}{3}e^{3t}$ b. $x_1 = e^{2t} - \frac{1}{3}e^{3t}$
      $x_2 = c_1 + 2c_2 e^{2t} - \frac{4}{3}e^{3t}$ $x_2 = 2e^{2t} - \frac{4}{3}e^{3t}$

15a. $x_1 = c_1 e^{2t} - c_2 \sin 3t + c_3 \cos 3t + te^{2t} - 1$
      $x_2 = \qquad\quad c_2 \cos 3t + c_3 \sin 3t$
      $x_3 = c_1 e^{2t} \qquad\qquad\qquad\quad + te^{2t}$
  b. $x_1 = 3e^{2t} - 2\sin 3t - \cos 3t + te^{2t} - 1$
      $x_2 = \qquad\quad 2\cos 3t - \sin 3t$
      $x_3 = 3e^{2t} \qquad\qquad\qquad\quad + te^{2t}$

17. $Q = -c_1 e^{-t}\sin t + c_2 e^{-t}\cos t$
     $I_2 = \quad c_1 e^{-t}\cos t + c_2 e^{-t}\sin t + e^{-t}$

19a. $A = 500,\ B = -2000e^{-t/10} + 2500,\ C = 2000e^{-t/10} - 1500 + 150t$
  b. Yes

21a. $p = -c_1 e^{-t/10} - \quad c_2 e^{-t/2}$
      $s = 500c_1 e^{-t/10} + 100c_2 e^{-t/2} + 5000$
  b. $p = 5e^{-5.2} - e^{-26} \approx .03$
      $s = -2500e^{-5.2} + 100e^{-26} + 5000 \approx 4986$
      $w = -3000e^{-5.2} + 600e^{-26} + 5000 \approx 4983$

Review Problems

1a. $x_1 = c_1 e^{3t}(\cos 2t - \sin 2t) + c_2 e^{3t}(\cos 2t + \sin 2t)$
     $x_2 = 2c_1 e^{3t}\cos 2t \qquad\quad + 2c_2 e^{3t}\sin 2t \qquad\quad + 2e^{3t}$
  b. $x_1 = -e^{3t}(\cos 2t + 5\sin 2t),\ x_2 = e^{3t}(4\cos 2t - 6\sin 2t + 2)$

3a. $x_1 = c_1(1 + 4t)e^{-t} - 4c_2te^{-t} + 1 + (t + 2t^2)e^{-t}$,
 $x_2 = 4c_1te^{-t} + c_2(1 - 4t)e^{-t} + 1 + 2t^2e^{-t}$
 b. $x_1 = 1 + (1 + 5t + 2t^2)e^{-t}$, $x_2 = 1 + (4t + 2t^2)e^{-t}$

5a. $x_1 = 6c_1e^{4t}$
 $x_2 = 4c_1e^{4t} + c_2e^{2t} + c_3e^{-2t} + e^{3t}$
 $x_3 = c_1e^{4t} \phantom{+ c_2e^{2t}} + c_3e^{-2t} + e^{3t}$
 b. $x_1 = 6e^{4t}$
 $x_2 = 4e^{4t} - 4e^{2t} - e^{-2t} + e^{3t}$
 $x_3 = e^{4t} \phantom{- 4e^{2t}} - e^{-2t} + e^{3t}$

7a. $x_1 = -2c_1e^{2t} \phantom{+ c_2e^{2t} + c_3e^{4t}} + 2te^{2t}$ b. $x_1 = (2t - 2)e^{2t}$
 $x_2 = \phantom{-2c_1e^{2t}} c_2e^{2t} + c_3e^{4t} + e^{4t}$ $x_2 = 8e^{2t} - e^{4t}$
 $x_3 = c_1e^{2t} \phantom{+ c_2e^{2t}} + c_3e^{4t} - te^{2t}$ $x_3 = (-t + 1)e^{2t} - 2e^{4t}$

9a. $x_1 = c_1e^{2t} \phantom{+ c_2e^{2t}} + c_3te^{2t} - \frac{1}{4}te^{-2t}$ b. $x_1 = \frac{5}{4}te^{2t} - \frac{1}{4}te^{-2t}$
 $x_2 = \phantom{c_1e^{2t}} c_2e^{2t} + c_3te^{2t} + \frac{1}{16}e^{-2t}$ $x_2 = \frac{5}{16}e^{2t} + \frac{5}{4}te^{2t} + \frac{1}{16}e^{-2t}$
 $x_3 = \phantom{c_1e^{2t} + c_2e^{2t} + } c_3e^{2t} - \frac{1}{4}e^{-2t}$ $x_3 = \frac{5}{4}e^{2t} - \frac{1}{4}e^{-2t}$

11a. $x_1 = c_1e^t - c_2te^t + 4e^{2t}$, $x_2 = -c_3e^{-t} + c_4e^{-t} + \frac{2}{3}e^{2t}$,
 $x_3 = c_2e^t + c_4e^{-t} + \frac{4}{3}e^{2t}$, $x_4 = c_3e^{-t} + \frac{2}{3}e^{2t}$
 b. $x_1 = 4e^{2t}$, $x_2 = \frac{13}{3}e^{-t} + \frac{2}{3}e^{2t}$, $x_3 = \frac{14}{3}e^{-t} + \frac{4}{3}e^{2t}$, $x_4 = \frac{1}{3}e^{-t} + \frac{2}{3}e^{2t}$

13a. $I_1 = c_1 - c_2e^{-4t} + (2t - \frac{1}{2})e^{-4t}$
 $I_2 = c_1 + c_2e^{-4t} - (2t + \frac{1}{2})e^{-4t}$
 b. $Q = c_1(1 + \frac{1}{2}t)e^{-t/2} + c_2te^{-t/2} + \frac{1}{2}t^2e^{-t/2}$,
 $I_2 = -\frac{1}{4}c_1te^{-t/2} + c_2(1 - \frac{1}{2}t)e^{-t/2} + (t - \frac{1}{4}t^2)e^{-t/2}$
 c. $Q = c_1e^{-t/2}\left(\sin\frac{t}{2} - \cos\frac{t}{2}\right) - c_2e^{-t/2}\left(\sin\frac{t}{2} + \cos\frac{t}{2}\right) + 4e^{-t/2}$
 $I_2 = c_1e^{-t/2}\cos\frac{t}{2} \phantom{+ c_2e^{-t/2}\sin\frac{t}{2}} + c_2e^{-t/2}\sin\frac{t}{2} - 2e^{-t/2}$
 d. $Q = c_1e^{-t/2}\left(\sin\frac{t}{2} + \cos\frac{t}{2}\right) + c_2e^{-t/2}\left(\sin\frac{t}{2} - \cos\frac{t}{2}\right) + 8e^{-t/2}$
 $I_3 = c_1e^{-t/2}\cos\frac{t}{2} \phantom{+ c_2e^{-t/2}\sin\frac{t}{2}} + c_2e^{-t/2}\sin\frac{t}{2} + 4e^{-t/2}$
 e. $Q_1 = 3c_1e^{-t} - c_2e^{-t/6} + 15$
 $Q_2 = 2c_1e^{-t} + c_2e^{-t/6}$

CHAPTER FOUR

Section 4.1

1a. Yes   b. No   c. No   d. Yes   e. No

Section 4.2

1a. Yes   b. Yes   c. No   d. No   e. No   f. Yes

3a. Yes   b. No   c. Yes   d. No   e. Yes   f. Yes

7a. Yes   b. No   c. No   d. Yes   e. Yes   f. No

9a. Yes   b. No   c. Yes

15a. $D(D - 1)(D - 2)$ b. $D(D^2 + 1)$ c. $(D - 1)(D + 1)$
   d. $(D - 1)(D + 1)$ e. D^3

Section 4.3

1a. $\begin{bmatrix} 1 \\ 1 \\ 0 \end{bmatrix}$ b. $\begin{bmatrix} 1 \\ 0 \\ 1 \end{bmatrix}, \begin{bmatrix} 0 \\ 1 \\ 1 \end{bmatrix}$ c. $\begin{bmatrix} 1 \\ 0 \\ 1 \end{bmatrix}, \begin{bmatrix} 1 \\ 1 \\ 0 \end{bmatrix}, \begin{bmatrix} 0 \\ 1 \\ 1 \end{bmatrix}$ d. $\begin{bmatrix} 1 \\ 1 \\ 2 \end{bmatrix}, \begin{bmatrix} 1 \\ -1 \\ 0 \end{bmatrix}$

e. $\begin{bmatrix} 1 \\ 1 \\ 0 \end{bmatrix}, \begin{bmatrix} 1 \\ 0 \\ 1 \end{bmatrix}, \begin{bmatrix} 0 \\ -1 \\ 1 \end{bmatrix}$ f. $\begin{bmatrix} 1 \\ -1 \\ 1 \end{bmatrix}, \begin{bmatrix} -1 \\ 1 \\ -1 \end{bmatrix}$

3a. $1, t, t^2, t^3, t^4$ b. $e^{2t}, te^{2t}, e^{-2t}, te^{-2t}$ c. $e^t, e^{-t}, \sin t, \cos t$

5a. No   b. Yes   c. No   d. Yes   e. No

7a. Yes   b. No   c. No   d. Yes

9a. Yes   b. Yes   c. No   d. Yes

Section 4.4

1a. (i) Yes   (ii) No   b. (i) Yes   (ii) Yes
 c. (i) No   (ii) No   d. (i) No   (ii) No
 e. (i) No   (ii) No

3a. (i) Yes   (ii) Yes   b. (i) Yes   (ii) No
 c. (i) No   (ii) No   d. (i) No   (ii) No

5a. (i) Yes   (ii) Yes   b. (i) Yes   (ii) Yes
 c. (i) No   (ii) No   d. (i) No   (ii) No

7a. $\begin{bmatrix} 1 \\ 1 \\ 0 \end{bmatrix}$ b. $\begin{bmatrix} 1 \\ 0 \\ 1 \end{bmatrix}, \begin{bmatrix} 0 \\ 1 \\ 1 \end{bmatrix}$ c. $\begin{bmatrix} 1 \\ 0 \\ 1 \end{bmatrix}, \begin{bmatrix} 1 \\ 1 \\ 0 \end{bmatrix}, \begin{bmatrix} 0 \\ 1 \\ 1 \end{bmatrix}$

d. $\begin{bmatrix} 1 \\ 1 \\ 2 \end{bmatrix}, \begin{bmatrix} 1 \\ -1 \\ 0 \end{bmatrix}$ e. $\begin{bmatrix} 1 \\ 1 \\ 0 \end{bmatrix}, \begin{bmatrix} 1 \\ 0 \\ 1 \end{bmatrix}$ f. $\begin{bmatrix} 1 \\ -1 \\ 1 \end{bmatrix}$

9a. $1, t, t^2, t^3, t^4$ b. $e^{2t}, te^{2t}, e^{-2t}, te^{-2t}$ c. $e^t, e^{-t}, \sin t, \cos t$

11. $\begin{bmatrix} 1 \\ 0 \\ 0 \\ \cdot \\ \cdot \\ \cdot \\ 0 \end{bmatrix}, \begin{bmatrix} 0 \\ 1 \\ 0 \\ \cdot \\ \cdot \\ \cdot \\ 0 \end{bmatrix}, \begin{bmatrix} 0 \\ 0 \\ 1 \\ \cdot \\ \cdot \\ \cdot \\ 0 \end{bmatrix}, \ldots, \begin{bmatrix} 0 \\ 0 \\ 0 \\ \cdot \\ \cdot \\ \cdot \\ 1 \end{bmatrix}$

13b and c. $M_{11}, M_{12}, \ldots, M_{1n}, M_{21}, \ldots, M_{mn}$, where M_{ij} is the $m \times n$ matrix whose only nonzero entry is a 1 in the ith row and the jth column.

15a. $1, \frac{3}{2}, 1$ b. $-2, 3, 1$ c. $-1, 3, 1$ d. $-2, 4, -1$

17a. $\frac{1}{2}, \frac{-1}{2}$ b. $\frac{1}{2}, \frac{1}{2}$ c. $\frac{1}{2}i, \frac{1}{2}i$ d. $\frac{1}{2} - \frac{1}{2}i, \frac{1}{2} + \frac{1}{2}i$

Section 4.5

1a. 1   b. 2   c. 3   d. 2   e. 2   f. 1

3a. 2   b. 1   c. 2   d. 1

5a. 2   b. 2   c. 2

7. $m + 1$ 9. mn (or $4mn$, if considered as a real vector space)

Section 4.6

1a. Yes   b. $\{0\}$ c. V

3a. Yes   b. $\{0\}$ c. R^3

5a. No

7a. Yes   b. The 3-vectors of the form $\begin{bmatrix} 0 \\ 0 \\ z \end{bmatrix}$ c. R^2

9a. No

11a. Yes   b. $\{0\}$ c. The 4-vectors of the form $\begin{bmatrix} x \\ y \\ z \\ x + y \end{bmatrix}$

13a. Yes   b. The functions of the form ce^t c. C_∞

15a. Yes   b. The functions of the form $c_1 + c_2 t + c_3 e^t$ c. C_∞

17a. Yes   b. The functions of the form $c_1 + c_2 t$ c. P

19a. Yes   b. The functions of the form $a_1 t$

  c. The 3-vectors of the form $\begin{bmatrix} x \\ x \\ y \end{bmatrix}$

21a. Yes   b. $\{0\}$ c. C_∞

23a. Yes   b. The matrices of the form $\begin{bmatrix} a & a \\ c & c \end{bmatrix}$ c. R^2

25a. No

Section 4.7

1a. Yes   b. Yes   c. Yes

3a. Yes   b. Yes   c. Yes

5a. No   b. Yes   c. No

7a. Yes   b. No   c. No

9a. Yes   b. Yes   c. Yes

11a. No   b. Yes   c. No

13a. Yes   b. Yes   c. Yes

15a. No   b. Yes   c. No

17a. Yes   b. Yes   c. Yes

19a. Yes   b. No   c. No

Section 4.8

1a. (i) 6   (ii) no   b. (i) 0 (ii) yes   c. (i) 0 (ii) yes
  d. (i) -2 (ii) no   e. (i) 0 (ii) yes   f. (i) 2 (ii) no

3a. (i) $\begin{bmatrix} \frac{5}{9} \\ \frac{10}{9} \\ -\frac{10}{9} \end{bmatrix}$ (ii) $\begin{bmatrix} \frac{4}{9} \\ \frac{17}{9} \\ \frac{19}{9} \end{bmatrix}$ b. (i) $\begin{bmatrix} \frac{5}{11} \\ \frac{15}{11} \\ \frac{5}{11} \end{bmatrix}$ (ii) $\begin{bmatrix} \frac{6}{11} \\ \frac{7}{11} \\ -\frac{27}{11} \end{bmatrix}$

  c. (i) $\begin{bmatrix} 0 \\ \frac{44}{25} \\ -\frac{33}{25} \end{bmatrix}$ (ii) $\begin{bmatrix} 1 \\ \frac{6}{25} \\ \frac{8}{25} \end{bmatrix}$ d. (i) $\begin{bmatrix} \frac{3}{4} \\ \frac{3}{4} \\ \frac{3}{4} \\ -\frac{3}{4} \end{bmatrix}$ (ii) $\begin{bmatrix} \frac{5}{4} \\ \frac{5}{4} \\ -\frac{3}{4} \\ \frac{7}{4} \end{bmatrix}$

5a. (i) $\sqrt{2}$ (ii) $\dfrac{1}{\sqrt{2}}$ b. (i) $\sqrt{\frac{2}{3}}$ (ii) $\sqrt{\frac{3}{8}}\,(t-1)$

  c. (i) $\sqrt{\frac{2}{7}}$ (ii) $\sqrt{\frac{7}{2}}t^3$ d. (i) 1 (ii) $\cos \pi t$

7a. $\begin{bmatrix} 0 \\ 3 \\ 0 \end{bmatrix}, \begin{bmatrix} 1 \\ 0 \\ -2 \end{bmatrix}$ b. $\begin{bmatrix} 0 \\ 1 \\ 0 \end{bmatrix}, \begin{bmatrix} 1/\sqrt{5} \\ 0 \\ -2/\sqrt{5} \end{bmatrix}$ c. $\mathbf{w} = 3\sqrt{5} \begin{bmatrix} 1/\sqrt{5} \\ 0 \\ -2/\sqrt{5} \end{bmatrix}$

9a. $\begin{bmatrix} 1 \\ 1 \\ 0 \end{bmatrix}, \begin{bmatrix} 1 \\ 2 \\ 0 \end{bmatrix}, \begin{bmatrix} 0 \\ 0 \\ 1 \end{bmatrix}$ b. $\begin{bmatrix} 1/\sqrt{2} \\ 1/\sqrt{2} \\ 0 \end{bmatrix}, \begin{bmatrix} -1/\sqrt{2} \\ 1/\sqrt{2} \\ 0 \end{bmatrix}, \begin{bmatrix} 0 \\ 0 \\ 1 \end{bmatrix}$

c. $\mathbf{w} = \dfrac{1}{\sqrt{2}} \begin{bmatrix} 1/\sqrt{2} \\ 1/\sqrt{2} \\ 0 \end{bmatrix} - \dfrac{1}{\sqrt{2}} \begin{bmatrix} -1/\sqrt{2} \\ 1/\sqrt{2} \\ 0 \end{bmatrix} + \begin{bmatrix} 0 \\ 0 \\ 1 \end{bmatrix}$

11a. $1, t, t^2 - \frac{1}{3}$ b. $\dfrac{1}{\sqrt{2}}, \sqrt{\dfrac{3}{2}}\, t, \sqrt{\dfrac{45}{8}}\, (t^2 - \frac{1}{3})$

c. $\mathbf{w} = \dfrac{2\sqrt{2}}{3} \dfrac{1}{\sqrt{2}} - \sqrt{\dfrac{8}{45}} \sqrt{\dfrac{45}{8}}\, (t^2 - \frac{1}{3})$

13a. $1, \cos \pi t, \sin \pi t$ b. $\dfrac{1}{\sqrt{2}}, \cos \pi t, \sin \pi t$ c. $\mathbf{w} = \sin \pi t$

17a. $\langle \mathbf{v}, \mathbf{w} \rangle = 3, \|\mathbf{v}\| = \sqrt{3}$ b. $\langle \mathbf{v}, \mathbf{w} \rangle = 0, \|\mathbf{v}\| = \sqrt{2}$

Review Problems

3b. **Ker**(L) is the solution space of $Lx = 0$. Its dimension is n.
   c. **Range**(L) = $C[a, b]$.

5a. Yes   b. No   c. Yes   d. No

7a. Yes   b. Yes   c. No   d. Yes

9a. No   b. No   c. Yes   d. No

11a. (i) Yes   (ii) Yes   (iii) Yes   b. (i) No   (ii) No   (iii) No
   c. (i) No   (ii) Yes   (iii) No   d. (i) Yes   (ii) No   (iii) No

13a. $\begin{bmatrix} 2 \\ -1 \end{bmatrix}$; one   b. $\begin{bmatrix} 1 \\ -2 \\ 1 \end{bmatrix}$; one   c. $\begin{bmatrix} -8 \\ 6 \\ 5 \\ 0 \end{bmatrix}, \begin{bmatrix} -3 \\ 1 \\ 0 \\ 5 \end{bmatrix}$; two

15a. 4   b. 8   c. 9

17a. Yes   b. No   c. No   d. Yes   e. Yes   f. No

19a. Yes   b. Yes   c. Yes   d. No

21a. (i) 2   (ii) C_∞ b. (i) 1   (ii) C_∞
   c. (i) 7   (ii) C_∞ d. (i) 0   (ii) C_∞^2
   e. (i) 2   (ii) C_∞^2 f. (i) 2   (ii) C_∞^2

23a. (i) No (ii) Yes (iii) No   b. (i) No (ii) Yes (iii) No
   c. (i) No (ii) Yes (iii) No   d. (i) Yes (ii) Yes (iii) Yes
   e. (i) No (ii) Yes (iii) No   f. (i) No (ii) Yes (iii) No

CHAPTER FIVE

Section 5.1

1a. $\begin{bmatrix} 0 & 0 & 2 \\ 1 & 1 & -1 \end{bmatrix}$ b. $\begin{bmatrix} \frac{3}{2} & \frac{1}{2} & \frac{1}{2} \\ \frac{1}{2} & -\frac{1}{2} & -\frac{1}{2} \end{bmatrix}$

c. $S\begin{bmatrix} x \\ y \\ z \end{bmatrix} = \begin{bmatrix} x - y \\ 2y + 3z \end{bmatrix}$ d. $R\begin{bmatrix} x \\ y \\ z \end{bmatrix} = \begin{bmatrix} 2x + 2y - 3z \\ -4x - 2y + 7z \end{bmatrix}$

3a. $\begin{bmatrix} 2 & -1 & 0 \\ 0 & 0 & 1 \\ 0 & 3 & 0 \end{bmatrix}$ b. $\begin{bmatrix} 1 & 0 & -1 \\ 4 & -1 & 1 \\ -1 & 1 & 2 \end{bmatrix}$

c. $S\begin{bmatrix} x \\ y \\ z \end{bmatrix} = \begin{bmatrix} x + z \\ 2y + 2z \\ -x \end{bmatrix}$ d. $R\begin{bmatrix} x \\ y \\ z \end{bmatrix} = \begin{bmatrix} y \\ -x + y \\ -x + 2z \end{bmatrix}$

5a. $\begin{bmatrix} 1 & 0 & 0 \\ 1 & -1 & 1 \\ -1 & 1 & 0 \end{bmatrix}$ b. $\begin{bmatrix} 1 & 0 & 0 \\ 1 & 0 & 1 \\ 0 & 1 & 1 \end{bmatrix}$

c. $S\begin{bmatrix} x \\ y \\ z \end{bmatrix} = \begin{bmatrix} x \\ x + z \\ y + z \end{bmatrix}$ d. $R\begin{bmatrix} x \\ y \\ z \end{bmatrix} = \begin{bmatrix} x \\ x - y + z \\ -x + y \end{bmatrix}$

7a. $\begin{bmatrix} 0 & 0 & 1 \\ 0 & 1 & 0 \\ 1 & 0 & 0 \end{bmatrix}$ b. $\begin{bmatrix} 1 & 0 & 0 \\ -1 & 1 & 1 \\ 2 & 0 & -1 \end{bmatrix}$

c. $S(a_0 + a_1t + a_2t^2) = (a_0 + a_2) + (2a_1 + 2a_2)t - a_0t^2$
d. $R(a_0 + a_1t + a_2t^2) = (-a_0 + 2a_1 + 3a_2) + (-a_0 + 2a_1 + a_2)t$
$+ (-a_1 + 2a_2)t^2$

9a. $\begin{bmatrix} 1 & 1 \\ 1 & -1 \end{bmatrix}$ b. $\begin{bmatrix} 1 + i & -1 + i \\ -2 & -1 - i \end{bmatrix}$ c. $S\begin{bmatrix} x \\ y \end{bmatrix} = \begin{bmatrix} ix \\ x + 2y \end{bmatrix}$

d. $R\begin{bmatrix} x \\ y \end{bmatrix} = \begin{bmatrix} -2y \\ 2y + (x + y)i \end{bmatrix}$

Section 5.2

1a. $ST\begin{bmatrix} x \\ y \end{bmatrix} = \begin{bmatrix} 2x \\ 2x \\ 2y \end{bmatrix}$ b. TS is not defined.

c. $_{\mathcal{A}}[S]_{\mathcal{C}} = \begin{bmatrix} 2 & -1 & 0 \\ 1 & 0 & 1 \\ 0 & 1 & 0 \end{bmatrix}$, $_{\mathcal{C}}[T]_{\mathcal{B}} = \begin{bmatrix} 3 & 3 \\ 4 & 2 \\ -1 & 1 \end{bmatrix}$, $_{\mathcal{A}}[S]_{\mathcal{C}}\,_{\mathcal{C}}[T]_{\mathcal{B}} = \begin{bmatrix} 2 & 4 \\ 2 & 4 \\ 4 & 2 \end{bmatrix}$

d. $_{\mathcal{A}}[ST]_{\mathcal{B}} = \begin{bmatrix} 2 & 4 \\ 2 & 4 \\ 4 & 2 \end{bmatrix}$

3a. $ST\begin{bmatrix} x \\ y \\ z \end{bmatrix} = \begin{bmatrix} 0 \\ 0 \\ 3x \end{bmatrix}$ b. TS is defined, but not equal to ST.

c. $_{\mathcal{A}}[S]_{\mathcal{C}} = \begin{bmatrix} 0 & 1 & 1 \\ 0 & 1 & 1 \\ 1 & -1 & -1 \end{bmatrix}$, $_{\mathcal{C}}[T]_{\mathcal{B}} = \begin{bmatrix} 3 & 0 & 0 \\ 0 & 0 & -1 \\ 0 & 0 & 1 \end{bmatrix}$,

$_{\mathcal{A}}[S]_{\mathcal{C}}\,_{\mathcal{C}}[T]_{\mathcal{B}} = \begin{bmatrix} 0 & 0 & 0 \\ 0 & 0 & 0 \\ 3 & 0 & 0 \end{bmatrix}$

d. $_{\mathcal{A}}[ST]_{\mathcal{B}} = \begin{bmatrix} 0 & 0 & 0 \\ 0 & 0 & 0 \\ 3 & 0 & 0 \end{bmatrix}$

5a. $ST(a_0 + a_1 t + a_2 t^2) = a_0 + a_1 t$
 b. TS is defined, but not equal to ST.

c. $_{\mathcal{A}}[S]_{\mathcal{C}} = \begin{bmatrix} 1 & 0 & -1 \\ 0 & 1 & -1 \\ 0 & 0 & 1 \end{bmatrix}$, $_{\mathcal{C}}[T]_{\mathcal{B}} = \begin{bmatrix} 1 & 0 & 0 \\ 0 & 1 & 0 \\ 0 & 0 & 0 \end{bmatrix}$, $_{\mathcal{A}}[S]_{\mathcal{C}}\,_{\mathcal{C}}[T]_{\mathcal{B}} = \begin{bmatrix} 1 & 0 & 0 \\ 0 & 1 & 0 \\ 0 & 0 & 0 \end{bmatrix}$

d. $_{\mathcal{A}}[ST]_{\mathcal{B}} = \begin{bmatrix} 1 & 0 & 0 \\ 0 & 1 & 0 \\ 0 & 0 & 0 \end{bmatrix}$

7a. $ST\begin{bmatrix} x \\ y \end{bmatrix} = \begin{bmatrix} 2y + x \\ 0 \end{bmatrix}$ b. TS is defined, but not equal to ST.

c. $_{\mathcal{A}}[S]_{\mathcal{C}} = \begin{bmatrix} -2i & -i \\ 0 & 0 \end{bmatrix}$, $_{\mathcal{C}}[T]_{\mathcal{B}} = \begin{bmatrix} 0 & 1 \\ 1 & 0 \end{bmatrix}$, $_{\mathcal{A}}[S]_{\mathcal{C}}\,_{\mathcal{C}}[T]_{\mathcal{B}} = \begin{bmatrix} -i & -2i \\ 0 & 0 \end{bmatrix}$

d. $_{\mathcal{A}}[ST]_{\mathcal{B}} = \begin{bmatrix} -i & -2i \\ 0 & 0 \end{bmatrix}$

9a. $x(t) - x(0)$ b. $x(t)$ c. No

11a. $D^2x - (t + 1)Dx + (t - 1)x$ b. $D^2x - (t + 1)Dx + tx$ c. No

15. $T\begin{bmatrix} x \\ y \end{bmatrix} = \begin{bmatrix} x \\ 0 \end{bmatrix}, S\begin{bmatrix} x \\ y \end{bmatrix} = \begin{bmatrix} 0 \\ y \end{bmatrix}$

Section 5.3

1a. $\begin{bmatrix} \frac{1}{2} & \frac{1}{2} \\ \frac{1}{2} & -\frac{1}{2} \end{bmatrix}$ b. $\begin{bmatrix} 1 & 0 & 1 \\ 1 & 1 & 0 \\ 1 & 0 & 0 \end{bmatrix}$ c. $\begin{bmatrix} \frac{3}{2} & \frac{1}{2} & \frac{1}{2} \\ \frac{1}{2} & -\frac{1}{2} & -\frac{1}{2} \end{bmatrix}$

d. $\begin{bmatrix} \frac{5}{2} & 1 & \frac{1}{2} \\ -\frac{5}{2} & -1 & \frac{1}{2} \end{bmatrix}$

3a. $\begin{bmatrix} 1 & 0 & 0 \\ 1 & -1 & 1 \\ -1 & 1 & 0 \end{bmatrix}$ b. $\begin{bmatrix} 1 & 0 & 0 \\ 1 & 0 & 1 \\ 0 & 1 & 1 \end{bmatrix}$ c. $\begin{bmatrix} 1 & 0 & -1 \\ 4 & -1 & 1 \\ -1 & 1 & 2 \end{bmatrix}$

d. $\begin{bmatrix} 1 & 1 & 1 \\ -2 & -1 & -3 \\ 1 & 1 & 3 \end{bmatrix}$

5. a, b, and c. $\begin{bmatrix} 1 & 0 & 0 \\ 1 & 0 & 1 \\ 0 & 1 & 1 \end{bmatrix}$ d. $\begin{bmatrix} 1 & 0 & 0 \\ 1 & 1 & 1 \\ 1 & 1 & 2 \end{bmatrix}$

7a. $\begin{bmatrix} 0 & 1 & 0 \\ 1 & -1 & 0 \\ -1 & 1 & 1 \end{bmatrix}$ b. $\begin{bmatrix} 1 & 1 & 0 \\ 1 & 0 & 0 \\ 0 & 1 & 1 \end{bmatrix}$ c. $\begin{bmatrix} 1 & 0 & 0 \\ -1 & 1 & 1 \\ 2 & 0 & -1 \end{bmatrix}$

d. $\begin{bmatrix} 2 & 2 & 2 \\ -1 & 0 & -1 \\ 0 & -1 & 1 \end{bmatrix}$

9a. $\begin{bmatrix} \frac{1}{2} - \frac{1}{2}i & \frac{1}{2} - \frac{1}{2}i \\ -\frac{1}{2} + \frac{1}{2}i & \frac{1}{2} + \frac{1}{2}i \end{bmatrix}$ b. $\begin{bmatrix} i & -1 \\ 1 & 1 \end{bmatrix}$ c. $\begin{bmatrix} 1 + i & -1 + i \\ -2 & -1 - i \end{bmatrix}$

d. $\begin{bmatrix} 1 & -i \\ 1 + i & 1 + i \end{bmatrix}$

Section 5.4

1. Yes   3. No   5. No

7. $\begin{bmatrix} 0 & \frac{1}{3} \\ \frac{1}{2} & 0 \end{bmatrix}$ 9. None   11. None

13. $\begin{bmatrix} \frac{1}{2} & 0 & -\frac{1}{2} & \frac{1}{2} \\ -\frac{1}{2} & -1 & -\frac{1}{2} & \frac{1}{2} \\ \frac{1}{2} & 0 & \frac{1}{2} & -\frac{1}{2} \\ 0 & -1 & -1 & 0 \end{bmatrix}$

15. $\begin{bmatrix} \frac{35}{2} & 1 & -\frac{11}{2} & -2 & -\frac{1}{2} \\ -1 & 0 & 1 & 0 & 0 \\ -\frac{7}{2} & 0 & -\frac{5}{2} & 1 & -\frac{1}{2} \\ 1 & 0 & 0 & 0 & 0 \\ -\frac{5}{2} & 0 & \frac{5}{2} & 0 & \frac{1}{2} \end{bmatrix}$

17a. $\begin{bmatrix} -2 & 0 & 2 \\ 0 & 1 & 0 \\ 1 & 0 & 1 \end{bmatrix}$ b. $\begin{bmatrix} -\frac{1}{4} & 0 & \frac{2}{4} \\ 0 & 1 & 0 \\ \frac{1}{4} & 0 & \frac{2}{4} \end{bmatrix}$ c. $\begin{bmatrix} -1 & 0 & 0 \\ 0 & 3 & 0 \\ 0 & 0 & 3 \end{bmatrix}$

19a. $\begin{bmatrix} 1 & 1 & 0 \\ 0 & 1 & -1 \\ -1 & 0 & 1 \end{bmatrix}$ b. $\begin{bmatrix} \frac{1}{2} & -\frac{1}{2} & -\frac{1}{2} \\ \frac{1}{2} & \frac{1}{2} & \frac{1}{2} \\ \frac{1}{2} & -\frac{1}{2} & \frac{1}{2} \end{bmatrix}$ c. $\begin{bmatrix} -\frac{3}{2} & -\frac{3}{2} & 3 \\ -\frac{3}{2} & \frac{5}{2} & 1 \\ -\frac{3}{2} & -\frac{1}{2} & 4 \end{bmatrix}$

21b. $\begin{bmatrix} -1 & 0 \\ 0 & -1 \end{bmatrix}, \begin{bmatrix} -1 & 0 \\ 0 & 1 \end{bmatrix}, \begin{bmatrix} 1 & 0 \\ 0 & -1 \end{bmatrix}$

27. The columns of a noninvertible matrix P do not form a basis for R^n.

Section 5.5

1a. $\begin{bmatrix} 1 & 0 \\ 0 & 2 \end{bmatrix}$ b. $\begin{bmatrix} 2 & 1 \\ 1 & 1 \end{bmatrix}$

3a. $\begin{bmatrix} 2 & 0 & 0 \\ 0 & 0 & 0 \\ 0 & 0 & -1 \end{bmatrix}$ b. $\begin{bmatrix} 1 & -1 & 0 \\ 1 & 1 & 0 \\ 0 & 0 & 1 \end{bmatrix}$

5a. $\begin{bmatrix} 1 & 0 & 0 \\ 0 & 1 & 0 \\ 0 & 0 & -2 \end{bmatrix}$ b. $\begin{bmatrix} 2 & -2 & -1 \\ 0 & 1 & 1 \\ 1 & 0 & 2 \end{bmatrix}$

7a. $\begin{bmatrix} i & 0 \\ 0 & -i \end{bmatrix}$ b. $\begin{bmatrix} -i & i \\ 1 & 1 \end{bmatrix}$

9a. A itself is in block diagonal form.   b. $\begin{bmatrix} 1 & 0 & 0 \\ 0 & 1 & 0 \\ 0 & 0 & 1 \end{bmatrix}$

11a. $\begin{bmatrix} 2 & 0 & 0 & 0 \\ 0 & 1 & 1 & 0 \\ 0 & 0 & 1 & 1 \\ 0 & 0 & 0 & 1 \end{bmatrix}$ b. $\begin{bmatrix} 1 & 1 & 0 & 0 \\ 1 & 0 & 1 & 0 \\ 1 & 0 & 0 & 1 \\ 1 & 0 & 0 & 0 \end{bmatrix}$

13a. A itself is in block diagonal form.   b. $\begin{bmatrix} 1 & 0 & 0 & 0 \\ 0 & 1 & 0 & 0 \\ 0 & 0 & 1 & 0 \\ 0 & 0 & 0 & 1 \end{bmatrix}$

15a. $\begin{bmatrix} -1 & 0 & 0 & 0 & 0 \\ 0 & -1 & 0 & 0 & 0 \\ 0 & 0 & 0 & 0 & 0 \\ 0 & 0 & 0 & 0 & 1 \\ 0 & 0 & 1 & 0 & 0 \end{bmatrix}$ b. $\begin{bmatrix} 0 & 0 & 1 & 1 & 0 \\ 1 & 0 & 0 & 0 & 1 \\ 0 & 1 & 1 & 0 & 0 \\ 0 & 0 & 0 & 1 & 0 \\ 0 & 0 & 0 & 0 & 1 \end{bmatrix}$

17a. $\begin{bmatrix} 1+i & 0 & 0 & 0 \\ i & 1+i & 0 & 0 \\ 0 & 0 & 1-i & 0 \\ 0 & 0 & -i & 1-i \end{bmatrix}$ b. $\begin{bmatrix} i & -i & -i & i \\ 0 & 1 & 0 & 1 \\ 2i & 0 & -2i & 0 \\ 2 & 0 & 2 & 0 \end{bmatrix}$

Section 5.6

1a. $\begin{bmatrix} 2 & 0 \\ 1 & 2 \end{bmatrix}$ b. $\begin{bmatrix} 1 & -1 \\ 0 & 1 \end{bmatrix}$ c. $\mathbf{x} = c_1 \begin{bmatrix} (1-t)e^{2t} \\ te^{2t} \end{bmatrix} + c_2 \begin{bmatrix} -e^{2t} \\ e^{2t} \end{bmatrix}$

3a. $\begin{bmatrix} -1 & 0 & 0 \\ 0 & 1 & 0 \\ 0 & 1 & 1 \end{bmatrix}$ b. $\begin{bmatrix} 1 & 0 & -1 \\ 2 & -1 & 0 \\ 0 & 2 & 0 \end{bmatrix}$

c. $\mathbf{x} = c_1 \begin{bmatrix} e^{-t} \\ 2e^{-t} \\ 0 \end{bmatrix} + c_2 \begin{bmatrix} -te^t \\ -e^t \\ 2e^t \end{bmatrix} + c_3 \begin{bmatrix} -e^t \\ 0 \\ 0 \end{bmatrix}$

5a. $\begin{bmatrix} -2 & 0 & 0 & 0 \\ 0 & -2 & 0 & 0 \\ 0 & 0 & 2 & 0 \\ 0 & 0 & 1 & 2 \end{bmatrix}$ b. $\begin{bmatrix} 1 & 0 & 0 & 0 \\ 0 & 1 & 1 & -2 \\ 0 & 0 & 1 & -6 \\ 0 & 0 & 0 & -4 \end{bmatrix}$

c. $\mathbf{x} = c_1 \begin{bmatrix} e^{-2t} \\ 0 \\ 0 \\ 0 \end{bmatrix} + c_2 \begin{bmatrix} 0 \\ e^{-2t} \\ 0 \\ 0 \end{bmatrix} + c_3 \begin{bmatrix} 0 \\ (1-2t)e^{2t} \\ (1-6t)e^{2t} \\ -4te^{2t} \end{bmatrix} + c_4 \begin{bmatrix} 0 \\ -2e^{2t} \\ -6e^{2t} \\ -4e^{2t} \end{bmatrix}$

7a. $\begin{bmatrix} 1 & 0 & 0 \\ 1 & 1 & 0 \\ 0 & 0 & 1 \end{bmatrix}$ b. $\begin{bmatrix} 1 & 0 & 1 \\ 0 & 1 & 1 \\ 0 & 1 & 0 \end{bmatrix}$ c. $\mathbf{x} = c_1 \begin{bmatrix} e^t \\ te^t \\ te^t \end{bmatrix} + c_2 \begin{bmatrix} 0 \\ e^t \\ e^t \end{bmatrix} + c_3 \begin{bmatrix} e^t \\ e^t \\ 0 \end{bmatrix}$

9a. $\begin{bmatrix} -1 & 0 & 0 & 0 \\ 0 & 1 & 0 & 0 \\ 0 & 1 & 1 & 0 \\ 0 & 0 & 0 & 1 \end{bmatrix}$ b. $\begin{bmatrix} 0 & 1 & 1 & 1 \\ 1 & 0 & 0 & -1 \\ 1 & 0 & 0 & 1 \\ 1 & 0 & -1 & 0 \end{bmatrix}$

c. $\mathbf{x} = c_1 \begin{bmatrix} 0 \\ e^{-t} \\ e^{-t} \\ e^{-t} \end{bmatrix} + c_2 \begin{bmatrix} (1+t)e^t \\ 0 \\ 0 \\ -te^t \end{bmatrix} + c_3 \begin{bmatrix} e^t \\ 0 \\ 0 \\ -e^t \end{bmatrix} + c_4 \begin{bmatrix} e^t \\ -e^t \\ e^t \\ 0 \end{bmatrix}$

11a. $\begin{bmatrix} 2 & 0 & 0 & 0 \\ 1 & 2 & 0 & 0 \\ 0 & 0 & 2 & 0 \\ 0 & 0 & 0 & 2 \end{bmatrix}$ b. $\begin{bmatrix} 1 & 0 & 0 & 1 \\ 0 & 0 & 1 & 0 \\ 0 & 1 & 0 & 1 \\ 0 & 1 & 0 & 0 \end{bmatrix}$

c. $\mathbf{x} = c_1 \begin{bmatrix} e^{2t} \\ 0 \\ te^{2t} \\ te^{2t} \end{bmatrix} + c_2 \begin{bmatrix} 0 \\ 0 \\ e^{2t} \\ e^{2t} \end{bmatrix} + c_3 \begin{bmatrix} 0 \\ e^{2t} \\ 0 \\ 0 \end{bmatrix} + c_4 \begin{bmatrix} e^{2t} \\ 0 \\ e^{2t} \\ 0 \end{bmatrix}$

13a. $\begin{bmatrix} 2i & 0 & 0 & 0 \\ 1 & 2i & 0 & 0 \\ 0 & 0 & -2i & 0 \\ 0 & 0 & 1 & -2i \end{bmatrix}$ b. $\begin{bmatrix} -i & -2 & i & -2 \\ 0 & -2i & 0 & 2i \\ -4i & 0 & 4i & 0 \\ 4 & 0 & 4 & 0 \end{bmatrix}$

c. $\mathbf{x} = c_1 \begin{bmatrix} -2t\cos 2t + \sin 2t \\ 2t\sin 2t \\ 4\sin 2t \\ 4\cos 2t \end{bmatrix} + c_2 \begin{bmatrix} -\cos 2t - 2t\sin 2t \\ -2t\cos 2t \\ -4\cos 2t \\ 4\sin 2t \end{bmatrix}$

$\qquad\qquad + c_3 \begin{bmatrix} -2\cos 2t \\ 2\sin 2t \\ 0 \\ 0 \end{bmatrix} + c_4 \begin{bmatrix} -2\sin 2t \\ -2\cos 2t \\ 0 \\ 0 \end{bmatrix}$

15a. $\begin{bmatrix} -1+i & 0 & 0 & 0 & 0 & 0 \\ 0 & -1-i & 0 & 0 & 0 & 0 \\ 0 & 0 & 2 & 0 & 0 & 0 \\ 0 & 0 & 1 & 2 & 0 & 0 \\ 0 & 0 & 0 & 1 & 2 & 0 \\ 0 & 0 & 0 & 0 & 1 & 2 \end{bmatrix}$

b. $\begin{bmatrix} 0 & 0 & 0 & 0 & 0 & 1 \\ 0 & 0 & 0 & 0 & 1 & 0 \\ 0 & 0 & 0 & 1 & 0 & 0 \\ 0 & 0 & 1 & 0 & 0 & 0 \\ -1-i & -1+i & 0 & 0 & 0 & 0 \\ 2 & 2 & 0 & 0 & 0 & 0 \end{bmatrix}$

c. $\mathbf{x} = c_1 \begin{bmatrix} 0 \\ 0 \\ 0 \\ 0 \\ e^{-t}(\sin t - \cos t) \\ 2e^{-t}\cos t \end{bmatrix} + c_2 \begin{bmatrix} 0 \\ 0 \\ 0 \\ 0 \\ -e^{-t}(\sin t + \cos t) \\ 2e^{-t}\sin t \end{bmatrix} +$

$\qquad c_3 \begin{bmatrix} \frac{1}{6}t^3 e^{2t} \\ \frac{1}{2}t^2 e^{2t} \\ te^{2t} \\ e^{2t} \\ 0 \\ 0 \end{bmatrix} + c_4 \begin{bmatrix} \frac{1}{2}t^2 e^{2t} \\ te^{2t} \\ e^{2t} \\ 0 \\ 0 \\ 0 \end{bmatrix} + c_5 \begin{bmatrix} te^{2t} \\ e^{2t} \\ 0 \\ 0 \\ 0 \\ 0 \end{bmatrix} + c_6 \begin{bmatrix} e^{2t} \\ 0 \\ 0 \\ 0 \\ 0 \\ 0 \end{bmatrix}$

Review Problems

1a. $\begin{bmatrix} 1 & -2 & 1 \\ 1 & -5 & -1 \\ 0 & 2 & 0 \end{bmatrix}$ b. $\begin{bmatrix} 2 & 2 & 0 \\ -7 & -6 & 1 \\ 5 & 3 & 0 \end{bmatrix}$

c. $\begin{bmatrix} 1 & 0 & 0 \\ 0 & -1 & 0 \\ 0 & 0 & -4 \end{bmatrix}$ d. $\begin{bmatrix} 1 & 0 & 0 \\ 0 & -1 & 0 \\ 0 & 0 & -4 \end{bmatrix}$

3a. $\begin{bmatrix} 0 & 1 & 0 & 0 \\ 0 & 0 & 2 & 0 \\ 0 & 0 & 0 & 3 \\ 0 & 0 & 0 & 0 \end{bmatrix}$ b. $\begin{bmatrix} 0 & 1 & -3 & -4 \\ 0 & 0 & 2 & 5 \\ 0 & 0 & 0 & 3 \\ 0 & 0 & 0 & 0 \end{bmatrix}$

c. $\begin{bmatrix} 0 & 1 & 0 & 0 \\ 0 & 0 & 2 & 0 \\ 0 & 0 & 0 & 3 \\ 0 & 0 & 0 & 0 \end{bmatrix}$ d. $\begin{bmatrix} 0 & 0 & 0 & 0 \\ 1 & 0 & 0 & 0 \\ 0 & 1 & 0 & 0 \\ 0 & 0 & 1 & 0 \end{bmatrix}$

5a. $\begin{bmatrix} 0 & 0 & 0 & 1 \\ 0 & 0 & 1 & 0 \\ 0 & 1 & 0 & 0 \\ 1 & 0 & 0 & 0 \end{bmatrix}$ b. $\begin{bmatrix} 1 & 0 & 0 & 0 \\ 0 & -1 & 0 & 0 \\ 0 & 0 & 1 & 0 \\ 0 & 0 & 0 & -1 \end{bmatrix}$

c. $\begin{bmatrix} 1 & 0 & 0 & 0 \\ 0 & 1 & 0 & 0 \\ 0 & 0 & -1 & 0 \\ 0 & 0 & 0 & -1 \end{bmatrix}$ d. $\begin{bmatrix} 1 & 0 & 0 & 0 \\ 0 & 1 & 0 & 0 \\ 0 & 0 & -1 & 0 \\ 0 & 0 & 0 & -1 \end{bmatrix}$

7a. $\begin{bmatrix} 2 & -2 & 2 \\ 0 & 1 & 1 \\ 0 & -1 & 3 \end{bmatrix}$ b. $\begin{bmatrix} 2 & 0 & 0 \\ 1 & 2 & 0 \\ 0 & 0 & 2 \end{bmatrix}$

c. $\mathbf{x} = c_1 \begin{bmatrix} 2te^{2t} \\ te^{2t} \\ (1+t)e^{2t} \end{bmatrix} + c_2 \begin{bmatrix} 2e^{2t} \\ e^{2t} \\ e^{2t} \end{bmatrix} + c_3 \begin{bmatrix} e^{2t} \\ 0 \\ 0 \end{bmatrix}$

9a. $\begin{bmatrix} 1 & 0 & 0 \\ 0 & 2+i & 0 \\ 0 & 0 & 2-i \end{bmatrix}$ b. $\begin{bmatrix} 1 & 0 & 0 \\ 0 & 2+i & 0 \\ 0 & 0 & 2-i \end{bmatrix}$

c. $\mathbf{x} = c_1 \begin{bmatrix} -e^t \\ -e^t \\ e^t \end{bmatrix} + c_2 \begin{bmatrix} e^{2t}\sin t \\ -e^{2t}\sin t \\ e^{2t}\cos t \end{bmatrix} + c_3 \begin{bmatrix} -e^{2t}\cos t \\ e^{2t}\cos t \\ e^{2t}\sin t \end{bmatrix}$

11a.
$$\begin{bmatrix} -1 & 0 & 0 & 0 & 0 & 0 \\ 0 & -1 & 0 & 0 & 0 & 0 \\ 0 & 0 & 1 & 0 & 0 & 0 \\ 0 & 0 & 1 & 1 & 0 & 1 \\ 0 & 0 & 0 & 0 & 1 & 1 \\ 0 & 0 & 0 & 0 & 0 & 1 \end{bmatrix}$$

b.
$$\begin{bmatrix} -1 & 0 & 0 & 0 & 0 & 0 \\ 0 & -1 & 0 & 0 & 0 & 0 \\ 0 & 0 & 1 & 0 & 0 & 0 \\ 0 & 0 & 1 & 1 & 0 & 0 \\ 0 & 0 & 0 & 0 & 1 & 0 \\ 0 & 0 & 0 & 0 & 1 & 1 \end{bmatrix}$$

c. $\mathbf{x} = c_1 \begin{bmatrix} 0 \\ 0 \\ 0 \\ 0 \\ e^{-t} \\ 0 \end{bmatrix} + c_2 \begin{bmatrix} 0 \\ 0 \\ -e^{-t} \\ 0 \\ 0 \\ e^{-t} \end{bmatrix} + c_3 \begin{bmatrix} 0 \\ 0 \\ e^t \\ te^t \\ 0 \\ 0 \end{bmatrix}$

$+ c_4 \begin{bmatrix} 0 \\ 0 \\ 0 \\ e^t \\ 0 \\ 0 \end{bmatrix} + c_5 \begin{bmatrix} -e^t \\ -te^t \\ 0 \\ te^t- \\ te^t \\ e^t \end{bmatrix} + c_6 \begin{bmatrix} 0 \\ -e^t \\ 0 \\ e^t \\ e^t \\ 0 \end{bmatrix}$

CHAPTER SIX

Section 6.1

1. $(D - .08)x = \begin{cases} 400, & 0 \le t < 2 \\ 500, & t \ge 2 \end{cases};$ $x(0) = 1000$

3. $(D^2 + 4)x = \begin{cases} 0, & 0 \le t < 2 \\ 32, & t \ge 2 \end{cases};$ $x(0) = -\dfrac{1}{2}, x'(0) = 0$

5. $\left(LD^2 + RD + \dfrac{1}{C}\right)Q = \begin{cases} t, & 0 \le t < 10 \\ 10, & t \ge 10 \end{cases};$ $Q(0) = Q'(0) = 0$

7a. $(D^3 + 4D^2 + 4D)x = 8; x(0) = x'(0) = 0, x''(0) = -8$
 b. $(D^3 + 4D^2 + 4D)x = 8; x(0) = x'(0) = 0, x''(0) = 8$

9. $DQ = -Q - I_2 + 10, DI_2 = Q - 3I_2; Q(0) = 10, I_2(0) = 0$

Section 6.2

1. $F(s) = \dfrac{1}{s - 4},\ s > 4$

3. $F(s) = \dfrac{2}{s^3},\ s > 0$

5. $F(s) = \dfrac{2}{s^2 + 4},\ s > 0$

7. $F(s) = \begin{cases} 3, & s = 0 \\ (1 - e^{-3s})/s, & s \neq 0 \end{cases}$

9. $F(s) = \dfrac{24}{s^5}$

11. $F(s) = \dfrac{2}{s^3} - \dfrac{7}{s} + \dfrac{s}{s^2 + 4}$

13. $F(s) = \dfrac{2}{s^3} + \dfrac{3}{s^2} + \dfrac{2}{s}$

15. $f(t) = e^{2t}$

17. $f(t) = \dfrac{1}{2} e^{t/2}$

19. $f(t) = \dfrac{1}{12} t^4$

21. $f(t) = \dfrac{1}{\sqrt{3}} \sin t\sqrt{3}$

23. No

27a. All except ii are of exponential order.
  b. All except i, v, vi are piecewise continuous on $[0,h]$.

Section 6.3

1. $\mathcal{L}[x] = \dfrac{-3}{s - 1}$

3. $\mathcal{L}[x] = \dfrac{1}{(s - 2)(s^2 - 1)}$

5. $\mathcal{L}[x] = \dfrac{s}{(s^2 + 9)(s^2 + 1)}$

7. $\mathcal{L}[f(t)] = \dfrac{s}{s^2 + \beta^2}$

9. $\mathcal{L}[f(t)] = \dfrac{2s + 3}{(s + 1)^2}$

11. $-\dfrac{1}{3} + \dfrac{1}{4} e^{-t} + \dfrac{1}{12} e^{3t}$

13. $-\dfrac{1}{2} \sin t + \dfrac{1}{4} e^{t} - \dfrac{1}{4} e^{-t}$

15. $-\dfrac{1}{4} \cos 2t + \dfrac{1}{4} t$

17. $x = -\dfrac{1}{3} + \dfrac{9}{4} e^{-t} + \dfrac{1}{12} e^{3t}$

19. $x = -\dfrac{1}{5} \cos 2t - \dfrac{1}{10} \sin 2t + \dfrac{1}{5} e^{t}$

21. $x = e^{t} + e^{2t}$

23. $x = 3 - \dfrac{1}{3} e^{t} + \dfrac{7}{2} e^{2t} - \dfrac{7}{6} e^{-2t}$

25. $x = -1 + \dfrac{1}{2} e^{t} + \dfrac{1}{2} e^{-t} + \cos t$

27. $x = 5e^{t-1} - 3t$

29. $x = \dfrac{1}{4} e^{\pi - t} - \dfrac{1}{4} e^{t - \pi} - \dfrac{1}{2} \sin t$

31. $x = 70 - 3t$

33. $x = 4 - \dfrac{1}{4} e^{-4t} + \dfrac{3}{4} e^{-2t}$

37a. $A_1 = p(m_1)/(m_1 - m_2) \cdots (m_1 - m_k)$

Section 6.4

1. $\dfrac{1}{(s - 3)^2}$

3. $\dfrac{12s}{(s^2 + 4)^2}$

5. $\dfrac{18(s^2 - 3)}{(s^2 + 9)^3}$

7. $\dfrac{2}{(s - m)^3}$

9. $\dfrac{2}{s^2 - 6s + 13}$

11. $\dfrac{6(s - 2)}{(s^2 - 4s + 13)^2}$

13. $\dfrac{1}{2s^2} + \dfrac{16 - s^2}{2(s^2 + 16)^2}$

15. $\dfrac{t^3 e^{-2t}}{6}$

17. te^{-3t}

19. $\frac{1}{9} - \frac{1}{9} e^{-3t} - \frac{1}{3} te^{-3t}$

21. $e^{-2t} \sin t$

23. $e^{-3t/2}\left(\cos \dfrac{t\sqrt{3}}{2} - \dfrac{\sqrt{3}}{3} \sin \dfrac{t\sqrt{3}}{2} \right)$

25. $x = \dfrac{t^4 e^t}{4}$

27. $x = -e^t + 3te^t + \frac{1}{20} t^5 e^t$

29. $x = \frac{1}{8} e^{-t} - \frac{1}{8} e^t \cos 2t + \frac{1}{8} e^t \sin 2t$

31. $x = \frac{1}{3} e^t - \frac{1}{3} e^{-t/2} \left(\cos \dfrac{t\sqrt{3}}{2} + \sqrt{3} \sin \dfrac{t\sqrt{3}}{2} \right)$

33. $Q = te^{-t}$

35. $x = -2e^{-3t/2} \cos \dfrac{t}{2} - 4e^{-3t/2} \sin \dfrac{t}{2}$

37. $x = 2t - 2te^{-2t}$

39a. $\displaystyle\int_0^t \mathcal{L}^{-1}\left[\dfrac{s}{(s^2 + 1)^2} \right] ds$ b. $-\frac{1}{2} t \cos t + \frac{1}{2} \sin t$

c. $\frac{1}{8} t \sin t - \frac{1}{8} t^2 \cos t$ d. $\frac{3}{8} \sin t - \frac{3}{8} t \cos t - \frac{1}{8} t^2 \sin t$

41. $x = 1 - 2t^2$

Section 6.5

1. $f(t) = u_3(t)(t - 3),\qquad \mathcal{L}[f(t)] = \dfrac{e^{-3s}}{s^2}$

3. $f(t) = u_\pi(t) \sin t,\qquad \mathcal{L}[f(t)] = \dfrac{-e^{-\pi s}}{s^2 + 1}$

5. $f(t) = u_1(t)e^t - u_2(t)e^t,\qquad \mathcal{L}[f(t)] = \dfrac{e^{-(s-1)}}{s - 1} - \dfrac{e^{-2(s-1)}}{s - 1}$

7. $f(t) = 3 - t + 2u_3(t)(t-3),\qquad \mathcal{L}[f(t)] = \dfrac{3}{s} - \dfrac{1}{s^2} + \dfrac{2e^{-3s}}{s^2}$

9. $f(t) = t - u_1(t) + u_2(t)(2 - t),\qquad \mathcal{L}[f(t)] = \dfrac{1}{s^2} - \dfrac{e^{-s}}{s} - \dfrac{e^{-2s}}{s^2}$

11. $f(t) = t - 2u_1(t) + u_3(t)(2 - t),\qquad \mathcal{L}[f(t)] = \dfrac{1}{s^2} - \dfrac{2e^{-s}}{s} - e^{-3s}\left(\dfrac{1}{s} + \dfrac{1}{s^2} \right)$

13. $u_4(t)e^{-4(t-4)}$ 15. $u_4(t)(t-4)e^{-4(t-4)}$

17. $\cos t + u_\pi(t)\sin(t-\pi) = \cos t - u_\pi(t)\sin t$

19. $u_1(t)\,[1 - e^{-(t-1)}]$ 21. $x = -5e^t + u_3(t)(2 - t + e^{t-3})$

23. $x = \frac{1}{2}e^{2t} + \frac{1}{2}e^{-2t} + u_2(t)[-\frac{1}{2} + \frac{1}{4}e^{2(t-2)} + \frac{1}{4}e^{-2(t-2)}]$

25. $x = u_2(t)\left[\dfrac{1}{2}e^{-t} - \dfrac{e^{-2}}{2}\cos(t-2) + \dfrac{e^{-2}}{2}\sin(t-2)\right]$

27. $x = \frac{2}{3}\sin t - \frac{1}{3}\sin 2t + u_\pi(t)(\frac{4}{3}\sin t + \frac{2}{3}\sin 2t)$

29. $x = \frac{1}{6}t^3 e^{-t} - \frac{1}{6}u_3(t)(t-3)^3 e^{-t}$

31a. $x = -\frac{1}{2}\cos 2t + u_2(t)[8 - 8\cos 2(t-2)]$
  b. $x(1) = -\frac{1}{2}\cos 2 \approx 0.2081,\ x(4) = -\frac{1}{2}\cos 8 - 8\cos 4 + 8 \approx 13.3019$

33. (i) a. $Q = u_2(t)[5 + 5e^{-2(t-2)} - 10e^{-(t-2)}]$
       $\quad - u_5(t)\,[5 + 5e^{-2(t-5)} - 10e^{-(t-5)}]$
    b. $Q(1) = 0,\ Q(3) = 5 + 5e^{-2} - 10e^{-1} \approx 1.9979,$
       $Q(6) = 5e^{-8} - 10e^{-4} - 5e^{-2} + 10e^{-1} \approx 2.8206$

    (ii) a. $Q = 5 + 5e^{-2t} - 10e^{-t} - u_2(t)[5 + 5e^{-2(t-2)} - 10e^{-(t-2)}]$
       $\quad + u_5(t)[5 + 5e^{-2(t-5)} - 10e^{-(t-5)}]$
    b. $Q(1) = 5 + 5e^{-2} - 10e^{-1} \approx 1.9979$
       $Q(3) = 5e^{-6} - 10e^{-3} - 5e^{-2} + 10e^{-1} \approx 2.5166$
       $Q(6) = 5 + 5e^{-12} - 10e^{-6} - 5e^{-8} + 10e^{-4} + 5e^{-2} - 10e^{-1}$
       $\quad \approx 2.1546$

35. $\mathcal{L}[f(t)] = e^{-ks}\sqrt{\pi/s}$

Section 6.6

1. $\frac{1}{20}t^5$ 3. $\frac{1}{4}t + \frac{1}{16}e^{-4t} - \frac{1}{16}$

5. $\frac{1}{2}t - \frac{1}{4}\sin 2t$ 7. $\frac{2}{3}\cos t - \frac{2}{3}\cos 2t$

11. $\frac{3}{4} - \frac{3}{4}\cos 2t$ 13. $\frac{3}{4}t - \frac{3}{16} + \frac{3}{16}e^{-4t}$

15. $\frac{1}{4}t\sin 2t$ 17. $\frac{1}{2}\sin t - \frac{1}{2}\cos t + \frac{1}{2}e^t$

19. $x = \frac{1}{2}\sin t - \frac{1}{2}t\cos t$ 21. $x = \frac{3}{2}\sin t - \frac{1}{2}t\cos t$

23. $x = 3u_1(t)[1 - \cos(t-1) - \frac{1}{2}(t-1)\sin(t-1)]$

25. $x = -3e^{-t}(\frac{3}{2}\sin t + \frac{1}{2}t\sin t - \frac{3}{2}t\cos t)$

29. $\frac{1}{2}u_a(t)(t-a)^2$

35d. (Exercise 12, Section 2.10) $H(t, u) = \dfrac{t}{1-u} - \dfrac{ue^{t-u}}{1-u}$

   (Exercise 13, Section 2.10) $H(t, u) = \dfrac{-u^2 t^{-1}}{u+1} + \dfrac{ue^{t-u}}{u+1}$

Section 6.7

1. $x = \frac{15}{4} e^{2t/3} - \frac{3}{4} e^{-2t/3}$
3. $x = 2e^{-t}(\cos 2t + \sin 2t)$

5. $x = e^{-2t}(t + \frac{1}{6} t^3)$

7. $x = e^t(-2 \sin 3t + \frac{2}{81} \cos 3t - \frac{2}{81} + \frac{1}{9} t^2)$

9. $x = \frac{1}{12} e^{2t} + \frac{1}{4} e^{-2t} - \frac{1}{3} e^{-t}$
    $+ u_1(t)(\frac{1}{20} e^{3t} + \frac{1}{5} e^{5-2t} - \frac{1}{4} e^{4-t} - \frac{1}{12} e^{2t} - \frac{1}{4} e^{4-2t} + \frac{1}{3} e^{3-t})$

11. $x = \sin t + u_1(t)[t - \sin(t - 1) - \cos (t - 1)]$

13. $x = 2 + t + (t - 2) e^t + u_2(t)[\frac{1}{2} (t - 2)^2 e^t + (10 - 3t)e^{2-t} - t - 2]$

15. $x = 1$
17. $x = -t + \frac{1}{2} e^t - \frac{1}{2} e^{-t}$

19. $x = \frac{1}{6} t^3 e^t + \frac{5}{2} t^4 e^t$

21a. $x = \frac{32}{125} - \frac{3}{500} \cos 5t\sqrt{5} + \frac{4}{125} u_3(t) [1 - \cos 5(t - 3)\sqrt{5}]$

Section 6.8

1. $x = 5e^{-t}, y = -5e^{-t}$

3. $x = 5e^{-t} + u_1(t)[\frac{1}{3} + \frac{1}{6} e^{3(t-1)} - \frac{1}{2} e^{-(t-1)}],$
    $y = -5e^{-t} + u_1(t)[-\frac{2}{3} + \frac{1}{6} e^{3(t-1)} + \frac{1}{2} e^{-(t-1)}]$

5. $x = e^t(t - t^2), y = t^2 e^t$

7. $x = 9e^{2t} - \cos 2t - \sin 2t, y = 3e^{2t} - 3 \cos 2t - 3 \sin 2t$

9. $x = 9 + 6t - 8e^t - e^{-2t}, y = 3 + 6t - 3e^{-2t}, z = -3 + 6t + 3e^{-2t}$

11a. $Q = -e^{-t} + 2e^{-5t/2}, I_2 = 2e^{-t} - e^{-5t/2}$
  b. $Q = -2e^{-t}(\cos t + \sin t), I_2 = 2e^{-t}(\cos t - \sin t)$
  c. $Q = e^{-2t}(1 - t), I_2 = e^{-2t}(2 - t)$
  d. $Q = \frac{15}{2} + \frac{5}{2} e^{-2t} + 5te^{-2t}, I_2 = \frac{5}{2} - \frac{5}{2} e^{-2t} + 5te^{-2t}$

13a. $A = 500, B = 2500 - 2000e^{-t/10},$
    $C = 150t - 1500 + 2000e^{-t/10} - 25u_1(t)(t - 1)$
  b. No

15. $p(52) = -e^{-26} + 5e^{-5.2} \approx .03,$
    $s(52) = 5000 + 100e^{-26} - 2500e^{-5.2} \approx 4986,$
    $w(52) = 5000 + 600e^{-26} - 3000e^{-5.2} \approx 4983$

17d. The same, with λ replaced by s.
  e. Partial fractions lead to expressions for the functions $x_i(t)$ as linear combinations of functions of the form $t^k e^{\lambda t}$, $t^j e^{\alpha t} \cos \beta t$, and $t^j e^{\alpha t} \sin \beta t$, where λ is a real root of multiplicity m_λ and $0 \le k < m_\lambda$, and $\alpha \pm \beta i$ are complex roots of multiplicity $m_{\alpha + \beta i}$ and $0 \le j < m_{\alpha + \beta i}$.

CHAPTER SEVEN

Section 7.1

1. $[D^2 + (10 + t)D + 1]Q = 10$ 3. $[10D^2 + (10 - t)D + 10]x = 0$

5a. $m = (100{,}000 - 100t)/32$ b. $F(t) = 100t$

  c. $[(1000 - t)D^2 - D]x = 32t$

Section 7.2

1. $|t| < \infty$ 3. Converges only when $t = 0$

5. $|t| < 3$ 7. $|t - 2| < 1$

9. $|2t - 1| < \infty$ 11. $\displaystyle\sum_{n=0}^{\infty} 2t^{2n}$

13. 1

15. $\displaystyle\sum_{k=1}^{\infty} k(3k + 1)t^{k-1} = \sum_{j=0}^{\infty} (j + 1)(3j + 4)t^{j}$

17. $\displaystyle\sum_{k=1}^{\infty} \frac{(3k - 12)}{k} t^{k}$

19. $1 + 2t + 2t^2 + \frac{4}{3} t^3 + \frac{2}{3} t^4 + \frac{4}{15} t^5 + \cdots$

21. $1 + t + t^4 + t^5 + \cdots$

23. $\displaystyle\sin\frac{t}{2} = \sum_{k=0}^{\infty} \frac{(-1)^k t^{2k+1}}{(2k + 1)!2^{2k+1}}, \qquad |t| < \infty$

25. $\displaystyle e^{-t} = \sum_{k=0}^{\infty} \frac{(-1)^k}{k!} t^{k}, \qquad |t| < \infty$

27. $\displaystyle\frac{1}{(1 + t)^2} = \sum_{j=0}^{\infty} (-1)^j(j + 1)t^{j}, \qquad |t| < 1$

29. $\displaystyle\frac{1}{1 + t^2} = \sum_{k=0}^{\infty} (-1)^k t^{2k}, \qquad |t| < 1$

Section 7.3

1a. $b_{j+2} = \dfrac{-(j + 4)b_j}{(j + 2)(j + 1)}, \qquad j = 0, 1, 2, \ldots$

  b. $x = b_0(1 - 2t^2 + t^4 - \ldots) + b_1(t - \frac{5}{6} t^3 + \frac{7}{24} t^5 - \ldots)$

3a. $b_2 = -\frac{1}{2} b_1$, and $b_{j+2} = \dfrac{-(j + 1)b_{j+1} - b_{j-1}}{(j + 2)(j + 1)}$ for $j = 1, 2, \ldots$

  b. $x = b_0(1 - \frac{1}{6} t^3 + \frac{1}{24} t^4 - \frac{1}{120} t^5 + \ldots)$
      $+ b_1(t - \frac{1}{2} t^2 + \frac{1}{6} t^3 - \frac{1}{8} t^4 + \frac{1}{20} t^5 - \ldots)$

5a. $b_2 = 2b_0 - 1$, $b_3 = \dfrac{b_1 + 1}{6}$, and $b_{j+2} = \dfrac{(4 - 3j)b_j}{(j + 2)(j + 1)}$ for $j = 2, 3, \ldots$

  b. $x = b_0(1 + 2t^2 - \frac{1}{3} t^4 + \ldots) + b_1(t + \frac{1}{6} t^3 - \frac{1}{24} t^5 + \ldots)$
      $+ (-t^2 + \frac{1}{6} t^3 + \frac{1}{6} t^4 - \frac{1}{24} t^5 + \ldots)$

7a. $b_2 = \dfrac{-b_0 - 10b_1 + 10}{2}$, and $b_{j+2} = \dfrac{-10b_{j+1} - b_j}{j + 2}$ for $j = 1, 2, \ldots$

  b. $x = 5t^2 - \frac{50}{3} t^3 + \frac{485}{12} t^4 - \ldots$

9a. $b_{j+2} = \dfrac{-10(j + 1)b_{j+1} + (j - 10)b_j}{10(j + 2)(j + 1)}, \quad j = 0, 1, 2, \ldots$

  b. $x = 1 - \frac{1}{2} t^2 + \frac{1}{6} t^3 - \frac{1}{120} t^4 + \ldots$

11a. $b_{j+2} = \dfrac{-j(j + 1)b_{j+1} + 4b_j}{(j + 2)(j + 1)}, \quad j = 0, 1, 2, \ldots$

  b. $x = b_0[1 + 2(t - 1)^2 - \frac{2}{3} (t - 1)^3 + (t - 1)^4 + \ldots]$
      $+ b_1[(t - 1) + \frac{2}{3} (t - 1)^3 - \frac{1}{3} (t - 1)^4 + \ldots]$

13a. $b_{j+2} = \dfrac{-b_j}{j + 2}, \quad j = 0, 1, 2, \ldots$

  b. $x = b_0 \displaystyle\sum_{m=0}^{\infty} \dfrac{(-1)^m t^{2m}}{2^m m!} + b_1 \displaystyle\sum_{m=0}^{\infty} \dfrac{(-1)^m t^{2m+1}}{1 \cdot 3 \cdot 5 \ldots (2m + 1)}$

15a. $b_{j+2} = \dfrac{(j + \mu + 1)(j - \mu)}{(j + 2)(j + 1)} b_j, \quad j = 0, 1, 2, \ldots$

17. $x = b_0 \displaystyle\sum_{m=0}^{\infty} \dfrac{(-1)^m 1 \cdot 4 \cdot 7 \ldots (3m - 2)}{(3m)!} t^{3m}$

    $+ b_1\left[t + \displaystyle\sum_{m=1}^{\infty} \dfrac{(-1)^m \cdot 2 \cdot 5 \ldots (3m - 1)}{(3m + 1)!} t^{3m+1} \right]$

19. $x = 1 - \frac{1}{6} t^3 + \frac{1}{20} t^5 + \ldots$

Section 7.4

1. (i) a. 1, 0, -0.5, 0.1666667, 0.0833333, $- 0.05$, -0.0055556,
        0.0079365, -0.0002976, -0.0008488, 0.0001146
    b. $x(0.1) \approx 0.9951745$; $x(1) \approx 0.7013492$
    c. $x(0.1) \approx 0.9951745$, $N = 5$; $x(1) \approx 0.7014008$, $N = 12$
    d. $x(0.1) \approx 0.995$, $x(1) \approx 0.701$

(ii) a. 0, 1, -0.5, -0.1666667, 0.1666667, 0, -0.0277778, 0.0039683,
0.0029762, -0.0007716, -0.0002205
  b. $x(0.1) \approx 0.09485$; $x(1) \approx 0.4781746$
  c. $x(0.1) \approx 0.09485$, $N = 5$; $x(1) \approx 0.4782679$, $N = 13$
  d. $x(0.1) \approx 0.095$; $x(1) \approx 0.478$

(iii) a. -2, 3, -0.5, -0.8333333, 0.3333333, 0.1, -0.0722222,
-0.0039683, 0.0095238, -0.0006173, -0.0008907
  b. $x(0.1) \approx -1.7057991$; $x(1) \approx 0.0318254$
  c. $x(0.1) \approx -1.7057995$, $N = 5$; $x(1) \approx 0.320099$, $N = 13$
  d. $x(0.1) \approx -1.706$, $x(1) \approx 0.032$

3. (i) a. 1, 0, 0, -0.6666667, 0, 0, 0.0888889, 0, 0, -0.0049383, 0
  b. $x(0.1) \approx 0.9993334$; $x(1) \approx 0.417284$
  c. $x(0.1) \approx 0.9993333$, $N = 5$; $x(1) \approx 0.3333333$, $N = 5$
  d. $x(0.1) \approx 0.999$; $x(1) \approx 0.417$

(ii) a. 0, 1, 0, 0, -0.3333333, 0, 0, 0.031746, 0, 0, -0.0014109
  b. $x(0.1) \approx 0.0999667$; $x(1) \approx 0.6970018$
  c. $x(0.1) \approx 0.1$, $N = 3$; $x(1) \approx 1$, $N = 3$
  d. $x(0.1) \approx 0.100$; $x(1) \approx 0.697$

(iii) a. -2, 3, 0, 1.3333333, -1, 0, -0.1777778, 0.0952381, 0, 0.0098765,
-0.0042328
  b. $x(0.1) \approx -1.6987668$, $x(1) \approx 1.2564373$
  c. $x(0.1) \approx -1.6987667$, $N = 6$; $x(1) \approx 1.2562523$, $N = 15$
  d. $x(0.1) \approx -1.699$; $x(1) \approx 1.256$

Section 7.5

1. $x = c_1 t + c_2 t^{-1}$ 3. $x = c_1 t + c_2 t \ln t$

5. $x = c_1 t^{-1/3} + c_2 t^{-1/3} \ln t$

7. $x = c_1 t^{-1} \cos (2 \ln t) + c_2 t^{-1} \sin (2 \ln t)$

9. $x = c_1 t^{\sqrt{3}} + c_2 t^{-\sqrt{3}}$ 11. $x = c_1 t^2 \cos (\ln t) + c_2 t^2 \sin (\ln t)$

13a. $R(r) = c_1 r^n + c_2 r^{-n}$ c. $u_n = r^n \cos n\theta$, $u_n^* = r^n \sin n\theta$
  d. $u(r, \theta) = r \cos \theta - 2r^3 \sin 3\theta$

21a. $p_0 \dfrac{d^2 x}{ds^2} + (q_0 - p_0) \dfrac{dx}{ds} + r_0 x = 0$

  b. If the indicial equation has distinct real roots, m_1 and m_2, then $x = c_1 e^{m_1 s} + c_2 e^{m_2 s}$. If there is only one root, m_1, then $x = c_1 e^{m_1 s} + c_2 s e^{m_1 s}$. If the roots are complex, $\alpha \pm \beta i$, then $x = c_1 e^{\alpha s} \cos \beta s + c_2 e^{\alpha s} \sin \beta s$.

Section 7.6

1. $x = c_1 t^{1/3}(1 - \frac{2}{5}t + \frac{1}{20}t^2 - \frac{1}{330}t^3 + \ldots)$
   $+ c_2 t^{1/3}(1 - 2t + \frac{1}{2}t^2 - \frac{1}{21}t^3 + \ldots)$

3. $x = c_1 t^{1/2}(1 - \frac{1}{6}t^2 + \ldots) + c_2 t^{1/2}(1 - \frac{1}{2}t^2 + \ldots)$

5. $x = c_1 t^{3/2}(1 - \frac{1}{10}t^2 + \ldots) + c_2 t^{-3/2}(1 + \frac{1}{2}t^2 + \ldots)$

7. $x = c_1 t(1 - \frac{1}{5}t - \frac{1}{70}t^2 + \frac{23}{1890}t^3 + \ldots)$
   $+ c_2 t^{-1/2}(1 + \frac{1}{4}t - \frac{17}{32}t^2 + \frac{121}{1152}t^3 + \ldots)$

9. $x = c_1 t \sum\limits_{k=1}^{\infty} \dfrac{t^k}{5 \cdot 7 \ldots (3 + 2k)} + c_2 t^{-1/2} \sum\limits_{k=0}^{\infty} \dfrac{t^k}{2^k k!}$

11. $x = c_1 t^{1/2} \sum\limits_{n=0}^{\infty} \dfrac{(-1)^n t^{2n}}{2^n(n!)1 \cdot 5 \ldots (4n + 1)}$

$\qquad + c_2 \left[1 + \sum\limits_{n=1}^{\infty} \dfrac{(-1)^n t^{2n}}{2^n(n!)3 \cdot 7 \ldots (4n - 1)} \right]$

13. $x = \sum\limits_{n=0}^{\infty} \dfrac{(-1)^n t^{2n}}{2^{2n}(n!)^2}$ 15. $x = t \sum\limits_{k=0}^{\infty} \dfrac{(-1)^k 2 t^k}{k!(k + 2)!}$

17. $x = -c_1 t\left[1 + \sum\limits_{k=1}^{\infty} \dfrac{t^k}{5 \cdot 7 \ldots (2k + 3)} \right] + c_2 |t|^{-1/2} \sum\limits_{k=0}^{\infty} \dfrac{t^k}{k! 2^k}$

19. $x = c_1(t - 1)\left[1 + \sum\limits_{k=1}^{\infty} \dfrac{(-1)^k (t - 1)^k}{5 \cdot 7 \ldots (2k + 3)} \right]$

$\qquad + c_2(t - 1)^{-1/2} \sum\limits_{k=0}^{\infty} \dfrac{(-1)^k (t - 1)^k}{2^k k!}$

21. $x = c_1(t + 1) \sum\limits_{k=0}^{\infty} \dfrac{(-1)^k 2 (t + 1)^k}{(k + 2)!} - c_2(t + 1)^{-1}t$

Section 7.7

13c. $J_{3/2}(t) = \sqrt{\dfrac{2}{\pi t}}\left[\dfrac{\sin t - t \cos t}{t} \right]; \quad J_{-3/2}(t) = -\sqrt{\dfrac{2}{\pi t}}\left[\dfrac{\cos t + t \sin t}{t} \right]$

17a.

t	$J_0(t)$	$J_1(t)$	t	$J_0(t)$	$J_1(t)$
0.5	0.9384766	0.2422689	5.0	−0.1776033	−0.3275530
1.0	0.7651910	0.4400499	5.5	−0.0068876	−0.3414492
1.5	0.5118315	0.5579109	6.0	0.1506638	−0.2767514
2.0	0.2238889	0.5767361	6.5	0.2600867	−0.1538123
2.5	−0.0484110	0.4970892	7.0	0.3000329	−0.0046954
3.0	−0.2600409	0.3390610	7.5	0.2663599	0.1351787
3.5	−0.3801323	0.1373984	8.0	0.1716420	0.2346672
4.0	−0.3971882	−0.0660519	8.5	0.0418914	0.2731083
4.5	−0.3205267	−0.2311232	9.0	−0.0903122	0.2452409

17b.

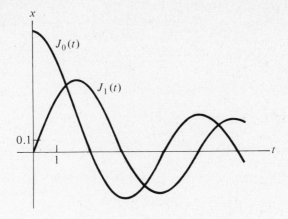

Section 7.8

1. $x = c_1 x_1(t) + c_2[x_1(t) \ln t + 2t^2 - \frac{3}{4} t^3 + \frac{11}{108} t^4 - \ldots)]$, where
   $x_1(t) = t - t^2 + \frac{1}{4} t^3 - \frac{1}{36} t^4 + \ldots$

3. $x = c_1 x_1(t) + c_2[x_1(t) \ln t - 16t^{-2} - 4 + \ldots]$, where
   $x_1(t) = t^2 - \frac{1}{12} t^4 + \ldots$

5. $x = c_1 x_1(t) + c_2[x_1(t) \ln t + \frac{2}{9} t^{2/3} - \frac{1}{108} t^{5/3} + \frac{11}{78732} t^{8/3} + \ldots]$, where
   $x_1(t) = t^{-1/3} - \frac{1}{9} t^{2/3} + \frac{1}{324} t^{5/3} - \frac{1}{26244} t^{8/3} + \ldots$

7. $x = c_1 t \displaystyle\sum_{k=0}^{\infty} \frac{2t^k}{(k+2)!} + c_2 t^{-1}(1 + t)$

9. $x = c_1(1 - t) + c_2\left[(1 - t) \ln t + \frac{5}{2} t - \displaystyle\sum_{k=2}^{\infty} \frac{(k+1)t^k}{2^k k(k-1)}\right]$

15d. $u(\rho, \phi) = \frac{2}{3} + \rho^2(\frac{1}{3} - \cos^2 \phi)$

Review Problems

1. $x = b_0\left[1 + \displaystyle\sum_{n=1}^{\infty} \frac{1 \cdot 3 \cdot 5 \ldots (2n-3)}{2^n n!} t^{2n}\right] + b_1 t$

3. $x = c_1 t^{1/4} \displaystyle\sum_{n=0}^{\infty} \frac{(-1)^n t^{2n}}{(n!)1 \cdot 5 \cdot 9 \ldots (4n+1)}$
   $+ c_2 t^{-1/4}\left[1 + \displaystyle\sum_{n=1}^{\infty} \frac{(-1)^n t^{2n}}{(n!)3 \cdot 7 \cdot 11 \ldots (4n-1)}\right]$
   $= k_1 J_{1/4}(t) + k_2 J_{-1/4}(t)$

5. $x = c_1 t^{2/3} + c_2 t^{2/3} \ln t$

7. $x = b_0 + b_1 \displaystyle\sum_{n=0}^{\infty} \frac{2^n t^{2n+1}}{n!(2n+1)}$

9. $x = c_1 t^2 \left[1 + \sum\limits_{k=1}^{\infty} \frac{(-1)^k 3^k (k+1)! t^k}{8 \cdot 11 \cdot 14 \ldots (3k+5)} \right]$

$+ c_2 t^{1/3} \left[1 + \sum\limits_{k=1}^{\infty} \frac{1 \cdot 4 \cdot 7 \ldots (3k-2) t^k}{3^k k!} \right]$

11. $x = c_1 x_1(t) + c_2 [x_1(t) \ln t - 2t^{-1} + 2 - \frac{4}{9} t^2 + \ldots]$, where

$x_1(t) = t \sum\limits_{k=0}^{\infty} \frac{2t^k}{(k+2)! k!}$

13. $x = c_1(t^{-1} - 5) + c_2[(t^{-1} - 5) \ln t + 16 - 5t - \frac{5}{3} t^2 - \frac{5}{12} t^3 - \frac{1}{20} t^4]$

15. $x = b_0(1 - \frac{1}{2} t^2 + \frac{1}{6} t^3 + \ldots) + b_1(t - \frac{1}{2} t^2 - \frac{1}{6} t^3 + \ldots)$

17. $x = c_1 t^{-1} + c_2 \left[t^{-1} \ln t - t^{-2} + t^{-2} \sum\limits_{k=2}^{\infty} \frac{t^k}{k!(k-1)} \right]$

19. $x = 1 + 2t + \sum\limits_{n=1}^{\infty} \frac{1 \cdot 3 \cdot 5 \ldots (2n-3)}{2^n n!} t^{2n}$

21a. $x = \sum\limits_{j=3}^{\infty} \frac{2t^j}{125 j (1000)^{j-3}}$ b. 5.3737

23. $x = b_0[1 - \frac{1}{2}(t-1)^2 + \frac{1}{3}(t-1)^3 - \frac{1}{8}(t-1)^4 + \ldots]$
$+ b_1[(t-1) - \frac{1}{2}(t-1)^2 + \frac{1}{8}(t-1)^4 + \ldots]$

Index

Abel's formula, 78
Accuracy, of series calculations, 537, 541–42
Airy equation, 528, 593–94
Algorithms for series calculations, 529–31, 535–38
Analytic function, 506
Annihilate, 97, 100
Annihilator, 108
Applications. *See* Bending beams; Compound interest; Diffusion; Electrical circuits; Laplace equation; Memorization; Mixing; Newton's law of cooling; Newton's second law of motion; Pendulum; Population models; Principle of Archimedes; Radioactive decay; Springs; Supply and demand; Temperature distribution; Twisted shaft; Vibrating drumhead.
Augmented matrix, 192

Basis, 296
  coordinates with respect to, 299
  orthogonal, 336
  orthonormal, 337
  standard, for R^n, 297
Bending beams, 130–33, 135–36
Bernoulli, Johannes, 13, 21, 573
Bernoulli equation, 24
Bessel, Friedrich Wilhelm, 573
Bessel equation, 450, 499, 551, 564–76, 580–82, 585–87, 591
Bessel functions,
  of the first kind, 564–76, 594
  of the second kind, 582, 587
Block diagonal form, 390
Block, Jordan, 397
Boole, George, 458
Boundary conditions, for the bending beam, 132

$C[a, b]$, 271, 331
C^n, 269, 270
C_∞^n, 266, 270

Calculator, HP-25 and TI-57, 529, 532–35
  programs for series calculations, 532–35, 538–40
Capacitance, 140
Capacitor, 139–40
Cauchy, Augustin Louis, 22
Cauchy-Euler equation, 109–10, 497, 501, 543–50, 551, 592
Change of bases, 371, 374
  -matrix, *See* Matrix, transition
Characteristic polynomial, 85, 185, 190
Characteristic value. *See* Eigenvalue.
Characteristic vector. *See* Eigenvector.
Charge, 139, 140
Chebyshev equation, 593
Circuits. *See* Electrical circuits.
Coefficients, of a linear o.d.e., 49
  of a linear system of o.d.e.'s, 147
  of a system of algebraic equations, 192
Coil, 139–40
Comb $(v_1, \ldots, v_n)$, 276
Complete list of solutions. *See* General solution.
Component
  parallel, 334
  orthogonal, 334
Compound interest, 8, 15, 24, 418, 462
Convergence,
  of power series, 505
  of series, 504
Convolution, 463–75
  and response to physical systems, 470–72
  formula, 464, 469–70
Cooling. *See* Newton's law of cooling.
Coordinate vector, 304, 354
Coordinates, 299
  with respect to two bases, 360
Corner entry, 196
Cramer's determinant test, 65
Cramer's rule, 71
Critically damped motion, 123
Current, 139
  mesh, 142

Damping coefficient, 44, 501
Determinants, calculation of, 63, 65–66, 70–71, 201

Determinant of coefficients, 63, 65
Diagonal
  block diagonal form, 390
  matrix, 387
Diagonalizable matrix, 387
Differential equation, 1
  general solution of, 3, 7
  linear, 17, 49
  nonlinear, 129–33
  order of, 4
  ordinary, 6
  partial, 6, 493–504
  solution of, 2, 7
  specific solution of, 3, 7
  system of, 6
Differential operator. See Linear differential oper-
    ator.
Differentials, 10, 13–14
Differentiation, of power series, 509
  of vector valued functions, 152, 155
Differentiation formula, first, 431, 433, 438, 439–40
  second, 447, 448–49
Diffusion, 145, 214, 261
Dimension, 306
Divergent series, 504

Eigenvalue, 181, 183, 215. See also Eigenvector
Eigenvector, 181, 183, 209, 215
  and solutions to differential systems, 181, 205,
    217–18
  complex, 215, 217
  generalized, 228, 238
    string of, 249, 398
Electrical circuits, 47, 139–45
  examples, 139–45, 154, 164–67, 187–89, 205–06,
    209–12, 215–17, 218–19, 238–39, 253–58
  exercises, 47, 92, 99, 109, 126, 145–46, 213,
    222, 235, 260, 262
Equidimensional equation. See Cauchy-Euler equa-
    tion.
Euler, Leonhard, 98, 116
Euler equation. See Cauchy-Euler equation.
Euler's formula, 94
Exact differential equation, 32
Existence and uniqueness of solutions,
  importance of theory, 3
  of first-order o.d.e.'s, 22–23
  of linear o.d.e.'s, 56, 61
  of linear systems of o.d.e.'s, 162–63
Expansion by minors, 65–66
Exponential order, function of, 427, 430
Exponential shift, 87. See also Shift formula.

Finite dimensional vector space, 306

Floating. See Principle of Archimedes.
Fourier, Jean Baptiste Joseph, 573
Frobenius, Ferdinand Georg, 551
Frobenius series, 552. See also Series.
  and solutions to o.d.e.'s, 551–90
Fuchs, Lazarus, 551
Function defined in pieces, 411–19, 451–52
Fundamental matrix, 170

Gamma function, 568–69, 572–73, 574–75
General solution, 3, 7
  of a first-order o.d.e., 21–22
  of a homogeneous linear o.d.e., 60, 74, 77, 82
    about an ordinary point, 516–28
    about a singular point, 543–90
    constant-coefficient case, 98
  of a homogeneous linear system of o.d.e.'s,
    162–64, 168, 177–78
    constant-coefficient case, 205, 217–18, 246–47
  of a nonhomogeneous linear o.d.e., 53, 108, 117,
    475–76
    about an ordinary point, 524–28
  of a nonhomogeneous linear system of o.d.e.'s,
    162, 250–51
Generalized eigenspace, 389
  dimension of, 392
Generalized eigenvectors, 228, 238, 388
  and solutions to differential systems, 228, 237,
    239–40, 241
  string of, 249, 398
Geometry and vectors, 156, 263–4, 267, 278–9,
    282. See also Parallel; Orthogonal.
Gram-Schmidt Orthogonalization Process, 337, 340
Graphing solutions of first-order o.d.e.'s, 26–31
Green's function, 474–5

Half-life, 15
Heat flow. See Temperature distribution.
Heaviside, Oliver, 458
Heaviside function. See Unit step function.
Hermite equation, 528
Homogeneous coefficients, 15
Homogeneous linear o.d.e, 18, 49, 55–61
  related, 18, 52
  with constant coefficients, 85–98
  with variable coefficients, 516–90
Homogeneous linear system of o.d.e.'s 147, 162–64,
    167–68
  related, 161
  with constant coefficients, 180–90, 203–47
Hooke, Robert, 43
Hooke's law, 43

Identity
  matrix, 157
  transformation, 370
Independence. *See* Linear independence.
Indicial equation, 543, 555
  and solutions to o.d.e.'s, 549, 562, 577, 583, 587
Inductance, 140
Initial conditions, 3, 7, 163
  and the Laplace transform, 431–42, 475–84
Inner product, 330
  space, 330
  standard, 330
Integrating factor, 25, 40
Interval of convergence, 505
Inverse
  Laplace transform, 425, 428–29. *See also* Laplace transform.
  of a matrix, 377. *See also* Matrix, invertible.
Invertible matrix, 377–86
Isocline, 26
Isomorphism, 328

Jordan block, 397
Jordan form, 397

Ker, 315
Kernel, 315
Kirchhoff, Gustav Robert, 139
Kirchhoff's laws, 139

Lagrange, Joseph Louis, 116, 573
Laguerre equation, 551, 591
Laplace, Marquis Pierre Simon de, 427
Laplace equation,
  two-dimensional, 493–97
    in polar coordinates, 495–97, 549–50
  three-dimensional, 493–94, 497–501
    in cylindrical coordinates, 499–501
    in spherical coordinates, 499–501, 592
Laplace transform, 419–20
  and initial conditions, 431–42, 475–84
  at nonzero time, 440
  convolution formula, 464, 469–70
  domain of, 427–28
  differentiation formula, first, 431, 433, 438, 439–40
    second, 447, 448–49
  inverse, 425, 428–29
  linearity of, 423, 425, 428
  of periodic function, 459–60, 462–63

shift formula, first, 443, 444
  second, 454, 455
solution of systems of o.d.e.'s, 484–91
table of formulas, 482–83
Legendre equation, 501, 528, 551, 591, 592
Legendre polynomials, 591–92
Leibniz, Gottfried Wilhelm von, 1
Lerch's theorem, 429
Linear combination, 56, 162, 276
Linear dependence. *See* Linear independence.
Linear differential equation, 17, 49
  coefficients of, 49
  existence and uniqueness of solutions, 56, 61
  general solution of, 21–22, 53, 60, 77, 82
  homogeneous, 17, 49, 55–61
    with constant coefficients, 85–98
  Laplace transform methods, 431–42, 475–84
  nonhomogeneous, 17
    undetermined coefficients, 99, 108
    variation of parameters, 18–19, 110–117
  normal, 49
  series methods, about an ordinary point, 516–42
    about a singular point, 551–90
  standard form, 18, 110, 114
Linear differential operator, 51
  multiplication of, 85–86, 91
Linear independence, of eigenvectors, 209, 214
  of Frobenius series, 564
  of functions, 79–85
  of initial vectors, 205, 217–18, 238, 239–40, 241, 246
  of solutions, to Cauchy-Euler equations, 550
    to homogeneous o.d.e.'s, 87, 89, 93, 94, 96
  of vectors, 172–78, 292–3
    and linear combinations, 295
    characterizations, 295
Linear system of o.d.e.'s, 147
  coefficients of, 147, 156
  existence and uniqueness of solutions, 162–63
  general solution of, 162–64, 168, 177–78
  homogeneous, 147, 156
    with constant coefficients, 180–90, 203–47
  Laplace transform methods, 485–90
  matrix form, 153
  nonhomogeneous, 250–59
  order of, 147
Linear transformation
  definition, 313
  identity, 370
  isomorphism, 328
  kernel of, 315
  matrix of, 349–60
  one-to-one, 321
  onto, 324
  product of, 363
  range of, 317
Liouville, Joseph, 22
Lipschitz, Rudolph, 22
LRC circuits. *See* Electrical circuits.

Matrix, 149–50. *See also* Vectors.
  algebra, 156–58, 233
  augmented, 192
  block diagonal. *See* normal forms.
  change-of-basis. *See* transition.
  characteristic polynomial of, 185
  diagonal, 387
    block. *See* normal forms.
  diagonalizable, 387, 393
  diagonalizing of a matrix, 387
  eigenvalues of, 181, 183, 215
  eigenvectors of, 181, 215
  entries of, 150
  equality, 150
  form of a system of o.d.e.'s, 153
  fundamental, 170
  generalized eigenspace, 389
  identity, 157
  inverse of, 377, 380
  invertible, 377–86
  Jordan form. *See* normal forms.
  normal forms:
    block diagonal, 390
    diagonal, 387
    Jordan, 397
  of a linear transformation, 349–62, 370–77
  of a product of transformations, 365
  powers of, 227
  product by a number, 151
  product with a vector, 151, 155
  product with another matrix, 225, 232–33, 365
  reduced, 196
  reduction of, 197
  row equivalent, 193
  similar, 383
  similarity, 383
  square of, 227
  sums, 151
  transition, 360, 376, 384
  triangular, 191, 407
  zero, 157
Memorization, 8, 16
Mesh current, 142
Method of Frobenius, 551–90
Minor, 65
Mixing, 8, 15, 441
Models. *See* Applications.
Moment, 128
Moment of inertia, 128

Newton, Sir Isaac, 1
Newton's law of cooling, 8, 24, 40–41, 418, 441, 484
Newton's second law of motion, 1
  exercises, 9, 24, 41, 418, 501, 528
  rotational form, 128, 135
  variable mass form, 501

Nonhomogeneous linear o.d.e., 18
  series methods, 524–28
  undetermined coefficients, 99–108
  variation of paramers, 18–19, 110–117
Nonhomogeneous linear systems of o.d.e.'s, 250–59
Norm, 332
Normal forms. *See* Matrix, normal forms.
Normal linear o.d.e., 49
Normalizing a basis, 337
Nullspace. *See* Kernel.
n-vector. *See* Vectors.

O.d.e. *See* Ordinary differential equation.
One-to-one, 321
  and kernel, 323
Onto, 324
Operator, Laplace transform, 419–20
  linear differential, 51
  multiplication of, 85–86, 91
Order. *See also* Exponential order, function of.
  of a Bessel function, 565, 567, 569, 570, 571, 573, 582, 587
  of an o.d.e., 4
  of a linear system of o.d.e.'s, 147
Ordinary differential equation, 6, 7
Ordinary point, 516
Orthogonal
  basis, 336
  component, 334
  vectors, 333
Orthonormal basis, 337
Overdamped motion, 122

Parallel
  component, 334
  vectors, 332
Particular solution. *See* Specific solution, of a differential equation.
Partial differential equation, 6, 493–504
Partial fractions, 438–39, 442
Partial sum of a series, 504
P.d.e. *See* Partial differential equation.
Pendulum, 129–30, 133–34
Periodic functions, 459–60, 462–63
Permeability, 145
Phase-amplitude form, 120
Picard, Emile, 22
Piecewise continuous function, 427–28, 429
Poisson, Simeon Denis, 573
Polynomials
  finding roots of, 91
  vector space of, 271, 280, 290
Population models, examples, 4–6, 12–13, 16–17, 411–13

exercises, 15, 16, 25, 145, 214, 462
Power series. *See also* Series.
Prices. *See* Supply and demand.
Principle of Archimedes, 46–47, 48, 99, 109, 119, 127, 418, 484, 501, 593
Principle of proportionality, 52. *See also* Linear transformation.
Principle of superposition, 51. *See also* Linear transformation.
Product. *See also* Matrix, product with another matrix.
of mappings, 363, 365, 368
Programs. *See* Calculator, programs.

R^n, 263
Radioactive decay, 8, 15, 145, 214, 260–61, 419, 490
Radius of convergence, 505
Range, 317
Ratio test, 505
Recurrence relation, 517
and series solutions, 526, 562
Reduced matrix, 196
Reduction of a matrix, 197
Reduction of order, 118
Regular singular point, 551
Related homogeneous o.d.e., 18, 52
Related homogeneous system, 161
Resistance, 140
Resistor, 139–40
Resonance, 126
Roots of polynomials, finding, 91
Roundoff error, 542
Row equivalent matrices, 193
Row operation, 193, 200
and determinants, 201
Row reduction, 193–201

Savings account. *See* Compound interest.
Saw-tooth function, 459, 462
Scalar, 265
Separable o.d.e., 13
Separation of variables, for o.d.e.'s, 9–14
for p.d.e.'s, 494–501
Series, Frobenius series, 552
methods, about an ordinary point, 516–42
about a regular singular point, 551–90
power series, 505
on programmable calculators, 529–42
review of, 504–14
Shift formula, first, 443, 444
second, 454, 455
Similar matrices, 383
Similarity matrix, 383

Simple pendulum. *See* Pendulum.
Simply connected, 38
Singular point, 516
regular, 551
Solution of an o.d.e., 2, 7
general, 3, 7
specific, 3, 7
Spanning set, definition, 282, 283
Specific solution, of a differential equation, 3, 7
Spring constant, 43
Springs, 43–46, 119–26
examples, 43–46, 94–95, 104–07, 119–26, 413–14, 417, 477–78, 480–81, 487–88
exercises, 47–48, 109, 119, 126–27, 134–35, 160, 222, 262, 418, 441, 450, 462, 484, 501, 528
Square wave, 415–16, 459, 462
Standard basis for R^n, 297
inner product, 330
Standard form, 18, 110, 114
Steady-state temperature. *See* Temperature distribution.
String of generalized eigenvectors, 249, 398
Subspace
basis for, 296
dimension of, 306
of a vector space, 272, 278
spanning set, 282, 283
trivial, 272
Supply and demand, 9, 24, 47, 92, 145, 261, 490
System of algebraic equations, 62–71, 191–201
System of differential equations, 6
equivalent to an o.d.e., 147–49
linear, 147
solution, by Laplace transforms, 484–91
by matrix methods, 139–262

$_\mathscr{C}[T]_\mathscr{B}$, 350, 355–6
$T_A(\mathrm{v})$, 349
Taylor series, 513–14
Temperature distribution. *See also* Laplace equation.
examples, 494–501
exercises, 501–02, 549–50, 592
Transformation. *See* Linear transformation.
Transition matrix, 360
Trivial subspace, 272
Turning moment, 131
Twisted shaft, 128, 133

$u_a(t)$. *See* Unit step function.
Uncoupled system of o.d.e.'s, 393
Underdamped motion, 121
Undetermined coefficients, 99–108
Unit step function, 451, 458–59
Laplace transform of, 453–54

$_\mathscr{B}[v]$, 354
Vandermonde determinant, 78–79
Variation of parameters, 21, 110–117, 250–59
Vector space, 265
  complex, 268
Vectors, 150. *See also* Eigenvector, Generalized eigenvector.
  and geometry, 156, 263–64, 267, 278–79, 282
  coordinate, 304, 354
  linear combination of, 276
  linear independence of, 292–93
  parallel, 332, 334
  product by a matrix, 151, 265
  orthogonal, 333, 334
  zero, 150, 265
Vector space, 265, 268. *See also* $C[a, b]$; C^n; C^n_∞; R^n; Polynomials, vector space of.
  basis of, 296
  complex, 268
  dimension of, 306
  inner product on, 330
  properties of, 261, 268, 271
  real, 265

  spanning set for, 282
  subspace, 272
Vector valued function, 150
  derivative of, 152
Viscous damping, 44
Voltage source, 139–40

Wronski, Count Hoëné, 74
Wronskian, 74, 166–67, 172
  tests, for independence of functions, 81
    for solutions of homogeneous o.d.e.'s, 74, 77
    for solutions of homogeneous systems, 164, 169

Zero matrix, 157
Zero vector, 265